ERRATA

Advances in Organometallic Chemistry, 13

The following entries for the Author Index on pp. 543-550 have been corrected and should be substituted for those in the index.

Advances in

ORGANOMETALLIC CHEMISTRY

VOLUME 13

CONTRIBUTORS TO THIS VOLUME

Joyce Y. Corey

Patrick M. Henry

D. Max Roundhill

W. E. Silverthorn

J. D. Smith

John S. Thayer

D. R. M. Walton

Advances in Organometallic Chemistry

EDITED BY

F. G. A. STONE

DEPARTMENT OF INORGANIC CHEMISTRY
THE UNIVERSITY
BRISTOL, ENGLAND

ROBERT WEST

DEPARTMENT OF CHEMISTRY
UNIVERSITY OF WISCONSIN
MADISON, WISCONSIN

VOLUME 13

1975

ACADEMIC PRESS New York · San Francisco · London
A Subsidiary of Harcourt Brace Jovanovich, Publishers

ACADEMIC PRESS, INC.
111 Fifth Avenue, New York, New York 10003

United Kingdom Edition published by
ACADEMIC PRESS, INC. (LONDON) LTD.
24/28 Oval Road, London NW1

LIBRARY OF CONGRESS CATALOG CARD NUMBER: 64-16030

ISBN 0-12-031113-5

PRINTED IN THE UNITED STATES OF AMERICA

Contents

Organometallic Chemistry: A Historical Perspective

JOHN S. THAYER

Arene Transition Metal Chemistry

W. E. SILVERTHORN

Organometallic Benzheterocycles

JOYCE Y. COREY

Organometallic Reactions Involving Hydro-Nickel, -Palladium, and -Platinum Complexes

D. MAX ROUNDHILL

Palladium-Catalyzed Organic Reactions

PATRICK M. HENRY

The Organometallic Chemistry of the Main Group Elements—A Guide to the Literature

J. D. SMITH and D. R. M. WALTON

List of Contributors

Numbers in parentheses indicate the pages on which the author's contributions begin.

JOYCE Y. COREY (139), *Department of Chemistry, University of Missouri—St. Louis, St. Louis, Missouri*

PATRICK M. HENRY (363), *Department of Chemistry, University of Guelph, Guelph, Ontario, Canada*

D. MAX ROUNDHILL (273), *Department of Chemistry, Washington State University, Pullman, Washington*

W. E. SILVERTHORN (47), *Department of Chemistry, Oregon State University, Corvallis, Oregon*

J. D. SMITH (453), *School of Molecular Sciences, University of Sussex, Brighton, England*

JOHN S. THAYER (1), *Department of Chemistry, University of Cincinnati, Cincinnati, Ohio*

D. R. M. WALTON (453), *School of Molecular Sciences, University of Sussex, Brighton, England*

Organometallic Chemistry: A Historical Perspective

JOHN S. THAYER

Department of Chemistry
University of Cincinnati
Cincinnati, Ohio

I

INTRODUCTION

A. Scope of Organometallic Chemistry

On encountering organometallic chemistry for the first time, one is likely to feel bewildered. What an array of compounds! What a variety of odd structures! What a range of reactions and applications! Above all, what energy and vigor researchers in this area show! How did it come about? How has organometallic chemistry evolved into such a thriving discipline? This article, although it does not cover the subject in detail, attempts to provide a historical survey that may help bring the present activity into perspective.

First of all, the very definition has varied. Even today not everybody is agreed as to what constitutes an organometallic compound. The most frequently heard definition is that an organometal is any species containing one or more metal-to-carbon bonds. All well and good, it would seem, until one realizes that boron, silicon, or arsenic are not true metals; yet their organic derivatives are very much a part of organometallic chemistry. Phosphorus and tellurium both have large numbers of organic compounds; are they organometals? Likewise, does *any* carbon atom bonded to a metal (however defined) yield an organometal? Most chemists would not consider $Fe(CN)_6^{2-}$ or $Hg(CNO)_2$ as organometals; yet they undoubtedly contain metal–carbon bonds. Binary metal–carbon compounds (the metal carbides) may or may not fall into this area, depending on the mood of the definer!

An early attempt to define the field appears on the opening page of the classic monograph by Krause and Von Grosse (*156*):

> In the category of "metal-organic or organo-metallic compounds" is included a series of versatile, reactive materials in which a metal is bonded directly through at least one primary valence to the carbon atom of a hydrocarbon radical. . . . Excluded by this definition are the numerous metal-containing organic compounds in which the organic portion is bonded to the metal through oxygen or another nonmetal such as nitrogen, phosphorus or sulfur. . . . Carbides, cyanides and metal carbonyls also do not belong to the class of organometallic compounds, because they contain no hydrocarbon radical. A completely sharp line, however, cannot be drawn between pure organic and metal-organic (such as Na_4C and $NaCH_3$) since there are also intermediate species such as Na_3CH and Na_2CH_2.

Some 35 years have passed since that definition was written. In the interim, there have been changes. Professor Gilman notes (*114a*):

> The more comprehensive definitions have grown to include metalloids and even what were considered only a few decades ago to be *non*metals. Furthermore, some are of the opinion that the so-called metals need not always be attached formally to carbon in order to have compounds with organometallic characteristics. An extreme case is the concept that carbon itself may partake of the characteristics of a metal; and on such a classification simple compounds such as ethane or benzene are in and of themselves, organometallic compounds.

Under all current definitions, the metal carbonyls (and their many derivatives), along with most metal carbides, have been incorporated into organometallic chemistry. In addition, such species as $Si(C_6F_5)_4$, $Mo(CNCH_3)_6$, and C_5Cl_5Tl are likewise organometals, even though the carbon atoms are not part of hydrocarbons in the strictest sense. Many of these changes reflect the great upheaval in the field since the preparation of ferrocene in 1951.

B. Nature of Organometallic Chemistry

Organometallic chemistry is the border area between the classic subdivisions of organic and inorganic chemistry. As such, it has benefited from, and contributed to, advances in both. But it has also suffered from an almost schizophrenic tendency toward division. Its several components originated separately and developed largely independently. Only within the last quarter-century have these components drawn together and organometallic chemistry become a coherent self-aware discipline. The four major components are (*a*) those compounds in which metal and carbon are linked only by σ-bonds—these are mostly derivatives of the representative elements; (*b*) the metal carbonyls and their derivatives; (*c*) those compounds in which unsaturated organic molecules are bonded to metals through electrons in π-bonds; (*d*) those compounds in categories *a–c* that show biological effects.

The dual nature of organometallic compounds was perhaps best expressed by C. A. Kraus (*154*):

> it is true that any compound containing carbon may properly be termed an organic compound, but it does not necessarily follow that all the reactions of such compounds are properly classed as organic reactions. . . . Organic chemists have long made use of the organic derivatives of the halogens and of oxygen, sulfur, nitrogen, etc., and most chemists consider that their reactions are typical of organic compounds. This, however, is not strictly true. All these compounds have an inorganic side, so to speak; they have properties that are primarily determined by their inorganic constituents and these properties, in many instances, are the very ones that render these compounds important. No sharp distinction can be drawn between the organic derivatives of most of these elements. The chemistry of the organic derivatives of oxygen or nitrogen, for example, cannot be differentiated from that of tin, germanium, or boron.

Thus the many reactions of organometallic compounds may involve organic group(s) only, inorganic ligand(s) only, the metal only, or the metal–carbon bond. Small wonder there is such diversity and such a vast literature!

Another feature of organometallic chemistry is that, perhaps more than any other area of chemistry, it has been dominated by a comparatively small number of men. It has also been strongly influenced by serendipity in that most of the major advances have come from experiments intended for quite different purposes. Once the initial report appears, there is a vast proliferation of research papers, frequently with exciting developments and quite unforeseen chemistry.

II
σ-BONDED ORGANOMETALS

A. Cacodyl

The first organometal to be reported was cacodyl [tetramethyldiarsine, $(CH_3)_4As_2$]. It was prepared almost as an afterthought by Cadet de Gassicourt, a Parisian military apothecary, in 1760 (*45*). Cadet was working on cobalt solutions for use as invisible inks. The two common ores of cobalt, smaltite and cobaltite, both contain arsenic, and arsenic trioxide was formed as a by-product. When he pyrolyzed this oxide with potassium acetate, Cadet got a red-brown liquid that fumed in air and gave off a terrible stench.

Other groups worked on "Cadet's fuming arsenical liquid" for three-quarters of a century after its initial report (*183*, *251*, *257*), but the definitive characterization came in a series of papers by Bunsen (*38–43*). He originally termed the liquid "alkarsin," but later it was named cacodyl from two Greek words meaning foul-smelling. Bunsen prepared numerous derivatives of cacodyl (*251*), including the cyanide, which he described as *diese schöne aber beispiellos giftige Verbindung*. He nearly died after tasting this compound—an experience which no doubt hastened the decline of tasting as a generally practiced analytical technique. Bunsen's research was done under primitive conditions and at considerable personal risk. Some experiments were done out of doors; during many indoors experiments, he was forced to breathe through long glass tubes reaching into the outside air. He lost part of the sight of one eye when a little cacodyl oxide exploded on a heated surface.

Cacodyl was later found to be tetramethyldiarsine (*251*), with the weakness of the arsenic–arsenic bond accounting for its high reactivity. But before this became known, cacodyl was believed to be a free radical; hence the "-yl" in the name. This belief contributed both to research on the theory of radicals, and, more importantly, to the further development of organometallic chemistry.

Cacodyl and its derivatives are now mostly of historical interest. One derivative, cacodylic acid [dimethylarsonic acid, $(CH_3)_2AsO_2H$], and its sodium salt, are commercially available and are widely used as herbicides.

B. Diethylzinc

The conversion of ammonium cyanate to urea by Wöhler (*202*) led to a rapid development of organic chemistry. Bunsen's work on cacodyl grew out of this and created its own line of research. The belief that cacodyl was an isolated free radical naturally implied that other such radicals might be isolable, and workers endeavored to confirm this idea. Bunsen himself continued research along this line for several years after his last paper on cacodyl, and trained a number of younger chemists in his laboratory at Marburg. Among these was the Englishman Edward Frankland. At Bunsen's suggestion, Frankland tried reacting ethyl iodide with zinc metal (*92*):

> The metal, either finely granulated or otherwise treated so as to expose a large surface, was then introduced and the open extremity of the tube was drawn out to the thickness of a straw; about an inch of this narrow tube was allowed to shrink up until a fine capillary bore was obtained. The narrow tube was then bent twice at right angles . . . the whole [tube] now being warmed, the open extremity was immersed in the iodide of ethyl, which, by the subsequent contraction of the enclosed air, was forced into the apparatus in the required quantity. . . . (After sealing the tube and heating to 150°) white crystals gradually encrust the zinc and glass, while a colorless, mobile liquid remains, equal in volume to only about half the iodide of ethyl employed and very different from that liquid in appearance . . . (after two hours) the reaction appearing to be complete, the tube was removed from the bath and allowed to cool. On, afterwards, breaking off its capillary extremity under water, about forty times its volume of gas was evolved, whilst the whole of the mobile fluid, above mentioned, disappeared. . . . On cutting off the upper portion of the tube and introducing distilled water, the white mass of crystals dissolved with brisk effervescence, occasioned by the evolution of a considerable quantity of a gas (similar to the above). . . . The solution of the crystals thus obtained possesses the properties of a solution of iodide of zinc and, with the exception of a trace of undecomposed iodide of ethyl, appeared to contain no organic substance.

This "mobile fluid" was diethylzinc, whereas the "white mass of crystals" contained both ethylzinc iodide and ZnI_2. Dimethylzinc was prepared in similar fashion (*93*):

> [Dimethylzinc] was subjected to distillation in an apparatus filled with dry hydrogen (!); a colorless, pellucid liquid, possessing a peculiarly penetrating and exceedingly nauseous odor, condensed in the receiver. This liquid spontaneously inflames on coming in contact with air or with oxygen, burning with a brilliant greenish-blue flame and forming dense clouds of oxide of zinc. . . . the vapour of this compound is highly poisonous, producing shortly after its incautious inhalation all the symptoms of poisoning by zinc. It decomposes water with as much violence as potassium. . . . the products of this decomposition are oxide of zinc and two equivalents of pure light carburretted hydrogen (CH_4).

A second description, equally graphic, further describes the hazards that Frankland and other chemists faced when working with such air-sensitive poisonous materials (*98*):

> Frankland says that when he first prepared zinc methyl and zinc ethyl in July 1849 in Bunsen's laboratory, he was trying the action of water on the residue in the tube used for preparing "methyl" from methyl iodide and zinc, when a "greenish-blue flame several feet long shot out of the tube . . . which diffused an abominable odor through the laboratory." Bunsen thought this was the smell of cacodyl, and told Frankland that he was "already irrecoverably poisoned." These forebodings were quelled in a few minutes by an examination of the black stain left upon porcelain by the flame, which proved to be zinc.

The isolation and characterization of these dialkylzinc compounds paved the way for the development of organometallic chemistry. Frankland and others quickly found that methyl and ethyl groups could be transferred from the zinc alkyls to other metals and metalloids. This will be discussed in the following section.

C. Organometals 1850–1900

During the years immediately following Frankland's report of diethylzinc, alkyl derivatives of many other metals were prepared. Table I gives a chronological account of how this occurred. Three synthetic methods received wide usage:

$$2CH_3I + 2Zn \longrightarrow (CH_3)_2Zn + ZnI_2 \qquad (1)$$

$$2SbCl_3 + 3(CH_3)_2Zn \longrightarrow 2(CH_3)_3Sb + 3ZnCl_2 \qquad (2)$$

$$(CH_3)_2Hg + Zn \longrightarrow (CH_3)_2Zn + Hg \qquad (3)$$

TABLE I

CHRONOLOGY OF METAL ALKYL REPORTS

Date	Metal	Compound	Worker	References
1840	Te	$(C_2H_5)_2Te$	F. Wöhler	*287*
1842	As	$(CH_3)_4As_2$	R. Bunsen[a]	38, 39
1849	Zn	$(C_2H_5)_2Zn$	E. Frankland	*92*
1850	Sb, Bi	$(C_2H_5)_3Sb$, $(C_2H_5)_3Bi$	C. Löwig	*170*
1852	Hg	$(CH_3)_2Hg$	E. Frankland	*94*
	Sn	$(C_2H_5)_4Sn$	E. Frankland[b]	*94*
1853	Pb	$(C_2H_5)_3PbI$	C. Löwig	*169*
1856	Cd	$(C_2H_5)_2Cd$	J. A. Wanklyn[c]	*272*
1857	P	$(CH_3)_3P$	A. Cahours[d]	*48*
1858	Na, K	C_2H_5Na, C_2H_5K	J. A. Wanklyn[e,f]	*273*
1860	B	$(CH_3)_3B$	E. Frankland	*99*
1863	Si	$(C_2H_5)_4Si$	C. Friedel and J. M. Crafts	*103, 104*
1864	Li	C_2H_5Li	E. Frankland[e]	*97*
1865	Al	$(C_2H_5)_3Al$	G. Buckton[c]	*36*
1866	Mg	$(C_2H_5)_2Mg$	J. A. Wanklyn[c]	*274*
1870	Tl	$(C_2H_5)_2TlI$	C. Hansen	*125*
1873	Be	$(C_2H_5)_2Be$	A. Cahours	*47*
1887	Ge	$(C_2H_5)_4Ge$	C. Winkler	*281*
1905	Ca	C_2H_5CaI	E. Beckman	*14*
1907	Au	$(C_2H_5)_2AuBr$	W. J. Pope	*210*
1909	Pt	$(CH_3)_3PtI$	W. J. Pope[g]	*211*
1914	Sr, Ba	C_2H_5SrI, C_2H_5BaI	G. Grüttner	*119*
1926	Rb, Cs	C_2H_5Rb, C_2H_5Cs	A. Von Grosse[e]	*269*
1932	Ga	$(C_2H_5)_3Ga$	L. M. Dennis	*72*
1934	In	$(CH_3)_3In$	L. M. Dennis	*73*

[a] First prepared in impure form by Cadet in 1760 (*45*).

[b] C. Löwig reported $(C_2H_5)_2Sn$ (probably impure) in 1852 (*168*).

[c] The reaction of C_2H_5I with Cd was first reported in 1853 by E. Schüler (*238*). Similar reactions with Mg and Al were reported in 1859 by Cahours (*46*) and by Hallwachs and Schafarik in 1859 (*123*). These latter authors also reported the preparation of $UO_2(C_2H_5)_2$.

[d] P. Thenard may have prepared methylphosphines in 1845 by the reaction of chloromethanes with calcium phosphide (*258, 259*). His father, L. J. Thenard, did some early research on Cadet's liquid (*257*).

[e] These compounds were not isolated but characterized by their reaction with diethylzinc. Pure organolithium (*230*) and organosodium (*232*) compounds were prepared by Schlenck and co-workers, while Gilman and Young prepared the first pure alkyl compounds of the heavier alkali metals (*114*).

[f] In 1855, Würtz prepared organosodium compounds as intermediates (*293*).

[g] Frankland reported in 1861 that $PtCl_4$ reacted with zinc alkyls, but no products were isolated (*95*).

These three methods, all developed by Frankland, remain the basis for most organometal syntheses to this day. In some cases, the direct reaction using metals such as Mg or Al actually gave decomposition products, usually hydrocarbons. Study of Table I reveals some interesting points:

1. Alkyl compounds were isolated first for the less active metals. Isolation of alkyl metals for elements such as Na or K required the development of chemical techniques; no derivatives of transition metals (except Pt and Au) could be reliably isolated, despite many attempts.

2. In most instances, the "peralkyls" were the first to be isolated. This arose primarily from their volatility, which enabled easy separation from the nonvolatile solid residues. Thallium is one exception; diethylthallium iodide was isolated in 1870, but the considerably less stable triethylthallium was not reported until 1930 (*118*).

3. In all cases, the first compounds reported were ethyl or methyl derivatives. Derivatives containing longer-chain alkyls usually proved harder to handle and in most cases were not synthesized until much later, if at all. Aryls came only after there had been some development in the chemistry of benzene; diphenylmercury was first reported in 1870 (*77*).

The rapid growth in the numbers and knowledge of organometallic compounds around 1850 enabled Frankland to use them in his formulation of the concept of valence; some of the ones he used are shown in Table II. From these and similar comparisons, he could make the following statement (*94*):

> When the formulae of inorganic chemical compounds are considered, even a superficial observer is struck with the general symmetry of their construction; the compounds of nitrogen, phosphorus, antimony, and arsenic especially exhibit the tendency of these elements to form compounds containing three or five equivalents of other elements, and it is in these proportions that their affinities are best satisfied. . . . Without offering any hypothesis regarding the cause of this symmetrical grouping of atoms, it is sufficiently evident . . . that such a tendency or law prevails, and that, no matter what the character of the uniting atoms may be, the combining power of the attracting element is always satisfied by the same number of these atoms.

This formulation was a significant step in the development of both inorganic and organic chemistry.

Twenty years after Frankland's initial paper, the chemistry of the alkyl–metal compounds had developed to such an extent that Mendeleev could use them in his formulation of the periodic law (*176*). He made the

TABLE II

COMPOUNDS CITED BY FRANKLAND IN THE STATEMENT ON VALENCE[a]

Frankland's formulation[b]		Current formulation	
Inorganic types	Organometallic derivatives	Inorganic	Organometallic
$As\left\{\begin{matrix}S\\S\end{matrix}\right\}$	$As\left\{\begin{matrix}C_2H_3\\C_2H_3\end{matrix}\right\}$ cacodyl	As_4S_4	$(CH_3)_4As_2$
$As\left\{\begin{matrix}O\\O\\O\end{matrix}\right\}$	$As\left\{\begin{matrix}C_2H_3\\C_2H_3\\O\end{matrix}\right\}$ oxide of cacodyl	As_4O_6	$(CH_3)_4As_2O$
$As\left\{\begin{matrix}O\\O\\O\\O\\O\end{matrix}\right\}$	$As\left\{\begin{matrix}C_2H_3\\C_2H_3\\O\\O\\O\end{matrix}\right\}$ cacodylic acid	As_4O_{10}	$(CH_3)_2AsO_2H$
$Zn\ O$	$Zn\ (C_2H_3)$ zincmethylium	ZnO	$(CH_3)_2Zn$
$Zn\left\{\begin{matrix}O\\O_x\end{matrix}\right\}$	$Zn\left\{\begin{matrix}C_2H_3\\O_x\end{matrix}\right\}$ oxide of Zincmethylium	ZnO	$(CH_3Zn)_2O$
$Sb\left\{\begin{matrix}O\\O\\O\end{matrix}\right\}$	$Sb\left\{\begin{matrix}C_4H_5\\C_4H_5\\C_4H_5\end{matrix}\right\}$ stibethine	Sb_2O_3	$(C_2H_5)_3Sb$
$Sb\left\{\begin{matrix}O\\O\\O\\O\\O\end{matrix}\right\}$	$Sb\left\{\begin{matrix}C_4H_5\\C_4H_5\\C_4H_5\\O\\O\end{matrix}\right\}$ binoxide of stibethine	Sb_2O_5	$(C_2H_5)_3SbO$
$Sb\left\{\begin{matrix}O\\O\\O\\O\\O\end{matrix}\right\}$	$Sb\left\{\begin{matrix}C_4H_5\\C_4H_5\\C_4H_5\\C_4H_5\\O\end{matrix}\right\}$ oxide of stibethylium	Sb_2O_5	$[(C_2H_5)_4Sb]_2O$
$Sn\ O$	$Sn\ (C_4H_5)$ stanethylium	SnO	$(C_2H_5)_2Sn$
$Sn\left\{\begin{matrix}O\\O\end{matrix}\right\}$	$Sn\left\{\begin{matrix}C_4H_5\\O\end{matrix}\right\}$ oxide of stanethylium	SnO_2	$(C_2H_5)_2SnO$
$Hg\left\{\begin{matrix}I\\I\end{matrix}\right\}$	$Hg\left\{\begin{matrix}C_2H_3\\I\end{matrix}\right\}$ Iodide of hydrargyromethylium	HgI_2	CH_3HgI

[a] References *94*, *253*.

[b] It should be remembered that Frankland used weights of 6 and 8 for C and O, respectively, in determining his formulas.

following specific statements involving organometals, namely about the known elements indium (Ae = ethyl), "Indium must be capable of forming, in the usual way, an isolable triethylindium, Ae_3In, since Ae_2Cd and Ae_4Sn exist; by analogy, this compound should boil at around 150°" [extr. b.p. 184° (*126*)], and cadmium, "Diethylcadmium is poorly known and worth investigation for many reasons. By analogy with the Zn and Hg compounds, this should boil at about 130° [observed: 64°/19.5 mm]. Research on Ae_3In and Ae_3Tl should throw new light on the regrettably poorly known Ae_3Al." He also noted that, regarding the then unknown element, eka-silicon (Es; now germanium), "A sharp difference between Es and Ti arises from the fact that Es, like Si and Sn, can form volatile organometallics such as Ae_4Es, while Ti, a member of an odd row of the system, forms no such compounds. By comparison to Sn and Si, the compound Ae_4Es should boil around 160° and have a density of 0.96" (observed: 163.5° and 0.99). More recent work indicates that germanium actually resembles silicon or tin less than Mendeleev had predicted (*226*), and in certain respects it is surprisingly similar to arsenic.

Tetraethyltitanium has still not been prepared. Ethyltitanium trichloride and tetramethyltitanium are known but are extremely unstable compounds (*64*). In fact, for over 50 years after Frankland's first paper, no stable σ-bonded organo derivatives of transition metals were isolated.

D. *Organometals and Organic Synthesis*

One result of Frankland's research was the realization that alkyl groups could be transferred *intact* from one atom to another. If the recipient atom were a metal, another organometallic formed; if the recipient were carbon or a nonmetal, a purely organic compound was the result. In 1855, Würtz (*293*) demonstrated this by reacting alkyl iodides with sodium metal to produce hydrocarbons. Frankland and his students performed a variety of syntheses using organozinc or organosodium species (*96*, *100*, *101*); Eq. (4) is typical of this work:

$$CH_3CO_2Et \xrightarrow{Na} NaCH_2CO_2Et \xrightarrow{EtI} CH_3CH_2CH_2CO_2Et \quad (4)$$

Butlerov reacted acyl halides with diethylzine to form tertiary alcohols (*44*)

$$CH_3C(:O)Cl + (CH_3)_2Zn \xrightarrow{H_2O} (CH_3)_3COH + Zn(OH)Cl \quad (5)$$

This reaction would usually be run in two stages: addition of the zinc alkyl followed by hydrolysis. Later workers found that it was unnecessary to isolate the zinc alkyl. Reaction could be carried out simply by adding the metal (zinc) and the alkyl halide to the organic compound being alkylated. The Reformatsky and Saytzeff reactions are examples of this approach. Barbier provided a major advance when he substituted magnesium for zinc in the synthesis of 2,6-dimethylhept-5-ene-2-ol (*12*):

$$(CH_3)_2C{=}CHCH_2\overset{\overset{\displaystyle O}{\|}}{C}CH_3 + CH_3I + Mg \xrightarrow{H_2O} (CH_3)_2C{=}CHCH_2\overset{\overset{\displaystyle OH}{|}}{C}(CH_3)_2 \quad (6)$$

This reaction would not proceed in the presence of zinc but went smoothly with magnesium. Barbier gave this problem to his student Victor Grignard.

The work and contribution of Grignard have been written up in exhaustive detail (*146*, *166*, *221*, *229*). The Grignard reaction is usually the first organometallic synthesis that a chemistry student encounters. (Table III lists some well-known syntheses involving organometals.) It is one of the most commonly used synthetic preparations. Grignard's first record of this reaction reads as follows (*166*):

> I have ascertained that, while heating Mg turnings with isobutyl iodide, the Mg is attacked very rapidly. Moreover, if, as soon as the attack has begun, anhydrous ether is introduced, the reaction continues very vigorously and Mg dissolves rapidly, simultaneously forming a brown deposit in very small amounts. If the iodide is replaced by the bromide, the reaction takes longer to begin but it is still quite vigorous and proceeds very rapidly in anhydrous ether. Diisobutylmagnesium has not been reported, and, for the moment, I am not bothering to isolate it, since, it does not seem necessary for my purpose.

TABLE III

Some Organic Syntheses Involving Organometals (*249*)

Barbier–Wieland degradation
Boudroux–Chichibabin synthesis
Bouveault aldehyde synthesis
Darzens synthesis
Friedel–Crafts Reaction
Gatterman–Koch reaction
Grignard reaction
Reformatsky reaction
Saytzeff reaction
Ullmann coupling reaction
Würtz coupling reaction
Würtz–Fittig reaction

Grignard's first paper on this reaction appeared in 1900, serving as a very appropriate opening for the new century (*116*). In it he stated:

> Methyl iodide attacks magnesium turnings only very slowly when cold, but if a little anhydrous ether is added, reaction occurs immediately and rapidly becomes very vigorous. It then becomes necessary to cool [the reaction vessel] and add excess ether. Under these conditions the dissolution of the magnesium proceeds rapidly to give a very fluid, slightly colored liquid with no appreciable deposit. If the ether is evaporated off, there remains a grayish mass, confusedly crystalline, which absorbs moisture very rapidly, gives off heat and falls into deliquescence. But the great advantage of this compound is that it is unnecessary to isolate it.

Since the Grignard reagent forms in ether solution, it can be reacted as a solution and, if desired, removed from the vessel in which it was formed. Furthermore, the solution is easier to handle than the air-inflammable zinc alkyls. A second major advantage is that the magnesium–carbon bond is more reactive than the zinc–carbon bond, making it a better alkyl transfer agent. Grignard concluded his first paper saying that "I am continuing the investigation into the application of these new organometallic compounds." And so he did—along with many others. Within 5 years, 200 additional papers had appeared. This number swelled to over 500 by 1908, and in excess of 6000 by the time of Grignard's death in 1935. Today, the number would likely be around 40,000 or more. For his work, Grignard shared the Nobel prize in chemistry for 1912 with Paul Sabatier.

E. σ-Bonded Organometals 1900–1950

The half-century after the discovery of the Grignard reagent proved a rich and rewarding one for organometallic chemistry. A number of chemists who worked during that period have written articles for *Advances in Organometallic Chemistry* (*28*, *111*, *133*, *187*, *226*, *298*). The availability of the Grignard reagent enabled a considerable expansion in the number of known organometallics, including the first stable alkyl derivatives of gold and platinum (see Table I). With the synthesis of $(CH_3)_3In$ (*73*), every nonradioactive main-group element was now known to form organo derivatives. Gradually the emphasis shifted from synthesis to reactions, structure, and bonding. Development turned out to be quite uneven. Organic compounds of magnesium received extensive research because of their synthetic utility;

organic compounds of arsenic, antimony, and mercury likewise became well-known because of medicinal applications (see Section VI). The first monographs on organoarsenic (*16*, *183*, *215*), antimony (*61*, *183*), magnesium (*229*, *234*), and mercury (*278*) were published. Organosilicon chemistry advanced substantially through the work of F. S. Kipping (*54*, *184*), and organolead chemistry was spurred by the development of gasoline additives (see below). However, unless some major use or application developed, organometallic compounds remained largely laboratory curiosities.

The continuing use of the Grignard reagent, for synthetic purposes, encouraged research on related compounds. Although alkyllithium compounds had been isolated by Schlenck and Holtz (*232*), they became important for synthesis only after an easy preparative method became available (*299*). Thereafter, they grew in importance and now run the Grignard reagent a close second in the number of syntheses for which they can be used. The work of Ziegler (*298*) and Gilman (*111*) also led to the development of organosodium and organopotassium compounds for synthetic use. Subsequent development of organocadmium species for ketone synthesis (*50*, *112*) gave chemists a greater range of activity. However, Grignard reagents remained the first choice. If greater reactivity was needed, organolithium compounds were available, or, in more extreme cases, organosodium compounds. For those reactions in which the Grignard reagent is too reactive, the alkyls of aluminum, zinc, or cadmium may be used. In principle, any organometal can exchange its organic group, and almost all of them do. During this period also a variety of reactions was studied in which the metal of an organolithium or organosodium compound undergoes exchange ("metalation") with hydrogen or halogen to form a new organometal (*111*).

Three σ-bonded organometals found widespread nonmedicinal commercial uses. These, and later developments, are covered in a collection of papers published by the American Chemical Society (*177*). First, in 1923, Thomas Midgeley and associates (*106*, *188*) developed the use of tetraethyllead as an additive to gasoline in order to control fuel combustion. The great growth in the number of automobiles and similar vehicles since that time has likewise led to vast expansion in the use of tetraethyllead as well as tetramethyllead. In recent years, these additives have come under heavy criticism as a source of lead pollution.

Then followed the development and widespread usage of silicon polymers. Although organosiloxy polymers had been known from the work of Friedel and Crafts and Kipping (*105*, *149*, *184*), they did not become commercially important until a convenient, large-scale preparative method for the precursor organohalosilanes appeared. The development of the direct process by Rochow and others (*223*, *226*) in the early 1940s opened up this area to commercial development, and the silicones are now used in a wide variety of applications that take advantage of their stability (both thermal and chemical) and impermeability to water. The direct process looks simple enough on paper:

$$2CH_3Cl + Si \xrightarrow[\text{Cu catalyst}]{300^\circ} (CH_3)_2SiCl_2 \qquad (7)$$

However, the catalyst must be prepared in a certain way, and the reaction actually yields many products containing one or more silicon atoms.

Finally, there is also the development for commercial use of aluminum alkyls as polymer catalysts (*298*, *300*). This arose initially from the observation that organo compounds of active metals, such as Li or K would add across olefinic double bonds repeatedly:

$$C_2H_5Li + CH_2{=}CH_2 \longrightarrow C_4H_9Li \xrightarrow{C_2H_4} C_6H_{13}Li \xrightarrow{\text{etc.}} C_nH_{2n+1}Li \qquad (8)$$

Long research into this reaction resulted in Al catalysts, plus certain other metals, that enabled the preparation of polyethylene and similar hydrocarbon polymers. Although the full commercial development of these processes did not come until after 1950, the groundwork was prepared earlier. K. Ziegler and G. Natta shared the 1963 Nobel Prize in chemistry for their work in this area (*185*, *297*).

The comparative recentness of the industrial development of organometallic chemistry may be illustrated by the following quotation (*110*, p. 577):

> Undoubtedly the greatest value of organometallic compounds is their laboratory use for synthesis. The most important technical application is that of tetraethyllead as an antiknock compound. It is doubtful that any other group of organic compounds combines at the same time an astonishingly high utility in the laboratory with an equally low usefulness in the works. With increasing reductions in the cost of metals and their salts and with newer solvents or with techniques to reduce the need of solvents, it is rather likely that industrial uses for organometallic compounds will expand.

A further indication of how things have changed since the preceding passage appeared in the 1940s can be seen in the recent statement by Gilman (*114a*):

> There was a time not long ago when the greatest value of organometallic compounds was their laboratory use for synthesis. The industrial or plant applications were, by contrast, almost extraordinarily minimal. Gradually this has undergone a marked change so that today the industrial applications are very pronounced. This applies not only to the organometallic compounds themselves, but also to "captive" uses in many operations, and especially to their applications as important and indispensable synthetic agents on scales never dreamed of about four decades ago. The huge quantities of such materials used in both batch and continuous processes is a rich compliment to those in industry who have developed techniques for the preparation, transport and handling, almost routinely, of such compounds considered heretofore as quite sensitive reagents.

Also during this period, the role of organometals in the formation of free radicals came under study. Frankland's original intention in reacting ethyl iodide with zinc was to produce free "ethyl" (*92*). The first evidence for the free methyl and ethyl radicals came from the work of Paneth and associates (*196–200*). They thermally decomposed tetramethyl (or -ethyl)lead and passed the vapors over mirrors of elements such as arsenic, zinc, and tellurium. By this method, they could isolate the methyl derivatives. For arsenic, antimony, and bismuth the product obtained would depend on whether or not the mirror itself was heated (*200*):

$$3CH_3\cdot + M \longrightarrow (CH_3)_3M \quad \text{cold mirror} \tag{9}$$
$$4CH_3\cdot + M \longrightarrow (CH_3)_4M_2 \quad \text{heated mirror} \tag{10}$$

In a similar manner, Pearson and Purcell (*205*) could obtain dimethyltellurium from photolysis of acetone if the Te mirror were not heated, but dimethylditellurium formed from a hot mirror. The Paneth technique can be used as a test for the presence of organic free radicals in the vapor phase. Other early work on organometals and free radicals is given in a book by Waters (*275*).

Comparable in significance, Gilman states (*114a*):

> is the classical and almost phenomenal intentional use of RM compounds for the study of the first rigorously established trivalent carbon compounds; then, later, the carbenes; and, finally, atomic carbon and other elements. Such free radical studies of "anomalously valenced" elements has been extended from carbon to a wide spectrum of elements (both metalloidal and metallic). Hand in hand with this has been the continuous and sound and fundamental studies on types of bonding, as well as atomic and other structural developments.

Moreover, during this period, there also appeared considerable unheralded but indispensable research that laid a partial foundation for the spectacular advances after 1950. Gilman notes further (*114a*):

> Among important contributions which helped not only lay the groundwork for later research but also provided complementary assistance in a variety of studies were:
>
> (1) determination of optimal conditions for the preparation of fundamental reagents such as RMgX, RLi, etc.
> (2) studies on the "permanence" or keeping qualities of these compounds.
> (3) an examination of the effects of various solvents or no solvent.
> (4) qualitative tests to help establish when an RM reagent was formed, and when a reaction was completed.
> (5) quantitative analyses of many important RM compounds.
> (6) studies on the relative reactivities of a wide variety of organometals (particularly valuable in establishing selective and preferential reactions, especially in polyfunctional substrates).
>
> It is appropriate to mention or to emphasize that these invaluable "groundwork studies" were carried out successfully without the aid of latterly developed and now indispensable instrumental techniques, such as chromatography and various spectrographic measurements.

Thus, by 1950, the σ-bonded organo derivatives of metals had an extensive, thriving, and growing chemistry. Many further developments followed, but these were strongly influenced by other areas of organometallic chemistry, which we shall now consider.

III

METAL CARBONYLS

A. Nickel Tetracarbonyl

By the latter part of the nineteenth century, the Solvay process for the production of sodium bicarbonate had become very important commercially. Ammonium chloride formed as a by-product in this reaction, and chemists sought ways to convert it to chlorine on a commercially feasible scale. The reaction vessels contained valves made of nickel, which corroded severely during many of these reactions. Mond carried out a detailed study on this corrosion, and found that carbon monoxide was the active agent (*180*):

> when a finely-divided nickel, such as is obtained by reducing nickel oxide by hydrogen at about 400°, is allowed to cool in a slow current of carbon monoxide, this

gas is very readily absorbed as soon as the temperature has descended to about 100°, and if the current of carbon monoxide is continued, or if this gas is replaced by a current of an inert gas (such as carbon dioxide, nitrogen, hydrogen, or even air) a mixture of gases is obtained, which contains upwards of 30 percent of nickel–carbon–oxide. . . . When these mixtures of gases are heated above 150°, their volume increases and nickel separates, which, according to the temperature, is more or less contaminated with carbon resulting from the action of the nickel upon the carbon monoxide generated. . . . Numerous experiments made to obtain similar compounds of carbon monoxide with other metals, notably with cobalt, iron, copper, and platinum have only led to negative results, although they were carried out at temperatures from 15° to 750°.

The reaction, in equation form, is

$$Ni(s) + 4CO(g) \longrightarrow Ni(CO)_4(g) \tag{11}$$

Despite Mond's pessimistic closing sentence, research continued on these compounds. In 1891, Berthelot (*18*) and Mond and Quincke (*182*) almost simultaneously synthesized $Fe(CO)_5$. Later that same year Mond and Langer reported $Fe_2(CO)_9$, although they gave an incorrect formulation (*181*). Some years later Dewar reported some reactions of nickel tetracarbonyl with inorganic (*74*) and organic (*75*) compounds, and corrected the formulation for $Fe_2(CO)_9$, as well as reporting $Fe_3(CO)_{12}$. Liebig had reacted potassium with CO as early as 1834 (*164*), getting a compound that he formulated as $(KCO)_n$. Later work showed this to be $K_6C_6O_6$, the potassium derivative of hexahydroxybenzene (*189*). More recently, the salts NiC_4O_4 and FeC_5O_5, isomeric to the carbonyls, have been reported (*276*). Other early work on the metal carbonyls and their derivatives is reported in a series of review articles by Trout (*263–266*).

B. *Developments in the Metal Carbonyls to 1950*

A substantial amount of work was done on metal carbonyls during the first half of the twentieth century. Derivatives containing ligands began to be reported. As a matter of fact, the platinum carbonyl chlorides were originally reported in 1868 (*239*). Table IV shows some of these arranged in chronological order. Numerous reviews on the carbonyls have appeared (*1*, *6*, *148*, *209*, *263–266*), to which the interested reader is referred for details. Extensive work on the carbonyls began about 1930 with Hieber and his group (*133*).

One point of great interest was the structure and bonding in these compounds. The effective atomic number rule, first formulated by

TABLE IV

TYPICAL METAL CARBONYLS AND DERIVATIVES (TO 1950)

Year	Compound	References
1868	$(PtCl_2CO)_2$	*239*
1890	$Ni(CO)_4$	*180*
1891	$Fe(CO)_5$	*18, 181*
1905	$Fe_2(CO)_9$, $Fe_3(CO)_{12}$	*76*
1930	$C_4H_6Fe(CO)_3$	*218*
1931	$H_2Fe(CO)_4$	*134*
1932	$NaHFe(CO)_4$	*85*
	$Fe(CO)_2(NO)_2$	*4*

Sidgwick (*245*), proved an extremely useful formalism for considering structural possibilities of metal carbonyls. Langmuir suggested that there might be double bonding between Ni and C in $Ni(CO)_4$ (*161*). Brockway and co-workers (*23–25*, *84*) carried out electron diffraction studies on various metal carbonyls and established that the metal–carbon bond was indeed shorter than expected for a purely single bond. They proposed resonance between two forms:

$$\mathrm{Ni{-}C{\equiv}O} \longleftrightarrow \mathrm{Ni{=}C{=}O}$$

Detailed pictures of the bonding in carbonyls are given in the books of Orgel (*194*), Pauling (*203*), and Zeiss (*59*, *222*). These ideas became fully developed only after 1950.

During this period, metal carbonyls first began to be used as catalysts. The Fischer–Tropsch synthesis was developed around 1925 (*21*),

$$n\mathrm{CO} + (2n + 1)\mathrm{H_2} \longrightarrow \mathrm{C}_n\mathrm{H}_{2n+2} + n\mathrm{H_2O} \tag{12}$$

but the role of metal carbonyls as catalytic intermediates was not recognized until later. The work of Reppe showed that $Ni(CO)_4$ could be used to generate carbonyl compounds from acetylenes (*66*, *219*, *220*). From this work would grow the great industrial use of the "oxo" reaction and related catalytic applications. For a while, $Fe(CO)_5$ was considered as a gasoline additive, but finally yielded to tetraethyllead (*188*).

IV

METAL COMPLEXES WITH UNSATURATED HYDROCARBONS

A. Zeise's Salt

The first olefin–metal complex to be isolated had the formula $K^+[C_2H_4PtCl_3]^-$ (*252*). This compound, reported in 1827 by W. C. Zeise, was prepared by heating $PtCl_2$–$PtCl_4$ with ethyl alcohol, evaporating, and treating the residue with aqueous KCl (*294*). The analogous salt $NH_4{}^+C_2H_4PtCl_3{}^-$ could be made by using NH_4Cl.

Zeise had formulated this salt as a compound of ethylene, namely $KCl \cdot PtCl_2 \cdot H_2O$. Liebig disagreed, considering it rather to contain an ethoxide group (*165*, *252*). Since the compound usually formed as a monohydrate and since the water of hydration could be removed fairly easily, analytical results could be used to justify either viewpoint. Zeise's view prevailed, and was verified by later work. Zeise's last paper (*295*), by an ironical juxtaposition, appeared right next to Bunsen's first paper on cacodyl (*39*)!

In 1868, Birnbaum prepared Zeise's salt directly by passing ethylene into H_2PtCl_6 (*22*). That same year $PtCl_2CO$ dimer was also reported (*239*). Birnbaum also prepared analogous compounds containing propylene and amylene. Two years later the bromo analog was reported (*60*). Similar complexes of copper (*19*), palladium (*147*), and iron (*218*) were also reported. The first complex of a metal with a nonaromatic polyene was $C_{10}H_{12}PtCl_2$, made from dicyclopentadiene and K_2PtCl_4 (*136*).

Anderson showed that ethylene could be displaced from Zeise's salt by various other ligands (*5*). However, the nature of the olefin–metal linkage was puzzling. Winstein and Lucas (*282*) proposed that the linkage in olefin–silver complexes involved resonance between three canonical forms:

$$\underset{Ag^+}{>C{=}C<} \longleftrightarrow \underset{\diagdown Ag}{>C{-}\overset{+}{C}<} \longleftrightarrow \underset{Ag \diagup}{>\overset{+}{C}{-}C<}$$

Complete understanding of bonding in these systems would not be developed until after the discovery of ferrocene.

B. Aromatic Complexes before 1950

The first complexes to be reported of an aromatic compound and a metal involved benzene and chromium. These were studied by Hein (*127–129*) over a period of almost 20 years and for a long time were believed to be σ-bonded phenylchromium compounds. Hein reacted $CrCl_3$ with phenylmagnesium bromide (*127*) and obtained a species formulated as $(C_6H_5)_nCrX$ (n = 3,4,5). These species could be reduced electrolytically to the neutral species $(C_6H_5)_nCr$. Attempts to extend this reaction to derivatives of benzene gave differing results; for example, p-BrC_6H_5MgBr and $CrCl_3$ gave compounds such as $(BrC_6H_4C_6H_4C_6H_4)_5CrBr$ and $(BrC_6H_4C_6H_4C_6H_4C_6H_4)_2CrOH$ (*131*). On the basis of magnetic moments, Klemm and Neuber (*152*) proposed that the oxidation state of the chromium was the same in all these compounds and that there were biphenyl linkages present.

Attempts to prepare aromatic derivatives of other transition metals generally failed; a typical reaction would be

$$MX_2 + 2ArMgX \longrightarrow M + Ar—Ar + 2MgX_2 \qquad (13)$$

Only the elements of Group IB gave isolable phenyl derivatives: C_6H_5Cu (*217*), $(C_6H_5Ag)_2 \cdot AgNO_3$ (*113, 157, 217*), $(C_6H_5)_nAuCl_{3-n}$ (n = 1,2) (*145*). These are σ-bonded compounds.

Silver salts had unusually and unexpectedly high solubility in aromatic solvents such as benzene. Andrews and Keefer (*7, 8*) found that a complex actually formed, and proposed an interaction between metal and hydrocarbon, similar to that proposed by Winstein and Lucas for olefins.

V

FERROCENE

In 1951 the compound bis(π-cyclopentadienyl)iron(II) was announced, and organometallic chemistry has not been the same since! This compound was prepared independently by two groups (*144, 178*). As happened with $Fe(CO)_5$, the reports on ferrocene came virtually simultaneously: Kealy and Pauson's paper was received August 7, 1951, while the paper by Miller *et al.* was received July 11, 1951. Both groups used synthetic methods well established by prior workers. Kealy and Pauson used the Grignard reagent:

$$FeCl_3 + C_5H_5MgBr \longrightarrow (C_5H_5)_2Fe \quad (14)$$

Cyclopentadienylmagnesium bromide has been known since 1914 (*117*) and potassium cyclopentadienide since 1901 (*260*). Considering that many attempts had been made to prepare organoiron compounds in this manner, it is surprising that ferrocene was not reported considerably earlier.

Miller *et al.* (*178*) reacted specially reduced iron metal with cyclopentadiene at 300° and atmospheric pressure:

$$Fe + 2C_5H_6 \longrightarrow (C_5H_5)_2Fe + H_2 \quad (15)$$

Both groups noted the unexpectedly great stability of this compound to heat and to air. They both proposed the σ-bonded structure shown in Fig. 1. The actual structure was first proposed by Wilkinson *et al.* (*280*) and confirmed by Fischer and Pfab (*90*), who also reported the existence of $(C_5H_5)_2Co^+$. X-Ray diffraction studies on ferrocene were also done by Dunitz and Orgel (*79*) and Eiland and Pepinsky (*82*). These three papers were received by the respective journals on June 20, July 5, and August 12, 1952! For their work in this area, Fischer and Wilkinson shared the 1973 Nobel Prize in chemistry (*239a*).

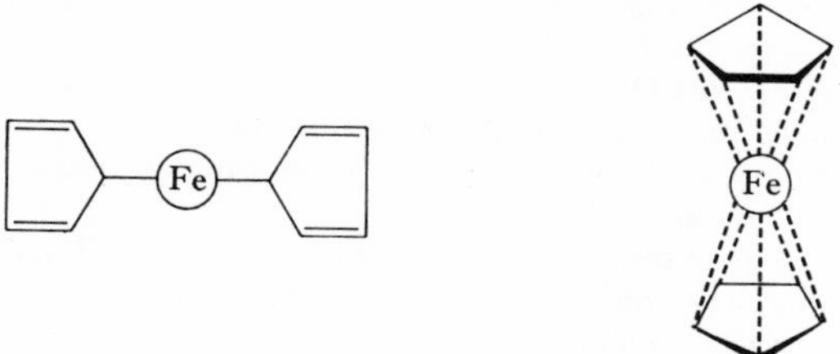

FIG. 1. Structures for ferrocene.

Also in 1952, Woodward *et al.* (*291*) reported that ferrocene would undergo a number of substitution reactions typical of aromatic compounds and that the rings had overall electrical neutrality—a fact that would have to be considered in drawing up any bonding picture. Finally, they proposed the name "ferrocene" itself. This has now become the commonly used term for $(C_5H_5)_2Fe$ and has spawned other names such as metallocene and cobaltocene.

VI

BIOLOGICAL EFFECTS OF ORGANOMETALS

A. Early Work

Ever since they were first reported, organometallic compounds have been known to have biological effects. The experiences of Frankland and Bunsen have already been mentioned. Two workers were poisoned, both fatally, by dimethylmercury in 1866 (*115*). Industrial development of metal carbonyls and tetraethyllead caused a number of deaths (*188*, *265*). In general, organoarsenic compounds have been most investigated for their toxic effects, with organomercury compounds a close second. Table V outlines the chronology of these developments, and Fig. 2 shows some representative compounds studied.

TABLE V

CHRONOLOGY OF ORGANOMETALLIC BIOCHEMISTRY

1760:	Cadet prepares cacodyl solution, notes biological effects.
1837:	Bunsen's first report on cacodyl. Over the next several years, he publishes several papers, including notes on the deleterious properties. About this time "arsenic rooms" began receiving the attention of Gmelin and others.
1849:	Frankland prepares $(CH_3)_2Zn$ and $(C_2H_5)_2Zn$. He observes that inhalation causes zinc poisoning.
1858:	Buckton notes irritating effects of alkyltin compounds on nasal passages.
1866:	First reported fatality from dimethylmercury poisoning.
1890:	Mond prepares $Ni(CO)_4$, which leads to the development of industrial uses of metal carbonyls and concomitant health problems.
1891:	Gosio determines that the arsenic of "arsenic rooms" exists in the form of an alkylarsenic compound.
1894:	Hofmeister suggests that "bismuth breath" is actually $(CH_3)_2Te$.
1908.	Ehrlich begins development of Salvarsan. Over the next several years his work in this area lays the foundation of chemotherapy and greatly extends the known chemistry of organo compounds of arsenic, antimony, and bismuth.
1914:	World War I begins. This conflict involves the use of Lewisite and other organoarsenicals as poison gases.
1923:	Development of tetraethyllead as gasoline additive. Its toxicity causes special handling measures to be developed.
1933:	Challenger reports generation of $(CH_3)_3As$ and $(CH_3)_2Te$ by molds. Subsequent work leads to formulation of "biological methylation."
1954:	Stalinon disaster in France. About the same time the first cases of Minimata disease (methylmercury poisoning) appear.
1961:	Coenzyme form of vitamin B_{12} containing Co—C bond is reported.
1967:	Discovery that methylmercury compounds can be generated by microorganisms from inorganic mercury by biological methylation.

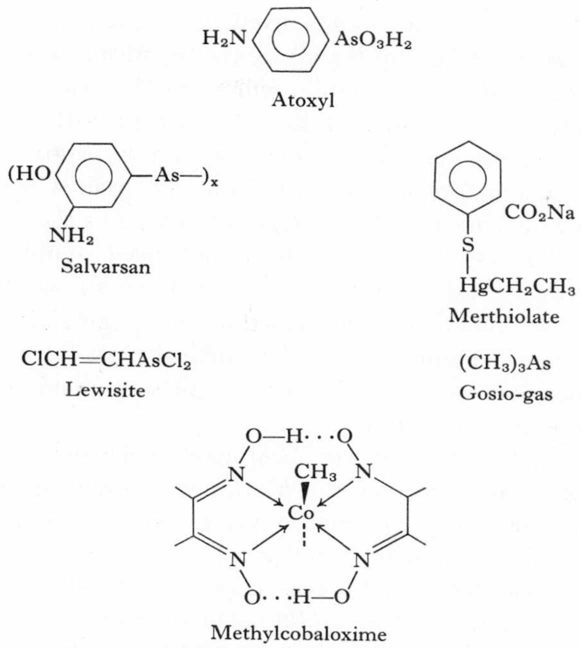

H_2N AsO_3H_2

Atoxyl

(HO —As—)$_x$

NH_2

Salvarsan

CO_2Na

S

$HgCH_2CH_3$

Merthiolate

$ClCH{=}CHAsCl_2$

Lewisite

$(CH_3)_3As$

Gosio-gas

O—H· · ·O

N CH_3 N

Co

N N

O· · ·H—O

Methylcobaloxime

FIG. 2. **Some biologically active organometals.**

Systematic investigation on the biological interactions of organometals began with the research of Paul Ehrlich (*135*), although some organomercury compounds had previously been used in the treatment of syphilis and various infectious diseases. Ehrlich laid the groundwork, both theoretical and practical, of chemotherapy. He found that the structure of Atoxyl contained a C—As bond rather than being C_6H_5-$NHAsO_3H_2$ as originally formulated. Atoxyl was very effective against various microorganisms; unfortunately, it was also very toxic. Ehrlich studied hundreds of derivatives or analogs of Atoxyl, and found that Salvarsan combined the highest effectiveness against microorganisms with the lowest toxicity toward humans. This is represented quantitatively in the *chemotherapeutic index*:

$$\text{chemotherapeutic index} = \frac{\text{maximum tolerated dose}}{\text{minimum effective dose}}$$

Ehrlich's research yielded compounds where this index was 5–10. Currently, the best drugs have values of several hundred.

Ehrlich's basic principle was *corpora non agunt nisi fixata* ("bodies i.e., molecules, do not act unless fixed"). A compound will not be effective unless it is taken up (fixed) by the organism to be acted upon. Minor modifications in composition will affect the degree to which the organometal will be taken up and, thus, the effect it will have. Salvarsan was effective against spirochetes (syphilis) and trypanosomes (African sleeping sickness and related ailments) but not against bacteria. Neosalvarsan (in which one amino hydrogen is replaced by the $—CH_2SO_3Na$ group) proved more effective due to its enhanced solubility in water. For many years these compounds had widespread usage. Although displaced by the sulfa drugs and penicillin, Salvarsan and Neosalvarsan even today retain certain specialized medicinal uses. Their importance led to an enormous amount of research into the organic chemistry of arsenic, antimony, and bismuth.

Organomercury compounds are the other class of widely used medicinal organometals. Mercurochrome and Merthiolate, originally developed for treating gonorrhea, now commonly serve as external antiseptics. Other organomercurials are used as diuretics.

Organoarsenicals were used as poison gases in World War I. The most notorious of these was Lewisite (*140*). During the years before World War II, researchers looked for antidotes and found the compound 2,3-dimercaptopropanol, now commonly known as British Anti-Lewisite or BAL (*190*). This compound has proven extremely effective in the treatment of lead and mercury poisoning.

B. Biological Methylation

During much of the last century, people were poisoned by arsenic not through any deliberate administration but because they lived in certain rooms. These "arsenic rooms" had three common factors: wallpaper using arsenic compounds for coloring; molds growing on them, enhanced by dampness; and an unpleasant smell. In 1839, Gmelin suggested that the smell was a volatile arsenic compound which caused the As poisoning, and, in 1874, Selmi proposed that this volatile compound was AsH_3 arising from the action of the mold. In 1891, Gosio began a study of the interaction of molds on media containing arsenious oxide. He found that some of them, including *Penicillium brevicaule* (now *Scopulariopsis brevicaulis*), generated a gas with a garliclike odor, even when only 1 ppm of arsenic was present. Gosio only partially characterized this gas. His student Biginelli found that this gas (now called

Gosio gas) formed a precipitate when passed into aqueous mercuric chloride. Both Gosio and Biginelli believed the gas to be $(C_2H_5)_2AsH$.

Definitive work on Gosio gas came from Challenger and his associates (*52, 53, 55, 56*). They identified the emitted gas as being solely trimethylarsine and found that it could be generated from a variety of arsenic compounds, including methylarsonic acid and cacodylic acid. They also found that compounds such as ethylarsonic acid gained methyl groups only, forming e.g., $C_2H_5As(CH_3)_2$, and that selenium or tellurium would also be converted to the volatile, malodorous dimethyl compounds. Mechanistic studies using ^{14}C labeling showed that methionine served as the primary source of methyl groups (*55, 57*). Challenger referred to this process as "biological methylation."

In 1948 researchers isolated a red crystalline solid from liver extracts; this solid could be used to treat pernicious anemia and became known as vitamin B_{12}. The full name of this compound is α-5,6-dimethylbenzimidazolylcyanocobamide, but it is usually referred to as cyanocobalamin. Vitamin B_{12} exists in a number of forms and has an extensive chemistry (*213*). One of these has a 5′-deoxyadenosyl group at the sixth coordination site of cobalt; the structure of this coenzyme was established by crystallography in 1961 (*162*) (Fig. 3). It was the first compound containing a metal–carbon bond to be found as a natural material! During biological methylation, the deoxyadenosyl group drops off, and the cobalt atom serves as a transfer agent. The intermediate compound, methylcobalamin, will serve to methylate a large number of species, and the mechanism of its action has been the subject of numerous investigations (*2, 37, 174, 247, 290*), especially in conjunction with methylmercury (see Section VI, C).

The complexity of vitamin B_{12} led to the preparation and investigation of model compounds, of which derivatives of bis(dimethylglyoximato)-cobalt(III), termed cobaloximes, have been the most extensively studied (*29, 235–237*). These derivatives, especially methylaquocobaloxime, have helped cast considerable light on the mechanisms involved in the biological action of vitamin B_{12}.

C. Methylmercury

All too often the interaction of organometals with humans has caused suffering and death. Many people died from the "arsenic rooms" before the cause was discovered. The use of a diethyltin diiodide preparation

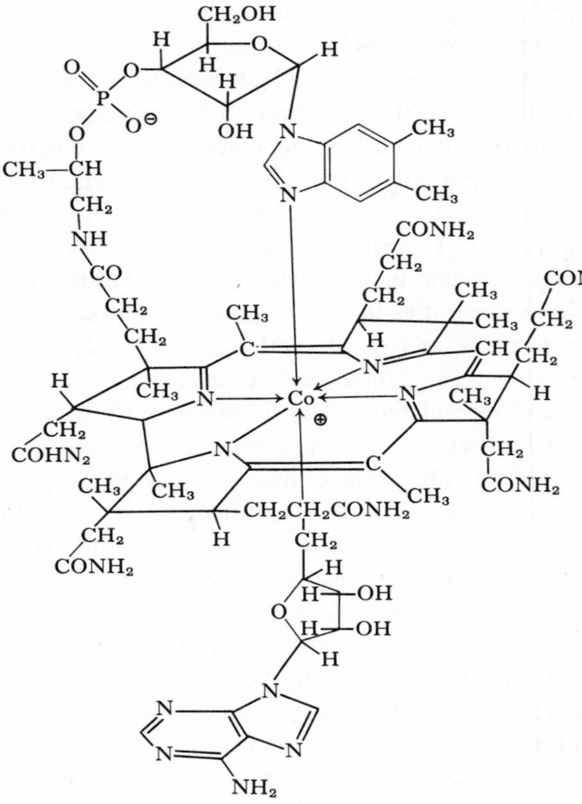

FIG. 3. Coenzyme form of vitamin B_{12}.

(Stalinon) in France in 1954 caused 102 deaths and left an equal number permanently injured (*207*). However, the most recent and most widespread of these tragedies is that caused by methylmercuric compounds (usually collectively termed methylmercury).

Organomercurials have been used as seed coatings to protect against mold. Methylmercuric dicyandiamide (Panogen) is one such preparation. Although the dangers in handling such preparations had long been recognized (*138*), their connection with dying from mercury poisoning in Sweden and the nervous affliction of people in the vicinity of Minimata Bay, Japan, was not immediately recognized. Both arose from methylmercury poisoning, as given in the report by Nelson (*186*).

Initially it seemed as if the methylmercury were coming solely from seed dressing (Sweden) or factory effluent (Japan). Yet while the use of Panogen and similar fungicides was discontinued, the level of methylmercury in organisms remained high. Furthermore, Westöö (*277*) found that methylmercury was the *only* form of organomercurial found in animal tissue. In a now classic paper, Jensen and Jernelöv (*142*) found that mercury metal and its inorganic compounds could be converted to CH_3Hg^+ or $(CH_3)_2Hg$ through the process of biological methylation by bacteria dwelling in the mud at the bottom of lakes. This work was confirmed and expanded by Wood *et al.* (*289*). In the years since these two papers, a substantial literature on all aspects of methylmercuric ion in biochemistry has appeared (*70*, *81*, *102*, *124*, *151*, *186*, *288*). More than anything else, the reports of the tragic deaths from methylmercury poisonings and the even more tragic cases of prenatal poisonings *in utero* (*11*, *186*) have made chemists, biochemists, biologists all aware of the role that organometallic compounds can play in bodily metabolism.

VII
COALESCENCE OF A DISCIPLINE

A. Recent Research Developments

1. *Instrumental Methods*

The enormous advances and changes in organometallic chemistry since the discovery of ferrocene would not have been possible had there not been a concomitant development of instrumental techniques and widespread availability of instruments. Infrared spectroscopy has long been known, but recent extensions in both theory and instrumentation have greatly expanded its applications. More recently, it has been complemented and supplemented by Raman spectroscopy. Nuclear magnetic resonance (NMR) spectroscopy, particularly for the hydrogen nucleus, has been an extremely important tool; much early work is reviewed in the article by Maddox *et al.* (*172*). In more recent years, nuclei such as ^{19}F, ^{13}C, ^{10}B, ^{11}B, ^{31}P, and a variety of others have also been studied by this technique. Other spectroscopic techniques applied to structure and bonding in organometallic compounds have been mass spectroscopy (*33*, *49*), electron spin resonance (ESR) spectroscopy, and

Mössbauer spectroscopy (*301*). All these are reviewed in the book by George (*108*).

Predominant among the nonspectroscopic instrumental methods is X-ray diffraction. This technique has been known for many years; for example, Brill used it to work out the structure of $Fe_2(CO)_9$ in 1927 (*32*). However, important and extensive application of this technique to organometallic chemistry came only when computer technology had been sufficiently advanced to speed up the computations involved. X-Ray diffraction is reviewed by Baenziger (*10*). Electron and neutron diffraction have also been used. Tsutsui has covered these various methods in a two-volume series (*267*).

2. *Transition Metal Compounds*

The synthesis of ferrocene opened up a new vista in organometallic chemistry. Within a decade, hundreds of cyclopentadienyl derivatives of transition metals had been prepared (*216*). Furthermore, the ferrocene research joined together the substantial but disjointed earlier research. Two groups independently prepared bis(π-benzene)chromium(0) (*87*, *88*, *296*) and proposed that Hein's earlier "polyphenylchromium" compounds were actually compounds containing benzene and/or biphenyl bonded to chromium in a manner analogous to ferrocene. Mixed compounds containing both aromatics and carbonyl groups could be prepared; $(\pi C_5H_5Fe)_2(CO)_4$ and π-$C_6H_6Cr(CO)_3$ were isolated in 1955 (*208*) and 1957 (*89*), respectively. In 1960 the compound π-$C_5H_5Mn(CO)_2(\pi$-$C_2H_4)$ was reported, one of the earliest species containing all three kinds of groups (*153*).

The existence and great stability of ferrocene and related compounds forced the development of bonding theory to explain these observations. In 1953, Chatt and Duncanson (*58*) explained the bonding in Zeise's salt on the basis of donation of π-electrons from ethylene to platinum, along with counterdonation of metal *d* electrons into the ethylene π^* orbitals (back-bonding). This two-way electron interchange created a linkage stronger than would have been expected *a priori*. This concept applied equally readily to metal carbonyls, and, with slight modifications, to aromatic complexes. Detailed descriptions of the bonding in ferrocene soon appeared (*80*, *141*, *179*). Understanding of the bonding in these species developed so quickly that in 1956 Longuet-Higgins and Orgel were able to predict the existence of stable metal complexes of the then unknown hydrocarbon cyclobutadiene (*167*). Shortly thereafter, two groups independently synthesized these complexes (*68*, *137*, *173*). Bonding theory has also been extended to cover those olefin complexes that exhibit unusual behavior (fluxional molecules, allenyl derivatives, etc.).

σ-Bonded organo-transition-metal compounds also came into greater prominence after 1950. Early work on these has been reviewed by Cotton (*67*). In 1952 and 1953, a series of organotitantium compounds were prepared and studied (*132*). It was soon found that the presence of π-bonding ligands had a stabilizing effect; thus compounds such as $(\pi\text{-}C_5H_5)_2Ti(C_6H_5)_2$, $CH_3Fe(CO)_2(\pi\text{-}C_5H_5)$, and $CH_3Mn(CO)_5$ have been isolated (*216*).

In recent years, techniques have advanced to the point where it is possible to prepare compounds such as $(t\text{-}C_4H_9)_4Cr$ (*158*) and $(CH_3)_6W$ (*244*) in pure form. As a result, the relationship between these compounds and corresponding ones of the representative elements has become better understood (*201*).

Organo-transition-metal compounds containing a variety of ligands on the metal have been prepared. Of these, the fluorocarbons perhaps deserve particular attention (*261*). The bonds between these and the metal are almost always σ-bonds, even for aromatic fluorocarbons, and they are appreciably more stable thermally than the corresponding hydrocarbons.

In recent years, more emphasis has been placed on compounds with metal–metal bonds (especially the cluster compounds) and on the mechanism of reactions. The observed role of transition metals in biochemistry and many industrial processes as catalysts (*51*, *65*) has also abetted research on them. The literature on these compounds has indeed become quite substantial; it has recently been compiled by Bruce (*34*, *35*).

3. *Representative Metal Compounds*

Since 1945, the organo compounds of the main group metals have been studied extensively, although they have been somewhat overshadowed by the spectacular postferrocene development of transition metal chemistry. They have also shared in the growth and refinement of instrumental techniques; in fact, probably the major part of research in this area over the last 20 years has emphasized structure, bonding, and mechanism over synthesis and chemical reactions.

One example of these compounds is trimethylaluminum. Long known to be associated, the compound's structure was not fully detailed until 1953 (*163*). The bonding description of this molecule, $(CH_3)_6Al_2$, involved the concept of three-center two-electron (electron-deficient) bonds, which was rapidly extended to a variety of other organometallics (*30*, *227*, *228*). Also from this compound came the concept of alkyl exchange. The expansion of NMR techniques led to extensive studies in this area (*191*).

With the continuing development of techniques, more and more studies of the organo derivatives of the most active metals have appeared. The true structure and nature of the Grignard reagent was the subject of numerous papers and workers (*9*, *271*). Organolithium compounds have become widely studied for both synthetic and structural reasons (*30*, *31*). Organosodium compounds, already known for many years (*292*), became important because of their synthetic utility and their ability to undergo exchange with hydrogen (*15*). Addition compounds between sodium metal and aromatic compounds, also known for many years (*17*, *231*, *292*), became better known as ESR spectroscopy developed. This reaction became steadily more extended, until numerous organometals, such as triphenylborane (*155*) and dodecamethylcyclohexasilane (*139*), were also found to form such anion radicals. They have been reviewed recently (*71*, *175*).

Although numerous advances in syntheses occurred during this period, two warrant special mention. First is the addition of boron–hydrogen bonds to olefins (hydroboration) (*26–28*):

$$\text{B—H} + \text{C=C} \longrightarrow \text{B—C—C—H} \tag{16}$$

This reaction opened up a new method for preparing boron–carbon bonds. In addition, because of the nature of the organoborane group, reaction (*16*) could be run reversibly, leading to stereospecific syntheses of olefins. This reaction will occur even for highly hindered olefins without skeletal rearrangement. The resulting organoboranes are highly versatile synthetic reagents (*28*). Related to these compounds are the more recently reported carboranes, in which one or more carbon atoms is incorporated into a boron cage structure (*193*). These compounds show exceptional thermal stability and have been much studied for their structural and bonding properties. Equation (16) is one specific example of a general synthesis (hydrometallation) that has been reported for a number of metals and metalloids.

Second are the ylids, specifically those of phosphorus (*143*). These were first reported by Wittig (*284*) when he tried to prepare pentamethylphosphorane:

$$(CH_3)_4P^+I^- + CH_3Li \longrightarrow (CH_3)_3P{=}CH_2 + CH_4 + LiI \qquad (17)$$

These compounds were found to react with ketones to form olefins (*262*, *285*, *286*), and are frequently termed Wittig reagents. Similar compounds have been prepared for a number of the elements in that part of the Periodic table. The carbon atoms of these compounds will form exceptionally stable bonds to transition metals (*233*).

Organometals containing metal–metal bonds have become more important in recent years. Of especial interest are those compounds with both representative and transition metals. These have been known for over 30 years (*130*), but have become well known only comparatively recently (*69*); they provide a link (both literally and figuratively) between the two areas of organometallic chemistry. The growing commercial interest in the silicones has spurred research in organosilicon chemistry, especially polysilane chemistry. Work in this area has been reviewed by Kumada (*159*). The disilane fraction from the direct synthesis of methylchlorosilanes contains a mixture of compounds of type $(CH_3)_x$-$Cl_{6-x}Si_2$, which are readily converted to hexamethyldisilane. This may then be converted to the chloride in a two-step synthesis (*160*):

$$(CH_3)_6Si_2 + H_2SO_4 \longrightarrow (CH_3)_5Si_2OSO_3H \xrightarrow{NH_4Cl} (CH_3)_5Si_2Cl \qquad (18)$$

The chloride then serves as starting point for other derivatives. The ready availability of the disilane fraction in large quantities has made disilanes and higher polysilanes easy to acquire in bulk, and they have been studied more than other compounds with metal–metal bonds.

An idea of the volume of organometallic chemical literature can be surmised from Table VI. It gives the number of references for the organo derivatives of the various elements for the years 1966 (*240*) and 1969 (*241*, *242*). Not all of the increase is necessarily due solely to growth of

TABLE VI

NUMBER OF ORGANOMETALLIC CHEMICAL REFERENCES

Group IA			Group IIA			Group IIIA			Group IVA			Group VA[a]		
Metal	1969	1966	Metal	1969	1966	Metal	1969	1966	Metal	1969	1966	Metal	1969	1966
Li	239	139	Be	11	13	B	283	235	Si	823	615	P	NL	NL
Na/K	46	14	Mg	237	82	Al	143	72	Ge	208	148	As	249	NL
Rb	—	—	Ca	4	2	Ga/In	20	8	Sn	537	207	Sb	103	15
Cs	—	—	Sr/Ba	—	—	Tl	27	13	Pb	82	71	Bi	25	3
	285	153		252	97		473	328		1650	1041		377	18

Group VIIB			Group VIII			Totals					
Metal	1969	1966	Metal	1969	1966	Group	1969	1966	Metal	1969	1966
Zn	54	34	Fe, Ru, Os	453	177	IB	41	17	Representative	3037	1665[b]
Cd	23	19	Co, Rh, Ir	145	93	IIIB	18	8			
Hg	240	95	Ni, Pd, Pt	287	103	IVB	74	41	Transition	2281[c]	850[d]
	317	148		885	373	VB	44	12		5318	2515
						VIB	291	130			
						VIIB	187	68			
							655	276			

[a] NL, not listed.
[b] Includes 28 under "general reference."
[c] Includes 424 under "general reference."
[d] Includes 53 under "general reference."

research, but much of it is, and the figures do give some indication of the relative interests.

4. *Biological Organometallics*

As previously noted (Section VI), there has been extensive work on the biological applications of organometallic compounds. The potentialities in this area were pointed out by Gilman in 1941 (*109*). Some aspects of this work are discussed in articles by Thayer (*254*, *255*, *255a*). Various specific aspects have been reviewed (*13*, *171*, *243*, *246*, *268*). The heavy metals such as Hg, Sn, and Pb have been studied primarily for their toxicity and physiological activity. Organotin compounds in particular have received attention for their numerous applications in situations where small quantities of compound are needed to control microorganisms (*171*) and have received widespread use as biocides. Recently, it has been found that the methyl derivatives of Tl, Pt(IV), and Au(III) are similar in their biological behavior to methylmercury and trimethyllead compounds (*256*).

Substantial research has also been performed on organo compounds of the metalloids. Silicon compounds particularly have played a major role, as witness the monograph by Voronkov (*270*) and the articles by Fessenden and Fessenden (*86*) and Garson and Kirchner (*107*). Silicones, because of their inertness to biological processes, have been used as heart valves, blood vessels, implants, ointments, etc. (*192*). Other organosilicon compounds are extremely active biologically (*270a*).

Many compounds of B, P, As, and Sb have medicinal importance because of their antibiotic properties. Table VII shows some representative applications of these compounds.

The methylmercury tragedy and the role of vitamin B_{12} have generated much current research as well. Pratt has reviewed the extensive chemistry of the latter (*213*, *214*). Methylcobalamin's role in methylation likewise received attention (*2*, *3*, *20*, *206*). The myriad aspects of the mechanism by which methylmercuric compounds and other organomercurials interact with biochemical processes are being studied by many groups of professionals.

The interaction of organometallic chemistry with biochemistry and the biological sciences has become substantial only in recent years. The impact is substantial and is increasing. More and more in the years to come will this aspect make itself felt on the discipline as a whole.

TABLE VII

SELECTED BIOLOGICAL APPLICATIONS OF ORGANOMETALS

Element	Compound	Use	Reference
B	$C_6H_5B(OH)_2$	Root growth stimulation	*279*
Si	$ClCH_2CH_2SiR_3$	Control of fruit ripening	*91*
Ge	$HO_2CCH_2CH_2GeO_{1.5}$	Treatment of hypertonia	*195*
P	$H_2NCH_2CH_2PO_3H_2$	Metabolic intermediate in certain marine organisms	*150*
P	$R_3P \cdot AuX$	Treatment of rheumatoid arthritis	*250*
As	$HO\text{-}C_6H_3(NO_2)\text{-}AsO_3H_2$	Enhancement of chick growth	*212*

B. Organometal Chemical Literature

The growth in organometallic chemistry, especially over the last quarter century, is reflected in its literature. All early papers, and the majority of present ones, appear in general journals. Since 1951, an increasing number have been in specialized journals. In 1963, the *Journal of Organometallic Chemistry* was founded and now publishes substantial numbers of papers annually. *Organometallic Chemistry Reviews* was founded in 1966 by Elsevier, and was later incorporated into the *Journal*. The hardcover series *Annual Surveys of Organometallic Chemistry* (also by Elsevier) likewise started in 1966, but after three volumes was incorporated into *Reviews* as Series B. More recently, two journals dealing with organometallics and synthesis have appeared: *Organometallics in Chemical Synthesis* (Elsevier, 1970; now discontinued), and *Synthesis in Inorganic and Metal-Organic Chemistry* (Dekker, 1971).

The classic reference book in the field is Krause and Von Grosse (*156*). It gives a fairly complete coverage of the literature in the field up to 1937. Since that time, a number of books covering the entire field have been published. Most of these will be discussed in the following section. Dub (*78*) has published a compilation of the literature between 1937 and 1964 for the transition metals and the elements of groups IV and V (except Si and P). The *Handbook of Organometallic Compounds* (*122*) lists a large number of organometals and their syntheses. Many, many review articles and monographs on the organic chemistry of individual elements have been published.

In 1963, Academic Press inaugurated the serial publication *Advances in Organometallic Chemistry*, edited by F. G. A. Stone and R. West. In the first twelve volumes of this series, 76 articles appeared; these break down into 30 on transition metals, 20 on representative metals, 7 on topics including both, 12 on structural-bonding-mechanistic aspects, and 7 historical articles.

C. Teaching Organometallic Chemistry

The first textbook containing any material on organometallics appeared in 1778. Volume III of "Elemens de Chymie Theorique et Pratique" (*120*) contains a small section on Cadet's fuming liquid and some work the authors had done on it. This set a pattern that was to last for a century and three-quarters. Organometallic compounds would appear as parts of textbooks on more general subjects. By and large, these compounds, especially the alkyls and aryls, were considered organic compounds. Interestingly enough, the metal carbonyls, when they became known, were considered inorganic. Textbooks on inorganic chemistry generally included a section on these compounds, usually in connection with bonding theory and coordination compounds. The alkyls would receive little or no mention. Organic textbooks gave the alkyls fuller coverage—especially the Grignard reagent—and generally emphasized their applications to synthesis.

After the synthesis of ferrocene, this situation changed. The inorganic aspect of organometallic chemistry received increasing emphasis, and detailed discussion of this subject began to appear in inorganic texts. Then, organometallic chemistry developed as a separate subject—first taught as special seminars, then as special courses, and finally as regular courses. Two books appeared, one by Coates in 1956 (*62a*) and the other by Rochow, Hurd, and Lewis in 1957 (*224*), that could serve as textbooks as well as reference books. The book by Coates has gone through three editions and has the distinction of being the only book on organometallics to do so. The relative changes give an indication of the changes in emphasis of material during the period. The third edition was published as two volumes, and no longer could be considered as a textbook. More recently, four other English-language texts have been published: two in hard covers (*83*, *204*) and two paperbacks (*63*, *225*). There is now ample material to give courses in the subject at almost any level.

D. Attitudes

The viewpoint of workers in organometallic chemistry has changed in recent years. Because this field does overlap both inorganic and organic chemistry, and given the sharp, almost hostile demarcation of those two disciplines in earlier years, there always existed a possibility that organometallic chemistry might be partitioned into two (or more) noninteracting areas. Much of its recent vigor comes from the fact that this has *not* happened. On the contrary, such a strong sense of coherence and synergic interaction has emerged that organometallic chemistry has not only retained its unity, but has actually brought the two larger disciplines into closer contact with each other. This has been helped by the fact that, over the last quarter-century, all internal demarcation lines in chemistry have tended to blur and also by the fact that physical chemistry, analytical chemistry, and biochemistry have also made contributions to organometallic chemistry. This latter has tended to weaken any dichotomy and added to the centripetal strength of the field. Two passages will illustrate this. The first is taken from the preface to Volume 1 of *Advances in Organometallic Chemistry* (*248*):

> During the past decade the study of organometallic chemistry has grown rapidly. Expansion of this area of endeavor has been fostered by the discovery of several classes of compounds possessing remarkable structures, by the development of valence theory to account both for the existence of these novel compounds and for the nature of carbon–metal bonds in general, and by the growing use of organometallics in industrial processes. Organometallic chemistry now seems well on its way toward establishing its identity as an important domain of science, representing a convergence of inorganic, organic, and physical chemistry where the respective disciplines can benefit by interaction with each other.

The second is from Gilman (*114a*):

> During much of its early development, organometallic chemistry had a special attraction for organic chemists. However, interests in this area have steadily broadened to such an extent that it is doubtful whether any other segment of chemistry has touched so intimately (and attracted the research interests of) *all* branches of chemistry, as well as allied sciences such as biology and engineering.

The selection of any year to mark the emergence of organometallic chemistry in its present form must be inherently subjective and arbitrary. Nonetheless, the year 1963 probably has a better claim than most others. In that year three firsts occurred: the first volume of *Advances in Organometallic Chemistry*; the first journal exclusively devoted to the subject; and the first international conference on organometals, held at Cincinnati, Ohio. This last has now become a biennial event under the auspices of IUPAC; the most recent conference (the sixth) was held at Amherst, Massachusetts, in the summer of 1973.

VIII
CONCLUSION

Unlike this article, the history of organometallic chemistry has no end. The momentum generated by the announcement of ferrocene continues as a potent driving force. Subsequent events have served to augment this vigor and alter the direction somewhat. The number of known compounds probably approximates 50,000 and is growing daily—in fact, almost hourly! Yet the development is rather uneven, as Tables VI and VII indicate. Many once-or-twice reported compounds have not had their physical properties determined. A majority of organometallic reactions still have not had their mechanisms determined. Much remains to be discovered on the role of substituents on the organic group in modifying the metal–carbon bond; note the striking differences, for example, between hydrocarbons and fluorocarbons. Biological aspects of organometals offer vast scope for further research. One fascinating aspect of reading the early literature is the many loose ends and unanswered questions that the authors bequeathed unto posterity. Another is the many reactions and preparations that were once termed impossible but which are now feasible and in some cases routine. Gilman notes (*114a*):

> The development of organometallic chemistry provided numerous hypotheses concerning the limits of preparation of these compounds. Gradually, the hypotheses have been narrowed down so that fewer and fewer metals are included in the groups which are said to be incapable of forming organometallic compounds. Some of these earlier hypotheses have been useful in pointing out those organometallic compounds which will be prepared with difficulty; other hypotheses have undergone revision necessitated by the preparation of organometallic compounds the existence of which had been denied. This sound prediction of a few decades ago has been essentially completely confirmed.

Every metallic or semimetallic element now seems capable of forming organo derivatives. Such derivatives have actually been reported for all metals except those that are radioactive with short half-lives. Even these last may ultimately yield; see the article by Gysling and Tsutsui (*121*). As a concluding illustration the following are some organometallic compounds that theory predicts to be possible and whose nonexistence remains a challenge to synthetic chemists: $(CH_3)_4Th$, $(C_6H_5)_2Ba$, $(C_8H_8)_2Md$, $(C_6F_6)_6Te$, $(CH_3)_3PdI$, $(CH_3)_5Pb_2Cl$, $(CH_3)_4AsAs(CH_3)_2$, and $(CH_3)_5Bi$.

Acknowledgments

I wish to thank the Editors of *Advances in Organometallic Chemistry* for their kind invitation to prepare this article, and for their permission, along with the permission of Academic Press, to reproduce the passage from the preface to Volume 1. I also wish to thank Professor Henry Gilman for his kind permission to quote the passage from his treatise on organic chemistry and for generously providing notes on organometallic chemistry in the period 1930–1950.

References

1. Abel, E. W., *Quart. Rev.* **17**, 133 (1963).
2. Abeles, R. H., *in* "Bioinorganic Chemistry" (R. Dessy, J. Dillard, and L. Taylor, eds.), pp. 346–364. Amer. Chem. Soc., Washington, D.C., 1971.
3. Agnes, G., Bendle, S., Hill, H. A. O., Williams, F. R., and Williams, R. J. P., *Chem. Commun.* 850 (1971).
4. Anderson, J. S., *Z. Anorg. Allgem. Chem.* **208**, 232 (1932).
5. Anderson J. S., *J. Chem. Soc.* 971 (1934).
6. Anderson, J. S., *Quart. Rev.* **1**, 331 (1947).
7. Andrews, L. J., and Keefer, R. M., *J. Amer. Chem. Soc.* **71**, 3644 (1949).
8. Andrews, L. J., and Keefer, R. M., *J. Amer. Chem. Soc.* **72**, 3113 (1950).
9. Ashby, E. C., *Quart. Rev.* **21**, 259 (1967).
10. Baenziger, N. C., *in* "Characterization of Organometallic Compounds" (M. Tsutsui, ed.), Part I, pp. 213–276. Wiley (Interscience), New York, 1969.
11. Bakir, F., Damluji, S. F., Amin-Zaki, L., Murtadha, M., Khalidi, A., Al-Rawi, N.Y., Tikriti, S., Dhahir, H. I., Clarkson, T. W., Smith, J. C., and Doherty, R. A., *Science* **181**, 230 (1973).
12. Barbier, P. A., *C. R. Acad. Sci.* **128**, 110 (1899).
13. Barnes, J. M., and Magos, L., *Organomet. Chem. Rev.* **3**, 137 (1968).
14. Beckman, E., *Ber.* **38**, 904 (1905).
15. Benkeser, R. A., Foster, D. J., and Sauve, D. M., *Chem. Rev.* **57**, 867 (1957).
16. Bertheim, A., "Handbuch der organischen Arsenverbindungen." Stuttgart, 1913.
17. Berthelot, M., *Ann. Chim. Phys.* **12**, 155 (1867).
18. Berthelot, M., *C. R. Acad. Sci.* **112**, 1343 (1891).
19. Berthelot, M., *Ann. Chim.* **23**, 32 (1901).

20. Bertilsson, L., and Neujahr, H. Y., *Biochemistry* **10**, 2805 (1971).
21. Bird, C. W., *Chem. Rev.* **62**, 283 (1962).
22. Birnbaum, K., *Annalen* **145**, 68 (1868).
23. Brockway, L. O., and Anderson, J. S., *Trans. Faraday Soc.* **33**, 1233 (1937).
24. Brockway, L. O., and Cross, P. C., *J. Chem. Phys.* **3**, 828 (1935).
25. Brockway, L. O., Ewens, R. V. G., and Lister, M. W., *Trans. Faraday Soc.* **34**, 1350 (1938).
26. Brown, H. C., "Hydroboration." Benjamin, New York, 1962.
27. Brown, H. C., "Boranes in Organic Chemistry." Cornell Univ. Press, Ithaca, New York, 1972.
28. Brown, H. C., *Advan. Organometal. Chem.* **11**, 1–20 (1973).
29. Brown, K. L., and Kallen, R. G., *J. Amer. Chem. Soc.* **94**, 1894 (1972).
30. Brown, T. L., *in Advan. Organometal. Chem.* **3**, 365–395 (1965).
31. Brown, T. L., *Accounts Chem. Res.* **1**, 23 (1968).
32. Brill, R., *Z. Kristallogr., Kristallgeometrie, Kristallphys., Kristallchem.* **65**, 85 (1927).
33. Bruce, M. I., *Advan. Organometall. Chem.* **6**, 273–334 (1968).
34. Bruce, M. I., *Advan. Organometal. Chem.* **10**, 274–346 (1972).
35. Bruce, M. I., *Advan. Organometal. Chem.* **11**, 447–471 (1973).
36. Buckton, G., and Odling, W., *Proc. Roy. Soc.* **14**, 19 (1865).
37. Burke, G. T., Mangum, J. H., and Brodie, J. D., *Biochemistry* **10**, 3079 (1971).
38. Bunsen, R. W., *Annalen* **24**, 271 (1837).
39. Bunsen, R. W., *Ann. Phys.* **40**, 219 (1837).
40. Bunsen, R. W., *Annalen* **31**, 175 (1839).
41. Bunsen, R. W., *Annalen* **37**, 1 (1841).
42. Bunsen, R. W. *Annalen* **42**, 14 (1842).
43. Bunsen, R. W., *Annalen* **46**, 1 (1843).
44. Butlerov, A., *Annalen* **144**, 39 (1867).
45. Cadet de Gassicourt, L. C., *Mem. Math. Phys.* **3**, 623 (1760).
46. Cahours, A., *Annalen* **114**, 227 (1859).
47. Cahours, A., *C. R. Acad. Sci.* **76**, 1383 (1873).
48. Cahours, A., and Hofmann, A. W., *Annalen* **104**, 29 (1857).
49. Cais, M., and Lupin, M. S. *Advan. Organometal. Chem.* **8**, 211–325 (1970).
50. Cason, J., *Chem. Rev.* **40**, 15 (1947).
51. Chalk, A. J., and Harrod, J. F., *Advan. Organometal. Chem.* **6**, 119–170 (1968).
52. Challenger, F., *Chem. Ind.* (*London*) 657 (1935).
53. Challenger, F., *Chem. Rev.* **36**, 315 (1945).
54. Challenger, F., *J. Chem. Soc.* **849** (1951).
55. Challenger, F., *Quart. Rev.* **9**, 255 (1955).
56. Challenger, F., Higginbottom, C., and Ellis, L., *J. Chem. Soc.* 95 (1933).
57. Challenger, F., Lisle, D. B., and Dransfield, P. B., *J. Chem. Soc.* 1760 (1954).
58. Chatt, J., and Duncanson, L. A., *J. Chem. Soc.* 2939 (1953).
59. Chatt, J., Pauson, P. L., and Venanzi, L. M., *in* "Organometallic Chemistry" (H. Zeiss, ed.), pp. 468–528. Reinhold, New York, 1960.
60. Chojnacki, C., *Z. Chem.* 419 (1870).
61. Christiansen, W. G., "Organic Derivatives of Antimony," *Amer. Chem. Soc.*, New York, 1925.
62. (a) Coates, G. E., "Organometallic Compounds," 1st ed. Methuen, London, 1956; (b) 2nd ed., 1960.
63. Coates, G. E., Green, M. L. H., Powell, P., and Wade, K., "Principles of Organometallic Chemistry." Methuen, London, 1968.

64. Coates, G. E., Green, M. L. H., and Wade, K., "Organometallic Chemistry," 3rd Ed. Methuen, London, 1968.
65. Collman, J. P., *Accounts Chem. Res.* **1**, 136 (1968).
66. Copenhaver, J., and Bigelow, M. H., "Acetylene and Carbon Monoxide Chemistry." Reinhold, New York, 1949.
67. Cotton, F. A., *Chem. Rev.* **55**, 551 (1955).
68. Criegee, R., and Schröder, G., *Angew. Chem.* **71**, 70 (1959).
69. Cundy, C. S., Kingston, B. M., and Lappert, M. F., *Advan. Organometal. Chem.* **11**, 253–330 (1973).
70. Dales, L. G., *Amer. Med.* **53**, 219 (1972).
71. DeBoer, E., *Advan. Organometal. Chem.* **2**, 115–155 (1964).
72. Dennis, L. M., and Patnode, W., *J. Amer. Chem. Soc.* **54**, 182 (1932).
73. Dennis, L. M., Work, R. W., Rochow, E. G., and Chamot, E. M., *J. Amer. Chem. Soc.* **56**, 1047 (1934).
74. Dewar, J., and Jones, H. O., *J. Chem. Soc.* **85**, 203 (1904).
75. Dewar, J., and Jones, H. O., *J. Chem. Soc.* **85**, 212 (1904).
76. Dewar, J., and Jones, H. O., *Proc. Roy. Soc., Ser. A* **76**, 558 (1905).
77. Dreher, E., and Otto, R., *Annalen* **154**, 94 (1870).
78. Dub, M., "Organometallic Compounds," Vol. 1. Springer-Verlag, Berlin and New York, 1966; Vol. 2, 1967; Vol. 3, 1968.
79. Dunitz, J. D., and Orgel, L. E., *Nature (London)* **171**, 121 (1953).
80. Dunitz, J. D., and Orgel, L. E., *J. Chem. Phys.* **23**, 954 (1955).
81. Dunlap, L., *Chem. Eng. News* **22** (July 5, 1971).
82. Eiland, P. F., and Pepinsky, R., *J. Amer. Chem. Soc.* **74**, 4971 (1952).
83. Eisch, J. J., "The Chemistry of Organometallic Compounds." Macmillan, New York, 1967.
84. Ewens, R. V. G., and Lister, M. W., *Trans. Faraday Soc.* **35**, 681 (1939).
85. Feigl, F., and Krumholz, P., *Monatsh. Chem.* **59**, 314 (1932).
86. Fessenden, R. J., and Fessenden, S. J., *Advan. Drug. Res.* **4**, 95–132 (1967).
87. Fischer, E. O., and Hafner, W., *Z. Naturforsch. B* **10**, 655 (1955).
88. Fischer, E. O., and Hafner, W., *Z. Anorg. Allgem. Chem.* **286**, 146 (1956).
89. Fischer, E. O., and Ofele, K., *Chem. Ber.* **90**, 2532 (1957).
90. Fischer, E. O., and Pfab, W., *Z. Naturforsch. B* **7** 378 (1952).
91. Foorg, W., and Fischer, H. P., *Chem. Abstr.* **77**, 5605 (1972).
92. Frankland, E., *J. Chem. Soc.* **2**, 263 (1849).
93. Frankland, E., *J. Chem. Soc.* **2**, 297 (1849).
94. Frankland, E., *Phil. Trans.* **142**, 417 (1852).
95. Frankland, E., *J. Chem. Soc.* **13**, 188 (1861).
96. Frankland, E., *Annalen* **126**, 109 (1863).
97. Frankland, E., *Proc. Roy. Inst.* **4**, 309 (1864).
98. Frankland, E., "Experimental Researches," p. 144. 1877.
99. Frankland, E., and Duppa, D. F., *Annalen* **115**, 319 (1860).
100. Frankland, E., and Duppa, D. F., *Annalen* **135**, 217 (1865).
101. Frankland, E., and Duppa, D. F., *J. Chem. Soc.* **20**, 102 (1867).
102. Friberg, L. T., and Vostal, J. J., "Mercury in the Environment." CRC Press, New York, 1972.
103. Friedel, C., and Crafts, J. M., *Annalen* **127**, 28 (1863).
104. Friedel, C., and Crafts, J. M., *Annalen* **136**, 203 (1865).
105. Friedel, C., and Crafts, J. M., *C. R. Acad. Sci.* **84**, 1392 (1877).

106. Garrett, A. B., *J. Chem. Educ.* **39**, 414 (1962).
107. Garson, L. R., and Kirchner, L. K., *J. Pharm. Sci.* **60**, 1113 (1971).
108. George, W. O., "Spectroscopic Methods in Organometallic Chemistry." Butterworths, London, 1970.
109. Gilman, H., *Science* **93**, 47 (1941).
110. Gilman, H., "Organic Chemistry: An Advanced Treatise," 2nd ed. Wiley, New York, 1943.
111. Gilman, H., *Advan. Organometal. Chem.* **7**, 1–52 (1968).
112. Gilman, H., and Nelson, J. F., *Rec. Trav. Chim.* **55**, 518 (1936).
113. Gilman, H , and Straley, J. M., *Rec. Trav. Chim.* **55**, 821 (1936).
114. Gilman, H., and Young, R. V., *J. Org. Chem.* **1**, 315 (1936).
114a. Gilman, H., private communication (1973).
115. Greco, A. R., *Riv. Neurol.* **3**, 515 (1930).
116. Grignard, V., *C. R. Acad. Sci.* **130**, 1322 (1900).
117. Grignard, V., and Courtot, C., *C. R. Acad. Sci.* **158**, 1763 (1914).
118. Groll, H. P., *J. Amer. Chem. Soc.* **52**, 2998 (1930).
119. Grüttner, G., Doctoral Dissertation (Berlin, 1914), p. 84.
120. Guyton de Morveau, L. B., Maret, H., and Durande, J. F., "Elemens de Chymie Theorique et Pratique." Dijons, 1778.
121. Gysling, H., and Tsutsui, M., *Advan. Organometal. Chem.* **9**, 361–395 (1970).
122. Hagihara, N., Kumada, M., and Okawara, R., "Handbook of Organometallic Compounds." Benjamin, New York, 1968.
123. Hallwachs, W., and Schafarik, A., *Annalen* **109**, 207 (1859).
124. Hammond, A. L., *Science* **171**, 788 (1971).
125. Hansen, C., *Ber.* **3**, 9 (1870).
126. Hartmann, H., and Lutsche, H., *Naturwissenschaften* **49**, 182 (1962).
127. Hein, F., *Ber.* **52**, 195 (1919).
128. Hein, F., *J. Prak. Chem.* **132**, 59 (1931).
129. Hein, F., *Z. Anorg. Chem.* **227**, 272 (1936).
130. Hein, F., and Pobloth, H., *Z. Anorg. Allgem. Chem.* **248**, 84 (1941).
131. Hein, F., and Späte, R., *Ber.* **59**, 751 (1926).
132. Herman, D. F., and Nelson, W. K., *J. Amer. Chem. Soc.* **75**, 3877 (1953).
133. Hieber, W., *Advan. Organometal. Chem.* **8**, 1–28 (1970).
134. Hieber, W., and Leutert, F., *Naturwissenschaften* **19**, 360 (1931).
135. Himmelweit, F., "The Collected Papers of Paul Ehrlich," Vol. 3: Chemotherapy. Pergamon, Oxford 1960.
136. Hofmann, K. A., and Von Narbutt, J., *Ber.* **41**, 1625 (1908).
137. Hübel, W., and Braye, E. H., *J. Inorg. Nucl. Chem.* **10**, 250 (1959).
138. Hunter, D., Bomford, R. R., and Russell, D. S., *Quart. J. Med.* **9**, 193 (1940).
139. Husk, G. R., and West, R., *J. Amer. Chem. Soc.* **87**, 3993 (1965).
140. Jackson, K. E., and Jackson, M. A., *Chem. Rev.* **16**, 439 (1935).
141. Jaffé, H. H., *J. Chem. Phys.* **21**, 156 (1953).
142. Jensen, S., and Jernelöv, A., *Nature (London)* **223**, 753 (1969).
143. Johnson, A. W., "Ylid Chemistry." Academic Press, New York, 1966.
144. Kealy, T. J., and Pauson, P. J., *Nature (London)* **168**, 1039 (1951).
145. Kharasch, M. S., and Isbell, H. S., *J. Amer. Chem. Soc.* **53**, 3053 (1931).
146. Kharasch, M. S., and Reinmuth, O., "Grignard Reactions of Nonmetallic Substances." Prentice-Hall, Englewood Cliffs, New Jersey, 1954.
147. Kharasch, M. S., Seyler, R. C., and Mayo, F. R., *J. Amer. Chem. Soc.* **60**, 882 (1938).

148. King, R. B., *Advan. Organometal. Chem.* **2**, 157–256 (1964).
149. Kipping, F. S., and Lloyd, L., *J. Chem. Soc.* **79**, 449 (1901).
150. Kittredge, J. S., and Roberts, E., *Science* **164**, 37 (1969).
151. Klein, D. M., *J. Chem. Educ.* **49**, 7 (1972).
152. Klemm, W., and Neuber, A., *Z. Anorg. Allgem. Chem.* **227**, 261 (1936).
153. Kögler, H. P., and Fischer, E. O., *Z. Naturforsch. B* **15**, 676 (1960).
154. Kraus, C. A., *J. Chem. Educ.* **6**, 1478 (1929).
155. Krause, E., *Ber.* **57**, 216 (1924).
156. Krause, E., and Von Grosse, A., "Die Chemie der Metall-Organischen Verbindungen." Borntraeger, Berlin, 1937.
157. Krause, E., and Wendt, B., *Ber.* **56**, 2064 (1923).
158. Kruse, W., *J. Organometal. Chem.* **42**, C39 (1972).
159. Kumada, M., and Tamao, K., *Advan. Organometal. Chem.* **6**, 19–117 (1968).
160. Kumada, M. Yamaguchi, M., Yamamoto, Y., Nakajima, J., and Shiina, K., *J. Org. Chem.* **21**, 1264 (1956).
161. Langmuir, I., *Science* **54**, 59 (1921).
162. Lenhert, P. G., and Crowfoot-Hodgkin, D., *Nature* (*London*) **192**, 937 (1961).
163. Lewis, P. H., and Rundle, R. E., *J. Chem. Phys.* **21**, 986 (1953).
164. Liebig, J. *Pogg. Ann.* **30**, 90 (1834).
165. Liebig, J., *Annalen* **23**, 12 (1837).
166. Locquin, R., *Bull. Soc. Chim. Fr.* 897 (1950).
167. Longuet-Higgins, H. C., and Orgel, L. E., *J. Chem. Soc.* 1969 (1956).
168. Löwig, C., *Annalen* **84**, 309 (1852).
169. Löwig, C., *Annalen* **88**, 318 (1853).
170. Löwig, C., and Schweizer, E., *Annalen* **75**, 315 (1850).
171. Luijten, J. G. A., *in* "Organotin Compounds" (A. K. Sawyer, ed.), Vol. 3, pp. 931–974. Dekker, New York, 1972.
172. Maddox, M. L., Stafford, S. L., and Kaesz, H. D., *Advan. Organometal. Chem.* **3**, 1–180 (1965).
173. Maitlis, P. M., *Advan. Organometal. Chem.* **4**, 95–143 (1966).
174. McBride, B. C., and Wolfe, R. S., *Biochemistry* **10**, 4312 (1971).
175. McClelland, B. J., *Chem. Rev.* **64**, 301 (1964).
176. Mendeleyeev, D., *Annalen Suppl.* **8**, 133–229 (1871).
177. "Metal-Organic Compounds." Amer. Chem. Soc., Washington, D.C., 1959.
178. Miller, S. A., Tebboth, J. A., and Tremaine, J. F., *J. Chem. Soc.* 632 (1952).
179. Moffitt, W., *J. Amer. Chem. Soc.* **76**, 3386 (1954).
180. Mond. L., *J. Chem. Soc.* **57**, 749 (1890).
181. Mond, L., and Langer, C., *J. Chem. Soc.* **59**, 1090 (1891).
182. Mond, L., and Quincke, F., *J. Chem. Soc.* **59**, 604 (1891).
183. Morgan, G. T., "Organic Derivatives of Arsenic and Antimony," Longmans, Green, London, 1918.
184. Müller, R., *J. Chem. Educ.* **42**, 41 (1965).
185. Natta, G., *Angew. Chem.* **76**, 553 (1964).
186. Nelson, N., *Environ. Res.* **4**, 1 (1971).
187. Nesmeyanov, A. N., *Advan. Organometal. Chem.* **10**, 1–78 (1972).
188. Nickerson, S. P., *J. Chem. Educ.* **31**, 560 (1954).
189. Nietski, R. H., and Benckiser, T., *Ber.* **18**, 499 (1885).
190. Oehme, F. W., *Clin. Toxicol.* **5**, 215 (1972).
191. Oliver, J. P., *Advan. Organometal. Chem.* **8**, 167–209 (1970).

192. Olson, K. J., *Toxicol. Appl. Pharmacol.* **21**, 12 (1972).
193. Onak, T., *Advan. Organometal. Chem.* **3**, 263–363 (1965).
194. Orgel, L. E., "An Introduction to Transition-Metal Chemistry," pp. 135–143. Methuen, London, 1960.
195. Pahk, U. S., *Chem. Abstr.* **78**, 148074 (1973).
196. Paneth, F. A., and Hofeditz, W., *Ber.*, **62**, 1335 (1929).
197. Paneth, F. A., and Lautsch, W., *Ber.* **64**, 2702 (1931).
198. Paneth, F. A., and Lautsch, W., *Ber.* **64**, 2708 (1931).
199. Paneth, F. A., and Lautsch, W., *J. Chem. Soc.* 380 (1935).
200. Paneth, F. A., and Loleit, H., *J. Chem. Soc.* 366 (1935).
201. Parshall, G. W., and Mrowca, J. J., *Advan. Organometal. Chem.* **7**, 157–209 (1968).
202. Partington, J. R., "A History of Chemistry," Vol. 4, pp. 259–260. Macmillan, London, 1964.
203. Pauling, L., "The Nature of the Chemical Bond," 3rd ed., pp. 331–336. Cornell Univ. Press, Ithaca, New York, 1960.
204. Pauson, P. L., "Organometallic Chemistry." Arnold, London, 1967.
205. Pearson, T. G., and Purcell, R. H., *J. Chem. Soc.* **1151** (1935).
206. Penley, M. W., Brown, D. G., and Wood, J. M., *Biochemistry* **9**, 4302 (1970).
207. P., H., *Brit. Med. J.* **1**, 515 (1958).
208. Piper, T. S., Cotton, F. A., and Wilkinson, G., *J. Inorg. Nucl. Chem.* **1**, 165 (1955).
209. Podall, H. E., *J. Chem. Educ.* **38**, 187 (1961).
210. Pope, W. J., and Gibson, C. S., *J. Chem. Soc.* **91**, 2061 (1907).
211. Pope, W. J., and Peachey, S. J., *J. Chem. Soc.* **95**, 571 (1909).
212. Prasad, S., Hairr, W. T., and Dallas, J. T., *Biol. Abstr.* **54**, 16603 (1972).
213. Pratt, J. M., "Inorganic Chemistry of Vitamin B_{12}." Academic Press, New York, 1971.
214. Pratt, J. M., and Craig, P. J., *Advan. Organometal. Chem.* **11**, 332–446 (1973).
215. Raiziss, G. W., and Gavron, J. L., "Organic Arsenical Compounds." Amer. Chem. Soc., New York, 1923.
216. Rausch, M. D., *in* "Werner Centennial," pp. 486–531. Amer. Chem. Soc., Washington, D.C., 1967.
217. Reich, R., *C. R. Acad. Sci.* **177**, 322 (1923).
218. Reihlen, H., Gruhl, A., Hessling, G., and Pfrengle, O., *Annalen* **482**, 161 (1930).
219. Reppe, W., "Neue Entwicklungen auf dem Gebiete der Chemie des Acetylens und des Kohlenoxyds." Springer-Verlag, Berlin and New York, 1949.
220. Reppe, W., *Annalen* **582**, 1 (1953).
221. Rheinboldt, H., *J. Chem. Educ.* **27**, 476 (1950).
222. Richardson, J. W., *in* "Organometallic Chemistry" (H. Zeiss, ed.), pp. 12–20. Reinhold, New York, 1960.
223. Rochow, E. G., "An Introduction to the Chemistry of the Silicones." Wiley, New York, 1946.
224. Rochow, E. G., Hurd, D. T., and Lewis, R. N., "The Chemistry of Organometallic Compounds." Wiley, New York, 1957.
225. Rochow, E. G., "Organometallic Chemistry." Reinhold, New York, 1964.
226. Rochow, E. G., *Advan. Organometal. Chem.* **9**, 1–19 (1970).
227. Rundle, R. E., *J. Amer. Chem. Soc.* **69**, 1327 (1947).
228. Rundle, R. E., *J. Chem. Phys.* **17**, 671 (1949).
229. Runge, F., "Organometallverbindungen." Stuttgart, 1944.
230. Schlenck, W., Appenrodt, J., Michael, A., and Thal, A., *Ber.* **47**, 473 (1914).

231. Schlenck, W., and Bergmann, E., *Annalen* **464**, 1 (1928).
232. Schlenck, W., and Holtz, J., *Ber.* **50**, 262 (1917).
233. Schmidbaur, H., *Abstr. Int. Conf. Organometal. Chem., 6th* P5 (1973).
234. Schmidt, J., "Die organischen Magnesiumverbindungen and ihre Anwendung zu Synthesen." Enke, Stuttgart, 1908.
235. Schrauzer, G. N., *Accounts Chem. Res.* **1**, 97 (1968).
236. Schrauzer, G. N., *in* "Bioinorganic Chemistry" (R. Dessy, J. Dillard and L. Taylor, eds.), pp. 1–20. Amer. Chem. Soc., Washington, D.C., 1971.
237. Schrauzer, G. N., and Sibert, J. W., *J. Amer. Chem. Soc.* **92**, 1022 (1970).
238. Schüler, E., *Annalen* **87**, 34 (1853).
239. Schützenberger, M. P., *Annales* **15**, 100 (1868).
239a. Seyferth, D., and Davison, A., *Science* **182**, 699 (1973).
240. Seyferth, D., and King, R. B., "Annual Surveys of Organometallic Chemistry," Vol. 3. Elsevier, Amsterdam, 1967.
241. Seyferth, D., and King, R. B., *Organometal. Chem. Rev.* **6**, 1–1202 (1970).
242. Seyferth, D., and King, R. B., *Organometal. Chem. Rev.* **7**, 175–306 (1971).
243. Shapiro, H., and Frey, F. W., "The Organic Compounds of Lead," pp. 17–23, 416–421. Wiley (Interscience), New York, 1968.
244. Shortland, A., and Wilkinson, G., *J. Chem. Soc. D* 318 (1972).
245. Sidgwick, N. V., "The Electronic Theory of Valence," p. 163. Oxford Univ. Press, London and New York, 1927.
246. Sijpesteijn, A. K., Luijten, J. G. A., and Van Der Kerk, G. J. M., *in* "Fungicides: An Advanced Treatise" (D. C. Torgeson, ed.), Vol. 2, pp. 331–366. Academic Press, New York, 1969.
247. Stadtman, T. C., *Science* **171**, 859 (1971).
248. Stone, F. G. A., and West, R., *Advan. Organometal.* **1**, vii (preface) (1963).
249. Surrey, A. R., "Name Reactions in Organic Chemistry." Academic Press, New York, 1954.
250. Sutton, B. M., McGusty, E., Walz, D. T., and DiMartino, M. J., *J. Med. Chem.* **15**, 1095 (1972).
251. Thayer, J. S., *J. Chem. Educ.* **43**, 594 (1966).
252. Thayer, J. S., *J. Chem. Educ.* **46**, 442 (1969).
253. Thayer, J. S., *J. Chem. Educ.* **46**, 764 (1969).
254. Thayer, J. S., *J. Chem. Educ.* **48**, 806 (1971).
255. Thayer, J. S., *J. Chem. Educ.* **50**, 390 (1973).
255a. Thayer, J. S., *J. Organometal. Chem.* **76**, 265 (1974).
256. Thayer, J. S., *Abstr. Int. Conf. Organometal. Chem., 6th* **253** (1973).
257. Thenard, L. J., *Annales* **52**, 54 (1804).
258. Thenard, P., *C. R. Acad. Sci.* **21**, 144 (1845).
259. Thenard, P., *C. R. Acad. Sci.* **25**, 892 (1847).
260. Thiele, J., *Ber.* **34**, 68 (1901).
261. Treichel, P. M., and Stone, F. G. A., *Advan. Organometal. Chem.* **1**, 143–220 (1964).
262. Trippett, S., *Pure Appl. Chem.* **9**, 255 (1964).
263. Trout, W. E., *J. Chem. Educ.* **14**, 453 (1937).
264. Trout, W. E., *J. Chem. Educ.* **14**, 575 (1937).
265. Trout, W. E., *J. Chem. Educ.* **15**, 77 (1938).
266. Trout, W. E., *J. Chem. Educ.* **15**, 113 (1938).
267. Tsutsui, M., "Characterization of Organometallic Compounds," Part I, Wiley (Interscience), New York, 1969; Part 2, 1971.

268. Ulfvarson, U., *in* "Fungicides: An Advanced Treatise" (D. C. Torgeson, ed.), Vol. 2, pp. 303–329. Academic Press, New York, 1969.
269. Von Grosse, A., *Ber.* **59**, 2646 (1926).
270. Voronkov, M. G., Zelchan, G. I., and Lukevitz, E. Y., "Silicon and Life: Biochemistry, Toxicology, and Pharmacology of Silicon Compounds." Zinatne Publ., Riga, 1971.
270a Voronkov, M. G., *Chem. Brit.* **9**, 411 (1973).
271. Wakefield, B. J., *Organometal. Chem. Rev.* **1**, 131 (1966).
272. Wanklyn, J. A., *J. Chem. Soc.* **9**, 193 (1857).
273. Wanklyn, J. A., *Annalen* **108**, 67 (1858).
274. Wanklyn, J. A., *Annalen* **140**, 353 (1866).
275. Waters, W. A., "The Chemistry of Free Radicals." Oxford Univ. Press (Clarendon), London and New York, 1946.
276. West, R., and Niu, H. Y., *J. Amer. Chem. Soc.* **85**, 2589 (1963).
277. Westöö, G., *Acta. Chem. Scand.* **20**, 213 (1966).
278. Whitmore, F. C., "Organic Compounds of Mercury." Amer. Chem. Soc., New York, 1921.
279. Wildes, R. A., and Neales, T. F., *J. Exp. Bot.* **20**, 591 (1969).
280. Wilkinson, G., Rosenblum, M., Whiting, M. C., and Woodward, R. B., *J. Amer. Chem. Soc.* **74**, 2125 (1952).
281. Winkler, C., *J. Prakt. Chem.* **36**, 177 (1887).
282. Winstein, S., and Lucas, H. J., *J. Amer. Chem. Soc.* **60**, 836 (1938).
283. Wittig, G., *Pure Appl. Chem.* **9**, 245 (1964).
284. Wittig, G., and Rieber, M., *Annalen* **562**, 177 (1949).
285. Wittig, G., and Schöllkopf, U., *Chem. Ber.* **87**, 1318 (1954).
286. Wittig, G., and Wetterling, M. H., *Annalen* **557**, 193 (1947).
287. Wöhler, F., *Annalen* **35**, 111 (1840).
288. Wood, J. M., *Environment* **14**, 33 (1972).
289. Wood, J. M., Kennedy, F. S., and Rosen, C. G., *Nature* (*London*) **220**, 173 (1968).
290. Wood, J. M., *Abstr. Int. Conf. Organometal. Chem., 6th*, P9 (1973).
291. Woodward, R. B., Rosenblum, M., and Whiting, M. C., *J. Amer. Chem. Soc.* **74**, 3458 (1952).
292. Wooster, C. B., *Chem. Rev.* **11**, 1 (1932).
293. Würtz, A., *Ann. Chim. Phys.* **44**, 279 (1855).
294. Zeise, W. C., *Ann. Phys.* **9**, 632 (1827).
295. Zeise, W. C., *Ann. Phys.* **40**, 234 (1837).
296. Zeiss, H., *in* "Organometallic Compounds" (H. Zeiss, ed.), pp. 380–425. Reinhold, New York, 1960.
297. Ziegler, K., *Angew. Chem.* **76**, 545 (1964).
298. Ziegler, K., *Advan. Organometal. Chem.* **6**, 1–17 (1968).
299. Ziegler, K., and Colonius, H., *Annalen* **479**, 135 (1930).
300. Ziegler, K., Holzkamp, E., Breil, H., and Martin, H., *Angew. Chem.* **67**, 427 (1955).
301. Zuckerman, J. J., *Advan. Organometal. Chem.* **9**, 22–134 (1970).

Arene Transition Metal Chemistry

W. E. SILVERTHORN

Department of Chemistry
Oregon State University
Corvallis, Oregon

I

INTRODUCTION

This review article concerns those complexes in which a transition metal is π-bonded to a six-membered aromatic hydrocarbon ring. The last comprehensive review of this area of chemistry covered the literature through January, 1965 (*461*). Since that time the number of papers in this area has expanded considerably. Recent X-ray structural investigations along with spectroscopic and theoretical studies have contributed significantly to the understanding of the bonding in arene metal complexes. Many new types of complexes have been synthesized including "electron-rich" monoarene complexes containing ligands other than carbon monoxide. New synthetic methods, such as metal vapor synthesis, have led to the preparation of arene complexes not accessible by conventional methods. Literature coverage is through December, 1973.

II

SOME GENERAL ASPECTS OF ARENE TRANSITION METAL CHEMISTRY

A. *Preparative Methods*

1. *Fischer–Hafner Method*

The widely applicable Fischer–Hafner method has been used to prepare arene complexes of most of the transition metals. The method involves the reaction of metal halide, aromatic hydrocarbon, aluminum trihalide, and aluminum metal, e.g. (*53*, *157b*)(M = Cr, V),

$$3MCl_3 + 2Al + AlCl_3 + 6 \text{ arene} \longrightarrow 3[(\text{arene})_2M][AlCl_4] \quad (1)$$

If reduction of the metal is not required, the reaction is carried out in the absence of aluminum metal, e.g. (*116*, *272*) (M = Fe, Ni),

$$MBr_2 + 2 \text{ arene} \xrightarrow{AlBr_3} [(\text{arene})_2M][AlBr_4]_2 \quad (2)$$

Bisarene metal complexes are generally obtained by this method; however, with metals of the early and late transition groups, monoarene complexes such as $(C_6H_6)Ti(AlCl_4)_2$, $[(Me_6C_6)NbCl_2]_2$, and $[(C_6H_6)Pd(AlCl_4)]_2$ have been isolated (*5*, *150*, *284*).

2. *Metal Vapor Synthesis*

The cocondensation of transition metal atoms and aromatic hydrocarbon at liquid nitrogen temperature leads directly to zero-valent arene complexes (*27*, *288*, *376*). Complexes, such as $(C_6H_6)_2Ti$ and $(C_6F_6)(C_6H_6)Cr$, that are not accessible by the Fischer–Hafner method have been prepared by this method.

3. *Cyclic Condensation*

Reaction of disubstituted acetylenes with transition metal compounds often leads to arene complexes (*93*, *154*, *245*, *416*, *460*), e.g. (M = Mn, Co),

$$MCl_2 + \text{aryl MgBr} \xrightarrow[\text{ii. } H_2O]{\text{i. MeC}\equiv\text{CMe}} [(Me_6C_6)_2M]^+ \quad (3)$$

The reaction may in some cases involve a metalocycle intermediate such as (*44*)

$$[(MeO)_3P]_2(CO)_2\overline{RuC(CF_3){=}C(CF_3)C(CF_3){=}C}CF_3 \xrightarrow{CF_3C\equiv CCF_3} [C_6(CF_3)_6]Ru(CO)_2[P(OMe)_3] \quad (4)$$

4. *Carbonyl Replacement*

The direct reaction between a metal carbonyl and an aromatic hydrocarbon often leads to the replacement of 2 or 3 carbonyl groups yielding arene complexes. Complexes of the type $[(\text{arene})V(CO)_4][V(CO)_6]$ (*50*), $(\text{arene})Cr(CO)_3$ (*327*), and $(\text{arene})Fe(CO)_3$ (*281*) have been prepared in this manner. Carbonyl metal halide complexes have also been observed to react with aromatic hydrocarbons in the presence of $AlCl_3$ yielding cationic arene derivatives, e.g. (*71*, *104*, *448*),

$$Mn(CO)_5Cl + \text{arene} \xrightarrow{AlCl_3} [(\text{arene})Mn(CO)_3]^+ \quad (5)$$

$$(\pi{-}C_4Ph_4)Co(CO)_2Br + \text{arene} \xrightarrow{AlCl_3} [(\text{arene})(\pi{-}C_4Ph_4)Co]^+ \quad (6)$$

Other preparative methods that have not been so widely applied are discussed in Sections IV to XI.

B. Reactions

1. *Arene Replacement*

Monoarene complexes often undergo arene replacement reactions with ligands such as arenes (arene exchange) (*283*, *300*, *337*), amines (*14*), phosphines (*67*, *466*), and halides (*83*, *441*) under fairly mild conditions. In contrast, the bisarene complexes are relatively unreactive. Bisarene chromium complexes have been observed to undergo arene exchange in the presence of $AlCl_3$ (*207*), but only in the case of bisarene molybdenum complexes has it proved possible to replace an arene with tertiary or ditertiary phosphine ligands (*65*, *182*). Under similar conditions, neither $(C_6H_6)_2Cr$ nor $(C_6H_6)_2V$ showed any tendency to react with phosphines (*65*, *373b*). However, a systematic study of the replacement reactions of bisarene transition metal complexes does not appear to have been carried out. Of particular interest are the bisarene complexes of the Group VIII transition metals containing an excess of electrons over the 18-electron configuration (see, for example, Table II). The presence of electrons in the antibonding e_{1g} orbital (Fig. 1, Section III) should lead to a weakening of the arene metal bond and a possible greater reactivity than observed for Group V and VI analogs.

2. *Nucleophilic Addition*

A widely observed reaction of cationic arene complexes is the addition of hydride, alkyl, or aryl anions to the arene ligand giving cyclohexadienyl derivatives (*51*, *154*, *185*, *243*, *251*). Other nucleophiles such as CN^-, OMe^-, and N_3^- have been employed, but the products are generally less stable (*185*, *432*). The following order of reactivity of π-complexes toward nucleophiles has been noted: cycloheptatriene > arene > C_4Ph_4 > C_5H_5 (*103*). However, the order will depend to some extent on the system (*17*).

3. *Catalysis*

Arene complexes have been found to be catalysts for the polymerization, hydrogenation, dismutation, and oxidative dimerization of olefins

(*166*, *181*, *270*, *411*). They have also proved useful in the study of nitrogen fixation (*189*, *429*) and the catalytic reduction of molecular oxygen to hydrogen peroxide (*192*). The ability of the arene ligand to stabilize electron-rich metal complexes and the large number of arene complexes covering a wide range of electronic and stereochemical environments (see Tables III to VI) ensure that this area of study will continue to be fruitful.

III
STRUCTURE AND BONDING

A. Bisarene Complexes

The parent compound bisbenzene chromium has been the subject of several structural analyses. The compound has a sandwich structure with the benzene rings in an eclipsed configuration. However, there has been considerable controversy as to whether the molecule possesses full D_{6h} symmetry or whether the true symmetry might be D_{3d} due to the presence of alternating short and long carbon–carbon bonds as in a Kékulé benzene structure (*461*). A recent low-temperature X-ray analysis (*246*) and a neutron diffraction study (*2*) on $(C_6H_6)_2Cr$ showed no evidence for any significant deviation from D_{6h} symmetry. Heat capacity measurements from 5° to 350°C are also consistent with D_{6h} symmetry in $(C_6H_6)_2Cr$ with rotation about the ring-to-metal bonds (*13*). However, another neutron diffraction study (*165*) reported D_{3d} symmetry with alternating C—C bond lengths of 1.406 and 1.424 Å and Cr—C bond lengths of 2.095 and 2.137 Å. Standard deviations, however, were between 0.03 and 0.05 Å. Electron diffraction studies have shown $(C_6H_6)_2Cr$ to possess D_{6h} symmetry in the gas phase with the C—C bond lengths equal to 1.423 ± 0.002 Å (*196*). Infrared studies also support D_{6h} symmetry for $(C_6H_6)_2Cr$ in the gas phase (*323*, *366*). The crystal structure of the Cr(I) compound $[(MeC_6H_5)_2Cr]I$ has also been determined and again there is no evidence for deviation from D_{6h} local symmetry (*383*).

The photoelectron spectra (PES) of several neutral bisarene–metal complexes have been obtained (Table I) and the spectra assigned on the basis of a simple MO model (*110*, *111*). An energy level diagram was deduced and is shown in Fig. 1. The predominantly ring e_{1u} and e_{1g}

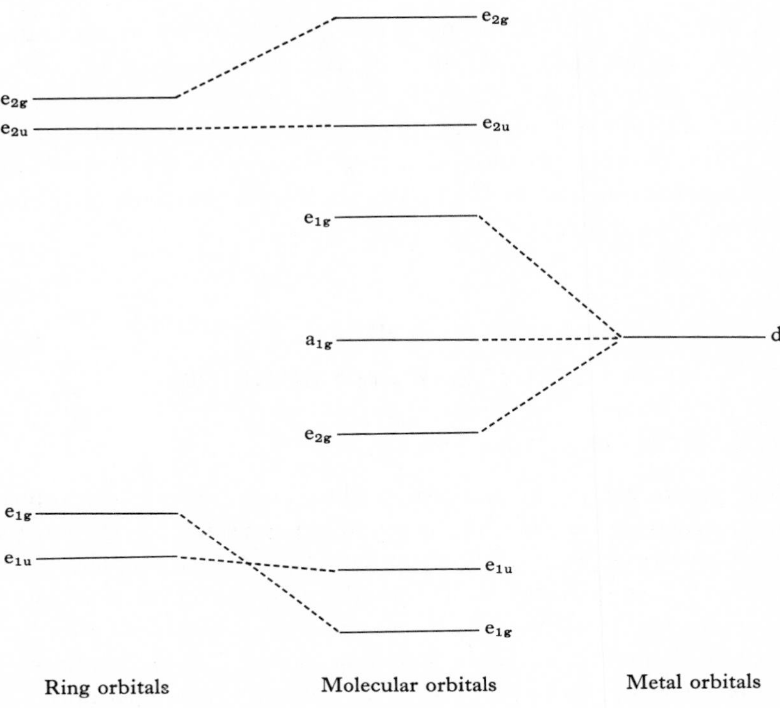

FIG. 1. Qualitative MO Scheme for bis-π-arene metal complexes.

MO's and the predominantly metal e_{2g} and a_{1g} MO's are completely filled in 18-electron bisarene complexes such as those of Cr(0) and Mo(0). The data (Table I) and the line widths of the observed PES bands showed that the a_{1g} orbital is largely nonbonding and lies approximately 1 eV higher in energy than the bonding e_{2g} orbital for the bisarene chromium and molybdenum complexes. A similar ordering of a_1 and e_2 levels was observed for the isoelectronic compound (C_6H_6)-$(\pi$-$C_5H_5)$Mn. The greater separation between the essentially ligand e_{1u} and e_{1g} orbitals for the bisarene molybdenum compounds as compared with the bisarene chromium compounds was attributed to a stronger arene ring–metal e_1 interaction in the former compounds. Also, the larger separation between the metal a_1 and e_2 orbitals observed for $(C_6H_6)_2Cr$ as compared to $(C_6H_6)(\pi$-$C_5H_5)$Mn was partially attributed to the effect of a change in the metal (*110*), i.e., the d-orbital energies are higher in chromium than in manganese, and there should, therefore,

TABLE I

PHOTOELECTRON SPECTRA OF SOME BISARENE–METAL COMPLEXES

Compound	Ionization energies(eV) (assignment)							
$(C_6H_6)_2Cr$	5.4	(a_{1g})	6.4	(e_{2g})	9.6	$(e_{1u} + e_{1g})$		—
$(MeC_6H_5)_2Cr$	5.24	(a_{1g})	6.19	(e_{2g})	9.16	(e_{1u})	9.53	(e_{1g})
$(1,3,5\text{-}Me_3C_6H_3)_2Cr$	5.01	(a_{1g})	5.88	(e_{2g})	8.90	$(e_{1u} + e_{1g})$		—
$(C_6H_6)_2Mo$	5.52	(a_{1g})	6.59	(e_{2g})	9.47	(e_{1u})	10.15	(e_{1g})
$(MeC_6H_5)_2Mo$	5.32	(a_{1g})	6.33	(e_{2g})	9.05	(e_{1u})	9.75	(e_{1g})
$(1,3,5\text{-}Me_3C_6H_3)_2Mo$	5.13	(a_{1g})	6.03	(e_{2g})	8.63	(e_{1u})	9.31	(e_{1g})
$(C_6H_6)(C_5H_5)Mn$	6.36	$(3a_1)$	6.72	$(1e_2)$				

be a stronger metal–ligand e_2 interaction in the chromium compound than in the manganese compound. The effect of replacing a π-C_5H_5 ligand with a π-C_6H_6 ligand has also been shown to lead to a greater metal–ligand e_2 interaction (*372*). A similar argument based on a change in the metal may account for the fact that $(C_6H_6)_2Ti$ is diamagnetic (see Table II), whereas the isoelectronic cation $[(1,3,5\text{-}Me_3C_6H_3)_2V]^+$ has two unpaired electrons, i.e., the a_{1g}–e_{2g} separation is greater than

TABLE II

MAGNETIC PROPERTIES OF SOME BISARENE COMPLEXES

Compound or cation	No. of unpaired electrons	μ eff Exptl. (BM)	μ eff Spin only (BM)	References
$(C_6H_6)_2Ti$	0	0	0	*27*
$(C_6H_6)_2V$	1	1.68 ± 0.08	1.73	*131*
$[(1,3,5\text{-}Me_3C_6H_3)_2V]^+$	2	2.80 ± 0.17	2.83	*131*
$(C_6H_6)_2Cr$	0	0	0	*148*
$(C_6H_6)_2Mo$	0	0	0	*148*
$[(C_6H_6)_2Cr]^+$	1	1.77	1.73	*148*
$[(1,3,5\text{-}Me_3C_6H_3)_2Fe]^{2+}$	0	0	0	*148*
$[(Me_6C_6)_2Fe]^+$	1	1.89	1.73	*149*
$(Me_6C_6)_2Fe$	2	3.08	2.83	*149*
$[(Me_6C_6)_2Co]^{2+}$	1	1.73 ± 0.05	1.73	*141*
$[(Me_6C_6)_2Co]^+$	2	2.95 ± 0.08	2.83	*141*
$(Me_6C_6)_2Co$	1	1.86	1.73	*142*
$[(Me_6C_6)_2Ni]^{2+}$	2	3.00 ± 0.09	2.83	*272*

the spin-pairing energy in the titanium compound but less than the spin-pairing energy in the vanadium compound. Molecular orbital calculations have shown that the a_{1g}–e_{2g} separation is approximately doubled on going from the cation $[(C_6H_6)_2Cr]^+$ to the isoelectronic compound $(C_6H_6)_2V$ (*8*, *9*).

The energy level ordering in Fig. 1 is in agreement with the PES results on the 17-electron compound $(1,3,5\text{-}Me_3C_6H_3)_2V$ (*111*) and is in accord with EPR evidence indicating that the ground states of bisarene vanadium(0) compounds and the isoelectronic cation $[(C_6H_6)_2Cr]^+$ are $^2A_{1g}$ (*8*, *333*, *347*).

The observed experimental ionization energies for the bisarene complexes are in reasonable agreement with detailed MO calculations (*8*, *95*, *372*).

A number of bisarene complexes are known which can be formulated as 19- or 20-electron compounds. The magnetic properties of these compounds (Table II) can be accounted for by addition of 1 or 2 electrons into the antibonding e_{1g} orbital of Fig. 1. Solution contact shift studies (*9*) on the cations $[(Me_6C_6)_2Fe]^+$ and $[(arene)_2Co]^+$ have shown that the unpaired electron(s) are in an e_{1g} orbital, in accord with Fig. 1, rather than in an essentially ligand e_{2u} orbital as had previously been proposed (*34*). The magnetic properties of $(Me_6C_6)_2Co$ can be accounted for by addition of 3 electrons to the e_{1g} orbital of Fig. 1, or it is possible that the complex is a 19-electron compound containing a nonplanar tetrahapto-coordinated ring as has been observed for $(Me_6C_6)_2Ru$ (*231*).

Although the bisarene complexes have been interpreted as being primarily "π-complexes," recent evidence suggests that the σ framework of the ring MO's plays an important role in the bonding as well. Extended Hückel calculations on metalocenes and bisarene complexes show that the σ framework of the ring orbitals is comparable in importance to the π framework in bonding to the metal (*8*, *16*). The proton hyperfine structure which has been observed in the EPR spectrum of bisarene complexes (*8*, *191*, *330*, *333*, *347*) can only be accounted for by extensive delocalization of the ring σ MO's (*7*, *8*). Nuclear magnetic resonance, contact shift studies also show the importance of σ delocalization in bisarene complexes (*9*).

Mass spectral studies on bisarene chromium derivatives have shown that the strength of the metal arene bond increases when electron-releasing substituents are present in the order: $Me_6C_6 \gg C_6H_5C_6H_5 \geq$ 1,3,5-$Me_3C_6H_3 > C_6H_6$ (*221*).

B. Arene Chromium Tricarbonyls and Related Compounds

An X-ray diffraction study of $(C_6H_6)Cr(CO)_3$ at room temperature established the staggered configuration (I) (*21*). There was no evidence of a threefold distortion of the ring. However, a recent X-ray and neutron diffraction analysis at 78°K showed that the C—C bonds which project on the Cr—CO bonds are lengthened by about 0.020 Å, with the remaining bonds being similar to free benzene (1.398 Å). Also, the H atoms are displaced from the benzene ring plane toward the Cr atom by

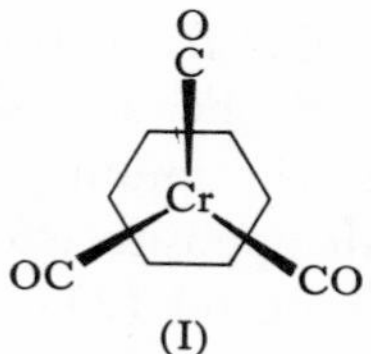

(I)

0.021(3) to 0.038(3) Å (*352*). This is the first established case of ring bond distortion in an arene chromium complex and points out the importance of accurate low-temperature studies in determining whether such distortions exist. The structures of a considerable number of substituted arene chromium tricarbonyl complexes have also been determined. Both staggered and eclipsed configurations have been observed, the configuration depending on the nature of the arene (see Section VI,F).

Molecular orbital calculations have been carried out on arene tricarbonyl complexes of Cr, Mo, and W (*37, 40, 56*). The calculated first ionization potential of $(C_6H_6)Cr(CO)_3$ was found (*40*) to be in good agreement with the experimental value of 7.39 eV as determined by mass spectroscopy (*344*). Calculations have also shown a net positive charge to reside on the arene ring in $(C_6H_6)Cr(CO)_3$ and $(NH_2C_6H_5)Cr(CO)_3$ (*40, 56*). The electron-withdrawing effect of the $Cr(CO)_3$ moiety has long been known from the chemistry of the arene chromium tricarbonyl complexes, for instance $(ClC_6H_5)Cr(CO)_3$ undergoes nucleophilic substitution by methoxide ion much more readily than does free chlorobenzene (*327*). Also, $(NH_2C_6H_5)Cr(CO)_3$ is observed to be a weaker base than free aniline, whereas the benzoic acid complex is a stronger acid than benzoic acid itself (*147*, *327*). Charge distributions comparable to that in $(C_6H_6)Cr(CO)_3$ have been observed for the Mo and W analogs and the results found to correlate well with the observed trends in dipole moments and IR spectra within the series Cr, Mo, W (*40*).

On the basis of overlap considerations, it has been suggested that the σ MO framework of the arene ring may play an important role in the metal–ligand bonding in arene chromium tricarbonyl complexes (*56*). Unfortunately, there are no MO data on arene chromium tricarbonyl complexes showing the distribution of charge on both the σ- and π-ring MOs. However, several recent chemical and spectroscopic studies on the effect of the $(CO)_3CrC_6H_5$ group as a substituent have clearly shown that the electron-withdrawing effect of the $Cr(CO)_3$ moiety is primarily transmitted through the σ-orbital framework of the ring (*62*, *194*, *252*). Previous studies on the NMR chemical shift of a hydrogen meta to various substituents in arene chromium tricarbonyl complexes showed that the substituents had an opposite effect on the chemical shift as compared to the free arene ligand (*258*, *279*, *286*). The NMR data suggest that the $Cr(CO)_3$ moiety may be modifying the ability of the ring σ-orbitals to transmit inductive effects. Further studies are necessary to determine the relative roles of σ- and π-ring orbitals in the bonding and transmission of substituent effects in arene metal tricarbonyl complexes.

Microcalorimetric studies on substituted arene chromium tricarbonyl complexes and cycloheptatriene chromium tricarbonyl showed that the ligand binding energy decreases in the order: $Me_6C_6 > MeC_6H_5 \sim C_6H_6 >$ cycloheptatriene (*74a*). The overall effect of replacing C_6H_6 by Me_6C_6 in the chromium tricarbonyl complexes was not as pronounced as in the bisarene chromium complexes (*221*).

Detailed vibrational assignments have been carried out on arene chromium and arene molybdenum tricarbonyl complexes (*78*, *79*). Splitting of the E band in the carbonyl region was observed for substituted benzene chromium tricarbonyl complexes but not in $(C_6H_6)Cr(CO)_3$ itself, showing that the concept of local symmetry (C_{3v}) is of very restricted validity when discussing the C—O stretching vibrations in such complexes (*78*).

IV

TITANIUM AND ZIRCONIUM COMPLEXES

A. Bisarene Complexes

Bisarene titanium complexes cannot be prepared by the Fischer–Hafner aluminum reduction method as reduction does not proceed beyond the Ti(II) stage. However, $(C_6H_6)_2Ti$ has recently been prepared in good yield by the cocondensation of benzene and titanium vapor at liquid nitrogen temperature (*27*). The red diamagnetic compound showed a parent ion peak in the mass spectrum and fragmented by successive loss of benzene. The observation of the ion $(C_6H_6)Ti^+$ in the mass spectrum of $(\pi\text{-}C_7H_7)(\pi\text{-}C_5H_5)Ti$ has led to the proposal that the parent ion $(C_6H_6)_2Ti^+$ is initially formed by a novel rearrangement of the ion $(\pi\text{-}C_7H_7)(\pi\text{-}C_5H_5)Ti^+$ (*421*). A similar conclusion has been arrived at for the mono and dimethyl derivatives $(\pi\text{-}C_7H_6R)(\pi\text{-}C_5H_4R)Ti$ (R = H or Me) (*422*).

B. Monoarene Complexes

The reaction of C_6H_6, $TiCl_4$, Al, and $AlCl_3$ leads to a product of stoichiometry $C_6H_6 \cdot TiCl_2 \cdot Al_2Cl_6$ for which structure (II) has been proposed (*92*, *284*, *302*, *337*). Toluene, 1,3,5-mesitylene, 1,2,4,5-tetramethylbenzene, and hexamethylbenzene analogs have been similarly prepared (*284*, *337*, *427*). The 1,2,4,5-tetramethylbenzene and hexamethylbenzene derivatives were also prepared by ligand exchange on (II) (*337*).

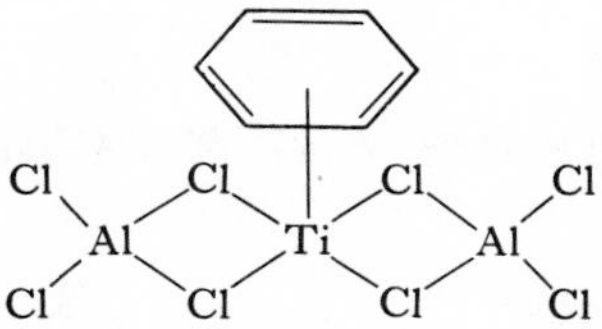

(II)

On heating (II) at 130°C under N_2, a product of stoichiometry C_6H_6-$(TiCl_2Al_2Cl_6)_3N$ was obtained (*429*). Hydrolysis of this product, which is presumably a nitride, gave a stoichiometric amount of ammonia.

Compound (II) is a catalyst for the polymerization of olefins (*284*, *454*, *465*) and the cyclotrimerization of butadiene (*426*).

When hexamethylbenzene, $TiCl_4$, Al, and $AlCl_3$ were heated together at 130°C for 6 hours and the resultant mixture carefully hydrolyzed at low temperature, a dark violet solid of stoichiometry $[Ti_3(Me_6C_6)_3Cl_6]Cl$ was obtained (*150*). The compound was observed to be a 1:1 electrolyte in MeOH. The cation is paramagnetic with one unpaired electron and is thought to have the structure (III) (M = Ti). The zirconium analog was prepared in a similar manner and is also thought to have the structure (III) (M = Zr).

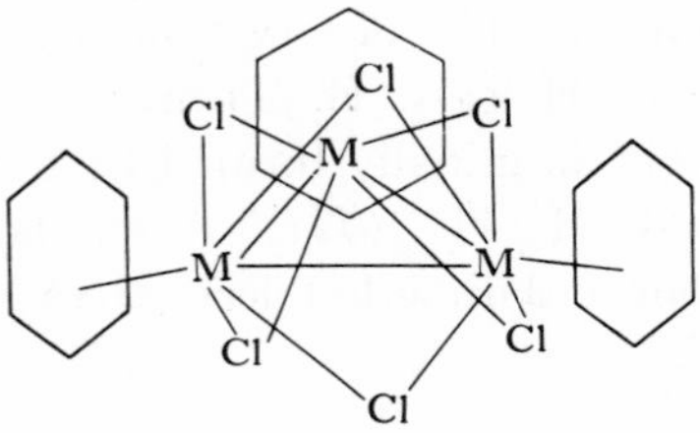

(III)

The reaction of 2-butyne, $TiCl_4$, and $Et_3Al_2Cl_3$ followed by hydrolysis yielded a product of stoichiometry $C_{12}H_{18}Ti(OH)Cl_2$. The same product was obtained when the 2-butyne was replaced by hexamethylbenzene. The compound was formulated as a π-complex of hexamethylbenzene which is supported by the observation of only one methyl resonance in the NMR (*433*).

V

VANADIUM, NIOBIUM, AND TANTALUM COMPLEXES

A. Bisarene Complexes

Bisbenzene vanadium and bismesitylene vanadium have been synthesized by the Fischer–Hafner method (*53*, *133*). The reaction is thought to proceed via formation of the bisarene vanadium cation, e.g.,

$$VCl_4 + Al + AlCl_3 + 2C_6H_6 \longrightarrow [(C_6H_6)_2V][AlCl_4] \tag{7}$$

which yields the neutral complex on hydrolysis. The hydrolysis step involves disproportionation of the bisarene vanadium cation to the neutral complex plus V(II) or V(III) depending on the pH of the solution (*53*):

$$2[(1,3,5\text{-}Me_3C_6H_3)_2V]^+ \xrightarrow[\text{Neutral pH}]{H_2O} (1,3,5\text{-}Me_3C_6H_3)_2V + V(II) + 2(1,3,5\text{-}Me_3C_6H_3) \quad (8)$$

$$3[(1,3,5\text{-}Me_3C_6H_3)_2V]^+ \xrightarrow[\text{Alkaline pH}]{H_2O} 2(1,3,5\text{-}Me_3C_6H_6)_2V + V(III) + 2(1,3,5\text{-}Me_3C_6H_3) \quad (9)$$

The cation $[(1,3,5\text{-}Me_3C_6H_3)_2V]^+$ has been isolated as the iodide salt by hydrolysis of the crude reaction mixture at 0°C in the presence of LiI (*53*). The cation may also be prepared from the neutral complex by oxidation with $V(CO)_6$ [Eq. (10)] (*53*). The same compound is obtained if $(1,3,5\text{-}Me_3C_6H_3)_2V$ is treated with CO under pressure [Eq. (11)] (*54a*).

$$(1,3,5\text{-}Me_3C_6H_3)_2V + V(CO)_6 \longrightarrow [(1,3,5\text{-}Me_3C_6H_3)_2V][V(CO)_6] \quad (10)$$

$$(1,3,5\text{-}Me_3C_6H_3)_2V \xrightarrow[\text{100 atm}]{CO} V(CO)_6 \xrightarrow{(1,3,5\text{-}Me_3C_6H_3)_2V} [(1,3,5\text{-}Me_3C_6H_3)_2V][V(CO)_6] \quad (11)$$

Compound $(1,3,5\text{-}Me_3C_6H_3)V(CO)_3$ was proposed as an intermediate in the reaction shown in Eq. (11) but was not isolated. Treatment of compound $[(1,3,5\text{-}Me_3C_6H_3)_2V][V(CO)_6]$ with LiI in tetrahydrofuran (THF) precipitated the cation as the iodide salt. Both the neutral and cationic bisarene vanadium complexes are paramagnetic having 1 and 2 unpaired electrons, respectively (*131*).

Bisbenzene vanadium has also been prepared by reacting phenyl magnesium bromide with $VCl_3(THF)_3$ or VCl_4 in ether solution (*263, 264*). Bisbiphenyl vanadium was also produced. Evidence was presented for formation of an unstable σ-phenyl V(II) complex which collapses to the bisarene complexes via a redox rearrangement. The bisarene complexes $(arene)_2V$ (arene = benzene, toluene, mesitylene) have been detected by EPR spectroscopy on reduction of $VO(acac)_2$ with Et_3Al or EtMgBr in the appropriate arene as solvent (*330, 333*). Bisnaphthalene and bisanthracene vanadium(0) have also been detected by EPR spectroscopy on reduction of VCl_3 in THF with 3 equivalents of sodium naphthalide or lithium anthracide, respectively (*334, 335*). With excess lithium naphthalide, the EPR signal was observed to disappear, possibly due to the formation of $(C_{10}H_8)_2V(-I)$.

The observation of the ion $(C_6H_6)V^+$ in the mass spectrum of $(\pi\text{-}C_7H_7)(\pi\text{-}C_5H_5)V$ (*296, 354*) has led to the proposal that the parent ion $(C_6H_6)_2V^+$ is initially formed by rearrangement of the ion $(\pi\text{-}C_7H_7)$-$(\pi\text{-}C_5H_5)V^+$ in a vibrationally or electronically excited state (*354*).

B. Arene Vanadium Carbonyl Complexes

Vanadium hexacarbonyl reacts with various arenes yielding complexes $[(\text{arene})V(CO)_4]^+[V(CO)_6]^-$ from which hexafluorphosphate or tetraphenylborate salts can be prepared by metathesis reaction (*50, 52*). The cation $[(C_6H_6)V(CO)_4]^+$ reacts with $NaBH_4$ to give the neutral cyclohexadienyl compound $(C_6H_7)V(CO)_4$ (*51*). Similar results are observed with methyl-substituted derivatives. The addition of the hydride ion was shown to occur preferentially at unsubstituted positions of the aromatic ring.

Treatment of the cation $[(1,3,5\text{-}Me_3C_6H_3)V(CO)_4]^+$ with NaI in THF gives the neutral compound $(1,3,5\text{-}Me_3C_6H_3)V(CO)_3I$ which reacts further with $NaBH_4$ yielding the hydride $(1,3,5\text{-}Me_3C_6H_3)V(CO)_3H$. Attempts to prepare the 17-electron compound $(1,3,5\text{-}Me_3C_6H_3)V(CO)_3$ have been unsuccessful. However, the anion $[(1,3,5\text{-}Me_3C_6H_3)V(CO)_3]^-$ is readily formed on treating $(1,3,5\text{-}Me_3C_6H_3)V(CO)_3H$ with aqueous sodium hydroxide (*83*).

C. Arene Metal Halide Complexes

The reaction of $NbCl_5$ or $TaCl_5$ with Me_6C_6, Al, and $AlCl_3$ yielded, after hydrolysis, ionic complexes of stoichiometry $[M_3(Me_6C_6)_3Cl_6]Cl$. In the case of Nb, the neutral compound $(Me_6C_6)_2Nb_2Cl_4$ was also obtained. On the basis of molecular weight determinations and the observed diamagnetism of the compounds, structures (III) (M = Nb, Ta) and (IV) were proposed for the cations and neutral complex, respectively (*150*). However, the cations have recently been observed to undergo oxidation by a variety of oxidizing agents to give cations of stoichiometry $[M_3(Me_6C_6)_3X_6]^{2+}$ (X = Cl, Br) (*254*). The observed diamagnetism of the cations requires an even number of metal atoms and the hexametallic formulation $[M_6(Me_6C_6)_6X_{12}]^{4+}$ was proposed. The authors suggested reformulation of (III) (M = Nb, Ta) as the dication which would give the overall 2-electron oxidation reaction,

$$[M_6(Me_6C_6)_6X_{12}]^{2+} \xrightarrow{-2e} [M_6(Me_6C_6)_6X_{12}]^{4+} \tag{12}$$

analogous to the known oxidation of $[(M_6Cl_{12})Cl_6]^{4-}$ to $[(M_6Cl_{12})Cl_6]^{2-}$ (M = Nb, Ta).[1]

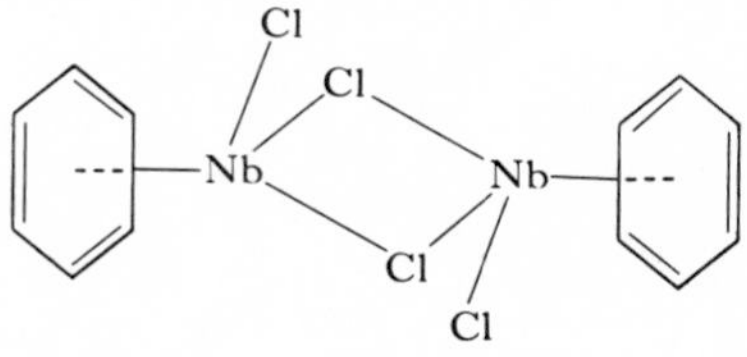

(IV)

VI

CHROMIUM, MOLYBDENUM, AND TUNGSTEN COMPLEXES

A. Preparation of the Bisarene Complexes

1. *Grignard Synthesis*

The Grignard method was employed by Hein (1919) in the first preparation and isolation of organochromium compounds (*204*). However, it was not until 1954 that the compounds prepared by Hein were recognized as being π-arene complexes (*455*, *458*). The reaction involves treatment of anhydrous $CrCl_3$ with PhMgBr in diethyl ether. The black pyrophoric material obtained yields, after hydrolysis under nitrogen, bis-π-biphenyl chromium (1–5%), π-benzene-π-biphenyl chromium (10–12%), and bis-π-benzene chromium (10–12%) (*224*). Complete removal of THF from the complex $Ph_3Cr(THF)_3$ also gives a similar black intermediate and approximately the same yield of bisarene complexes after hydrolysis (*415*). The exact nature of the black intermediate material is unknown, however, it is thought to contain the bisarene π-complexed radicals (V) and (VI) along with bis-π-biphenyl chromium (*197*, *224*). This hypothesis is in agreement with deuterolysis experiments in which deuterium is incorporated only into the benzene rings (*456*) and also with the observation that only bis-π-biphenyl chromium can be extracted prior to hydrolysis (*197*). It is interesting to note that a paramagnetic species has been prepared by reacting the cation $[(C_6H_6)_2Cr]^+$ with a strong base for which the structure $(\pi\text{-}C_6H_6)(\pi\text{-}C_6H_5)Cr$ was suggested on the basis

[1] Structure III has recently been confirmed for the cation $[Nb_3(Me_6C_6)_3Cl_6]^+$ in the solid state [M. R. Churchill and S. W.-Y. Chang, *J. Chem. Soc., Chem. Commun.*, 248 (1974)].

of EPR evidence (*107*)—such a structure is closely related to that of compound (VI).

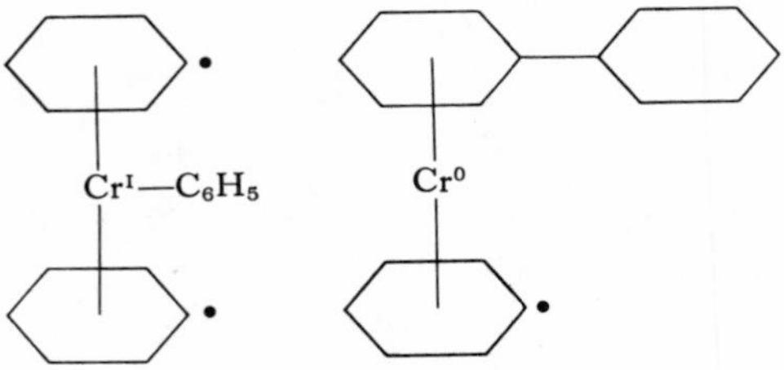

$Cr^{I}—C_6H_5$ Cr^{0}

(V) (VI)

Compounds (p-tolyl)$_3$Cr·4THF, (m-tolyl)$_3$Cr·3THF, (p-biphenyl)$_3$-Cr·3THF, and (mesityl)$_3$Cr·3THF have also been shown to yield mixtures of π-complexes containing both coupled and uncoupled aromatic ligands, after removal of THF and hydrolysis (*386*, *413*). Similarly, (benzyl)$_3$Cr·3THF yielded π-2-benzyltoluene-π-toluene chromium in which an orthophenyl hydrogen has been transferred to a CH_2 group (*174*) or π-bibenzyl-π-toluene chromium (*413*), depending on the reaction conditions. However, little or no π-complex was obtained from ($PhCHCH_2$—)$_3$Cr·3THF, (PhCH═CH—)$_3$Cr·3THF, or (PhC≡C—)$_3$Cr·3THF and this was attributed to the lower proximity to the metal of the aromatic group during homolytic bond cleavage as compared to the phenyl and benzyl derivatives (*413*).

Treatment of the black intermediate obtained from the reaction between $CrCl_3$ and phenyl Grignard in diethyl ether with CO_2, followed by hydrolysis and acidification, yielded the cation $[(\pi\text{-}C_6H_5COOH)(\pi\text{-}C_6H_5C_6H_5)Cr]^+$. No $[(\pi\text{-}C_6H_5COOH)_2Cr]^+$ was isolated, presumably due to its decomposition into free benzoic acid under the conditions employed (*43*).

2. *Cyclic Condensation Reactions*

Trialkyl and triaryl Cr(III) compounds are effective in promoting the cyclic trimerization of disubstituted acetylenes to aromatic hydrocarbons (*222*, *457*). In the reaction of Ph_3Cr·3THF with 2-butyne, the bisarene complexes of hexamethylbenzene and 1,2,3,4-tetramethylnaphthalene are obtained, as well as the free aromatic hydrocarbons (*222*). It is clear that a phenyl substituent has been incorporated in the formation of the coordinated and uncoordinated 1,2,3,4-tetramethylnaphthalene. Although the presumed intermediate in the cyclotrimerization reaction, $Ph_3Cr(MeC{\equiv}CMe)_3$, has not been isolated, isotopic studies (*442*, *443*) are consistent with such an intermediate and rule out a cyclobutene type of intermediate, as has been proposed (*455*) to account for the formation of the 1,2,3,4-tetramethylnaphthalene.

3. *Fischer–Hafner Synthesis*

Bisarene cations of Cr, Mo, and W are best synthesized by the reaction of aromatic hydrocarbon, metal halide, aluminum trihalide and aluminum metal at elevated temperature. Aromatic hydrocarbons, such as benzene, toluene, mesitylene, and hexamethylbenzene have been employed (Table III). The method is not applicable, however, when the aromatic hydrocarbon is itself reactive under Friedel–Crafts conditions. The bisarene cations are readily reduced to the neutral compounds by aqueous dithionite or, in the case of Mo and W, by alkaline disproportionation, e.g. (*152*),

$$6[(C_6H_6)_2Mo]^+ + 8OH^- \longrightarrow 5(C_6H_6)_2Mo + MoO_4^{2-} + 4H_2O + 2C_6H_6 \quad (13)$$

The yield of bisarene complex prepared under Fischer–Hafner conditions appears to decrease in the order Cr > Mo > W for a given arene, the yield in the case of $(C_6H_6)_2W$ being only 2% (*152*). The neutral bisarene complexes are readily oxidized back to their cations by iodine (*152*) and, in the case of Cr, by oxygen (*192*) or $AlCl_3$ (*208*):

$$3(\text{arene})_2Cr + 4AlCl_3 \longrightarrow 3[(\text{arene})_2Cr]^+[AlCl_4]^- + Al \quad (14)$$

Several modifications of the aluminum method are known. The cation $[(MeC_6H_5)_2Cr]^+$ has been prepared by the reduction of $Cr(acac)_3$ with R_3Al (R = Et, *i*-Bu) or $(i\text{-Bu})_2AlH$ in toluene (*333*). Bisarene cation complexes of Cr, Mo, and W have been prepared directly by the reaction of aromatic hydrocarbon, $AlCl_3$, HCl, and the metal powder (*70*).

4. *Arene Exchange Synthesis*

Bisarene chromium compounds have been shown to undergo reversible exchange with aromatic hydrocarbons in the presence of aluminum trihalides (*207*). The exchange reaction has been used to improve the preparation of the bisbenzene chromium cation via the more readily prepared bismesitylene chromium cation (*157b*).

TABLE III

BISARENE–METAL COMPLEXES

Compound	References
$(C_6H_6)_2Ti$	*27*
$(C_6H_6)_2V$	*133, 263, 264*
$(1,3,5\text{-}Me_3C_6H_3)_2V$	*53*
$(C_6H_5C_6H_5)_2V$	*264*
$(C_{10}H_8)_2V$ naphthalene	*334, 335*
$(C_{14}H_{10})_2V$ anthracene	*335*
$[(1,3,5\text{-}Me_3C_6H_3)_2V]^+$	*53, 54a, 133*
$(C_6H_6)_2Cr$	*129, 130, 157b, 223, 407, 408, 413, 436, 456*
$(C_6H_5D)_2Cr$	*121, 456*
$(MeC_6H_5)_2Cr$	*70, 148, 376, 437*
$(EtC_6H_5)_2Cr$	*376, 423*
$(i\text{-}PrC_6H_5)_2Cr$	*288*
$(C_6H_5C_6H_5)_2Cr$	*158, 459*
$(C_{12}H_{12})_2Cr$ tetralin	*148*
$(4\text{-}ClC_6H_4C_6H_5)_2Cr$	*46*
$(C_6H_5COOMe)_2Cr$	*121–123*
$(FC_6H_5)_2Cr$	*376*
$(ClC_6H_5)_2Cr$	*376*
$(Me_3SiC_6H_5)_2Cr$	*105*
$(1,3\text{-}(i\text{-}Pr)_2C_6H_4)_2Cr$	*288*
$(4,4'\text{-}MeC_6H_4C_6H_4Me)_2Cr$	*386*
$(1,4\text{-}F_2C_6H_4)_2Cr$	*376*
$(1,3,5\text{-}Me_3C_6H_3)_2Cr$	*148, 157b, 228*
$(1,3,5\text{-}Et_3C_6H_3)_2Cr$	*70*
$(Me_6C_6)_2Cr$	*148, 445*
$(1,2,3,4\text{-}(C_6H_5)_4C_4H_6)Cr$	*412*
$(C_6H_6)(C_6H_5C_6H_5)Cr$	*212*
$(C_6H_6)(C_6H_5CHO)Cr$	*123*
$(C_6H_6)(C_6H_5CH_2OH)Cr$	*123*
$(C_6H_6)(C_6H_5COMe)Cr$	*123*
$(C_6H_6)(C_6H_5CH(OH)Me)Cr$	*123*
$(C_6H_6)(C_6H_5COOMe)Cr$	*122, 123*
$(C_6H_6)[(C_6H_5)_2CO]Cr$	*122*
$(C_6H_6)[(C_6H_5)_2CHOH]Cr$	*123*
$(C_6H_6)[(C_6H_5)_3COH]Cr$	*123*
$(C_6H_6)(F_6C_6)Cr$	*288*
$[(C_6H_6)_2Cr]^+$	*129, 130, 132, 148, 163, 205, 207, 212, 223, 452, 453, 456*
$[(C_6H_6)(C_6H_5D)Cr]^+$	*106*
$[(MeC_6H_5)_2Cr]^+$	*386*
$[(i\text{-}PrC_6H_5)_2Cr]^+$	*288*
$[(C_6H_5C_6H_5)_2Cr]^+$	*148, 158, 206–212, 215, 216, 274, 351, 386, 456, 458, 459*

TABLE III—*Continued*

Compound	References
$[(C_{12}H_{12})_2Cr]^+$ tetralin	*148, 207*
$[(4\text{-}ClC_6H_4C_6H_5)_2Cr]^+$	*46, 213*
$[(1,3\text{-}(i\text{-}Pr)_2C_6H_4)_2Cr]^+$	*288*
$[(C_6H_5C_6H_4C_6H_4C_6H_5)_2Cr]^+$	*386*
$[(4,4'\text{-}MeC_6H_4C_6H_4Me)_2Cr]^+$	*386*
$[(1,2,3,4\text{-}(C_6H_5)_4C_4H_6)Cr]^+$	*412*
$[(1,3,5\text{-}Me_3C_6H_3)_2Cr]^+$	*207, 413*
$[(3,3',5,5'\text{-}Me_2C_6H_3C_6H_3Me_2)_2Cr]^+$	*70*
$[(C_6H_6)(C_6H_5C_6H_5)Cr]^+$	*205, 206, 210–212, 214, 456, 459*
$[(C_6H_5D)(C_6H_5C_6H_5)Cr]^+$	*456*
$[(MeC_6H_5)(C_6H_5CH_2CH_2C_6H_5)Cr]^+$	*413*
$[(MeC_6H_5)(4,4'\text{-}MeC_6H_4C_6H_4Me)Cr]^+$	*386*
$[(MeC_6H_5)(2\text{-}(C_6H_5CH_2)C_6H_4Me)Cr]^+$	*174*
$[(4\text{-}ClC_6H_4C_6H_5)(C_6H_5C_6H_5)Cr]^+$	*213*
$[(C_6H_5COOH)(C_6H_5C_6H_5)Cr]^+$	*43*
$[(C_6H_5COOMe)(C_6H_5C_6H_5)Cr]^+$	*43*
$[(C_{10}H_8)_2Cr]^+$ naphthalene	*335*
$(C_6H_6)_2Mo$	*70, 152, 159*
$(MeC_6H_5)_2Mo$	*70, 189*
$(1,3,5\text{-}Me_3C_6H_3)_2Mo$	*189*
$[(C_6H_6)_2Mo]^+$	*70, 116, 152*
$[(MeC_6H_5)_2Mo]^+$	*70*
$(C_6H_6)_2W$	*152*
$[(C_6H_6)_2W]^+$	*152*
$[(2\text{-}Me_2C_6H_4)_2W]^+$	*70*
$[(1,3,5\text{-}Me_3C_6H_3)_2W]^+$	*70*
$[(Me_6C_6)_2Mn]^+$	*416*
$[(C_6H_6)(Me_6C_6)Mn]^+$	*154*
$[(C_6H_6)_2Tc]^+$	*23, 24, 153, 336*
$[(Me_6C_6)_2Tc]^+$	*153*
$[(C_6H_6)_2Re]^+$	*156, 161*
$[(1,3,5\text{-}Me_3C_6H_3)_2Re]^+$	*161*
$[(Me_6C_6)_2Re]^+$	*156*
$(C_6H_6)_2Fe$	*407*
$(Me_6C_6)_2Fe$	*149*
$[(Me_6C_6)_2Fe]^+$	*149*
$[(C_6H_6)_2Fe]^{2+}$	*217, 414*
$[(MeC_6H_5)_2Fe]^{2+}$	*217, 414*
$[(o, m, p\text{-}Me_2C_6H_4)_2Fe]^{2+}$	*414*
$[(1,3,5\text{-}Me_3C_6H_3)_2Fe]^{2+}$	*116, 148, 217*
$[(1,2,4,5\text{-}Me_4C_6H_2)_2Fe]^{2+}$	*414*
$(1,3,5\text{-}Me_3C_6H_3)_2Ru$	*148*
$(Me_6C_6)_2Ru$	*124*

Continued

TABLE III—*Continued*

Compound	References
$[(C_6H_6)_2Ru]^{2+}$	*124, 243*
$[(1,3,5\text{-}Me_3C_6H_3)_2Ru]^{2+}$	*117*
$[(Me_6C_6)_2Ru]^{2+}$	*124*
$[(C_{10}H_8)_2Ru]^{2+}$ naphthalene	*125*
$[(1,3,5\text{-}Me_3C_6H_3)_2Os]^{2+}$	*127*
$(Me_6C_6)_2Co$	*142*
$(C_5H_5BR)_2Co(II)$ (R = Me, C_6H_5, OMe, OH)	*220*
$[(Me_6C_6)_2Co]^+$	*141, 416, 460*
$[(Me_6C_6)_2Co]^{2+}$	*141*
$[(Me_6C_6)_2Rh]^+$	*141*
$[(Me_6C_6)_2Rh]^{2+}$	*141*
$[(Me_6C_6)_2Ni]^{2+}$	*272*

5. *Reduction of a Metal Halide by an Arene Radical Anion*

The reduction of $CrCl_3$ with lithium naphthalide in THF ($LiC_{10}H_8$-to-Cr ratio = 5:1), followed by hydrolysis in the presence of oxygen, leads to the formation of the cation $[(\pi\text{-}C_{10}H_8)_2Cr]^+$ as evidenced by EPR spectroscopy. At higher ratios of $LiC_{10}H_8$ to Cr, EPR signals were observed due to more highly reduced species, possibly $[(C_{10}H_8)_2Cr]^-$ and $[(C_{10}H_8)_2Cr]^{3-}$ (*334, 335*). The reaction of $CrCl_3$, biphenyl, and lithium metal in a 1:2:3 molar ratio in THF followed by addition of KI gives $[(\pi\text{-}C_6H_5\text{-}C_6H_5)_2Cr]I$ in 60% yield (*274*).

6. *Metal Vapor Synthesis*

Cocondensation of metal vapor and organic ligands at liquid nitrogen temperature has proved to be a very useful method for preparing organometallic compounds of the first-row transition elements (*409*), as exemplified by the synthesis of bisbenzene chromium (*408*):

$$Cr(vap) + 2C_6H_6 \xrightarrow{-178°C} (C_6H_6)_2Cr \tag{15}$$

The method is particularly valuable for the preparation of bisarene compounds for which the Fischer–Hafner method is unsuitable. It has thus been possible to prepare for the first time compounds $(ClC_6H_5)_2Cr$,

$(FC_6H_5)_2Cr$, $(p\text{-}F_2C_6H_4)_2Cr$, $(C_6F_6)(C_6H_6)Cr$, $(i\text{-}PrC_6H_5)_2Cr$, and $(m\text{-}(i\text{-}Pr)_2C_6H_4)_2Cr$ (*288*, *376*). The extension of metal vapor synthesis to the second- and third-row transition metals is difficult because of the much lower volatility of these metals. However, bisarene molybdenum complexes have recently been synthesized by electron beam evaporation techniques (*27*).

B. Reactions of the Bisarene Complexes

1. *Arene Replacement Reactions*

Bisarene chromium compounds have proven to be rather inert toward arene replacement by nonaromatic ligands. However, reactions do occur at high temperature between $(C_6H_6)_2Cr$ and good π-acceptor ligands such as CO (*145*), PF_3 (*260*), and tripyridyl (*26*) yielding $Cr(CO)_6$, $Cr(PF_3)_6$, and $Cr(tripyridyl)_2$, respectively. No reaction is observed with diphosphines (*65*). Replacement of the arene ring occurs much more readily in the case of $(C_6H_6)_2Mo$. Thus, $Mo(PF_3)_6$ is prepared under milder conditions than is $Cr(PF_3)_6$ (*261*), whereas diphosphines react at 150°C yielding compounds of the type $Mo(diphos)_3$ (*65*). With some monodentate alkyl and aryl phosphines, replacement of only one ring occurs yielding compounds of the type $(C_6H_6)Mo(PR_3)_3$ under mild conditions (*182*) (see Section VI,G). Bisbenzene molybdenum also reacts with allyl chloride at room temperature yielding the dimer $[(C_6H_6)Mo(\pi\text{-}C_3H_5)Cl]_2$ (*189*).

2. *Metalation*

Bisbenzene chromium has been metalated with amylsodium (*121–123*) or *n*-butyl lithium in the presence of tetramethylethylenediamine (*106*). The presence of one metal atom on the ring appears to activate strongly the molecule toward further metalation. Thus, on treatment of $(C_6H_6)_2Cr$ with *n*-butyl lithium–tetramethylethylenediamine, followed by hydrolysis with D_2O in the presence of oxygen, the cations $[(C_{12}H_{12-n}D_n)Cr]^+$ (n = 1, 2, 3, or 4) were obtained (*106*). Similarly, treatment of $(C_6H_6)_2Cr$ with amyl sodium followed by carbonation and esterification led to the compounds $[C_{12}H_{12-n}(COOMe)_n]Cr$ (n = 1, 2, 3, or 4) (*123*). Compounds $(RC_6H_5)(C_6H_6)Cr$ [R = CH_2OH,

CH(OH)Me, and $C(OH)Ph_2$] were obtained from the reaction of $(C_{12}H_{12-n}Na_n)Cr$ with formaldehyde, acetaldehyde, and benzophenone, respectively after chromatographic separation from the higher-substituted derivatives (*123*). Compound $(Me_3SiC_6H_5)_2Cr$ has been prepared by the reaction of $(C_6H_5Li)_2Cr$ with Me_3SiCl (*105*).

The compounds $(RC_6H_5)_2Cr$ (R = H, Me, Et) and the corresponding cations have been shown to undergo isotopic exchange with EtOD in the presence of sodium ethoxide. The cation exchanged at a greater rate than the neutral compound, whereas alkyl groups caused a reduction in the rate of exchange. Under similar conditions, the free aromatic molecules do not undergo hydrogen exchange, thus demonstrating the increased kinetic C—H acidity in the coordinated aromatic molecule (*266*).

3. *Formation of Charge Transfer Compounds*

Bisbenzene chromium reacts with good π-acceptor Lewis acids to form complexes $(C_6H_6)_2Cr^+ \cdot L^-$ (L = tetracyanoethylene, trinitrobenzene, *p*-quinone, chloranil) in which electron transfer from the $(C_6H_6)_2Cr$ to the Lewis acid has taken place. The complexes are best described as bisbenzene chromium cation salts of radical anions (*163*). The crystal structure of one such compound $[(MeC_6H_5)_2Cr]^+ (TCNQ)^-$ (TCNQ = 7,7,8,8-tetracyanoquinodimethane) has been determined and consists of stacks of TCNQ anions and bisbenzene chromium cations with interplanar spacings of 3.42 Å (*371*).

4. *Decomposition*

Decomposition of $[(MeC_6H_5)_2Cr]I$ proceeds via formation of $(MeC_6H_5)_2Cr$, CrI_2, and toluene at about 200°C. The $(MeC_6H_5)_2Cr$ thus formed further decomposes to chromium metal and toluene at about 320°C (*193*). The thermal decomposition of the tetraphenyl borate salts of some alkyl-substituted bisbenzene chromium cations at 195°–210°C gave the free organic ligands and a compound of stoichiometry $CrBPh_4$ (*339*). The catalytic activity of bisarene compounds, such as in ethylene polymerization (*410*, *411*), may be due to the formation of highly reactive metal atoms via thermal decomposition.

5. *Catalysis*

Bisbenzene chromium has been found to be a catalyst for the polymerization of ethylene at 200°–250°C. Chromium metal was postulated as the active catalyst in the system (*410, 411*). The polymerization of ethylene by bisarene chromium(I) salts in the presence of $(i\text{-Bu})_3Al$ has also been studied (*406*). The catalytic activity was found to be a function of both the arene and the anion present. When bisarene chromium complexes are air oxidized in water, hydrogen peroxide is produced:

$$2(\text{arene})_2Cr + O_2 + 2H_2O \longrightarrow 2[(\text{arene})_2Cr]OH + H_2O_2 \qquad (16)$$

The reaction can be made catalytic by addition of $Na_2S_2O_4$ which reduces the bisarene chromium cation formed back to the neutral species (*192*).

C. *Mixed Sandwich Complexes*

The diamagnetic cations $[(C_6H_6)(\pi\text{-}C_5H_5)M(CO)]^+$ (M = Mo, W) have been prepared by reacting $(\pi\text{-}C_5H_5)M(CO)_3Cl$ with benzene in the presence of $AlCl_3$ (*135*). Hydride reduction of $[(C_6H_6)(\pi\text{-}C_5H_5)Mo(CO)]PF_6$ yielded the paramagnetic compound $(C_6H_6)(\pi\text{-}C_5H_5)Mo$, whereas hydride reduction of the tungsten analog yielded instead the diamagnetic hydride $(\pi\text{-}C_5H_5)(1,3\text{-}C_6H_8)W(CO)H$ (*136, 137*). The hydride $(C_6H_6)(\pi\text{-}C_5H_5)W(H)$ has been prepared in low yield through reaction of 1,3-cyclohexadiene, WCl_6, $(i\text{-Pr})MgBr$, and cyclopentadiene. Under similar conditions with $MoCl_5$ in place of WCl_6, only $(C_6H_6)(\pi\text{-}C_5H_5)Mo$ was obtained (*434*). The chromium compound $(C_6H_6)(\pi\text{-}C_5H_5)Cr$ has been prepared directly from the reaction of $CrCl_3$ with a 1:1 mixture of PhMgBr and C_5H_5MgBr or NaC_5H_5 (*120, 134*). Attempts to acetylate the benzene ring in $(C_6H_6)(\pi\text{-}C_5H_5)Cr$ in the presence of $AlBr_3$ lead instead to expansion of the benzene ring giving the methyl cycloheptatriene cation $[(\pi\text{-}C_7H_6Me)(\pi\text{-}C_5H_5)Cr]^+$. With benzyl chloride in place of acetyl chloride in the foregoing reaction, the cation $[(\pi\text{-}C_7H_6Ph)(\pi\text{-}C_5H_5)Cr]^+$ was obtained (*120*). The parent cation $[(\pi\text{-}C_7H_7)(\pi\text{-}C_5H_5)Cr]^+$ was prepared by reacting $(C_6H_6)(\pi\text{-}C_5H_5)Cr$, C_7H_8, and $AlCl_3$, followed by hydrolysis (*118, 119*). Diamagnetic cations of the type $[(C_6H_5R)(\pi\text{-}C_7H_7)Mo]^+$ (R = H, Me) have recently been prepared by treating the dimers $[(C_6H_5R)Mo(\pi\text{-}C_3H_5)Cl]_2$ with C_7H_8 in the presence of $EtAlCl_2$ (*17*). The cations were shown to undergo arene replacement reactions readily with ligands such as tertiary phosphines and acetonitrile yielding complexes of the type $[(\pi\text{-}C_7H_7)MoL_3]^+$. The benzene derivative was shown to undergo hydride attack giving the neutral cyclohexadienyl complex (C_6H_7)-$(\pi\text{-}C_7H_7)Mo$.

D. *Preparation of the Arene Tricarbonyl Complexes*

The parent compounds $(C_6H_6)M(CO)_3$ (M = Cr, Mo, W) were first prepared by the direct reaction between benzene and the metal hexacarbonyl (*146*, *301*, *326*, *327*, *387*):

$$C_6H_6 + M(CO)_6 \longrightarrow (C_6H_6)M(CO)_3 + 3CO \tag{17}$$

However, yields are generally low, and high temperatures and long reaction times are required. The method is more useful for the preparation of substituted benzene derivatives and compounds containing a condensed ring system (Table IV). Milder reaction conditions and higher yields often result on replacing the metal hexacarbonyl with a trisubstituted derivative such as (diglyme)$Mo(CO)_3$ (*64*), $(CH_3CN)_3W(CO)_3$ (*255*), (4-methylpyridine)$_3Cr(CO)_3$ (*331*), or $(NH_3)_3Cr(CO)_3$ (*349*). Compounds that are inaccessible by the metal hexacarbonyl route, such as $(BrC_6H_5)Cr(CO)_3$ and $(IC_6H_5)Cr(CO)_3$ (*331*), have been prepared by this method. Benzene chromium tricarbonyl has recently been prepared in almost quantitative yield by the reaction of $Cr(CO)_6$ with benzene in the presence of 2-picoline, presumably via intermediates of the type (2-picoline)$_nCr(CO)_{6-n}$ (n = 1–3) (*348*). Benzene chromium tricarbonyl has also been prepared in high yield by the reaction of 1-methoxycyclohexa-1,4-diene with $Cr(CO)_6$. More complex analogs also readily aromatized with loss of methanol yielding adducts such as $(MeOC_6H_5)$-$Cr(CO)_3$ and (1,4-dihydronapthalene)$Cr(CO)_3$ (*29*). A novel route to $(C_6H_6)Cr(CO)_3$ has been observed in the reaction between sodium cyclopentadienide and the cation $[(\pi\text{-}C_7H_7)Cr(CO)_3]^+$ in which a contraction of the tropylium ring occurs to give the complex while the C_5H_5 ring expands giving free benzene (*298*, *299*). A similar reaction was observed with the cation $[(\pi\text{-}C_7H_7)Mo(CO)_3]^+$ yielding (C_6H_6)-$Mo(CO)_3$ (*298*).

The complete mechanism of formation of arene–metal–tricarbonyl complexes from an arene and a metal hexacarbonyl is uncertain. A

TABLE IV

ARENE–CARBONYL COMPLEXES

Compound	References
$(1,3,5\text{-}Me_3C_6H_3)V(CO)_3I$	*83*
$(1,3,5\text{-}Me_3C_6H_3)V(CO)_3H$	*83*
$[(1,3,5\text{-}Me_3C_6H_3)V(CO)_3]^-$	*83*
$[(C_6H_6)V(CO)_4]^+$	*52*
$[(MeC_6H_5)V(CO)_4]^+$	*52*
$[MeOC_6H_5)V(CO)_4]^+$	*50*
$[(1,2\text{-}Me_2C_6H_4)V(CO)_4]^+$	*50*
$[(1,4\text{-}Me_2C_6H_4)V(CO)_4]^+$	*52*
$[(1,2,3\text{-}Me_3C_6H_3)V(CO)_4]^+$	*50*
$[(1,2,4\text{-}Me_3C_6H_3)V(CO)_4]^+$	*50*
$[(1,3,5\text{-}Me_3C_6H_3)V(CO)_4]^+$	*52*
$[(1,2,3,4\text{-}Me_4C_6H_2)V(CO)_4]^+$	*51*
$[(1,2,3,4\text{-}Me_4C_6H_2)V(CO)_4]^+$	*50*
$[(Me_6C_6)V(CO)_4]^+$	*50*
$[(C_{10}H_8)V(CO)_4]^+$ naphthalene	*50*
$(C_6H_6)Cr(CO)_3$	*29, 151, 145, 146, 301, 326, 327, 348*
$(MeC_6H_5)Cr(CO)_3$	*146, 218, 301, 326, 327*
$(EtC_6H_5)Cr(CO)_3$	*235*
$(i\text{-}PrC_6H_5)Cr(CO)_3$	*235, 244*
$(Ph(CH_2)_2C_6H_5)Cr(CO)_3$	*36, 108*
$(Ph(CH_2)_3C_6H_5)Cr(CO)_3$	*75*
$(Ph(CH_2)_4C_6H_5)Cr(CO)_3$	*75*
$(HOCH_2C_6H_5)Cr(CO)_3$	*300, 327*
$(HO(CH_2)_2C_6H_5)Cr(CO)_3$	*236*
$(HOOCCH_2C_6H_5)Cr(CO)_3$	*326*
$(EtOOCCH_2C_6H_5)Cr(CO)_3$	*327*
$(HCOCH_2C_6H_5)Cr(CO)_3$	*76*
$(MeCOCH_2C_6H_5)Cr(CO)_3$	*76*
$(PhCOCH_2C_6H_5)Cr(CO)_3$	*76*
$(Me_3SnCH_2C_6H_5)Cr(CO)_3$	*203*
$(MeCH(OH)C_6H_5)Cr(CO)_3$	*236*
$(CH_2{:}CHC_6H_5)Cr(CO)_3$	*348*
$(PhCH{:}CHC_6H_5)Cr(CO)_3$	*36, 108, 96*
$(RCH{:}CHC_6H_5)Cr(CO)_3$ (R = 4-biphenyl, 4p-terphenyl, α-naphthyl, *p*-cyanophenyl, 4-styrylphenyl, 4-phenethylphenyl)	*96*
$(Ph(CH)_4C_6H_5)Cr(CO)_3$	*47, 282*
$(Me_3CC_6H_5)Cr(CO)_3$	*327, 348*
$(1,1\text{-}(C_6H_5)_2CHCH_3)Cr(CO)_3$	*75*
$(4\text{-}MeC_6H_4C_6H_5)Cr(CO)_3$	*38*
$(4\text{-}FC_6H_4C_6H_5)Cr(CO)_3$	*38*

Continued

TABLE IV—*Continued*

Compound	References
$(HOOCC_6H_5)Cr(CO)_3$	*146, 300, 304*
$(MeOOCC_6H_5)Cr(CO)_3$	*146, 300, 326*
$(HCOC_6H_5)Cr(CO)_3$	*76, 96*
$(MeCOC_6H_5)Cr(CO)_3$	*76, 109, 327, 355*
$(PhCOC_6H_5)Cr(CO)_3$	*76*
$(FC_6H_5)Cr(CO)_3$	*327*
$(ClC_6H_5)Cr(CO)_3$	*146, 300, 327, 328*
$(BrC_6H_5)Cr(CO)_3$	*331*
$(IC_6H_5)Cr(CO)_3$	*331, 348*
$(HOC_6H_5)Cr(CO)_3$	*300*
$(MeOC_6H_5)Cr(CO)_3$	*29, 146, 238, 286, 327*
$(H_2NC_6H_5)Cr(CO)_3$	*146, 300, 326, 327, 328*
$(MeHNC_6H_5)Cr(CO)_3$	*327*
$(Me_2NC_6H_5)Cr(CO)_3$	*300*
$(MeCONHC_6H_5)Cr(CO)_3$	*328*
$(MeSC_6H_5)Cr(CO)_3$	*279*
$(Me_3SiC_6H_5)Cr(CO)_3$	*244, 348, 370*
$(Me_3GeC_6H_5)Cr(CO)_3$	*244, 370*
$(Me_3SnC_6H_5)Cr(CO)_3$	*203, 244, 370*
$(Me_2PhSnC_6H_5)Cr(CO)_3$	*203*
$(Ph_3SnC_6H_5)Cr(CO)_3$	*276*
$(Ph_2BC_6H_5)Cr(CO)_3$	*84*
$(Ph_2PC_6H_5)Cr(CO)_3$	*348*
$(2\text{-}C_6H_5CH_2C_4H_3NH)Cr(CO)_3$ 2-benzylpyrole	*73*
$(2\text{-}C_6H_5C_5H_4N)Cr(CO)_3$ 2-phenylpyridine	*348*
$(2,6\text{-}(C_6H_5)_2C_5H_3N)Cr(CO)_3$	*85*
$(2,4,6\text{-}(C_6H_5)_3C_5H_2N)Cr(CO)_3$	*85*
$(1,2\text{-}Me_2C_6H_4)Cr(CO)_3$	*146, 301, 327*
$(1,3\text{-}Me_2C_6H_4)Cr(CO)_3$	*146, 301, 327*
$(1,4\text{-}Me_2C_6H_4)Cr(CO)_3$	*146, 168, 301, 327*
$(1,3\text{-}(i\text{-}Pr)_2C_6H_4)Cr(CO)_3$	*235*
$(1,4\text{-}(i\text{-}Pr)_2C_6H_4)Cr(CO)_3$	*244*
$(1,3\text{-}(t\text{-}Bu)_2C_6H_4)Cr(CO)_3$	*235*
$(1,4\text{-}(t\text{-}Bu)_2C_6H_4)Cr(CO)_3$	*235, 244*
$(p\text{-}C_6H_4)$ and $(p\text{-}C_6H_4)Cr(CO)_3$ bridged by $(CH_2)_m$ and $(CH_2)_n$ ($m = 1, n = 8$–12; $m = 2, n = 2, 3$; $m = 3, n = 4$; $m = 4, n = 4$–6; $m = 5, n = 5, 6$; $m = 6, n = 6$; $m = 9, n = 9$; $m = 10, n = 10$; $m = 12, n = 12$)	*75*

TABLE IV—*Continued*

Compound	References
(4,4′-$MeC_6H_4C_6H_4Me)Cr(CO)_3$	*38*
(2,2′-$FC_6H_4C_6H_4F)Cr(CO)_3$	*38*
(4,4′-$FC_6H_4C_6H_4F)Cr(CO)_3$	*38*
$(MeC_6H_4R)Cr(CO)_3$	
[R = 4-(*i*-Bu), 4-$CH(OEt)_2$	*168, 283*
= 2-F, 3-F, 4-F, 4-Cl, 4-Br	*168, 239*
= 4-OH, 2-OMe, 4-OMe	*168, 327*
= 2-NH_2, 3-NH_2, 4-NH_2, 2-NMe_2, 4-NMe_2	*168, 327*
= 4-SMe, 4-$Si(Me)_3$]	*168*
$(HOCH_2C_6H_4R)Cr(CO)_3$	
(R = 2-Me, 3-Me, 2-OMe, 3-NMe_2)	*77*
$(HCOC_6H_4R)Cr(CO)_3$	
(R = 4-Me, 4-Cl, 2-OMe, 4-OMe)	*77, 168*
$(HONCHC_6H_4R)Cr(CO)_3$	
[R = H (benzaldoxime)	*76*
= 4-Me, 4-Cl, 2-OMe, 4-OMe]	*77*
$(MeCOC_6H_4R)Cr(CO)_3$	
[R = 2-Me, 4-Me	*218, 279, 355*
= 2-Et, 3-Et, 3-(*i*-Pr), 4-(*i*-Pr), 4-(*t*-Bu), 2-OMe]	*233*
$(HOOCC_6H_4R)Cr(CO)_3$	
(R = 2-Me, 3-Me, 4-Me	*77, 112*
= 4-F	*239*
= 2-OEt, 3-OEt, 2-OPr, 3-OPr	*77, 239*
= 2-OMe, 4-OMe, 2-SMe	*278*
= 2-NMe_2, 3-NMe_2)	*278*
$(MeOOCC_6H_4R)Cr(CO)_3$	
(R = 2-Me, 3-Me, 4-Me	*77, 168, 257, 278*
= 2-F, 3-F, 4-F, 2-Cl, 3-Cl	*239, 257*
= 2-OMe, 4-OMe, 2-OEt, 3-OEt, 2-OPr, 3-OPr, SMe	*77, 239, 257*
= 4-NH_2	*257*
= 3-NMe_2	*77, 239*
= 2-COOMe, 4-COOMe)	*392*
$(NH_2C_6H_4R)Cr(CO)_3$	
(R = 2-F, 4-F, 2-Cl, 3-Cl, 4-Cl)	*239*
(1,4-$F_2C_6H_4)Cr(CO)_3$	*389*
(1,4-$Cl_2C_6H_4)Cr(CO)_3$	*389*
(1,2-$(MeO)_2C_6H_4)Cr(CO)_3$	*286*
(1,3-$(MeO)_2C_6H_4)Cr(CO)_3$	*286*
(1,4-$(MeO)_2C_6H_4)Cr(CO)_3$	*286*
(1,4-$(HOCH_2CH_2OOC)_2C_6H_4)Cr(CO)_3$	*392*
(1,4-$(Me_3Si)_2C_6H_4)Cr(CO)_3$	*244*
(1,4-$(Me_3Sn)_2C_6H_4)Cr(CO)_3$	*203*
(1,4-$(Me_3Si)C_6H_4NMe_2)Cr(CO)_3$	*168*
(1,3,5-$Me_3C_6H_3)Cr(CO)_3$	*146, 238, 326–328*

Continued

TABLE IV—*Continued*

Compound	References
$(1,3,5\text{-}(t\text{-Bu})_3C_6H_3)Cr(CO)_3$	*325*
$(1,3,5\text{-}Ph_3C_6H_3)Cr(CO)_3$	*86*
$(2,4,6\text{-}Ph_3C_5H_2P)Cr(CO)_3$ 2,4,6-triphenyl-1-phosphabenzene	*87*
$(1,2,3\text{-}(MeO)_3C_6H_3)Cr(CO)_3$	*286*
$(1,2,4\text{-}(MeO)_3C_6H_3)Cr(CO)_3$	*286*
$(1,3,5\text{-}(MeO)_3C_6H_3)Cr(CO)_3$	*286*
$(2,3\text{-}Me_2C_6H_3COOH)Cr(CO)_3$	*112*
$(2,5\text{-}Me_2C_6H_3COOH)Cr(CO)_3$	*112*
$(1,2,3,4\text{-}Me_4C_6H_2)Cr(CO)_3$	*346*
$(1,2,3,5\text{-}Me_4C_6H_2)Cr(CO)_3$	*346*
$(1,3,5\text{-}Me_3C_6H_2COOEt)Cr(CO)_3$	*151*
$(1,4\text{-}(t\text{-Bu})_2\text{-}2,5\text{-}(MeO)_2C_6H_2)Cr(CO)_3$	*235*
$(Me_5C_6H)Cr(CO)_3$	*346*
$(Me_6C_6)Cr(CO)_3$	*146*
$[(C_6H_5)_4C_5O]Cr(CO)_3$ tetraphenylcyclopentenone	*36*
$(C_9H_8)Cr(CO)_3$ indane	*138, 256*
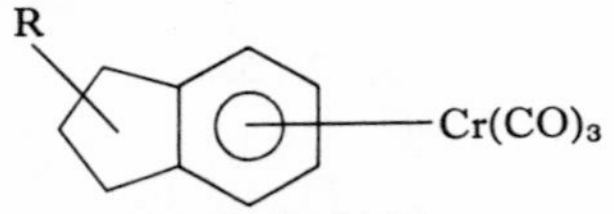 R; $Cr(CO)_3$	
cis- and *trans-*R-Indanes	
[R = 1-Me, 2-Me, 1-(*i*-Pr), 2-(*i*-Pr)	*175, 178*
= 1-CH_2OH, 2-CH_2OH, 1-CH_2OOCMe	*178, 236*
= 1-CH_2CN, 1-CN, 2-CN	*175, 178*
= 1-OH, 2-OH, 1-OOCMe,	*175, 178, 236, 237*
= 1-COOMe, 2-COOMe,	*175, 178, 236*
= 1-SO_2CH_2Me]	*178*
$(C_{13}H_{16}O)Cr(CO)_3$ 5-acetyl-*trans*-1,3-dimethylindane	*176*
$(C_{13}H_{16}O)Cr(CO)_3$ 6-acetyl-*trans*-1,3-dimethylindane	*176*
$(C_{10}H_{12})Cr(CO)_3$ tetralin	*301, 326, 327*
$(C_{10}H_{12}O)Cr(CO)_3$ 1-hydroxytetralin	*226*
$(C_{10}H_{10}O)Cr(CO)_3$ 1-tetralone	*226*
$(C_{10}H_{10})Cr(CO)_3$ 1,2-dihydronaphthalene	*226*
$(C_{10}H_{10})Cr(CO)_3$ 1,4-dihydronaphthalene	*29*
$(C_{10}H_8)Cr(CO)_3$ naphthalene	*90, 146, 300*
$(C_{10}H_9N)Cr(CO)_3$ 1-aminonaphthalene	*58*
$(R_2C_{10}H_6)Cr(CO)_3$	
[R_2 = 1,4-Me_2, 2,3-Me_2, 1,4-F_2, 1,4-$(OMe)_2$, 2,3-$(OMe)_2$, 1,4-$(NMe_2)_2$, 1,4-$(COOMe)_2$]	*91*
$(1,2,3,4\text{-}Me_4C_{10}H_4)Cr(CO)_3$	*91*
$(1,4,6,7\text{-}Me_4C_{10}H_4)Cr(CO)_3$	*91*
$(C_{12}H_8)Cr(CO)_3$ acenaphthylene	*256*
$(C_{12}H_{10})Cr(CO)_3$ biphenylene	*348*
$(C_{13}H_{10})Cr(CO)_3$ fluorene	*138*

TABLE IV—*Continued*

Compound	References
$(C_{14}H_{10})Cr(CO)_3$ anthracene	*90, 139, 446*
$(C_{14}H_{10})Cr(CO)_3$ phenanthrene	*90, 139, 256*
$(C_{14}H_{12})Cr(CO)_3$ dihydroanthracene	*139*
$(C_{14}H_{12})Cr(CO)_3$ 9,10-dihydrophenanthrene	*295*
$(C_{15}H_{12})Cr(CO)_3$ 1-methylphenanthrene	*38*
$(C_{16}H_{10})Cr(CO)_3$ pyrene	*90, 256*
$(C_{16}H_{10})Cr(CO)_3$ fluoranthene	*90*
$(C_{16}H_{10})Cr(CO)_3$ benzo[*b*]fluorene	*90*
$(C_{16}H_{12})Cr(CO)_3$ dibenzo[*a,e*]cyclooctene	*297*
$(C_{18}H_{12})Cr(CO)_3$ chrysene	*90, 139*
$(C_{18}H_{12})Cr(CO)_3$ benz[*a*]anthracene	*90*
$(C_8H_6O)Cr(CO)_3$ benzofuran	*128*
$(C_{12}H_8O)Cr(CO)_3$ dibenzofuran	*128*
$(C_{16}H_{10}O)Cr(CO)_3$ benzo[*b*]naphtho[2,3-*d*]furan	*128*
$(C_8H_7N)Cr(CO)_3$ indole	*128*
$(C_{12}H_9N)Cr(CO)_3$ carbazole	*128*
$(C_{13}H_9N)Cr(CO)_3$ benzo[*h*]quinoline	*128*
$(C_{13}H_9N)Cr(CO)_3$ benzo[*f*]quinoline	*128*
$(C_{17}H_{11}N)Cr(CO)_3$ benzo[*a*]acridine	*128*
$(C_8H_6S)Cr(CO)_3$ benzo[*b*]thiophene	*128*
$(C_{12}H_8S)Cr(CO)_3$ dibenzothiophene	*128*
$(C_{16}H_{10}S)Cr(CO)_3$ benzo[*b*]naptho[2,1-*d*]thiophene	*128*
$(C_6H_5C_6H_5)[Cr(CO)_3]_2$	*108*
$(C_6H_5CH_2C_6H_5)[Cr(CO)_3]_2$	*75, 108*
$[C_6H_5(CH)_4C_6H_5][Cr(CO)_3]_2$	*47*
$[(C_6H_5(CH)_4C_6H_5)Cr(CO)_3]Fe(CO)_3$	*47, 282*
$[(C_6H_5(CH)_4C_6H_5)\{Cr(CO)_3\}_2]Fe(CO)_3$	*47*
$(C_6H_5(CH)_2$-(*p*-C_6H_5)-$(CH)_2C_6H_5)[Cr(CO)_3]_2$	*96*
$([C_6H_5(CH)_2$-(4,4′-$C_6H_4C_6H_4$)-$(CH)_2C_6H_5][Cr(CO)_3]_2$	*96*
$(CH_2)_m$ (*p*-C_6H_4) (*p*-C_6H_4)$[Cr(CO)_3]_2$ $(CH_2)_n$ ($m = 4, n = 5; m = 6, n = 6$)	*75*
$[1,3,5\text{-}(C_6H_5)_3C_6H_3][Cr(CO)_3]_2$	*86*
$(1,3,5\text{-}(C_6H_5)_3C_6H_3)[Cr(CO)_3]_3$	*86*
$(2,6\text{-}(C_6H_5)_2C_5H_3N)[Cr(CO)_3]_2$ 2,6-diphenylpyridine	*85*
$(2,4,6\text{-}(C_6H_5)_3C_5H_2N)[Cr(CO)_3]_2$	*85*
$(2,4,6\text{-}(C_6H_5)_3C_5H_2N)[Cr(CO)_3]_3$	*85*

Continued

TABLE IV—*Continued*

Compound	References
$(C_6H_5—NH—C_6H_5)[Cr(CO)_3]_2$	*108*
$(C_6H_5HgC_6H_5)[Cr(CO)_3]_2$	*348*
$\{(C_6H_5)_2P(n\text{-Bu})\}[Cr(CO)_3]_2$	*348*
$\{(C_6H_5)_2PPh)\}[Cr(CO)_3]_2$	*348*
$\{(C_6H_5)_2SnMe_2\}[Cr(CO)_3]_2$	*203*
$(C_{12}H_8)[Cr(CO)_3]_2$ biphenylene	*348*
$(C_{16}H_{12})[Cr(CO)_3]_2$ dibenzo[*a,e*]cyclooctene	*297*
$(C_{16}H_{10}O)[Cr(CO)_3]_2$ benzo[*b*]naptho[2,3-*d*]furan	*128*
$(C_6H_6)Cr(CO)_3\cdot(1,3,5\text{-}(NO_2)_3C_6H_3)$	*227*
$(MeC_6H_5)Cr(CO)_3\cdot(1,3,5\text{-}(NO_2)_3C_6H_3)$	*162, 227, 229, 259*
$(MeOC_6H_5)Cr(CO)_3\cdot(1,3,5\text{-}(NO_2)_3C_6H_3)$	*162, 227, 229, 259*
$(Me_2NC_6H_5)Cr(CO)_3\cdot(1,3,5\text{-}(NO_2)_3C_6H_3)$	*162, 227, 229, 259*
$(C_6H_6)Cr(CO)_2L$	
[L = PPh_3, PEt_3, $P(n\text{-Bu})_3$, $P(C_6H_{11})_3$	*394, 401*
= $P(C_5H_4FeC_5H_5)_3$	*305*
= C_2H_4, cyclopentene, PhC⋮CPh, maleic anhydride	*226, 396*
= CPhOMe (phenylmethoxycarbene)	*25*
= C_5H_5N, quinoline, cyclohexylisonitrile, pyrrolidine	*394, 395, 402*
= N_2	*369*
= $OSMe_2$	*394*
= $S(CH_2)_4$, $OS(CH_2)_4$, $OS(OCH_2)_2$, $OSPh_2$, SO_2]	*402*
$(MeC_6H_5)Cr(CO)_2(PPh_3)$	*273*
$(MeC_6H_5)Cr(CO)_2(CPhOMe)$ phenylmethoxycarbene	*25*
$(MeOC_6H_5)Cr(CO)_2(C_4H_2O_3)$ maleic anhydride	*226*
$(Me_2NC_6H_5)Cr(CO)_2(PPh_3)$	*273*
$(1,4\text{-}Me_2C_6H_4)Cr(CO)_2(CPhOMe)$ phenylmethoxycarbene	*25*
$(1,4\text{-}(MeOOC)_2C_6H_4)Cr(CO)_2L$	
[L = PPh_3, PMe_3, PEt_3, $P(n\text{-Bu})_3$, $P(C_6H_{11})_3$, $As(C_6H_{11})_3$	*394, 401*
= MeCN, PhCN, $PhNH_2$ cyclohexylisonitrile	*393, 395*
= C_5H_5N, $C_5H_{11}N$, quinoline, pyrrolidine	*393, 394, 402*
= PhC⋮CPh, maleic anhydride	*226, 396*
= $S(CH_2)_4$, $OS(CH_2)_4$, $OS(OCH_2)_2$, $OSPh_2$]	*402*
$(1,3,5\text{-}Me_3C_6H_3)Cr(CO)_2L$	
[L = PPh_3, PMe_3, PEt_3, $P(n\text{-Bu})_3$, $P(CH_2Ph)_3$	*394, 401*
= C_2H_4, cyclopentene	*140, 396*
= PhC⋮CPh, maleic anhydride, $C_2(COOEt)_2$	*226, 396*
= CPhOMe (phenylmethoxycarbene)	*25*
= $S(CH_2)_4$, $OS(OCH_2)_2$, $OSMe_2$, SO_2	*390, 394*
= C_5H_5N, quinoline]	*394, 395*
$(Me_6C_6)Cr(CO)_2L$	
[L = PPh_3, PEt_3, $P(n\text{-Bu})_3$, $P(OEt)_3$	*395, 401*
= C_2H_4, PhC⋮CPh, $C_2(COOEt)_2$, maleic anhydride	*226, 396*
= maleic acid, fumaric acid, endic anhydride	*15*
= N_2	*369*
= quinoline, $OSMe_2$	*395*
= $OS(CH_2)_4$, $OS(OCH_2)_2$, $OSPh_2$, SO_2]	*402*

TABLE IV—*Continued*

Compound	References
$[(Me_6C_6)Cr(CO)_2]_2N_2$	*369*
$[(C_6H_5)_2AsCH_2AsPh_2]Cr(CO)_2$	*357*
$[\{(C_6H_5)_3P\}Cr(CO)_2]_2$	*31*
$[\{(3\text{-}MeC_6H_4)_3P\}Cr(CO)_2]_2$	*31*
$[\{(4\text{-}MeC_6H_4)_3P\}Cr(CO)_2]_2$	*31*
$(C_6H_6)Cr(CO)_2(SiCl_3)H$	*240*
$[(C_6H_6)Cr(CO)_2CN]^-$	*157a*
$[(1,3,5\text{-}Me_3C_6H_3)Cr(CO)_3(HgCl)]^+$	*102*
$(C_6H_6)Mo(CO)_3$	*21, 146, 387*
$(MeC_6H_5)Mo(CO)_3$	*387*
$(i\text{-}PrC_6H_5)Mo(CO)_3$	*235*
$(t\text{-}BuC_6H_5)Mo(CO)_3$	*235*
$(FC_6H_5)Mo(CO)_3$	*387*
$(MeOC_6H_5)Mo(CO)_3$	*307*
$(Me_2NC_6H_5)Mo(CO)_3$	*341*
$(Me_3SnC_6H_5)Mo(CO)_3$	*203*
$(1,2\text{-}Me_2C_6H_4)Mo(CO)_3$	*341, 346*
$(1,3\text{-}Me_2C_6H_4)Mo(CO)_3$	*341, 346*
$(1,4\text{-}Me_2C_6H_4)Mo(CO)_3$	*168, 387, 391*
$(1,4\text{-}MeC_6H_4CHMe_2)Mo(CO)_3$	*283*
$(1,4\text{-}MeC_6H_4NMe_2)Mo(CO)_3$	*168*
$(1,4\text{-}MeC_6H_4SiMe_3)Mo(CO)_3$	*168*
$(1,4\text{-}Me_2NC_6H_4SiMe_3)Mo(CO)_3$	*168*
$(1,4\text{-}(Me_3Si)_2C_6H_4)Mo(CO)_3$	*168*
$(1,2,3\text{-}Me_3C_6H_3)Mo(CO)_3$	*346*
$(1,2,4\text{-}Me_3C_6H_3)Mo(CO)_3$	*346*
$(1,3,5\text{-}Me_3C_6H_3)Mo(CO)_3$	*146, 327, 439*
$(1,2,3,4\text{-}Me_4C_6H_2)Mo(CO)_3$	*346*
$(1,2,3,5\text{-}Me_4C_6H_2)Mo(CO)_3$	*346*
$(1,2,4,5\text{-}Me_4C_6H_2)Mo(CO)_3$	*346*
$(Me_5C_6H)Mo(CO)_3$	*346*
$(Me_6C_6)Mo(CO)_3$	*341*
$(C_{12}H_8)Mo(CO)_3$ biphenylene	*64*
$(C_{12}H_8)[Mo(CO)_3]_2$	*64*
$(1,3,5\text{-}Me_3C_6H_3)Mo(CO)_2(C_2H_4)$	*140*
$[(C_6H_5)_2AsCH_2As(C_6H_5)_2]Mo(CO)_2$	*357*
$(C_6H_6)Mo(CO)(PPh_3)_2$	*189*
$(MeC_6H_5)Mo(CO)(PPh_3)_2$	*189*
$[(1,3,5\text{-}Me_3C_6H_3)Mo(CO)_3I]^+$	*379*
$[(1,3,5\text{-}Me_3C_6H_3)Mo(CO)_3HgCl]^+$	*102*
$[(1,3,5\text{-}Me_3C_6H_3)Mo(CO)_3HgBr]^+$	*102*
$[(Me_6C_6)Mo(CO)_3Cl]^+$	*384*
$(C_6H_6)W(CO)_3$	*146, 255*
$(MeC_6H_5)W(CO)_3$	*255, 388*

Continued

TABLE IV—*Continued*

Compound	References
$(MeOC_6H_5)W(CO)_3$	*307, 343*
$(H_2NC_6H_5)W(CO)_3$	*343*
$(Me_2NC_6H_5)W(CO)_3$	*343*
$(MeOOCC_6H_5)W(CO)$	*343*
$(1,2\text{-}Me_2C_6H_4)W(CO)_3$	*346*
$(1,4\text{-}Me_2C_6H_4)W(CO)_3$	*168, 255, 388*
$(1,4\text{-}MeC_6H_4CHMe_2)W(CO)_3$	*86*
$(1,4\text{-}MeC_6H_4NMe_2)W(CO)_3$	*168*
$(1,4\text{-}Me_2NC_6H_4SiMe_3)W(CO)_3$	*168*
$(1,4\text{-}(Me_3Si)_2C_6H_4)W(CO)_3$	*168*
$(1,2,3\text{-}Me_3C_6H_3)W(CO)_3$	*346*
$(1,2,4\text{-}Me_3C_6H_3)W(CO)_3$	*346*
$(1,3,5\text{-}Me_3C_6H_3)W(CO)_3$	*146, 255*
$(1,2,3,4\text{-}Me_4C_6H_2)W(CO)_3$	*346*
$(1,2,3,5\text{-}Me_4C_6H_2)W(CO)_3$	*346*
$(1,2,4,5\text{-}Me_4C_6H_2)W(CO)_3$	*346*
$(Me_5C_6H)W(CO)_3$	*346*
$(Me_6C_6)W(CO)_3$	*384*
$[(Me_6C_6)W(CO)_3Cl]^+$	*380, 385*
$[(Me_6C_6)W(CO)_3I]^+$	*379*
$(C_6H_6)Mn(CO)_2CN$	*430, 431*
$(1,3,5\text{-}Me_3C_6H_3)Mn(CO)_2CN$	*72, 430, 431*
$(1,3,5\text{-}Me_3C_6H_3)Mn(CO)_2(CNBF_3)$	*430*
$(1,3,5\text{-}Me_3C_6H_3)Mn(CO)_2(NCO)$	*14*
$(Me_6C_6)Mn(CO)_2(NCO)$	*14*
$(Me_6C_6)Mn(CO)_2(CONH_2)$	*14*
$(Me_6C_6)Mn(CO)_2[CONH(C_6H_{11})]$	*14*
$[(C_6H_6)Mn(CO)_3]^+$	*447, 448*
$[(MeC_6H_5)Mn(CO)_3]^+$	*448*
$[(1,3,5\text{-}Me_3C_6H_3)Mn(CO)_3]^+$	*71, 448*
$[(Me_6C_6)Mn(CO)_3]^+$	*448*
$[(C_{10}H_8)Mn(CO)_3]^+$ naphthalene	*448*
$[(1,3,5\text{-}Me_3C_6H_3)Mn(CO)_2(CNEt)]^+$	*430*
$[(1,3,5\text{-}Me_3C_6H_3)Mn(CO)_2(CNCPh_3)]^+$	*430*
$[(Me_6C_6)Fe(CO)_2]_2$	*115*
$(1\text{-}CH_2CHC_{10}H_7)Fe(CO)_3$ 1-vinylnaphthalene	*281*
$(2\text{-}CH_2CHC_{10}H_7)Fe(CO)_3$ 2-vinylnaphthalene	*81*
$(C_{14}H_{10})Fe(CO)_3$ anthracene	*281*
$(C_{18}H_{12})Fe(CO)_3$ benz[*a*]anthracene	*22*
$(C_{16}H_{12}O)Fe(CO)_3$ 9-acetylanthracene	*281*
$(C_{13}H_9N)Fe(CO)_3$ acridine	*22*
$(C_{17}H_{11}N)Fe(CO)_3$ benz[*c*]acridine	*22*
$(C_{12}H_8N_2)Fe(CO)_3$ phenazine	*22*
$(C_{16}H_{10}N_2)Fe(CO)_3$ benzo[*a*]phenazine	*22*

TABLE IV—*Continued*

Compound	References
$(C_{20}H_{12}N_2)Fe(CO)_3$ dibenzo[*a*,*c*]phenazine	*22*
$(1,3\text{-}(C_2H_3)_2C_6H_4)[Fe(CO)_3]_2$*m*-divinylbenzene	*282*
$(1,4\text{-}(C_2H_3)_2C_6H_4)[Fe(CO)_3]_2$ *p*-divinylbenzene	*282*
$(C_{10}H_{12})[Fe(CO)_3]_2$ 3,α-dimethylstyrene	*424*
$(C_{12}H_8)[Fe_2(CO)_5]$acenaphthylene	*253*
$(C_{18}H_{12})[Fe_2(CO)_6]$ naphthacene	*22*
$[(CF_3)_6C_6]Ru(CO)_2[P(OMe)_3]$	*44*
$(MeC_6H_5)Ru_6(C)(CO)_{14}$	*241*
$(1,3\text{-}Me_2C_6H_4)Ru_6(C)(CO)_{14}$	*241*
$(1,3,5\text{-}Me_3C_6H_3)Ru_6(C)(CO)_{14}$	*241*
$(C_6H_6)Co_4(CO)_9$	*247*
$(MeC_6H_5)Co_4(CO)_9$	*247*
$(MeOC_6H_5)Co_4(CO)_9$	*247*
$(Me_2C_6H_4)Co_4(CO)_9$	*247*
$(1,3,5\text{-}Me_3C_6H_3)Co_4(CO)_9$	*247*
$(C_8H_{12})Co_4(CO)_9$ tetrahydronaphthalene	*365*
$(C_6H_6)Co_3(CPh)(CO)_6$	*359*
$(MeC_6H_5)Co_3(CPh)(CO)_6$	*359*
$(1,2\text{-}Me_2C_6H_4)Co_3(CPh)(CO)_6$	*359*
$(1,2\text{-}Me_2C_6H_4)Co_3(CMe)(CO)_6$	*359*
$(1,3\text{-}Me_2C_6H_4)Co_3(CPh)(CO)_6$	*359*
$(1,4\text{-}Me_2C_6H_4)Co_3(CMe)(CO)_6$	*359*
$(1,4\text{-}Me_2C_6H_4)Co_3(CPh)(CO)_6$	*359*
$(1,4\text{-}Me_2C_6H_4)Co_3(CF)(CO)_6$	*359*
$(1,3,5\text{-}Me_3C_6H_3)Co_3(CPh)(CO)_6$	*359*
$(1,3,5\text{-}Me_3C_6H_3)Co_3(CMe)(CO)_6$	*358, 359*
$(1,3,5\text{-}Me_3C_6H_3)Co_3(CF)(CO)_6$	*359*
$[(C_6H_6)_3Co_3(CO)_2]^+$	*66, 114*
$[(MeC_6H_5)_3Co_3(CO)_2]^+$	*66*

kinetic study of the reaction of $Mo(CO)_6$ with arenes has established that the initial step is an S_N1 dissociation of the $Mo(CO)_6$ into $Mo(CO)_5$ plus CO (*438*). This step might then be followed by further dissociation or more likely the formation of (arene)$Mo(CO)_5$. Infrared evidence has been presented for the formation of unstable compounds of the type (arene)$W(CO)_5$ on irradiation of $W(CO)_6$ with arenes at −80°C (*385*). When the arene was naphthalene, the compound decomposed at −10°C to give $(C_{10}H_8)W(CO)_3$. This observation supports a mechanism in which a compound of the type (arene)$M(CO)_5$ is first formed in the reaction between an arene and a metal hexacarbonyl followed by intramolecular displacement of CO to give (arene) $M(CO)_3$. This mechanism has, however, been disputed (*327*).

Of considerable interest is the position of attachment of the $Cr(CO)_3$ moiety in benzenoid compounds containing more than one aromatic nuclei. In fused ring systems, such as anthracene, phenanthrene or benzanthrene, the $Cr(CO)_3$ moiety is always bonded to a terminal ring (*90*). Terminal-ring bonding has also been observed in more complex systems where a benzene ring is separated from a naphthalene system by a heterocycle ring as in benzo[*a*]acridine (*128*) and benzo[*b*]naptho-[2,3-*d*]furan (*90*). Terminal-ring bonding in arene–metal–tricarbonyl derivatives has been attributed to the terminal ring being the ring of lowest bond localization energy (*324*). The position of attachment of the $Cr(CO)_3$ moiety in substituted naphthalene and biphenyl derivatives has also been studied. In a series of naphthalene derivatives in which only one ring was substituted, the $Cr(CO)_3$ moiety was found to be bonded to either ring with about equal probability when the substituents were methyl groups but to only the unsubstituted ring when the substituents were OMe, NMe_2, COOMe, or F (*91*). Thus, factors other than the electron density of the naphthalene rings appear to be controlling the position of attachment of the $Cr(CO)_3$ moiety. In contrast, the $Cr(CO)_3$ moiety appears to bond solely to the ring of highest electron density in biphenyl derivatives (*38*).

In general, it has not proved possible to attach more than one $Cr(CO)_3$ moiety to a fused ring system, although a few exceptions have been noted (*128*, *297*, *348*). However, many complexes containing more than one $Cr(CO)_3$ moiety are known with nonfused multiring systems, including the bistricarbonyl chromium complexes of biphenyl (*108*), 1,4-diphenylbutadiene (*47*), diphenyldimethyltin (*203*), diphenylmercury, and triphenylphosphine (*348*). An X-ray study has established a trans configuration for the $Cr(CO)_3$ moieties in $(C_6H_5C_6H_5)[Cr(CO)_3]_2$ (*4*, *74b*). In the case of 1,3,5-triphenylbenzene, complexes of the type $[(C_6H_5)_3C_6H_3][Cr(CO)_3]_n$ ($n = 1$, 2, or 3) have been isolated. With $n = 1$, the $Cr(CO)_3$ moiety is bonded to the central ring, whereas with $n = 2$ or 3, the $Cr(CO)_3$ moieties are bonded to the terminal rings only (*86*). The molecule 2,4,6-triphenylpyridine has also been observed to form complexes of the type $[(C_6H_5)_3C_5H_2N][Cr(CO)_3]_n$ ($n = 1$, 2, or 3). However, in this case the $Cr(CO)_3$ moieties were found to be bonded to the phenyl rings only (*85*). It is interesting to note in this regard that there appears to be no example of a pyridine ring π-bonded to a transition metal.

E. Reactions of the Arene Tricarbonyl Complexes

1. *Arene Replacement Reactions*

The exchange of one arene ligand for another in arene tricarbonyl compounds has been of some synthetic value. For example, $(Me_6C_6)W(CO)_3$ has been prepared by reacting $(1,4\text{-}(i\text{-}Pr)_2C_6H_4)W(CO)_3$ with Me_6C_6 (*283*), whereas $(MeSC_6H_5)Cr(CO)_3$ (*279*) and $(C_6H_5HgC_6H_5)[Cr(CO)_3]_2$ (*348*) have been prepared by treating $(C_6H_6)Cr(CO)_3$ with the respective arenes. Thiophene has also been shown to exchange with the benzene in $(C_6H_6)Cr(CO)_3$ yielding $(\pi\text{-}C_4H_4S)Cr(CO)_3$ (*280*).

The kinetics of the exchange reaction have been studied by the use of ^{14}C-tagged arenes (*397–400, 403*):

$$\text{(arene)M(CO)}_3 + {}^*\text{arene} \rightleftharpoons ({}^*\text{arene})\text{M(CO)}_3 + \text{arene} \qquad (18)$$

In the reaction between ^{14}C-tagged benzene and $(C_6H_6)Cr(CO)_3$, the rate was found to be second order in $(C_6H_6)Cr(CO)_3$ and approximately one-third order in benzene. This result was interpreted in terms of two rate-determining steps—a fast step that is second order in $(C_6H_6)Cr(CO)_3$ and a slow step that is first order in both $(C_6H_6)Cr(CO)_3$ and benzene (*398, 461*). In the second-order step the formation of a bimolecular activated complex is thought to be rate-determining followed by a fast series of reactions leading to the ^{14}C-tagged product as follows:

$$2(C_6H_6)Cr(CO)_3 \xrightarrow{\text{Slow}} (OC)_3Cr\cdots(C_6H_6)\cdots Cr(CO)_3\cdots(C_6H_6)$$

$$\downarrow \text{Fast} \qquad (19a)$$

$$({}^*C_6H_6)Cr(CO)_3 \xleftarrow[\text{Fast}]{{}^*C_6H_6} Cr(CO)_3 + (C_6H_6)Cr(CO)_3 + C_6H_6$$

The step considered to lead to first-order dependence in both $(C_6H_6)Cr(CO)_3$ and benzene is depicted in Eq. (19b). Similar conclusions were arrived at for substituted benzene complexes of Cr, Mo, and W (*400*). The rate of exchange for a given arene was found to increase in the order Mo > W > Cr (*400*).

$$(C_6H_6)Cr(CO)_3 + {}^{*}C_6H_6 \xrightarrow{\text{Slow}} (C_6H_6)\cdots Cr(CO)_3 \cdots ({}^{*}C_6H_6) \xrightarrow{\text{Fast}} C_6H_6 + ({}^{*}C_6H_6)Cr(CO)_3 \qquad (19b)$$

Phosphine and phosphite ligands react with arene tricarbonyl complexes of Cr, Mo, and W yielding complexes of the type *fac*-$(PR_3)_3M(CO)_3$ (*32*, *261*, *341–343*, *466*). Kinetic studies reveal that the reaction proceeds by an S_N2 mechanism (*341–343*, *466*) which is thought to involve the stepwise displacement of the arene via tetrahapto- and dihapto-coordinated arene intermediates (*466*):

$$(C_6H_6)Mo(CO)_3 \xrightarrow{L} (C_6H_6)Mo(CO)_3L \xrightarrow{L} (C_6H_6)Mo(CO)_3L_2 \xrightarrow{L} L_3Mo(CO)_3 \qquad (20)$$

The arene ligand can also be replaced by several anions leading to anionic complexes of the type $[(CO)_3MX_3M(CO)_3]^{3-}$ (M = Mo, W; X = F, Cl, Br, I, OH, SCN, or N_3) (*441*). The reaction of dipyridyl and tripyridyl with $(C_6H_6)Cr(CO)_3$ causes replacement of all of the ligands yielding $Cr(dipyridyl)_3$ and $Cr(tripyridyl)_2$, respectively (*26*). When arene chromium tricarbonyl complexes are irradiated in methanol, the ligands are evolved quantitatively yielding $Cr(OMe)_3$ (*35*).

2. *Carbonyl Replacement Reactions*

Replacement of one carbonyl ligand in arene chromium tricarbonyl complexes has been achieved with a variety of ligands yielding complexes of the type (arene)$Cr(CO)_2L$. The reaction is normally carried out under UV irradiation:

$$(\text{arene})Cr(CO)_3 + L \xrightarrow[\text{Solvent}]{\text{uv}} (\text{arene})Cr(CO)_2L + CO \qquad (21)$$

In some cases, yields are improved by irradiation of the arene chromium tricarbonyl complex in a donor solvent such as THF until 1 equivalent of CO has been evolved followed by addition of the ligand L (*401*). Ligands, such as phosphines (*394*, *401*), amines (*394*), nitriles (*393*), sulfides, sulfoxides (*390*, *402*), olefins, acetylenes (*15*, *396*), hydrazine, and molecular nitrogen (*369*) have been employed (see Table IV). A molybdenum complex (1,3,5-$Me_3C_6H_3$)$Mo(CO)_2(C_2H_4)$ has also been similarly prepared (*140*). The reaction of $(C_6H_6)Cr(CO)_3$ with sodium cyanide yielded the anion $[(C_6H_6)Cr(CO)_2CN]^-$ (*157a*). A new type of hydride complex $(C_6H_6)Cr(CO)_2(SiCl_3)H$ has been prepared by UV irradiation of $(C_6H_6)Cr(CO)_3$ and $SiCl_3H$ (*240*).

Chromium hexacarbonyl reacts with $Ph_2AsCH_2AsPh_2$ yielding a compound of stoichiometry $Cr(CO)_2(Ph_2AsCH_2AsPh_2)$ (*357*) and with triarylphosphines yielding complexes of the type $[Cr(CO)_2L]_2$ [L = PPh_3, P(*m*-tolyl)$_3$, and P(*p*-tolyl)$_3$] (31). The arsine and triphenylphosphine compounds have been shown by X-ray analysis to possess the structures (VII) and (VIII), respectively, in which a phenyl group is

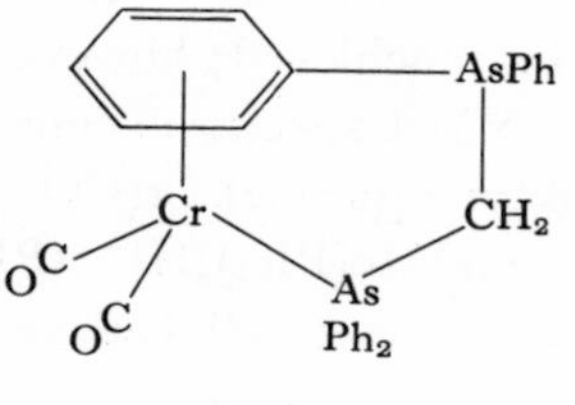

(VII)

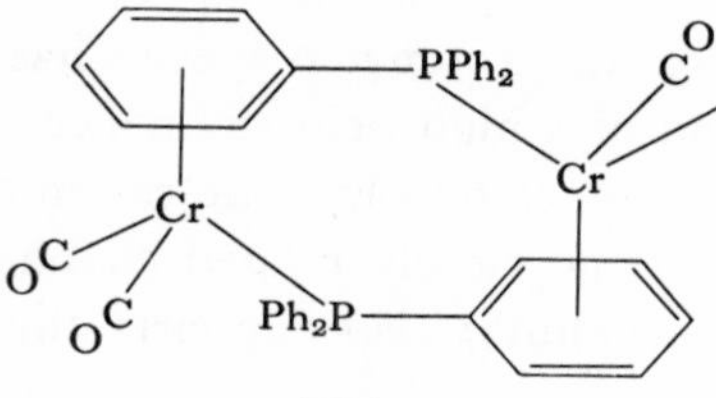

(VIII)

π-bonded to the chromium atom (*356*, *357*).

The replacement of one carbonyl group appears to stabilize the remaining carbonyl groups toward replacement since no disubstituted compounds have been observed.

3. *Oxidation and Protonation Reactions*

Cations of the type $[(Me_6C_6)M(CO)_3Cl]^+$ have been prepared by oxidation of the compounds $(Me_6C_6)M(CO)_3$ (M = Mo, W) with

$SbCl_5$ in dichloromethane solution (*380*, *384*). Iodine oxidation of $(Me_6C_6)W(CO)_3$ yielded the cation $[(Me_6C_6)W(CO)_3I]^+$ for which structure (IX) was determined by X-ray analysis (*379*). Complexes

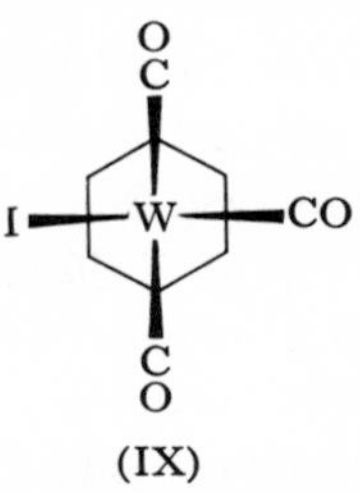

(IX)

$(1,3,5\text{-}Me_3C_6H_3)M(CO)_3$ (M = Cr, Mo) have been shown to react with $HgCl_2$ giving complexes of stoichiometry $(1,3,5\text{-}Me_3C_6H_3)M(CO)_3\text{-}(HgCl_2)_x$ (x = 1 or 2). For x = 2, the compounds were formulated as the HgCl adduct $[(1,3,5\text{-}Me_3C_6H_3)M(CO)_3(HgCl)]^+(HgCl_3)^-$; for x = 1, the compounds were formulated as the neutral $HgCl_2$ adducts $(1,3,5\text{-}Me_3C_6H_3)M(CO)_3HgCl_2$. However, the formulation $[(1,3,5\text{-}Me_3C_6H_3)M(CO)_3HgCl]^+Cl$ could not be excluded (*102*).

Arene chromium tricarbonyl complexes can be protonated by strong acids such as $BF_3 \cdot H_2O—CF_3CO_2H$ (*82*, *267*) and FSO_3H (*271*). Protonation is enhanced by electron-releasing groups on the arene ring and by substitution of one of the carbonyl ligands by PPh_3 (*267*, *273*). Isolation of the protonated species has not been achieved; however, the observation of a high field resonance in the NMR spectra is consistent with the presence of a hydride cation of the type $[(\text{arene})\,Cr(CO)_3H]^+$. Isolation of the closely related cations $[(\text{arene})Mo(PR_3)_3H]^+$ (PR_3 = tertiary phosphine) also supports this formulation (*182*) (see Section VI, G).

4. *Ion–Molecule Reactions*

Ions of the type $[(\text{arene})_2Cr_2(CO)_3]^+$ and $[(\text{arene})_2Cr_3(CO)_6]^+$ have been detected in the mass spectra of several arene chromium tricarbonyl complexes (*172*, *173*). The products were shown to arise through the following ion–molecule reactions:

$$(\text{arene})Cr(CO)_3 + [(\text{arene})Cr(CO)_3]^+ \longrightarrow [(\text{arene})_2Cr_2(CO)_3]^+ + 3CO \quad (22)$$

$$[(\text{arene})_2Cr_2(CO)_3]^+ + (\text{arene})Cr(CO)_3 \longrightarrow [(\text{arene})_2Cr_3(CO)_6]^+ + \text{arene} \quad (23)$$

5. *Formation of Charge Transfer Complexes*

Several arene chromium tricarbonyl complexes form 1:1 adducts with Lewis acids such as tetracyanoethylene and 1,3,5-trinitrobenzene (TNB) (*162*, *227*, *229*, *259*). The TNB adducts have been isolated as crystaline solids and the structure of the anisole derivative determined by X-ray analysis (*57*, *229*). The plane of the TNB ring was found to be parallel to the anisole ring with an average separation of 3.41 Å. This is a somewhat larger separation than that observed in the charge transfer complexes of TNB with aromatic molecules, and the increased separation was attributed to the strong electron-withdrawing capacity of the tricarbonyl chromium moiety which decreases the π-electron donor capacity of the anisole molecule (*229*).

6. *Metalation Reactions*

The lithiated derivative $(LiC_6H_5)Cr(CO)_3$ has been prepared in high yield by the reaction of $[(CO)_3Cr](C_6H_5HgC_6H_5)[Cr(CO)_3]$ with *n*-butyl lithium. Complexes, such as 2-phenylpyridine chromium tricarbonyl and $(Ph_2PC_6H_5)Cr(CO)_3$, which are not otherwise obtainable were prepared by the reaction of the lithiated derivative with pyridine and Ph_2PCl, respectively (*348*). Benzene chromium tricarbonyl has been metalated by treatment with *n*-butyl lithium in THF and after carbonation yielded π-benzoic acid chromium tricarbonyl (*304*).

Deuterium–hydrogen exchange in $(C_6H_6)Cr(CO)_3$ and $(MeC_6H_5)Cr(CO)_3$ has been shown to take place in ethanol solution in the presence of sodium ethoxide, whereas no exchange is observed in the uncomplexed arenes (*265*).

7. *Electrophilic Substitution Reactions*

The only electrophilic substitution of arene chromium tricarbonyl complexes so far achieved is Friedel–Crafts acetylation. Benzene and substituted benzene chromium tricarbonyls undergo this reaction under mild conditions giving the corresponding acetyl-substituted complexes (*109*, *176*, *218*, *233*, *234*, *355*). Substituent and conformational effects play an important role in directing the position of acetylation in arene chromium tricarbonyl complexes (*176*, *218*, *233*, *234*).

8. *Nucleophilic Substitution Reactions*

Halogen-substituted arene chromium tricarbonyl complexes undergo nucleophilic substitution by alkoxide ions at a considerably enhanced rate over the free arene (*39*, *327*, *444*). The effect of the chromium tricarbonyl moiety on the rate of substitution of the arene is approximately equal to that of a *p*-nitro group (*39*). Treatment of ethylbenzene chromium tricarbonyl with *tert*-butyl lithium followed by hydrolysis and decomposition of the resulting complexes with Ce(IV) yielded *m*- and *p*-ethyl-*tert*-butylbenzene along with some unsubstituted ethylbenzene (*55*). The reaction represents a novel nucleophilic displacement of a hydride ion and contrasts with the metalation reaction observed with *n*-butyl lithium (*304*).

9. *Catalysis*

Arene chromium tricarbonyl complexes are selective hydrogenation catalysts for the 1,4-addition of hydrogen to diolefins (*48*, *49*, *166*, *167*). Electron-withdrawing substituents on the arene enhance the rate of reaction (*48*), presumably due to a weakening of the arene–chromium–bond which is thought to be partially or completely broken by the incoming diolefin (*49*, *166*). Toluene tungsten tricarbonyl is an effective catalyst for the disproportionation of olefins. The reaction is thought to involve initially displacement of the arene forming a trisolefin tungsten tricarbonyl complex (*270*).

F. Structures of the Arene Chromium Tricarbonyl Complexes

The crystal structures of a considerable number of arene chromium tricarbonyl complexes have been determined. The structures can be broken into two categories—those in which the carbonyl chromium vectors adopt a staggered configuration with respect to the carbon–carbon bonds of the arene ring (X) and those in which the carbonyl chromium vectors adopt an eclipsed configuration with respect to the carbon atoms of the arene ring (XI). The tricarbonyl chromium complexes of benzene (*21*, *352*), hexamethylbenzene (*20*), acetylbenzene (*100*), the *o*- and *m*-toluate anions (*45*), and *exo*-2-acetoxybenzonorbornene (*275*), display the staggered configuration. In the last example, the staggered configuration was attributed to a steric interaction between the chromium tricarbonyl moiety and a norbornene hydrogen (*275*).

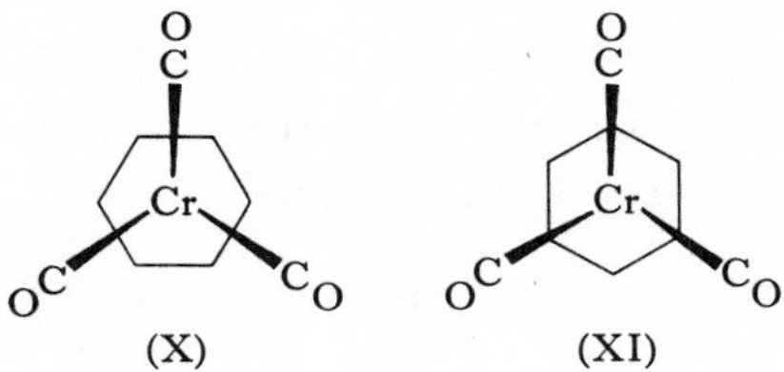

(X) (XI)

Staggered configurations have also been observed for the tricarbonyl chromium complexes of phenanthrene (*294*, *295*), 9,10-dihydrophenanthrene (*293*, *295*), anthracene (*202*), naphthalene (*262*), and 1-aminonaphthalene (*58*). The eclipsed configuration has been observed for the tricarbonyl complexes of anisole (*57*, *229*), toluidine (*60*, *61*), methylbenzoate (*59*), *o*-methoxyacetylbenzene, *o*-hydroxyacetylbenzene (*101*), 2-methoxy-[1-hydroxy-ethyl]benzene (*99*), and 2-methyl-[1-hydroxy-1-phenylpropyl]benzene (*97*). It is apparent that the orientation of the chromium tricarbonyl moiety is in many cases controlled by the substituents on the ring to which it is coordinated, and this has been attributed to mesomeric electron repulsion or withdrawal by the substituents (*374*).

Nuclear magnetic resonance experiments indicate that both staggered and eclipsed conformations of substituted arene chromium tricarbonyls exist in solution (*177*, *235*). The preferred conformations are attributed to both steric and electronic effects.

G. Monoarene Complexes Containing Ligands Other Than Carbon Monoxide

Bisbenzene molybdenum reacts with tertiary phosphine or phosphite ligands with loss of one molecule of benzene yielding the neutral complexes $(C_6H_6)Mo(PR_3)_3$ [$PR_3 = P(OPh)_3, P(OMe)_3, PPh_2OMe, PPh_2Me$, and $PPhMe_2$] (*182*). The reaction does not proceed with PPh_3 presumably for steric reasons nor with the very basic phosphines PEt_3 and PMe_3. The phosphine complexes are very electron-rich and react with 1 or 2 equivalents of acid yielding stable hydride cations that have been isolated as their PF_6 salts (*180*, *182*, *186*) [see Eq. (24a)]. The reaction is readily reversed by adding a base such as sodium hydroxide. The dihydrides represent the first example of the isolation of a compound in which two protons have been added to the same atom of a neutral molecule. Low-temperature NMR studies on the dication $[(C_6H_6)Mo(PEt_3)_3H_2]^{2+}$ are consistent with the presence of 2 inequivalent hydrogen atoms on the metal as shown in (XII) (*186*).

$$(C_6H_6)Mo(PR_3)_3 \underset{-H^+}{\overset{+H^+}{\rightleftharpoons}} [(C_6H_6)Mo(PR_3)_3H]^{1+} \underset{-H^+}{\overset{+H^+}{\rightleftharpoons}} [(C_6H_6)Mo(PR_3)_3H_2]^{2+} \quad (24a)$$

Another type of reaction of bisarene molybdenum complexes has been observed with allyl chloride (*188*, *189*) (arene = C_6H_6, MeC_6H_5, 1,3,5-$Me_3C_6H_3$):

$$(arene)_2Mo + C_3H_5Cl \longrightarrow \tfrac{1}{2}[(arene)Mo(\pi\text{-}C_3H_5)Cl]_2 \quad (24b)$$

The benzene derivative has been shown by X-ray analysis to possess the dimeric structure (XIII) (*54b*). In contrast, reaction between $(C_6H_6)_2Mo$ and allyl acetate gave the monomeric compound $(C_6H_6)Mo(\pi\text{-}C_3H_5)$-$O_2CMe$ (*180*). The dimers were shown to react with tertiary phosphines giving monomeric complexes of the type (arene)$Mo(\pi\text{-}C_3H_5)(PR_3)Cl$ (*188*, *189*). Reduction of the PPh_3 derivatives (arene)$Mo(\pi\text{-}C_3H_5)$-$(PPh_3)Cl$ (arene = C_6H_6, MeC_6H_5, 1,3,5-$Me_3C_6H_3$) with sodium borohydride in the presence of excess PPh_3 led to the neutral dihydride derivatives (arene)$Mo(PPh_3)_2H_2$ in high yield (*188*, *189*). Under similar conditions, the less bulky phosphine derivatives yielded the tris-phosphine compounds (arene)$Mo(PR_3)_3$ described in the foregoing as well as these compounds where PR_3 = PEt_3 and PMe_3 (*186*, *188*, *189*). When the reduction was carried out in the absence of free phosphine, the dihydride complexes (arene)$Mo(PR_3)_2H_2$ were always obtained (*189*). The neutral dihydride complexes were shown to react reversibly with dinitrogen yielding monomeric dinitrogen derivatives [Eq. 25], except in

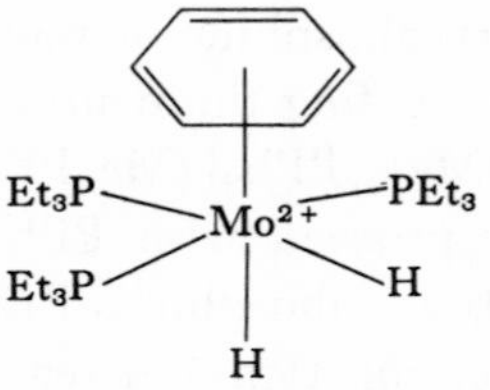

(XII)

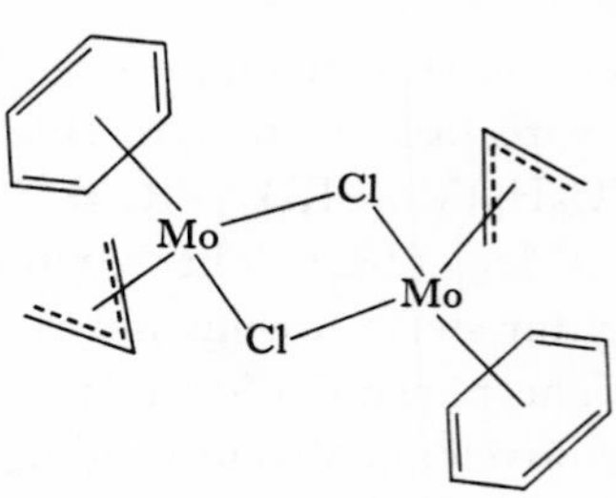

(XIII)

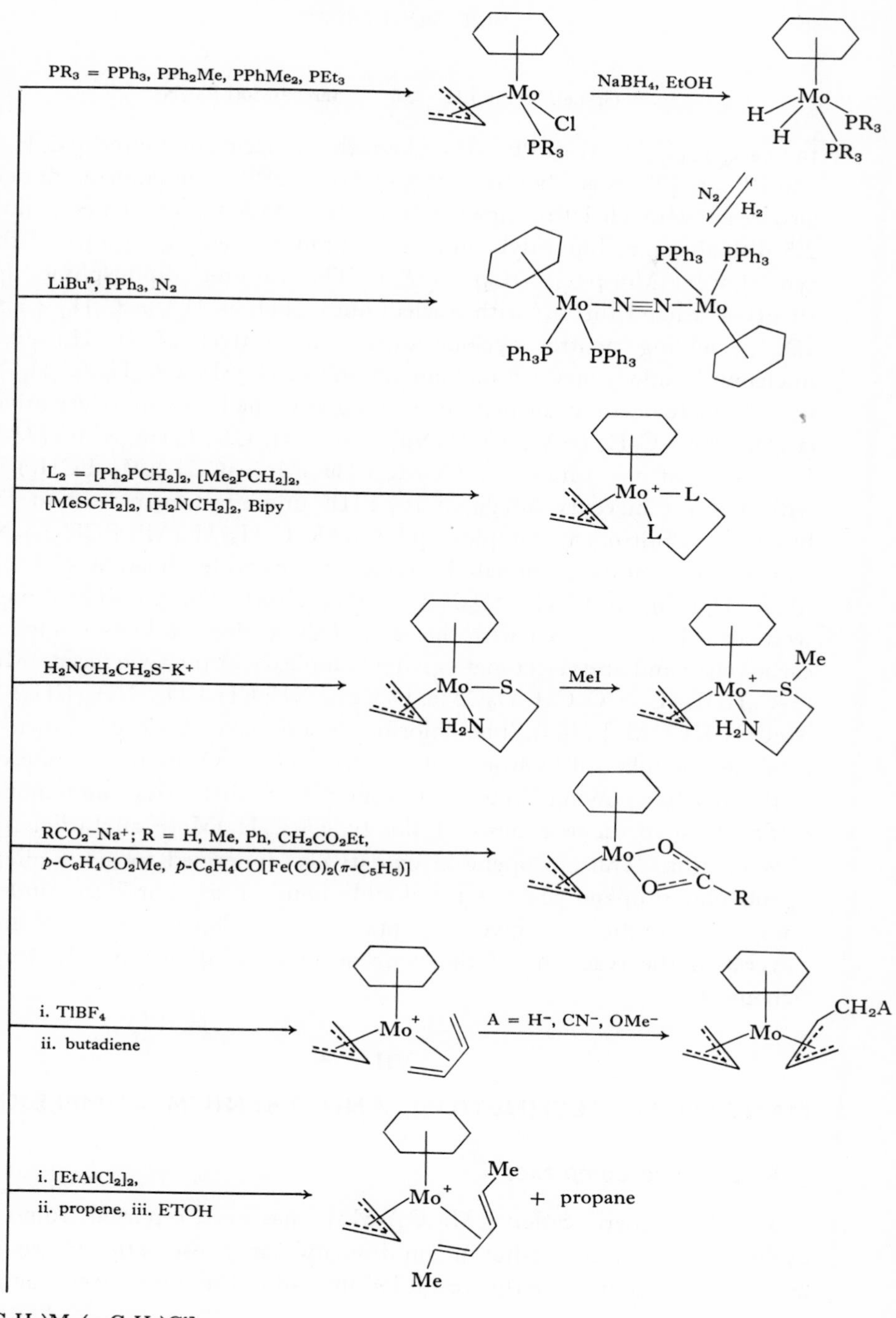
PR3 = PPh3, PPh2Me, PPhMe2, PEt3
Mo
Cl
PR3
NaBH4, EtOH
H
Mo
H
PR3
PR3
N2
H2
LiBun, PPh3, N2
PPh3 PPh3
Mo—N≡N—Mo
Ph3P PPh3
L2 = [Ph2PCH2]2, [Me2PCH2]2,
[MeSCH2]2, [H2NCH2]2, Bipy
Mo+—L
L
H2NCH2CH2S−K+
Mo—S
H2N
MeI
Me
Mo+—S
H2N
RCO2−Na+; R = H, Me, Ph, CH2CO2Et,
p-C6H4CO2Me, p-C6H4CO[Fe(CO)2(π-C5H5)]
Mo
O
O
C
R
i. TlBF4
ii. butadiene
Mo+
A = H−, CN−, OMe−
CH2A
Mo
i. [EtAlCl2]2,
ii. propene, iii. ETOH
Mo+
Me
+ propane
Me
[(C6H6)Mo(π-C3H5)Cl]2

Scheme 1

$$\text{(arene)Mo(PR}_3)_2\text{H}_2 \underset{\text{H}_2}{\overset{\text{N}_2}{\rightleftharpoons}} \text{(arene)Mo(PR}_3)_2\text{N}_2 \tag{25}$$

the case of $(C_6H_6)Mo(PPh_3)_2H_2$ where the dimeric compound $[(C_6H_6)$-$Mo(PPh_3)_2]_2N_2$ was obtained instead (*188*, *189*). The chloride dimers also react with chelating ligands, such as butadiene, ethylenediamine, 2,5-dithiahexane, bipyridyl, and diphosphines, yielding cations of the type $[(\text{arene})Mo(\pi\text{-}C_3H_5)L_2]^+$ (*185*). The cationic diphosphine derivatives reacted further with nucleophiles, such as H^-, $n\text{-}C_4H_9^-$, and CN^-, yielding neutral cyclohexadienyl derivatives (*185*). However, nucleophilic attack on the butadiene cation $[(C_6H_6)Mo(\pi\text{-}C_3H_5)(C_4H_6)]^+$ was shown to occur at the olefinic ligand giving the bis-π-allyl derivatives $(C_6H_6)Mo(\pi\text{-}C_3H_5)(\pi\text{-}C_3H_4CH_2Nu)$ (Nu = H, CN, OMe, SMe) (*184*). Reduction of the cation $[(1,3,5\text{-}Me_3C_6H_3)Mo(\pi\text{-}C_3H_5)(Me_2PCH_2)_2]^+$ with sodium mercury amalgam in THF under nitrogen lead to the binuclear dinitrogen complex $[(1,3,5\text{-}Me_3C_6H_3)Mo(Me_3PCH_2)_2]_2N_2$ which was readily protonated giving the hydride dication $\{[(1,3,5\text{-}Me_3C_6H_3)Mo(Me_2PCH_2)_2H]_2N_2\}^{2+}$ (*190*, *373a*). The chloride dimers were also shown to react with the salts of three electron ligands such as carboxylate and acetylacetonate giving complexes of the type (arene)Mo-$(\pi\text{-}C_3H_5)L$ [L = O_2CH, O_2CMe, O_2CPh, $NH_2CH_2CO_2$, $NH_2(CH_2)_2S$, MeCOCHCOMe] (*183*). The chloride dimers have also been shown to catalyze a number of hydrocarbon reactions (*181*). Of particular interest is the reaction among dimer, ethyl aluminum dichloride, and mono-olefins to give diene cations of the type $[(C_6H_6)Mo(\pi\text{-allyl})(\text{diene})]^+$ plus paraffin. Thus, propene is oxidatively dimerized to a hexadiene ligand plus propane whereas the double bond of *cis*-pent-2-ene undergoes redistribution to give a penta-1,3-diene ligand plus pentane. Several of the reactions of the benzene chloride dimer are shown in Scheme 1.

VII
MANGANESE, TECHNETIUM, AND RHENIUM COMPLEXES

A. Bisarene Complexes

The diamagnetic cation $[(Me_6C_6)_2Mn]^+$ has been reported from the cyclic condensation of 2-butyne on diphenylmanganese (*416*). However, an attempt to reproduce this result led instead to the mixed arene cation

$[(C_6H_6)(Me_6C_6)Mn]^+$ (*154*). Bisarene cations of technetium and rhenium with arenes such as benzene, mesitylene, and hexamethylbenzene have been synthesized by the Fischer–Hafner method (*153, 156, 161, 336*). The bisbenzene technetium cation has also been formed by neutron bombardment of bisbenzene molybdenum (*23, 24*):

$$(C_6H_6)_2Mo \xrightarrow{n,\gamma} (C_6H_6)_2\,{}^{99}Mo \xrightarrow{\beta^-} (C_6H_6)_2\,{}^{99}Tc^+ \qquad (26)$$

Cation $[(C_6H_6)(Me_6C_6)Mn]^+$ was shown to react with $LiAlH_4$ giving the neutral cyclohexadienyl complex $(C_6H_7)(Me_6C_6)Mn$ (*154*). The cations $[(C_6R_6)_2M]^+$ (M = Tc, Re; R = H, Me) were also shown to react with $LiAlH_4$ yielding neutral cyclohexadienyl complexes (*153, 156, 243*). When cations $[(Me_6C_6)_2M]^+$ (M = Tc, Re) were heated with molten sodium or lithium under vacuum, the ring-coupled cyclohexadienyls (XIV) were obtained as sublimates (*153, 156*). In the case of rhenium, a monomer was obtained by collecting the sublimate at

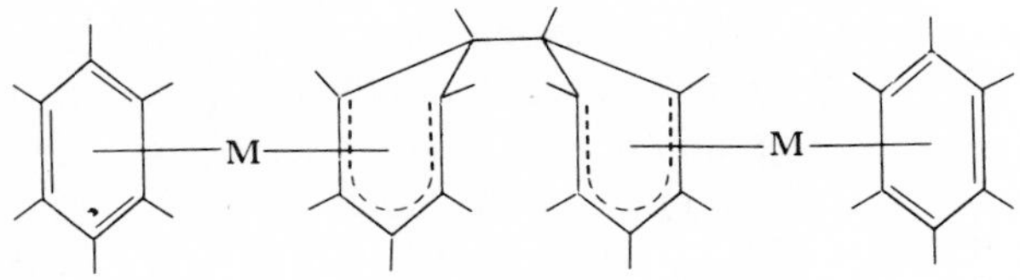

(M = Tc, Re)

(XIV)

−196°C. The compound dimerized irreversibly on warming above −196°C, yielding (XIV) (M = Re) (*156*). The cyclohexadienyl complexes of rhenium were readily oxidized to the cation $[(Me_6C_6)_2Re]^+$ (*156*).

B. *Mixed Sandwich Complexes*

The diamagnetic complex $(C_6H_6)(\pi\text{-}C_5H_5)Mn$ has been prepared by treating C_5H_5MgBr with $MnCl_2$ in the presence of benzene (*89*) and by reacting $(\pi\text{-}C_5H_5)_2Mn$ with a phenyl Grignard reagent (*155*). In the latter reaction, $(C_6H_5C_6H_5)(\pi\text{-}C_5H_5)Mn$ was the main product, with a smaller amount of the bimetallic compound $(C_6H_5C_6H_5)[Mn(\pi\text{-}C_5H_5)]_2$ also being produced. The benzene derivative is isomorphous with $(C_6H_6)(\pi\text{-}C_5H_5)Cr$. The rhenium compound $(C_6H_6)(\pi\text{-}C_5H_5)Re$ has been prepared by a Grignard synthesis (*160*):

$$ReCl_5 + i\text{-}PrMgBr + C_5H_5MgBr + 1,3,\text{-cyclohexadiene} \xrightarrow{\text{ether}} (C_6H_6)(\pi\text{-}C_5H_5)Re + (\pi\text{-}C_5H_5)_2ReH \quad (27)$$

The neutral complex was oxidized to the paramagnetic cation $[(C_6H_6)(\pi\text{-}C_5H_5)Re]^+$ by HCl and reacted with acetyl chloride in the presence of $AlBr_3$ giving $(MeCOC_6H_5)(\pi\text{-}C_5H_5)Re$ and $(C_6H_6)(\pi\text{-}C_5H_4COMe)Re$ in approximately equal amounts (*160*). This contrasts with the analogous manganese complex which does not give a stable cation and which undergoes a ring expansion reaction with acetyl chloride giving the methylcycloheptatrienyl cation $[(\pi\text{-}C_7H_6Me)(\pi\text{-}C_5H_5)Mn]^+$ (*155*).

C. Carbonyl Complexes

Manganese pentacarbonyl chloride reacts with arenes in the presence of $AlCl_3$ yielding cations $[(\text{arene})Mn(CO)_3]^+$. Arenes such as benzene, toluene, mesitylene, hexamethylbenzene, and naphthalene have been employed (*71*, *447*, *448*). The compounds are diamagnetic and, thus, isoelectronic with the arene chromium tricarbonyl derivatives. The cations react with many nucleophiles such as $C_6H_5^-$ (*243*), H^- (*447*, *448*), N_3^-, OMe^-, PPh_2^-, NCS^- (*432*), and CN^- (*431*, *432*), yielding neutral cyclohexadienyl derivatives. Although the stability of many of the adducts was low, stable complexes of the nucleophiles $C_6H_5^-$, H^-, and CN^- were isolated. The *exo*-cyano group can be removed from the complexes $(C_6H_{6-n}Me_nCN)Mn(CO)_3$ by treatment with $[Ph_3C]^+$, $[Et_3O]^+$, or H^+, regenerating the cations $[(C_6H_{6-n}Me_n)Mn(CO)_3]^+$. Furthermore, treatment of the cyanide adducts with Ce(IV) in H_2SO_4 gave the free cyanoarene uncontaminated with $C_6H_{6-n}Me_n$ (*432*). On refluxing the neutral cyano derivative $(1,3,5\text{-}Me_3C_6H_3CN)Mn(CO)_3$ in aqueous suspension, the neutral arene complex $(1,3,5\text{-}Me_3C_6H_3)Mn(CO)_2CN$ was obtained (*72*, *431*). The cyano group in the arene complex reacts with Lewis acids such as BF_3, Et^+, and Ph_3C^+, yielding complexes of the type $(1,3,5\text{-}Me_3C_6H_3)Mn(CO)_2(CNBF_3)$ and $[(1,3,5\text{-}Me_3C_6H_3)Mn(CO)_2(CNX)]^+$ (X = Et, Ph_3C) (*430*). Evidence was also presented for formation of the protonated derivative $[(1,3,5\text{-}Me_3C_6H_3)Mn(CO)_2(CNH)]^+$ on dissolving the neutral cyano complex in water.

The arene manganese tricarbonyl cations undergo a reversible reaction with amines giving neutral carboxamido derivatives (*14*):

$$[(\text{arene})\text{Mn(CO)}_3]^+ + 2\text{RNH}_2 \rightleftharpoons (\text{arene})\text{Mn(CO)}_2\text{CONHR} + \text{RNH}_3^+ \quad (28)$$

Stable derivatives were isolated where arene = Me_6C_6 and R = H or C_6H_{11}. The tricarbonyl cation could be regenerated by treating the carboxamido derivatives with CCl_3CO_2H. When hydrazine was treated with the tricarbonyl cations, the unstable carbazoyl intermediates (arene)-$Mn(CO)_2CONH_2NH_2$ were detected which reacted further to give the more stable isocyanato complexes, (arene)$Mn(CO)_2NCO$ (arene = 1,3,5-$Me_3C_6H_3$, Me_6C_6).

In contrast to the preceding carbonyl group reactions, cations $[(C_6H_6)Mn(CO)_3]^+$ and $[(Me_6C_6)Re(CO)_3]^+$ reacted with primary amines with loss of the arene group to give derivatives of the type $[(RNH_2)_3M(CO)_3]^+$ (*14*). Arene replacement reactions have also been observed on treatment of the cations with diethylenetriamine (*1*) and some tridentate phosphines (*67*). The arene manganese tricarbonyl cations have been observed to react with halide ions yielding complexes of the type $Mn_2(CO)_8X_2$ (X = Br, I) (*83*).

VIII

IRON, RUTHENIUM, AND OSMIUM COMPLEXES

A. Bisarene Complexes

Diamagnetic bisarene cations of the type $[(\text{arene})_2Fe]^{2+}$ (arene = benzene, toluene, mesitylene, hexamethylbenzene) have been prepared by the direct reaction of $FeCl_2$ or $FeBr_2$ with the arene in the presence of $AlCl_3$ (*116*, *148*, *149*, *217*, *414*). The yields can be considerably improved by employing $FeCl_3$ in the reaction which is reduced to highly reactive $FeCl_2$ by the arene solvent (*217*). The stability of the complexes has been shown to increase with increasing methyl substitution of the ring (*414*). The ruthenium cations $[(\text{arene})_2Ru]^{2+}$ (arene = benzene, mesitylene, hexamethylbenzene, naphthalene) have been prepared by the Fischer–Hafner method starting from $RuCl_3$ (*117*, *124*, *125*). The osmium cation $[(Me_3C_6H_3)_2Os]^{2+}$ has also been reported (*127*).

Sodium dithionite reduction of $[(Me_6C_6)_2Fe]^{2+}$ gave the paramagnetic Fe(I) cation $[(Me_6C_6)_2Fe]^+$ (*149*). Electron spin resonance studies have shown a loss of axial symmetry in the Fe(I) cation and this was attributed to a Jahn–Teller distortion due to the presence of an unpaired electron in a degenerate orbital (*34*) (see Section III,A). Complete reduction to the neutral paramagnetic (μ_{eff} = 3.08 BM) complex $(Me_6C_6)_2Fe$ was achieved with alkaline dithionite (*149*). The complex is sensitive to light, air, and temperature. The benzene analog has been synthesized by cocondensation of benzene and iron metal at −196°C, but the complex explodes on warming to −50°C (*407*). An excess of electrons over the 18-electron configuration as observed for the bisarene Fe(0) and Fe(I) complexes has also been observed for the bisarene complexes of other Group VIII metals such as Co(II) and Ni(II) (*141*, *272*) but not for metals in the preceding groups. This effect must be due in part to the relatively high positive charges and increased nuclear charges of the Group VIII complexes which should lead to a lowering in energy of the antibonding, primarily metal-like e_{1g} orbitals which contain the excess electrons (see Fig. 1, Section III,A).

In contrast to the case of the iron analog, the ruthenium cation $[(Me_6C_6)_2Ru]^{2+}$ yielded a diamagnetic complex $(Me_6C_6)_2Ru$ on reduction (*124*). Solution NMR studies (*124*) and an X-ray analysis (*230*, *231*) established structure (XV). The ruthenium atom thus adopts an 18-electron configuration by coordination of a nonplanar tetrahaptohexamethylbenzene ligand. The nonplanar Me_6C_6 ligand bears some resemblance to the $(CF_3)_6C_6$ ligand in $[(CF_3)_6C_6](\pi\text{-}C_5H_5)Rh$ (see Section IX,B). In both cases the ring is tilted at an angle of approximately 45° and the C_5–C_6 distances [see structures (XV) and (XXIX)] are very close to that of a normal double bond. The bond distances C_1–C_6, C_2–C_3, C_3–C_4, and C_4–C_5 are also comparable in the two compounds. However, there is a significant difference in the C_1–C_2 distances in the two compounds. In the case of $[(CF_3)_6C_6](\pi\text{-}C_5H_5)Rh$, the near equality of the C_1–C_2 and C_3–C_4 distances (~1.50 Å) leads to a diene-type description of the bonding of the $(CF_3)_6C_6$ group to the metal atom (see Section IX,B). However, the much shorter C_1–C_2 distance (1.415 Å) in $(Me_6C_6)_2Ru$ is not characteristic of diene-type bonding (*3*, *69*).

The cation $[(C_6H_6)_2Ru]^{2+}$ has been shown to react with 2 equivalents of phenylithium giving the neutral phenylcyclohexadienyl derivative $(C_6H_6Ph)_2Ru$. Reduction with $LiAlH_4$ gave the cyclohexa-1,3-diene derivative $(C_6H_6)(C_6H_8)Ru$ as the main product along with some $(C_6H_7)_2Ru$ (*243*).

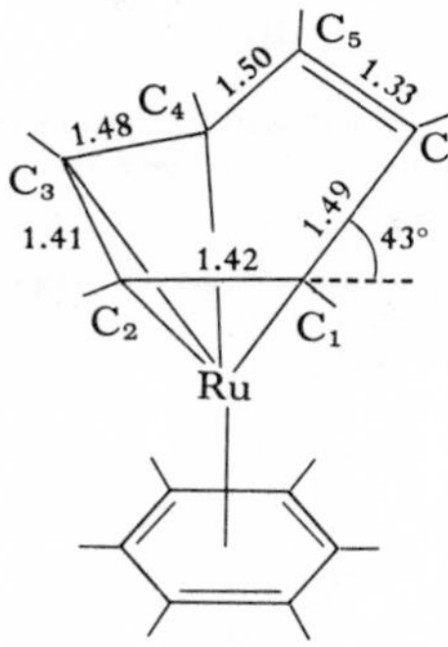

(XV)

B. Mixed Sandwich Complexes

1. *Preparation*

One of the cyclopentadienyl ligands in ferrocene can readily be replaced by benzene in the presence of $AlCl_3$ giving the benzene cyclopentadienyl cation (*310*, *311*):

$$(\pi\text{-}C_5H_5)_2Fe + C_6H_6 \xrightarrow{AlCl_3} [(C_6H_6)(\pi\text{-}C_5H_5)Fe]^+ + C_5H_5 \quad (29)$$

The reaction also occurs with a wide variety of substituted arenes (Table V). Electron-donating substituents attached to either the cyclopentadienyl or the arene ring have been shown to enhance the reaction, whereas electron-accepting substituents hinder the reaction (*316*). The following intramolecular exchange reaction has also been observed (*268*):

$$\xrightarrow{BF_3\cdot Et_2O} \quad (30)$$

The cations $[(\text{arene})(\pi\text{-}C_5H_5)Fe]^+$ (arene = benzene, mesitylene) have also been prepared by the reaction of $(\pi\text{-}C_5H_5)Fe(CO)_2Cl$ with the arene in the presence of $AlCl_3$ (*71*, *187*). A ruthenium analog $[(C_6H_6)$-$(\pi\text{-}C_5H_5)Ru]^+$ has been prepared by the reaction of $[(C_6H_6)RuCl_2]_2$

TABLE V

MIXED SANDWICH COMPLEXES

Compound	References
$(C_6H_6)(C_5H_5)Cr$	*120, 134*
$(C_6H_6)(C_5H_5)Mo$	*136, 137, 434*
$[(C_6H_6)(C_5H_5)Mo(CO)]^+$	*135*
$[(C_6H_6)(C_7H_7)Mo]^+$	*17*
$[(MeC_6H_5)(C_7H_7)Mo]^+$	*17*
$(C_6H_6)(C_5H_5)W(H)$	*434*
$[(C_6H_6)(C_5H_5)W(CO)]^+$	*135*
$(C_6H_6)(C_5H_5)Mn$	*89, 155*
$(C_6H_6)(C_5H_4Me)Mn$	*71*
$(C_6H_5C_6H_5)(C_5H_5)Mn$	*89, 155*
$(C_6H_5C_6H_5)[Mn(C_5H_5)]_2$	*155*
$(C_6H_6)(C_5H_5)Re$	*160*
$(C_6H_6)(C_5H_4COMe)Re$	*160*
$(C_6H_5COMe)(C_5H_5)Re$	*160*
$[(C_6H_6)(C_5H_5)Re]^+$	*160*
$(C_6H_6)(C_5H_5)Fe$	*321*
$[(C_6H_6)(C_5H_5)Fe]^+$	*187, 310, 311, 316*
$[(C_6H_6)(C_5H_4R)Fe]^+$	
[R = Et	*18, 306, 316*
= *n*-Pr, *i*-Pr, $PhCH_2$	*18*
= COOH, COOEt, COMe	*306, 322*
= CN	*316*
= Cl	*306, 313, 314*
= OMe, OEt	*306, 313, 315*
= SPh	*306, 313*
= $C_6H_4(CO)_2N$ (phthalimide)	*313*
= NH_2	*306, 312, 313*
= NHCOMe	*306, 313*
= NHMe, NHEt	*314*
= NHC_5H_{11}, NHC_5H_{10} (piperidine)]	*306, 313*
$[(RC_6H_5)(C_5H_5)Fe]^+$	
[R = Me	*18, 251, 306, 309*
= Et	*18, 306, 375*
= *n*-Pr, *i*-Pr, *t*-Bu	*18*
= Ph, $PhCH_2$	*18, 269, 309*
= 3-FC_6H_4, 4-FC_6H_4, CN	*306*
= F	*306, 375*
= Cl	*248, 306, 309*
= Br	*248*
= OMe, OEt, OPh	*249, 306, 308, 315*
= SPh, S(*n*-Bu)	*308*
= NH_2	*306, 308, 312*
= $NHCOCH_3$, $C_6H_4(CO)_2N$]	*306, 308, 309*

TABLE V—*Continued*

Compound	References
$[(RC_6H_4Me)(C_5H_5)Fe]^+$	
(R = *o*-, *m*-, *p*-Me	*251, 306, 309*
= *p*-Et, *p*-CN	*306, 375*
= *o*-, *m*-, *p*-Cl; *o*-, *m*-, *p*-Br; *m*-, *p*-F	*248, 306, 322, 375*
= *o*-, *m*-, *p*-OMe	*249, 306, 375*
= *p*-SMe	*375*
= *p*-NH_2, *p*-NHCOMe)	*375*
$[(RC_6H_4R')(C_5H_5)Fe]^+$	
(R = R′ = *o*-, *m*-, *p*-Cl	*248, 375*
R = R′ = *p*-OMe	*249*
R = Cl, R′ = *m*-, *p*-CN; *p*-$CONH_2$	*375*
R = Cl, R′ = *m*-, *p*-F; *m*-, *p*-OMe	*375*
R = COOH, R′ = *m*-, *p*-Cl; *p*-OEt	*322, 375*
R = COOH, R′ = *m*-, *p*-SPh; *p*-SO_2Ph)	*322, 375*
$[(RC_6H_5)(R'C_5H_4)Fe]^+$	
(R = Cl, R′ = Et	*322*
R = Cl, R′ = CN, COOH, $CONH_2$)	*375*
$[(1,3,5\text{-}Me_3C_6H_3)(C_5H_5)Fe]^+$	*71, 187, 251, 310, 311, 316*
$[(1,3,5\text{-}Me_3C_6H_3)(C_5H_4Et)Fe]^+$	*311*
$[(1,3,5\text{-}Me_3C_6H_3)(C_5H_4COMe)Fe]^+$	*310, 311, 316*
$[(1,2,3\text{-}Me_3C_6H_3)(C_5H_5)Fe]^+$	*251*
$[(1,2,4\text{-}Me_3C_6H_3)(C_5H_5)Fe]^+$	*251*
$[(1,2,3,5\text{-}Me_4C_6H_2)(C_5H_5)Fe]^+$	*251*
$[(Me_5C_6H)(C_5H_5)Fe]$	*251*
$[(Me_6C_6)(C_5H_5)Fe]^+$	*251*
$[(Me_6C_6)(C_5H_4Me)Fe]^+$	*250*
$[(C_{10}H_8)(C_5H_5)Fe]^+$ naphthalene	*309*
$[(C_{12}H_{12})(C_5H_5)Fe]^+$ tetralin	*310, 311*
$[(C_{13}H_{10})(C_5H_5)Fe]^+$ fluorene	*269, 309*
$[(C_9H_8)(C_9H_7)Fe]^+$ indene, indenyl	*268*
$[(C_9H_{10})(C_9H_{11})Fe]^+$ indane	*268*
$[(C_6H_6)(C_5H_4NH_3)Fe]^{2+}$	*314*
$[(C_6H_6)(C_5H_4N_2)Fe]^{2+}$	*314*
$[(C_6H_5C_6H_5)\{Fe(C_5H_5)\}_2]^{2+}$	*269*
$[(C_6H_5CH_2C_6H_5)\{Fe(C_5H_5)\}_2]^{2+}$	*269*
$[(C_{13}H_{10})\{Fe(C_5H_5)\}_2]^{2+}$ fluorene	*269*
$(\pi\text{-}C_6H_5BPh_3)(C_5H_5)Ru(II)$	*198*
$[(C_6H_6)(C_5H_5)Ru]^+$	*463*
$(C_5H_5BR)(C_5H_5)Co(II)$ (R = Me, Ph, OMe, OH)	*220*
$[(C_5H_5BPh)(C_5H_5)Co(III)]^+$	*219*
$[(C_6H_6)(C_4Ph_4)Co]^+$ tetraphenylcyclobutadiene	*104, 277*
$[(C_6H_6)\{C_4(C_6H_4Me)_4\}Co]^+$	*104*

Continued

TABLE V—*Continued*

Compound	References
$[(MeC_6H_5)(C_4Ph_4)Co]^+$	*104, 277*
$[(n\text{-}BuC_6H_5)(C_4Ph_4)Co]^+$	*103, 104*
$[(PhC_6H_5)(C_4Ph_4)Co]^+$	*104*
$[(NH_2C_6H_5)(C_4Ph_4)Co]^+$	*104*
$[(p\text{-}Me_2C_6H_4)(C_4Ph_4)Co]^+$	*104*
$[(1,3,5\text{-}Me_3C_6H_3)(C_4Ph_4)Co]^+$	*104, 277*
$[(C_6H_6)(C_5H_5)Co]^{2+}$	*123*
$\{(CF_3)_6C_6\}(C_5H_5)Rh$	*93*
$\{(COOMe)_6C_6\}(C_5Me_5)Rh$	*245*
$[(p\text{-}Me_2C_6H_4)(C_5Me_5)Rh]^{2+}$	*440*
$[(Me_6C_6)(C_5Me_5)Rh]^{2+}$	*440*
$\{(COOMe)_6C_6\}(C_5Me_5)Ir$	*245*
$[(C_6H_6)(C_5Me_5)Ir]^{2+}$	*440*
$[(Me_5C_6H)(C_5Me_5)Ir]^{2+}$	*440*
$[(Me_6C_6)(C_5Me_5)Ir]^{2+}$	*440*

with thallium(I) cyclopentadienide (*463*). The neutral Ru(II) complex $(\pi\text{-}C_5H_5)RuBPh_4$ has been reported from the reaction of $(\pi\text{-}C_5H_5)Ru(PPh_3)_2Cl$ with sodium tetraphenyl boron (*198*). The compound was shown to contain a π-bonded phenyl group derived from the tetraphenyl boron anion.

The effect of cyclopentadienyl or benzene ring substituents on the proton NMR shifts, polarographic half-wave potentials, and Mössbauer quadrupole splittings in the iron compounds have been shown to correlate well with the Hammett–Taft σ_p parameters of the substituents (*303*, *306*, *420*).

2. *Reaction with Nucleophiles*

The halogen-substituted cations $[(ClC_6H_5)(\pi\text{-}C_5H_5)Fe]^+$ and $[(C_6H_6)(\pi\text{-}C_5H_4Cl)Fe]^+$ were shown to exchange readily their halogens for nucleophiles such as MeO^-, PhO^-, $n\text{-}BuS^-$, PhS^-, NH_2^-, NHR^-, and $C_6H_4(CO)_2N^-$ (phthalimide) (*308*, *313*, *315*, *375*). The benzene chlorine was found to be approximately 3 times as reactive as the cyclopentadienyl chlorine and much more reactive than in free chlorobenzene (*315*). Both cations reacted with ammonia in an autoclave yielding the corresponding amine complexes (*312*). The cation $[(C_6H_6)(C_5H_4NH_2)Fe]^+$ was shown to be a stronger base than the cation $[(NH_2C_6H_5)(C_5H_5)Fe]^+$, indicating a higher electron density in the cyclopentadienyl ring than in the benzene ring. Both of the amine complexes are weak bases, being weaker than ferrocenylamine and aniline, respectively (*312*).

Cations $[(\text{arene})(C_5H_5)Fe]^+$ (arene = benzene or a substituted benzene) react with hydride ion to give neutral cyclohexadienyl derivatives. With halogen-substituted arenes the reaction was shown to occur preferentially at positions ortho to the halogen (*248*), whereas with methoxy substituted arenes, the reaction occurred preferentially at the position meta to the methoxy group (*249*). Methyl groups were shown to influence the mode of reaction only insofar as hydride addition to an unsubstituted arene carbon atom is preferred (*251*). The reaction of the methanide anion to give neutral methylcyclohexadienyl complexes was subject to the same directive influences as for the hydride reaction. However, with the cation $[(Me_6C_6)(C_5H_5)Fe]^+$, addition occurred at the cyclopentadienyl ring giving the neutral methylcyclopentadiene derivative $(Me_6C_6)(C_5H_5Me)Fe$ (*250*).

3. *Arene Displacement Reactions*

The parent cation $[(C_6H_6)(C_5H_5)Fe]^+$ reacts with nucleophiles, such as hydroxide, alkoxide, or amide ions with loss of benzene and the formation of ferrocene (*317*). A similar reaction is induced by reducing agents such as sodium mercury amalgam and sodium naphthalide (*317*, *319*). Ultraviolet light and electrochemical reduction were also effective in bringing about the reaction (*18*, *320*), which is thought to involve initial reduction to the neutral species $(\text{arene})(\pi\text{-}C_5H_5)Fe$ followed by decomposition to ferrocene and free arene. This mechanism is supported by a recent publication that describes the isolation of the neutral compound $(C_6H_6)(\pi\text{-}C_5H_5)Fe$. The compound is reported to be stable in THF solution at $-78°C$ but decomposes on warming to give ferrocene. The neutral compound also readily exchanges with naphthalene and with CO or $P(OPh)_3$ giving $(C_{10}H_8)(\pi\text{-}C_5H_5)Fe$ and $[(\pi\text{-}C_5H_5)FeL_2]_2$ [L = CO, $P(OPh)_3$], respectively (*321*). Similar results have been observed on reduction of $[(C_{10}H_8)(\pi\text{-}C_5H_5)Fe]^+BF_4^-$ with sodium mercury amalgam in the presence of benzene or carbon monoxide (*318*). The cation $[(C_6H_6)(\pi\text{-}C_5H_5)Fe]^+$ also reacts with β-diketones such as acetylacetone yielding complexes of the type $(\pi\text{-}C_5H_5)Fe(\beta\text{-diketonate})_2$ (*338*, *350*).

C. Carbonyl Complexes

A compound of stoichiometry $[(Me_6C_6)Fe(CO)_2]_2$ has been prepared by reacting $Fe(CO)_5$ with hexamethyldewarbenzene and is thought to have structure (XVI) (*115*). Monomeric complexes of the type (arene)Fe-$(CO)_2$ have not yet been reported, but the PF_3 analogs $(C_6H_6)Fe(PF_3)_2$ (*288*) and $(MeC_6H_5)Fe(PF_3)_2$ (*377*) have been prepared by cocondensation of iron metal vapor, PF_3, and the arene at −196°C.

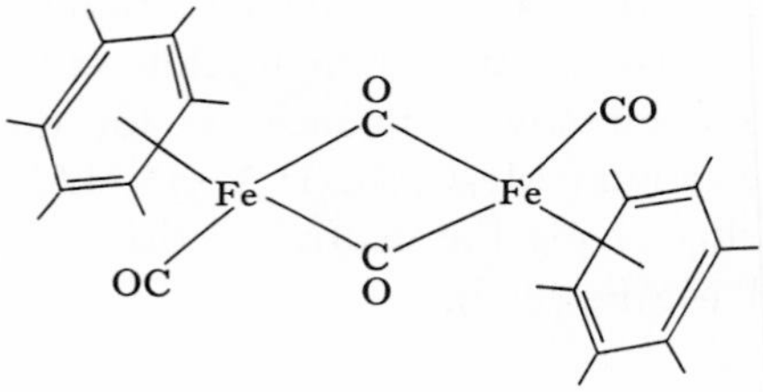

(XVI)

The reaction of $Fe(CO)_5$ or $Fe_2(CO)_9$ with vinyl-substituted arenes leads to adducts in which one or two $Fe(CO)_3$ moieties are bound to the aromatic molecule The structures of several of the adducts have been determined, including those of *p*-divinylbenzene (XVII), *m*-divinylbenzene (XVIII), 1-vinylnaphthalene (XIX), and 2-vinylnaphthalene (XX) (*81*) In all cases, the bonds a_1, a_2, and a_3 are approximately equal, averaging 1.42 Å whereas bond c has a length typical of that of an ethylenic bond (1.32 Å). The bond lengths a_1, a_2, and a_3 are similar to those in butadiene iron tricarbonyl (*290*), showing that the aromaticity of the ring coordinated to the $Fe(CO)_3$ moiety is largely destroyed. The ring not involved in bonding in the napththalene derivatives retains its aromaticity. Diene-type complexes have also been obtained by the irradiation of $Fe(CO)_5$ with styrenes (*424*, *425*). With 3,α-dimethylstyrene, two isomers were obtained for which structures (XXI) and (XXII) were proposed. Condensed ring systems such as anthracene, 9-acetylanthracene (*281*), and benz[*a*]anthracene(*22*) have been shown

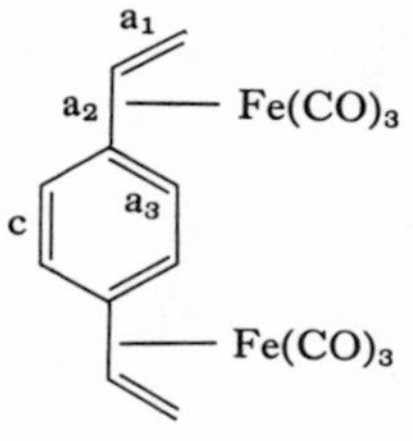

(XVII) 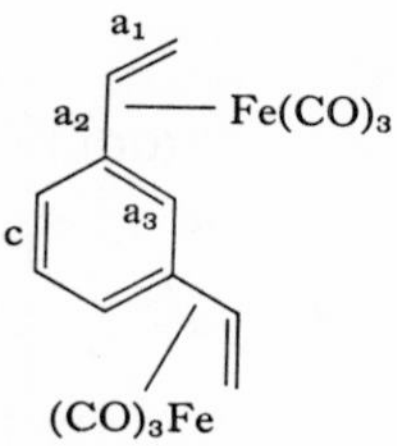 (XVIII)

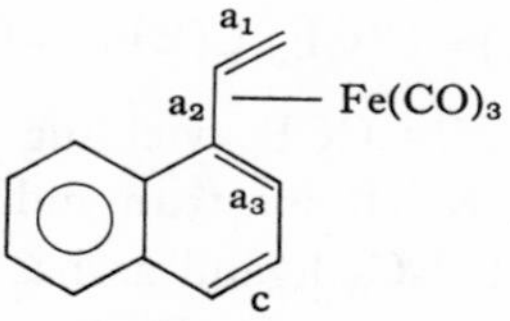

(XIX) 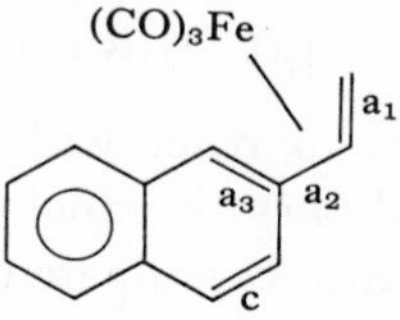 (XX)

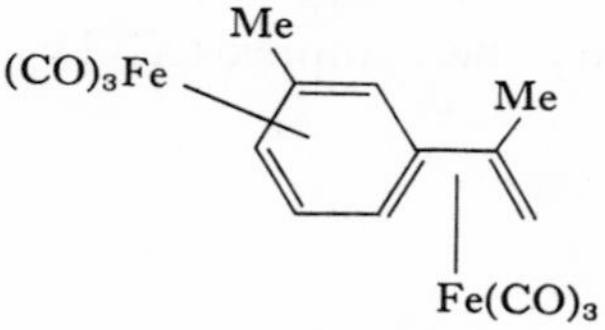

(XXI) 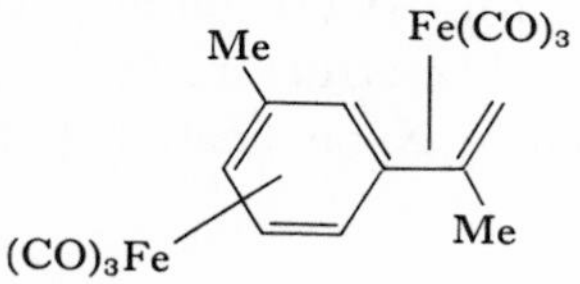 (XXII)

to form complexes in which an $Fe(CO)_3$ moiety is bonded to a terminal ring. The observed hydrogen–hydrogen coupling constants in the anthracene complex show that there is fixation of a "butadiene" unit in the ring bonded to the $Fe(CO)_3$ moiety with enhanced π-delocalization for the residual "naphthalene" unit (*195*). The position of attachment of the $Fe(CO)_3$ moiety in arene complexes has been correlated with the position of lowest bond localization energy for the fixation of a butadiene unit (*324*).

The reaction between $Fe_3(CO)_{12}$ and naphthacene lead to a new type of binuclear complex $(C_{18}H_{12})[Fe_2(CO)_6]$ for which structure (XXIII) was proposed (*22*). The condensed aromatics acridine, phenazine, 4-benz[*c*]acridine, 5-benzo[*a*]phenazine, and dibenzo[*a,c*]phenazine were shown to react with $Fe_3(CO)_{12}$ to give complexes in which an $Fe(CO)_3$ moiety is bound to a terminal ring directly adjacent to the heterocycle ring (*22*).

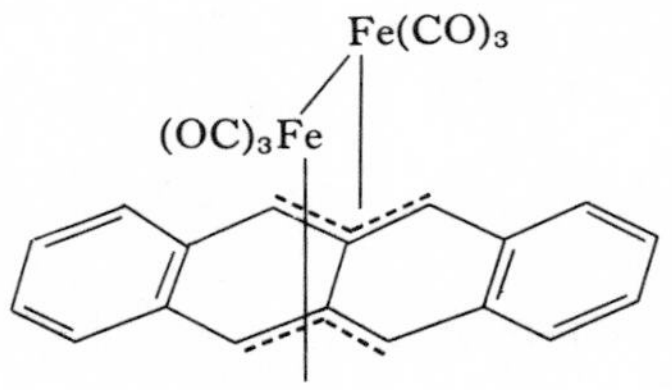

(XXIII).

The metallocyclopentadiene compound

$$[(MeO)_3P]_2(CO)_2\overline{RuC(CF_3){=}C(CF_3)C(CF_3){=}C}CF_3$$

has been shown to react with $CF_3C{\equiv}CCF_3$ yielding $[(CF_3)_6C_6]Ru(CO)_2[P(OMe)_3]$ (*44*) for which the NMR spectrum indicated the presence of a tetrahapto-coordinated $(CF_3)_6C_6$ ligand as has been observed in the isoelectronic complex $[(CF_3)_6C_6](\pi\text{-}C_5H_5)Rh$ (*68*).

Compound $Ru_3(CO)_{12}$ reacts with arenes under reflux, yielding complexes $Ru_6C(CO)_{14}$(arene) (arene = toluene, *m*-xylene, mesitylene) (*241*, *242*). The structure of the mesitylene complex (XXIV) has been determined by X-ray analysis (*285*).

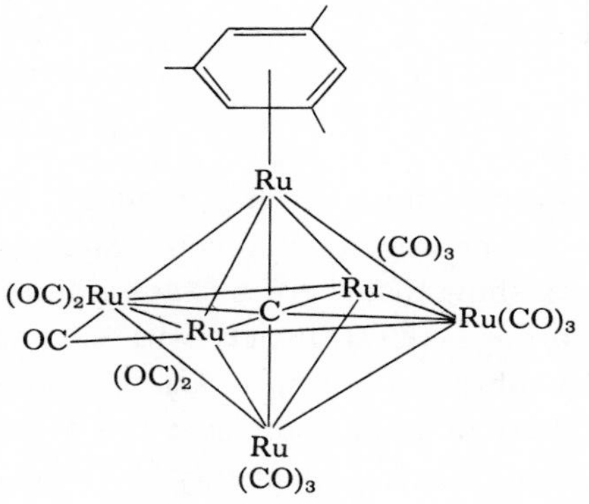

(XXIV)

D. Monoarene Complexes Containing Ligands Other Than Carbon Monoxide

The benzene cyclohexa-1,3-diene complexes $(C_6H_6)(1,3\text{-}C_6H_8)M$ (M = Fe, Ru, Os) have been prepared by reacting the metal trichlorides with *i*-PrMgBr in the presence of $1,3\text{-}C_6H_8$ (*143*, *144*). The iron and osmium reactions were carried out under UV irradiation. Formation of the iron complex is thought to occur via the following reaction:

$$Fe(i\text{-}Pr)_3 + 2(1,3\text{-}C_6H_8) \xrightarrow[35^\circ]{h\nu} (C_6H_6)(1,3\text{-}C_6H_8)Fe + \tfrac{3}{2}C_3H_8 + \tfrac{3}{2}C_3H_6 + H_2 \quad (31)$$

The ruthenium complex has also been prepared by $LiAlH_4$ reduction of the cation $[(C_6H_6)_2Ru]^{2+}$ (*243*), whereas the iron complex has recently been prepared by cocondensation of iron metal vapor and 1,3-C_6H_8 at $-196°C$ (*377*). Cocondensation of iron, toluene, and butadiene at $-196°$ yields $(MeC_6H_5)Fe(C_4H_6)$ (*377*). The PF_3 complexes $(C_6H_6)Fe(PF_3)_2$ (*288*) and $(MeC_6H_5)Fe(PF_3)_2$ (*377*) were prepared by cocondensation of iron, PF_3, and the arene at $-196°C$.

Benzene solutions of ferric chloride contain the adduct $(C_6H_6)FeCl_3$. All evidence was consistent with the adduct possessing C_{3v} symmetry in solution (*451*).

Ruthenium and osmium trihalides have been found to react with cyclohexa-1,3-diene in ethanol solution at 80° to give insoluble polymeric compounds of the type $[(C_6H_6)MX_2]_n$ (X = Cl, I). The polymeric compounds reacted with phosphines yielding the dimeric derivatives $[(C_6H_6)Ru\{P(n\text{-}Bu)_3\}Cl_2]_2$ and $[(C_6H_6)Os(PPh_3)I_2]_2$ (*449*, *450*). A soluble form of the ruthenium complex has recently been prepared by carrying out the reaction between $RuCl_3$ and 1,3-C_6H_8 in an ethanol–water mixture. The bromide, iodide, and thiocyanate analogs were prepared by metathetical replacement of the chloride ion. Molecular weight measurements showed that the complexes were dimeric, and structure (XXV) was proposed (*462*, *464*). Complexes $[(arene)RuCl_2]_2$ (arene = toluene, *p*-xylene, *p*-methylcumene, anisole) have been reported from the reaction of $RuCl_3$ with substituted cyclohexa-1,3- or 1,4-dienes (*28*). Addition of sodium tetraphenyl boron to the mother liquors from the preparation of $[(C_6H_6)RuCl_2]_2$ yielded the chloride-bridged cation $[(C_6H_6)(1,3\text{-}C_6H_8)RuClRu(1,3\text{-}C_6H_8)(C_6H_6)]^{3+}$. When HgCl was added in place of sodium tetraphenyl boron, the Hg_3Cl_2-bridged cation $[(C_6H_6)(1,3\text{-}C_6H_8)Ru{-}Cl{-}Hg{-}Hg{-}Hg{-}Cl{-}Ru(1,3\text{-}C_6H_8)(C_6H_6)]^{4+}$ was obtained (*462*).

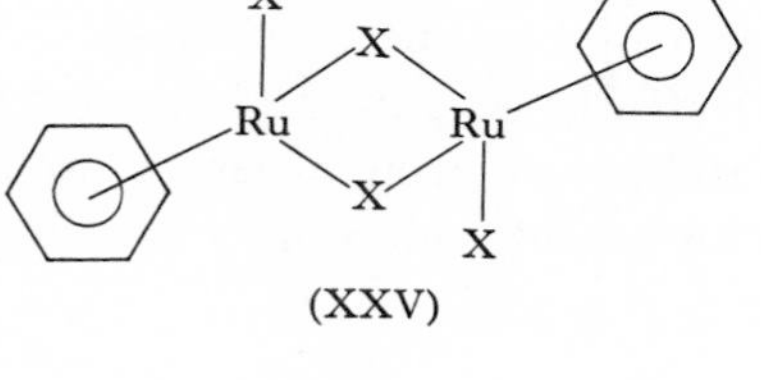

(XXV)

The ruthenium dimers have been shown to react with tertiary phosphines yielding monomeric derivatives of the type (arene)Ru(PR_3)Cl_2 (*28*, *462*, *464*) (Table VI). An X-ray analysis of the derivatives (C_6H_6)-Ru(PPh_2Me)Cl_2 and [1,4-(*i*-Pr)C_6H_4Me]Ru(PPh_2Me)Cl_2 showed both molecules to have the staggered structure (XXVI) (*28*). A small degree of bending of the ring about C_3 and C_6 was observed in both compounds.

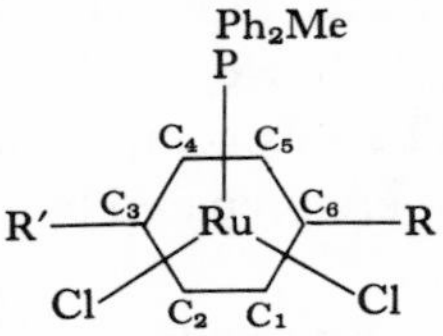

Ph₂Me P; C_4 C_5; R′—C_3 Ru C_6—R; Cl C_2 C_1 Cl (R = R′ = H; R = Me, R′ = *i*-Pr)

(XXVI)

The ring dihedral angles are 5° for the benzene compound and 2° for the *p*-methylcumene compound. The ruthenium carbon distances occurred as one set of four equivalent short bonds (Ru—C_3 through Ru—C_6; average = 2.20 Å) and one set of equivalent long bonds (Ru—C_1 and Ru—C_2; average = 2.26 Å). The two long bonds are trans to the phosphine, and the authors suggested that the asymmetry is a consequence of the trans bond weakening property of the tertiary phosphine.

Solutions of [(C_6H_6)RuCl_2]$_2$ in coordinating solvents, such as dimethyl sulfoxide (DMSO) or acetonitrile appear to contain complexes of the type (C_6H_6)RuCl_2 (solvent) (*332*, *462*). The addition of $HgCl_2$ to the solution gives rise to cations of the type [(C_6H_6)RuCl (solvent)$_2$]$^+$ and [(C_6H_6)Ru(solvent)$_3$]$^{2+}$. Nuclear magnetic resonance evidence also indicates the formation of the cations [(C_6H_6)RuCl(D_2O)$_2$]$^+$ and [(C_6H_6)Ru(D_2O)$_3$]$^{2+}$ on dissolving [(C_6H_6)RuCl_2]$_2$ in D_2O (*462*, *464*).

Dimethyl sulfoxide solutions of [(C_6H_6)RuCl_2]$_2$ were shown to react with nucleophiles such as H^-, OH^-, or CN^-. Nuclear magnetic resonance studies showed the presence of cyclohexadienyl complexes, but the compounds were too unstable to be isolated, and the identity of the other ligands coordinated to the ruthenium atom is uncertain (*462*, *464*).

The polymeric form of benzene ruthenium dichloride has catalytic behavior similar to Ru(PPh_3)$_3$$Cl_2$ in the hydrogenation of olefins (*332*). Similar catalytic activity has been observed for the dimer [(C_6H_6)-RuCl_2]$_2$ in dimethylformamide solution. On the basis of kinetic studies,

TABLE VI

MONOARENE COMPLEXES CONTAINING LIGANDS OTHER THAN CARBON MONOXIDE

Compound	References
$(C_6H_6)Ti(AlCl_4)_2$	*92, 284, 302, 337*
$(C_6H_6)Ti(AlBr_4)_2$	*427*
$(MeC_6H_5)Ti(AlCl_4)_2$	*284, 427*
$(1,3,5\text{-}Me_3C_6H_3)Ti(AlCl_4)_2$	*284, 337*
$(1,2,4,5\text{-}Me_4C_6H_2)Ti(AlCl_4)_2$	*337*
$(Me_6C_6)Ti(AlCl_4)_2$	*337*
$(Me_6C_6)TiCl_2(OH)$	*433*
$(C_6H_6)[Ti(AlCl_4)_2]_3N$	*429*
$[(Me_6C_6)_3Ti_3Cl_6]^+$	*150*
$[(Me_6C_6)_3Zr_3Cl_6]^+$	*150*
$(Me_6C_6)_2Nb_2Cl_4$	*150*
$[(Me_6C_6)_3Nb_3Cl_6]^+$	*150*
$[(Me_6C_6)_6Nb_6Cl_{12}]^{4+}$	*254*
$[(Me_6C_6)_6Nb_6Br_{12}]^{4+}$	*254*
$[(Me_6C_6)_3Ta_3Cl_6]^+$	*150*
$[(Me_6C_6)_6Ta_6Cl_{12}]^{4+}$	*254*
$[(Me_6C_6)_6Ta_6Br_{12}]^{4+}$	*254*
$(C_6H_6)Cr(PF_3)_3$	*288*
$(1,3,5\text{-}Me_3C_6H_3)Cr(PF_3)_3$	*288*
$(F_6C_6)Cr(PF_3)_3$	*288*
$(C_6H_6)Mo[P(OPh)_3]_3$	*182*
$(C_6H_6)Mo[P(OMe)_3]_3$	*182*
$(C_6H_6)Mo[PPh_2(OMe)]_3$	*182*
$(C_6H_6)Mo(PPh_2Me)_3$	*182*
$(C_6H_6)Mo(PPhMe_2)_3$	*182*
$(C_6H_6)Mo(PEt_3)_3$	*186*
$(MeC_6H_5)Mo(PPh_2Me)_3$	*189*
$(1,3,5\text{-}Me_3C_6H_3)Mo(PPh_2Me)_3$	*189*
$(1,3,5\text{-}Me_3C_6H_3)Mo(PMe_3)_3$	*186*
$[(C_6H_6)Mo(PPh_3)_2]_2N_2$	*188, 189*
$(MeC_6H_5)Mo(PPh_3)_2N_2$	*189*
$(MeC_6H_5)Mo(PPh_2Me)_2N_2$	*189*
$(1,3,5\text{-}Me_3C_6H_3)Mo(PPh_3)_2N_2$	*189*
$[(1,3,5\text{-}Me_3C_6H_3)Mo(Me_2PCH_2)_2]_2N_2$	*190, 373a*
$(C_6H_6)Mo(PPh_3)_2H_2$	*188, 189*
$(MeC_6H_5)Mo(PPh_3)_2H_2$	*188, 189*
$(MeC_6H_5)Mo(PPh_2Me)_2H_2$	*189*
$(1,3,5\text{-}Me_3C_6H_3)Mo(PPh_3)_2H_2$	*189*
$[(C_6H_6)Mo(\pi\text{-}C_3H_5)Cl]_2$	*188, 189*
$[(C_6H_6)Mo(\pi\text{-}C_3H_4Me)Cl]_2$	*181*
$[(C_6H_6)Mo(\pi\text{-}C_3H_4Et)Cl]_2$	*181*
$[(C_6H_6)Mo(\pi\text{-}MeC_3H_3Me)Cl]_2$	*181*

Continued

TABLE VI—*Continued*

Compound	References
$[(MeC_6H_5)Mo(\pi\text{-}C_3H_5)Cl]_2$	*188, 189*
$[(1,3,5\text{-}Me_3C_6H_3)Mo(\pi\text{-}C_3H_5)Cl]_2$	*189*
$(C_6H_6)Mo(\pi\text{-}C_3H_5)(PPh_3)Cl$	*188, 189*
$(C_6H_6)Mo(\pi\text{-}C_3H_5)(PPh_2Me)Cl$	*188, 189*
$(C_6H_6)Mo(\pi\text{-}C_3H_5)(PPhMe_2)Cl$	*188, 189*
$(MeC_6H_5)Mo(\pi\text{-}C_3H_5)(PPh_3)Cl$	*188, 189*
$(MeC_6H_5)Mo(\pi\text{-}C_3H_5)(PPh_2Me)Cl$	*188, 189*
$(1,3,5\text{-}Me_3C_6H_3)Mo(\pi\text{-}C_3H_5)(PPh_2Me)Cl$	*189*
$(C_6H_6)Mo(\pi\text{-}C_3H_5)_2$	*180, 184*
$(C_6H_6)Mo(\pi\text{-}C_3H_5)(\pi\text{-}1\text{-}MeC_3H_4)$	*180, 184*
$(C_6H_6)Mo(\pi\text{-}C_3H_5)[\pi\text{-}1\text{-}(CH_2OMe)C_3H_4]$	*180, 184*
$(C_6H_6)Mo(\pi\text{-}C_3H_5)[\pi\text{-}1\text{-}(CH_2CN)C_3H_4]$	*180, 184*
$(C_6H_6)Mo(\pi\text{-}C_3H_5)[\pi\text{-}1\text{-}(CH_2SMe)C_3H_4]$	*184*
$(MeC_6H_5)Mo(\pi\text{-}C_3H_5)_2$	*184*
$(C_6H_6)Mo(\pi\text{-}C_3H_5)O_2R$ carboxylate [R = CH, CMe, CPh, p-$CC_6H_4CO_2Me$, CCH_2CO_2Et, $CCH_2Fe(CO)_2(\pi\text{-}C_5H_5)$, *p*-$CC_6H_4COFe(CO)_2(\pi\text{-}C_5H_5)$, $CCH_2Mo(CO)_3(\pi\text{-}C_5H_5)$]	*183*
$[(C_6H_6)Mo(\pi\text{-}C_3H_5)]_2\text{-}\mu\text{-}C_2O_4$	*183*
$(C_6H_6)Mo(\pi\text{-}C_3H_5)MeCO\cdot CH\cdot COMe$	*183*
$(C_6H_6)Mo(\pi\text{-}C_3H_5)MeCO\cdot CH\cdot COEt$	*183*
$(C_6H_6)Mo(\pi\text{-}C_3H_5)NH_2CH_2CO_2$	*180, 183*
$(C_6H_6)Mo(\pi\text{-}C_3H_5)NH_2(CH_2)_2S$	*180, 183*
$(MeC_6H_5)Mo(\pi\text{-}C_3H_5)NH_2CH_2CO_2$	*183*
$(MeC_6H_5)Mo(\pi\text{-}C_3H_5)NH_2(CH_2)_2S$	*183*
$[(C_6H_6)Mo(\pi\text{-}C_3H_5)(C_4H_6)]^+$ butadiene	*180, 181*
$[(C_6H_6)Mo(\pi\text{-}C_3H_5)(1,4\text{-}Me_2C_4H_4)]^+$	*181*
$[(C_6H_6)Mo(\pi\text{-}C_3H_4Et)(1\text{-}MeC_4H_5)]^+$	*181*
$[(C_6H_6)Mo(\pi\text{-}MeC_3H_3Me)(1\text{-}MeC_4H_5)]^+$	*181*
$[(MeC_6H_5)Mo(\pi\text{-}C_3H_5)(C_4H_6)]^+$	*180*
$[(C_6H_6)Mo(\pi\text{-}C_3H_5)L_2]^+$ [L_2 = $(Me_2PCH_2)_2$, $(Ph_2PCH_2)_2$, $(MeSCH_2)_2$, $(H_2NCH_2)_2$, bipyridyl]	*180, 185*
$[(MeC_6H_5)Mo(\pi\text{-}C_3H_5)(NH_2CH_2)_2]^+$	*185*
$[(MeC_6H_5)Mo(\pi\text{-}C_3H_5)(o\text{-}(NH_2)_2C_6H_4)]^+$	*185*
$[(C_6H_6)Mo(PPh_2Me)_3H]^+$	*182*
$[(C_6H_6)Mo(PPhMe_2)_3H]^+$	*182*
$[(C_6H_6)Mo(PEt_3)_3H]^+$	*186*
$[(1,3,5\text{-}Me_3C_6H_3)Mo(PMe_3)_3H]^+$	*186*
$[(MeC_6H_5)Mo(PPh_3)_2\text{-}N_2\text{-}(Me_2PCH_2)_2Fe(C_5H_5)]^+$	*189*
$[(C_6H_6)Mo(PPh_2Me)_3H_2]^{2+}$	*180, 186*
$[(C_6H_6)Mo(PPhMe_2)_3H_2]^{2+}$	*180, 186*
$[(C_6H_6)Mo(PEt_3)_3H_2]^{2+}$	*186*
$[(1,3,5\text{-}Me_3C_6H_6)Mo(PMe_3)_3H_2]^{2+}$	*186*

TABLE VI—*Continued*

Compound	References
$[\{(1,3,5\text{-}Me_3C_6H_3)Mo(Me_2PCH_2)_2H\}_2N_2]^{2+}$	*190, 373a*
$(Me_6C_6)Mn(C_6H_7)$	*154*
$(C_6H_6)Tc(C_6H_7)$	*153*
$(Me_6C_6)Tc(Me_6C_6H)$	*153*
$(C_6H_6)Re(C_6H_7)$	*156, 243*
$(Me_6C_6)Re(Me_6C_6H)$	*156*
$(Me_6C_6)_2Re_2(Me_6C_6 \cdot C_6Me_6)$	*156*
$(C_6H_6)Fe(PF_3)_2$	*288*
$(MeC_6H_5)Fe(PF_3)_2$	*377*
$(C_6H_6)Fe(1,3\text{-}C_6H_8)$	*143, 377*
$(MeC_6H_5)Fe(C_4H_6)$	*377*
$(C_6H_6)Ru(1,3\text{-}C_6H_8)$	*144, 243*
$(C_6H_6)Ru(\pi\text{-}C_3H_5)Cl$	*463*
$[(C_6H_6)RuCl_2]_n$ $(n > 2)$	*449*
$[(C_6H_6)RuX_2]_2$	
(X = Cl, Br, I, SCN)	*462, 464*
$[(MeC_6H_5)RuCl_2]_2$	*28*
$[(MeOC_6H_5)RuCl_2]_2$	*28*
$[(1,4\text{-}Me_2C_6H_4)RuCl_2]_2$	*28*
$[(1,4\text{-}MeC_6H_4CHMe_2)RuCl_2]_2$	*28*
$(C_6H_6)RuCl_2(PR_3)$	
$[PR_3 = PPh_3, PPhMe_2, PEt_3, P(OPh)_3, P(OEt)_3$	*462, 464*
$= PPh_2Me, P(OMe)_3$	*28, 462, 464*
$= P(n\text{-}Bu)_3, P(n\text{-}C_8H_{17})_3]$	*28*
$(C_6H_6)RuCl_2(AsPh_3)$	*462, 464*
$(C_6H_6)RuCl_2(Ph_2PCH_2PPh_2)$	*462*
$(C_6H_6)RuCl_2(Ph_2P(CH_2)_4PPh_2)$	*462*
$(C_6H_6)RuBr_2(PPh_3)$	*462*
$(C_6H_6)RuI_2(PPh_3)$	*462*
$(C_6H_6)Ru(SCN)_2(PPh_3)$	*462*
$[(C_6H_6)RuCl_2\{P(n\text{-}Bu)_3\}]_2$	*449*
$(1,4\text{-}MeC_6H_4CHMe_2)RuCl_2(PPh_2Me)$	*28*
$(1,3,5\text{-}Me_3C_6H_3)RuCl_2[P(n\text{-}Bu)_3]$	*28*
$(C_6H_6)Ru(Me)Cl(PPh_3)$	*463*
$(C_6H_6)Ru(Ph)Cl(PPh_3)$	*463*
$(C_6H_6)RuCl_2(OSMe_2)$	*332*
$(C_6H_6)RuHCl(OSMe_2)$	*332*
$[(C_6H_5PPh_2)RuH(PPh_3)_2]^+$	*364*
$[(C_6H_6)Ru(C_{18}H_{18}N_8B)]^+$ tetrakis-(1-pyrazolyl) borate	*353*
$[(C_6H_6)RuCl(CH_3CN)_2]^+$	*462, 464*
$[\{(C_6H_6)Ru(1,3\text{-}C_6H_8)\}_2Cl]^{3+}$	*462*
$[\{(C_6H_6)Ru(1,3\text{-}C_6H_8)\}_2Hg_3Cl_2]^{4+}$	*462*
$(C_6H_6)Os(1,3\text{-}C_6H_8)$	*144*
$[(C_6H_6)OsCl_2]_n$	*450*

Continued

TABLE VI—*Continued*

Compound	References
$[(C_6H_6)OsI_2]_n$	*450*
$[(C_6H_6)OsI_2(PPh_3)]_2$	*450*
$(C_6H_5BPh_3)Rh(PPh_3)_2$	*367*
$(C_6H_5BPh_3)Rh[P(OR)_3]_2$	
(R = Me, *i*-Pr, *n*-Bu, *i*-Bu)	*329*
$(C_6H_5BPh_3)Rh(C_2H_4)_2$	*367*
$(C_6H_5BPh_3)Rh(diene)$	
(diene = nbd, C_4H_6, 2,3-$Me_2C_4H_4$, C_5H_6, 1,3-C_6H_8, 1,5-hexadiene, 1,5-cod, 1,3,5,7-cyclooctatetraene)	*367*
$[(C_6H_6)Rh\{P(OPh)_3\}_2]^+$	*368*
$[(C_6H_6)Rh(C_2H_4)_2]^+$	*179, 368*
$[(C_6H_6)Rh(diene)]^+$	
(diene = nbd, 1,5-hexadiene, 1,3-C_6H_8, 1,5-cod)	*179, 368*
$[(MeC_6H_5)Rh(nbd)]^+$	*179*
$[(MeC_6H_5)Rh(1,4\text{-}cod)]^+$	*179*
$[(HOC_6H_5)Rh(nbd)]^+$	*179*
$[(MeOC_6H_5)Rh(nbd)]^+$	*179*
$[(1,3\text{-}Me_2C_6H_4)Rh(C_2H_4)_2]^+$	*179*
$[(1,3\text{-}Me_2C_6H_4)Rh(nbd)]^+$	*179*
$[(1,3\text{-}Me_2C_6H_4)Rh(1,5\text{-}cod)]^+$	*179*
$[(1,3,5\text{-}Me_3C_6H_3)Rh\{P(OPh)_3\}_2]^+$	*179*
$[(1,3,5\text{-}Me_3C_6H_3)Rh(C_2H_4)_2]^+$	*179*
$[(1,3,5\text{-}Me_3C_6H_3)Rh(diene)]^+$	
(diene = nbd, 1,3-C_6H_8, 1,5-hexadiene, 1,5-cod)	*179, 368*
$[(Me_6C_6)Rh\{P(OPh)_3\}_2]^+$	*368*
$[(Me_6C_6)Rh(C_2H_4)_2]^+$	*368*
$[(Me_6C_6)Rh(diene)]^+$	
(diene = nbd, 1,3-C_6H_8, 1,5-hexadiene, 1,5-cod)	*179, 368*
$[(Me_6C_6)RhCl(C_5H_5N)]^{2+}$	*30*
$[(Me_6C_6)_2Rh_2Cl_2]^{4+}$	*30*
$(C_6H_5BPh_3)Ir(1,5\text{-}cod)$	*367*
$(C_6H_5BPh_3)Ir(1,5\text{-}hexadiene)$	*367*
$[(C_6H_6)Ir(cyclooctene)_2]^+$	*368*
$[(C_6H_6)Ir(1,5\text{-}cod)]^+$	*368*
$[(1,3,5\text{-}Me_3C_6H_3)Ir(cyclooctene)_2]^+$	*368*
$[(1,3,5\text{-}Me_3C_6H_3)Ir(1,5\text{-}cod)]^+$	*368*
$[(Me_6C_6)Ir(cyclooctene)_2]^+$	*368*
$[(Me_6C_6)Ir(1,5\text{-}cod)]^+$	*368*
$[(CF_3)_6C_6]Ni(1,5\text{-}cod)$	*41*
$[(CF_3)_6C_6]Ni(1,5\text{-}cod)$	*41*
$[(CF_3)_6C_6]NiL_2$	
[L = $AsPhMe_2$, PPh_2Me, PPh_3, $P(OMe)_3$, $P(OCH_2)_3CMe$]	*41*
$[(CF_3)_6C_6]Ni_2[P(OMe)_3]_4$	*41*
$[(CF_3)_6C_6]Ni_2[P(OCH_2)_3CMe]_4$	*41*

TABLE VI—*Continued*

Compound	References
$[(CF_3)_6C_6]Pt(PEt_3)_2$	*42*
$[(C_6H_6)Pd(AlCl_4)]_2$	*5*
$[(C_6H_6)Pd(Al_2Cl_7)]_2$	*5, 6*
$(C_6H_6)Pd(H_2O)(ClO_4)$	*80*
$(C_6H_6)Cu(AlCl_4)$	*382, 417, 418*
$(C_6D_6)Cu(AlCl_4)$	*382*
$(C_6H_6)Cu(OSO_2CF_3)_2$	*94, 381*
$(MeC_6H_5)Cu(AlCl_4)$	*287*
$(MeC_6H_5)_2Cu(AlCl_4)$	*287*
$(C_6H_6)Ag(AlCl_4)$	*419*
$(C_6D_6)Ag(AlCl_4)$	*382*
$(C_6H_6)Ag(ClO_4)$	*362, 378*
$(C_6H_5CH_2CH_2C_6H_5)Ag(ClO_4)$	*405*
$(C_6H_{11}C_6H_5)_2Ag(ClO_4)$	*199, 200*
$(1,2\text{-}Me_2C_6H_4)Ag(ClO_4)$	*19*
$(1,2\text{-}Me_2C_6H_4)_2Ag(ClO_4)$	*19, 404, 405*
$(1,3\text{-}Me_2C_6H_4)Ag(ClO_4)$	*19*
$(1,3\text{-}Me_2C_6H_4)_2Ag(ClO_4)$	*19, 405*
$(1,4\text{-}Me_2C_6H_4)_2Ag(ClO_4)$	*19, 405*
$(1,3,5\text{-}Me_3C_6H_3)Ag(ClO_4)$	*19*
$(1,3,5\text{-}Me_3C_6H_3)_2Ag(ClO_4)$	*405*
$(C_9H_8)Ag(ClO_4)$ indane	*361*
$(C_{10}H_8)[Ag(ClO_4)]_4 \cdot 4H_2O$ naphthalene	*201, 405*
$(C_{12}H_8)Ag(ClO_4)$ acenaphthylene	*340, 360*
$(C_{12}H_8)Ag(BF_4)$	*340*
$(C_{12}H_{10})Ag(ClO_4)$ acenaphthene	*340, 360, 405*
$(C_{12}H_{10})Ag(BF_4)$	*340*
$(C_{14}H_{10})[Ag(ClO_4)]_4 \cdot H_2O$ anthracene	*201*

a benzene ruthenium hydride complex has been proposed as the catalytic species (*225*). Furthermore, the reaction of aqueous $(C_6H_6)RuCl_2$-(DMSO) with hydrogen in the presence of triethylamine yielded a hydride that, on the basis of its NMR spectrum, was formulated as the compound $(C_6H_6)RuCl(H)(DMSO)$ (*332*).

Treatment of the dimer $[(C_6H_6)RuCl_2]_2$ with R_2Hg (R = Me, Ph) followed by the addition of PPh_3 yielded complexes of the type (C_6H_6)-$RuCl(R)(PPh_3)$. With $(C_3H_5)_4Sn$, the π-allyl complex $(C_6H_6)Ru$-$(\pi\text{-}C_3H_5)Cl$ was obtained (*463*). The allyl complex is isoelectronic with $[(C_6H_6)Mo(\pi\text{-}C_3H_5)Cl]_2$ (*189*), and in many respects the chemistry of the Ru(II) complexes described in this section resembles that of the Mo(II) complexes described in Section VI,G.

Zero-valent ruthenium complexes (arene)Ru$(PR_3)_2$ (PR_3 = tertiary phosphine), isoelectronic with the molybdenum complexes (C_6H_6)Mo-$(PR_3)_3$ (*182*), have not yet been reported. However, the hydride cation (XXVII), isoelectronic with the molybdenum hydride cations $[(C_6H_6)$-Mo$(PR_3)_3H]^+$ (*182*), has been obtained from the dissociation of [RuH-$(PPh_3)_4]PF_6$ in dichloromethane solution (*364*).

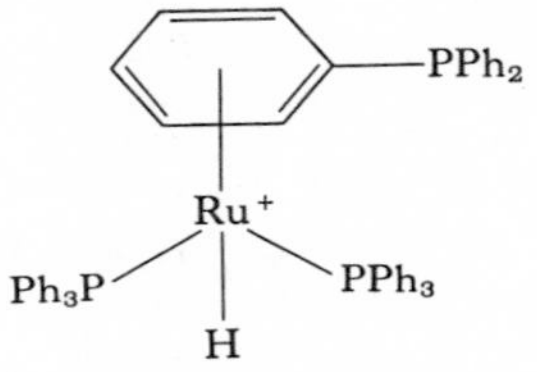

(XXVII)

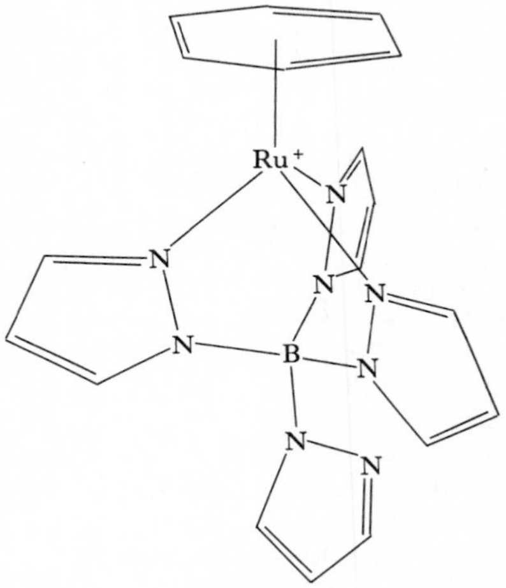

(XXVIII)

A new type of arene Ru(II) complex $[(C_6H_6)Ru\{B(C_3H_3N_2)_4\}]PF_6$ (XXVIII) has been recently prepared and its structure determined by X-ray analysis (*353*). The benzene ring is staggered with respect to the RuN_3 moiety. No distortion of the benzene carbon–carbon bonds from the mean value of 2.20(2) Å was observed.

Reduction of *trans*-$RuCl_2[Me_2PCH_2)_2]_2$ by arene radical anions such as sodium naphthalide has been shown to lead to a series of hydridoaryl complexes of the type Ru(H)(aryl)$[(Me_2PCH_2)_2]_2$ (*63*). The hydride structure of the napthalene derivative has been shown in solution (*63*) and in the solid state (*232*). However, on the basis of the chemistry of the complexes, the following equilibrium was proposed (*63*), between a hydride and a small amount of a dihapto-coordinated arene-Ru(0) complex:

$$\text{Ru(H)(aryl)}[(Me_2PCH_2)_2]_2 \rightleftharpoons \text{Ru}^0\text{(arene)}[(Me_2PCH_2)_2]_2 \tag{32}$$

Arene complexes may also occur as intermediates in the formation of other aryl hydride complexes such as $(\pi\text{-}C_5H_5)_2Mo(H)(aryl)$ (aryl = phenyl, *p*-tolyl) (*171*).

IX

COBALT, RHODIUM, AND IRIDIUM COMPLEXES

A. *Bisarene Complexes*

No cations of the type $[(arene)_2M]^{3+}$ (M = Co, Rh, Ir) which would be isoelectronic with the corresponding Fe(II) and Ru(II) complexes have been isolated, although cation $[(Me_6C_6)_2Co]^{3+}$ is thought to arise in the disproportionation of $[(Me_6C_6)_2Co]^{2+}$ to $[(Me_6C_6)_2Co]^+$ (*141*). Apparently bisarene cations with a charge greater than 2 are unstable.

Cation $[(Me_6C_6)_2Co]^{2+}$ has been prepared by reacting $CoCl_2$ with hexamethylbenzene in the presence of $AlCl_3$ (*141*). Cation $[(Me_6C_6)_2Rh]^{2+}$ has also been prepared by the Fischer–Hafner method starting from $RhCl_3$ (*141*). Both of the cations are paramagnetic with one unpaired electron. Addition of aluminum metal to the reaction mixture in the case of cobalt lead to the cation $[(Me_6C_6)_2Co]^{1+}$. The cation is paramagnetic with two unpaired electrons which can be accounted for by accommodating 2 electrons in the e_{1g} orbital (Section III, A; Fig. 1). However, a diamagnetic analog has been reported from the condensation of dimethylacetylene on dimesityl cobalt(II) (*416*, *460*). A diamagnetic Rh(I) cation $[(Me_6C_6)_2Rh]^+$ has also been obtained by reduction of the cation $[(Me_6C_6)_2Rh]^{2+}$ with Zn/HCl (*141*). The observed diamagnetism of the Co(I) and Rh(I) cations suggests that they may be 18-electron compounds containing a nonplanar tetrahapto-coordinated Me_6C_6 ligand as has been found for $(Me_6C_6)_2Ru$ (see Section VIII,A) (*230*). Reduction of the paramagnetic cations $[(Me_6C_6)_2Co]^+$ or $[(Me_6C_6)_2Co]^{2+}$ with sodium in liquid ammonia gave the neutral complex $(Me_6C_6)_2Co$ (*142*). The complex was found to be paramagnetic with one unpaired electron. The magnetic properties can be accounted for by accommodating 3 electrons in the e_{1g} orbital (Section III,A; Fig. 1) or, alternatively, by considering the complex to be a 19-electron compound with one nonplanar tetrahapto-coordinated Me_6C_6 ligand. In support of the latter possibility is the observed dipole moment of 1.78 D which excludes a centrosymmetric structure. A Jahn–Teller distortion of the molecule has also been proposed to account for the observed dipole moment (*325*).

Paramagnetic Co(II) borabenzene complexes $(C_5H_5BR)_2Co$ (R = Br, Me, Ph) have been synthesized by treating $(\pi\text{-}C_5H_5)_2Co$ with $RBBr_2$. Hydrolysis or methanolysis of the bromo derivative led to complexes $(C_5H_5BOH)_2Co$ and $(C_5H_5BOMe)_2Co$, respectively (*220*).

B. *Mixed Sandwich Complexes*

The Co(III) cation $[(C_6H_6)(\pi\text{-}C_5H_5)Co]^{2+}$ has been prepared by hydrogen abstraction on $(\pi\text{-}C_5H_5)(1,3\text{-}C_6H_8)Co$ with triphenylmethyl tetrafluoroborate (*126*). Cations $[(\text{arene})(\pi\text{-}C_5Me_5)M]^{2+}$ (M = Rh, Ir; arene = benzene, *p*-xylene, pentamethylbenzene, hexamethylbenzene) have been prepared by the reaction of $(\pi\text{-}C_5Me_5)M(OCOCF_3)_2(H_2O)$ with the arene in trifluoracetic acid (*440*). The rhodium and iridium cations undergo hydride attack giving cyclohexadienyl cations of the type $[(C_6H_{7-n}Me_n)(\pi\text{-}C_5Me_5)M]^+$ (n = 0, 5, 6), the unsubstituted ring positions being preferentially attacked. The arene ligand in the cation $[(1,4\text{-}Me_2C_6H_4)(\pi\text{-}C_5Me_5)Rh]^{2+}$ was readily displaced by DMSO giving the cation $[(\pi\text{-}C_5Me_5)Rh(DMSO)_3]^{2+}$.

The Co(III) borabenzene cation $[(C_5H_5BPh)(\pi\text{-}C_5H_5)Co]^+$ has been prepared by reaction of $(\pi\text{-}C_5H_5)_2Co$ with $PhBBr_2$ (*219*). The NMR spectrum supports a sandwich structure for the complex with the borabenzene ring coordinated to the cobalt. Neutral paramagnetic borabenzene complexes of the type $(C_5H_5BR)(\pi\text{-}C_5H_5)Co$ (R = Br, Me, Ph) have also been obtained from the reaction of $(\pi\text{-}C_5H_5)_2Co$ with $RBBr_2$. The bromo derivative yielded $(C_5H_5BOH)(\pi\text{-}C_5H_5)Co$ on hydrolysis (*220*).

The neutral Rh(I) complex $[(CF_3)_6C_6](\pi\text{-}C_5H_5)Rh$ has been prepared by reacting $CF_3C{\equiv}CCF_3$ with $(\pi\text{-}C_5H_5)Rh(CO)_2$ (*93*). An X-ray analysis has revealed a nonplanar tetrahapto-coordinated $(CF_3)_6C_6$ ligand (XXIX) (*68*). The C_5–C_6 distance of 1.31 Å is very close to that of a normal double bond. Bonds C_1–C_2 and C_3–C_4 average 1.50 Å which is very close to the accepted value of 1.51 Å for a single bond, whereas C_2–C_3 is significantly shorter at 1.42 Å. The squeezing together of carbon atoms C_1 and C_4, separation = 2.56 ± 0.2 Å as compared to 2.80 Å in a regular benzene ring, led the authors to suggest that carbon atoms C_1 and C_4 form strong σ-bonds to the formally

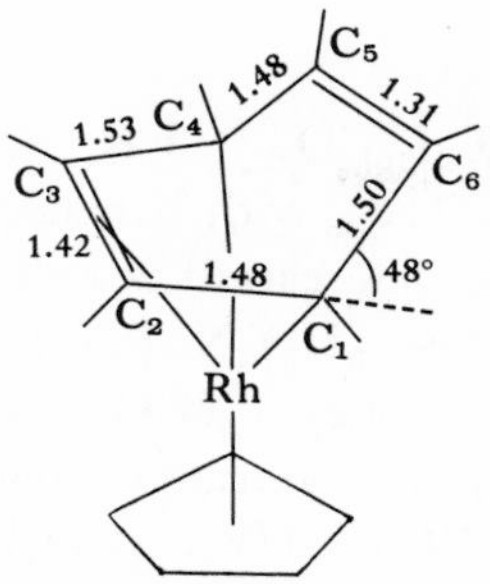

(XXIX) 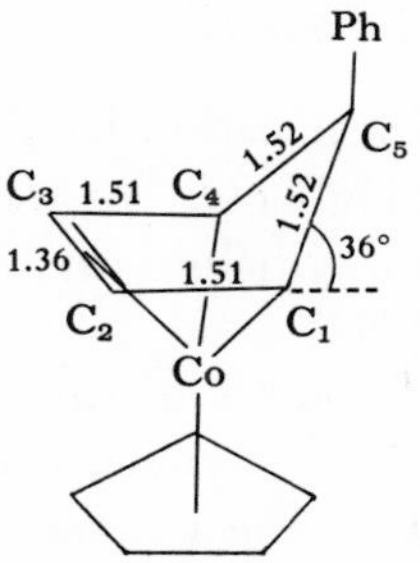 (XXX)

Rh(III) ion, whereas carbon atoms C_2 and C_3 form a normal π-bond to the metal. The bonding of the $(CF_3)_6C_6$ ring in $[(CF_3)_6C_6](\pi\text{-}C_5H_5)Rh$ is in many respects comparable to the phenylcyclopentadiene ring in the compound $(\pi\text{-}C_5H_5)(C_5H_5Ph)Co$ (XXX). In this case, bonds C_1–C_2 and C_3–C_4 average 1.51 Å whereas C_2–C_3 is considerably shorter at 1.36(3) Å (*69*). The bonding of the $(CF_3)_6C_6$ ligand in $[(CF_3)_6C_6]$-$(\pi\text{-}C_5H_5)Rh$ can thus be described as being of the "diene" type. The bending of the $(CF_3)_6C_6$ ring allows the rhodium atom to adopt the stable 18-electron configuration just as has been observed for $(Me_6C_6)_2Ru$ (*230*). Compounds $[(MeOOC)_6C_6](\pi\text{-}C_5Me_5)M$ (M = Rh, Ir) have been prepared by a cyclic condensation reaction (*245*):

$$(\pi\text{-}C_5Me_5)M(OAc)_2(H_2O) + MeOOCC{\equiv}CCOOMe + H_2 \longrightarrow [(MeOOC)_6C_6](\pi\text{-}C_5Me_5)M + 2HOAc \quad (33)$$

The NMR spectrum showed the presence of a tetrahapto-coordinated $(MeOOC)_6C_6$ ligand. The rhodium compound showed fluxional NMR behavior on heating to 150°C in d_6 DMSO which was reversed on cooling.

Diamagnetic arene-π-cyclobutadiene cations of the type $[(arene)(\pi\text{-}C_4Ph_4)Co]^+$ have been prepared by the reaction of $(\pi\text{-}C_4Ph_4)Co(CO)_2Br$ with a variety of arenes in the presence of $AlCl_3$ (*104, 277*) (see Table V). The benzene derivative reacts with the nucleophiles H^- or $n\text{-}Bu^-$, yielding the neutral cyclohexadienyl derivatives $(C_6H_6R)(\pi\text{-}C_4Ph_4)Co$ (*103*). Treatment of the derivative where R = H with *N*-bromosuccinimide gave back the benzene cation, whereas the derivative for which R = *n*-Bu gave the cation $[(n\text{-}BuC_6H_5)(\pi\text{-}C_4Ph_4)Co]^+$.

C. Carbonyl Complexes

The polynuclear cation $[(C_6H_6)_3Co_3(CO)_2]^+$ has been reported from the reaction of $Hg[Co(CO)_4]_2$ or $Co_2(CO)_8$ with benzene in the presence of an aluminum trihalide (*66*, *114*). The cation is believed to have the structure (XXXI) analogous to $(\pi\text{-}C_5H_5)_3Ni_3(CO)_2$ (*289*). Neutral polynuclear complexes of the type (arene)$Co_4(CO)_9$ (arene = benzene, toluene, anisole, *p*-xylene, mesitylene, tetrahydronaphthalene) have been prepared by the reaction of $(RC{\equiv}CH)Co_2(CO)_6$ (R = H or Ph) with norbornadiene in the appropriate aromatic solvent or in some cases by simply warming $Co_4(CO)_{12}$ with the arene (*247*, *365*). The compounds are believed to have the structure (XXXII) derived from that of $Co_4(CO)_{12}$ (*435*) by replacement of three apical CO groups by the arene. A normal coordinate analysis has been carried out on several of

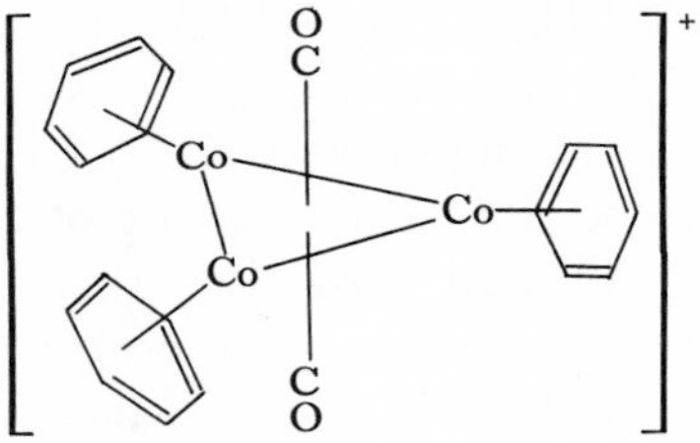

(XXXI)

the derivatives and the carbonyl frequencies assigned and the force constants calculated (*365*). Polynuclear complexes of the type $RCCo_3(CO)_6$(arene) (R = Me, Ph, F; arene = benzene, toluene, *o*-, *m*-, *p*-xylene, mesitylene) have been prepared by the reaction of $RCCo_3(CO)_9$

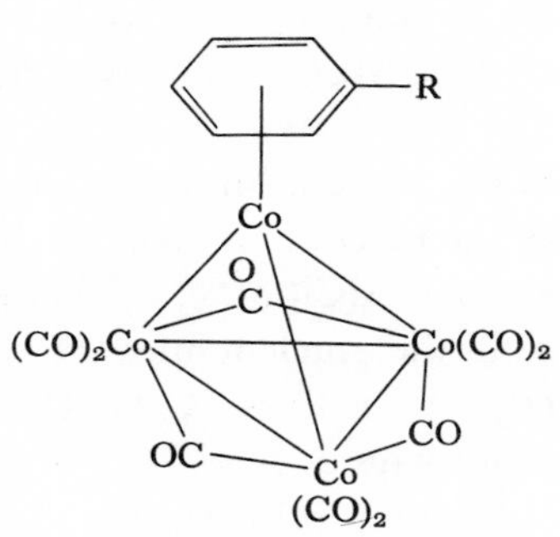

(XXXII)

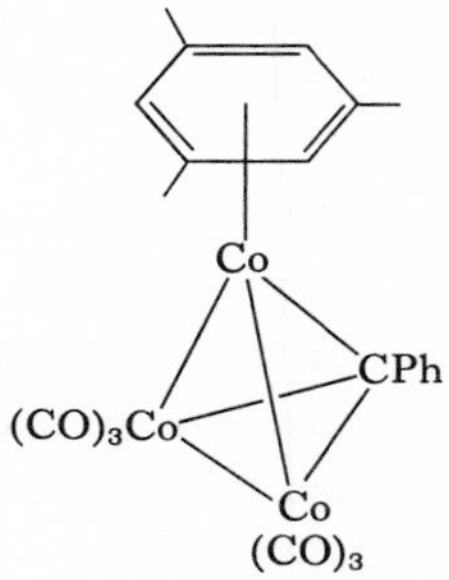

(XXXIII)

with the arene at elevated temperatures (*358*, *359*). The complex where R = Ph and arene = mesitylene has been shown by X-ray analysis to have the structure (XXXIII) (*33*, *88*). The structure can be thought of as being derived from the parent compound $MeCCo_3(CO)_9$ by replacement with mesitylene of the three CO groups attached to one of the cobalt atoms.

D. *Monoarene Complexes Containing Ligands Other Than Carbon Monoxide*

Treatment of the diene cations $[M(diene)_2]^+$ (M = Rh, Ir) with an arene gives cations of the type $[(arene)M(diene)]^+$ (*179*, *368*) (see Table VI). The reaction occurred more readily in the sequence: $Me_6C_6 >$ 1,3,5-$Me_3C_6H_3 >$ 1,3-$Me_2C_6H_4 > MeC_6H_5 > C_6H_6$ (*179*). Similarly, the cations $[RhL_4]^+$ [L = C_2H_4, $P(OPh)_3$] reacted with arenes yielding the cations $[(arene)RhL_2]^+$. Cations $[(Me_6C_6)RhL_2]^+$ (L_2 = norbornadiene, 1,5-cyclooctadiene) were also prepared by treating the complexes $Rh(L_2)(acac)$ with $[Ph_3C]^+BF_4^-$ in the presence of hexamethylbenzene (*179*). The arene ligand in the cations $[(arene)Rh(diene)]^+$ could be displaced by various π-acceptor or σ-donor ligands yielding cations of the type $[Rh(diene)L_2]^+$ (*179*). Cation $[Rh(nbd)(1,3\text{-}C_6H_8)]^+$ is an effective catalyst for the disproportionation of 1,3-cyclohexadiene into benzene and cyclohexene. The disproportionation reaction occurred much more slowly however with 1,4-cyclohexadiene. The mechanism shown in Eq. (34) favoring the 1,3-reaction was proposed (*179*).

Neutral Rh(I) complexes containing a π-bonded phenyl group derived from a tetraphenylboron anion have been described (*329*, *367*). The compounds are of the type RhL_2BPh_4 [L = C_2H_4, PPh_3, $P(OR)_3$] and $Rh(diene)BPh_4$ (see Table VI). The Ir(I) complexes Ir(1,5-cod)BPh_4 and Ir(1,5-hexadiene)BPh_4 were also described. The compounds are clearly closely related to the Rh(I) and Ir(I) cationic complexes already described. The structure of the compound $Rh[P(OMe)_3]_2BPh_4$ (XXXIV) prepared by allowing a solution of $[Rh\{P(OMe)_3\}_5]^+BPh_4^-$ to stand in air, has been determined by X-ray analysis. The bonded phenyl ring is distorted into a boat configuration and this is thought to be due to either packing forces or possibly to a slight localization of the bonding molecular orbitals on the carbon atoms 2, 3, 5, and 6 (*329*).

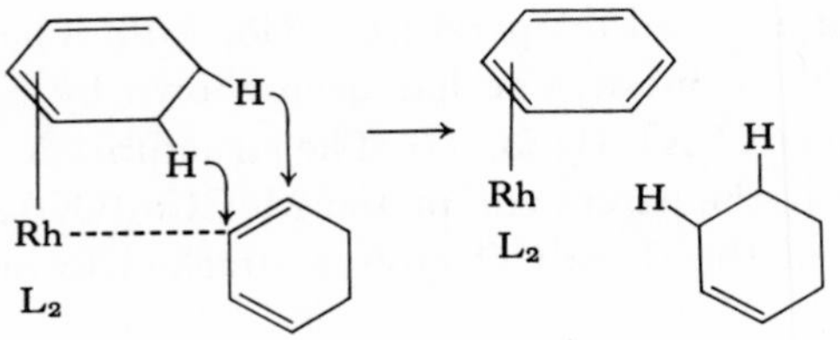
H
H
Rh
L2
Rh
L2
H
H

(34)

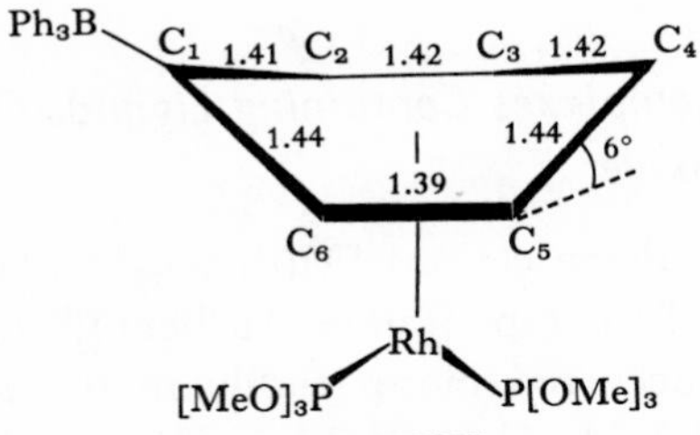
Ph3B
C1
1.41
C2
1.42
C3
1.42
C4
1.44
1.44
6°
1.39
C6
C5
Rh
[MeO]3P
P[OMe]3

(XXXIV)

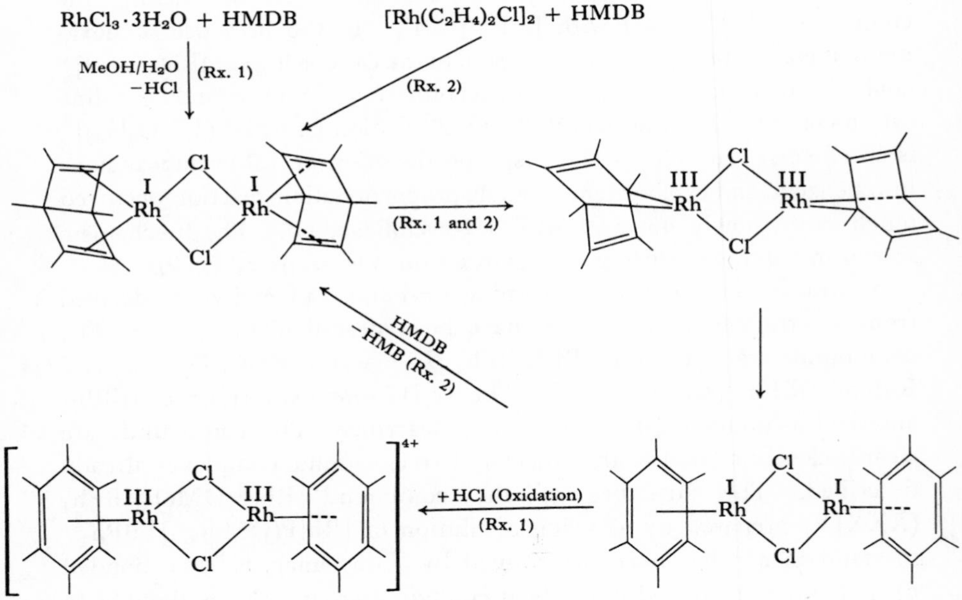
RhCl3·3H2O + HMDB
[Rh(C2H4)2Cl]2 + HMDB
MeOH/H2O
−HCl
(Rx. 1)
(Rx. 2)
I
Rh
Cl
Cl
I
Rh
(Rx. 1 and 2)
III
Rh
Cl
Cl
III
Rh
HMDB
−HMB (Rx. 2)
4+
III
Rh
Cl
Cl
III
Rh
+HCl (Oxidation)
(Rx. 1)
I
Rh
Cl
Cl
I
Rh

Scheme 2

The reaction of hexamethyldewarbenzene (HMDB) with rhodium compounds gives hexamethylbenzene (HMB) complexes. Thus, the reaction between HMDB and the cations $[Rh(diene)_2]^+$ (diene = norbornadiene, 1,5-cyclooctadiene) yielded cations $[(Me_6C_6)Rh(diene)]^+$ (*179*), whereas the reaction between HMDB and $RhCl_3 \cdot 3H_2O$ yielded a complex formulated as the chloride-bridged dimer $[(Me_6C_6)RhCl_2Rh(Me_6C_6)]^{4+}(Cl^-)_4$. The latter compound reacted with pyridine yielding the cation $[(Me_6C_6)RhCl(C_5H_5N)]^{2+}$. No reaction was observed between hexamethylbenzene and $RhCl_3 \cdot 3H_2O$ which was taken to indicate formation of a HMDB intermediate in the reaction between HMDB and $RhCl_3 \cdot 3H_2O$ (*30*). Furthermore, the dimer $[(HMDB)RhCl]_2$ catalyzes the formation of HMB from HMDB (*428*). Intermediates in which HMDB has oxidatively added to the metal atom are thought to occur in the reactions between HMDB and Rh(I) complexes (*179*). The reactions between HMDB and the rhodium chloride compounds discussed in the foregoing can be described by Scheme 2.

X

NICKEL, PLATINUM, AND PALLADIUM COMPLEXES

A. Bisarene Complexes

The only bisarene compound reported to date is the Ni(II) cation $[(Me_6C_6)_2Ni]^{2+}$ prepared by reaction of $NiBr_2$ with hexamethylbenzene in the presence of $AlBr_3$ (*272*). The cation is paramagnetic with 2 unpaired electrons making it isoelectronic with the corresponding Fe(0) and Co(I) complexes. When the reaction was carried out in the presence of aluminum metal, an olive brown mass was obtained which appeared to contain the Ni(I) cation $[(Me_6C_6)_2Ni]^+$. However, attempts to separate the Ni(I) cation from the reaction mixture were unsuccessful, and only the Ni(II) cation along with nickel metal were obtained. The following disproportionation reaction was suggested to account for these observations:

$$2[(Me_6C_6)_2Ni]^+ \longrightarrow [(Me_6C_6)_2Ni]^{2+} + Ni(0) + 2Me_6C_6 \qquad (35)$$

B. Monoarene Complexes

The reaction among $PdCl_2$, $AlCl_3$, Al, and benzene has been shown to afford two types of π-complexes, depending on the amount of $AlCl_3$ used (*5, 6*). When the ratio of $PdCl_2$ to $AlCl_3$ was 1:1, diamagnetic crystals of the compound $[(C_6H_6)Pd(AlCl_4)]_2$ were obtained. When the ratio of $PdCl_2$ to $AlCl_3$ was 1:1.6, the diamagnetic compound $[(C_6H_6)$-$Pd(Al_2Cl_7)]_2$ was obtained. Both compounds were decomposed on addition of THF, depositing metallic Pd and $PdCl_2$ in equivalent amounts. An X-ray analysis showed the compounds to have the unusual binuclear sandwich structures (XXXV) and (XXXVI). In (XXXV) it appears that

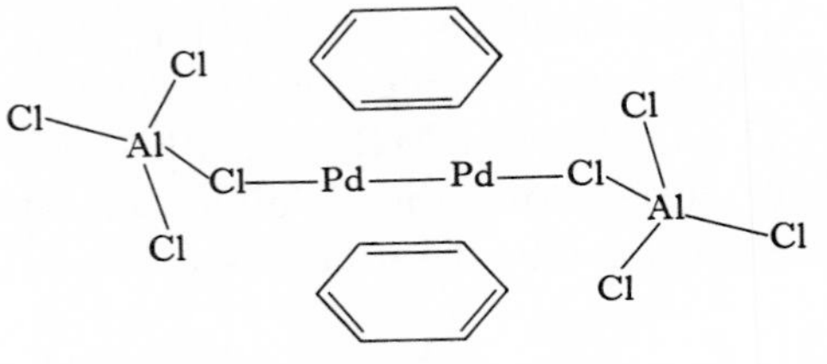

(XXXV)

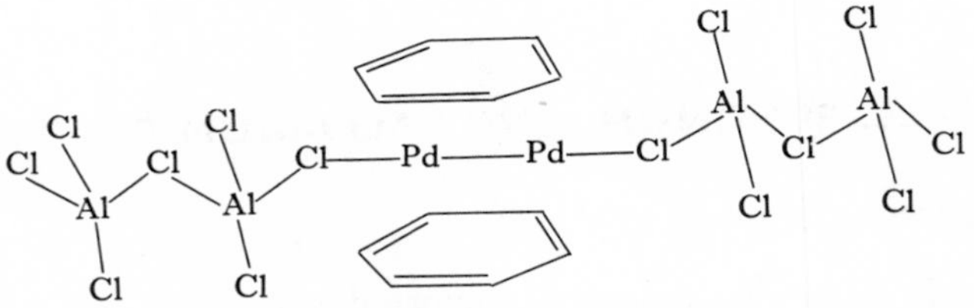

(XXXVI)

four of the carbon atoms in each ring are bonded to two Pd atoms. The observed deviation from planarity of the benzene rings by 7° was taken as evidence for reduced aromaticity. The benzene rings can thus be described as bonding in a diene fashion with the Pd atoms adopting a 16-electron configuration as is common for low-valent Pd complexes. Relatively wide thermal oscillations of the benzene rings in (XXXVI) precluded any definite conclusion to be drawn about the benzene-to-metal coordination (*5*).

The neutral complexes $[(CF_3)_6C_6]Ni(cod)$ (cod = 1,5-cyclooctadiene) and $[(CF_3)_6C_6]Ni_2(cod)_2$ have been prepared by the cyclic condensation of $CF_3C{\equiv}CCF_3$ on $Ni(cod)_2$ (*41*). The cod ligand in $[(CF_3)_6C_6]Ni(cod)$

could be displaced by tertiary phosphines and arsines such as PPh_2Me, PPh_3, and $AsPhMe_2$ yielding complexes of the type $[(CF_3)_6C_6]NiL_2$. The derivative for which L = $AsPhMe_2$ was also prepared by the reaction of $CF_3C{\equiv}CCF_3$ with $Ni(AsPhMe_2)_4$, whereas that for which L = PPh_3 was also prepared by the reaction between $(CF_3)_6C_6$ and $(PPh_3)_2$-$Ni(C_2H_4)$. Compound $[(CF_3)_6C_6]_2Ni(cod)_2$ reacted with tertiary phosphite ligands yielding complexes of the type $[(CF_3)_6C_6]Ni_2L_4$ [L = $P(OMe)_3$, $P(OCH_2)_3CMe$] for which the structure (XXXVII) was proposed. On the basis of the ^{19}F NMR spectrum, free rotation of the L_2Ni—NiL_2 system about an axis perpendicular to the plane of the ring and the Ni—Ni bond was proposed. The monomeric complexes $[(CF_3)_6C_6]NiL_2$ were also shown to be fluxional by ^{19}F NMR, and structure (XXXVIII) was proposed in which the $(CF_3)_6C_6$ ligand is bonded to an Ni(II) atom by two σ-bonds as has been observed in simple fluoroolefin complexes. This proposal is supported by a recent X-ray structure (XXXIX) on the closely related Pt compound $[(CF_3)_6C_6]Pt(PEt_3)_2$, prepared by the reaction of $(CF_3)_6C_6$ with $Pt(PEt_3)_3$ (*42*). However, the Pt atom is bonded to two adjacent carbon atoms in the ring rather than in the 1–4

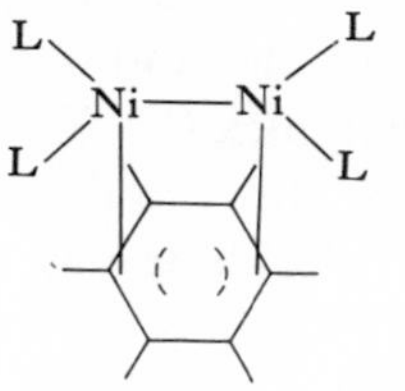

(XXXVII)

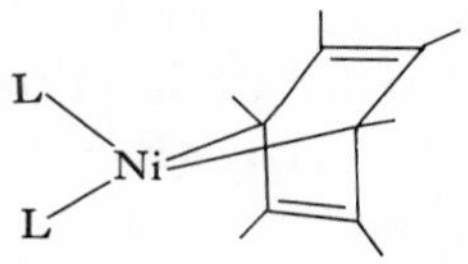

(XXXVIII)

fashion suggested for compounds $[(CF_3)_6C_6]NiL_2$. The aromaticity of the ring has been destroyed as can be seen from the alternating single- and double-bond distances. The Pt compound showed fluxional ^{19}F

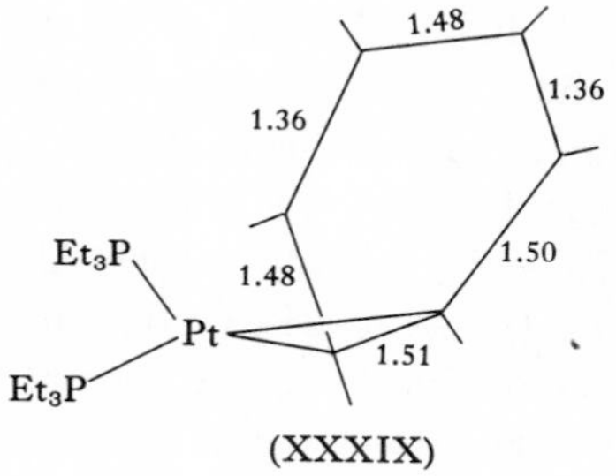

(XXXIX)

NMR behavior down to $-90°C$ where the single triplet resonance ($J_{PF} = 3.0$ Hz) became two unresolved multiplets. The low activation energy for the intramolecular rearrangement of the Ni and Pt complexes is in sharp contrast to the compound $[(CF_3)_6C_6](\pi\text{-}C_5H_5)Rh$ which has been shown to contain a tetrahapto-coordinated $(CF_3)_6C_6$ ligand (*68*).

Palladium(II) acetate has been found to be a catalyst for the dimerization of benzene to biphenyl in perchloric acid–acetic acid solutions (*80*). Kinetic studies suggested the formation of an intermediate complex between a benzene molecule and a Pd(II) ion, and the following reaction sequence was proposed:

$$C_6H_6 + Pd^{2+} \longrightarrow [(C_6H_6)Pd]^{2+} \xrightarrow{-H^+} [(C_6H_5)Pd]^+ \longrightarrow \tfrac{1}{2}(C_6H_5)_2 + Pd^+ \quad (36)$$

The Pd(I) ion generated also appears to form a benzene complex in solution as addition of acetic anhydride to the reaction mixture precipitated an explosive, paramagnetic Pd(I) complex formulated as $(C_6H_6)Pd(H_2O)(ClO_4)$. Palladium(II) acetate and the styrene complex $[(PhCH{=}CH_2)PdCl_2]_2$ have been shown to catalyze the reaction between benzene and styrene in acetic acid yielding stilbene (*169*, *170*). The reactions are also likely to involve an intermediate complex between benzene and palladium. Furthermore, the addition of benzene or toluene to the σ-bonded olefin in the complex $(Cl_2C{=}CH)Pd(PPh_3)_2Cl$ to give $Cl_2C{=}CHC_6H_4R$ (R = H or Me) requires the presence of Ag(I) ion, presumably to remove the chloride ion so that the aromatic can coordinate to the Pd(II) atom (*291*).

XI

COPPER AND SILVER COMPLEXES

The reaction among CuCl, $AlCl_3$, and benzene gives a compound of stoichiometry $(C_6H_6)Cu(AlCl_4)$ (*417*). The crystal structure has been determined (*418*), and the coordination about the Cu atom is shown in (XL). The structure is made up of sheets of $(C_6H_6)Cu(AlCl_4)$ units, the sheets being cross-linked by Cu—Cl bonds. The two short Cu—Cl bonds, Cu—Cl_3 and Cu—Cl_4, arise from one $AlCl_4$ unit in the sheet, whereas the longer bond, Cu—Cl_1 arises from another $AlCl_4$ unit in an adjacent sheet. The Cu atom sits above one edge of the benzene ring but is not equidistant from the two nearest carbon atoms, the Cu–C

$C_1–C_2 = 1.25 \pm 0.04$ Å
$C_2–C_3 = 1.41 \pm 0.05$ Å
$C_3–C_4 = 1.27 \pm 0.04$ Å
$C_4–C_5 = 1.40 \pm 0.05$ Å
$C_5–C_6 = 1.29 \pm 0.04$ Å
$C_1–C_6 = 1.37 \pm 0.04$ Å

(XL)

distances being 2.15 Å and 2.30 Å. The plane generated by the atoms Cu, C_3, and C_4, make an angle of 95° with the plane of the benzene ring. Similar features have been observed in the bonding of Ag(I) to various arenes as is discussed in the following. Although the bond distances in (XL) alternate in length around the benzene ring and suggest a cyclohexatriene system, the authors caution that the errors are sufficiently large that the variation may not be real. The benzene ring is planar within experimental error which is in sharp contrast to the arene ring in $[(CF_3)_6C_6]Pt(PEt_3)_2$ (XXXIX) (*42*).

The existence of a solid 1:1 and a liquid 2:1 toluene–$Cu(AlCl_4)$ molecular complex has been shown by vapor pressure–phase composition studies (*287*). The results indicated that the toluene molecules in the 2:1 complex are not as strongly bonded to the Cu(I) atom as in the 1:1 complex. The toluene could be removed from the 1:1 complex by heating at 50° to 60°C at a pressure of 10^{-4} mm, yielding pure $CuAlCl_4$.

The crystalline air-sensitive complex $(C_6H_6)[Cu(OSO_2CF_3)]_2$ has been isolated from the reaction of trifluoromethanesulfonicanhydride with Cu(I) oxide in benzene (*381*). The complex was stable to 100°C when heated in a sealed evacuated tube, the benzene being released quantitatively only above 120°C. The structure has been determined and consists of infinite chains of $Cu(SO_3CF_3)$ units cross-linked in sheets by the benzene molecules (*94*). The benzene–Cu(I) coordination is shown in (XLI). The structure was not sufficiently well-resolved to observe

$Cu_1–C_1 = 2.30$ Å
$Cu_1–C_2 = 2.12$ Å
$Cu_2–C_3 = 2.09$ Å
$Cu_2–C_4 = 2.12$ Å

(XLI)

any significant deviations of carbon–carbon distances from their value in free benzene. Rapid and dynamic arene-exchange equilibria were observed when the solid was mixed with various alkyl-substituted benzenes. The relative stabilities for a wide range of substituted benzenes were determined from which it was concluded that geometry and the size of the ring substituents predominantly dictate the stability ordering rather than the π-basicity of the aromatic ring (*94*). The benzene complex has also been shown to react with diene, triene, and tetraene ligands yielding cations of the type $[CuL]^+$ and $[CuL_2]^+$ (*381*).

Arene complexes of Cu(II) have been prepared by adsorption of an arene on the interlamellar surfaces of the Cu(II)-exchanged layer aluminosilicate, montmorillonite. Complexes of benzene, methyl-substituted benzenes, biphenyl, naphthalene, and anthracene have been studied (*292*, *345*, *363*). Two types of complexes were formed depending on the degree of dehydration of the montmorillonite. The complexes formed under conditions of moderate dehydration (Type 1) were shown by IR spectroscopy to contain a planar aromatic ring. The bonding between the arene and the Cu(II) is believed to be similar to that in arene–silver perchlorate complexes. Complex formation was also observed between arenes and Ag(I)-exchanged montmorillonites (*345*). A second type of complex (Type 2) was formed under conditions of extreme dehydration of the montmorillonite. Electron spin resonance studies suggested that a π-electron has been transferred from the arene to the Cu(II) giving a Cu(I) species plus an arene radical cation of the type (XLII). Addition of water regenerated the Type 1 complexes (*363*).

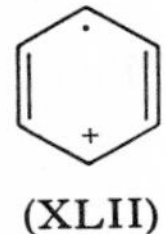

(XLII)

The existence of donor–acceptor complexes between aromatic molecules and Ag(I) has been known for some time (*10–12*). Many arene–Ag(I) complexes have been obtained from silver perchlorate and an aromatic hydrocarbon. Complexes containing various ratios of aromatic hydrocarbon to silver perchlorate have been observed including 1:1, 1:2, 2:1, and 1:4 complexes (see Table VI). Crystal structures have been obtained for several of the complexes. The 1:1 complex $(C_6H_6)Ag(ClO_4)$ (XLIII) consists of chains of alternating benzene molecules and silver atoms (*362*). The arene–silver atom coordination in another type of 1:1 complex, (acenaphthene)Ag(ClO_4) is shown in

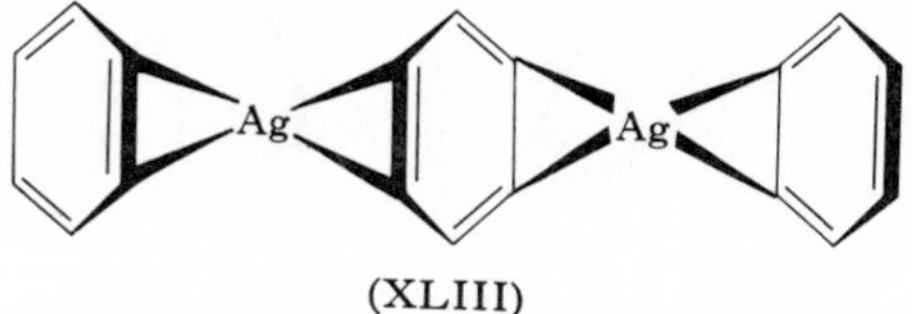

(XLIII)

(XLIV) (*360*). Another 1:1 complex, (indene)Ag(ClO_4) has been shown to have a structure similar to (XLIV) with one of the two silver atoms bridging two five-membered rings (*361*). In the complex (acenaphthylene)Ag(ClO_4) (XLV), the silver atoms are also observed to bridge two rings, but, in this case, each silver atom is bound to two aromatics which are further bound to two other silver atoms, giving a zigzagging infinite chain. The silver atoms are further bridged by perchlorate groups (*360*). The silver atom–arene coordination for the 2:1 complex

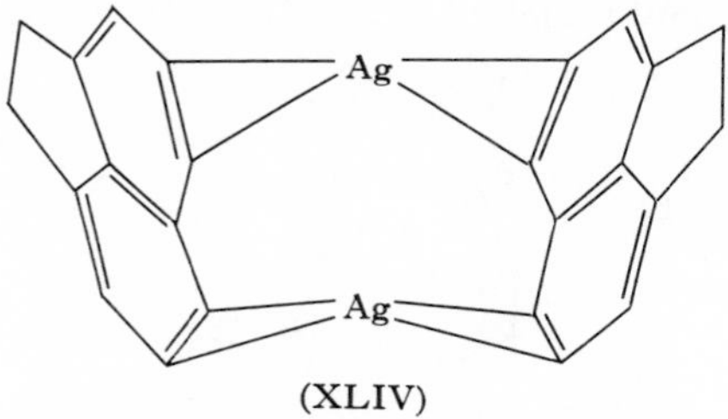

(XLIV)

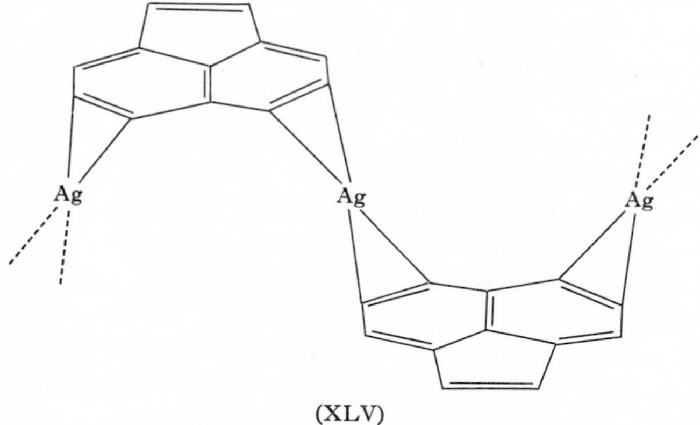

(XLV)

(*m*-xylene)$_2$Ag(ClO_4) is shown in (XLVI) (*405*). Silver atom–arene

coordination similar to that in (XLVI) has been observed in (*o*-xylene)$_2$-

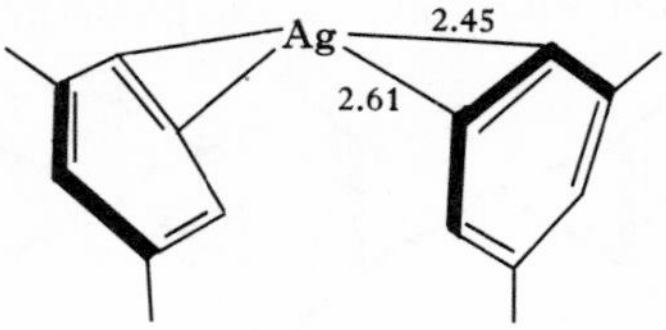

(XLVI)

Ag(ClO_4) (*404*) and (cyclohexylbenzene)$_2$Ag(ClO_4) (*199*, *200*). In the 1:4 complex, (anthracene)[Ag(ClO_4)]$_4 \cdot H_2O$ (XLVII), two silver atoms are coordinated to each of the outer rings (*201*). Similar coordination was observed for (naphthalene)[Ag(ClO_4)]$_4 \cdot 4H_2O$ (*201*).

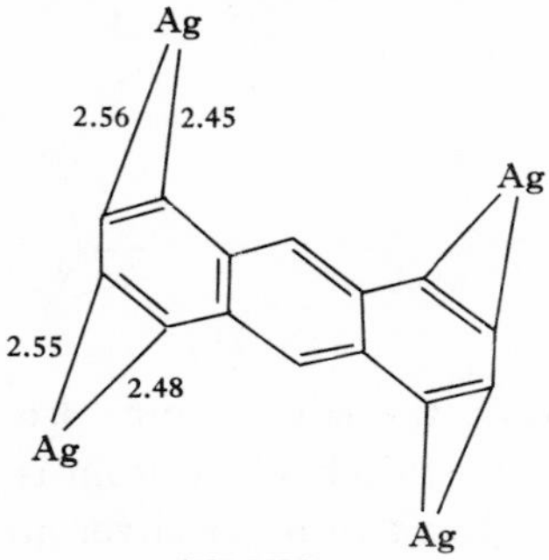

(XLVII)

The crystal structures of all the arene–silver perchlorate complexes can be described as layer structures (*404*). The layers or sheets are composed of silver perchlorate and arene units that are alternately stacked to make the crystal structure. The silver atoms lie above one edge of the arene group to which they are bonded, forming a short bond and a long bond to two adjacent carbon atoms. The remainder of the silver atom coordination polyhedron is completed by silver–oxygen interactions with the perchlorate groups. The data in Table VII summarize the silver–carbon bond lengths for most of the reported structures. In all cases except that of (naphthalene)[Ag(ClO_4)]$_4 \cdot 4H_2O$, which has been described as a chlatherate (*201*), the short silver–carbon bond length averages about 2.47 Å. The next-nearest silver–carbon bond length varies over wide limits from 2.51 to 2.92 Å. It has, thus, been concluded that the dominant interaction between the silver atom and the aromatic ligand is with the nearest carbon atom, whereas the other silver–carbon distance may depend on a number of factors such as molecular packing, nature of anion, and other structural details (*405*). It is also noteworthy that within experimental limits the structures of the aromatic ligands in the complexes determined to date are not significantly different from what would be expected in the uncomplexed molecules.

Anhydrous $AgBF_4$ has also been reported to form complexes with aromatic hydrocarbons, and these are generally more stable than the $AgClO_4$ analogs (*340*).

Compound $(C_6H_6)Ag(AlCl_4)$ has been prepared and its structure determined (Table VII) (*419*). The IR spectra of solid $(C_6H_6)Ag(AlCl_4)$, $(C_6D_6)Ag(AlCl_4)$, and the Cu(I) analogs have been studied in the range 4000–33 cm^{-1}. A metal–carbon vibrational frequency was observed near 100 cm^{-1} in all of the complexes (*382*).

Silver(I) ions have been observed to form complexes with benzene in aqueous solution. Nuclear magnetic resonance and solubility studies of aqueous silver nitrate–benzene mixtures established the presence of the cations $[(C_6H_6)Ag]^+$ and $[(C_6H_6)Ag_2]^+$ (*10*, *164*).

TABLE VII

Ag–C DISTANCES IN ARENE–Ag(I) COMPLEXES

Compound	Ag–C (Å)	Ag–C′ (Å)	References
(Benzene) $AgClO_4$	2.496(6)	2.634(6)	*362*
(Bibenzyl) $AgClO_4$	2.48(2)	2.72(2)	*405*
(Indene) $AgClO_4$	2.47	2.76	*361*
(Acenaphthene) $AgClO_4$	2.48, 2.44	2.51, 2.51	*360*
(Benzene) $AgAlCl_4$	2.47(6)	2.92(7)	*419*
(*m*-Xylene)$_2$ $AgClO_4$	2.45(2)	2.61(2)	*405*
(*o*-Xylene)$_2$ $AgClO_4$	2.44, 2.49	2.53, 2.57	*404*
(Cyclohexylbenzene)$_2$ $AgClO_4$	2.48(1)	2.67(1)	*199*, *200*
(Anthracene) $[AgClO_4]_4 \cdot H_2O$	2.48, 2.45	2.55, 2.56	*201*
(Naphthalene) $[AgClO_4]_4 \cdot 4H_2O$	2.60, 2.62	2.61, 2.63	*201*

REFERENCES

1. Abel, E. W., Bennett, M. A., and Wilkinson, G., *J. Chem. Soc.* 2323 (1959).
2. Albrecht, G., Forster, E., Sippel, D., Eichkorn, F., and Kurras, E., *Z. Chem.* **8**, 311 (1968).
3. Alcock, N. W., *Chem. Commun.* 177 (1965).

4. Allegra, G., *Atti Accad. Naz. Lincei, Cl. Sci. Fis. Mat. Natur. Rend.* **31**, 399 (1961).
5. Allegra, G., Casagrande, G. T., Immirzi, A., Porri, L., and Vitulli, G., *J. Amer. Chem. Soc.* **92**, 289 (1970).
6. Allegra, G., Immirzi, A., and Porri, L., *J. Amer. Chem. Soc.* **87**, 1394 (1965).
7. Anderson, S. E., and Drago, R. S., *J. Amer. Chem. Soc.* **91**, 3656 (1969).
8. Anderson, S. E., and Drago, R. S., *Inorg. Chem.* **11**, 1564 (1972).
9. Anderson, S. E., and Drago, R. S., *J. Amer. Chem. Soc.* **92**, 4244 (1970).
10. Andrews, L. J., and Kefer, R. M., *J. Amer. Chem. Soc.* **71**, 3644 (1949).
11. Andrews, L. J., and Kefer, R. M., *J. Amer. Chem. Soc.* **72**, 3113, (1950).
12. Andrews, L. J., and Kefer, R. M., *J. Amer. Chem. Soc.* **72**, 5034 (1950).
13. Andrews, J. T. S., and Westrum, E. F., *J. Organometal. Chem.* **17**, 293 (1969).
14. Angelici, R. J., and Blacik, L. J., *Inorg. Chem.* **11**, 1754 (1972).
15. Angelici, R. J., and Busetto, L., *Inorg. Chem.* **7**, 1935 (1968).
16. Armstrong, A. T., Carroll, D. G., and McGlynn, S. P., *J. Chem. Phys.* **47**, 1104 (1967).
17. Ashworth, E. A., Green, M. L. H., and Knight, J., *J. Chem. Soc., Chem. Commun.* **5**, (1974).
18. Astruc, D., Dabard, R., and Laviron, E., *C. R. Acad. Sci., Ser. C*, **269**, 608 (1969).
19. Avinur, P., and Eliezer, I., *Anal. Chem.* **42**, 1317 (1970).
20. Bailey, M. F., and Dahl, L. F., *Inorg. Chem.* **4**, 1298 (1965).
21. Bailey, M. F., and Dahl, L. F., *Inorg. Chem.* **4**, 1314 (1965).
22. Bauer, R. A., Fischer, E. O., and Kreiter, C. G., *J. Organometal. Chem.* **24**, 737 (1970).
23. Baumgärtner, F., Fischer, E. O., and Zahn. U., *Naturwissenschaften* **48**, 478 (1961).
24. Baumgärtner, F., Fischer, E. O., and Zahn, U., *Chem. Ber.* **94**, 2198 (1961).
25. Beck, H. J., Fischer, E. O., and Kreiter, C. G., *J. Organometal. Chem.* **26**, C41 (1971).
26. Behrens, H., Meyer, K., and Müller, A., *Z. Naturforsch. B* **20**, 74 ((1965).
27. Benfield, F. W. S., Green, M. L. H., Ogden, J. S., and Young, D., *J. Chem. Soc. Chem. Commun.* **866** (1973).
28. Bennett, M. A., Robertson, G. B., and Smith, A. K., *J. Organometal. Chem.* **43**, C41 (1972).
29. Birch, A. J., Cross, F. E., and Fitton, H., *Chem. Commun.* 366 (1965).
30. Booth, B. L., Haszeldine, R. N., and Hill, M., *Chem. Commun.* 1118 (1967).
31. Bowden, J. A., and Colton, R., *Aust. J. Chem.* **26**, 43 (1973).
32. Bowden, J. A., Colton, R., and Commons, C. J., *Aust. J. Chem.* **26**, 655 (1973).
33. Brice, M. D., Dellaca, R. J., and Penfold, B. R., *Chem. Commun.* 72 (1971).
34. Brintzinger, H., Palmer, G., and Sands, R. H., *J. Amer. Chem. Soc.* **88**, 623 (1966).
35. Brown, D. A., Cunningham, D., and Glass, W. K., *Chem. Commun.* 306 (1966).
36. Brown, D. A., Hargaden, J. P., McMullen, C. M., Gogan, N., and Sloan, H., *J. Chem. Soc.* 4914 (1963).
37. Brown, D. A., and McCormack, C. G., *Chem. Commun.* 383 (1967).
38. Brown, D. A., and Raju, J. R., *J. Chem. Soc., A* 1617 (1966).
39. Brown, D. A., and Raju, J. R., *J. Chem. Soc., A* 40 (1966).
40. Brown, D. A., and Rawlinson, R. M., *J. Chem. Soc., A* 1534 (1969).
41. Browning, J., Cundy, C. S., Green, M., and Stone, F. G. A., *J. Chem. Soc., A* 448 (1971).
42. Browning, J., Green, M., Penfold, B. R., Spencer, J. L., and Stone, F. G. A., *J. Chem. Soc. Chem. Commun.* 31 (1973).
43. Burger, T. F., and Zeiss, H., *Chem. Ind.* (*London*) 183 (1962).

44. Burt, R., Cooke, M., and Green, M., *J. Chem. Soc. A* 2981 (1970).
45. Bush, M. A., Dullforce, T. A., and Sim, G. A., *Chem. Commun.* 1491 (1969).
46. Bush, R. W., and Snyder, H. R., *J. Org. Chem.* **25**, 1240 (1960).
47. Cais, M., and Feldkimel, M., *Tetrahedron Lett.* **13**, 444 (1961).
48. Cais, M., Frankel, E. N., and Rejoan, *Tetrahedron Lett.* **16**, 1919 (1968).
49. Cais, M., and Rejoan, A., *Inorg. Chim. Acta* **4**, 509 (1970).
50. Calderazzo, F., *Inorg. Chem.* **4**, 223 (1965).
51. Calderazzo, F., *Inorg. Chem.* **4**, 429 (1966).
52. Calderazzo, F., *Inorg. Chem.* **3**, 1207 (1964).
53. Calderazzo, F., *Inorg. Chem.* **3**, 810 (1964).
54a. Calderazzo, F., and Cini, R., *J. Chem. Soc.* 818 (1965).
54b. Cameron, T. S., Prout, C. K., and Rees, G. V., personal communication.
55. Card, R. J., and Trahanovsky, W. S., *Tetrahedron Lett.* **39**, 3823 (1973).
56. Carroll, D. G., and McGlynn, S. P., *Inorg. Chem.* **7**, 1285 (1968).
57. Carter, O. L., McPhail, A. T., and Sim, G. A., *J. Chem. Soc. A* 822 (1966).
58. Carter, O. L., McPhail, A. T., and Sim, G. A., *J. Chem. Soc. A* 1866 (1968).
59. Carter, O. L., McPhail, A. T., and Sim, G. A., *J. Chem. Soc. A* 1619 (1967).
60. Carter, O. L., McPhail, A. T., and Sim, G. A., *J. Chem. Soc. A* 228 (1967).
61. Carter, O. L., McPhail, A. T., and Sim, G. A., *Chem. Commun.* 212 (1966).
62. Ceccon, A., and Biserni, G. S., *J. Organometal. Chem.* **39**, 313 (1972).
63. Chatt, J., and Davidson, J. M., *J. Chem. Soc.* 843 (1965).
64. Chatt, J., Guy, R. G., and Watson, H. R., *J. Chem. Soc.* 2332 (1961).
65. Chatt, J., and Watson, H. R., *Proc. Chem. Soc.* 243 (1960).
66. Chini, P., and Ercoli, R., *Gazz. Chim. Ital.* **88**, 1170 (1958).
67. Chiswell, B., and Venanzi, L. M., *J. Chem. Soc. A*, 417 (1966).
68. Churchill, M. R., and Mason, R., *Proc. Roy. Soc. Ser. A* **292**, 61 (1966).
69. Churchill, M. R., and Mason, R., *Proc. Roy. Soc. Ser. A* **279**, 191 (1964).
70. Closson, R. D., U.S. Patent 3,115,510; *Chem. Abstr.* **60**, P 6867d.
71. Coffield, T. H., Sandel, V., and Closson, R. D., *J. Amer. Chem. Soc.* **79**, 5826 (1957).
72. Coffield, T. H., Sandel, V., and Closson, R. D., *134th Meeting Amer. Chem. Soc. Abstr.* 58-P (1958).
73. Coleman, K. J., Daives, C. S., and Gogan, N. J., *Chem. Commun.* 1414 (1970).
74a. Connor, J. A., Skinner, H. A., and Virmani, Y., *J. Chem. Soc., Faraday Trans. I* **69**, 1218 (1973).
74b. Corradini, P., and Allegra, G., *J. Amer. Chem. Soc.* **82**, 2075 (1960).
75. Cram, D. J., and Wilkinson, D. I., *J. Amer. Chem. Soc.* **82**, 5721 (1960).
76. Dabard, R., Fourani, P., and Besancon, J., *C. R. Acad. Sci., Ser. C* **260**, 2833 (1965).
77. Dabard, R., and Meyer, A., *C. R. Acad. Sci., Ser. C* **264**, 903 (1967).
78. Davidson, G., and Riley, E. M., *Spectrochim. Acta, Part A* **27**, 1649 (1971).
79. Davidson, G., and Riley, E. M., *J. Organomet. Chem.* **19**, 101 (1969).
80. Davidson, J. M., and Triggs, C., *J. Chem. Soc. A* 1324 (1968).
81. Davis, R. E., and Pettit, R., *J. Amer. Chem. Soc.* **92**, 716 (1970).
82. Davison, A., McFarlane, W., Pratt, L., and Wilkinson, G., *J. Chem. Soc.* 3653 (1962).
83. Davison, A., and Reger, D. L., *J. Organometal. Chem.* **23**, 491 (1970).
84. Deberitz, J., Dirscherl, K., and Nöth, H., *Chem. Ber.* **106**, 2783 (1973).
85. Deberitz, J., and Nöth, H., *J. Organometal. Chem.* **61**, 271 (1973).

86. Deberitz, J., and Nöth, H., *J. Organometal. Chem.* **55**, 153 (1973).
87. Deberitz, J., and Nöth, H., *Chem. Ber.* **103**, 2541 (1970).
88. Dellaca, R. J., and Penfold, B. R., *Inorg. Chem.* **11**, 1855 (1972).
89. Denning, R. G., and Wentworth, R. A. D., *J. Amer. Chem. Soc.* **88**, 4619 (1966).
90. Deubzer, B., Fischer, E. O., Fritz, H. P., Kreiter, C. G., Kriebitzsch, N., Simmons, H. D., and Willeford, B. R., *Chem. Ber.* **100**, 3084 (1967).
91. Deubzer, B., Fritz, H. P., Kreiter, C. G., and Öfele, K., *J. Organometal. Chem.* **7**, 289 (1967).
92. DeVries, H., *Rec. Trav. Chim.* **81**, 359 (1962).
93. Dickson, R. S., and Wilkinson, G., *Chem. Ind.* (*London*) 1432 (1963).
94. Dines, M. B., and Bird, P. H., *Chem. Commun.* 12 (1973).
95. Domrachev, G. A., and Vyshinskii, N. N., *Dokl. Akad. Nauk SSSR* **194**, 583 (1970).
96. Drefahl, G., Hörhold, H., and Kühne, K., *Chem. Ber.* **98**, 1326 (1965).
97. Dusausoy, Y., Besancon, J., and Protas, J., *C. R. Acad. Sci., Ser. C* **274**, 774 (1972).
98. Dusausoy, Y., Protas, J., and Besancon, J., *J. Organometal. Chem.* **59**, 281 (1973).
99. Dusausoy, Y., Protas, J., Besancon, J., and Tirouflet, J., *C. R. Acad. Sci., Ser. C* **271**, 1070 (1970).
100. Dusausoy, Y., Protas, J., Besancon, J., and Tirouflet, J., *C. R. Acad. Sci., Ser. C* **270**, 1792 (1970).
101. Dusausoy, Y., Protas, J., Besancon, J., and Tirouflet, J., *Acta Crystallogr., Sect. B* **29**, 469 (1973).
102. Edgar, K., Johnson, B. F. G., Lewis, J., and Wild, S. B., *J. Chem. Soc. A* 2851 (1968).
103. Efraty, A., and Maitlis, P. M., *Tetrahedron Lett.* **34**, 4025 (1966).
104. Efraty, A., and Maitlis, P. M., *J. Amer. Chem. Soc.* **89**, 3744 (1967).
105. Elschenbroich, C., *J. Organometal. Chem.* **22**, 677 (1970).
106. Elschenbroich, C., *J. Organometal. Chem.* **14**, 157 (1968).
107. Elschenbroich, C., Gerson, F., and Heinzes, J., *Z. Naturforsch. B* **27**, 312 (1972).
108. Ercoli, R., Calderazzo, F., and Alberola, A., *Chim. Ind.* (*Milan*) **41**, 975 (1959).
109. Ercoli, R., Calderazzo, F., and Mantica, E., *Chim. Ind.* (*Milan*) **41**, 404 (1959).
110. Evans, S., Green, J. C., and Jackson, S. E., *J. Chem. Soc., Faraday Trans. II* **68**, 249 (1972).
111. Evans, S., Green, J. C., and Jackson, S. E., *J. Chem. Soc., Dalton Trans.*, 304 (1974).
112. Falk, H., and Schlögl, K., *Monatsh. Chem.* **99**, 578 (1968).
113. Falk, H., Schlögl, K., and Steyrer, W., *Monatsh. Chem.* **97**, 1029 (1966).
114. Fischer, E. O., and Beckert, O., *Angew. Chem.* **70**, 744 (1958).
115. Fischer, E. O., Berngruber, W., and Kreiter, C. G., *J. Organometal. Chem.* **14**, P25 (1968).
116. Fischer, E. O., and Böttcher, R., *Chem. Ber.* **89**, 2397 (1956).
117. Fischer, E. O., and Böttcher, R., *Z. Anorg. Allg. Chem.* **291**, 305 (1957).
118. Fischer, E. O., and Breitschaft, S., *Angew. Chem., Int. Ed. Engl.* **2**, 44 (1963).
119. Fischer, E. O., and Breitschaft, S., *Chem. Ber.* **99**, 2905 (1966).
120. Fischer, E. O., and Breitschaft, S., *Chem. Ber.* **99**, 2213 (1966).
121. Fischer, E. O., and Brunner, H., *Z. Naturforsch. B* **16**, 406 (1961).
122. Fischer, E. O., and Brunner, H., *Chem. Ber.* **95**, 1999 (1962).
123. Fischer, E. O., and Brunner, H., *Chem. Ber.* **98**, 175 (1965).
124. Fischer, E. O., and Elschenbroich, C., *Chem. Ber.* **103**, 162 (1970).

125. Fischer, E. O., Elschenbroich, C., and Kreiter, C. G., *J. Organometal. Chem.* **7**, 481 (1967).
126. Fischer, E. O., and Fischer, R. D., *Z. Naturforsch. B* **16**, 556 (1961).
127. Fischer, E. O., and Fritz, H. P., *Angew. Chem.* **73**, 353 (1961).
128. Fischer, E. O., Goodwin, H. A., Kreiter, C. G., Simmons, H. D., Sonogashira, K., and Wild, S. B., *J. Organometal. Chem.* **14**, 359 (1968).
129. Fischer, E. O., and Hafner, W., *Z. Naturforsch. B* **10**, 665 (1955).
130. Fischer, E. O., and Hafner, W., *Z. Anorg. Allg. Chem.* **286**, 146 (1956).
131. Fischer, E. O., Joos, G., and Meer, W., *Z. Naturforsch. B* **13**, 456 (1958).
132. Fischer, E. O., and Kögler, H. P., *Angew. Chem.* **68**, 426 (1956).
133. Fischer, E. O., and Kögler, H. P., *Chem. Ber.* **90**, 250 (1957).
134. Fischer, E. O., and Kögler, H. P., *Z. Naturforsch. B* **13**, 197 (1958).
135. Fischer, E. O., and Kohl, F. J., *Z. Naturforsch. B* **18**, 504 (1963).
136. Fischer, E. O., and Kohl, F. J., *Angew. Chem.* **76**, 98 (1964).
137. Fischer, E. O., and Kohl, F. J., *Chem. Ber.* **98**, 2134 (1965).
138. Fischer, E. O., and Kriebitzsch, N., *Z. Naturforsch. B* **15**, 465 (1960).
139. Fischer, E. O., Kriebitzsch, N., and Fischer, R. D., *Chem. Ber.* **92**, 3214 (1959).
140. Fischer, E. O., and Kuzel, P., *Z. Naturforsch. B* **16**, 475 (1961).
141. Fischer, E. O., and Lindner, H. H., *J. Organometal. Chem.* **1**, 307 (1964).
142. Fischer, E. O., and Lindner, H. H., *J. Organometal. Chem.* **2**, 222 (1964).
143. Fischer, E. O., and Müller, J., *Z. Naturforsch. B* **17**, 776 (1962).
144. Fischer, E. O., and Müller, J., *Chem. Ber.* **96**, 3217 (1963).
145. Fischer, E. O., and Öfele, K., *Chem. Ber.* **90**, 2532 (1957).
146. Fischer, E. O., and Öfele, K., *Z. Naturforsch. B* **13**, 458 (1958).
147. Fischer, E. O., Öfele, K., Essler, H., Fröhlich, W., Mortensen, P., and Semmlinger, W., *Chem. Ber.* **91**, 2763 (1958).
148. Fischer, E. O., and Piesberger, U., *Z. Naturforsch. B* **11**, 758 (1956).
149. Fischer, E. O., and Röhrscheid, F., *Z. Naturforsch. B* **17**, 483 (1962).
150. Fischer, E. O., and Röhrscheid, F., *J. Organometal. Chem.* **6**, 53 (1966).
151. Fischer, E. O., and Rühle, H., *Z. Anorg. Allg. Chem.* **341**, 137 (1965).
152. Fischer, E. O., Scherer, F., and Stahl, H. O., *Chem. Ber.* **93**, 2065 (1960).
153. Fischer, E. O., and Schmidt, M. W., *Chem. Ber.* **102**, 1954 (1969).
154. Fischer, E. O., and Schmidt, M. W., *Chem. Ber.* **100**, 3782 (1967).
155. Fischer, E. O., and Schmidt, M. W., *Chem. Ber.* **99**, 2213 (1966).
156. Fischer, E. O., and Schmidt, M. W., *Chem. Ber.* **99**, 2206 (1966).
157a. Fischer, E. O., and Schneider, R. J. J., *J. Organometal. Chem.* **12**, P27 (1968).
157b. Fischer, E. O., and Seeholzer, J., *Z. Anorg. Allg. Chem.* **312**, 244 (1961).
158. Fischer, E. O., and Seus, D., *Chem. Ber.* **89**, 1809 (1956).
159. Fischer, E. O., and Stahl, H. O., *Chem. Ber.* **89**, 1805 (1956).
160. Fischer, E. O., and Wehner, H. W., *Chem. Ber.* **101**, 454 (1968).
161. Fischer, E. O., and Wirzmuller, A., *Chem. Ber.* **90**, 1725 (1957).
162. Fitch, J. W., and Lagowski, J. J., *J. Organometal. Chem.* **5**, 483 (1966).
163. Fitch, J. W., and Lagowski, J. J., *Inorg. Chem.* **4**, 864 (1965).
164. Foreman, M. I., Gorton, J., and Foster, R., *Trans. Faraday Soc.* **66**, 2120 (1970).
165. Forster, E., Albrecht, G., Durselen, W., and Kurras, E., *J. Organometal. Chem.* **19**, 215 (1969).
166. Frankel, E. N., and Butterfield, R. O., *J. Org. Chem.* **34**, 3930 (1969).
167. Frankel, E. N., Selke, E., and Glass, C. A., *J. Org. Chem.* **34**, 3936 (1969).
168. Fritz, H. P., and Kreiter, C. G., *J. Organometal. Chem.* **7**, 427 (1967).

169. Fujiwara, Y., Moritani, I., and Matsuda, M., *Tetrahedron* **24**, 4819 (1968).
170. Fujiwara, Y., Moritani, I., Matsuda, M., and Teranishi, S., *Tetrahedron Lett.* **5**, 633 (1968).
171. Giannotti, C., and Green, M. L. H., *Chem. Commun.* 1114 (1972).
172. Gilbert, J. R., Leach, W. P., and Miller, J. R., *J. Organometal. Chem.* **56**, 295 (1973).
173. Gilbert, J. R., Leach, W. P., and Miller, J. R., *J. Organometal. Chem.* **30**, C41 (1971).
174. Glocking, F., Sneeden, R. P. A., and Zeiss, H., *J. Organometal. Chem.* **2**, 109 (1964).
175. Gracey, D. E. F., Henbest, H. B., Jackson, W. R., and McMullen, C. H., *Chem. Commun.* 566 (1965).
176. Gracey, D. E. F., Jackson, W. R., and Jennings, W. B., *Chem. Commun.* 366 (1968).
177. Gracey, D. E. F., Jackson, W. R., Jennings, W. B., Rennison, S. C., and Spratt, R., *Chem. Commun.* 231 (1966).
178. Gracey, D. E. F., Jackson, W. R., McMullen, C. H., and Thompson, N., *J. Chem. Soc. B* 1197 (1969).
179. Green, M., and Kuc, T. A., *J. Chem. Soc., Dalton Trans.* 832 (1972).
180. Green, M. L. H., Knight, J., Mitchard, L. C., Roberts, G. G., and Silverthorn, W. E., *Chem. Commun.* 1619 (1971).
181. Green, M. L. H., Knight, J., Mitchard, L. C., Roberts, G. G., and Silverthorn, W. E., *J. Chem. Soc., Chem. Commun.* 987 (1972).
182. Green, M. L. H., Mitchard, L. C., and Silverthorn, W. E., *J. Chem. Soc. A* 2929 (1971).
183. Green, M. L. H., Mitchard, L. C., and Silverthorn, W. E., *J. Chem. Soc., Dalton Trans.* 1403 (1973).
184. Green, M. L. H., Mitchard, L. C., and Silverthorn, W. E., *J. Chem. Soc., Dalton Trans.* 1952 (1973).
185. Green, M. L. H., Mitchard, L. C., and Silverthorn, W. E., *J. Chem. Soc., Dalton Trans.* 2177 (1973).
186. Green, M. L. H., Mitchard, L. C., and Silverthorn, W. E., *J. Chem. Soc., Dalton Trans.* 1361 (1974).
187. Green, M. L. H., Pratt, L., and Wilkinson, G., *J. Chem. Soc.* 989 (1960).
188. Green, M. L. H., and Silverthorn, W. E., *Chem. Commun* 557 (1971).
189. Green, M. L. H., and Silverthorn, W. E., *J. Chem. Soc., Dalton Trans.* 301 (1973).
190. Green, M. L. H., and Silverthorn, W. E., *J. Chem. Soc., Dalton Trans.* in press.
191. Gribov, B. G., Kozyrkin, B. I., Krivospitskii, A. D., and Chirkin, G. K., *Dokl. Akad. Nauk. SSSR* **193**, 91 (1970).
192. Gribov, B. G., Mozzhukkin, D. D., Suskina, I. A., and Salamatin, B. A., *Dokl. Akad. Nauk. SSSR* **196**, 586 (1971).
193. Gribov, B. G., Traukin, N. N., Talrina, G. M., Rumyantseva, V. P., Salamatin, B. A., Kozyrkin, B. I., and Pashinkin, A. S., *Dokl. Akad. Nauk. SSSR* **187**, 330 (1969).
194. Gubin, S. P., and Khandkarova, V. S., *J. Organometal. Chem.* **22**, 449 (1970).
195. Günther, H., Wenzl, R., Klose, H., *Chem. Commun.* 605 (1970).
196. Haaland, A., *Acta Chem. Scand.* **19**, 41 (1965).
197. Hähle, J., and Stolze, G., *Z. Naturforsch. B* **19**, 1081 (1964).
198. Haines, R. J., and DuPreez, A. L., *J. Amer. Chem. Soc.* **93**, 2820 (1971).
199. Hall, E. A., and Amma, E. L., *Chem. Commun.* 622 (1968).
200. Hall, E. A., and Amma, E. L., *J. Amer. Chem. Soc.* **93**, 3167 (1971).
201. Hall, E. A., and Amma, E. L., *J. Amer. Chem. Soc.* **91**, 6538, (1969).
202. Hanic, F., and Mills, O. S., *J. Organometal. Chem.* **11**, 151 (1968).

203. Harrison, P. G., Zuckerman, J. J., Long, T. V., Poeth, T. P., and Willeford, B. R., *Inorg. Nucl. Chem. Lett.* **6**, 627 (1970).
204. Hein, F., *Chem. Ber.* **52**, 195 (1919).
205. Hein, F., and Eisfeld, K., *Z. Anorg. Allg. Chem.* **292**, 162 (1957).
206. Hein, F., and Fischer, K. W., *Z. Anorg. Allg. Chem.* **288**, 279 (1956).
207. Hein, F., and Kartte, K., *Z. Anorg. Allg. Chem.* **307**, 22 (1960).
208. Hein, F., and Kartte, K., *Z. Anorg. Allg. Chem.* **307**, 52 (1960).
209. Hein, F., and Kartte, K., *Z. Anorg. Allg. Chem.* **307**, 89 (1960).
210. Hein, F., and Kartte, K., *Monatsber. Deut. Akad. Wiss., Berlin* **2**, 185 (1960).
211. Hein, F., Kleinert, P., and Jehn, W., *Naturwissenschaften* **44**, 34 (1956).
212. Hein, F., Kleinert, P., and Kurras, E., *Z. Anorg. Allg. Chem.* **289**, 229 (1957).
213. Hein, F., and Kleinwächter, K., *Monatsber. Deut. Akad. Wiss., Berlin* **2**, 610 (1960).
214. Hein, F., and Kurras, E., *Z. Anorg. Allg. Chem.* **290**, 179 (1957).
215. Hein, F., and Reinert H., *Chem. Ber.* **93**, 2089 (1960).
216. Hein, F., and Scheel, H., *Z. Anorg. Allg. Chem.* **312**, 264 (1961).
217. Helling, J. F., Rice, S. L., Braitsch, D. M., and Mayer, T., *Chem. Commun.* 931 (1971).
218. Herberich, G. E., and Fischer, E. O., *Chem. Ber.* **95**, 2803 (1962).
219. Herberich, G. E., Greiss, G., and Heil, H. F., *Angew. Chem., Int. Ed. Engl.* **9**, 805 (1970).
220. Herberich, G. E., Greiss, G., Heil, H. F., and Müller, J., *Chem. Commun.* 1328 (1971).
221. Herberich, G. E., and Müller, J., *J. Organometal. Chem.* **16**, 111 (1969).
222. Herwig, W., Metlesics, W., and Zeiss, H., *J. Amer. Chem. Soc.* **81**, 6203 (1959).
223. Herwig, W., and Zeiss, H., *J. Amer. Chem. Soc.* **79**, 6561 (1957).
224. Herwig, W., and Zeiss, H., *J. Amer. Chem. Soc.* **81**, 4798 (1959).
225. Hinze, A. G., *Rec. Trav. Chim. Pays-Bas* **92**, 542 (1973).
226. Hirberhold, M., and Jablonski, C., *J. Organometal. Chem.* **14**, 457 (1968).
227. Huttner, G., and Fischer, E. O., *J. Organometal. Chem.* **8**, 299 (1967).
228. Huttner, G., Fischer, E. O., and Elschenbroich, C., *J. Organometal. Chem.* **3**, 330 (1965).
229. Huttner, G., Fischer, E. O., Fischer, R. D., Carter, O. L., McPhail, A. T., and Sim, G. A., *J. Organometal. Chem.* **6**, 288 (1966).
230. Huttner, G., and Lange, S., *Acta Crystallogr., Sect. B.* **28**, 2049 (1972).
231. Huttner, G., Lange, S., and Fischer, E. O., *Angew. Chem., Int. Ed. Engl.* **10**, 556 (1971).
232. Ibekwe, S. D., Kilbourn, B. T., Raeburn, U. A., and Russell, D. R., *Chem. Commun.* 433 (1969).
233. Jackson, W. R., and Jennings, W. B., *J. Chem. Soc. B* 1221 (1969).
234. Jackson, W. R., and Jennings, W. B., *Chem. Commun.* 824 (1966).
235. Jackson, W. R., Jennings, W. B., Rennison, S. C., and Spratt, R., *J. Chem. Soc. B* 1214 (1969).
236. Jackson, W. R., and McMullen, C. H., *J. Chem. Soc.* 1170 (1965).
237. Jackson, W. R., McMullen, C. H., Spratt, R., and Bladon, P., *J. Orgometal. Chem.* **4**, 392 (1965).
238. Jackson, W. R., Nicholls, B., and Whiting, M. C., *J. Chem. Soc.* 469 (1960).
239. Jaouen, G., Tchissambou, L., and Dabard, R., *C. R. Acad. Sci., Ser. C* **274**, 654 (1972).
240. Jetz, W., and Graham, W. A. G., *J. Amer. Chem. Soc.* **91**, 3375 (1969).

241. Johnson, B. F. G., Johnston, R. D., and Lewis, J., *J. Chem. Soc. A* 2865 (1968).
242. Johnson, B. F. G., Johnston, R. D., and Lewis, J., *Chem. Commun.* 1057 (1967).
243. Jones, D., Pratt, L., and Wilkinson, G., *J. Chem. Soc.* 4458 (1962).
244. Jula, T. F , and Seyferth, D., *Inorg. Chem.* **7**, 1245 (1968).
245. Kang, J. W., Childs, R. F., and Maitlis, P. M., *J. Amer. Chem. Soc.* **92**, 720 (1970).
246. Keulen, K., and Jellinek, F., *J. Organometal. Chem.* **5**, 490 (1966).
247. Khand, I. U., Knox, G. R., Pauson, P. L., and Watts, W. E., *Chem. Commun.* 36 (1971).
248. Khand, I. U., Pauson, P. L., and Watts, W. E., *J. Chem. Soc. C* 2261 (1968).
249. Khand, I. U., Pauson, P. L., and Watts, W. E., *J. Chem. Soc. C* 116 (1969).
250. Khand, I. U., Pauson, P. L., and Watts, W. E., *J. Chem. Soc. C* 2024 (1969).
251. Khand, I. U., Pauson, P. L., and Watts, W. E., *J. Chem. Soc. C* 2257 (1968).
252. Khandkarova, V. S., Gubin, S. P., and Kvasov, B. A., *J. Organometal. Chem.* **23**, 509 (1970).
253. King, R. B., *J. Amer. Chem. Soc.* **88**, 2075 (1966).
254. King, R. B., Braitsch, D. M., and Kapoor, P. N., *J. Chem. Soc., Chem. Commun.* 1072 (1972).
255. King, R. B., and Fronzaglia, A., *Inorg. Chem.* **5**, 1837 (1966).
256. King, R. B., and Stone, F. G. A., *J. Amer. Chem. Soc.* **82**, 4557 (1960).
257. Klopman, G., and Calderazzo, F., *Inorg. Chem.* **6**, 977 (1967).
258. Klopman, G., and Noack, K., *Inorg. Chem.* **7**, 579 (1968).
259. Kobayashi, H., Kobayashi, M., and Kaizu, Y., *Bull. Chem. Soc. Jap.* **46**, 3109 (1973).
260. Kruck, T., *Chem. Ber.* **97**, 2018 (1964).
261. Kruck, T., and Prasch, A., *Z. Naturforsch. B* **19**, 669 (1964).
262. Kunz, V., and Norwacki, W., *Helv. Chim. Acta* **50**, 1052 (1967).
263. Kurras, E., *Angew. Chem.* **72**, 635 (1960).
264. Kurras, E., *Z. Anorg. Chem.* **351**, 268 (1967).
265. Kursanov, D. N., Setkina, V. N., Baranetskaya, N. K., and Anisimov, K. N., *Izv. Akad. Nauk. SSSR, Ser. Khim.* 1622 (1968).
266. Kursanov, D. N., Setkina, V. N., and Gribov, B. G., *J. Organometal. Chem.* **37**, C35 (1972).
267. Kursanov, D. N., Setkina, V. N., Petrovskil, P. V., Zdanovich, V. I., Baranetskaya, N. K., and Rubin, I. D., *J. Organometal. Chem.* **37**, 339 (1972).
268. Lee, C. C., Sutherland, R. G., and Thomson, B. J., *Chem. Commun.* 1071 (1971).
269. Lee, C. C., Sutherland, R. G., and Thomson, B. J., *Chem. Commun.* 907 (1972).
270. Lewandos, G. S., and Pettit, R., *J. Amer. Chem. Soc.* **93**, 7087 (1971).
271. Lillya, C. P., and Sahatjian, R. A., *Inorg. Chem.* **11**, 889 (1972).
272. Lindner, H. H., and Fischer, E. O., *J. Organometal. Chem.* **12**, P18 (1968).
273. Lokshin, B. V., Zdanovich, V. I., Baranetskaya, N. K., Setkina, V. N., and Kursanov, D. N., *J. Organometal. Chem.* **37**, 331 (1972).
274. Lühder, K., *Z. Chem.* **9**, 31 (1969).
275. Luth, H., Taylor, I. F., and Amma, E. L., *Chem. Commun.* 1712 (1970).
276. Magomedov, G., Syrkin, V., Frenkel, A., Medvedeva, V., and Morozova, L., *Zh. Obshch. Khim.* **43**, 804 (1973).
277. Maitlis, P. M., and Efraty, A., *J. Organometal. Chem.* **4**, 175 (1965).
278. Mandelbaum, A., Neuwith, Z., and Cais, M., *Inorg. Chem.* **2**, 902 (1963).
279. Mangini, A., and Toddei, F., *Inorg. Chim. Acta* **2**, 8 (1968).
280. Mangini, A., and Toddei, F., *Inorg. Chim. Acta* **2**, 12 (1968).
281. Manuel, T. A., *Inorg. Chem.* **3**, 1794 (1964).

282. Manuel, T. A., Stafford, S. L., and Stone, F. G. A., *J. Amer. Chem. Soc.* **83**, 3597 (1961).
283. Manuel, T. A., and Stone, F. G. A., *Chem. Ind.* (*London*) 231 (1960).
284. Martin, H., and Vohwinkel, F., *Chem. Ber.* **94**, 2416 (1961).
285. Mason, R., and Robinson, W. R., *Chem. Commun.* 468 (1968).
286. McFarlane, W., and Grim, S. O., *J. Organometal. Chem.* **5**, 147 (1966).
287. McVicker, G. B., *Inorg. Chem.* **11**, 2485 (1972).
288. Middleton, R., Hull, J. R., Simpson, S. R., Tomlinson, C. H., and Timms, P. L., *J. Chem. Soc., Dalton Trans.* 120 (1973).
289. Mills, O. S., and Hock, A. A., *in* "Advances in the Chemistry of the Coordination Compounds" (S. Kerschner, ed.), p. 640. Macmillan, New York, 1961.
290. Mills, O. S., and Robinson, G., *Acta Crystallogr.* **16**, 758 (1963).
291. Moritani, I., Fugiwara, Y., and Danno, S., *J. Organometal. Chem.* **27**, 279 (1971).
292. Mortland, M. M., and Pinnavaia, T. J., *Nature Phys. Sci.* **229**, 75 (1971).
293. Muir, K. W., and Ferguson, G., *J. Chem. Soc. B* 476 (1968).
294. Muir, K. W., and Ferguson, G., *J. Chem. Soc. B* 467 (1968).
295. Muir, K. W., Ferguson, G., and Sim, G. A., *Chem. Commun.* 465 (1966).
296. Müller, J., and Goser, P., *J. Organometal. Chem.* **12**, 163 (1968).
297. Müller, J., Goser, P., and Elian, M., *Angew. Chem., Int. Ed. Engl.* **8**, 374 (1969).
298. Munro, J. D., and Pauson, P. L., *Proc. Chem. Soc.* 267 (1959).
299. Munro, J. D., and Pauson, P. L., *J. Chem. Soc.* 3479 (1961).
300. Natta, G., Calderazzo, F., and Santambrogio, E., *Chim. Ind.* (*Milan*) **40**, 1003 (1958).
301. Natta, G., Ercoli, R., and Calderazzo, F., *Chim. Ind.* (*Milan*) **40**, 287 (1958).
302. Natta, G., Mazzanti, G., and Pregaglia, G., *Gazz. Chim. Ital.* **89**, 2065 (1959); *Tetrahedron* **8**, 86 (1960).
303. Nesmeyanov, A. N., Denisovich, L. I., Gubin, S. P., Vol'kenau, N. A., Sirotkina, E. I., and Bolesova, I. M., *J. Organometal. Chem.* **20**, 169 (1969).
304. Nesmeyanov, A. N., Kolobova, N. E., Anisimov, K. N., and Makarov, Yu. V., *Izv. Akad. Nauk. SSSR, Ser. Khim.* 2665 (1968).
305. Nesmeyanov, A. N., Kursanov, D. N., Setkina, V. N., Vil'chevskaya, V. D., Barametskaya, N. K., Krylova, A. I., and Gluschchenko, L. A., *Dokl. Akad. Nauk. SSSR* **199**, 1336 (1971).
306. Nesmeyanov, A. N., Leshchova, I. F., Ustynyuk, Yu. A., Sirotkina, Ye. I., Bolesova, I. N., Isaeva, L. S., and Vol'kenau, N. A., *J. Organometal. Chem.* **22**, 689 (1970).
307. Nesmeyanov, A. N., Rybin, L. V., Kakanovich, V. S., Krivykh, V. V., and Rybinskaya, M. I., *Izv. Akad. Nauk. SSSR, Ser. Khim.* 2090 (1973).
308. Nesmeyanov, A. N., Vol'kenau, N. A., and Bolesova, I. N., *Dokl. Akad. Nauk. SSSR* **175**, 606 (1967).
309. Nesmeyanov, A. N., Vol'kenau, N. A., and Bolesova, I. N., *Dokl. Akad. Nauk. SSSR* **166**, 607 (1966).
310. Nesmeyanov, A. N., Vol'kenau, N. A., and Bolesova, I. N., *Dokl. Akad. Nauk. SSSR* **149**, 615 (1963).
311. Nesmeyanov, A. N., Vol'kenau, N. A., and Bolesova, I. N., *Tetrahedron Lett.* **25**, 1725 (1963).
312. Nesmeyanov, A. N., Vol'kenau, N. A., Bolesova, I. N., and Isaeva, L. S., *Izv. Akad. Nauk. SSSR, Ser. Khim.* 2416 (1968).
313. Nesmeyanov, A. N., Vol'kenau, N. A., and Isaeva, L. S., *Dokl. Akad. Nauk. SSSR* **176**, 106 (1967).

314. Nesmeyanov, A. N., Vol'kenau, N. A., and Isaeva, L. S., *Dokl. Akad. Nauk. SSSR* **183**, 606 (1968).
315. Nesmeyanov, A. N., Vol'kenau, N. A., Isaeva, L. S., and Bolesova, I. N., *Dokl. Akad. Nauk. SSSR* **183**, 834 (1968).
316. Nesmeyanov, A. N., Vol'kenau, N. A., and Shiloutseva, L. S., *Dokl. Akad. Nauk. SSSR* **160**, 1327 (1965).
317. Nesmeyanov, A. N., Vol'kenau, N. A., and Shiloutseva, L. S., *Izv. Akad. Nauk. SSSR, Ser. Khim.* 726 (1969).
318. Nesmeyanov, A. N., Vol'kenau, N. A., and Shiloutseva, L. S., *Izv. Akad. Nauk. SSSR, Ser. Khim.* 1206 (1970).
319. Nesmeyanov, A. N., Vol'kenau, N. A., and Shiloutseva, L. S., *Dokl. Akad. Nauk. SSSR* **190**, 354 (1970).
320. Nesmeyanov, A. N., Vol'kenau, N. A., and Shiloutseva, L. S., *Dokl. Akad. Nauk. SSSR* **190**, 857 (1970).
321. Nesmeyanov, A. N., Vol'kenau, N. A., Shiloutseva, L. S., and Petrakova, V. A., *J. Organometal. Chem.* **61**, 329 (1973).
322. Nesmeyanov, A. N., Vol'kenau, N. A., Sirotkina, E. I., and Deryabin, V. V., *Dokl. Akad. Nauk. SSSR* **177**, 1110 (1967).
323. Ngai, L. H., Stafford, F. E., and Schäfer, L., *J. Amer. Chem. Soc.* **91**, 48 (1969).
324. Nicholson, B. J., *J. Amer. Chem. Soc.* **88**, 5156 (1966).
325. Nicholson, B. J., and Longuet-Higgins, H. C., *Mol. Phys.* **9**, 461 (1965).
326. Nicholls, B., and Whiting, M. C., *Proc. Chem. Soc.* 152 (1958).
327. Nicholls, B., and Whiting, M. C., *J. Chem. Soc.* 551 (1959).
328. Nicholls, B., and Whiting, M. C., *J. Chem. Soc.* 469 (1959).
329. Nolte, M. J., Gafner, G., and Haines, L. M., *Chem. Commun.* 1406 (1969).
330. Nozawa, Y., and Tokeda, M., *Bull. Chem. Soc. Jap.* **42**, 2431 (1969).
331. Öfele, K., *Chem. Ber.* **99**, 1732 (1966).
332. Ogata, I., Iwata, R., and Ikeda, Y., *Tetrahedron Lett.* **34**, 3011 (1970).
333. Olivé, G. Henrici, and Olivé, S., *Z. Phys. Chem.* (*Frankfurt am Main*) **56**, 223 (1967).
334. Olivé, G. Henrici, and Olivé, S., *J. Organometal. Chem.* **9**, 325 (1967).
335. Olivé, G. Henrici, and Olivé, S., *J. Amer. Chem. Soc.* **92**, 4831 (1970).
336. Palm, C., Fischer, E. O., and Baumgärtner, F., *Tetrahedron Lett.* **6**, 253 (1962).
337. Pasynkiewicz, S., Giezynski, R., and Zierzgowski, S. D., *J. Organometal. Chem.* **54**, 203 (1973).
338. Pavlycheva, A. V., Domrachev, G. A., Razuvaev, G. A., and Suvorova, O. N., *Dokl. Akad. Nauk. SSSR* **184**, 105 (1969).
339. Patukhov, G. G., and Koksharova, A. A., *Uch. Zap. Gork. Gos. Pedagog. Inst.* **266**, (1970), (C. A. 76, 25398).
340. Peyronel, G., Vezzosi, I. M., and Buffagni, S., *Gazz. Chim. Ital.* **97**, 845 (1967).
341. Pidcock, A., Smith, J. D., and Taylor, B. W., *J. Chem. Soc. A* 872 (1967).
342. Pidcock, A., Smith, J. D., and Taylor, B. W., *Inorg. Nucl. Chem. Lett.* **4**, 467 (1968).
343. Pidcock, A., Smith, J. D., and Taylor, B. W., *J. Chem. Soc. A* 1604 (1969).
344. Pignataro, S., and Lossing, F. P., *J. Organometal. Chem.* **10**, 531 (1967).
345. Pinnavaia, T. J., and Mortland, M. M., *J. Phys. Chem.* **75**, 3957 (1971).
346. Price, J. T., and Sorensen, T. S., *Can. J .Chem.* **46**, 515 (1968).
347. Prins, R., and Reinders, F. J., *Chem. Phys. Lett.* **3**, 45 (1969).
348. Rausch, M. D., *Pure Appl. Chem.* **30**, 523 (1972).
349. Rausch, M. D., Moser, G. A., Zaiko, E. J., and Lipman, A. L., *J. Organometal. Chem.* **23**, 185 (1970).

350. Razuvaev, G. A., Domrachev, G. A., Suvorava, O. N., and Abakumova, L. G., *J. Organometal. Chem.* **32**, 113 (1971).
351. Razuvaev, G. A., Sorokin, Y. A., and Domrachev, G. A., *Proc. Acad. Sci. USSR* **111**, 1264 (1956).
352. Rees, B., and Coppens, P., *J. Organometal. Chem.* **42**, C102 (1972).
353. Restivo, R. J., and Ferguson, G., *Chem. Commun.* 847 (1973).
354. Rettig, M. F., Stout, C. D., Klug, A., and Farnham, P., *J. Amer. Chem. Soc.* **92**, 5100 (1970).
355. Riemschneider, R., Becker, O., and Franz, K., *Monatsch. Chem.* **90**, 571 (1959).
356. Robertson, G. B., and Whimp, P. O., *J. Organometal. Chem.* **60**, C11 (1973).
357. Robertson, G. B., Whimp, P. O., Colton, R., and Rix, C. J., *Chem. Commun.* 573 (1971).
358. Robinson, B. H., and Spencer, J., *Chem. Commun.* 1480 (1968).
359. Robinson, B. H., and Spencer, J. L., *J. Chem. Soc. A* 2045 (1971).
360. Rodesiler, P. F., and Amma, E. L., *Inorg. Chem.* **11**, 388 (1972).
361. Rodesiler, P. F., Hall Griffith, E. A., and Amma, E. L., *J. Amer. Chem. Soc.* **94**, 761 (1972).
362. Rundle, R. E., and Goring, J. H., *J. Amer. Chem. Soc.* **72**, 5337 (1950).
363. Rupert, J. P., *J. Phys. Chem.* **77**, 784 (1973).
364. Sanders, J. R., *J. Chem. Soc., Dalton Trans.* 743 (1973).
365. Sbrignadello, G., and Marcati, F., *J. Organometal. Chem.* **46**, 357 (1972).
366. Schafer, L., and Southern, J. F., *J. Organometal. Chem.* **24**, C13 (1970).
367. Schrock, R. R., and Osborn, J. A., *Inorg. Chem.* **9**, 2339 (1970).
368. Schrock, R. R., and Osborn, J. A., *J. Amer. Chem. Soc.* **93**, 3089 (1971).
369. Sellman, D., and Maisel, G., *Z. Naturforsch. B* **27**, 465 (1972).
370. Seyferth, D., and Alleston, D. L., *Inorg. Chem.* **2**, 417 (1963).
371. Shibaeva, R. P., Atovmyan, L. O., and Rozenberg, L. P., *Chem. Commun.* 649 (1969).
372. Shustorovich, E. M., and Dyatkina, M. E., *Dokl. Akad. Nauk. SSSR* **131**, 215 (1960).
373a. Silverthorn, W. E., *Abstr. 6th Int. Conf. Organometal. Chem., Amherst*, 1973.
373b. Silverthorn, W. E., unpublished results.
374. Sim, G. A., *Annu. Rev. Phys. Chem.* **18**, 57 (1967).
375. Sirotkina, E. I., Nesmeyanov, A. N., and Vol'kenau, N. A., *Izv. Akad. Nauk. SSSR, Ser. Khim.* 1524 (1969).
376. Skell, P. S., Smith, D. L. Williams, and McGlinchey, M. J., *J. Amer. Chem. Soc.* **95**, 3337 (1973).
377. Smith, D. L. Williams, Wolf, L. R., and Skell, P. S., *J. Amer. Chem. Soc.* **94**, 4042 (1972).
378. Smith, H. G., and Rundle, R. E., *J. Amer. Chem. Soc.* **80**, 5075 (1958).
379. Snow, M. R., Pauling, P., and Stiddard, M. H. B., *Aust. J. Chem.* **22**, 709 (1969).
380. Snow, M. R., and Stiddard, M. H. B., *Chem. Commun.* 580 (1965).
381. Solomon, R. G., and Kochi, J. K., *J. Chem. Soc. Chem. Commun.* 559 (1972).
382. Sourisseau, C., and Pasquier, B., *Spectrochim. Acta., Part A* **26**, 1279 (1970).
383. Starovskii, O. V., and Struchkov, Y. T., *Dokl. Akad. Nauk. SSSR* **135**, 620 (1961).
384. Stiddard, M. H. B., and Townsend, R. E., *J. Chem. Soc. A* 2355 (1969).
385. Stolz, I. W., Haas, H., and Sheline, R. K., *J. Amer. Chem. Soc.* **87**, 716 (1965).
386. Stolze, G., and Hahle, J., *J. Organometal. Chem.* **5**, 545 (1966).
387. Strohmeier, W., *Chem. Ber.* **94**, 3337 (1961).

388. Strohmeier, W., *Z. Naturforsch. B* **17**, 566 (1962).
389. Strohmeier, W., *Z. Naturforsch. B* **17**, 627 (1962).
390. Strohmeier, W., Guttenberger, J. F., and Müller, F. J., *Z. Naturforsch. B* **22**, 1091 (1967).
391. Strohmeier, W., Hahgoub, A. E., and VonHobe, D., *Z. Phys. Chem.* (*Frankfurt am Main*) **35**, 253 (1962).
392. Strohmeier, W., and Hartmann, P., *Z. Naturforsch. B* **18**, 506 (1963).
393. Strohmeier, W., and Hellmann, H., *Z. Naturforsch. B* **18**, 769 (1963).
394. Strohmeier, W., and Hellmann, H., *Chem. Ber.* **96**, 2859 (1963).
395. Strohmeier, W., and Hellmann, H., *Chem. Ber.* **97**, 1877 (1964).
396. Strohmeier, W., and Hellmann, H., *Chem. Ber.* **98**, 1598 (1965).
397. Strohmeier, W., and Von Hobe, D., *Z. Naturforsch. B* **18**, 981 (1963).
398. Strohmeier, W., and Mittnacht, H., *Z. Phys. Chem.* (*Frankfurt am Main*) **29**, 339 (1961).
399. Strohmeier, W., and Mittnacht, H., *Chem. Ber.* **93**, 2085 (1960).
400. Strohmeier, W., and Müller, R., *Z. Phys. Chem.* (*Frankfurt am Main*) **40**, 85 (1964).
401. Strohmeier, W., and Müller, F. J., *Chem. Ber.* **102**, 3608 (1969).
402. Strohmeier, W., Popp, G., and Guttenberger, J. F., *Chem. Ber.* **99**, 165 (1966).
403. Strohmeier, W., and Staricco, E. H., *Z. Phys. Chem.* (*Frankfurt am Main*) **38**, 315 (1963).
404. Taylor, I. F., and Amma, E. L., *Chem. Commun.* 1442 (1970).
405. Taylor, I. F., Hall, E. A., and Amma, E. L., *J. Amer. Chem. Soc.* **91**, 5745 (1969).
406. Tazima, Y., and Yuguchi, S., *Bull. Chem. Soc. Jap.* **39**, 2534 (1966).
407. Timms, P. L., *Chem. Commun.* 1033 (1969).
408. Timms, P. L., *J. Chem. Educ.* **49**, 782 (1972).
409. Timms, P. L., *Advan. Inorg. Chem. Radiochem.* **14**, 121–171 (1972).
410. Tsutsui, M. Aryoshi, J., Koyano, T., and Levy, M. N., *Advan. Chem. Ser.* **70**, 266 (1968).
411. Tsutsui, M., and Koyano, T., *J. Polym. Sci. Part A-1* **5**, 683 (1967).
412. Tsutsui, M., and Levy, M. N., *Proc. Chem. Soc.* 117 (1963).
413. Tsutsui, M., and Levy, M. N., *Z. Naturforsch. B* **21**, 823 (1966).
414. Tsutsui, M., and Zeiss, H., *Naturwissenschaften* **44**, 420 (1957).
415. Tsutsui, M., and Zeiss, H., *J. Amer. Chem. Soc.* **81**, 1367 (1959).
416. Tsutsui, M., and Zeiss, H., *J. Amer. Chem. Soc.* **83**, 825 (1961).
417. Turner, R. W., and Amma, E. L., *J. Amer. Chem. Soc.* **85**, 4046 (1963).
418. Turner, R. W., and Amma, E. L., *J. Amer. Chem. Soc.* **88**, 1877 (1966).
419. Turner, R. W., and Amma, E. L., *J. Amer. Chem. Soc.* **88**, 3243 (1966).
420. Turta, K. I., Stukan, R. A., Gol'danskii, V. I., Vol'kenau, N. A., Sirotkina, E. I., Bolesova, I. N., Isaeva, L. S., and Nesmeyanov, A. N., *Teor. Eksp. Khim.* **7**, 486 (1971); (*Chem. Abstr.* **76**, 19852V).
421. Van Oven, H. O., and De Liefde Meijer, H. J., *J. Organometal. Chem.* **23**, 159 (1970).
422. Verkouw, H. T., and Van Oven, H. O., *J. Organometal. Chem.* **59**, 259 (1973).
423. Vertyulina, L. N., Domrachev, G. A., Korshunov, I. A., and Razuvaev, G. A., *J. Gen. Chem. USSR* **33**, 285 (1963).
424. Victor, R., Ben-Shoshan, R., and Sarel, S., *Chem. Commun.* 1680 (1970).
425. Victor, R., Ben-Shoshan, R., and Sarel, S., *Tetrahedron Lett.* **49**, 4257 (1970).
426. Vohwinkel, F., *Trans. N.Y. Acad. Sci.* **26**, 446 (1964).
427. Vohwinkel, F., and Preusser, G., U.S. Patent 3,322,803; (*Chem. Abstr.* **64**, P55762).

428. Volger, H. C., and Hogeveen, H., *Rec. Trav. Chim. Pays-Bas* **86**, 830 (1967).
429. Volpin, M. E., Ilatouskaya, M. A., Kosyakova, L. V., and Shur, V. B., *Dokl. Akad. Nauk. SSSR* **180**, 103 (1968).
430. Walker, P. J. C., and Mawby, R. J., *J. Chem. Soc. A* 3006 (1971).
431. Walker, P. J. C., and Mawby, R. J., *Inorg. Chem.* **10**, 404 (1971).
432. Walker, P. J. C., and Mawby, R. J., *Chem. Commun.* 330 (1972).
433. Weber, J., Ring, R., Hochmuth, U., and Franke, W., *Ann. Chim.* **10**, 681 (1965).
434. Wehner, W. H., Fischer, E. O., and Muller, J., *Chem. Ber.* **103**, 2258 (1970).
435. Wei, C. H., *Inorg. Chem.* **8**, 2384 (1969).
436. Weiss, E., *Z. Anorg. Allg. Chem.* **287**, 236 (1956).
437. Weiss, E., and Fischer, E. O., *Z. Anorg. Allg. Chem.* **286**, 142 (1956).
438. Werner, H., and Prinz, R., *J. Organometal. Chem.* **5**, 79 (1966).
439. Werner, R. P. M., and Coffield, T. H., *in* "Advances in the Chemistry of the Coordination Compounds" (S. Kerschner, ed.), p. 535. Macmillan, New York, 1961.
440. White, C., and Maitlis, P. M., *J. Chem. Soc. A* 3322 (1971).
441. White, J. F., and Farona, M. F., *J. Organometal. Chem.* **37**, 119 (1972).
442. Whitesides, G. M., and Ehmann, W. J., *J. Amer. Chem. Soc.* **90**, 804 (1968).
443. Whitesides, G. M., and Ehmann, W. J., *J. Amer. Chem. Soc.* **91**, 3800 (1969).
444. Whiting, M. C., U.S. Patent 3,225,071; *Chem. Abstr.* **64**, 6694h.
445. Wilke, G., and Kröner, M., *Angew. Chem.* **71**, 574 (1959).
446. Willeford, B. R., and Fischer, E. O., *Naturwissenschaften* **51**, 38 (1964).
447. Winkhaus, G., *Z. Anorg. Allg. Chem.* **319**, 404 (1963).
448. Winkhaus, G., Pratt, L., and Wilkinson, G., *J. Chem. Soc.* 3807 (1961).
449. Winkhaus, G., and Singer, H., *J. Organometal. Chem.* **7**, 487 (1967).
450. Winkhaus, G., Singer, H., and Kricke, M., *Z. Naturforsch. B* **21**, 1109 (1966).
451. Work, R. A., and McDonald, R. L., *Inorg. Chem.* **12**, 1936 (1973).
452. Yagubskii, E. B., Khidekel, M. L., Shchegolev, I. F., Buravov, L. I., Gribov, B. G., and Makova, M. K., *Izv. Akad. Nauk SSSR, Ser. Khim. Fiz.* **9**, 2124 (1968).
453. Yamazaki, H., Yamaguchi, M., and Hagihara, N., *Mem Inst. Sci. Ind. Res. Osaka Univ.* **20**, 107 (1963), (*Chem. Abstr.* **60**, 13320b; *Nippon Kagaku Zasshi* **81**, 819 (1960), *Chem. Abstr.* **56**, 1588i).
454. Zakharkin, L. I., and Akhmedov, V. M., *Azerb. Khim. Zh.* **2**, 58 (1969).
455. Zeiss, H., in "Organometallic Chemistry" (H. Zeiss, ed.), pp. 380–425. Reinhold New York, 1960.
456. Zeiss, H., and Herwig, W., *Justus Liebigs Ann. Chem.* **606**, 209 (1957).
457. Zeiss, H., and Herwig, W., *J. Amer. Chem. Soc.* **80**, 2913 (1958).
458. Zeiss, H. H., and Tsutsui, M., *126th Meeting Amer. Chem. Soc., Abstr.* p. 29–0 (1954).
459. Zeiss, H. H., and Tsutsui, M., *J. Amer. Chem. Soc.* **79**, 3062 (1957).
460. Zeiss, H. H., and Tsutsui, M., U.S. Patent 3,187,013 (*Chem. Abstr.* 63, 9992d).
461. Zeiss, H., Wheatley, P. J., and Winkler, H. J. S., "Benzenoid-Metal Complexes." Ronald Press, New York, 1966.
462. Zelonka, R. A., and Baird, M. C., *Can. J. Chem.* **50**, 3063 (1972).
463. Zelonka, R. A., and Baird, M. C., *J. Organometal. Chem.* **44**, 383 (1972).
464. Zelonka, R. A., and Baird, M. C., *J. Organometal. Chem.* **35**, C43 (1972).
465. Zimmer, H. J., Verfahrenstechnik Fr. 1,460,233 (*Chem. Abstr.* **67**, P 11928A).
466. Zingales, F., Chiesa, A., and Basolo, F., *J. Amer. Chem. Soc.* **88**, 2707 (1966).

Organometallic Benzheterocycles

JOYCE Y. COREY

Department of Chemistry
University of Missouri—St. Louis
St. Louis, Missouri

I

INTRODUCTION

Heterocyclic derivatives are cyclic compounds containing at least two different kinds of atoms in the ring. The majority of the known derivatives contain carbon and one or more nonmetal heteroatoms. The extensive

TABLE I

KNOWN "UNSATURATED" HETEROCYCLES OF THE MAIN GROUP ELEMENTS[a]

System	B	Al	Si	Ge	Sn	Pb	As	Sb	Bi	Se	Te
M (3-membered ring)	A*	—	A*	A*	—	—	—	—	—	—	—
M (4-membered ring)	—	—	A, B	—	—	—	—	—	—	—	—
M (5-membered ring)	—	A	A, B	A, B	—	—	B	—	—	—	—
M (5-membered ring)	B	B	A, B	A	—	—	B	—	—	—	—
M (5-membered ring)	A, C	A, B, C	A, B, C	A, C	A, B, C	C	A, B, C	A, C	C	A, B, C	A, B, C
M (6-membered ring)	A, B	B	A, B	—	—	—	B	—	—	—	—
M (6-membered ring)	—	—	A, B	A	—	—	A	—	—	—	—

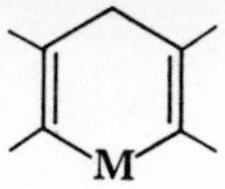	A, C	—	A, C	A, C	A, C	—	C	A, C	—	B, C	—
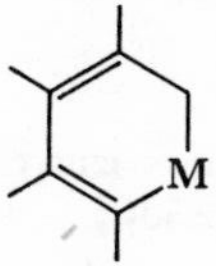	B	—	A, B, C	A	A	—	—	A	—	B	—
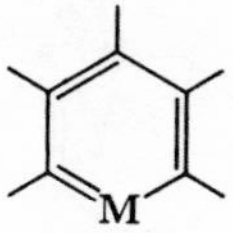	A, C*	—	C*	—	—	—	A, C	A	A	A, B	—
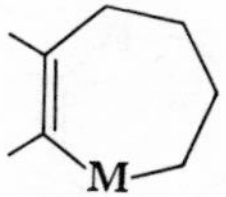	—	—	B	—	—	—	—	—	—	B	—
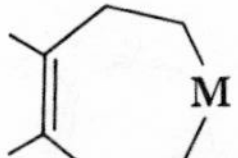	—	—	B	—	B	—	—	—	—	—	—
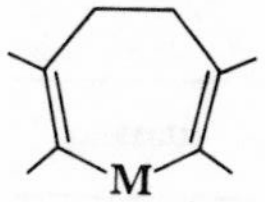	A, C	—	C	A, C	A, C	C	A, C	—	—	C	—
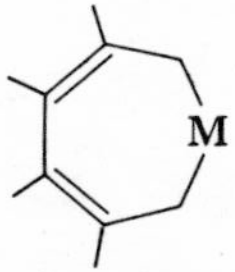	—	—	A*, C*	A	—	—	C	—	—	C	—

Continued

TABLE I—*Continued*

System	B	Al	Si	Ge	Sn	Pb	As	Sb	Bi	Se	Te
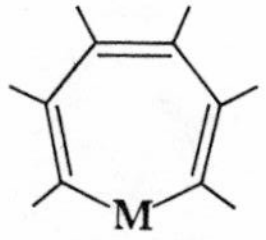	A*, C	—	C, D	C	—	—	D	—	—	—	—
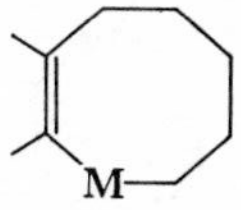	—	—	B	—	—	—	—	—	—	—	—
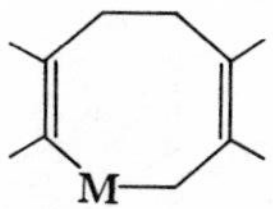	—	—	C	—	—	—	—	—	—	—	—

[a] *Speculative; A, unsaturated monocycle; B, monobenzo derivative of A; C, dibenzo derivative of A; D, tribenzo derivative of A. Benzene fused at the double-bond position.

literature on heterocycles containing nonmetal atoms stems partly from their importance in metabolism of living cells but also from extensive practical usages such as for drugs, pesticides, and dyes (*4*). Although it has been slower to develop, the study of heterocycles containing a metal(loid)[1] heteroatom has grown rapidly since the mid-1950s. The majority of the difficulties associated with the study of heterocyclic derivatives containing a metal atom has centered about the problems of formation. Several of these problems have recently been resolved, particularly those associated with the generation of analogs of aromatic derivatives.

This review will be concerned with the synthesis, reactions, and structural characteristics of unsaturated heterocycles that contain a single metal atom from Group III, IV, V, or VI and no other ring atoms except carbon. The benzheterocycles may be considered derivatives of unsaturated organometallic heterocycles with the benzene nucleus fused to the position formerly occupied by the double bond. The range of presently known, unsaturated heterocycles and benzene-annelated derivatives is given in Table I for the main group atoms as a function of ring size.

Methods for generation of unsaturated organometallic heterocycles are limited in number. The fact that the numbers of benzene-annelated derivatives exceeded the simple unsaturated monocycles in the past was a result of lack of methods of synthesis of the latter compounds. In the sections that follow, the details for synthesis of unsaturated heterocycles are developed starting with the monocycles and followed by derivatives with increased benzene annelation. The individual sections are divided by size of the heteroatom ring that is formed in the reaction, rather than on the basis of the particular metal–carbon bond generated. Synthetic methods will be divided into those that permit a direct synthesis of the heterocycle (formation of both ring bonds in one step) and those that involve an indirect synthesis (formation of one metal–carbon bond in a separate step followed by ring closure for the second M—C ring bond). Because the aromatic ring provides a site of reactivity not present in the intermediates for generation of unsaturated monocycles, some of the methods outlined are applicable only for synthesis of benzheterocycles. Ring closure by electrophilic aromatic substitution is the principal

[1] Hereafter the term organometal or metal will be used to include both metal and metalloid.

difference in generation of benzheterocycles as compared to the monocycles. General features of the reactivity of organometallic heterocycles are included at the ends of each of the sections, and preceding each section are comments on heterocyclic nomenclature.

Results of spectroscopic studies of organometallic heterocycles are treated in the last section. These studies provide insights into the problems of (*a*) stereochemistry and conformation of ring systems containing a metal–carbon bond compared to their organic analogs, (*b*) stereochemical nonrigidity of the heterocycles in solution, (*c*) "ionic" chemistry in the gas phase, and (*d*) aromaticity in ring systems that contain 6π (10π, 14π) electrons and a heteroatom with empty *d* orbitals.

Previous reviews and monographs have contained sections on heterocyclic chemistry of individual main group elements, cf. B (*190*), Si (*107*), Ge (*214*), Sn (*274*), Pb (*314*), and As, Sb, Bi (*230*); however, the emphasis of this review is on the heterocyclic systems and not on the chemistry of the individual element.

II
HETEROCYCLOALKENES, -DIENES, AND -TRIENES

Two different nomenclature systems are currently used for naming heterocyclic derivatives. Unfortunately the systems do not seem to be internally consistent and crossover within heteroatom families occurs. In general, for monocyclic derivatives the ring index system is used for Groups V and VI, and the parent carbocycle method is used for Groups III and IV. A brief description of the two methods follows and an illustration of the application of both methods is given in Table II. The reader should consult the *Chemical Abstract Index Guide to the Index of Ring Systems* for specific ring sizes and heteroatom in searching for a particular system.

In the ring index method, heterocycles are named according to Rule B-1 (*287*). The base name for a heterocyclic monocycle is generated by the combination of a prefix from column A of Table III and a suffix for the ring size from column B. Systems with unsaturation *less* than that corresponding to the maximum number of noncumulative double bonds require additional prefixes, e.g., dihydro, to denote the position of saturation. The heteroatom is always numbered as the one-position in the *monocyclic* system.

In the second method in general use, the prefix from Column A, Table III, is added to the name of the parent homocyclic derivative. In some cases (particularly the larger rings) the parent homocycle is that

TABLE II

Examples of Nomenclature Used for Heterocyclic Derivatives

Ring system	M	Ring index	Parent carbocycle
 M	Ge	Germirene[a]	Germacyclopropene
 M	Si	Siletine	1-Silacyclobut-2-ene[a]
 M	As	2-Arsolene[b]	Arsacyclopent-2-ene
 M	Sb	Stibole[b]	1-Stibacyclopenta-2,4-diene
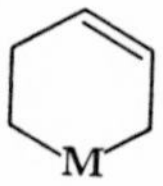 M	Ge	1,2,5,6-Tetrahydrogermanin	1-Germacyclohexa-3-ene[a]
 M	Si	1,6-Dihydrosilicin	1-Silacyclopenta-2,4-diene[c]
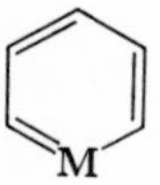 M	As	Arsenin[c]	Arsabenzene[b]
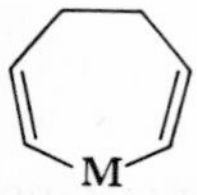 M	Sn	4,5-Dihydrostannepin[b]	4,5-Dihydro-1-stanna-2,6-cycloheptadiene[c]

[a] Only name used.
[b] Preferred for Groups V and VI.
[c] Preferred for Groups III and IV.

derivative which contains the maximum number of conjugated or multiple bonds. If the heterocyclic system has *less* than the maximum, then hydro prefixes and/or *H* are added before the heteroatom prefix. Examples of the application of both of these methods of nomenclature are given in Table II with an indication of their general use.

TABLE III

PREFIXES AND SUFFIXES USED IN NAMING METAL HETEROCYCLIC SYSTEMS

A (Prefix)		B (Suffix)	
Element		No. atoms	Unsaturated system
B	Bora	3	-irene
Al	Alumina	4[a]	-ete
Si	Sila	5[a]	-ole
Ge	Germa	6	-in[b]
Sn	Stanna	7	-epin
Pb	Plumba	8	-ocin
As	Arsa		
Sb	Stiba		
Bi	Bisma		
Se	Selena		
Te	Tellura		

[a] For four- and five-membered rings containing a single site of unsaturation, when it is theoretically possible for the presence of more than one noncumulative double bond, other endings are used: 4, -etine; 5, -oline or -olene.

[b] The "a" of the prefix is dropped and replaced by an element base name, cf., german-, arsen-, etc.

A. Heterocycloalkenes

1. *Heterocyclopropenes and Heterocyclobutenes*

Claims for the synthesis of small unsaturated ring systems containing a metalloid atom have an interesting history. In the early 1960s, a series of papers appeared describing the products of reaction of Group IV analogs of carbene with tolan (*197*, *208*, *346*, *350–353*):

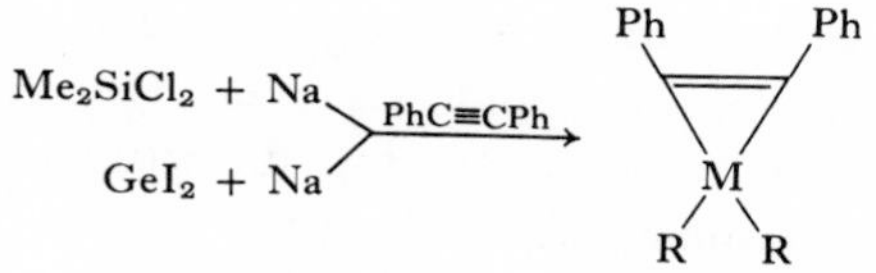

$$Me_2SiCl_2 + Na,\ GeI_2 + Na \xrightarrow{PhC\equiv CPh} \text{(I): three-membered ring, } C(Ph)=C(Ph) \text{ bridged by } MR_2 \quad (1)$$

The reaction products were formulated as monomeric three-membered ring compounds. Cryoscopic molecular weights determined by the Rast method and vibrational spectra for (I) ($MR_2 = GeMe_2, GeI_2$) appeared to be consistent with the monomer formulation (*197*, *208*, *346*, *349–353*). Derivatives of (I) had very high melting points (> 200°C) and were unreactive toward Br_2–CCl_4 (*348*), $KMnO_4$–acetone (*348*, *358*) and H_2–Pt (*348*, *358*). If the monomer formulation is a correct one, the metallirenes exhibit a rather unexpected stability.

The molecular weight of the silirine derivative (I) ($MR_2 = SiMe_2$) was redetermined by vapor phase osmometry in benzene and gave a value twice that of the monomeric formulation (*358*); isopiestic molecular weights for the germanium derivatives also gave dimer values (*174*). A mass-spectrometric investigation of germanium derivatives of (I) at inlet temperatures of 100° to 110°C showed molecular ions that were twice the molecular weight of the monomer, and from detailed analysis of the data a dimer structure with nonadjacent metal atoms was suggested (*175*):

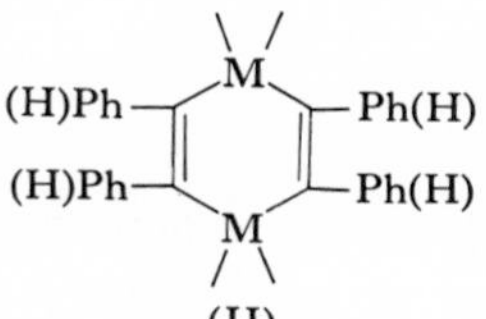

$$\text{(II): six-membered ring } M(R_2)\text{–}C(Ph(H))=C(Ph(H))\text{–}M(R_2)\text{–}C(Ph(H))=C(Ph(H))$$

(II)

The dimeric structure for (II) was confirmed by solid-state structural studies [$MR_2 = GeCl_2$ (*354*), GeI_2 (*347*), $GeMe_2$ (*354*), $SiMe_2$ (*347*)].

Although the structure of germirine compounds appeared to be settled, a controversy concerning the nature of germirine compounds in the gas phase may still exist (*345*). A recent electron diffraction investigation of the parent germirene (no ring substituents), i.e., $C_2H_2GeX_2$ ($X = CH_3$, Cl, I), reported results that seem to be at variance with the ring-substituted derivatives of (II) (dimer) (*345*). When the methyl or iodo derivatives are evaporated at low temperatures, the diffraction patterns are not consistent with the dimer formulation, structure (II):

the methyl derivative appears to be monomeric and the iodo derivative appears to be a mixture of dimer and monomer (*345*). A possibility exists that the parent compounds (i.e., no ring substituents) are structurally different from the ring-substituted derivatives or that a dimer-to-monomer structural change takes place in passing from the solid state to the gaseous state. Further investigations of the parent derivatives of (I) will be necessary before this structural dichotomy can be resolved. The high melting points and chemical inertness of the ring-substituted derivatives are certainly more consistent with structure (II) (dimer) than structure (I) (monomer). For comparison, both the saturated and the unsaturated four-membered rings containing either silicon or germanium are highly susceptible to ring-cleavage reactions (*95*). Three-membered rings should exhibit a similar reactivity.

Recently, another approach to the problem of heterocyclopropene synthesis has resulted in a claim for the generation of a phosphirene derivative similar to that of the controversial silirene and germirene (I) from reaction of a dibromide precursor with 1,5-diazabicyclo[4.3.0]non-5-ene (DBN) (*189*):

$$(\text{PhCHBr})_2\text{P(=O)Ph} \xrightarrow{\text{DBN}} \text{PhC}{=}\text{CPh}\ (\text{ring with } \text{P(=O)Ph}) \qquad (2)$$

The high reactivity of this system is demonstrated by the fact that the cyclic ring system is cleaved by sodium hydroxide to give (*cis*-1,2-diphenylvinyl)phosphinic acid and is thermally unstable giving tolan at 120°C (*189*). The extension of this synthetic route to other phosphorus or other heteroatom derivatives has not yet been tested.

A third synthetic route that has been successfully exploited in the formation of five- and six-membered heterocycles has been studied in an attempted synthesis of a borirene derivative. Reaction of phenylboron dibromide with tolan gives a mixture of products resulting from both haloboration and phenylboration (*109*). A Wurtz coupling reaction of the former derivative could generate a borirene derivative:[2]

[2] THF, tetrahydrofuran.

$$\text{Ph(PhBrB)C{=}C(Br)Ph} \xrightarrow[-77^\circ\text{C}]{\text{Li/THF}} \left[\text{PhB}\text{-cyclopropene (Ph, Ph)}\right] \begin{cases} \xrightarrow{CO_2} \text{No carboxylic acid derivative of } cis\text{-stilbene} \\ \xrightarrow[\text{Workup}]{\text{Acetolysis}} \text{PhCH{=}CHPh} \\ \xrightarrow{\text{Heat}} \text{Decomposition} \end{cases} \quad (3)$$

The chemical reactivity of the product species indicates the presence of *cis*-vicinal carbon–metal bonds (*109*). Unfortunately the chemical facts are consistent with either a monomer structure (I) or dimer structure (II).

Lack of three-membered rings containing a metalloid heteroatom is frequently attributed to the ring strain associated with replacement of a —CH_2— by the longer carbon–metal bond. A successful synthesis of a saturated silicon three-membered ring has been accomplished, however (*202*). Although the system exhibits a high chemical reactivity, the problems associated with generation of small rings may be solved (at least for silicon), and renewed interest in small, unsaturated ring systems is expected. The possible existence of 1,1-dimethylgermirene and the claim for formation of a phosphirene ring suggests that long carbon–metal bonds may not be a deterrent to small ring formation for metalloids.

Only one example of a cyclobutene derivative containing a metalloid heteroatom has been reported (*130*). The silicon derivative (III) was prepared by the following route:[3]

$$\text{PhCH(Br)(CH}_2)_2\text{SiPh}_2\text{Br} \xrightarrow{\text{Mg/THF}} \text{2-Ph-1,1-Ph}_2\text{-silacyclobutane} \xrightarrow{\text{NBS}} \text{2-Br-2-Ph-1,1-Ph}_2\text{-silacyclobutane} \xrightarrow{\text{PhMgBr}} \underset{\text{(III)}}{\text{2-Ph-1,1-Ph}_2\text{-silacyclobut-2-ene}} \quad (4)$$

The indicated ring closure to form the saturated four-membered ring is the only successful method that now exists for formation of such compounds in reasonable yield (*95*). In this case, ring closure is solvent-dependent, and the reaction does not take place in ether. Since a standard dehydrobromination method using a mixture of KOAc–HOAc and

[3] NBS, *N*-bromosuccinimide.

EtOH resulted in ring cleavage, dehydrobromination was effected by reaction with phenylmagnesium bromide. The overall yield of (III) from the saturated ring was 15%.

2. *Heterocyclopentenes and Heterocyclohexenes*

Two isomeric species are possible for five- and six-membered heterocycloalkenes and both are found for each of the ring systems. For structural types (IV), the unsaturation is adjacent (α) to the heteroatom, and in structure (V) the unsaturation is one carbon removed from the heteroatom (β).

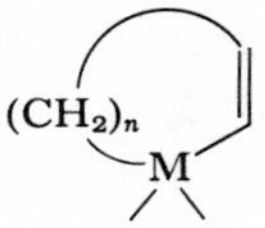

$(CH_2)_n$ M (n = 2, 3) (IV)

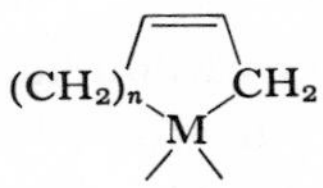

$(CH_2)_n$ CH_2 M (n = 1, 2) (V)

Three basic approaches have been developed for the formation of these unsaturated systems. Two of these involve direct ring-formation reactions and generally give only one of the isomers. The third method involves reaction of a preformed ring system and yields both of the possible isomers.

a. *Ring-Formation Reactions.* A convenient route to compounds of structural type (IV) takes advantage of the fact that hydrometallation of terminal alkynes, HC≡CR, affords ready access to reagents with unsaturation adjacent (α) to the metal heteroatom. If the alkyne contains a chlorinated side chain, a Wurtz coupling reaction may be effected:

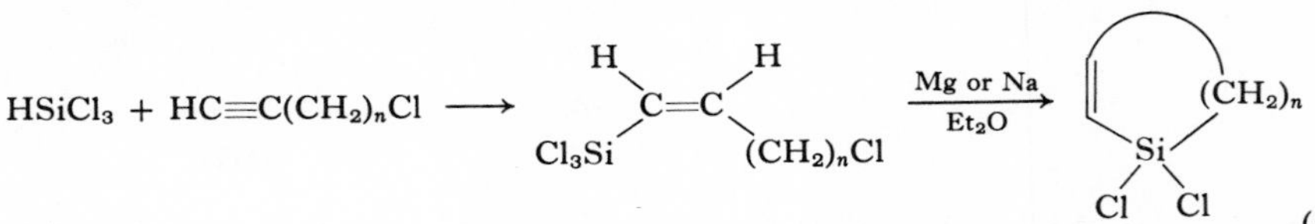

$$HSiCl_3 + HC{\equiv}C(CH_2)_nCl \longrightarrow \underset{Cl_3Si\quad\quad (CH_2)_nCl}{\overset{H\quad\quad H}{C{=}C}} \xrightarrow[Et_2O]{Mg\ or\ Na} \text{cyclo-}[CH{=}CH(CH_2)_nSiCl_2] \quad (5)$$

n = 2 (*32*)
n = 3 (*31*)

Although the monoaddition product of acetylene and Cl_3GeH has been reported (*255*) and the above type of ring closure has been utilized for formation of saturated four-membered Ge ring systems (*95*), the combination has not been utilized to form germanium ring systems with unsaturation α to the Ge atom.

The majority of the known compounds of type (V) with the unsaturation β to the heteroatom are five-membered rings. Silacyclopentenes are formed by the reaction of a suitable diene with an alkali metal to generate a transient organometallic reagent which is then trapped with a difunctional silicon compound (*359*):

$$Me_2SiCl_2 + \text{CH}_2\!=\!\text{CH}\!-\!\text{CH}\!=\!\text{CH}_2 \xrightarrow[\text{THF}]{\text{Na}} \text{1,1-(CH}_3)_2\text{-silacyclopent-3-ene} \qquad (6)$$

No other organometallic halides have been used to make derivatives of the intermediate butadiene dianion in the reaction depicted in Eq. (6), but similar reactions have been demonstrated for five-membered diene systems [see Eq. (12)]. The reactions can also be effected by lithium metal but the yields are considerably lower than the approximate 40% yields reported with sodium (*359*).

Heterocyclopentenes are also formed in the 1,4-addition of a presumed "divalent" Group IV intermediate with a diene. It is postulated that solutions of $HGeCl_3$ in dioxane contain the reactive species $GeCl_2$ that can be trapped with butadiene (*188*), i.e.,

$$HGeCl_3 + \text{dioxane} \longrightarrow GeCl_2\cdot\text{dioxane}$$

$$GeCl_2\cdot\text{dioxane} + \text{CH}_2\!=\!\text{CH}\!-\!\text{CH}\!=\!\text{CH}_2 \longrightarrow \text{1,1-Cl}_2\text{-germacyclopent-3-ene} \qquad (7)$$

Also GeI_2 has been used to generate germacyclopentenes. Reactions with methyl-substituted butadienes usually provide higher yields of heterocyclic derivatives.

The derivatives of heterocyclopentenes (V) prepared by these last methods are listed in Table IV.

b. *Synthesis from a Saturated Heterocycle.* Halogenation of a saturated ring system followed by dehydrohalogenation has been utilized as an entry into silacycloalkene systems. For unsubstituted ring systems the usual halogenating agent is sulfuryl chloride (in a 1:1 molar ratio) with or without peroxide catalysts, but chlorine gas has also been used (*117*). In all cases studied thus far, a mixture of monochlorinated isomers is

TABLE IV

DIRECT SYNTHESIS OF [1-metallacyclopent-3-ene, $M(R_1)(R_2)$]

Reacting olefin	Reacting organometallic	Metal	Solvent	Product R_1	Product R_2	References
$BrH_2CCH{=}CHCH_2Br$	Me_2SiCl_2	Mg	Et_2O	Me	Me	*235*
$ClCH_2CH{=}CHCH_2Cl$	—	Cu/Si	gas (300°)	Cl	Cl	*275, 288*
butadiene (structure)	$Me_3Al \cdot OEt_2$	Li, Na	THF, Me_2O, Et_2O	Me	Me[a]	*206*
	$Me_2Si(Cl)OMe$	Na[b]	THF	Me	Me	*359*
	Me_2SiCl_2	Na[b]	THF	Me	Me	*359*
	Me_2SiCl_2	Mg	$(Me_2N)_3PO$	Me	Me	*106*
	$Me_3Si_2Cl_3(F_3)$	—	[c]	Me	Cl(F)	*16*[d]
	Si_2Cl_6	—	[c]	Cl	Cl	*16, 374*
	$HGeCl_3$	—	Neat(−60°C)	Cl	Cl	*253, 254, 256*
	$HGeCl_3$	—	Dioxane	Cl	Cl	*188*
	$HGeBr_3$	—	Et_2O	Br	Br	*124*
	$GeBr_2$	—	Acetone	Br	Br	*94, 124*
	GeI_2	—	Heptane[e]	I	I	*250*
isoprene (CH_3-substituted butadiene, structure)	$Me_2SiCl(OMe)$	Na	THF	Me	Me	*359*

	$Me_4Si_2(OMe)_2$	—	[b]	Me	Me	*16*
	$HGeCl_3$	—	Neat(−60°C)	Cl	Cl	*254, 256*
	GeI_2	—	Neat[f]	I	I	*249*
	$EtClGe(H)OMe$	—	[g]	Et	Cl	*248*
H_3C CH_3 (2,3-dimethylbutadiene structure)	$Me_2Si(Cl)OMe$	Na	THF	Me	Me	*359*
	Me_2SiCl_2	Mg	$(Me_2N)_3PO$	Me	Me	*106*
	Ph_2SiCl_2	Mg	$(Me_2N)_3PO$	Ph	Ph	*106*
	$Me_4Si_2(OMe)_2$	—	[b]	Me	Me	*16*
	$Me_2Si_2(OMe)_4$	—	[b]	Me	Me	*16*
				Me	OMe	*16*
	GeI_2	—	Neat[f]	I	I	*249*
	$R_1EtGe(OMe)H$	—	[g]	Et	Cl, OMe	*248*
	$R_1PhGe(OMe)H$	—	[g]	Ph	X[h]	*248*

[a] Product is anionic.
[b] Considerably reduced yields if Li is substituted for Na.
[c] Gas phase reaction.
[d] Mixture of 2-ene and 3-ene.
[e] 100°C—no reaction in absence of solvent at room temperature.
[f] Mixed at room temperature followed by heating.
[g] Sealed tube reaction.
[h] X = F, Cl, Br, I, H, OMe.

TABLE V

SILACYCLOALKENES FROM DEHYDROHALOGENATION

System	Alkene product[a]	Dehydrohalogenation method			Reference
		Quinoline	$FeCl_3$	Pyrolysis	
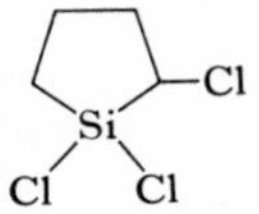 Cl; Si; Cl Cl	(IV)	67%	Trace	18%	*33*
	(V)	—	Trace	Trace	*33*
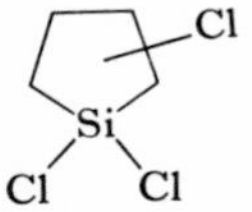 Cl; Si; Cl Cl	(IV)	Trace	16%	36%	*33*
	(V)	Trace	13%	7%	*33*
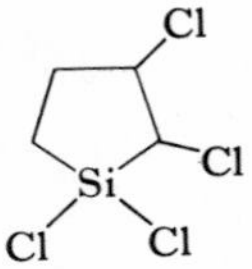 Cl; Cl; Si; Cl Cl	(IV)	50%	—	—	*265*
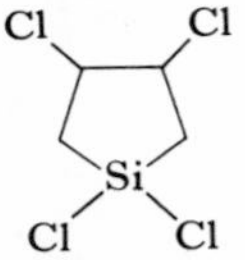 Cl Cl; Si; Cl Cl	(V)	—	—	11%	*265*
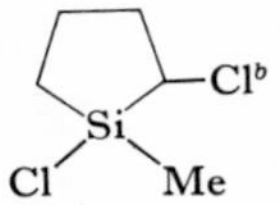 Cl[b]; Si; Cl Me	(IV)	—	—	53%	*265*
	(V)	—	—	~0%	
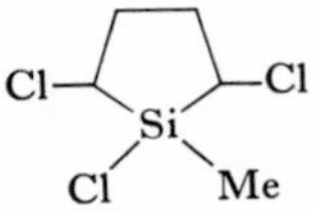 Cl Cl; Si; Cl Me	(IV)	—	—	~75%	*265*
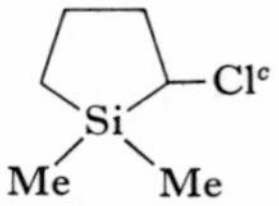 Cl[c]; Si; Me Me	(IV)	~100%[d]	—	—	*117*
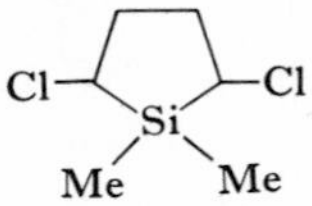 Cl Cl; Si; Me Me	(IV)	—	—	40–50%	*117*
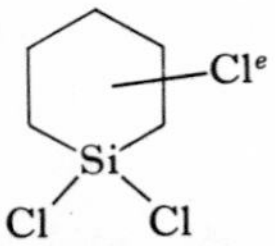 Cl[e]; Si; Cl Cl	(IV)	6%[f]	4%	11%	*31*
	(V)	15%[f]	11%	19%	*31*

obtained and physical separation of the isomers is quite tedious (*33*). However, it is possible to take advantage of the chemical reactivity of the β-chloro isomer which undergoes β-elimination (ring opening) in the presence of reagents such as $FeCl_3$ and thus obtain the pure α-chloro isomer. The dehydrohalogenation can be achieved by pyrolysis or chemically with iron(III) chloride or quinoline. The most detailed study has been associated with the chlorination of 1,1-dichlorosilacyclopentane and dehydrochlorination of the product mixture (*32*) as well as each of the pure isomers (*33*). As might be expected, chlorination of 1,1-dichlorosilacyclohexane affords each of the three possible monochlorinated products (*34*).

Halogenation of 1,1-dimethylsilacyclopentane gives α-chloromethyl-1-methylsilacyclopentane as the major product (*117*). Therefore, practically speaking, the synthetic method is limited to those cyclic derivatives with at least one Si—Cl linkage. Although it would seem reasonable to synthesize the 1,1-dichlorosilacyclopentene and convert to a 1,1-dialkylsilacyclopentene with RMgX, this latter reaction has been reported to give products of ring opening (*257*).

Results of dehydrohalogenation studies are summarized in Table V. Conditions for chlorination with sulfuryl chloride have been reported only for silicon heterocycles. Ring dehydrohalogenation reactions of cyclic germanium compounds have been reported to yield tars (*310*).

If one of the ring hydrogens is replaced by a phenyl group, specific benzylic bromination will take place at the phenyl-bearing carbon with NBS. Dehydrobromination can then be effected with a variety of reagents. This approach has been successful not only in generation of silacyclobutene [Eq. (4)] but also silacyclopentene (*130*).

Br, Ph / Si (Ph)(Ph) —(*130*)→ Ph / Si (Ph)(Ph) (8)

---

[a] Ring-opening products are usually observed but are not included in this table.

[b] Separated from 3-chloro isomer by heating with $FeCl_3$.

[c] Mixed with 2-chloro and α-chloromethyl isomers; the latter does not react with quinoline.

[d] Estimate from 2-chloro isomer.

[e] Mixture contains 23% 2-chloro, 47% 3-chloro, and 20% 4-chloro.

[f] Excess quinoline.

Other reactions involving heterocycloalkane to heterocycloalkene transformations have met with quite limited success. Dehydrogenation reactions of saturated silacyclopentane derivatives with Pt/C or with an Al–Cr–K oxide catalyst yield complex mixtures of unsaturated cyclic products with one and two double bonds (*140*, *272*). Dehydration or solvolysis reactions lead to ring cleavage (*356*); for instance, solvolysis of 4-tosyl-1,1-dimethyl-1-silacyclohexane affords low yields of 1,1-dimethyl-1-silacyclohex-3-ene and several ring-opening products (*356*).

c. *Miscellaneous Methods*. The direct reaction of CuSi with $ClCH_2CH{=}CHCH_2Cl$ affords low yields of 1,1-dichlorosilacyclopent-3-ene (*275*). An attempt to prepare 1,1-dichlorosilacyclopent-3-ene from the dilithio reagent of $ClCH_2CH{=}CHCH_2Cl$ failed (*116*), but 60% cyclized product was obtained from the reaction of Me_2SiCl_2 and $BrCH_2CH{=}CHCH_2Br$ with magnesium (*235*). A novel ring-closure reaction has produced a boracyclohexene (*122*):

$$(CH_2{=}CHCH_2)_3B + RC{\equiv}CH \longrightarrow$$

$$(CH_2{=}CHCH_2)_2B\overset{H}{C}{=}\overset{H}{C}{-}\overset{H}{C}{-}\overset{R}{C}{=}CH_2 \longrightarrow \text{[1-}(CH_2CH{=}CH_2)\text{-3-R-5-}(CH_2CH{=}CH_2)\text{-boracyclohex-3-ene ring]} \quad (9)$$

d. *Large Rings*. Reaction of 1-bromo-6-chlorohex-3-ene with lithium followed by reaction with tributylborate is reported to give 1-butoxy-boracyclohept-3-ene (*49*). An alkyne derivative of an eleven-membered Si heterocycle is also reported (*248a*).

3. *Reactions of Heterocycloalkenes*

In studying the reactivity of metallocycloalkenes there are two general questions to be considered: (*a*) what influence does the replacement of a $—CH_2—$ group in a cycloalkene by an $=MR_x$ group have on the organic chemistry of the double bond or on the structural characteristics of the system? and (*b*) how is the chemistry of the metal atom altered as a result of incorporation into a cyclic ring? To investigate either of these questions it is necessary to establish conditions under which the chemical properties of the ring system may be observed without ring cleavage.

If the heteroatom of a cyclic system contains a metal–halogen bond, several organometallic transformation reactions are possible, such as reduction with lithium aluminum hydride or reaction with an organolithium or Grignard reagent. Do these standard reactions occur without ring cleavage? Reduction of the five-membered ring isomers of silacyclopentene with lithium aluminum hydride has been reported (*32*). The reduction was carried out on the product mixture obtained from pyrolysis of a 4:1 mixture of 2-chloro- and 3-chlorosilacycloalkanes. Less than 50% of mixed reduced product was obtained, but the distribution of the hydride product isomers appears to be approximately the same as the starting isomer ratio.

Reaction of 1,1-dichlorosilacyclopent-2-ene with RMgX (*257*, *288*), 1,1-dichlorosilacyclohex-2-ene with PhLi or PhMgBr (*31*), and 1,1-dichlorogermacyclopent-3-ene with RMgX (*249*, *253*, *254*, *256*) have been reported. The substitution reactions of the silicon compound appear to proceed normally, although an account of ring opening with EtMgX exists (*257*). Reactions of 1,1-dichlorogermacyclopent-3-ene with Grignard reagents are invariably accompanied by extensive ring-cleavage products. The percent ring cleavage may be reduced by lowering the temperature of the reaction.

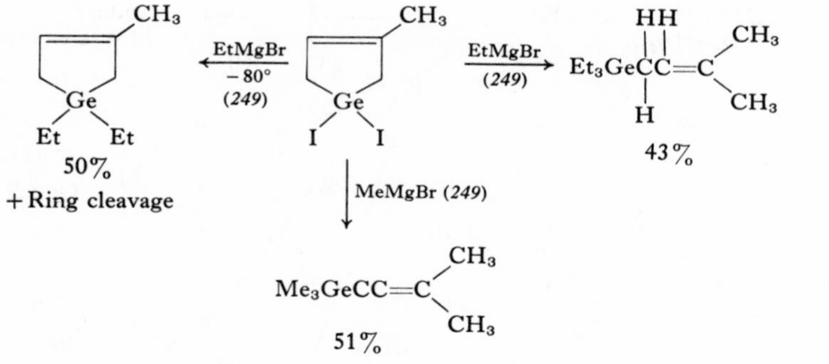

CH_3 CH_3 HH CH_3

EtMgBr, −80°, (*249*) ← | → EtMgBr (*249*) $Et_3GeCC{=}C(CH_3)_2$ (10)

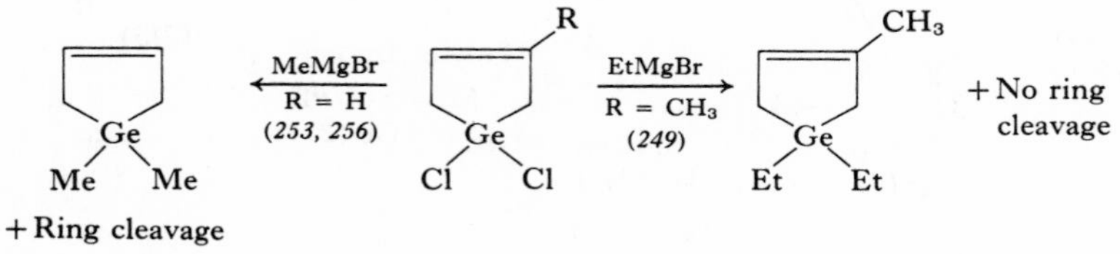

Ge (Et, Et) | Ge (I, I) | H

50% + Ring cleavage | 43%

↓ MeMgBr (*249*)

$Me_3GeCC{=}C(CH_3)_2$

51%

R CH_3

MeMgBr, R = H, (*253, 256*) ← | → EtMgBr, R = CH_3, (*249*) + No ring cleavage (11)

Ge (Me, Me) | Ge (Cl, Cl) | Ge (Et, Et)

+ Ring cleavage

Halogen exchange reactions of the type M–Cl to M–F (with SbF_3) have been reported for Si (*64*), and of M–I to M–Br or M–Cl (with AgCl or AgBr) for germanium (*249*). These reactions apparently produce little or no ring-cleavage products.

Heterocycloalkenes containing the maximum number of organic groups tend to be far less chemically reactive toward carbon–metal bond scission. One notable exception exists. The smallest known, unsaturated, heterocyclic ring system, 1,1-diphenylsilacyclobut-2-ene, is highly reactive. Ring opening occurs on reaction with $AgNO_3$–EtOH, or Li, or on contact with basic alumina (*130*).

Several standard organic reactions have been carried out at the carbon atoms including both C—H substitution reactions and addition reactions to the multiple bond. In the majority of the reactions, no ring-cleavage products are recorded. The methylene group adjacent to a multiple bond is susceptible to free radical bromination. Free radical bromination of (IV) ($MR_2 = SiCl_2$) is reported for both five- and six-membered rings (*31*, *32*). Allylic bromination of germacyclopent-3-ene has been reported for (V) when $MR_2 = GeMe_2$ but not when $MR_2 = GeBr_2$ (*310*).

Several reactions involving the multiple bond have been observed.

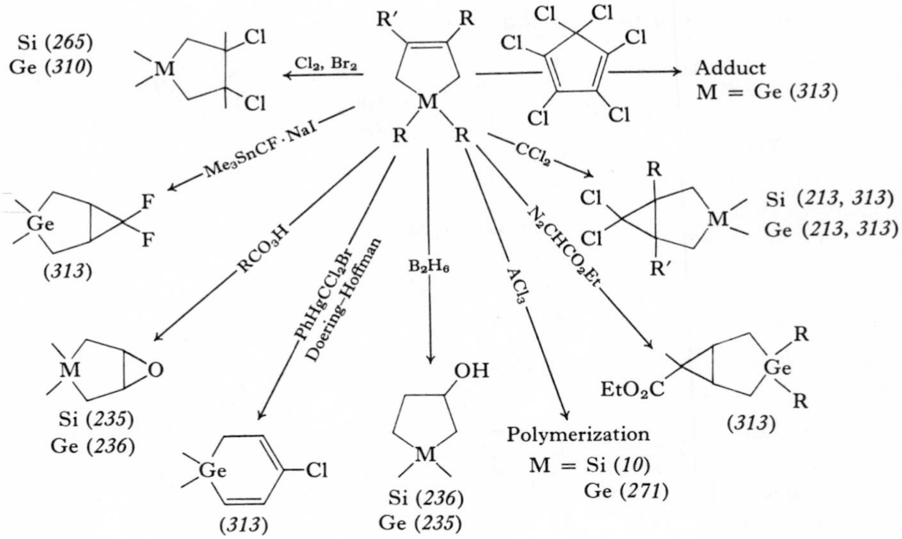

Scheme 1

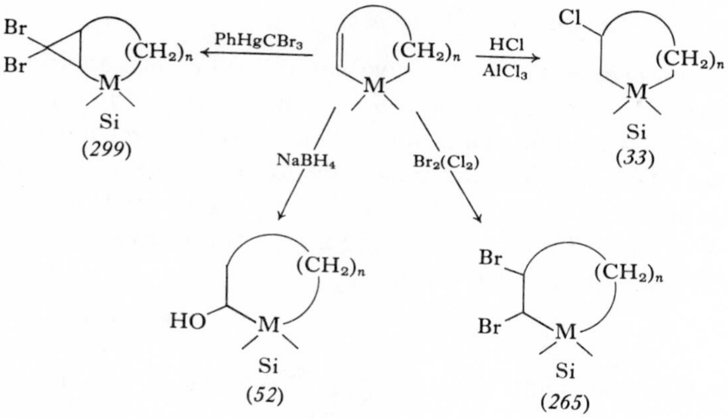

Scheme 2

These reactions are summarized in Schemes 1 and 2 and indicate the variety of organic transformations that may be carried out with heterocycloalkenes without ring cleavage.

B. *Heterocyclic Dienes*

Four-membered rings containing two double bonds would require a $p_\pi - p_\pi$ multiple bond that would include the heteroatom. The importance of this type of bonding in compounds containing a main group metal remains unresolved. Recent experiments in organosilicon chemistry suggest that reactive silicon–carbon double-bonded species may be generated from silacyclobutanes in photolysis experiments (*21*, *42*). This represents a preliminary report, however, and the generality of the reaction or of the method is yet to be tested.

The majority of the known heterocyclic dienes are pentadienes, largely as a result of the discovery that alkynes react with alkali metals and form dianions via dimer addition of the acetylene anion radical. A method that has become useful for the generation of larger cyclic dienes is the addition of an organometallic dihydride to a suitable dialkyne. Synthetic approaches to six-membered dienes tend to be unique for each system, and no general method has been developed. Some of the six-membered heterocyclic dienes have been converted to trienes for the study of systems that may be aromatic.

1. *Heterocyclic Derivatives of Cyclopentadiene*

a. *Ring-Formation Reactions.* Two slightly different conditions have been reported for the generation of the dianion from the dimerization of the anion radical of tolan and subsequent reaction to form heterocyclic derivatives (*47*, *132*). The intermediate lithium reagent, $Li_2C_4Ph_4$, is insoluble in ether, the solvent in which it is generated, and may be dissolved by the addition of THF (*132*) or 1,2-dimethoxyethane (*47*) before reaction with a metal dihalide:

$$PhC{\equiv}CPh + Li \xrightarrow{Et_2O} Li_2C_4Ph_4 \xrightarrow[(47)]{R_nMCl_2} \text{2,3,4,5-tetraphenyl-}C_4M(R)_n \; \text{(VI)} \qquad (12)$$

Dimerization of a mixed acetylene, $RC{\equiv}CR'$, may generate three different dianions on reaction with lithium. Addition of R_2SiCl_2 to this dimer-addition product has been shown to generate all three silole isomers (*54*). 1,4-Dilithiobutadiene intermediates have also been generated by the halogen–metal exchange reaction.

$$Ph\text{–}C(Br){=}CH\text{–}CH{=}C(Br)\text{–}Ph \xrightarrow{BuLi} Ph\text{–}C(Li){=}CH\text{–}CH{=}C(Li)\text{–}Ph \xrightarrow[(17,\,24)]{Me_2MCl_2} \text{2,5-diphenyl-}C_4H_2MMe_2 \qquad (13)$$

A second method that may be useful for formation of cyclic dienes is the addition of a metal dihydride to a substituted diyne. The approach has been reported for $PhAsH_2$ (*241*):

$$R'_xMH_2 + RC{\equiv}C\text{—}C{\equiv}CR \xrightarrow[(241)]{} \text{2,5-}R_2C_4H_2MR'_x \qquad (14)$$

This synthetic route has been useful for the formation of six-membered Sn heterocyclic dienes (see Section II, B, 2) but evidently has not been studied for formation of five-membered heterocycles by other metalloids that are known to undergo hydride addition to multiple bonds.

In a novel reaction, *syn*-disilanedihydrides have been shown to react in a catalyzed addition to alkynes to produce silacyclopentadienes (*282*):

$$HMe_2SiSiMe_2H + RC{\equiv}CR' \xrightarrow[90^\circ]{NiCl_2\cdot(PEt_3)_2} H_2SiMe_2 + \text{(2,5-R'-3,4-R-1,1-Me}_2\text{-silacyclopentadiene)} \quad (15)$$

b. *Synthesis from Saturated Heterocycles.* Halogenation–dehydrohalogenation reactions similar to those outlined in Section II, A, 2 (see Table V) as well as high-temperature dehydrogenation reactions have also been used to generate heterocyclodiene derivatives (*140*, *272*). Recently, a more convenient route that involves reaction of the saturated heterocycle with dichlorodicyanoquinone (DDQ) has been reported:

$$\text{2,5-Ph}_2\text{-1,1-Me}_2\text{-silacyclopentane} \xrightarrow[\text{16 hr, } C_6H_6\ (20)]{DDQ} \text{2,5-Ph}_2\text{-1,1-Me}_2\text{-silacyclopentadiene} \quad (16)$$

The heterocyclic derivatives of cyclopentadiene prepared by the ring-closure reactions are listed in Table VI.

TABLE VI

HETEROCYCLIC DERIVATIVES OF CYCLOPENTADIENE PREPARED BY RING-CLOSURE REACTIONS

System	Metal	R_1	R_2	m.p./b.p.	Method	%	References
Metallacyclopentadiene (M with R_1, R_2)	Te	—	—	−36°	A	69	*229*
				[a]	A	37	*25*
3-CH_3-metallacyclopentadiene (M with R_1, R_2)	Al	H	S[b]	—	B	—	*48*
		Me	S[b]	—	B	—	*48*
		Bu	S[b]	—	B	—	*48*
2,5-R'-metallacyclopentadiene (M with R_1, R_2)	As	Ph	—	81°/0.5 mm[c]	C	33	*241*
	Te	—	—	107°[d]	A	25	*229*
		—	—	94°[e]	A	100	*229*

Continued

TABLE VI—*Continued*

System	Metal	R_1	R_2	m.p./b.p.	Method	%	References
2,5-Ph₂C₄H₂M(R₁)(R₂)	Si	Me	Me	132°–133°	D	58	*17*
		Me	H		D		*145*
	Ge	Me	Me	128°	D	65	*24*
	Sn	Me	Me	119°–121°	D	38.5	*17*
	As	Ph	—	186.5°–187.5°	C	83	*241*
$Ph_4C_4M(R_1)(R_2)$	B	Ph	—	175d[f]	E	66	*47*
	Tl	Ph	—	—	E	—	*169, 170*
	Si	H	Cl	212°–216°	E	—	*145*
		Ph	Cl	177°–178°	E	—	*301*
				181°–183°	E	70[g]	*93*
		Ph	H	200°	E	—	*300*
		Ph	Ph	191°	E	50	*47*
				188°–191°	E	68	*133*
		Me	Cl	193°–194°	E	—	*301*
				194°–195°	E	58	*93*
		Me	H	225°–226°	E	—	*300*
		Me[h]	Me[h]	181°–182°	E[i]	72.5	*132*
				178°–179°	H	56	*282*
	Ge	Cl	Cl	197°–199°	E	70[g]	*93*
		Me	Me	179°–181°	E	—	*168*
				183°–184°	E	—	*368*
		Ph	Cl	210°–211°	E	47	*93*
		Ph	Ph		E[j]		*169, 204*
		Ph	Ge(Ph)C_4Ph_4	235°–236°	E	25	*93*
	Sn	Me	Me	192°–193°	E	67	*205, 368*
		Vi	Vi	158°–159°	E	69	*205*
		Ph	Ph	173°–174°	E	<20	*205*
				174°	E	40	*47*
	Pb	Ph	Ph	153°–154°	E	15	*368*
	As	Ph	—	214°	E	93	*205*
		Cl	—	182°–184°[k]	E	—	*205*

Continued

TABLE VI—*Continued*

System	Metal	R_1	R_2	m.p./b.p.	Method	%	References
	Sb	Ph	—	160°	E	52	*205*
				162°–170°d	E	21	*47*
	Se	—	—	183°–184°	E	65	*47*
					F	—	*169*
					G	98	*47*
	Te	—	—	239.5°	E	56	*47*
					G	82	*47*
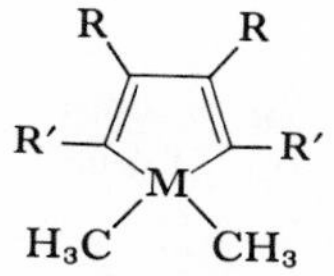 2,5-R′, 3,4-R, 1,1-dimethyl metallole (H_3C, CH_3 on M)	Si	Ph	Me	—	H	44	*282*
	Si	Et	Et	—	H	95	*282*
	Si	Me	Me	—	H	100	*282*
	Si	Bu	Bu	—	H	100	*282*
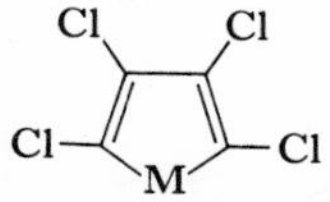 Tetrachloro metallole (Cl, Cl, Cl, Cl; M)	Te	—	—	49°	I	14	*228*
	Se	—	—	35°	I	32	*228*

Key to method: (A) Addition of Na_2Te to diyne; (B) direct reaction of Al and isoprene; (C) addition of $PhAsH_2$ to diyne; (D) reaction of $LiPh(CH)_4PhLi$ with R_xMCl_y; (E) reaction of $Li_2Ph_4C_4$ with R_xMCl_y; (F) photolysis of $Fe_2(CO)_7(PhC_2Ph)_2$; (G) from IPh=CPhCPh=CPhI + Na_2M; (H) reaction of $(HMe_2Si)_2$ with RC≡CR′ in presence of $NiCl_2(PEt_3)_2$; (I) Cl_6C_4 + M.

[a] Isolated by column chromatography.

[b] Coordinated with solvent molecules: tetrahydrofuran, dioxane, Et_2O, *N*-methylpyrolidine.

[c] R′ = Me; also reported were R_1 = Ph, R′ = -p-ClC_6H_4, -p-$CH_3C_6H_4$, α-naphthyl.

[d] R′ = CH_2OH.

[e] R′ = CMe_2OH.

[f] When prepared from exchange reaction with the tin heterocycle, Ph_5C_4B has m.p. 120°(d) (*111*).

[g] Based on assumed 80% yield, $Li_2C_4Ph_4$.

[h] Other alkyl derivatives reported R_1 = R_2 = Et, Pr, Bu, CH_2Ph (*296*).

[i] Reported as a by-product in reaction of $Li_2C_4Ph_4$ with $(ClMe_2Si)_2$ (*266*).

[j] $Ph_4C_4GePh_2$, m.p. 198–199°, prepared from $Ph_4C_4Ge(Ph)Cl$ + PhLi (*93*).

[k] Impure.

2. *Heterocyclic Derivatives of Cyclohexadiene and Cycloheptadiene*

a. *Ring-Formation Reactions.* The direct synthesis of cyclic dienes of intermediate size has not been well explored. The only method that may have general utility is M—H addition to a diyne. This reaction has been exploited mainly in the synthesis of tin heterocycles (*15*, *216*, *278*, *315*), but one germanium heterocycle has been prepared by this route (*278*).

$$(CH_2)_x(C\equiv CH)_2 + H_2SnR_2 \xrightarrow[(278)]{} (CH_2)_x(CH{=}CH)_2SnR_2 \qquad (17)$$

$$(x = 1, 2)$$

Recently, this approach has been utilized to generate 4,5-dihydroarsepins (*238*). This synthetic method is restricted to the formation of cyclic dienes where both double bonds are α to the heteroatom.

Two reactions have been reported that suggest possible synthetic routes to other isomers of metalloid cyclohexadienes and cycloheptadienes:

$$Me_2Si(Cl)CH_2Cl + LiPhC{=}CPh{-}CPh{=}CPhLi \xrightarrow[(243)]{THF} \text{1,1-dimethyl-2,3,4,5-tetraphenyl-1-silacyclohexa-2,4-diene (ring: } CH_2, CPh{=}CPh, CPh{=}CPh, Si(CH_3)_2) \qquad (18)$$

$$HGeCl_3 \xrightarrow{Dioxane} GeCl_2 \cdot dioxane \xrightarrow[(270)]{CH_2{=}CH{-}CH{=}CH{-}CH{=}CH_2} \text{1,1-dichlorogermacyclohepta-3,5-diene } (GeCl_2) \qquad (19)$$

Equation (19) is related to the 1,4-addition reactions reported for divalent Group IV analogs of "carbene" in the preparation of five-membered heterocycles containing a multiple-bond β to the heteroatom [see Eq. (7)]. If a method is developed for the generation of organometallic carbene analogs, reactions with trienes may provide a convenient route to seven-membered ring dienes.

b. *Synthetic Routes Involving Heterocyclic Derivatives.* Although direct methods are more desirable for generation of dienes, it is also possible to produce as follows the dienes from the saturated heterocycle:

$$\text{(1,1-dichloro-1-silacyclohex-2-ene)} \xrightarrow{\text{NBS}} \text{(4-bromo derivative)} \xrightarrow[(31)]{560^\circ} \text{(1,1-dichloro-1-silacyclohexa-2,4-diene)}$$

$$\text{(1,1-dimethyl-1-silacyclohex-2-en-4-one)} \xrightarrow[(357)]{\text{SeO}_2} \text{(1,1-dimethyl-1-silacyclohexa-2,5-dien-4-one)} \qquad (20)$$

One of the reactions that has been particularly useful in the generation of heterocyclic derivatives of Groups III and V is the exchange reaction with a tin heterocycle (normally derived by the direct reaction of a tin dihydride with a 1,3 diyne). The successful generation of Group V and III analogs of benzene relies on this synthetic route (*11–13*, *15*):

$$\text{(}R_2Sn\text{ heterocycle)} + AsCl_3 \xrightarrow[(11)]{} \text{(}AsCl\text{ heterocycle)} + R_2SnCl_2 \qquad (21)$$

The limitations of the exchange process have not yet been defined, and the overall approach appears to be limited by the availability of unsaturated diyne or diene for the formation of the tin heterocycle. Although exchange reactions between Group IV organometallics are known, preliminary attempts to exchange an R_2Sn grouping in a heterocyclic derivative with R_2SiCl_2 have not yet been successful (see Section IV, E).

The known heterocyclic derivatives of six- and seven-membered rings are listed in Table VII.

3. *Reactions of Heterocyclic Dienes*

a. *Reactions at the Heteroatom.* Standard organometallic reactions may be carried out at the heteroatom without ring cleavage. These reactions closely parallel those reported for heterocycloalkenes in Section II, A, 3.

The reduction of a metal–halogen bond with $LiAlH_4$ has been reported for both five- and six-membered heterocyclic diene systems of

TABLE VII

HETEROCYCLIC CYCLOHEXADIENES AND CYCLOHEPTADIENES

Ring substituents	M	R_1	R_2	%	m.p.	Method	References
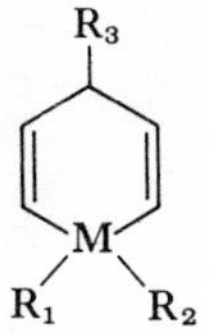 None	B	Ph	—	50	—	A	*15*
	Sn	Bu	Bu	40	—	B	*15*
	As	Cl	—	—	—	A	*11*
	Sb	Cl	—	—	115°–117°	A	*12*
$R_3 = O$	Bi	Cl	—	75	144°–145°	A	*13*
None	Si	Me	Me	~55	[a]	C	*357*
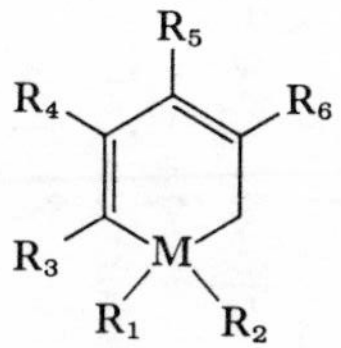	Si	Cl	Cl	78	[a]	D	*31*
	Si	Me	Me	—	—	E[b]	*21*
	Si	Ph	Ph	—	[a]	E	*31*
$R_3 = R_4 = R_5 = R_6 = Ph$	Si	Me	Cl	50–73	182°–183°	F	*146*
		Me	H	—	141°–142°	G	*146*
		Me	Me	87	161°–162°	F	*243*
				—	158°–159°	E	*146*
		Me	Ph	—	139°–140°	E	*146*
$R_5 = Cl; R_3 = R_4 = R_6 = H$	Ge	Et	Et	50	115°/15 mm	H	*313*
$R_5 = Cl; R_3 = R_4 = H; R_6 = CH_3$	Ge	Et	Et	10	[a]	H	*313*
$R_5 = Cl; R_3 = H; R_4 = R_6 = CH_3$	Ge	Et	Et	38	[a]	H	*313*

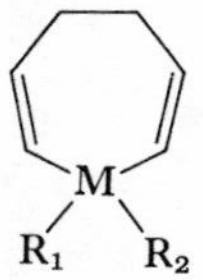	None	B	Cl	—	—	—	—	*315*
		B	Ph	—	—	—	—	*217*
		Ge	Bu	Bu	2	100°–105°/0.5 mm	B	*278*
		Sn	Ph	Ph	12	147°–150°/0.3 mm	B	*278*
			Me	Me	16	30°–32°/0.2 mm	B	*278*
			Et	Et	20	74°–76°/5 mm	B	*278*
			Pr	Pr	17	64°–65°/0.02 mm	B	*278*
			Bu	Bu	28	90°–92°/0.08 mm	B	*278*
		As	Ph	—	17	68°–70°/0.01 mm	B	*238*
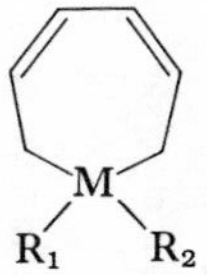	None	Ge	Cl	Cl	30–40	—	I	*270*
		Ge	Me	Me	—	—	E	*270*

Key to method: (A) exchange reaction with tin heterocycle; (B) hydride addition to diyne; (C) oxidation of saturated ketone; (D) bromination–dehydrobromination by pyrolysis; (E) reaction of RLi or ArLi with exocyclic M–Cl of preformed diene; (F) ring expansion reaction from cyclopentadiene derivative; (G) $LiAlH_4$ reduction of exocyclic M–Cl; (H) carbene insertion into five-membered cyclopentadiene derivative. Doering–Hoffman method; (I) 1,6-cycloaddition of $GeCl_2$.

[a] Separated by vapor phase chromatography.

[b] Modification.

both silicon and germanium (*92*, *145*, *146*, *301*). Normal reactions of metal–halogen bonds with organolithium reagents in heterocyclic hexadienes is possible (*93*, *146*). Reaction of a mixture of 1,1-dichlorosilacyclo-2,4-hexadiene, 1,1-dichlorosilacyclohex-2-ene, and 1,1-dichlorosilacyclohex-3-ene in a 7:1:1 ratio with PhMgBr gave the corresponding diphenyl derivatives in an approximate 1:1:1 ratio (*31*). It is not clear whether ring cleavage or reduction of diene occurs. An example of a replacement reaction of an exocyclic M—C bond in the presence of an organolithium reagent has also appeared (*368*):

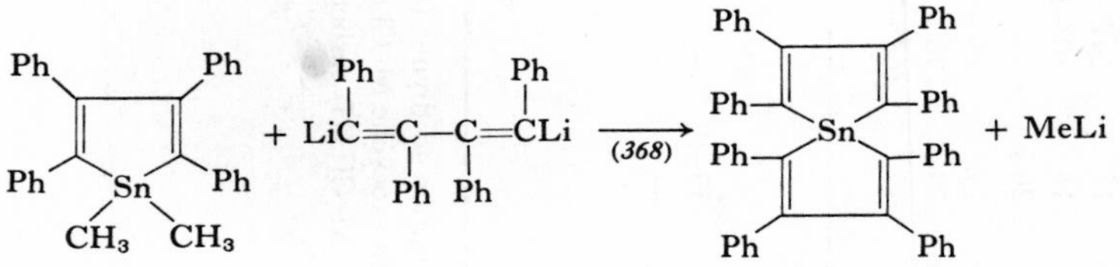

Ph Ph Ph Ph Sn CH$_3$ CH$_3$ + LiC(Ph)=C(Ph)—C(Ph)=C(Ph)Li $\xrightarrow{(368)}$ spiro-Sn(C$_4$Ph$_4$)$_2$ + MeLi (22)

The Si—Cl bond in pentaphenylsilacyclopentadiene can be converted to $SiNMe_2$ or to the siloxane successfully (*301*).

The chemical reactivity of several heterocyclic derivatives is closely associated with a reactivity one might expect for open-chain unsaturated derivatives. The boron heterocycles are sensitive to oxygen or hydrogen peroxide (*111*, *169*) and so apparently is 1,1,2,5-tetraphenylsilacyclopentadiene (*17*, *246*). These reactions probably lead to ring cleavage but the products have not been characterized. Vinyl–metal bonds are frequently cleaved by electrophilic reagents and the heterocycles are no exception.

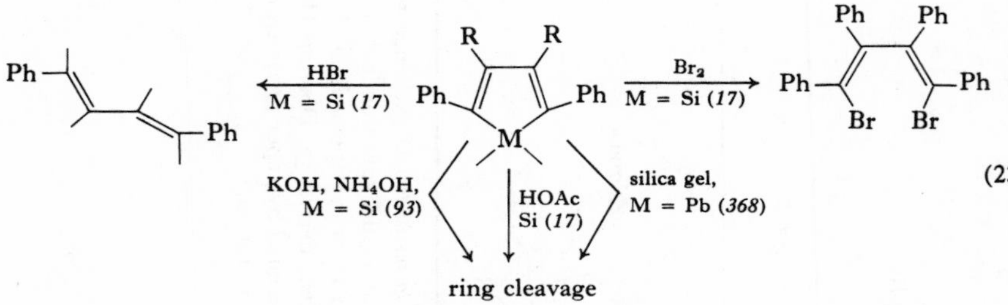

(23)

HBr, M = Si (*17*); Br$_2$, M = Si (*17*); KOH, NH$_4$OH, M = Si (*93*); HOAc Si (*17*); silica gel, M = Pb (*368*); ring cleavage

The structure of the product dienes was deduced from their NMR spectra. Both the boron and lead derivatives decompose when heated (*111*, *368*).

Lower-valent organometallic derivatives of Groups V and VI undergo oxidative addition reactions to produce the higher-valent state. These reactions can also be effected with heterocyclic compounds.

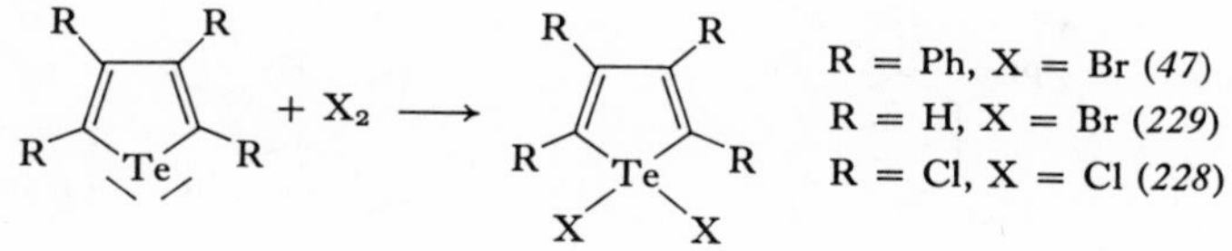

R = Ph, X = Br (*47*)
R = H, X = Br (*229*) (24)
R = Cl, X = Cl (*228*)

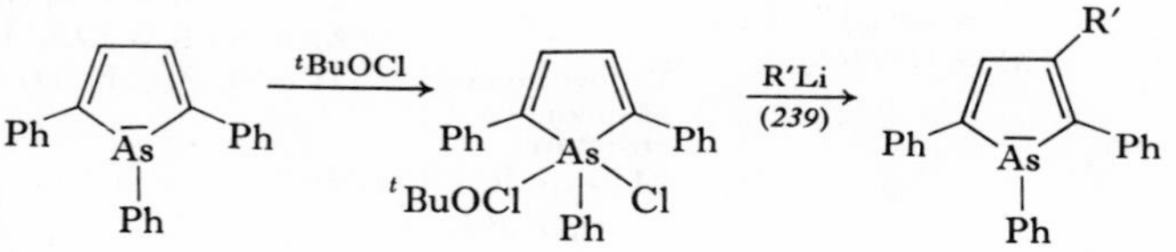

(25)

This latter ring-substitution reaction accompanying the reduction to trivalent arsenic is a unique reaction. The cleavage of 1-arylarsoles by alkali metals and subsequent reactions of the anion have been described (*46*, *239*). The reaction provides a convenient method of converting an exocyclic phenyl group to an alkyl substituent:

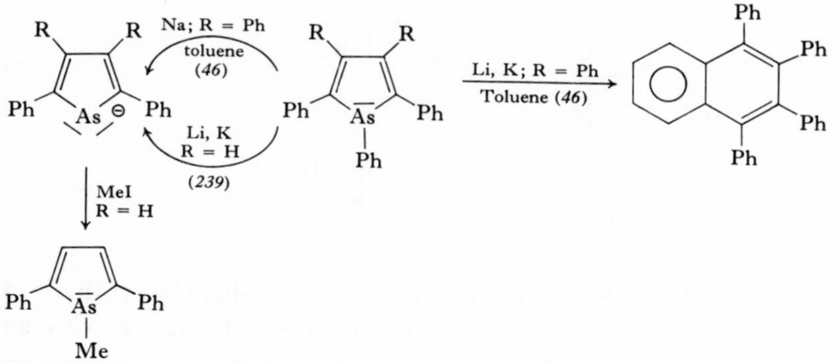

(26)

b. *Reactions of the Multiple Bonds.* Many of the reactions of the heterocyclic dienes involve Diels–Alder type additions; these are summarized in Schemes 3 and 4.

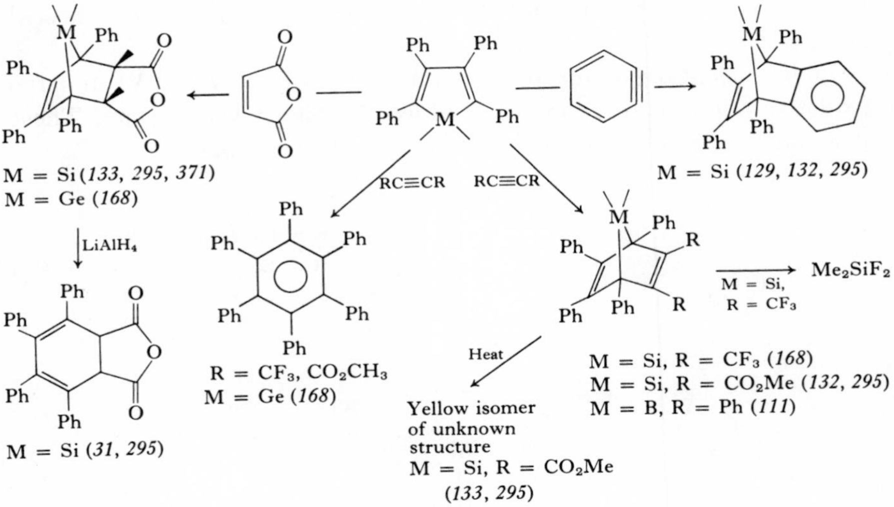

M = Si (*133, 295, 371*)
M = Ge (*168*)

M = Si (*31, 295*)

R = CF_3, CO_2CH_3
M = Ge (*168*)

M = Si (*129, 132, 295*)

$\xrightarrow[M\,=\,Si,\ R\,=\,CF_3]{}$ Me_2SiF_2

M = Si, R = CF_3 (*168*)
M = Si, R = CO_2Me (*132, 295*)
M = B, R = Ph (*111*)

Heat

Yellow isomer of unknown structure
M = Si, R = CO_2Me (*133, 295*)

Scheme 3

Reaction studies have been extended to six-membered cyclic dienes to generate bicyclooctane derivatives:

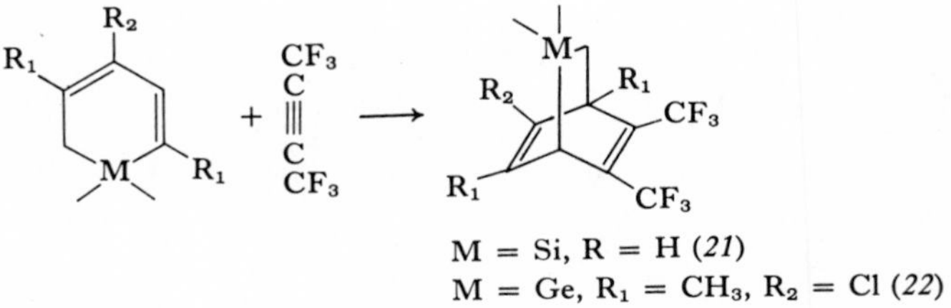

$$\text{(27)}$$

M = Si, R = H (*21*)
M = Ge, R_1 = CH_3, R_2 = Cl (*22*)

Analogs of norbornadiene with heteroatom bridges may be synthesized in theory by reaction of a heterocyclopentadiene derivative with an acetylene (Schemes 3 and 4). These analogs of norbornadiene may be sources of "silene" or "germene" intermediates. The reaction of heterocyclopentadienes with acetylenes yields isolable adducts if the heteroatom is Si but not if the heteroatom is Ge. The intermediate "dimethylgermene" may be trapped by excess alkyne reagent (*168*):

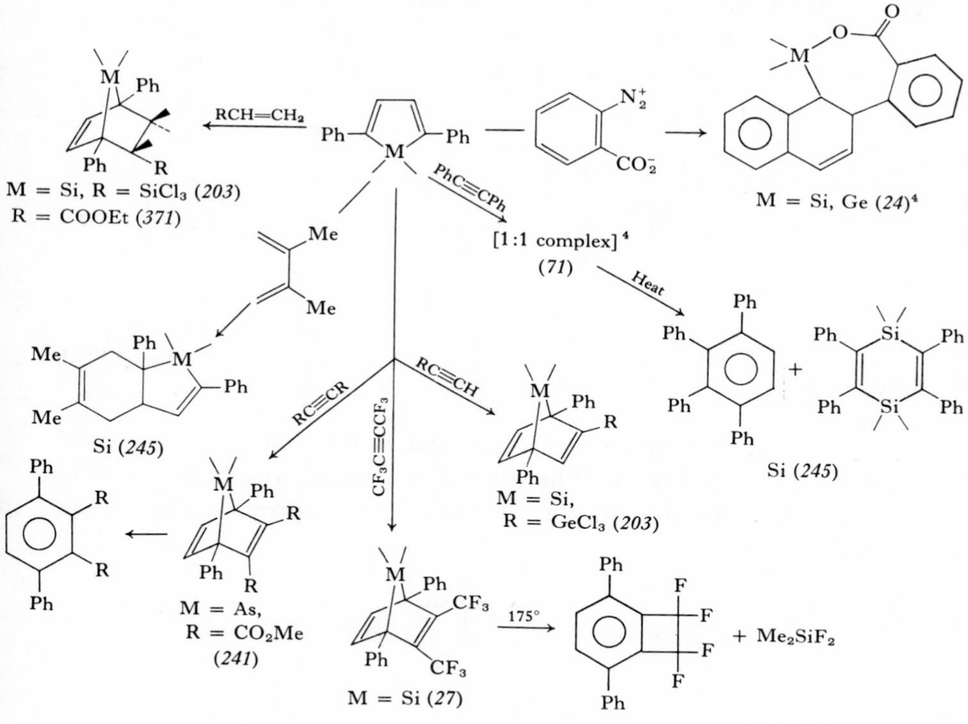

M = Si, R = $SiCl_3$ (*203*)
R = COOEt (*371*)

M = Si, Ge (*24*)[4]

[1:1 complex][4]
(*71*)

Si (*245*)

Si (*245*)

M = Si,
R = $GeCl_3$ (*203*)

M = As,
R = CO_2Me
(*241*)

M = Si (*27*)

$+ Me_2SiF_2$

Scheme 4

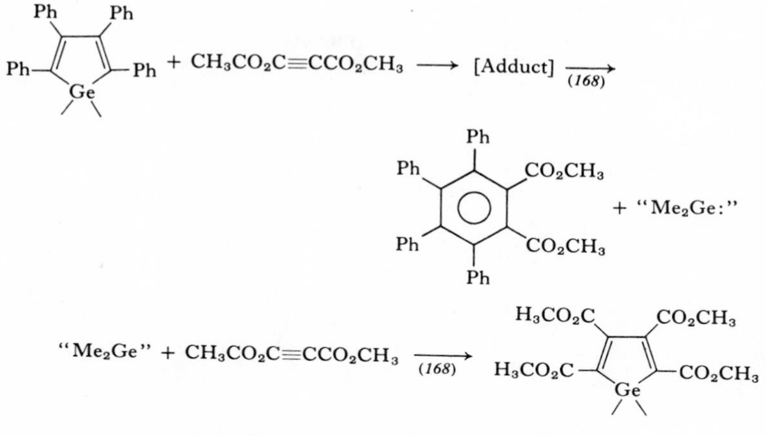

$$\text{Ph}_4\text{C}_4\text{GeMe}_2 + CH_3CO_2C{\equiv}CCO_2CH_3 \longrightarrow [\text{Adduct}] \xrightarrow{(168)} \text{Ph}_4C_6(CO_2CH_3)_2 + \text{“}Me_2Ge\text{:”} \tag{28}$$

$$\text{“}Me_2Ge\text{”} + CH_3CO_2C{\equiv}CCO_2CH_3 \xrightarrow{(168)} (H_3CO_2C)_4C_4GeMe_2$$

[4] Structure confirmed by solid-state study (*24, 71*).

The same general approach has been utilized to generate an "$M{=}CH_2$" species from the two-membered bridge system. The presence of $M{=}CH_2$ was inferred from the isolated product, 1,3-dimetallocyclobutane.

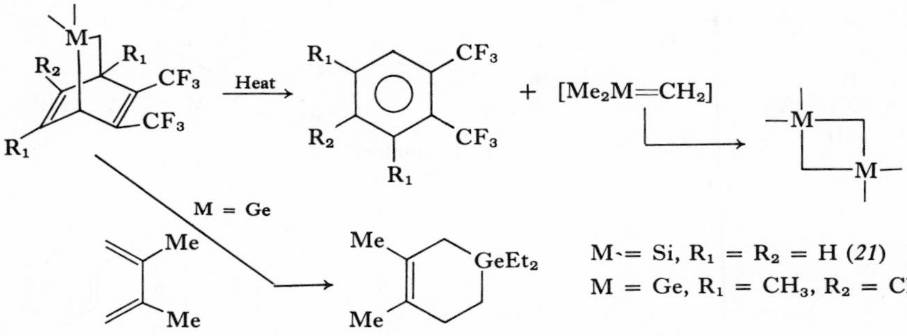

$M = Si$, $R_1 = R_2 = H$ (*21*)
$M = Ge$, $R_1 = CH_3$, $R_2 = Cl$ (*22*)

(29)

A few reactions have been reported that involve one of the multiple bonds of the diene system. The silacyclopentadiene system reacts with 1 mole of bromine to generate a mixture of brominated cyclopentene derivatives:

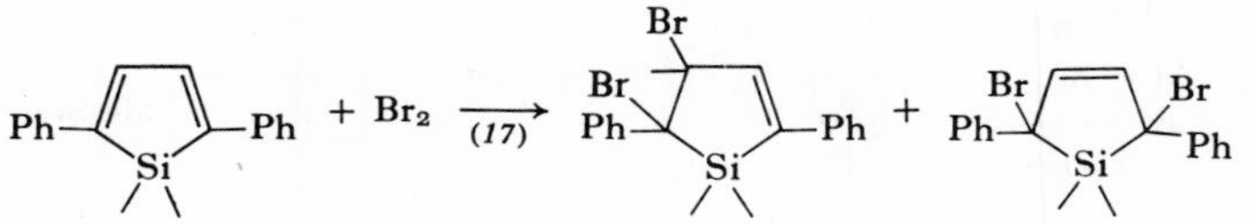

(30)

Unsaturated heterocyclic metalloid derivatives should be good candidates for photochemical studies. As yet, this area has been little explored. Preliminary reports of the photochemical reactivity of two silacyclopentadiene derivatives have appeared:

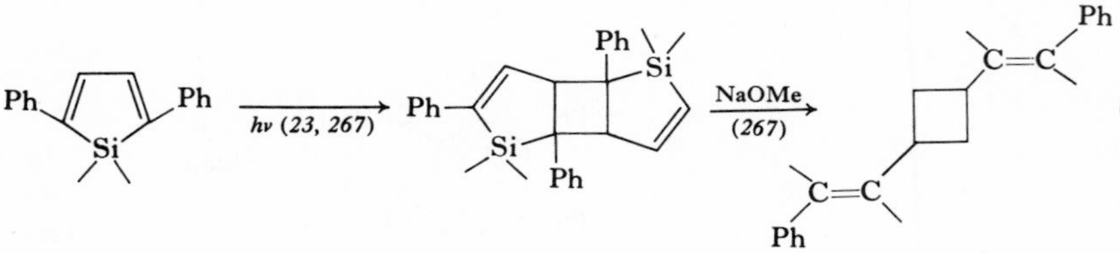

(31)

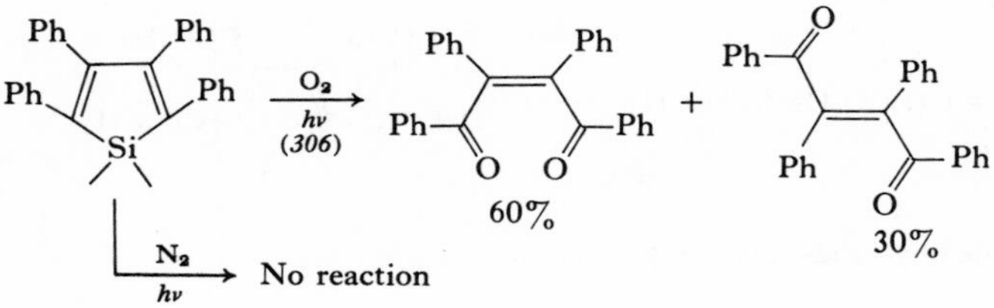

(32)

Dimer formation was also reported for the photolysis of the germanium heterocycle [Eq. (31)] (*23*) but not for tellurophene. Lack of reactivity in this latter compound was ascribed to aromaticity of the heterocycle. Tellurophene also undergoes electrophilic substitution at the α position in formylation, acetylation, and trifluoroacetylation reactions (*121*).

Photolysis of tin heterocycles will probably result in reaction occurring at the metal site. This type of reaction has been exhibited by 1,1-diphenyl-4,5-dihydrostannepin:

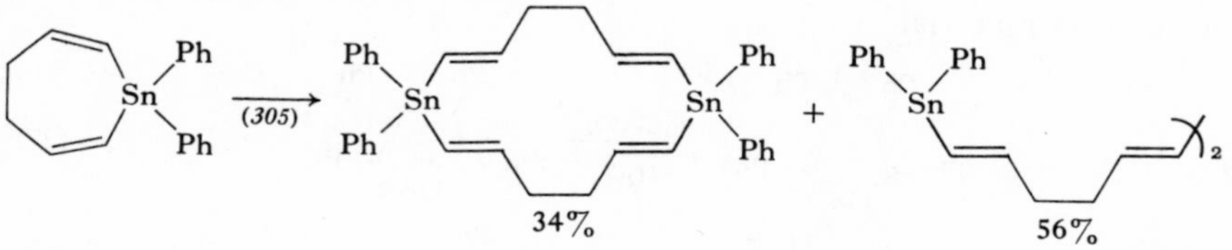

(305)

34%

56%

(33)

Some heterocyclic cyclopentadiene derivatives form π complexes with metal carbonyls (*47*, *55*, *373*, *384*):

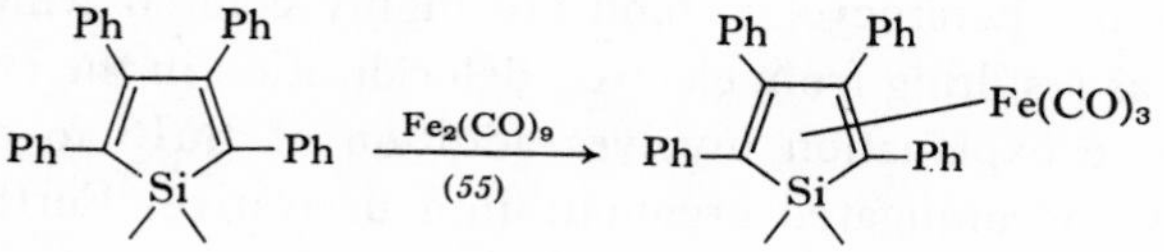

$Fe_2(CO)_9$ (*55*) $Fe(CO)_3$ (34)

C. Heterocyclic Trienes

1. *Synthesis of Trienes*

To date, metallocyclotrienes are, without exception, generated from the appropriate diene. Derivatives containing a Group III or IV heteroatom have been obtained in solution and their actual existence is speculative at this time. The derivatives of Si and B are generated by reaction of the cyclic diene with butyllithium:

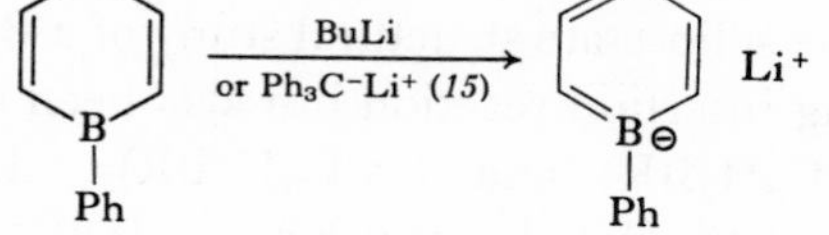

BuLi or $Ph_3C^-Li^+$ (*15*) Li^+ (35)

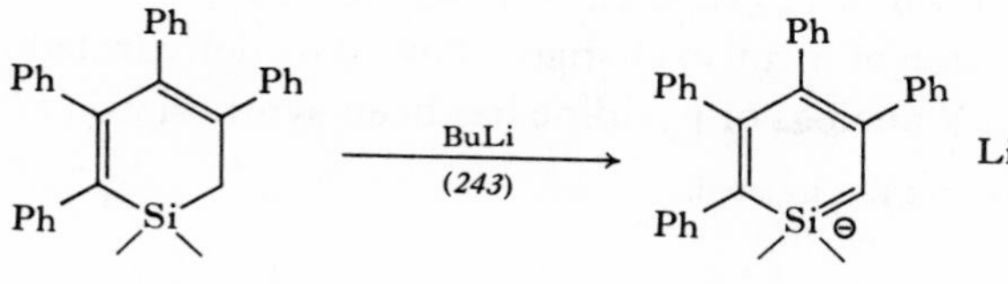

BuLi (*243*) Li^+ (36)

Deuterolysis of the boron derivative with acetic acid-d_1 gives *cis*-1,2-pentadiene which has incorporated three deuterium atoms. Thus, chemical evidence appears to support a borabenzene structure. If delocalization of the negative charge generated at the position α to the Si atom into the *d* orbitals on the silicon occurs, an "aromatic system" may result. The spectral properties of this system are discussed in a later section. Formation of germacyclopentadiene anion may also involve formation of a 6π-electron system. Protonation of the anion generates the starting material.

Ph Ph Ph Ph Ge Ph H $\underset{H_2O}{\overset{BuLi\ (92)}{\rightleftarrows}}$ Ph Ph Ph Ph Ge$^{\ominus}$ Ph (37)

Although the corresponding deprotonation of the Si analog occurs, reaction of the anion with water does not regenerate the starting hydride (*92*). The solutions of the heterocyclic anion are highly colored which could be interpreted as resulting from electron delocalization in the cyclic system. An alternative explanation involves addition of BuLi to the multiple bond to give a conjugated organolithium derivative. Further study is required before the alternatives can be resolved.

A novel approach to derivatives of a boron heterocycle isoelectronic to benzene was reported in the reaction of phenylboron dihalide with bis-(cyclopentadienyl)cobalt followed by reaction with tin tetrabromide:

Co $\xrightarrow[\text{Toluene PhBX}_2\ (161)]{-50°C}$ Co H BPhX $\xrightarrow{SnBr_4}$ Co$^+$ B—Ph X$^-$ (38)

The presence of a six-membered ring containing a boron heteroatom was confirmed by the solid-state structural study of a B—OMe derivative (*171*). The foregoing insertion reaction has also been shown to generate neutral $Co(C_5H_5)(C_5H_5BR)$ and $Co(C_5H_5BR)_2$, depending on the stoichiometric ratio of $(C_5H_5)_2$ Co and $RBCl_2$ (*160*).

In a combination of metal exchange followed by dehydrohalogenation, a series of Group V analogs of pyridine has been synthesized (*11–13*, *379*):[5]

[5] Bismuth derivative not isolated.

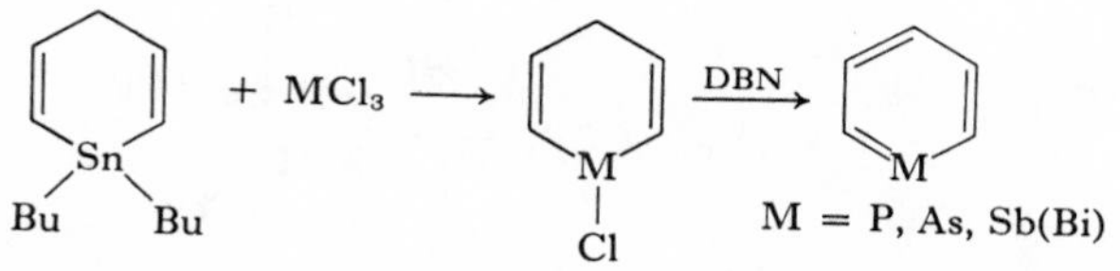

(39)

$M = P, As, Sb(Bi)$

The spectral properties of the metallobenzenes will be discussed in Section VI.

A general synthesis of substituted arsabenzenes has been formulated from the arsacyclopentadiene derivative. The method requires generation of a carbene in the α position on the metal side chain followed by insertion of the carbene into an As—C ring bond (*240*):

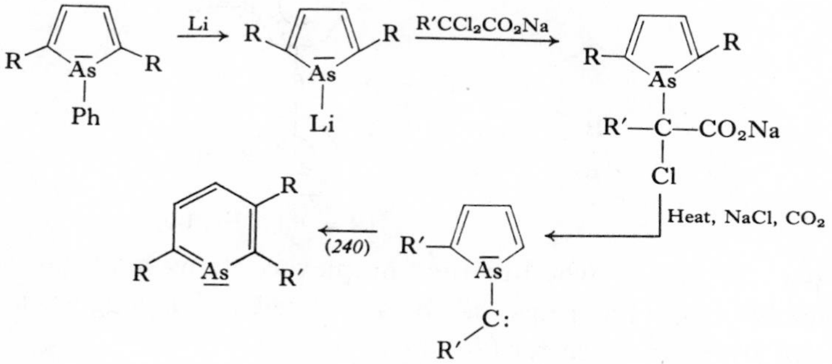

(40)

2. *Reactions of Trienes*

The silacyclohexadiene anion system can be characterized by reaction with Me_3SiCl or a ring-coupling reaction can be effected in presence of $PhICl_2$:

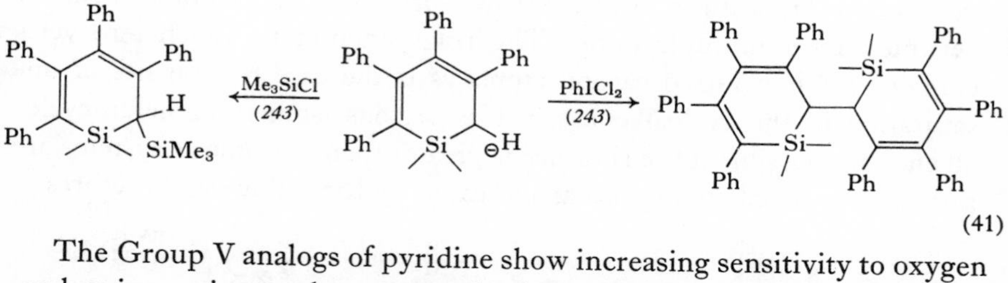

(41)

The Group V analogs of pyridine show increasing sensitivity to oxygen and an increasing tendency toward polymerization from phosphabenzene to bismabenzene (*11–13*). The bismabenzene derivative could not be isolated, but its presence was inferred by trapping the product with $CF_3{\equiv}CF_3$:

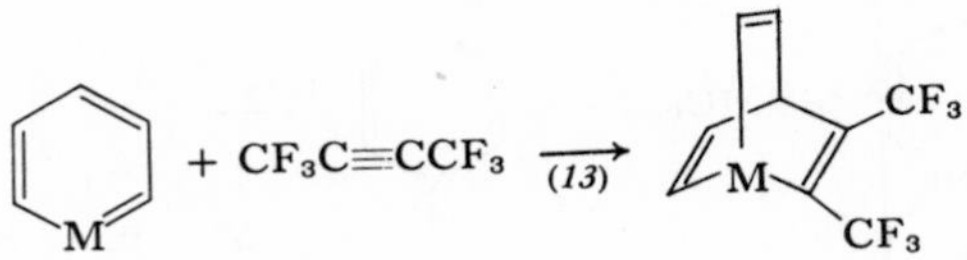

$$\text{(M-benzene)} + CF_3C{\equiv}CCF_3 \xrightarrow{(13)} \text{(M-barrelene with } CF_3, CF_3\text{)} \quad (42)$$

Ring-substituted arsabenzenes also will react with alkynes, RC≡CR (R = CF_3, CN, CO_2Et) to form arsabarrelene derivatives [Eq. (42)] (*237*).

Sandwich compounds have been reported from the reaction of borabenzene anions with iron(II) chloride.

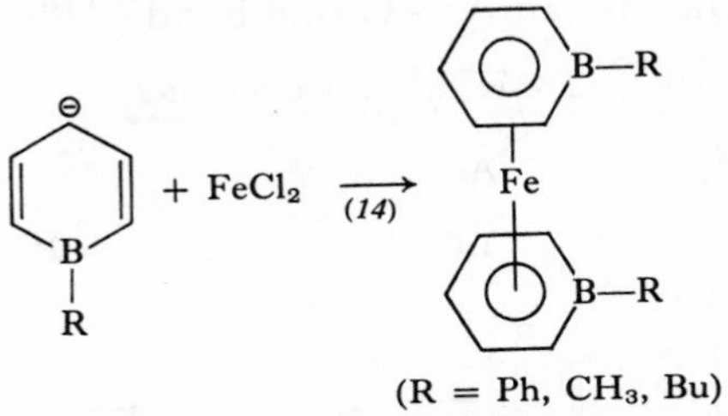

$$\text{(R–B borabenzene anion)} + FeCl_2 \xrightarrow{(14)} (\text{R–B}C_5H_5)_2Fe \quad (43)$$

(R = Ph, CH_3, Bu)

It is not yet known whether the complex contains a planar or bent borabenzene ring. The rings may be acetylated which suggests a certain degree of aromatic character (*14*).

III

BENZHETEROCYCLES

The benzo derivatives are similar to the heterocycloalkenes discussed in Section II, with a benzene ring fused to the position formerly occupied by a multiple bond. The ring junction in benzheterocycles contains sp^2 hybridized carbon atoms as is the case for the site of unsaturation in the cycloalkenes. Just as various isomers of heterocycloalkenes are possible if the ring size is greater than 4, isomeric systems are possible for benzheterocycles as indicated by the following structures:

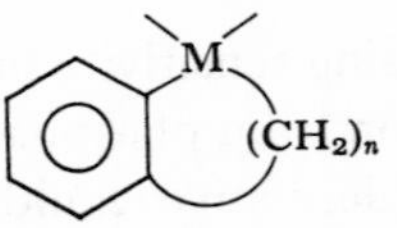

(known: $n = 1 \rightarrow 5$)
(VII)

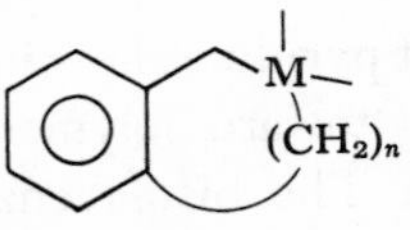

(known: $n = 1 \rightarrow 3$)
(VIII)

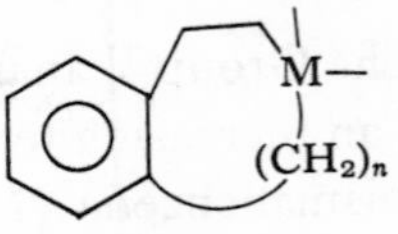

(known: $n = 2$)
(IX)

The nomenclature of the benzheterocycles is closely related to that described for the heterocycloalkenes. Examples of the nomenclature for benzheterocycles are given in Table VIII. As described in Section II, a base name could be generated for a heterocycloalkene by combination of an appropriate prefix that contained the identity of the heteroatom and a suffix that was specific for the ring size. For benzheterocycles the base name of the heterocyclene is preceded by the prefix benz or benzo. The position of the heteroatom is determined by the lowest number generated if the position adjacent (*ortho*) to the aromatic ring is labeled the 1-position as indicated in structures (VIIIa) and (VIIIb):

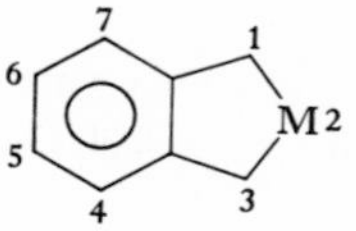

(VIIIa) 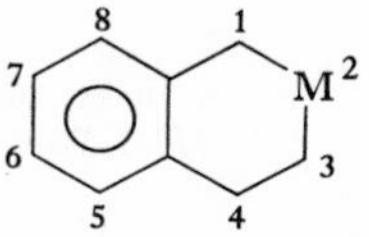 (VIIIb)

Saturation at position 1 is designated by 1*H*. Systems with unsaturation less than that corresponding to the maximum number of noncumulative double bonds require additional prefixes, e.g. dihydro, tetrahydro.

A few metalloid derivatives are named according to the parent heterocycle, e.g., (VII), $n = 2$, are metalloindanes and (VII), $n = 3$, are metallochroman derivatives or metallonaphthalene derivatives (hydro prefixes are required to denote saturation). Thus, compounds of structural type (VIII) are metal derivatives of isoindane or isochroman.

The methods used to generate benzheterocycles are closely related to those already outlined for the cycloalkene derivatives. Direct methods involve formation of both ring bonds by reaction of an appropriate aromatic derivative with a difunctional organometalloid. In the second approach, an aromatic derivative with a heteroatom in a side chain is prepared in a separate step, followed by the ring-closure reaction. This latter method usually involves electrophilic substitution on the aromatic nucleus, and accounts for the frequency of the two-step synthetic approach which is less commonly employed in synthesis of cycloalkenes.

A. Benzocyclobutene Derivatives

The only metalloid derivative of benzocyclobutene contains a Si heteroatom. This is not surprising since the only derivative of a four-membered heterocyclobutene is the silacyclobutene discussed in Section

TABLE VIII

NOMENCLATURE OF BENZHETEROCYCLES

System	Base name unsaturated derivative[a]	Prefix for saturated derivative
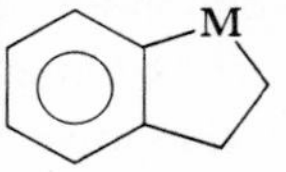	1*H*-1-Benzometallole[b]	2,3-Dihydro
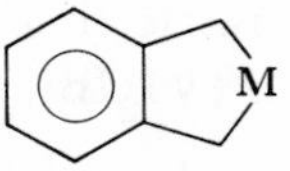	1*H*-2-Benzometallole[c]	2,3-Dihydro
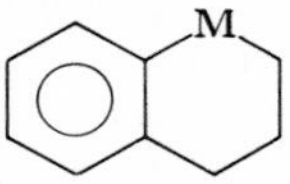	1-Benzometallin[d] 2*H*-1-Benzometallin	1,2,3,4-Tetrahydro 3,4-Dihydro
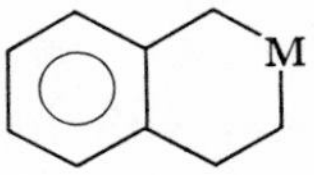	2-Benzometallin 1*H*-2-Benzometallin	1,2,3,4-Tetrahydro 3,4-Dihydro
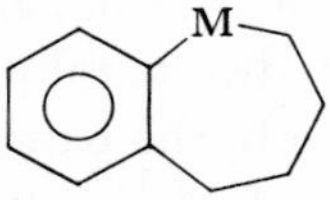	1*H*-1-Benzometallepin	2,3,4,5-Tetrahydro
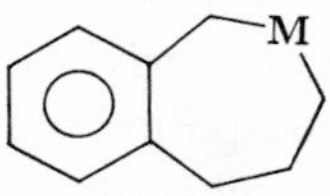	1*H*-2-Benzometallepin	2,3,4,5-Tetrahydro
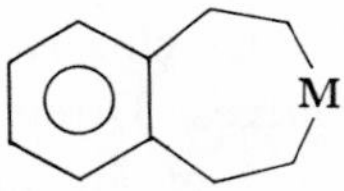	1*H*-3-Benzometallepin	2,3,4,5-Tetrahydro
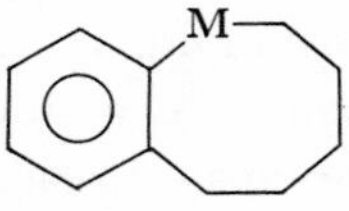	1-Benzometallocin 2*H*-1-Benzometallocin	1,2,3,4,5,6-Hexahydro 3,4,5,6-Tetrahydro
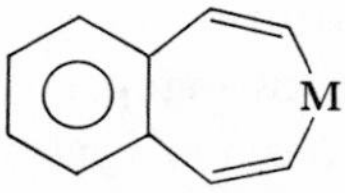	3*H*-3-Benzometallepin	—

[a] Unsaturation at carbon atoms only. Thus maximum number of multiple bonds is not possible and 1*H* prefixes are required. Benzo may be replaced by benz.

[b] Silaindane is preferred for Si; arsindoline is preferred for As.

[c] 2-Silaindane preferred for Si; isoarsindoline preferred for As.

[d] 1-Silanaphthalene preferred for Si; Arsinoline preferred for As.

II. The benzocyclobutene derivative can be prepared by direct reaction of *o*-bromobenzyl bromide with R_2SiCl_2 in the presence of magnesium (*108*, *129*). The small membered ring system has also been synthesized by ring closure with formation of one ring bond:

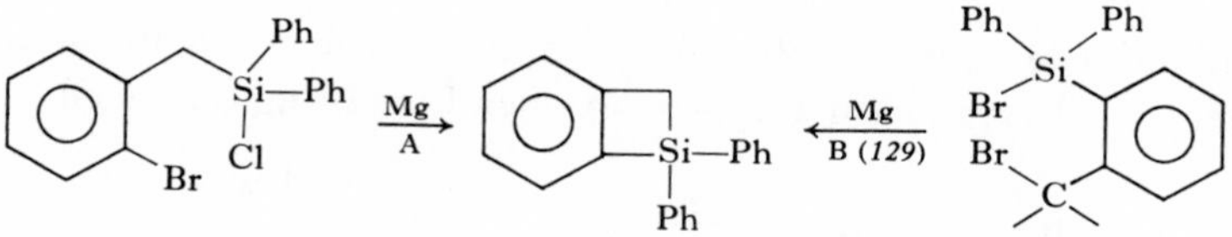

(44)

Method B is similar to the ring-closure reaction by Wurtz coupling, as indicated in Eq. (5) for five- and six-membered heterocycloalkenes.

B. Derivatives of Benzocyclopentene and Benzocyclohexene

1. *Metalloindanes and Metallochromans*

There are few examples of direct one-step reactions for generation of benzheterocycles with the heteroatom bonded directly to the aromatic ring. The two known methods are unique to the heteroatom. Aralkyl derivatives of boron will react at high temperatures to eliminate an alkene and cyclize to form a five-membered ring:

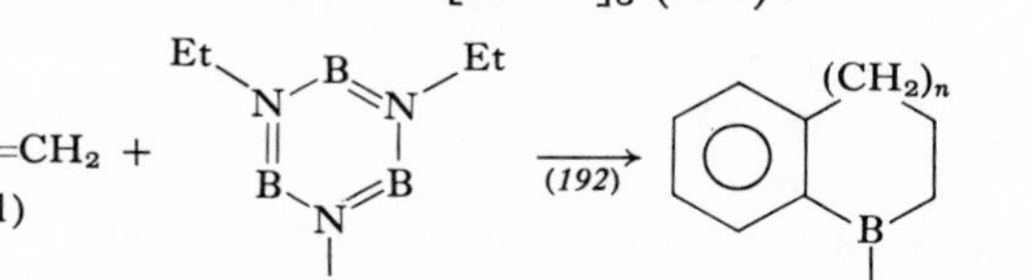

$$i\text{-}Bu_3B + Ph(Me)C{=}CH_2 \xrightarrow[(192)]{188^\circ} \text{(1-}Bu^i\text{-2-methylboraindane)} \quad (45)$$

Cyclic organoboron derivatives are also obtained when phenyl-substituted alkenes are heated with $[EtNB]_3$ (*192*):

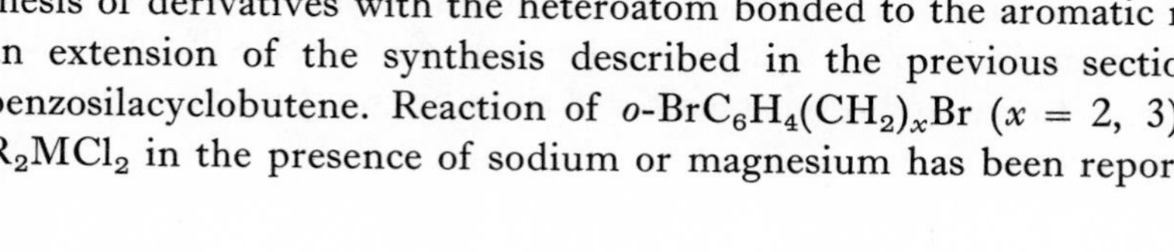

$$Ph(CH_2)_nCH{=}CH_2 + [EtNB]_3 \xrightarrow{(192)} \text{benzoborinane with } (CH_2)_n \text{ and } B\text{-}CH_2CH_2(CH_2)_nPh \quad (n = 0, 1) \quad (46)$$

The only other one-step route that has been employed for the synthesis of derivatives with the heteroatom bonded to the aromatic ring is an extension of the synthesis described in the previous section for benzosilacyclobutene. Reaction of $o\text{-}BrC_6H_4(CH_2)_xBr$ ($x = 2, 3$) with R_2MCl_2 in the presence of sodium or magnesium has been reported to yield derivatives of (VII) [M = Si, x = 2 (*108, 137, 138*); M = Si, x = 3 (*136–138*); M = Sn, x = 3 (*138*)].

Although the remaining routes parallel the direct methods described in the foregoing, isolation of an appropriate organometal reagent prior to the final ring-closure step is required. Pyrolysis of aralkylboron derivatives at temperatures greater than 200°C will result in the formation of cyclic derivatives if a five- or six-membered ring can be generated:

$$[Ph(CH_2)_nCH_2]_3B \xrightarrow[(192,\ 195)]{>200°C} \text{(benzo-fused ring: } (CH_2)_n \text{, B–}(CH_2)_nCH_2Ph) \quad (n = 1, 2) \tag{47}$$

Silicon heterocycles with different ring sizes have been synthesized by variations of the organometallic Wurtz coupling reaction. The method requires synthesis of intermediate organosilicon derivatives that can involve several steps.

$$o\text{-}ClC_6H_4(CH_2)_nSi(H)R_1R_2 \xrightarrow[(137)]{Na} \text{(benzo-fused ring: } (CH_2)_n \text{, }SiR_1R_2) \quad (n = 2 \rightarrow 5) \tag{48}$$

Not unexpectedly, yields of ring-cyclized product decrease sharply in going from the seven- to the eight-membered ring system. Further studies of the coupling reaction are reported for both five- and six-membered ring systems (*108, 138, 139*). Ring closure may also be accompanied by elimination of an organic group from the heteroatom:

$$o\text{-}ClC_6H_4(CH_2)_3SiPh_3 \xrightarrow[\text{Toluene } (139)]{Na} \text{(benzo-fused six-membered ring, }SiPh_2) \tag{49}$$

This type of silicon–carbon cleavage during ring formation is not without precedent. Spiro derivatives have been observed from this elimination reaction.

The standard organic reaction of aluminum chloride-catalyzed intramolecular cyclization by attack on a benzene nucleus has been used to generate both six- and seven-membered rings but not the five-membered ring of silicon.

$$Ph_3Si(CH_2)_n\overset{O}{\overset{\|}{C}}OH \xrightarrow[\text{Nitrobenzene } (361)]{AlCl_3} \begin{cases} \xrightarrow{n=1} Ph_3SiOH \\ \xrightarrow{n=2,3} \text{[benzo-fused ring: } Ph_2Si-(CH_2)_n-C(=O)\text{]} \end{cases} \tag{50}$$

A similar reaction of an arsenic derivative failed (*234*). It should be noted that many Ar—M bonds (including Si) are cleaved in the presence of $AlCl_3$ and, therefore, the reaction method is somewhat restrictive. Ring closure of the alcohols, $Ph_3Si(CH_2)_nCH_2OH(n = 2)$, in the presence of a Lewis acid in nitrobenzene was unsuccessful due to cleavage of Si—Ph bonds (*361*). Intramolecular cyclization involving electrophilic attack on a benzene nucleus has been used more successfully for formation of heterocycles of arsenic:

$$Ph(CH_2)_n\overset{O}{\overset{\|}{As}}(R)—OH \xrightarrow[2.\ SO_2\ (164,\ 177)]{1.\ H_2SO_4} \text{[benzo-fused ring: } As(R)-(CH_2)_n\text{]} \xleftarrow[(57,\ 333)]{AlCl_3} Ph(CH_2)_nAsMeCl \quad (n = 2, 3) \tag{51}$$

The arsenic acids have no structural parallel in the heavier elements of Group IV (Si, Ge, Sn, Pb) but appear to function as a carboxylic acid would in ring-closure reactions. The contrast in behavior of an As—Cl bond toward electrophilic attack on a benzene ring versus Si—Cl reactivity should be noted. No clear-cut examples of a Friedel-Crafts reaction involving a silicon–chlorine bond have been reported (*107*).

The only exchange reaction reported for synthesis of a derivative with the heteroatom bonded to the benzene ring is for an aluminum heterocycle:

$$\text{[3-methyl-1-propylboraindane: } B(Pr), CH_3\text{]} \xrightarrow[(191)]{AlPr_3} \left[\text{3-methyl-1-propylaluminaindane: } Al(Pr), CH_3\right]_2 \tag{52}$$

2. *Isomers of Metalloindanes and Metallochromans*

The isometalloindanes of arsenic and silicon and isometallochroman derivatives of silicon may be synthesized in a one-step process (Scheme 5).

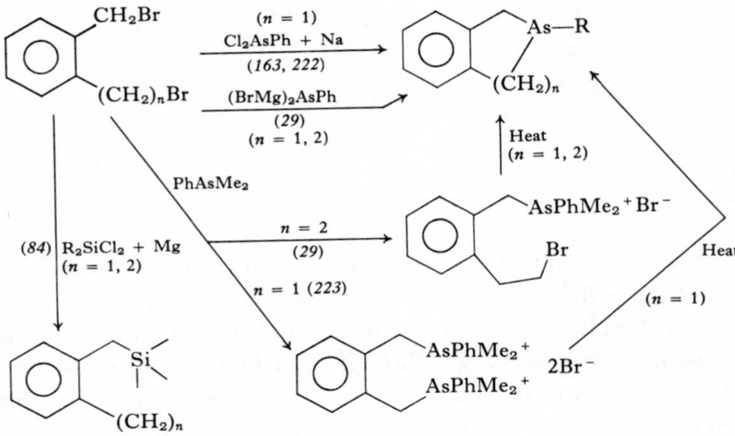

Scheme 5

A possible route to benzheterocycles containing a tin heteroatom is suggested by the observation that benzylbromide reacts with elemental tin to form $(PhCH_2)_2SnBr_2$ (*318*). It is possible that reaction of tin with *o*-xylene dibromide could generate a 1*H*-2-benzstannole. Although the reaction of tin with *o*-$(BrCH_2)_2C_6H_4$ is exothermic, only high-melting solids and gums were obtained and no isolable product consistent with a monomer formulation was observed (*75*).

The remaining derivatives are prepared by indirect routes and all contain silicon. These compounds require synthesis of an aromatic derivative with an Si atom in a side chain before the ring-closure reaction takes place. Ring closure requires the presence of an $AlCl_3$ catalyst (*84, 268*).

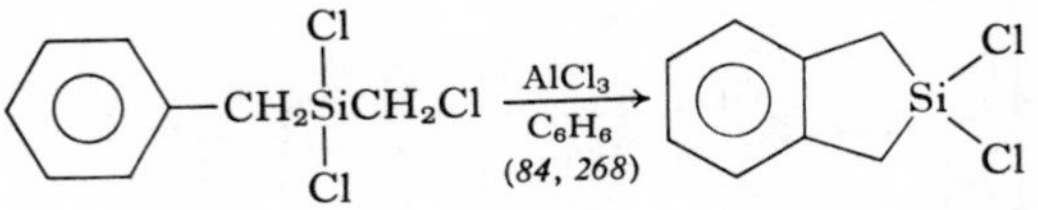

$$C_6H_5CH_2SiCl_2CH_2Cl \xrightarrow[C_6H_6\ (84,\ 268)]{AlCl_3} \text{2,2-dichloro-2-silaindan} \qquad (53)$$

$$C_6H_5CH_2Si(Cl)(R)CH{=}CH_2 \xrightarrow[C_6H_6\ (7,\ 8,\ 84,\ 87,\ 88,\ 341,\ 369)]{AlCl_3} \text{2-R-2-chloro-2-silatetralin} \qquad (54)$$

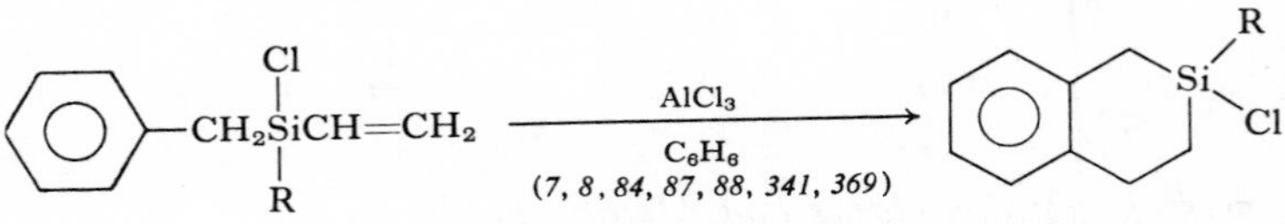

A variety of derivatives of (VIII) with various substituents on the aromatic ring have also been reported.

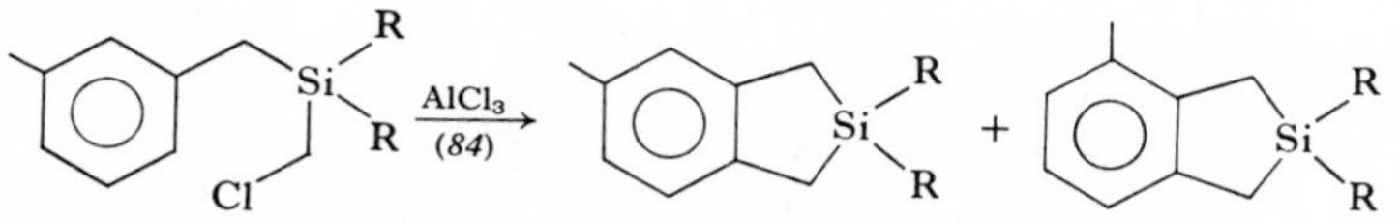

(55)

The general applicability of this route to other heteroatom systems has not been tested.

C. Benzheterocycles Containing a Seven- or Eight-Membered Ring

Three possible isomeric forms exist for benzheterocycles containing seven- or eight-membered rings, as depicted by structures (VII), (VIII), and (IX). Most of these heterocycles contain an Si heteroatom, but derivatives of (IX) with several different heteroatoms are known. Only one benzheterocycle with an eight-membered ring has been reported [(IX), M = Si].

The majority of the derivatives with benzene fused to a seven-membered ring are of structural type (IX). Since these may be viewed as symmetrical derivatives, the most convenient synthetic routes are those of the direct reaction of an appropriately substituted aryl derivative with a bifunctional organometallic.

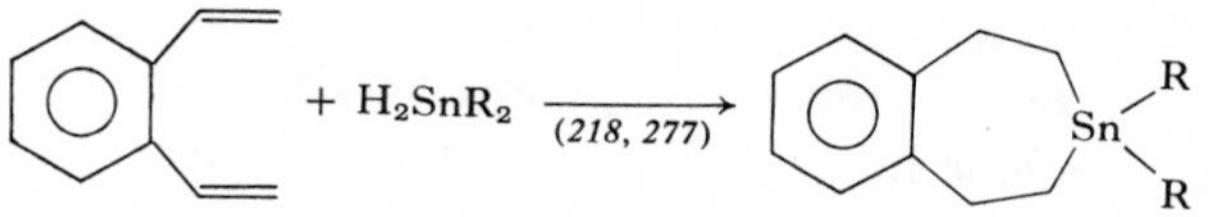

$+ H_2SnR_2 \xrightarrow{(218, 277)}$ (56)

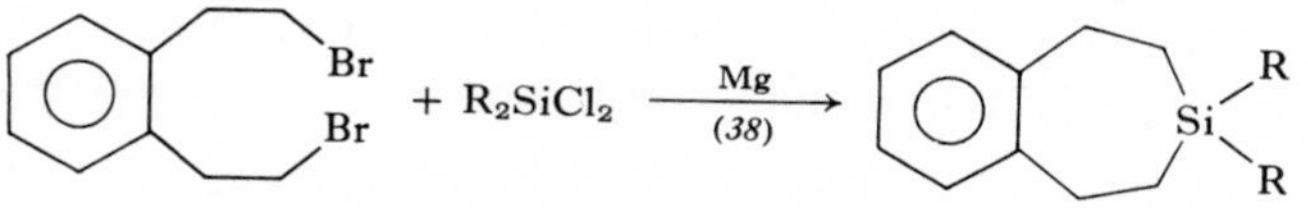

$+ R_2SiCl_2 \xrightarrow[(38)]{Mg}$ (57)

The remaining methods require formation of an organometallic intermediate prior to the ring-closure reaction. Ring closure via electrophilic attack on the aromatic ring has been demonstrated for the formation of each of the seven-membered ring isomers.

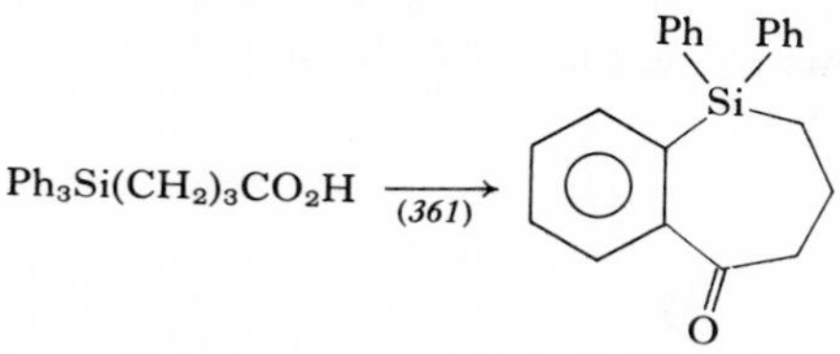

$Ph_3Si(CH_2)_3CO_2H \xrightarrow{(361)}$ (58)

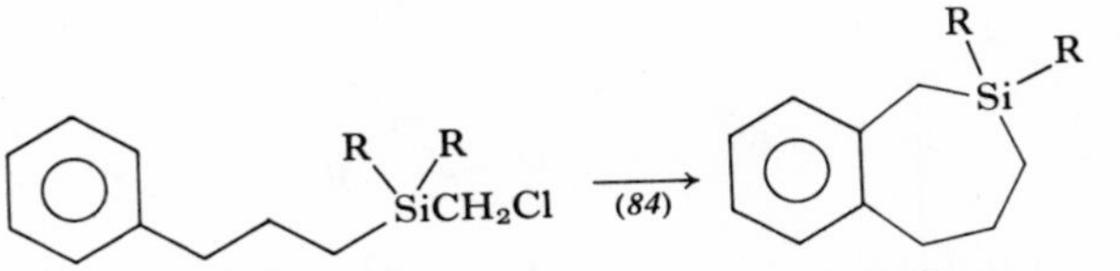

The derivatives of structural type (VII) may also be prepared by organometallic Wurtz coupling reactions as was the case for smaller ring sizes.

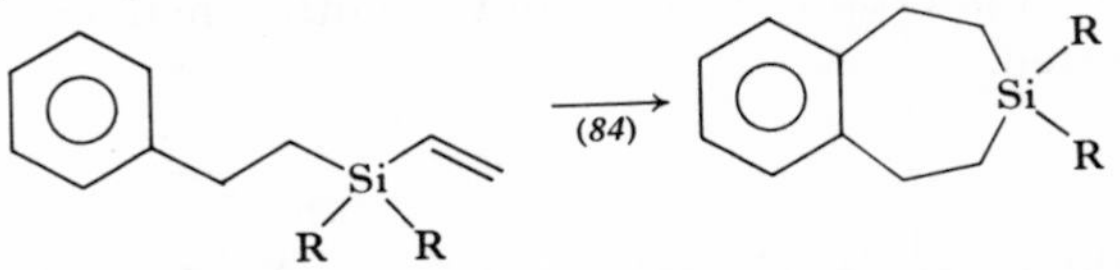

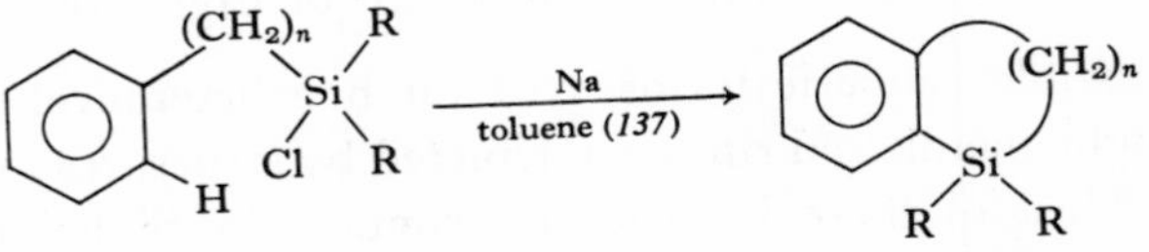

The majority of the ring formation and/or ring-closure reactions of heterocycloalkenes involve reactions at the heteroatom. Ring closure at a site remote from the heteroatom is generally employed only in the formation of large ring systems. This is also typical of the synthetic routes used to generate heterocycloalkanes. Ring closure at the metalloid atom is restricted to the few methods that are currently available for metal–carbon bond formation. A greater versatility of synthetic design is available in organic heterocyclic chemistry. As can be seen by reactions such as those depicted in Eqs. (50), (53), and (55), it is possible to obtain benzheterocycles by more standard organic reactions involving C—C bond formation. If the metal–carbon bonds are not cleaved by the reagent utilized for C—C bond formation, replacement of $>$C by $>MR_n$ at a site remote from the ring-closure reaction does not appear to alter the course of standard organic ring-closure reactions. Thus a greater flexibility of synthetic routes may be possible. Successful approaches to electrophilic ring closure may require appropriate substitution on the aromatic nucleus to facilitate the desired reaction.

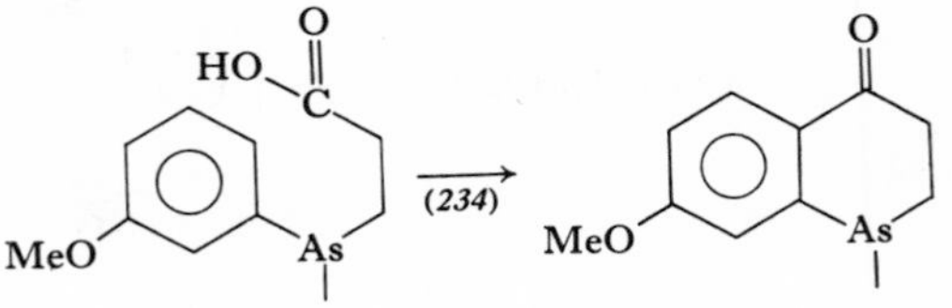

Ring closure of the acid derivative does not take place in absence of methoxy substitution. The limitation then becomes the synthesis of the intermediate aromatic derivative with a suitably substituted side chain.

Derivatives of the benzheterocycles are listed in Table IX.

D. Benzheterocycles Containing Additional Multiple Bonds: Derivatives of Cycloalkadienes and -trienes

Most of the derivatives of benzheterocycloalkenes containing additional multiple bonds have been synthesized by techniques that are not applicable generally but specific to a single heteroatom. Derivatives of Groups III and V are generated by addition cyclization reactions, derivatives of silicon are generated by bromination–dehydrobromination of saturated heterocycles, and tin derivatives are formed by hydride addition to aryl diynes which are then used, in turn, to generate heterocycles of boron by the exchange process. Applications of these synthetic approaches are outlined below as a function of ring size.

1. *Derivatives of Indene*

Addition of an aryl organometallic reagent to an alkyne followed by thermolysis has been used to form heterocyclic analogs of indene:

$Ph_3Al + PhC{\equiv}CPh \xrightarrow[(112,\ 113)]{200^\circ}$ Ph, Ph, Al, Ph $\xrightarrow{I_2}$ Ph, Ph, I, I (63)

$PhAsCl_2 + PhC{\equiv}CH \xrightarrow[(99)]{}$ Cl, As, Cl, Ph $\xrightarrow{\text{Heat}}$ Cl, As, H, Ph (64)

The structure of the intermediate benzaluminole was deduced from the degradation product with iodine.

TABLE IX

BENZHETEROCYCLES

System	R_3	R_1	R_2	M	m.p. or b.p.	Method	%	References
	H	Ph	Ph	Si	75°–76°	A	4.8	*129*
				Si	74°–75°	A	28	*129*
				Si	73°–75°	B	27	*129*
	H	Me	Me	Si	73°/17 mm	B	35	*108*
	H	$PhCH_2CH_2$	—	B	175°/11 mm	C	90	*192, 195*
	H	H	—	B[a]	132°	D	—	*195*
	CH_3	${}^{i}Bu$	—	B[b]	109°/11 mm	E(D)	89(90)	*192, 195*
	CH_3	H	—	B[a]	119°	D	—	*195*
	H	Pr	—	Al[b]	52°	F	—	*193*
	CH_3	Pr	—	Al[a,b,c]	103°	F	—	*193*
	H	Ph	H	Si	125°–126°/1.5 mm	G	54	*137*
	H	Ph	Ph	Si	62°–63°	G	62	*137*
					61°–63°	D	52	*138*
	H	Me	Me	Si	83°/15 mm	B	15	*108*
	H	Me	—	As	112°–113°/15 mm	H	—	*333*
	H	Ph	—	As	126°–128°/0.6 mm	I	76	*96*
						H	17	*96*
	H	Cl	Cl	Si[d]	109°–110°/13.5 mm	J	75	*84*
	H	Ph	OMe	Si	135°/0.25 mm	D	18	*84*
	H	Ph	Me	Si	122°–123°/0.6 mm	D	—	*84*
				Si	125°/1.2–1.3 mm	B	9	*84*
	H	Ph	Ph	Si	62°–63°	D	—	*84*
	H	Me	Me	Ge	—	—	—	*289*
	H	Me	Me	Sn	—	—	—	*148*
	H	Ph	—	As[e]	136°–138°/0.3 mm	K	19	*222, 223*
						L	80	*29*

H	Cl	—	As	73°–74°	D	—	*223*
H	I	—	As	107°–108°	D	—	*223*
H	CH_3	—	As	115°/17 mm	D	—	*223*
					K	4	*222*
H	O	OH	As	144°	D	—	*223*
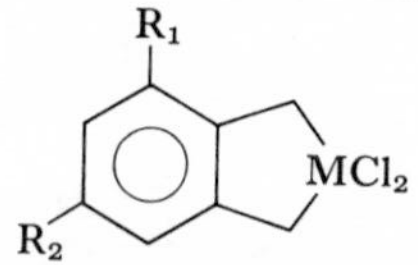 H	CH_3	H[f]	Si	115°–120°/12 mm	J	40–45	*84*
H	H	CH_3[f]	Si	115°–120°/12 mm	J	40–45	*84*
H	CH_3	CH_3	Si	130°–135°/11 mm	J	31	*84*
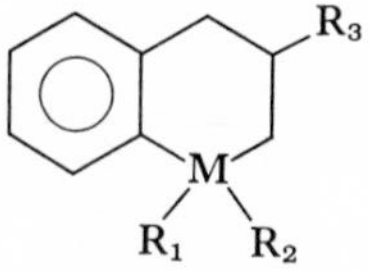 H	Bu^t	—	B	119°/11 mm	C	—	*192, 195*
H	H	—	B	103°–104°	D	—	*195*
H	Pr	—	Al	—	F	—	*193*
CH_3	Et	—	Al	85°	F	—	*193*
H	Me	H	Si	106°–106.5°/21 mm	B	33	*138*
H	Me	Me	Si	53°/0.5 mm	A	35	*108*
H	Ph	H	Si	143°–142.5°/2.5 mm	G	79	*137*
H	Ph	Me	Si	110°–112°/0.002 mm	G	24	*137*
H	Ph	Ph	Si[g]	76°–77.5°	G	44	*137*
				75°–77°	G	9.8	*139*
				78°–79.5°	B	54	*138*
H	Ph	Ph	Sn	70°–71°	B	9.4	*138*
H	Cl	—	As	22°	D	—	*297*
H	Me	—	As	140°/14 mm	H	—	*57*
H	Ph	—	As	Oil	D	75	*164*
					I	6	*164*
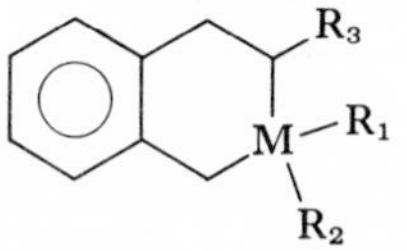 H	Cl	Cl	Si[d,h]	79°–81°/0.7 mm	M	55	*84*
H	Me	Me	Si	227°–229°	M	22	*341*
H	Ph	Ph	Si	72°–73°	D	83	*84*
H	Ph	Me	Si	170°–175°/14–15 mm	D	—	*84*
				180°–185°/14 mm	B	46	*84*
H	Ph	OMe	Si	145°/1.2 mm	D	62	*84*
H	Ph	H	Si	112°–113°/0.05 mm	D	70	*84*

Continued

TABLE IX—*Continued*

System	R_3	R_1	R_2	M	m.p. or b.p.	Method	%	References
	H	Cl	—	As	157°/14 mm	D	—	*163*
		Br	—	As	173°–176°/12 mm	D	36	*164*
		I	—	As	Oil	D	—	*29*
		Me	—	As	131°/18 mm	K	16	*163*
		Ph	—	As	110°–112°/0.01 mm	K	31	*163*
					127°–132°/0.1 mm	L	54	*29*
						C	8	*29*
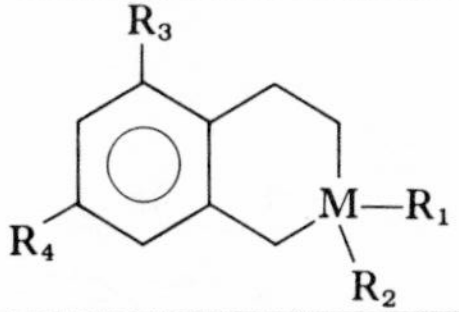	$R_3 = CH_3, R_4 = H$	Cl	Cl	Si[d]	118°–122°/14 mm	M	25	*84*
		α-Naph	OMe	Si	93.5°–94° mm	B	75	*84*
	$R_3 = CH_3, R_3 = CH_3$	Cl	Cl	Si[d]	117°–121°/14 mm	M	40	*84*
		α-Naph	OMe	Si	210°–211°/0.7 mm	B	78	*84*
	$R_3 = CH_3, R_3 = CH_3$	Cl?	Cl	Si[d]	130°–135°/14 mm	K	29	*84*
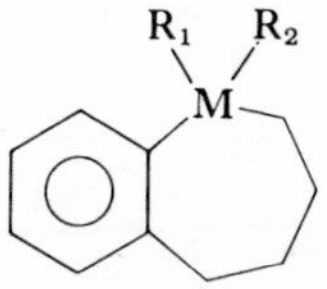	H	Ph	Ph	Si	151°–153°	G	56	*137*
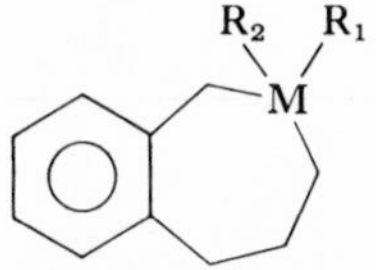	H	Cl	Cl	Si[d]	135°–145°/15 mm	A	25–35	*84*
	H	Ph	Ph	Si	185°–190°/0.3 mm	J	77	*84*

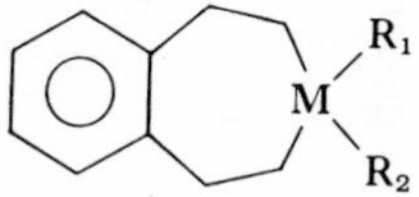	H	Cl	Cl	Si	125°–150°/11 mm[d,i]	M	15–20	*84*
					69°–70°	B	63	*38*
	H	H	H	Si	104°/12 mm	D	90	*38*
	H	Me	Me	Si	117°/12 mm	B	38	*38*
	H	Ph	Ph	Si	154°	B	58	*38*
	H	—	—	Si	157°–158°	B	—	*84*
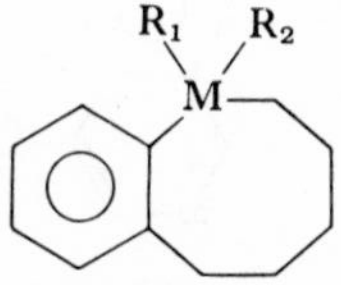	H	Ph	Ph	Si	154.5°–155.5°	G	5	*137*

Key to method: (A) Metal coupling reaction; (B) Grignard reaction; (C) thermolytic ring closure; (D) reaction of exocyclic bond of heterocycle; (E) reaction of R_3B with olefin; (F) exchange of boron heterocycle with AlR_3; (G) Wurtz coupling of Ar—Cl and H—M; (H) Friedel–Crafts reaction of As—Cl; (I) cyclization of arsinic acid; (J) $AlCl_3$ ring closure of M—CH_2Cl; (K) reaction of dibromide with R_xMCl_y in presence of sodium; (L) reaction of dibromide with $RAs(MgBr)_2$; (M) ring closure of alkene.

[a] Dimer.

[b] Derivatives where R_1 = Et, $CH_2CH(Ph)CH_3$ also reported (*193*).

[c] Forms 1:1 adduct with pyridine, m.p. 93.5°.

[d] Derivatives with $R_1 = R_2$ = OMe also reported.

[e] Derivatives with R_1 = Aryl also reported.

[f] Mixture of two isomers.

[g] Derivative with R_2 = o-$CH_3C_6H_4$ also reported.

[h] Also reported are the diacetate and dihydroxide.

[i] Mixed with bibenzyl.

The arsindole system has also been generated by other routes involving a similar thermolytic procedure or by a Friedel–Crafts reaction.

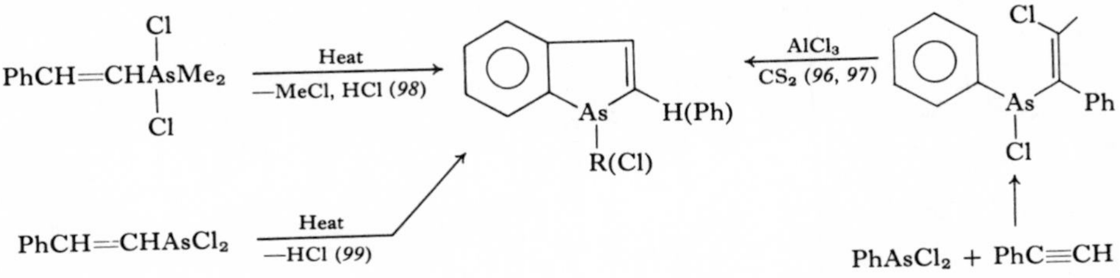

PhCH=CHAsMe$_2$Cl$_2$ —Heat, —MeCl, HCl (*98*)→ arsindole (As–R(Cl), H(Ph)) ←$AlCl_3$, CS_2 (*96, 97*)— PhAs(Cl)C(Cl)=CHPh ← $PhAsCl_2$ + PhC≡CH

PhCH=CHAsCl$_2$ —Heat, —HCl (*99*)→ arsindole

(65)

Diphenylacetylene has been used as a starting point for the generation of metalloindenes of Si (*294*), Sn (*294*), and As (*45*). Reaction of BuLi–tetramethylethylenediamine (TMED) with diphenylacetylene results in addition metallation to produce a dilithio organometallic that will react with a metal dihalide derivative:

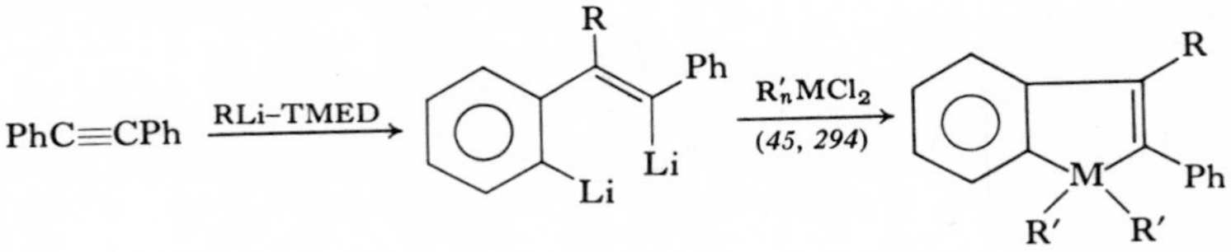

PhC≡CPh —RLi–TMED→ o-LiC$_6$H$_4$C(R)=C(Li)Ph —R'_nMCl_2 (*45, 294*)→ metalloindene (MR'_n, R, Ph) (66)

M = As; n = 1
M = Si, Sn; n = 2

Benzheterocycles of selenium have been prepared by ring-closure reactions involving C—C bond formation (*70, 251, 252, 334*):

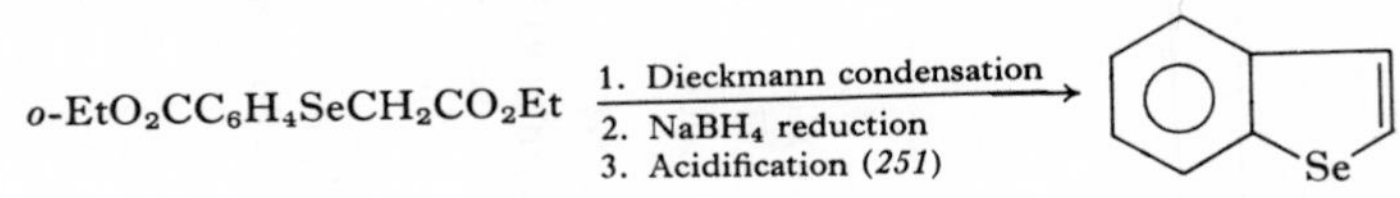

o-$EtO_2CC_6H_4SeCH_2CO_2Et$ —1. Dieckmann condensation; 2. $NaBH_4$ reduction; 3. Acidification (*251*)→ benzo[b]selenophene (Se) (67)

2. *Derivatives of Isochromene and Naphthalene*

The cycloaddition technique has also been employed for the generation of a six-membered ring system (*285*):

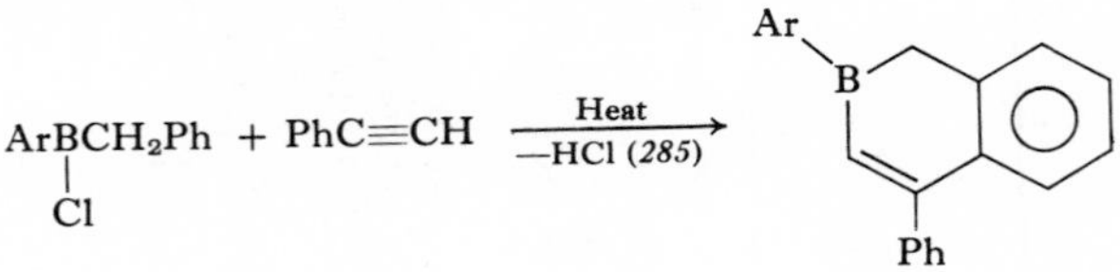

ArB(Cl)CH$_2$Ph + PhC≡CH —Heat, —HCl (*285*)→ 2-Ar-4-Ph-boranaphthalene (B–Ar, Ph) (68)

The difference in reactivity of an As–Cl [reaction (65)] and a B–Cl bond toward the same reagent, PhC≡CH, is noteworthy.

The approach to the diene system of silicon is related to the halogenation–dehydrohalogenation reactions discussed previously. The presence of two benzylic positions introduces additional complications.

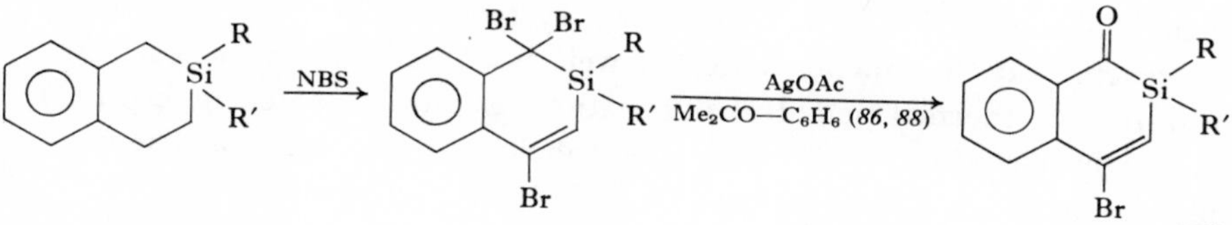
R
Si
R′
NBS
Br
Br
R
Si
R′
Br
AgOAc
Me₂CO—C₆H₆ (86, 88)
O
R
Si
R′
Br

(69)

Metallo derivatives of naphthalene have not yet been reported. Stable phosphanaphthalene derivatives can be generated by a thermolysis reaction from a phosphorus(III) derivative:

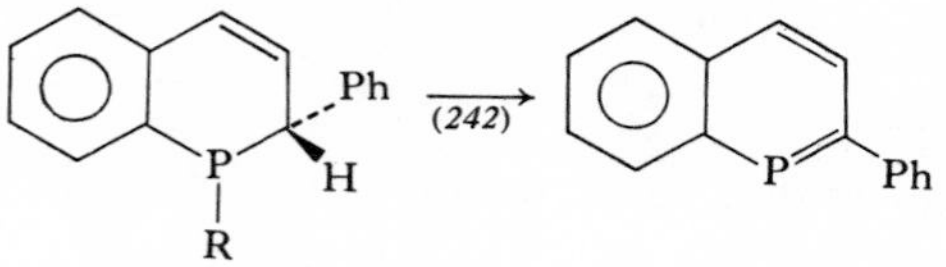
Ph
P
H
R
(242)
P
Ph

(70)

It is unlikely that this method could be extended to compounds containing a Group IV heteroatom. An attempt at chemical dehydrochlorination of an arsenic derivative was unsuccessful *(297)*.

3. *Derivatives of Cycloheptatriene*

Silicon derivatives have been generated by bromination followed by dehydrobromination either chemically or thermolytically. An unexpected metal extrusion reaction occurs during the bromination

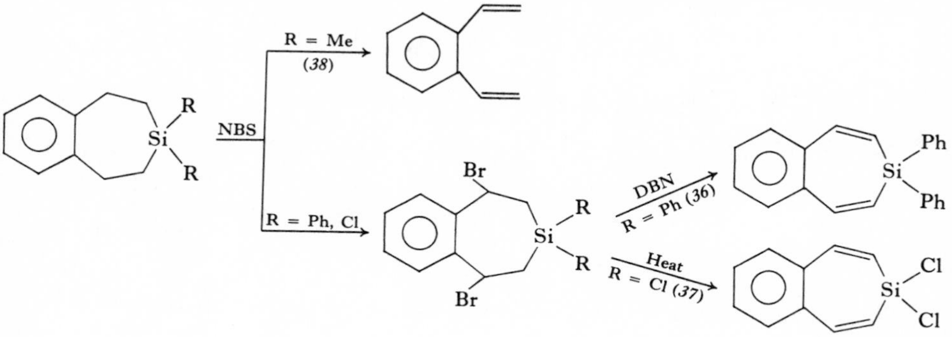
R
Si
R
NBS
R = Me
(38)
R = Ph, Cl
Br
Br
R
Si
R
DBN
R = Ph (36)
Heat
R = Cl (37)
Ph
Si
Ph
Cl
Si
Cl

(71)

process when the silicon exocyclic substituents are methyl groups (*38*). The reaction has not yet been studied for higher alkyl group substituents on Si, but the bromination reaction indicated in Eq. (69) for the six-membered heterocycle gives low yields of brominated product when R = α-Naph, R′ = Me, and higher yields when R′ is Et or Pr. Chemical dehydrohalogenation with a base for a system with Si—Cl bonds leads to reaction at the silicon center as well as the carbon center [see Eq. (102)]. The thermal dehydrohalogenation reaction depicted in Eq. (71) is closely related to the thermal dehydrohalogenation reactions discussed in Section II, A, 2 for synthesis of cycloalkene derivatives.

Metal-hydride addition reactions can be used to generate triene derivatives directly:

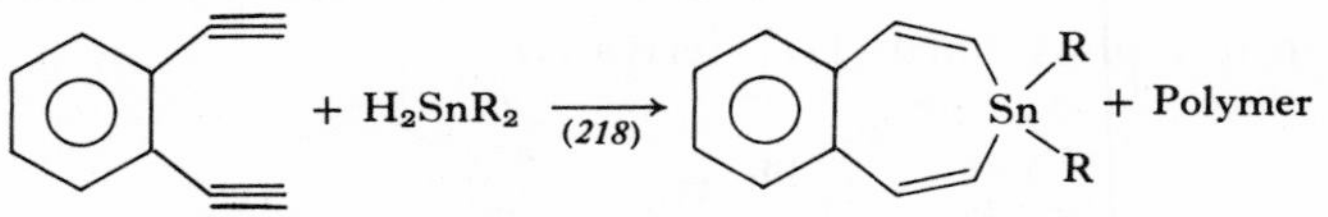

$$+ H_2SnR_2 \xrightarrow{(218)} \text{Sn(R)(R)} + \text{Polymer} \qquad (72)$$

The disadvantage of this route is the required multistep synthesis of *o*-diethynylbenzene.

The benzostannepins can be utilized in exchange reactions to generate benzoborepins:

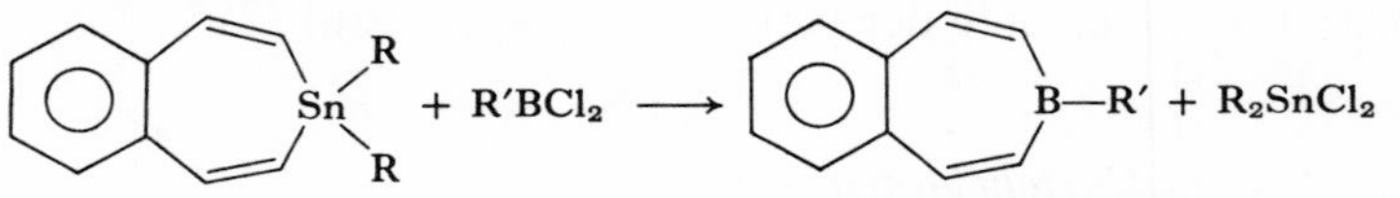

$$\text{Sn(R)(R)} + R'BCl_2 \longrightarrow \text{B—R}' + R_2SnCl_2 \qquad (73)$$

R′ = Ph (*217*)
R′ = OH (*18*)[6]

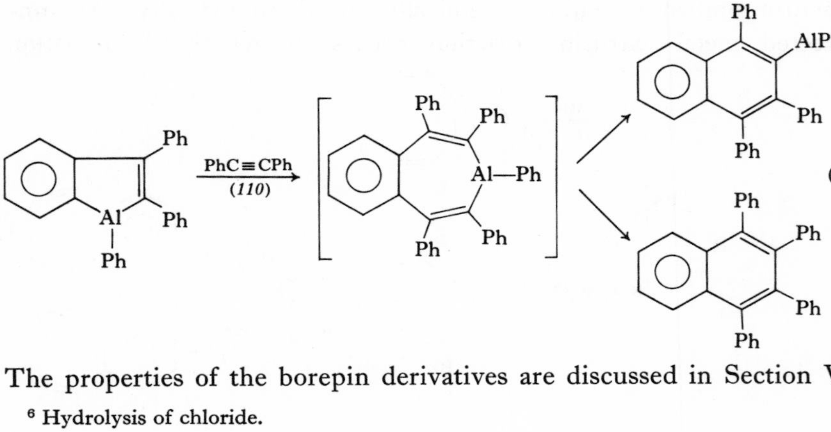

$$\text{Al—Ph} \xrightarrow[(110)]{PhC{\equiv}CPh} [\text{Al—Ph}] \rightarrow \text{AlPh}_2 \;/\; \text{Ph} \qquad (74)$$

The properties of the borepin derivatives are discussed in Section VI.

[6] Hydrolysis of chloride.

An attempt to generate an aluminepin by ring expansion of a benz-aluminole was unsuccessful (*110*).

E. Reactivity of Benzheterocycles

1. *Reactions at the Heteroatom*

Simple substitution reactions at the heteroatom have been effected without ring cleavage. Most of these reactions are general reactions depicting the chemistry that is typical of individual M—Cl bonds. The 1*H*-2-benzosilepin system provides a good illustrative example of the variation of heteroatom substitution possibilities.

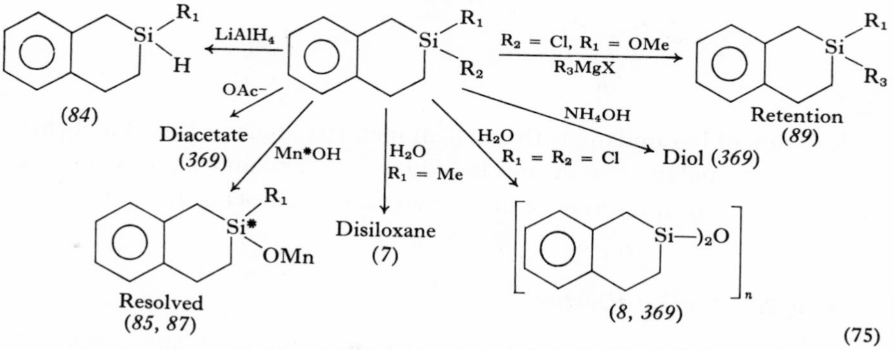

(75)

Reactions of other heterocycles have also been reported and proceed by a similar route, i.e., hydrolysis of B—Cl in 3*H*-benzoborepin to B—OH (*18*) and reduction of $SiCl_2$ in 1,2,4,5-tetrahydro-3*H*-benzosilepin with $LiAlH_4$ (*36*).

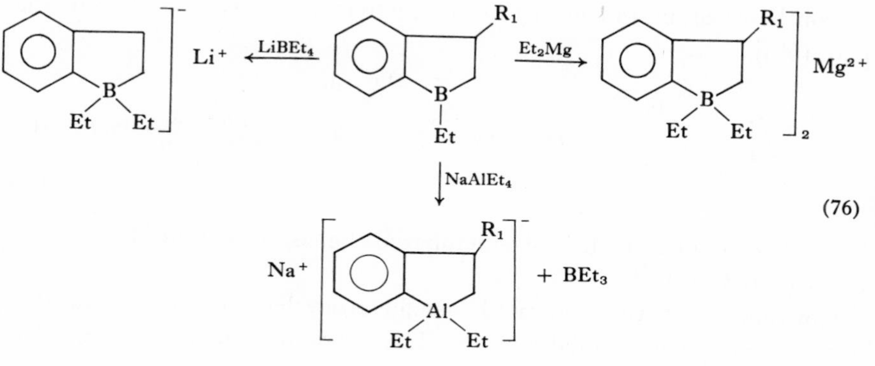

(76)

Reactions that are consistent with the chemistry of Group III (electrophilic) and Group V (nucleophilic) are also possible. The benzheterocycles of aluminum are dimeric but react with bases to form complexes (*193*). Complex anions of Group III have been produced in a variety of reactions including a metal exchange (*190*). Commonly, derivatives of Group V are isolated by formation of the quaternary salt (usually as the methiodide) (*230*). Reaction of an organic dibromide, however, can be accompanied by replacement of an As—Me bond and formation of a spiro derivative (*325*).

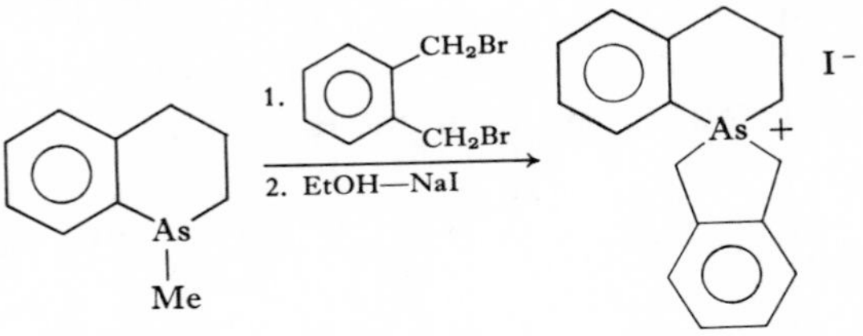

1. $C_6H_4(CH_2Br)_2$; 2. EtOH—NaI → As$^+$ I$^-$ (spiro) — As—Me (77)

Reaction of boraindane with an alkali metal produces a 1:1 adduct that is thought to polymerize in solution (*141*), and 2-silaindane also appears to polymerize in the presence of K even at low temperatures (*269*).

2. *Ring Substitution Reactions*

Few ring substitution reactions have been reported and these are predominantly benzylic brominations such as that indicated in Eq. (71) for the generation of benzosilepin systems. Multibromination of benzsilacyclopentane has been effected, but attempts to hydrolyze the tetrabromo derivative to the diketone resulted in the expulsion of the silicon heteroatom, in contrast to the reactivity of the six-membered ring [Eq. (69)].

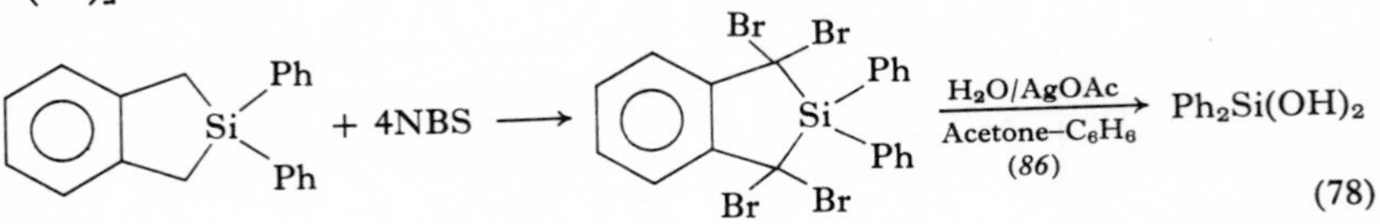

$$\text{Si(Ph)}_2 \text{ (benzsilacyclopentane)} + 4\text{NBS} \longrightarrow \text{Br}_4\text{-derivative, SiPh}_2 \xrightarrow[\text{Acetone–C}_6\text{H}_6\ (86)]{\text{H}_2\text{O/AgOAc}} \text{Ph}_2\text{Si(OH)}_2 \quad (78)$$

Several reactions of the six-membered benzsilyl ketone have been reported (*86*, *88*).

Conversion of the saturated benzheterocycles to the unsaturated derivative has had limited success. These few reactions were recorded in Section III, D. The arsindoline derivative is unreactive toward ring reactions (*177*).

$$\text{Arsindoline oxide (As–Ph, As=O)} \xleftarrow[SeO_2]{KMnO_4} \text{1-phenylarsindoline} \xrightarrow{\text{tetrachloro-}p\text{-benzoquinone}} \text{no reaction}; \quad \xrightarrow{NBS} \text{no reaction}; \quad \xrightarrow{Pd/C} \text{no reaction} \qquad (79)$$

Arsindoles have been successfully generated by other routes [see Eq. (65)].

3. *Ring-Cleavage Reactions*

Thermal and chemical stability toward ring-cleavage reactions have not been extensively studied. Reactions with strong nucleophiles have been reported to yield ring-cleavage products, and benzosilacyclobutane is cleaved by water (*108*). Halogens also cleave carbon–metal bonds.

$$\text{2,2-dimethyl-2-silaindane} \xrightarrow[EtOH\ (118)]{EtONa} \text{o-xylene} + Me_2Si(OEt)_2 \qquad (80)$$

$$\text{2,2-dimethyl-2-silatetralin} \xrightarrow[EtOH\ (118)]{EtONa} o\text{-}MeC_6H_4CH_2CH_2SiMe_2OEt \qquad (81)$$

$$\text{3,3-}R_2\text{-3-metallabenzazepine (M)} \xrightarrow[M\ =\ Sn\ (218)]{I_2} \text{diiodo product (I, I)} \qquad (82)$$

Benzheterocycles with an aromatic metal bond are susceptible to reagents that cleave this bond, e.g., 1-silaindane is cleaved in both acid and alkali media (*108*). Thermolysis reactions usually result in metal extrusion:

$$\text{3,3-dichloro-3-silabenzazepine (}SiCl_2\text{)} \xrightarrow[(37)^{7}]{500^\circ} \text{naphthalene} \qquad (83)$$

[7] A similar reaction is reported for benz[6,7]suberan (*65*).

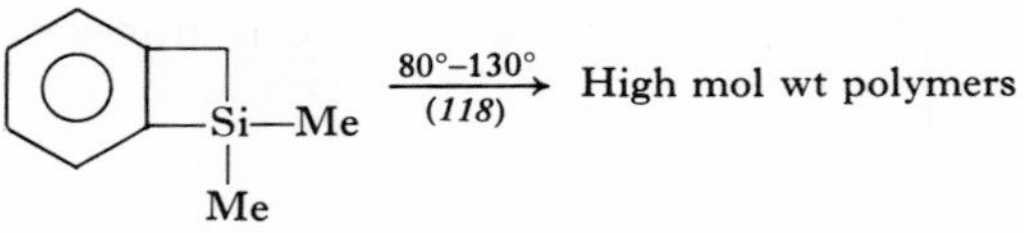

$\xrightarrow[(118)]{80°–130°}$ High mol wt polymers (84)

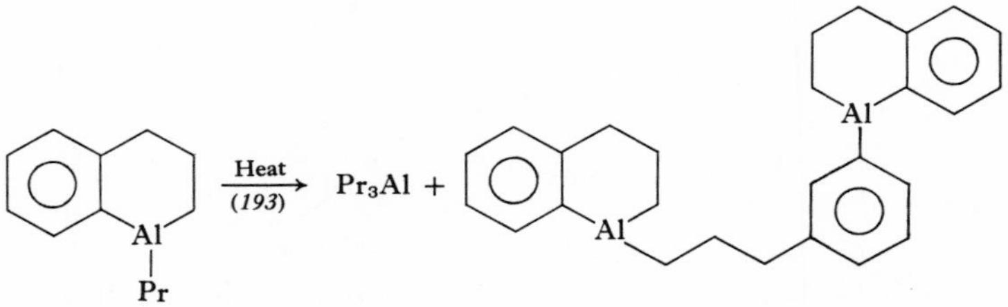

$\xrightarrow[(193)]{\text{Heat}}$ Pr_3Al + (85)

IV
DIBENZHETEROCYCLES

The dibenzheterocycles may be considered derivatives of heterocyclic dienes (Section II, B); the simplest member would contain a central five-membered ring. Several isomers are possible for the larger ring systems, but the majority of the known derivatives are those that are "symmetrical," i.e., contain a plane that bisects the molecule and passes through the heteroatom.

The first dibenzheterocyclic derivatives were named for the parent hydrocarbon system. A slow conversion to the ring index method is taking place, and the nomenclature now in use requires a prefix of dibenz or dibenzo preceding the base name of the heterocycloalkene. If the ring containing the heteroatom has six or more members, the position of the fused benzene rings are indicated by letters (the heteroatom is always lettered *a*). The lettering and numbering schemes respectively, for a central seven-membered ring are as follows:

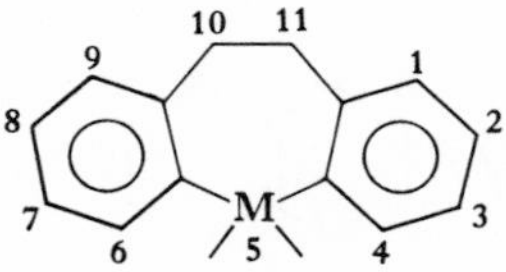

(Xa)

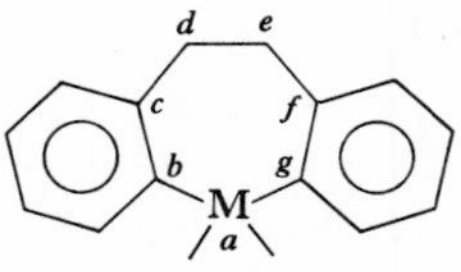

(Xb)

Saturation at the heteroatom is indicated by *H* (in the above example, 5*H*) and saturated carbons by "hydro" prefixes. As in the systems described in previous sections, common names are often used, particularly for heteroatom derivatives of dihydroanthracene. The nomenclature is briefly discussed for each of the systems in the following sections.

The dominant method of synthesis for the dibenz derivatives is the reaction of a dilithio organometallic with a metal dihalide. The dibenzheterocycles are listed in Table X.

A. Dibenzometalloles

Historically, several names have been used to describe structures of the type indicated as (XI). Originally the derivatives were named for the parent hydrocarbon, fluorene, and were numbered the same as fluorene. Thus, the heteroatom was position 9 and if $MR_1R_2 = AsPh$ [for (XI)], the derivative would be 9-phenyl-9-arsafluorene. Currently, the ring index system is used with the following numbering:

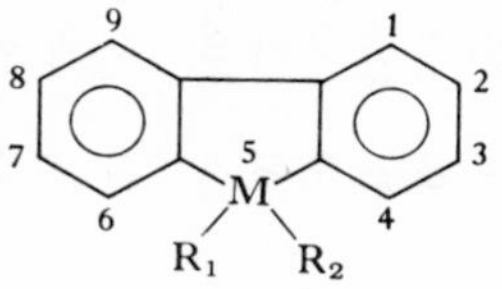

(XI)

Compound (XI), $MR_1R_2 = AsPh$, would be 5-phenyldibenzarsole (5*H* is not included). This latter nomenclature should eventually prevail and derivatives of (XI) are thus dibenzometalloles.

The majority of the known dibenzheterocycles are analogs of fluorene. This is no doubt related to the ease of synthesis. The discovery of the ready coupling of *o*-dibromobenzene provides a convenient entry into the heterocyclic analogs of fluorene through standard organometallic reactions. The method has the advantage of a one-step synthesis for generation of the desired heterocycle. Other methods have also been employed such as thermolytic ring closure, exchange reactions, and insertion into biphenylene, but these generally are a less desirable substitute for the *o*,*o*′-dilithiobiphenyl plus organometallic dihalide.

TABLE X

DIBENZHETEROCYCLES

System	R_1	R_2	M	b.p./mm or m.p.	Method	%	Other derivatives[a]	References
R_1–M–R_2	H	—	B	107°[b]	A	—	—	*194*
	H	py	B	117°–118.5°	B	66	—	*340*
	Cl	—	B	52°[c]	A	—	—	*194*
	Et	—	B	112°/0.3 mm	B	~65	^{i,n}Pr, ^{t}Bu	*194*
	Ph	—	B	118°	B	55	—	*194*
	$M(CO)_4PPh_3$	—	B	>140°	A	52	—	*276*
	$Co(diphos)_2$	—	B	186°	A	79	—	*309*
	Et	—	Al	230°[b]	C	—	—	*191*
	Ph	—	Al	225–230°[b,d]	B	—	—	*112*
	Cl	Cl	Si	108°–110°/0.01 mm	D	38[e]	—	*134*
					B	18	—	*66, 68, 69*
	Cl	Me	Si	63°–68°	D	83[e]	Dodecyl, Et	*134*
	Cl	Ph	Si	69°–72°	D	73	CH_2Ph[e]	*134*
	H	Ph	Si	58°–60°	A	88[e]	CH_2Ph	*135*
	Me	Me	Si	58°–59°	A	70	—	*134*
					E	54	—	*135*
					B	18	—	*90*
	Et	Et	Si	60°–61°	D	23	Dodecyl	*134*
	Ph	Ph	Si	147°–148°	D	25	—	*144*
					A	74	—	*134*
	Me	O	Si[f]	125°–127°	A	77	Dodecyl	*135*
	Ph	O	Si[f]	203°–204°	A	21	—	*135*
	Me	Si	Si[g]	180°–182°	F	43	Dodecyl	*135*
	Ph	Ph	Ge	152°–3°	D	75	—	*144*

Me	Me	Sn	123°–4°	D	—	—	*173*
Et	Et	Sn	73°	D	57	^{n}Bu, C_6H_{11}	*127*
Ph	Ph	Sn	141.5°	D	58	*p*-$CH_3C_6H_4$, *p*-PhC_6H_4	*127, 173*
			136°–137°	G	30–37	—	*126*
			—	H	—	—	*123*
Ph	Ph	Pb	136°–137.5°	G	30	—	*126*
O	OH	As	290°	I	—	—	*3*
O	—	As[f]	178°	A	—	—	*3*
Cl	—	As	161°	B	—	—	*3*
				J	43	Br[h]	*125*
				K	20	—	*41*
I	—	As	166°–167°	A	—	—	*3*
Me	—	As	40°–41	A	—	—	*153*
			46°	D	60	—	*153*
			41°–45°	B	69	—	*150*
Ph	—	As	88°	D	—	—	*152*
Cl	—	Sb	204.5°–207°	I	16	—	*262, 364*
Cl	—	Sb	204°–206°	C	53	—	*156*
I	—	Sb	222°	A	—	—	*262*
			212°–218°	A[h]	80	—	*156*
Me	—	Sb	57°	A	—	—	*262*
Ph	—	Sb	101°–102°	A	65(80)	Biph. naph.	*156, 364*
			101°	D	50	—	*153*
I	—	Bi	250°(d)	C	63	—	*156*
Ph	—	Bi	167°–168°	D	72	—	*364*
—	—	Te	96°–97°	D[i]	28	—	*158*
—	—	Te	93°–94°	D[i]	54	—	*158*
—	—	Te	—	C	82	—	*158*
Cl	Cl	Te	328°–330°	L	84	I	*158*

Continued

TABLE X—*Continued*

System	R_1	R_2	M	b.p./mm or m.p.	Method	%	Other derivatives[a]	References
9,10-dihydro-10-M-anthracene (M bearing R_1, R_2)	Cl	—	B	84°	C	70	—	*180, 182*
				95°–97.5°	A	95	[j]	*337*
	O	—	B[f]	232°–235°	A	96[e]	[j]	*337*
	$OCH_2CH_2NH_2$	—	B	170°–176°	D[k]	71	[j]	*337*
				170°–175°	B	61	—	*340*
	Me	—	B	116°/0.2 mm	C	49	—	*180*
	Ph	—	B	108°	C	93	—	*180*
	Mes	—	B	131°–134°	A	87	[j]	*338*
	H	H	Si		A	—	—	*67*
	Cl	Cl	Si	83°	D	37	—	*180*
					B	—	—	*67*
	Cl	Me	Si	128°/3 mm	D	50	OEt,[i] Bu	*180*
	Cl	Ph	Si	85°	D	40	Mes	*180*
	H	Ph	Si	163°/1 mm	D	39	—	*180*
	Me	Me	Si	102°/2 mm	D	61	—	*180*
	Cl	Me	Ge	75°	D	60	—	*180*
	Me	Me	Sn	64°	D	55	—	*180*
	O	OH	As	235°–236°	I	~100	—	*142*
	Cl	—	As	114°–115°	A	90	[j]	*142*
				97°	C	76	—	*180*
	Me	—	As	91°	C	72	—	*180*
	Ph	—	As	74°–75°	A	98	—	*176*
	O	OH	Sb	—	I	—	—	*263*
	Cl	—	Sb	105°	A	—	—	*263*
	—	—	—	89°	C	88	—	*180, 380*
	I	—	Sb	160°–162°	A	—	—	*263*
	Me	—	Sb	101°	A	—	—	*263*

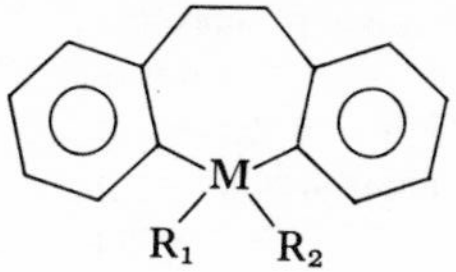	$OCH_2CH_2NH_2$		B	195°–196°	D	42^e	$OCH_2CH_2NMe_2$ (*102*)	*215*
	O	—	B^f	145°–146°	A	96	—	*215*
	OBu		B	150°–75°/0.1 mm	A	65	—	*102*
	Me	Me	Si	115°–118°/0.03 mm	D	48	—	*26, 63, 80*
	Ph	Cl	Si	182°–86°/0.05 mm	D	56	—	*74*
	Ph	H	Si	65°–67°	D	40	—	*74*
	Ph	OCH_3	Si	79°–81°	A	40	—	*80*
	Ph	O	Si^f	219°–221°	A	—	—	*75*
	Ph	Ph	Si	170°–172°	D	5	—	*80*
	Ph	Me	Si	100°–102°	D	18	—	*74*
	Me	Me	Ge	110°–116°/0.03 mm	D	35	—	*81*
	Ph	Ph	Ge	165°–168°	D	33	—	*81*
	H	H	Sn	—	A	—	—	*196*
	Cl	Cl	Sn	106°–106.5°	D	36	—	*196*
	Me	Me	Sn	109°–120°/0.02 mm	D	33	—	*81*
				130°–135°	A	73	—	*196*
	Ph	Ph	Sn	138°–141°	D	23	—	*81*
				136°–137°	A	83	—	*196*
				146°–147°	—	—	—	
	Me	Me	Pb	163°–165°/0.3 mm	D	25^e	—	*81*
	Ph	Ph	Pb	163°–164°	D	1.4	—	*81*
	Ph	—	As	59°–59.5°	D, M	75	—	*233*
				78.5–79°	—	—	—	*233*
	Cl	—	Sb	138°	C	—	—	*380*
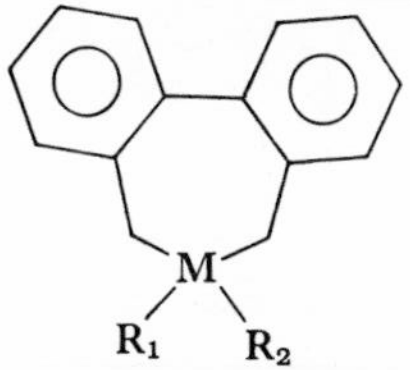	I	—	As	117°–117.5°	A	—	—	*30*
	Ph	—	As	118°–118.5°	P	45^e	—	*30*
	—	—	Se	65°–66°	Q	~100	—	*332*

Continued

TABLE X—*Continued*

System	R_1	R_2	M	b.p./mm or m.p.	Method	%	Other derivatives[a]	References
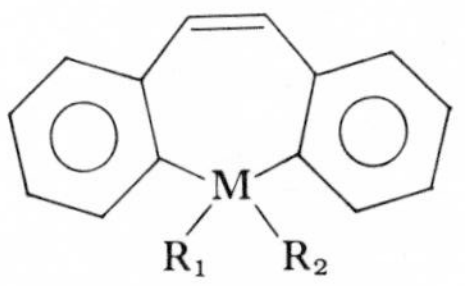	$OCH_2CH_2NH_2$	—	B	222°–226°	N	—	—	*336*
	Me	Me	Si	110–123°/0.2 mm[e]	N	36	—	*81*
					N	24	—	*26*
					O	23	—	*63*
	Ph	Ph	Si	195°–197°	N	40	—	*80*
	Me	Me	Ge		N	16	—	*81*
	Ph	Ph	Ge	193°–194.5°	N	6.3	—	*81*
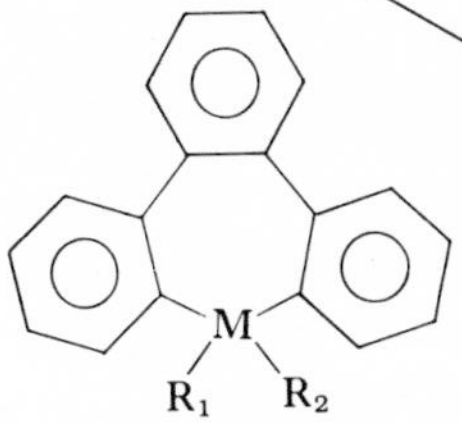	Me	Me	Si	225°–226.5°	D	3.5	—	*77*
	Ph	—	As	159°–160°	—	—	—	*232*

Key to method: (A) Reaction of exocyclic bond(s) of heterocycle; (B) thermolytic ring closure; (C) exchange reaction; (D) reaction of dilithio reagent or diGrignard with R_xMCl_2; (E) reaction of heterocyclic disilane with lithium; (F) coupling of M—Cl; (G) reaction of dilithio reagent with R_xMCl; (H) insertion of Ph_2Sn into biphenylene; (I) ring closure of arsonic acid or stibonic acid; (J) ring closure of acid in presence of PX_3; (K) reaction of biphenyl, $AsCl_3$, and $AlCl_3$; (L) halogenation; (M) thermolysis of methiodide adduct; (N) halogenation–dehydrohalogenation, halogenation–dehalogenation; (O) dehydrogenation; (P) R'_n $PhAs(MgBr)_2$ with organic dibromide; (Q) R'_n. K_2Se with organic dibromide.

[a] Additional derivatives of the heterocycle with exocyclic substituents of same type prepared by the same method. Types of exocyclic substituents; R = H, X, alkyl, aryl.

[b] Dimer.

[c] Etherate also reported to dissociate at 82°.

[d] Etherate, m.p. 132°–134°.

[e] Impure.

[f] Siloxane; diarsine oxide boron anhydride.

[g] Disilane.

[h] Derivatives with substituents on benzene ring also reported; As (*101*, *125*, *162*), Sb (*62*, *156*).

[i] From $TeCl_4$; from $TeCl_2$.

[j] 10-Phenyl derivatives; B (*339*), As (*343*).

[k] Followed by reaction with ethanolamine.

1. *Reactions of 1,1′-Dilithiobiphenyl*

The low-temperature generation of *o,o*′-dibromobiphenyl from BuLi and *o*-$Br_2C_6H_4$ can be effected in yields of up to 75% (*376*). The reactions of *o,o*′-dilithiobiphenyl with various organometallic halides (Scheme 6) provide an example of the utility of dilithio reagents in heterocyclic compound formation. Reaction of organoboron derivatives

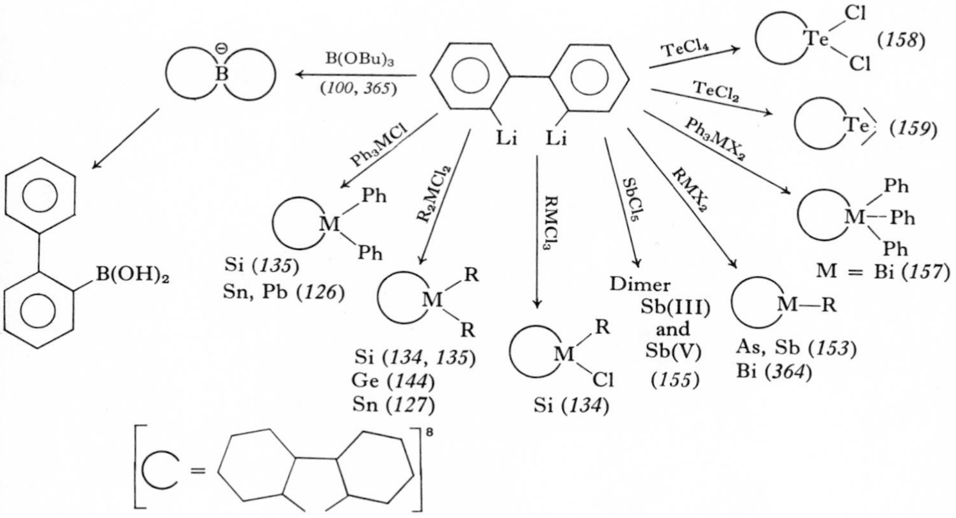

Scheme 6

yields the usual four-coordinate boron species generated in reactions with organolithium reagents. Thermolysis reactions are utilized to produce the neutral boron heterocycle (*194*, *112*). The silafluorene system may be generated from *o,o*′-dilithiobiphenyl even when the reacting halide is of the type R_3SiCl, which requires cleavage of R—M during the ring-closure reaction (*360*). This process has been observed in the formation of five-membered heterocyclic derivatives of As (*362*). Reaction of the lithium reagents with $RMCl_3$ (M = Group IV atom) provides fluorene derivatives with a reactive M—Cl bond that can be converted into a variety of other derivatives. If dilithiobiphenyl reacts with MCl_4 or MCl_5, spiro derivatives are formed (*134*, *135*).

[8] This notation is used in succeeding equations.

2. *Miscellaneous Methods Not Employing an Organolithium Reagent*

A second method utilized in the approach to dibenzometallole systems involves thermal ring closure of an appropriately substituted aromatic derivative. The method requires the synthesis of an unsymmetrically substituted organometallic derivative prior to the ring closure. Often, these syntheses involve several steps. The thermolytic reactions are summarized in Eq. (86).

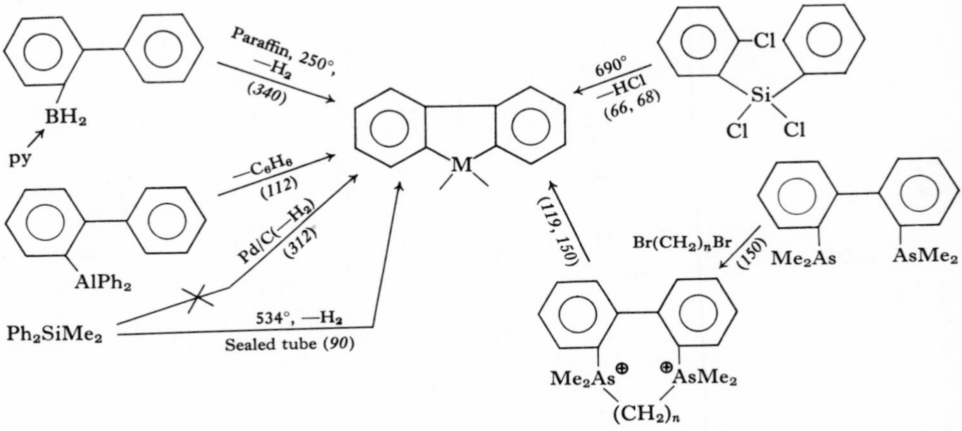

BH_2 ← py; Paraffin, 250°, $-H_2$ (*340*)

$AlPh_2$; $-C_6H_6$ (*112*)

Ph_2SiMe_2; Pd/C($-H_2$) (*312*); 534°, $-H_2$, Sealed tube (*90*)

M

Cl, Si, Cl, Cl; 690°, $-HCl$ (*66, 68*)

Me_2As, $AsMe_2$; $Br(CH_2)_nBr$ (*150*)

$Me_2As^{\oplus}$ $^{\oplus}AsMe_2$ $(CH_2)_n$ $Br^{\ominus}$ $Br^{\ominus}$; (*119, 150*)

(86)

The dehydrogenation reaction of Ph_2SiMe_2 with Pd/C catalyst at 400° to 500° under hydrogen flow afforded biphenyl as the only isolated reaction product (*312*). A later successful investigation of the thermal dehydrogenation of Ph_2SiMe_2 involved reaction in a sealed evacuated tube at similar temperatures (*90*). The metallole system is evidently thermally stable but reacts in presence of a metal catalyst.

Exchange reactions have been reported for formation of tellurium and aluminum benzheterocycles:

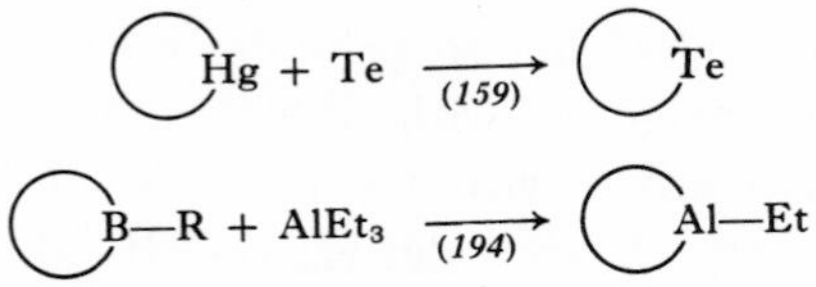

$$\text{Hg} + \text{Te} \xrightarrow[(159)]{} \text{Te}$$

$$\text{B—R} + \text{AlEt}_3 \xrightarrow[(194)]{} \text{Al—Et} \qquad (87)$$

The boron-to-aluminum exchange reaction appears to be the most useful method for synthesis of aluminum heterocycles. Dibenzotellurole has also been reported from the reaction of tellurium with the following compound (*286*):

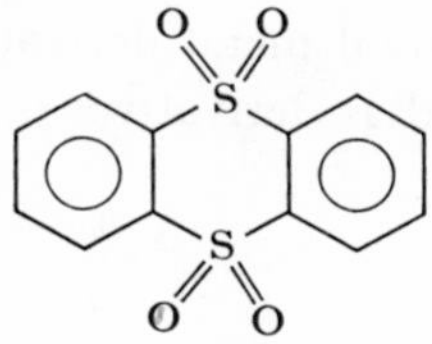

(XII)

In an interesting variation of a metalloid insertion reaction, biphenylene was found to react with selenium as well as with $Ph_{10}Sn_5$ at 275°C. No reaction was observed with tellurium.

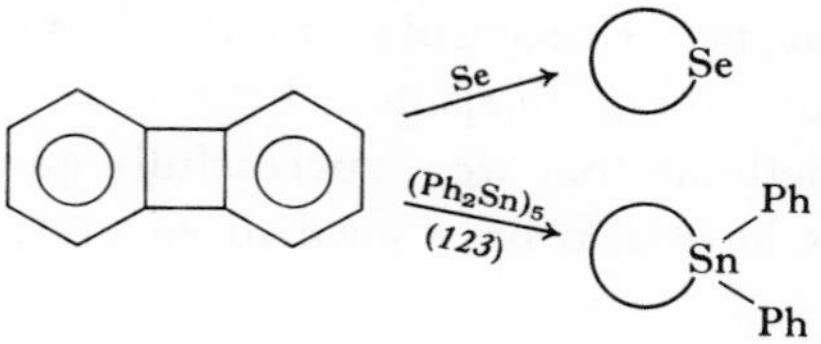

Se; $(Ph_2Sn)_5$ (*123*); Se; Sn; Ph; Ph (88)

The utility of routes requiring electrophilic attack on an aromatic nucleus, used so extensively in the synthesis of benzheterocycles, has been little explored as a method for generating dibenzometalloles. The only systems for which this route has been reported are those that contain a Group V heteroatom:

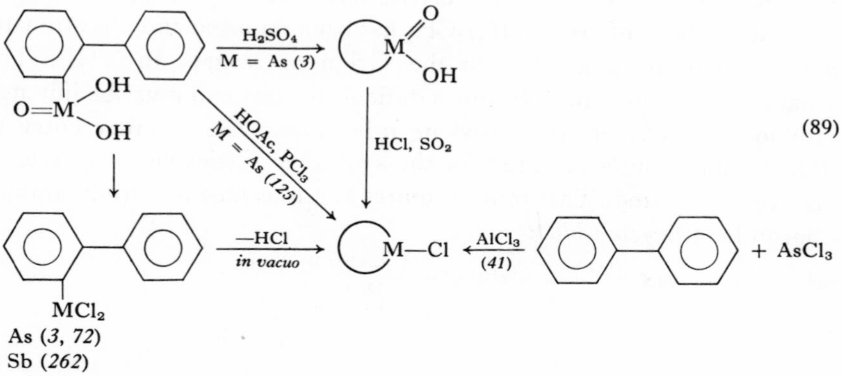

H_2SO_4, M = As (*3*); O; M; OH; OH; O=M; OH; HOAc, PCl_3, M = As (*125*); HCl, SO_2 (89)

—HCl, *in vacuo*; M—Cl; $AlCl_3$ (*41*); + $AsCl_3$

MCl_2
As (*3, 72*)
Sb (*262*)

The derivatives of dibenzometalloles (XI) are listed in Table X.

B. Dibenzometallins

Although arsenic derivatives of (XIII) have been known since the early 1930s, derivatives containing other metal heteroatoms have only recently been prepared. Several metal derivatives, particularly those of silicon and boron, are named as derivatives of anthracene:

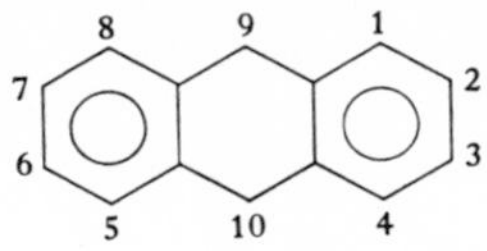

(XIIIa)

Numbering scheme: anthracene

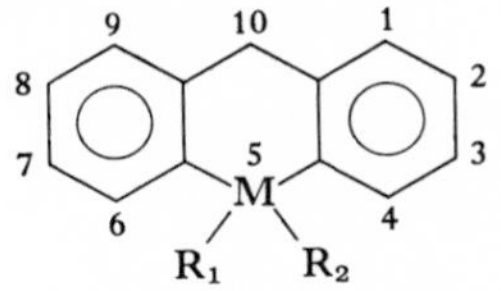

(XIIIb)

Ring index numbering scheme: heterocycle

If $MR_1R_2 = SiMe_2$ for (XIII), the derivative would be 9.9-dimethyl-9,10-dihydro-9-silaanthracene or, according to the Ring Index rules, 5,5-dimethyl-5,10-dihydrodibenzo[*b*,*e*]silicin.[9]

In general, the methods that were successfully employed to generate the fluorene analogs have also been used to generate the derivatives of metalloanthracenes.

1. *Heterocyclic Analogs of 9,10-Dihydroanthracene*

Since the methylene protons of diphenylmethane are slightly acidic, halogen–metal exchange reactions of $(o\text{-}BrC_6H_4)_2CH_2$ with BuLi are complicated by metallation at the methylene position. The organometallic reagent of choice is the diGrignard reagent. A simple, straightforward synthesis of $(o\text{-}ClC_6H_4)_2CH_2$ has been reported and this aromatic chloride may be converted to the Grignard reagent in THF. The reaction of the aromatic chloride is difficult to start and must be initiated with iodine (*180*) or with ethylene dibromide (*75*). Another entry to aromatic dibromides required for the synthesis of tricyclic ring systems has been suggested. This route requires the intermediate production of a boron heterocycle (*340*):

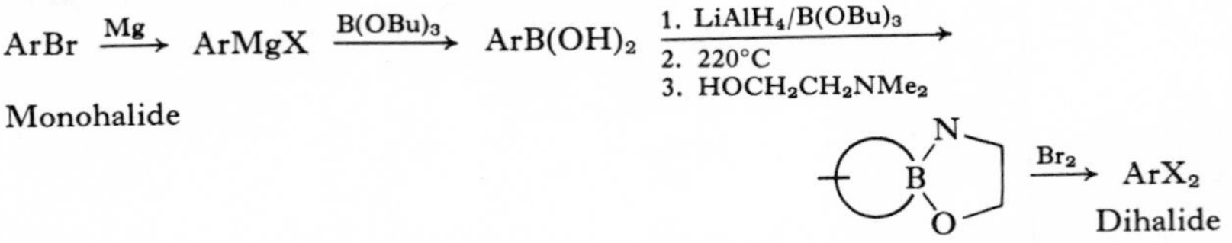

$$\underset{\text{Monohalide}}{ArBr} \xrightarrow{Mg} ArMgX \xrightarrow{B(OBu)_3} ArB(OH)_2 \xrightarrow[\substack{2.\ 220^\circ C \\ 3.\ HOCH_2CH_2NMe_2}]{1.\ LiAlH_4/B(OBu)_3} \text{[boron heterocycle]} \xrightarrow{Br_2} \underset{\text{Dihalide}}{ArX_2} \tag{90}$$

[9] The arsenic derivative of (XIII), $MR_1R_2 = AsPh$ has been called 5-phenyl-5,10-dihydroacridarsine.

The reaction of the diGrignard reagent with both group III and IV reagents have been studied:

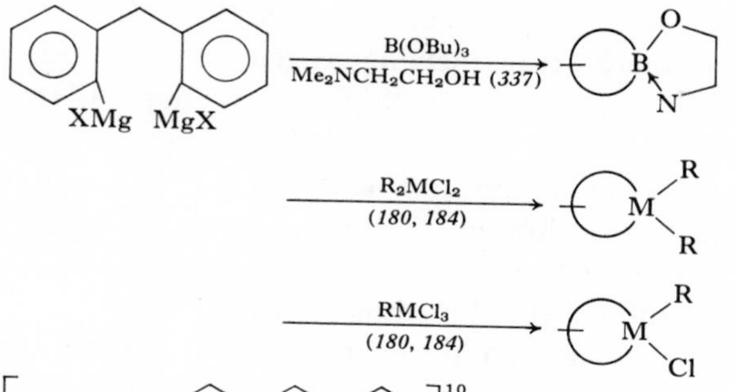

XMg MgX

$B(OBu)_3$
$Me_2NCH_2CH_2OH$ (*337*)

R_2MCl_2
(*180, 184*)

(91)

$RMCl_3$
(*180, 184*)

[10]

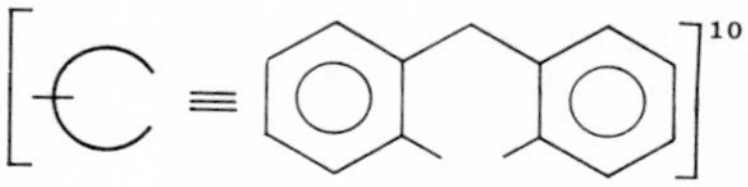

The two disadvantages of the Grignard method, i.e., synthesis of the aromatic dibromide and the difficulty of initiating the reaction of an aryl chloride with magnesium, have led to a search for other routes to

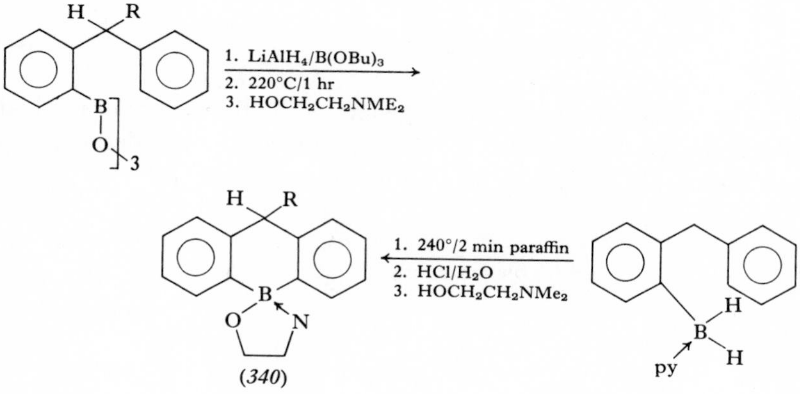

1. $LiAlH_4/B(OBu)_3$
2. 220°C/1 hr
3. $HOCH_2CH_2NME_2$

1. 240°/2 min paraffin
2. HCl/H_2O
3. $HOCH_2CH_2NMe_2$

(*340*)

(92)

220°C/8 hr
(*340*)

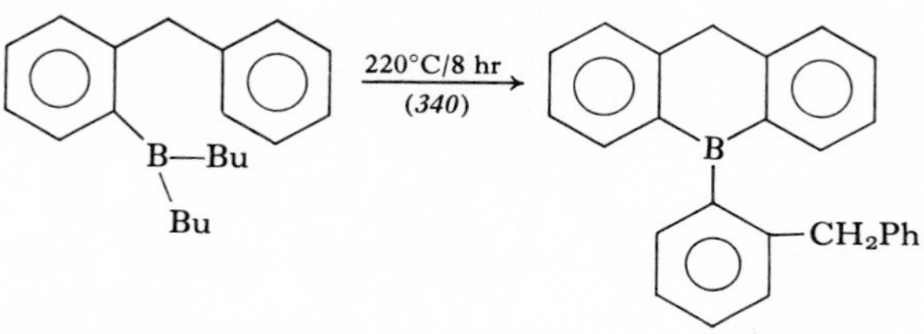

[10] This notation is used in succeeding equations.

the heterocyclic anthracene derivatives. Recently, a detailed report of thermolytic ring closure as a method for generation of boraanthracenes has been published. The monobromides, ArBr, required for the synthesis of the intermediate organometallic reagent are more readily accessible than the dibromides required for the Grignard route. All three thermolytic routes depicted in Eq. (92) were studied, but the preferred route is through the air-stable pyridine aryl borane. Although 9,9-dichloro-9,10-dihydro-9-silaanthracene has been prepared in the pyrolysis of *o*-$MeC_6H_4Si(C_6H_5)Cl_2$ at 670° to 680°C, this method is probably less desirable than the Grignard method (*67*).

In a reaction analogous to the generation of the arsafluorene derivative, electrophilic aromatic substitution of arsinic acids may be used to generate arsaanthracene:

H R' / As R O OH → H R' / As R O (*343*) $\xrightarrow[R\ =\ OH]{SO_2/HCl}$ H R' / As Cl (93)

Since the tin heterocycle has been generated by the Grignard method, the exchange reaction can be utilized to convert the tin derivative to either a Group III derivative or a Group V derivative.

Sn $\xrightarrow[(182)]{RBX_2}$ B—R ; $\xrightarrow[(185)]{AsCl_3}$ As—Cl (94)

The generation of 14π-electron systems, derivatives of anthracene, are discussed in Section IV, E.

2. *Derivatives of Dihydrophenanthrene: 5,6-Dihydro-5-Dibenzometallins*

Very few heterocyclic derivatives of dihydrophenanthrenes (XIV) are known.

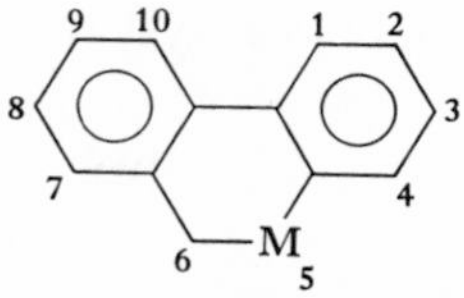

(XIV)

The Grignard method of synthesis requires the starting dibromide, *o*-$BrCH_2C_6H_4C_6H_4Br$-*o*. This approach to the synthesis of dibenzometallins has not been reported; a difficulty in the reaction of the dibromide with magnesium is the possibility of coupling to form fluorene. The reported syntheses involve an indirect approach. The 5,6-dihydro-5-dibenzometallins may be regarded as derivatives of a bridged biphenyl. The arsenic compound is synthesized through electrophilic attack by a side-chain substituent of a monosubstituted biphenyl at the *ortho* position of the remaining unsubstituted benzene ring. The arsanthridine derivatives were isolated as methiodide adducts (*72*, *73*):

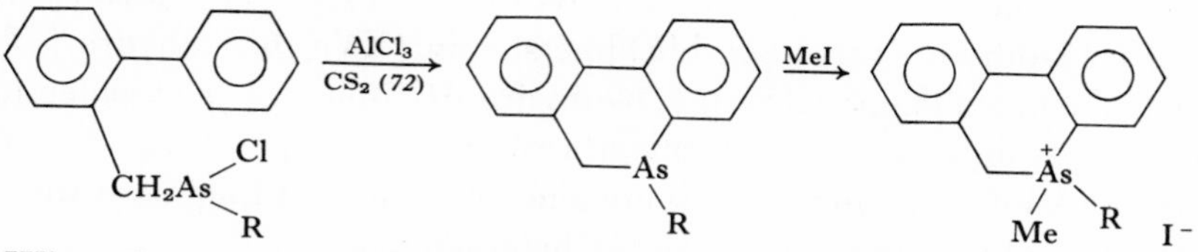

(95)

When R = Ph, the methiodide adduct occurs in two solid forms: the first form is converted irreversibly to a second form upon scratching or crushing.

A silicon derivative has been prepared by an extension of the ring-expansion reaction. Although the method has the disadvantage of requiring prior synthesis of a heterocyclic derivative, the ring-expansion reaction is a facile process (*83*):

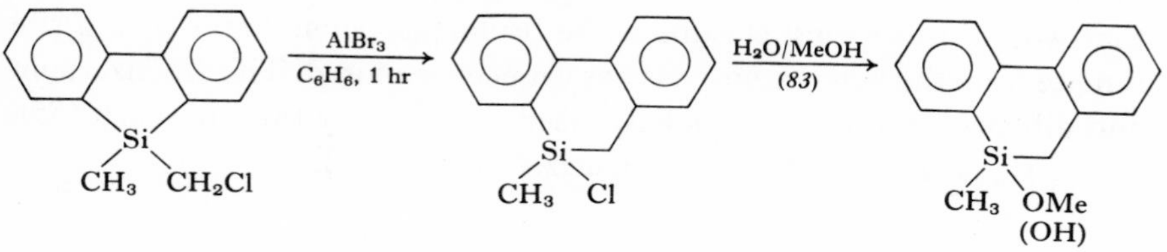

(96)

In principle, this ring-expansion reaction should proceed for any metalloid heterocycle with a chloromethyl substituent. However, no system other than one with a silicon heteroatom has been studied.

Traces of moisture in the $AlBr_3$ catalyst result in ring cleavage after the ring expansion occurs.

C. *Dibenzometallepins*

Four heterocyclic isomers of dibenzometallepins are possible:

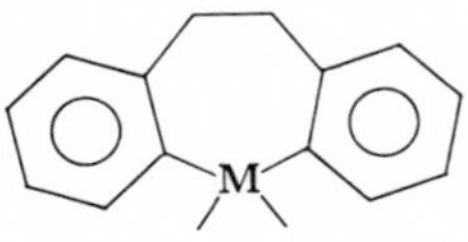

(XVa) 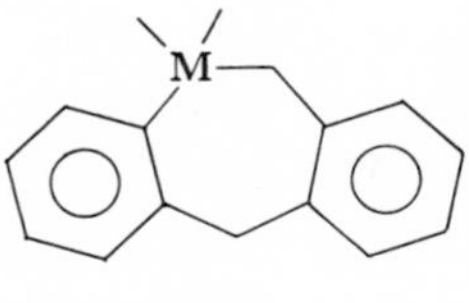 (XVb) 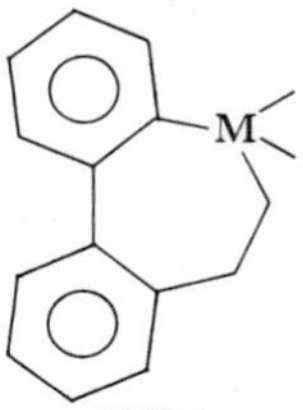 (XVc) 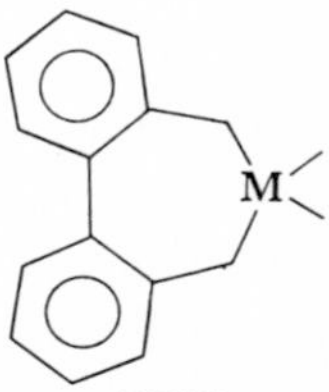 (XVd)

It is not surprising that all but one of the known heterocyclic metallepin derivatives are those associated with the symmetrical structures (XVa) and (XVd). The Ring Index system describes each of the preceding isomers as follows: (XVa), 10,11-dihydro-5*H*-dibenzo[*b,f*]metallepin; (XVb), 6,11-dihydro-5*H*-dibenzo[*b,e*]metallepin; (XVc), 6,7-dihydro-5*H*-dibenzo[*b,d*]metallepin; (XVd), 5,7-dihydro-6*H*-dibenzo[*c,e*]metallepin. The majority of the known derivatives are of type (XVa), but the bridged biphenyl (XVd) is known for arsenic and selenium. All four derivatives are known only for oxygen and sulfur heteroatoms.

1. *10,11-Dihydro-5H-Dibenzo[b,f]metallepins*

The sole method of synthesis that has been developed thus far involves the formation of a diGrignard or dilithio reagent from *o,o'*-dibromobibenzyl. The dibromo derivative is prepared by the coupling of *o*-bromobenzylbromide with phenyllithium (*215*). An alternative route that would utilize ring closure at the bridging carbon atoms (i.e., a site remote from the heteroatom) has not yet been explored. The dibenzazepin and dibenzoxepin derivatives have been prepared by this approach (*35*). The Grignard reactions are summarized in Eq. (97).

2. *5,7-Dihydro-6H-dibenzo[c,e]metallepins*

The most convenient way of viewing these derivatives is as bridged biphenyls. The starting dibromide in the synthesis of these symmetrical

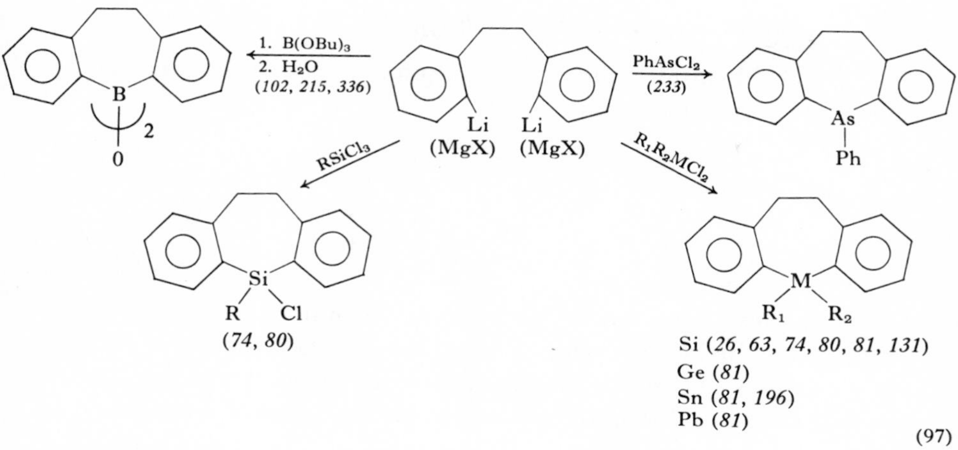

(97)

derivatives is *o,o'*-bis(bromomethyl)biphenyl which can be prepared from diphenic acid. Formation of the diGrignard reagent from this compound would provide a direct route into the heterocyclic system (XVd). However, reaction of *o,o'*-bis(bromomethyl)biphenyl with magnesium results in ring closure and generation of dihydrophenanthrene (*147*). An attempt to trap the organic Grignard reagent by reaction of the organic dibromide with dimethyldichlorosilane in the presence of magnesium gave small quantities of the silicon heterocycle, but attempts to separate and purify the 6,6-dimethyl-5,7-6*H*-dibenzo[*c,e*]silepin have been unsuccessful (*79*).

The heterocyclic derivatives that have been generated incorporate other reactions of *o,o'*-bis(bromomethyl)biphenyl.

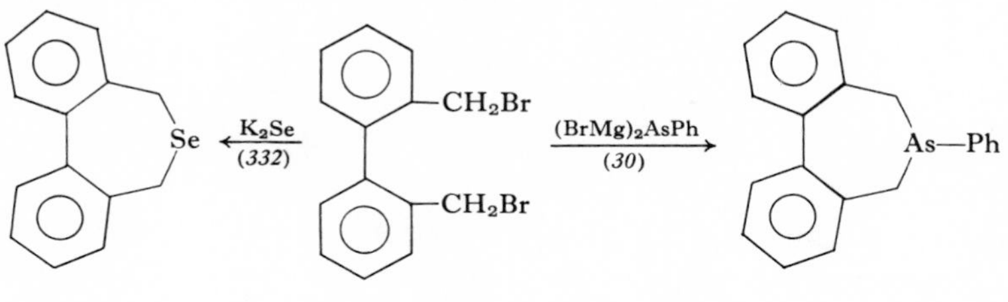

(98)

The reaction of the arsenic diGrignard is similar to that reported for benzheterocycles of arsenic (Scheme 5).

3. *6,11-Dihydro-5H-dibenzo[b,e]metallepin*

The 6,11-dihydro-5*H*-dibenzo[*b*,*e*]silepin system has been prepared by the ring-expansion reaction:

$$\xrightarrow[C_6H_6\ (83)]{AlCl_3} \quad \xrightarrow{LiAlH_4} \qquad (99)$$

Si, CH_3, CH_2Cl; Si, Me, Cl; Si, Me, H

Unsymmetrical heterocyclic derivatives containing large rings will probably require indirect synthetic routes due to the difficulty of obtaining the appropriate organic dibromides and the difficulty associated with coupling reactions during the generation of the Grignard reagent.

D. *Dibenzometallocins*

There are six possible isomers of dibenzometallocins. Reactions from Grignard or dilithio reagents are possible, but yields would probably be small. Only one derivative could practicably be prepared via the Grignard route, i.e.,

$$\xrightarrow[2.\ R_2MCl_2]{1.\ Mg} \qquad (100)$$

Br, Br; M, R_1, R_2

although a study of this reaction has not yet been reported.

Only one dibenzocyclooctane derivative, 5-methyl-5-chloro-5,6,10,11-tetrahydrodibenzo[*b*,*f*]silocin has been prepared at this writing. Derivatives of this system are prepared by the ring-expansion reaction (*83*):

$$\xrightarrow[C_6H_6\ (82)]{AlCl_3\ \text{or}\ AlBr_3} \qquad (101)$$

Si, Me, CH_2Cl; Si, CH_3, Cl

Alternative reactions leading to the formation of large ring heterocycles will probably require ring closure at sites remote from the heteroatom. Large-membered cycloalkenes containing silicon and germanium have been prepared by this approach.

E. Dibenzheterocycles Containing Additional Multiple Bonds—the 14π-Electron System

1. *Derivatives of Anthracene*

Two methods have evolved for the generation of 14π-electron, "aromatic" heterocyclic derivatives of anthracene. In the first approach neutral derivatives are obtained by a 9,10-HCl elimination from the dihydroanthracene derivative (B, Si, P, As); in the second method, anions are generated by proton abstractions from the neutral saturated heterocycle (B, Si).

Transannular HCl eliminations involving the heteroatom [cf. Eq. (102)] have achieved only moderate success. Although reactions of boron and silicon heterocycles with amine bases appear to proceed, monomeric products have not yet been isolated. The results are summarized in Eq. (102).

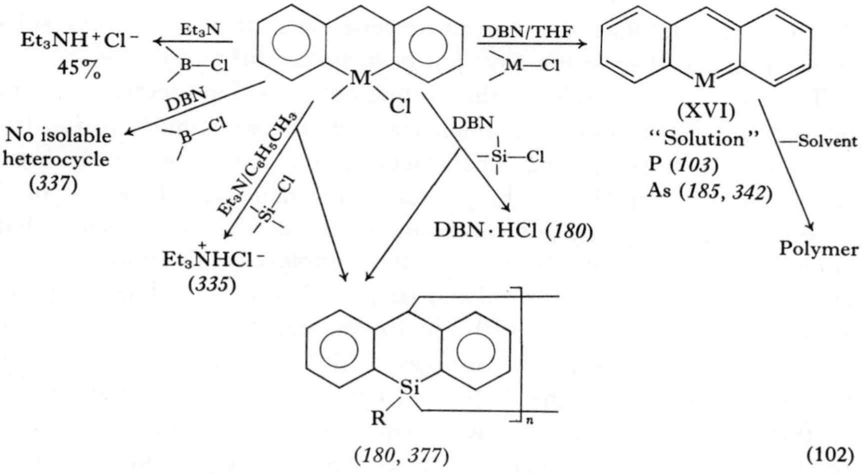

$Et_3NH^+Cl^-$ 45% ← Et_3N / B—Cl

DBN / B—Cl → No isolable heterocycle (*337*)

$Et_3N/C_6H_5CH_3$ / Si—Cl → $Et_3\overset{+}{N}HCl^-$ (*335*)

DBN / Si—Cl → DBN·HCl (*180*)

DBN/THF / M—Cl → (XVI) "Solution" P (*103*) As (*185, 342*) —Solvent → Polymer

R—Si [ ]$_n$ (*180, 377*)

(102)

Arsaanthracene appears to be unstable as the monomer since removal of the solvent from the HCl elimination reaction produced only polymeric material. The presence of arsaanthracene was inferred from spectral evidence (Section VI) and on the basis of adduct formation with maleic anhydride.

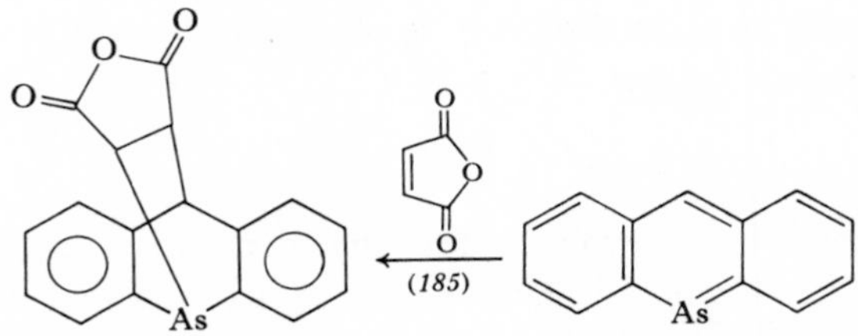

(185) (103)

Stable derivatives of (XVI) which are apparently monomers have been isolated for both P (*104*) and As (*343*) if a phenyl group is substituted for hydrogen in the 10-position. Since this report is preliminary, the assignment of the "anthracene" structure is only tentative. A recent X-ray structural study has shown that arsanthrene is a dimer and not the monomer (*178*):

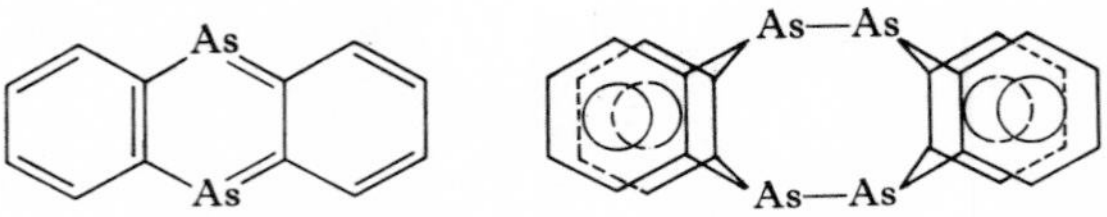

Attempts to stabilize the boraanthracene derivative by phenyl substitution in the 10-position have not been successful (*339*).

The second approach to the formation of a 14π-electron system involves the generation of an anion via proton abstraction from the 10-position with butyllithium. The reactions that have been studied are summarized in Eq. (104). The products indicated were those proposed in the original reports. Interpretation of these reactions is somewhat complicated. In alkane solvents, addition complexes of the boron heterocycle (i.e., quaternary boron anions) are probably formed. The exception appears to be the 9-mesityl-9,10-dihydro-9-boraanthracene system. Lack of complex formation in this case may be due to steric hindrance at the boron heteroatom by the bulky mesityl substituent. If the proton abstraction reaction is run in the presence of ethereal solvents, butane is produced and the precipitated salts contain solvated lithium cations. Support for lithiation in the 10-position is provided by hydrolysis studies with D_2O. Reaction with D_2O regenerates the neutral heterocycle with deuterium incorporated in the 10-position. The UV spectral data of solutions of the heterocyclic anions show similarities to the neutral 14π-electron systems but the nature of the electron distribution, i.e., localized vs. nonlocalized, remains unanswered. The state of aggregation of these organolithium reagents and the possible effect of

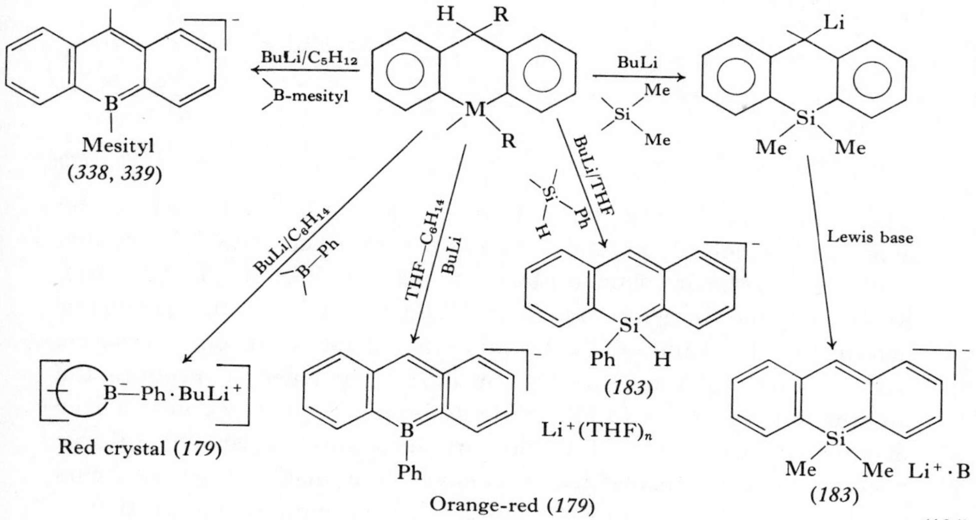

(104)

aggregation on the spectral data obtained need to be considered before the questions of electron delocalization can be answered. Spectral data for the 14π-electron systems are given in Section VI, F.

2. *Derivatives of Dibenzocycloheptatriene*

The impetus behind the generation of a 14π-electron system in which the central ring contains seven members is the formal analogy to the tropylium ion. Derivatives of (XVII) have been generated by traditional organic reactions at the ethylene bridge. Historically, the borepin derivative was the first of this class to be generated.

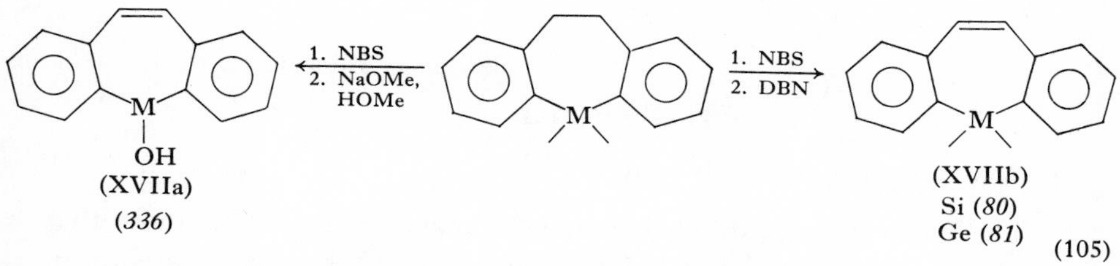

(105)

Other methods have also been developed for the generation of a double bond in the 10,11-position:

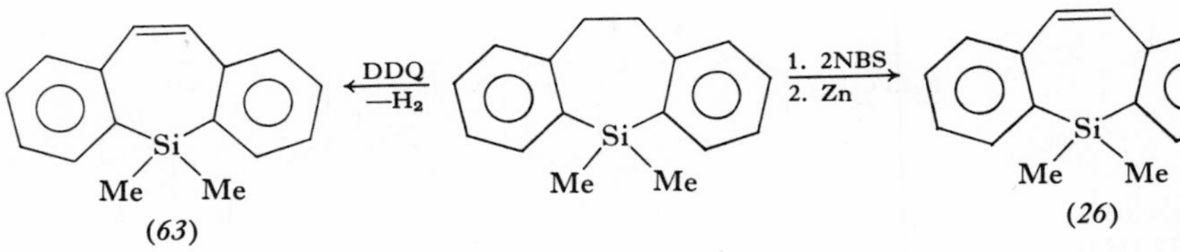

(63) ←DDQ / $-H_2$ — 1. 2NBS, 2. Zn → (26) (106)

There are several limitations on these methods with regard to the heteroatom involved. Reaction of the tin heterocycle with NBS results in cleavage of the aromatic–tin bonds to give *o,o'*-dibromobibenzyl (*81*). Reaction of the tin heterocycle with DDQ resulted only in products of heteroatom elimination (*81*). Attempts to generate a silicon heterocycle with a functional group on the Si atom have failed to produce the unsaturated system (XVII). Dehydrobromination reactions of the silane afford highly colored reaction products but no isolable identifiable monomer. It is probable that the dehydrobromination reaction occurs not only at the ethylene bridge but also by transannular elimination.

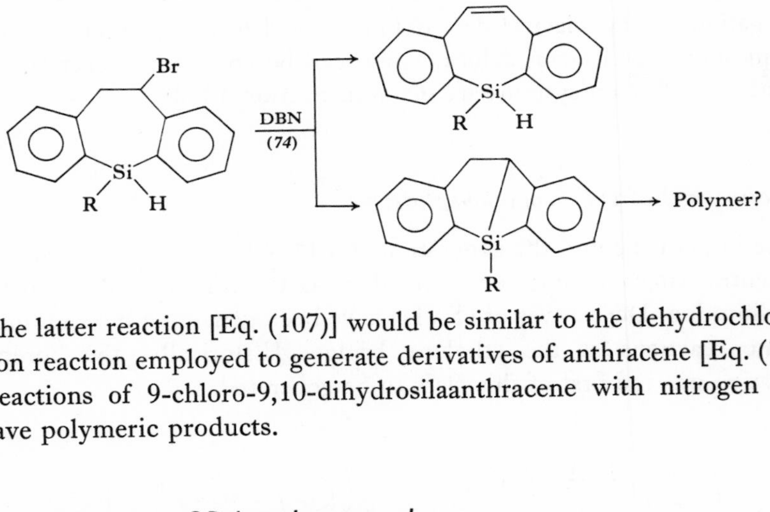

DBN (74) → Polymer? (107)

The latter reaction [Eq. (107)] would be similar to the dehydrochlorination reaction employed to generate derivatives of anthracene [Eq. (102)]. Reactions of 9-chloro-9,10-dihydrosilaanthracene with nitrogen bases gave polymeric products.

F. Reactions of Dibenzheterocycles

1. *Heteroatom Substitution Reactions*

The most versatile dibenzheterocyclic reagents have been those that contain an M—Cl (X) bond. Only the Group IV heterocycles with an M—Cl bond can be produced in a direct synthetic process (such as reaction of the appropriate dilithio derivative with $RMCl_3$). The chloro-substituted derivatives of boron are frequently prepared from the heterocyclic ethanolamine ester but have also been generated by cleavage of a boron–carbon bond. The Group V chloro derivatives may be generated by reduction of the arsinic acid or by cleavage of a carbon–metal bond:

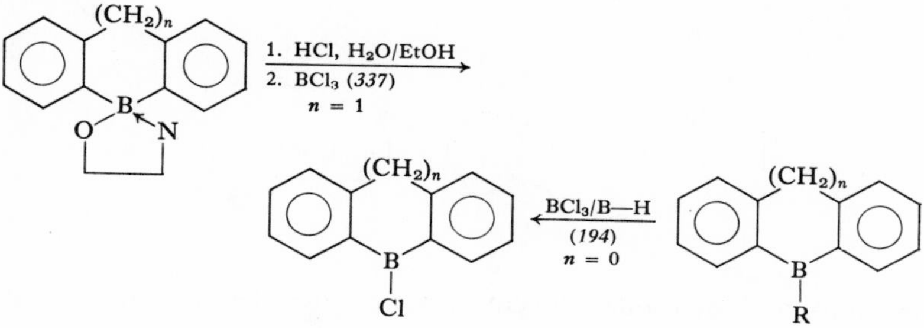

$(CH_2)_n$; B ; O ; N

1. HCl, H_2O/EtOH
2. BCl_3 (*337*)
$n = 1$

$(CH_2)_n$; B ; Cl

BCl_3/B—H
(*194*)
$n = 0$

$(CH_2)_n$; B ; R

(108)

$(CH_2)_n$; M ; O ; OH

SO_2/HCl
(*142*)
$n = 1$

$(CH_2)_n$; M ; Cl

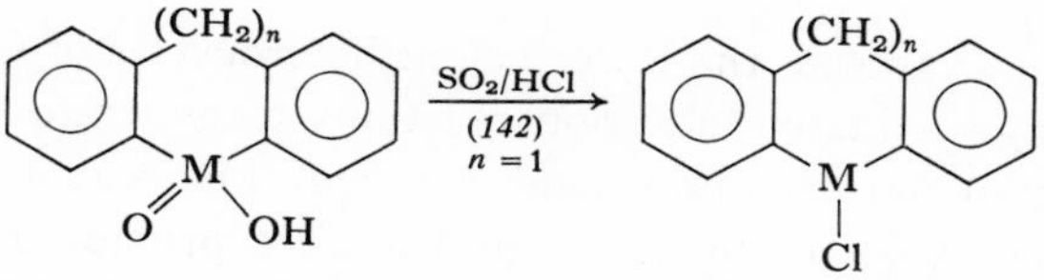

(109)

Reactions of the M—Cl bond are quite varied and include hydride reduction, metal coupling, and hydrolysis. The various types of reactions that have been carried out at the M—Cl center are summarized in

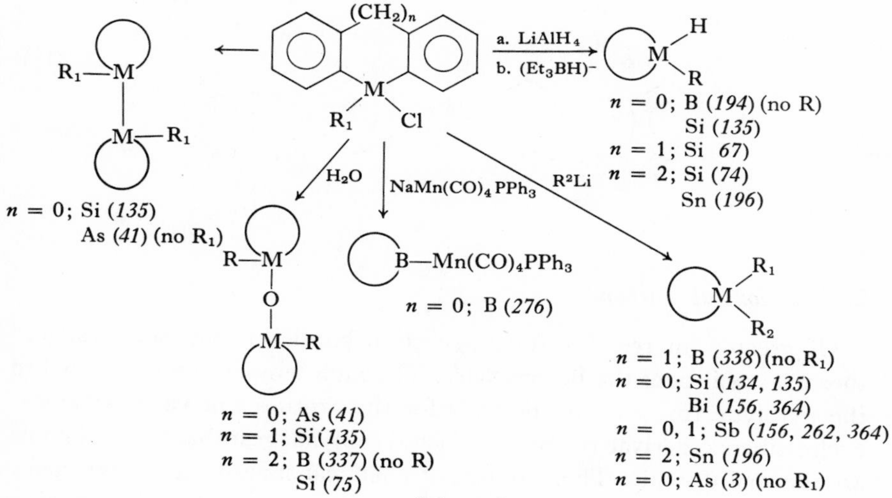

$(CH_2)_n$; M ; R_1 ; Cl

R_1—M ; M—R_1
$n = 0$; Si (*135*)
As (*41*) (no R_1)

a. $LiAlH_4$
b. $(Et_3BH)^-$

M ; H ; R
$n = 0$; B (*194*) (no R)
Si (*135*)
$n = 1$; Si (*67*)
$n = 2$; Si (*74*)
Sn (*196*)

H_2O

R—M ; O ; M—R
$n = 0$; As (*41*)
$n = 1$; Si (*135*)
$n = 2$; B (*337*) (no R)
Si (*75*)

$NaMn(CO)_4PPh_3$

B—$Mn(CO)_4PPh_3$
$n = 0$; B (*276*)

R^2Li

M ; R_1 ; R_2
$n = 1$; B (*338*) (no R_1)
$n = 0$; Si (*134, 135*)
Bi (*156, 364*)
$n = 0, 1$; Sb (*156, 262, 364*)
$n = 2$; Sn (*196*)
$n = 0$; As (*3*) (no R_1)

Scheme 7

Scheme 7. Substitution reactions at pentavalent Group V derivatives may result in reduction.

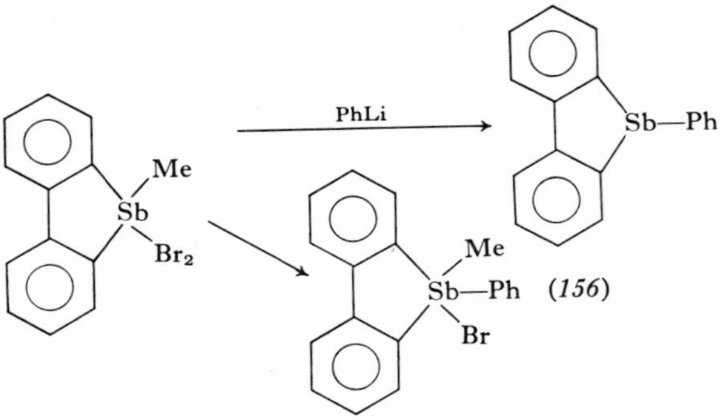

PhLi

Me Sb Br_2 → Sb—Ph

Me Sb—Ph (*156*) Br

(110)

The reactions and methods of synthesis are included in the data in Table X.

As might be expected, the lower valent heterocycles of Groups V and VI will undergo oxidative addition reactions to generate derivatives of higher oxidation states without ring cleavage. The reaction of halogens with the heterocycles of tellurium and arsenic provide an example in contrast to the chemistry of Groups III and IV, as depicted in Scheme 8.

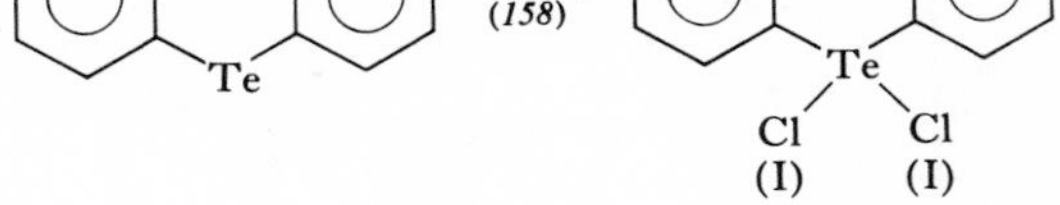

Sb R —Cl_2 (*156*)→ Sb Cl Cl R

(111)

Te —$Cl_2(I_2)$ (*158*)→ Te Cl (I) Cl (I)

2. *Reactions at Carbon*

Of greater interest for future reaction studies is the chemical and thermal stability of the heterocycles. The high temperatures involved in the pyrolytic ring-closure methods for the synthesis of various heterocyclic dibenzo derivatives [Eqs. (86) and (92)] suggest that these systems are thermally stable. The problem of chemical stability in the presence of halogens, hydrogen halides, acids, oxidizing agents, and reducing agents is important as it will limit the study of the reactivity of the heterocyclic systems. Since the known heterocyclic isomer systems are actually analogs of Ar_2MX_n, the question is related to the stability of the phenyl–metal bond in the presence of the foregoing list of reagents. Electrophilic reagents tend to cleave phenyl–metal bonds (*107*, *214*, *274*). This has been observed for heterocyclic systems as well. Both ring C—M bonds tend to be cleaved in the presence of halogens, but reactions of hydrogen halides frequently cause cleavage of only one of the ring carbon–metal bonds (Scheme 8). Dehydrogenation of the dibenzometalle-

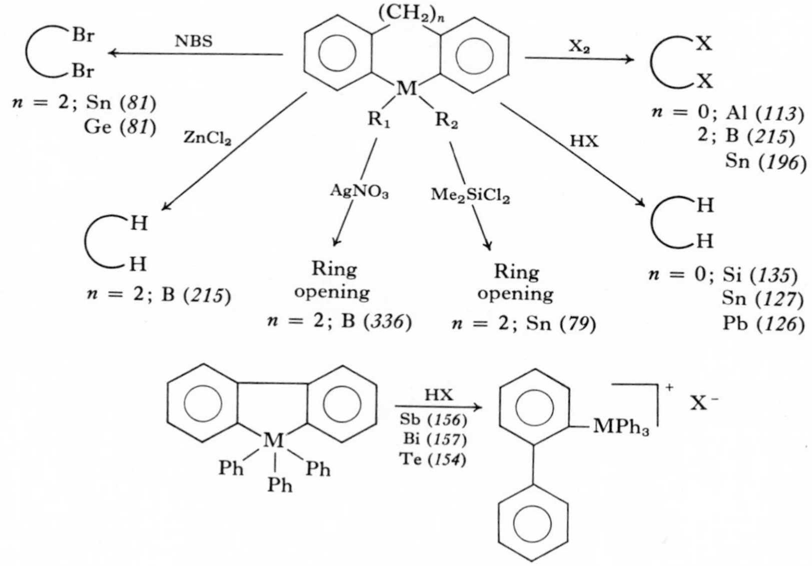

$(CH_2)_n$ NBS X_2 ZnCl$_2$ HX AgNO$_3$ Me$_2$SiCl$_2$

Br, Br: $n = 2$; Sn (*81*), Ge (*81*)

X, X: $n = 0$; Al (*113*); 2; B (*215*), Sn (*196*)

H, H: $n = 2$; B (*215*)

Ring opening $n = 2$; B (*336*)

Ring opening $n = 2$; Sn (*79*)

H, H: $n = 0$; Si (*135*), Sn (*127*), Pb (*126*)

HX: Sb (*156*), Bi (*157*), Te (*154*) → MPh_3^+ X^-

Scheme 8

pins also results in ring-cleavage reactions:

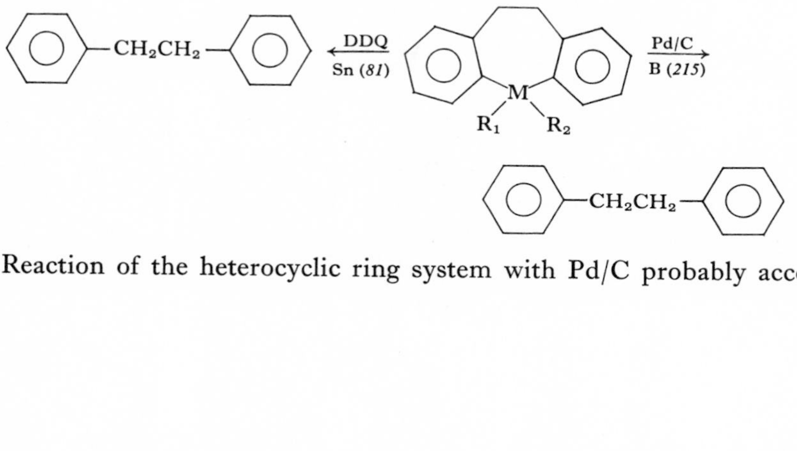

$C_6H_5CH_2CH_2C_6H_5$ ← DDQ, Sn (*81*) — M, R_1, R_2 — Pd/C, B (*215*) → $C_6H_5CH_2CH_2C_6H_5$ (112)

Reaction of the heterocyclic ring system with Pd/C probably accounts for the lack of success of cylodehydrogenation as a route to heterocyclic synthesis [Eq. (86)].

Reactions of oxidizing agents with dibenzheterocycles have also been reported. Arsenic acids are stable in sulfuric acid and can be nitrated without ring cleavage:

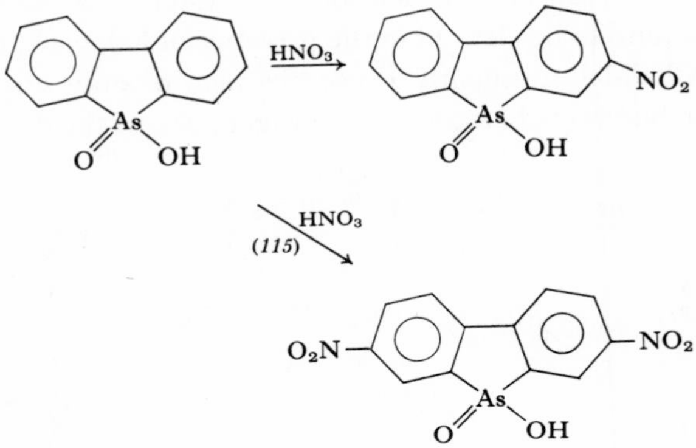

HNO_3 (113)

HNO_3 (*115*)

The nitro derivatives can be reduced to amines and subsequent diazatization reactions may afford other ring-substituted derivatives.

Oxidation of the methylene bridge of heterocyclic derivatives of 9,10-dihydroanthracene have also been successful:

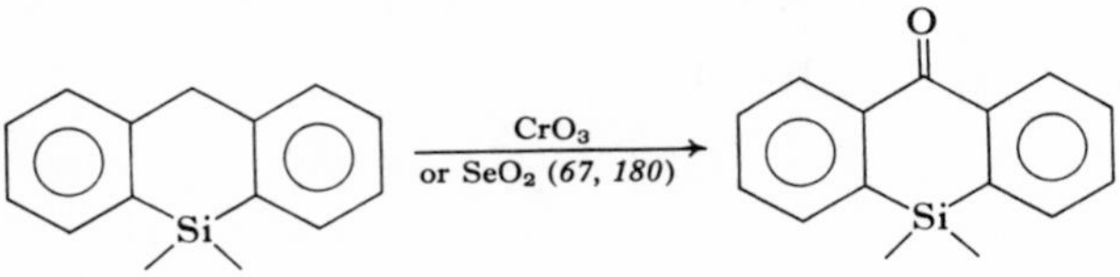

CrO_3 or SeO_2 (*67, 180*) (114)

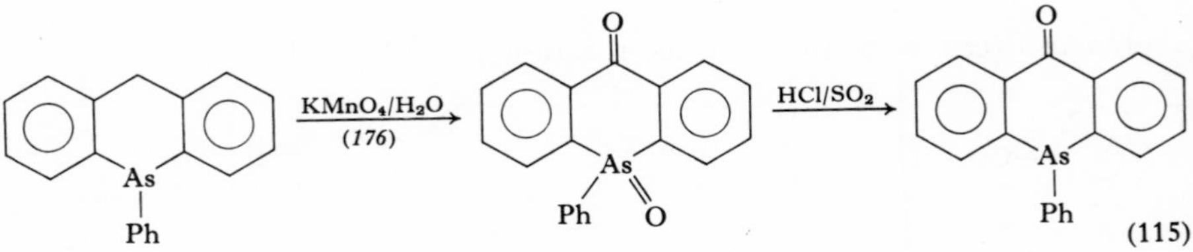

$KMnO_4/H_2O$ (*176*) HCl/SO_2 (115)

Reaction of boron heterocycles with hydrogen peroxide, however, give only ring-cleavage products (*215*). The derivatives of arsenic(V) are usually prepared through the tosylamine method:

$$\text{As—Ph} \xrightarrow[(363)]{} \text{As(=NTS)Ph} \xrightarrow{\text{PhLi}} \text{AsPh}_3 \quad (116)$$

Other reactions involving addition to the heteroatom have been reported for Group V and VI analogs. Frequently heterocycles of Group V are isolated and purified by conversion to the methiodide salt. The selenepin has been reported to react with 2 moles of methyl iodide where commonly the observed reactions are in a 1:1 ratio (*332*).

$$\text{Se} + \text{MeI} \xrightarrow[(332)]{} [\text{SeMe}_2]^{2+} \; 2\text{I}^- \quad (117)$$

V

TRIBENZOMETALLEPINS AND TETRABENZOMETALLONINS

Although large ring systems pose interesting structural and stereochemical problems, few heterocycles have been reported that contain a metal or metalloid heteroatom. Those that have been prepared are generally obtained by an indirect route, and only derivatives containing an arsenic or a silicon atom have been reported.

Formation of the tribenzoarsepin derivative involves thermolysis of a quaternary arsonium salt in a reaction similar to that which generates arsafluorene derivatives [Eq. (86)]. As for the dibenzheterocycles, an organic dihalide provides the starting point for the synthesis. Reaction of *o*-bromoiodobenzene with magnesium in ether at 0°C followed by reaction with dimethyliodoarsine provides a disubstituted terphenyl derivative. Addition of 1,3-dibromopropane forms the cyclic quaternary salt (*232*):

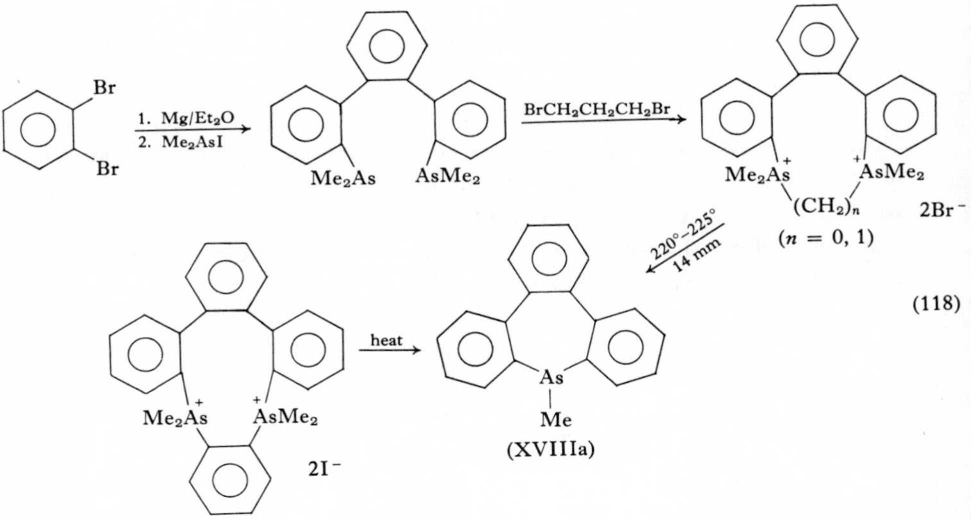

Br Br — 1. Mg/Et_2O, 2. Me_2AsI → Me_2As $AsMe_2$ — $BrCH_2CH_2CH_2Br$ → $Me_2\overset{+}{As}$ $\overset{+}{As}Me_2$ $(CH_2)_n$ $2Br^-$ (n = 0, 1) — 220°–225° 14 mm → As Me (XVIIIa)

$Me_2\overset{+}{As}$ $\overset{+}{As}Me_2$ $2I^-$ — heat → As Me (XVIIIa)

(118)

The reaction of the product produced from magnesium and *o*-bromoiodobenzene, in principle, should not be limited to a metal halide. It is possible that reaction with a metal dihalide may result in the direct formation of the tribenzometallepin system. The reaction will be complicated by formation of disubstituted products which may form in preference to the ring closure. Recent work suggests that low yields of ring-closure product can be obtained from *o,o'*-dilithioterphenyl. The *o,o'*-dilithioterphenyl is generated by an indirect route involving prior formation of a mercury heterocycle:

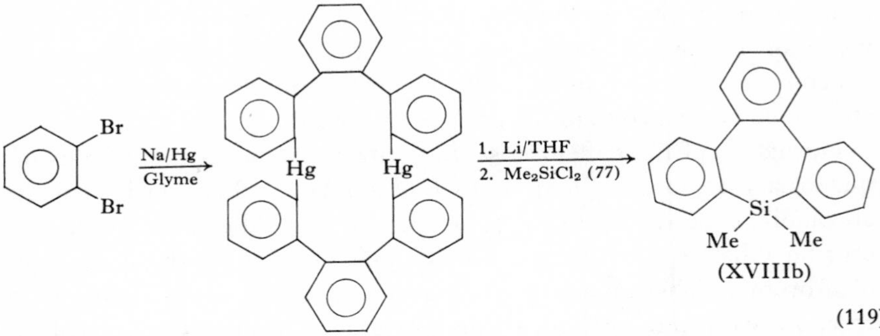

Br Br — Na/Hg, Glyme → Hg Hg — 1. Li/THF, 2. Me_2SiCl_2 (77) → Si Me Me (XVIIIb)

(119)

The success of the lithiation reaction is variable and seems to depend on the purity and crystalline form of the mercury heterocycle. The yields of the Si heterocycle are only about 3.5%. Products other than triphenylene and the recovered starting material have not been identified in the reaction products (*77*). The 9,9-dimethyl-9*H*-tribenzo[*b*,*d*,*f*]silepin has also been generated in trace quantities in the reaction of phenylmethyl methoxychlorosilane with sodium in toluene (*6*). Structural properties of the silepin system are discussed in Section VI. In an unusual ring closure reaction at a site remote from the heteroatom, derivatives of XVIIIb have been prepared by reaction of bis(o-phenylethynylphenyl)dimethylsilane with RC≡CR in the presence of a rhodium catalyst (*381*, *382*).

The nine-membered tetrabenzoarsonin system is produced in an interesting thermolysis reaction of an arsenic(V) derivative (*232*):

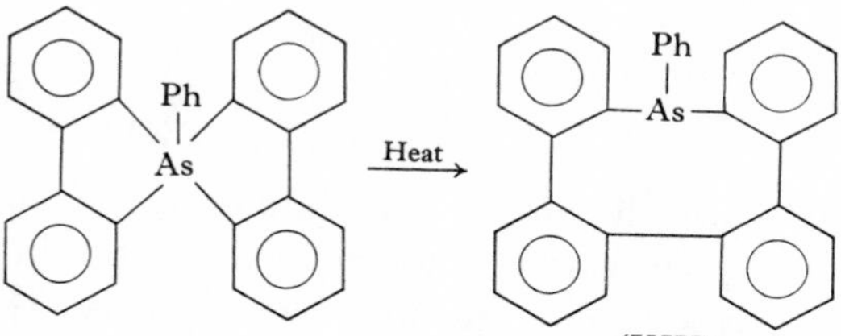

Ph As —Heat→ Ph As (120)

(XIX)

The reaction converts the arsenic(V) spiro derivative to an arsenic(III) heterocycle. This synthetic approach would be limited to those metal heteroatoms that can exist in two different stable oxidation states.

VI
STRUCTURE, STEREOCHEMISTRY, AND SPECTRA

Structural aspects of heterocyclic systems that contain a metal heteroatom have been little explored. Such a wealth of data is available for carbocycles that these will be used as initial structural models for the organometallic systems. The questions related to the effect of replacing a —CH_2— group in a carbocycle by a metal atom are many-sided. In the next few paragraphs the structures of carbocycles and their benzene-annelated derivatives will be briefly reviewed, followed by a discussion of the problems introduced, or changes that may be expected, upon heteroatom replacement. Then the data observed for organometallic systems will be discussed in terms of the points raised in these sections.

A. Structures of Carbocycles

Five-membered rings containing only one multiple bond are generally nonplanar (puckered) but cyclopentadiene is planar. All larger rings regardless of degree of unsaturation are nonplanar with the exception of benzene. Conformations of rings with six or more members are usually variations of a chair or boat form. These conformations for a six-membered ring are as follows:

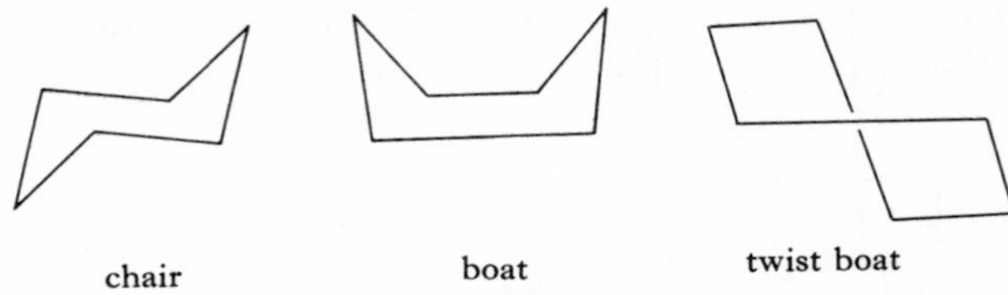

chair boat twist boat

The larger the ring size the greater the number of ring conformers that must be considered. Benzcarbocycles may exist either in chair or boat forms. The following are some possible forms of benzcycloheptane:

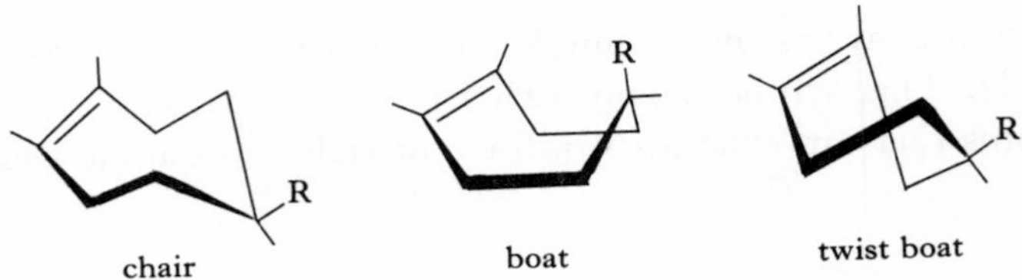

chair boat twist boat

Additional multiple bonds in the benzcarbocycle tend to favor the boat or twisted boat form.

Dibenzcarbocycles are of two general types, i.e., derivatives of biphenyl or those in which the benzene rings are separated by one or more —CH_2— groups. These two structural types are shown below for a system with a central six-membered ring:

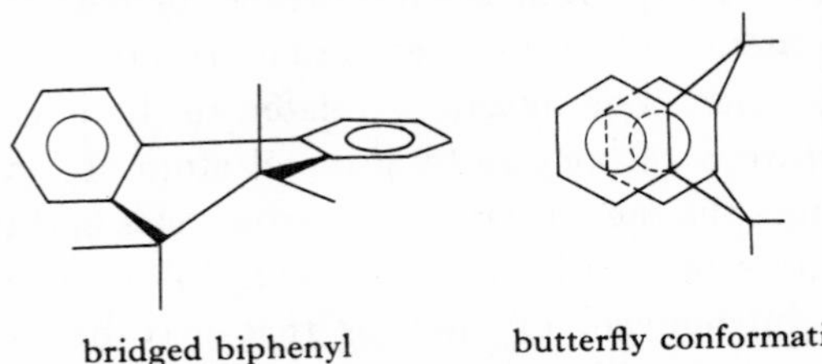

bridged biphenyl butterfly conformation

Tricyclic systems with a central ring size of six or seven members exhibit a boat conformation (butterfly conformation), and with an eight-membered ring either the chair or boat (basket) conformations:

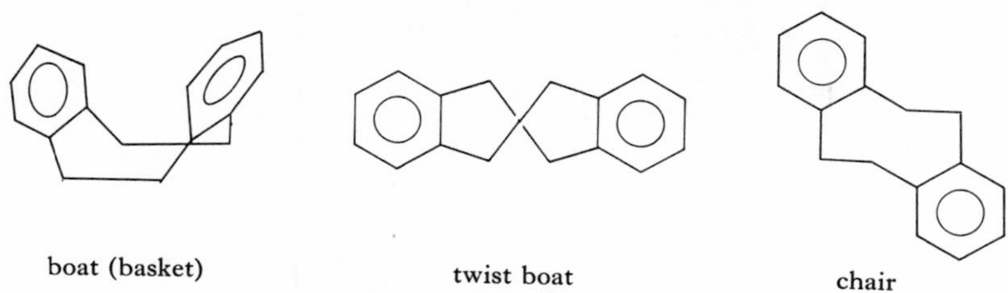

boat (basket) twist boat chair

B. Heteroatom Replacement: Possible Effects

Two interrelated effects may result when a —CH_2— group in a carbocycle is replaced by a heteroatom. The first is related to a possible change in structure as compared to the carbocycle and the second is related to the unique structural features of the heteroatom itself as compared to carbon. In Table XI are listed some of the properties associated with the heteroatom that are considered in the structural arguments here and in later sections.

As can be seen from Table XI, replacement of a —CH_2— group of a carbocyclic system with a metal heteroatom results in the introduction of two long-ring bonds as well as altering the internal ring angles. In a ring system the C—M—C bond angles are normally less than the C—C—C ring angles. If the carbocycle is used as a model for an organometallic heterocyclic system, several structural effects of heteroatom replacement should be considered. These include: (*a*) a possible change in ground-state conformation as a result of replacing —CH_2— by —MH_x—; (*b*) possible alteration of ground-state conformation by substituents at the metal heteroatom or at a ring carbon (may only be important in larger-ring systems); (*c*) substituent preference for axial or equatorial positions; and (*d*) change in barriers for conformational inversion processes for heterocyclic systems as compared to the carbocyclic analogs and effects on conformational equilibria by substituents on the heteroatom.

The unique structural features associated with the heteroatom may also play an important role in determining the ground-state structure of the heterocycle. Group IV atoms are most closely related to carbon in

TABLE XI

PHYSICAL PROPERTIES OF MAIN GROUP ELEMENTS

Heteroatom	χ_P [a]	$(M—C_{calc}(Å))$ [b]	$(M—C)_{obs}(Å)$ [c]	C—M—C [c] angle	Derivative	References
B	2.0	1.58	1.56(ED) [d]	120	Me_3B	*219*
Al	1.5	2.03	1.97(X) [e]	123	$(Me_3Al)_2$	*355*
Si	1.8	1.88	1.89(ED)	109	Me_4Si	*316*
Ge	1.8	1.93	1.98(ED)	109	Me_4Ge	*51*
Sn	1.8	2.11	2.18(ED)	109	Me_4Sn	*51*
Pb	1.8		2.20(ED)	109	Me_4Pb	*366*
As	2.0	1.94	1.98(ED)	96	Me_3As [f]	*320*
			1.96(MW)	96	Me_3As	*220*
Sb	1.9	2.13	2.20(ED)	100	$(CF_3)_3Sb$	*43*
Bi	1.9		2.24(X)	94	Ph_3Bi	*149*
			2.27(ED)	97	Me_3Bi	*372*
Se	2.4	1.92	1.93(X)	106	$(p\text{-}CH_3C_6H_5)_2Se$ [g]	*40*
Te	2.1	2.12	2.05(X)	101	$(p\text{-}CH_3C_6H_5)_2Te$	*39*

[a] Electronegativity values from L. Pauling, "Nature of the Chemical Bond," Cornell Univ. Press, Ithaca, New York (1960).
[b] Calculated from the Stevenson Schomaker Equation: $r_C + r_M - 0.09|\chi_M - \chi_C|$.
[c] For standard deviations, see original articles.
[d] Method of measurement given in parentheses: ED = electron diffraction; MW = microwave; X = X-ray.
[e] Data for terminal methyls only.
[f] In $(CF_3)_3As$: C—As is 2.05 Å and C—As—C is 100° (*43*).
[g] In $Me_3Se^+I^-$: C—Se = 1.96Å, C—Se—C = 98.5° (*165*).

that they normally exist in a tetrahedral, four-coordinate environment. Group III heteroatoms, however, normally exist in a trigonal, three-coordinate environment, and Group V heteroatoms in a pyramidal, three-coordinate environment. Additional properties of the heteroatom that may influence the structure adopted by the heterocycle are discussed below.

1. *Replacement of Carbon by Another Group IV Atom*

Experimental data as well as calculations (*284, 330*) reveal two important differences between silicon–carbon bonds and carbon–carbon bonds, i.e., torsional barriers are small and gauche interactions unimportant for the Si—C bond when compared to the C—C bond. Low torsional barriers may imply a more flexible system (lower barriers to ring inversion), and small gauche interactions may imply lack of substituent preference or low substituent preference when compared to the parent carbocycle. Although calculations for heavier Group IV heteroatoms have not been published, it is probable that results will be similar to those obtained for Si.

2. *Replacement of Carbon (or Group IV Atom) by a Group V Atom*

Although bond lengths of Group IV and V atoms are similar, the C—M—C bond angles are usually smaller for Group V than those observed for Group IV. In addition, there is a lone pair present on the Group V heteroatom. This lone pair may introduce a more subtle aspect of substituent preference in comparison with ring systems that contain a Group IV atom. A direct comparison between heterocyclic systems containing a Group IV atom or a Group V atom may be questionable and should be made with care. However, the tetrahedral quaternary salts of Group V heteroatomic systems could be comparable to the Group IV derivatives if electronic effects are unimportant.

3. *Replacement of Carbon by a Group III Element*

Borane derivatives of the type R_2BH are dimeric except when R is a very bulky group (*273*). This precludes a direct comparison between boracyclanes and parent carbocycles or other heteroatom, —MH_x— systems. Since R_3B systems are monomeric, replacement of —CHR— in a carbocycle with —BR— would have the effect of introducing a trigonal center (but a trigonal center with long bonds) and, thus, a more appropriate model for this system may be a carbocycle in which the —CH_2— has been replaced by a C═O or >C═CH_2 unit. The anionic salts of Group III are, however, structurally related to the four-coordinate Group IV derivatives and the four-coordinate quaternary salts of Group V. If carbocycles are used as model compounds, the effect of heteroatom replacement may more appropriately involve those heteroatoms that are in tetrahedral environments.

C. Solid- and Gas-State Structures

Structural data for cycloalkenes and related heterocycles are included in Table XII. All metal heteroatoms considered here except boron have larger single bond radii than carbon. From the limited amount of data available, it appears that introduction of a heteroatom with a larger single bond radius than carbon tends to flatten a five- or six-membered ring system (see cyclopentene and silacyclopentene, 1,4-disila-2,5-cyclohexadiene and 1,4-disila-2,5-cyclohexadiene). No structural information is yet available for benzheterocycles containing a metal atom. For comparison, however, replacement of one of the —CH_2— groups of tetrahydronaphthalene with an S═O group (S—X and Si—C bonds are of comparable length) results in a change from the half-chair exhibited by the hydrocarbon to a boat conformation for the heterocycle. This change, however, may be due to the presence of the *cis*-1,4-dimethyl substituents in the sulfur heterocycle. The tricyclic systems with a central ring size of six or seven members and containing a metal heteroatom exhibit a boat conformation, sometimes referred to as the butterfly conformation (or basket conformation for eight-membered rings).

At this time it is not clear whether the conformational changes that occur in the solid state for benzheterocycles, as compared to their hydrocarbon analogs, are a result of ring substitution or crystal forces or are due to the replacement of —CH_2— by a heteroatom with longer bond lengths. Further structural data are required, preferably for unsubstituted derivatives before an assessment of these factors can be made. For tricyclic systems containing central rings with eight or more members, the conformation present in the solid state may not be exclusively present in solution. Dibenzo[*a*,*e*]cyclooctane has the chair conformation in the solid state, but the low-temperature NMR spectral data have been interpreted in terms of a chair $\rightleftharpoons$ boat equilibration near $-72°C$ and equilibration of equivalent conformations of the boat family at $-115°C$ (*91*, *261*).

D. Solution Structure: NMR

Nuclear magnetic resonance spectroscopy is a useful tool for studying conformational equilibria in solution as well as providing information concerning "aromatic character" in 6π-, 10π-, and 14π-electron systems. Several types of conformational changes for ring systems have been studied through the vehicle of variable temperature NMR spectroscopy. Examples of such studies include (1) ring inversion, i.e., boat $\rightleftharpoons$ boat or chair $\rightleftharpoons$ chair processes, (2) equilibria between two different conformations, i.e., chair $\rightleftharpoons$ boat, and (3) substituent preference. Although considerable time and effort have been expended on the study of conformations of cycloalkanes, cycloalkenes, and organic heterocycles of varying ring sizes, few studies include comparative data for organometallic heterocycles.

The results that have been obtained for organic systems may be used as a beginning model for consideration of phenomena that could be observed for organometallic heterocycles. Aspects of conformational equilibria for six-membered ring systems are usually included in standard organic texts, and features of seven-membered ring systems have been reviewed (*326*). Scattered papers on eight-membered ring systems have also appeared recently (*9*, *56*, *91*, *261*, *311*). Data for organometallic heterocycles are still forthcoming, but we shall attempt here to compare data already available and to discuss pertinent results from organic heterocycles that may be of interest in studies of organometallic benzheterocycles.

TABLE XII

STRUCTURES OF UNSATURATED CYCLOALKENES AND HETEROCYCLOALKENES

Heterocycle	M, X	Geometry of ring system	Twist angle or dihedral angle	Geometry at heteroatom	Method	Reference
5-Membered rings:						
M	—CH_2—	Puckered	—	Tetrahedral	IR	*200*
	—SiH_2—	Planar[a]	0°	Tetrahedral	IR	*198*
	—$GeMe_2$—	Planar	—	Tetrahedral	Raman	*207, 209*
M	—SiH_2—	Planar	—	Tetrahedral	IR	*199*
M	—CH_2—	Planar	0°	Tetrahedral	Microwave	*308*
	—Se—	Planar	0°	—	IR	*331*
					Microwave	*53*
	As[b]	Planar	0°	Trigonal	X-ray	*1*
M	—AsPh—	Planar	—	Pyramidal	X-ray	*304*
	—Se—	~Planar	0.4° and 1.2°[c]	—	X-ray	*166*
6-Membered rings:						
M	—CH_2—	Half-chair[d]	—	Tetrahedral	X-Ray	*283, 383*

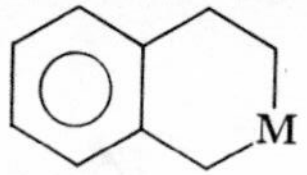	—CH_2—	Half-chair[e]	—	Tetrahedral	X-Ray	*278*
	—SO—	Boat[f]	—	Pyramidal	X-Ray	*290*
	—CH_2—	Nonplanar	17.5°	Tetrahedral	Microwave	*59*
		Nonplanar	—	Tetrahedral	ED	*280*
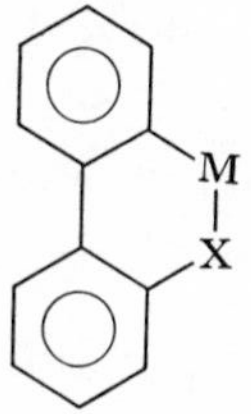	—$(R_1R_2)Si$—O[g]	Twisted	—	Tetrahedral	X-Ray	*143*
	—(R_1R_2P)—CH_2—[h]	Twisted	—	Tetrahedral	X-Ray	*221*
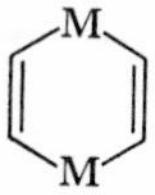	—CH_2—	Planar	—	Tetrahedral	R, IR	*128, 321*
	—CH_2—	Boat	159°	Tetrahedral	ED	*280*
	—GeI_2—	Planar	—	Tetrahedral	X-Ray	*347*
	—$SiPh_2$—[i]	Slight chair	—	Tetrahedral	X-Ray	*347*
	—CH_2— —CH_2—	Folded boat	145°	Tetrahedral	X-Ray	*375*
	—As(Me)— —As(Me)—	Folded boat	117°	Pyramidal	X-Ray	*178*
	—SO			Pyramidal	X-Ray	*172*
	—$CH(CH_3)$— —O—	Folded boat	—	Tetrahedral		
	—O— CDPh	Planar	180°	—	X-Ray	*167*
	B—Mes				X-Ray[m]	*339*

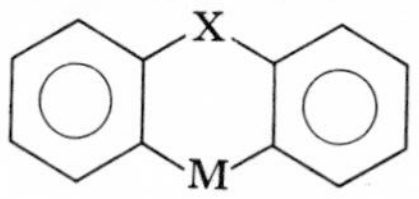

Continued

TABLE XII—*Continued*

Heterocycle	M, X	Geometry of ring system	Twist angle or dihedral angle	Geometry at heteroatom	Method	Reference
[6-membered ring containing M]	—B(R)[j]	Nearly planar[k]	—	Trigonal	X-Ray	*171*
	—As—[l]	Planar	—	—	X-Ray	*303*
7-Membered rings:						
[dibenzo 7-membered ring containing M]	—GePh$_2$—	Folded boat	156°	Tetrahedral	X-Ray	*78*
[7-membered ring containing M]	—CH$_2$—	Boat	—	Tetrahedral	Microwave	*60*
					ED	*329*

Ring	X	Conformation		Geometry at M	Method	Ref.
8-Membered rings:						
Dibenzo ring with X and M	—CH_2— —CH_2— —S— —NR—	Chair Basket	— —	Tetrahedral Pyramidal	X-Ray X-Ray	*19* *367*
 Cyclooctadiene ring with M	—CH_2—	Chair and boat	—	Tetrahedral	ED	*151*

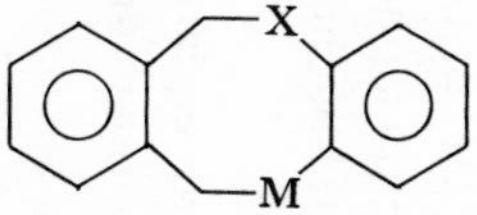

[a] Silacyclopent-2-ene is calculated also to be planar.
[b] In complex $Ph_4C_4AsMn(CO)_3$.
[c] Dihedral angle between five- and six-membered rings.
[d] 3,4,5,6-Tetrachlorocyclohexene.
[e] Tetrachloronaphthalene.
[f] *cis*-1,4-Dimethylisothiochroman-2,2-dioxide.
[g] R_1 = Me; R_2 = *p*-BrC_6H_4—.
[h] R_1 = OH; R_2 = O.
[i] $(Ph_2SiC_2Ph_2)_2$.
[j] In borabenzene complex of cobalt.
[k] 2,3,6-Triphenylarsenin.
[l] Mean torsion angle is 5.3°.
[m] In progress.

1. *Conformation and Substituent Preference*

Not unexpectedly the systems that have received the most attention are derivatives of cyclohexane containing a silicon or phosphorus heteroatom. Although cyclohexane is a saturated system, the results of these studies should provide insight for future work on unsaturated systems. Until future studies provide data to the contrary, it will be assumed that Si provides the model for Group IV heteroatom behavior and P a model for Group V behavior. The first aspects to be considered are associated with ground-state geometry and substituent preference.

The calculated ground-state structure for silacyclohexane is the chair conformer (*330*). Although no solid-state structural work has been published that confirms this assignment, an analysis of 1-methyl-1-*p*-bromophenyl-4-*tert*-butylsilacyclohexane is in progress (*302*). In the solid state, the conformation of 1-phenylphosphorinanone and its dimethyl ketal as well as *trans*-1-methyl-4-*tert*-butyl-4-phosphorinanol (*227*) is the chair form somewhat flattened relative to cyclohexane and with lower torsion angles (*225*, *226*). It is probably reasonable to assume a chair conformation for the saturated heterocyclohexane, independent of the heteroatom.

The question of substituent preference at the heteroatom for equatorial or axial positions remains ambiguous at this time, again due to the paucity of experimental data. Calculations for methylsilacyclohexane indicate that either the axial conformation is slightly more stable (*284*) or that the axial and equatorial isomers are of equal energy (*330*). Similar calculations also show that a *t*-butyl substituent on Si favors the equatorial position but only by 1.6 kcal/mole. In contrast a *t*-butyl group on cyclohexane favors the equatorial position by 5.5 kcal/mole. Substituents on phosphorus in phosphacyclohexane appear to favor the axial position: (*a*) a hydrogen substituent is $\geq 90\%$ axial at $-50°C$ (*201*); (*b*) for a methyl group the axial form predominates (2:1) at room temperature and the equatorial form (2:1) at $-130°C$ (*114*); (*c*) a phenyl group is presumably axial at room temperature in 1-phenyl-4-phosphorinanone (only one conformer observed in solution) (*226*).

In cyclohexane chemistry, a *t*-butyl group that occupies an equatorial position can function as a holding group. Substitution of a *t*-butyl group in the 4-position relative to the heteroatom has been successfully utilized to prepare conformationally stable derivatives of both silacyclohexane and phosphacyclohexane. The derivatives may be separated into cis and trans isomers:

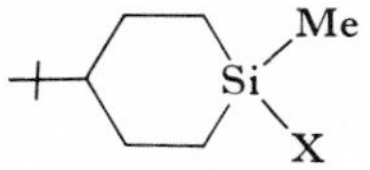

(*302*)
(XX)

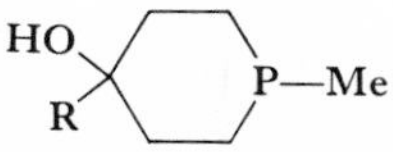

(*291, 317*)
(XXI)

As was the case for 1-methylphosphacyclohexane, the methyl group in (XXI) demonstrates little configurational preference, and, in a cis–trans pair, configurational differences occur at P rather than at C—4.

The foregoing results suggest three generalizations that may be relevant in studies on other ring systems. First, the carbocycle provides a tentative structural model for the heteroatomic system. Second, substituent effects at the metal heteroatom will be small. One of the most commonly used substituents is the methyl group, and experimental data thus far indicate that the free-energy difference between axial M—Me and equatorial M—Me is very small ($\lesssim 0.5$ kcal/mole). Third, substitution of a ring C—H with a bulkier C—R group in the heterocycle should produce an effect related to that found for the parent carbocycle, i.e., a possible holding group.

The only unsaturated heterocycles for which a structure has been calculated by the method of molecular mechanics are silacyclopent-2-ene and silacyclopent-3-ene where all conformers minimized to the planar form (*330*). The far-IR spectra are consistent with a planar conformation (*198*, *199*).

2. *Conformational Inversion*

The systems for which the greatest amount of data are available are the cyclohexane derivatives. The chair $\rightleftharpoons$ chair inversion process is depicted as follows:

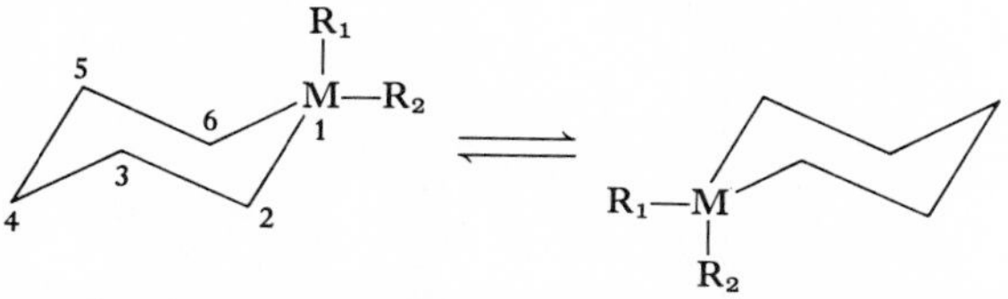

Substituents on the heteroatom convert from eq $\rightleftharpoons$ ax during the inversion process. The barriers to this inversion process for various heteroatom

TABLE XIII

ACTIVATION PARAMETERS FOR CHAIR ⇌ CHAIR INVERSION FOR HETEROCYCOLALKANES

MR_1R_2	Resonance	T_c(°C)	$\Delta G^{\ddagger}$(kcal/mole)	E_a	Solvent	Reference
CMe_2	^{13}C NMR	—36	10.5	—	Neat	*58*
$SiMe_2$	$H—C_{3,4,5}$	−162	5.5^a	6.1	$CBrF_3$	*58*
N—Me	—	−28	—	14.4	CD_3OD	*201b*
P—Me	^{31}P NMR	−75	9.2^b	—	ViCl	*114*
S	$H—C_2$	−81	9.4^a	$11.6^{b,c}$	$CD_3OD/CHClF_2$	*201b,c*
Se	$H—C_2$	−105	8.2^a	11.2	$CHClF_2$	*201c*
Te	$H—C_2$	−119	7.3^a	—	$CHClF_2/CHCl_2F$	*201c*
SO	$H—C_2$	−70	$10.1^{a,d}$	14.2	CH_2Cl_2	*201a–c*
SeO	$H—C_2$	−102	$8.3^{a,d}$	8.2	$CHClF_2$	*201c*
SO_2	$H—C_4$	−63	10.3^b	14.9	CH_2Cl_2	*201a–c*
SeO_2	$H—C_4$	−133	6.7^b	—	$CH_2Cl_2/CHClF_2$	*201c*

[a] Line-shape analysis.
[b] Coalescence temperature method.
[c] E_a reported for −93°C in CH_2Cl_2 (*201b*).
[d] Weighted average of both isomers.

systems are given in Table XIII. The compounds may be grouped according to the environment of the heteroatom. The Group IV derivatives and the sulfones are tetrahedral four-coordinate, and the Group V derivatives and sulfoxides contain three-coordinate pyramidal heteroatoms. The majority of the E_a values for ring inversion fall in the range of 10 to 15 kcal/mole expected for cyclohexane derivatives. A notable exception is the 1,1-dimethylsilacyclohexane derivative. It has been suggested that the origin of the low barrier may be the long C—Si bonds that tend to flatten the ring in the vicinity of the heteroatom and produce a system more coplanar than cyclohexane. If ring inversion proceeds through a semiplanar transition state, the ground-state structure of silacyclohexane is closer to this transition state and, thus, the lower barrier (*58*). Calculations show that the best planar arrangement of four atoms in 1,1-dimethylsilacyclohexane is the 6-1-2-3 set (in contrast to the 2-3-4-5 set for 1,1-dimethylcyclohexane) (*330*). It is probable that similar arguments could be used for the remaining group IV heteroatoms, and these would also be expected to have low barriers to ring inversion. The C—S and C—Se bond distances in the Group VI metallocyclohexanes should be similar to that of C—Si, yet the inversion barriers are greater than that of the silicon derivative. Because the heteroatom–carbon bond distances are so similar, the nonbonded interactions might be expected to be closely related for the sulfones and dimethylsilacyclohexane and the inversion barriers similar for the two systems. However, the chair $\rightleftharpoons$ chair inversion for the sulfone may not proceed through a planar transition state that involves the heteroatom or, possibly, the torsional barriers are high in spite of the long S—C bond lengths. A recent report of a conformational analysis study of Group VI heterocycles indicates that torsional barriers for sulfones should be similar to those for carbon (*201c*). Whether silicon (and other Group IV heteroatoms) is unique with low ring inversion barriers or whether silacyclohexane derivatives are anomalous requires future study of a variety of compounds. Several difficulties attend the study of organometallic heterocycles. Even under conditions of slow exchange in 1,1-dimethylsilacyclohexane, there was no detectable chemical shift difference for the presumed nonequivalent *gem*-dimethylsilyl group (*58*). The persistence of a singlet $Si(CH_3)_2$ in variable temperature NMR studies may not necessarily mean ring flexibility. Conclusions based on equivalence at the metal heteroatom should be made with caution. In a study of bridged biphenyls, it has been shown that the apparent magnetic equivalence of diasteromeric protons is a solvent-dependent phenomenon (*259*). Unfortunately, chemical shifts of Si—Me are not particularly sensitive to solvent media (*293*). Therefore, studies of ring inversion phenomena in organometallic heterocycles will require a study of exchanging nuclei at carbon atoms as well.

Benzene annelation significantly increases the barrier to ring inversion processes. Saturated seven-membered rings can invert by pseudorotation (pseudorotation barrier is ~2 kcal/mole) which has not been measured by NMR techniques. The $\Delta G^{\ddagger}$ for the ring inversion process for the benzene-annellated derivatives of cycloheptane (which probably exists in a chair form) was determined to be 10.9 kcal/mole at $T_c = -57°C$ (*186*, *298*, *322*). Equilibria in tricyclic organic derivatives have been studied more extensively. Several different inversion processes have been reported and depend on structural type. The following are examples of ring inversion processes.

a. Boat ⇌ boat:

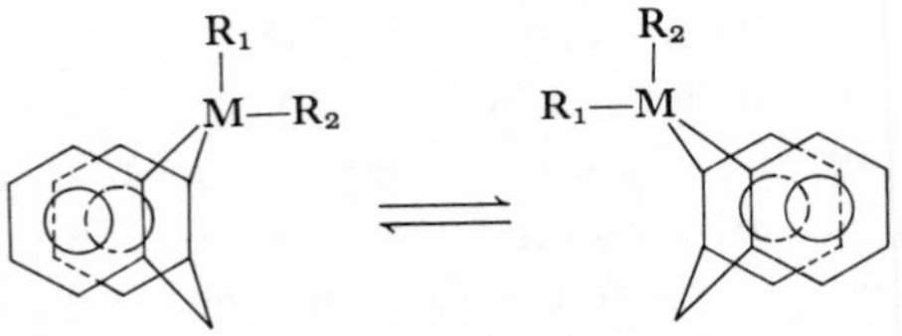

(121)

b. Biphenyl inversion:

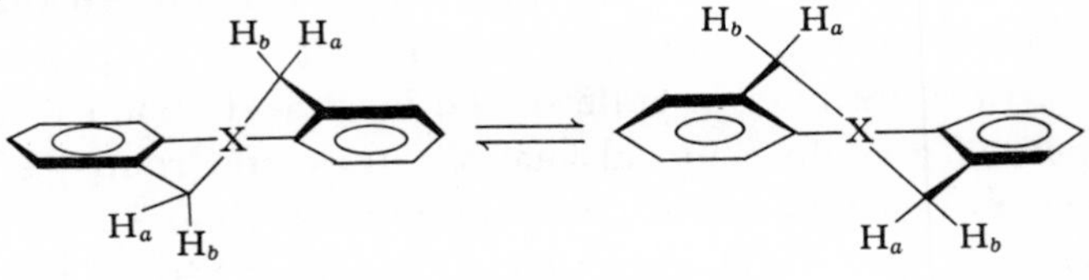

(122)

c. Chair ⇌ boat:

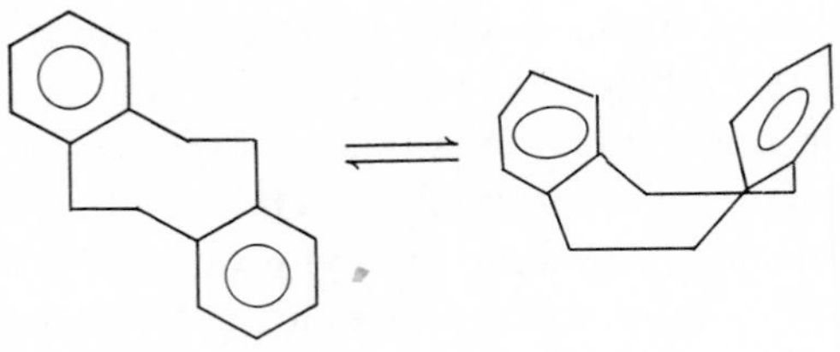

(123)

Representative data for tricyclic derivatives containing a pyramidal or tetrahedral heteroatom are given in Table XIV. As can be seen from Table XIV derivatives of anthracene appear to be quite flexible, as well as the tricycles containing a central seven-membered ring. The *N*-acetyl group apparently functions as a "locking" group for the 10,11-dihydrodibenzazepin derivative (*2*). Introduction of a multiple bond in the 10,11-position of the dibenz[*b,f*]heteroepins and substituents in the 10-position raise the barriers to ring inversion (*279*). The increase in the barrier for Groups VI and VII is undoubtedly due to repulsion with the *peri*-hydrogen on the benzene in the transition state for ring inversion.

Dibenzheterocycles containing central rings with eight members are more complex because several conformational processes may occur in solution including B $\rightleftharpoons$ B, C $\rightleftharpoons$ C, B $\rightleftharpoons$ C, and B $\rightleftharpoons$ TB. The conformation in the solid state may not be exclusively present in solution (*91*). Dibenzocycloocta-1,5-diene exhibits the chair conformation in the solid state but exhibits both C $\rightleftharpoons$ B and B $\rightleftharpoons$ B transformations in solution

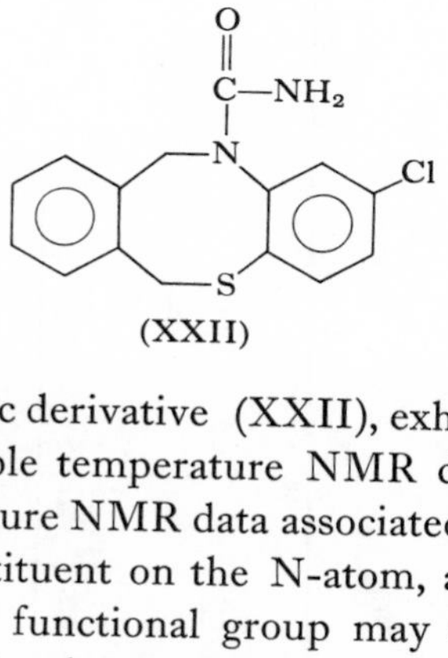

(XXII)

(*91, 261*). The heterocyclic derivative (XXII), exhibits the basket form in the solid state, but variable temperature NMR data were not reported (*367*). The room temperature NMR data associated with the 6,11-protons are a function of the substituent on the N-atom, and the more complex spectrum with an acetyl functional group may imply a locking effect similar to that observed for the azepine derivative.

It is noteworthy that in the tribenzcycloheptatriene series, the replacement of a —CH_2— group with an $>SiMe_2$ group does *not* result in a decrease in the barrier to ring inversion in direct contrast to the previously discussed silacyclohexane system (*77*). There now exist two extremes in ring flexibility for silicon compounds. It is necessary to study other heterocyclic derivatives before it will be possible to evaluate the effect on ring flexibility with heteroatom replacement.

Other pertinent results from conformational studies of tricycles that may be of interest in future work on organometallic systems concerns

TABLE XIV

CONFORMATIONAL INVERSION IN TRICYCLIC DERIVATIVES CONTAINING PYRAMIDAL OR TETRAHEDRAL HETEROATOMS

System	MR_1R_2	T_c(°C)[a]	$\Delta G^{\ddagger}$(kcal/mole)	Comments	Reference
	CH_2	< -60	—	$B \rightleftharpoons B$	*319*
	$SiMe_2$	< -40	—	$B \rightleftharpoons B$	*180*
	AsMe	[b,c]	—	See section on pyramidal inversion.	*260*
	SO	—	—	Spectrum remains unchanged from $-90°$ to $+35°$.	*324*
	SO_2	—	—	Low-temperature work precluded by lack of solubility	*324*
	CH_2	< -90	9[d]	—	*281*
	CH_2	<R.T.[b]	—	—	*264*
	$SiMe_2$	< -94	—	—	*63*
	NMe	< -100	—	—	*2*
	N—Ac	112	19.9	$B \rightleftharpoons B$	*2*
	CH_2	<R.T.	—	—	*259*
	$C(CO_2Et)_2$	9	14	Biphenyl inversion	*259*
	$N(Me_2)^+$	-1	13.4	Biphenyl inversion	*323*
	SO_2	87.5	18.2	Biphenyl inversion	*323*
	AsMe	—	—	Model calculations suggest a rigid structure.	*231*

Structure	M			Process	Ref.
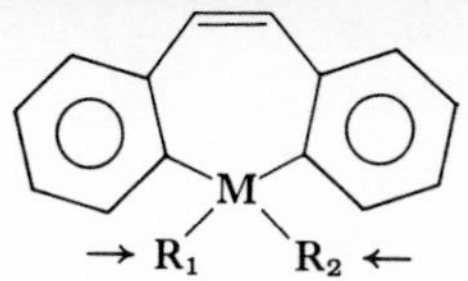 [structure: M, → R_1, R_2 ←]	CH_2	−90	9.0	B ⇌ B	*279*
	$C(OCH_3)_2$	30	15	B ⇌ B	*328*
	$SiMe_2$	< −90	8.5[e]	—	*63*
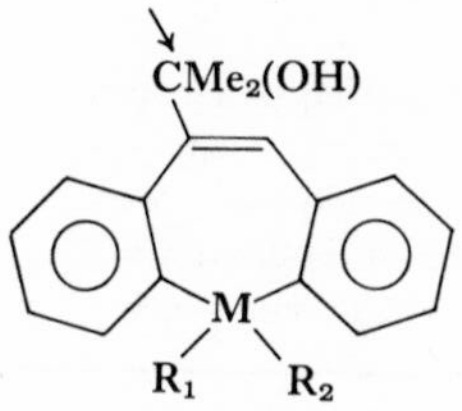 [structure: ↘ $CMe_2(OH)$, M, R_1, R_2]	CH_2	44	17.5	B ⇌ B	*279*
	NEt	116	21.7	B ⇌ B	*279*
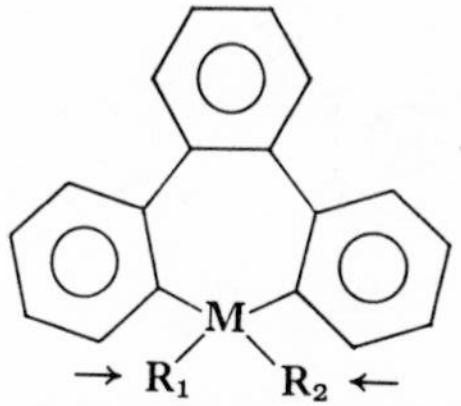 [structure: M, → R_1, R_2 ←]	CH_2	202	24	B ⇌ B	*279*
	$C(OCH_3)_2$	> 180	> 23	B ⇌ B	*327*
	$SiMe_2$	> 200	> 23 (45.4)[e]	B ⇌ B	*77*
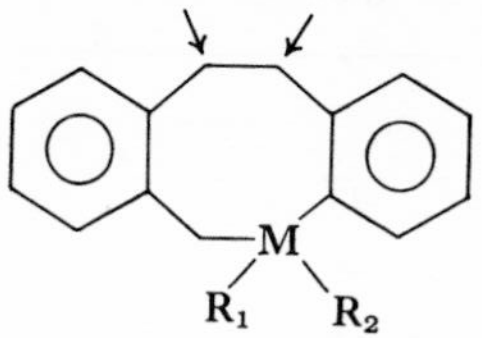 [structure: ↘ ↙, M, R_1, R_2]	CH_2	−72	10.2 ± 1	C ⇌ B	*91, 261*
		−115	7.5 ± 1	Boat inversion	*261*
	SiMe(OH)	< −60		Line broadening at −50°	*83*
	NAc	—	11.1	($NCOCH_3$) C ⇌ B (−77° to 56°)	*91*
			20.1	(N—CH_2) B ⇌ B (+88° to +113°)	*91*

Continued

TABLE XIV—*Continued*

System	MR_1R_2	T_c(°C)[a]	$\Delta G^{\ddagger}$(kcal/mole)	Comments	Reference
[structural formula: dibenzo ring system with two $M(R_1)(R_2)$ groups and CH_3; arrow at R_1]	SO_2	—	11.9	$C \rightleftharpoons C^*$ (−55° to 5°)	*91*

[a] Coalescence of protons indicated by arrows in structural formulas.
[b] Only room temperature (R.T.) results were reported.
[c] Calculated barrier in 5,10-dihydroarsanthrene is ~6–7 kcal (*260*).
[d] Calculated barrier.
[e] Calculated barrier (*370*).

substituent preferences. Structure (XXIII) shows substituent positions for anthracene:

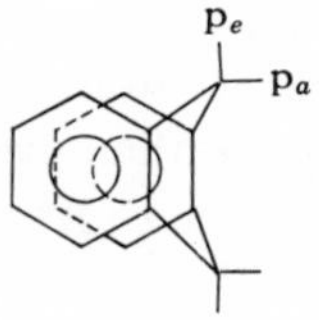

p_e p_a

(XXIII)

The preferred orientation of the alkyl substituent in 9-alkyl-9,10-dihydroanthracenes is the pseudoaxial position (Me and Ph substituents occupy a pseudoequatorial position to some extent) (*50*), and the alkyl groups occupy the diaxial positions in *cis*-9,10-dihydroanthracene derivatives (*292*). The favored conformation of the 10-substituted 10,11-dihydrodibenz[*b,f*]oxepins is that in which the 10-substituent is quasi-equatorial with respect to the seven-membered ring (*187*). In a study of the *cis*- and *trans*-thioxanthenes, the lone pair was found to prefer the *axial* position (*324*):

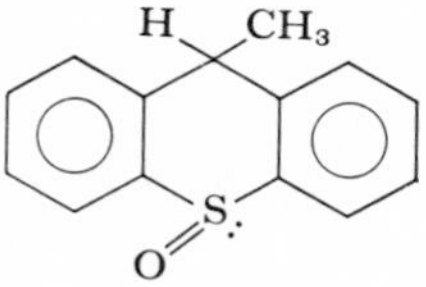

H CH_3 S O

(XXIV)

It has been suggested that substituent preference for the axial position in 9,10-dihydroheteroanthracene derivatives is related to repulsions with hydrogens *peri* to the substituent. If heteroatoms with longer C—M bonds are present the substituents will extend further away from the *peri*-hydrogens and repulsive interactions should decrease. Thus organometallic heterocycles may not exhibit a substituent preference.

3. *Pyramidal Inversion*

Heterocycles that contain a pyramidal heteroatom may exhibit pyramidal inversion as well as the conformational inversion processes outlined in the previous section. Pyramidal inversion barriers have been measured for a number of phosphorus and arsenic derivatives and have been shown to be considerably higher than their nitrogen analogs (*258*).

Isolation of optically active phenoxarsenes (an analog of 9,10-dihydroanthracene) in the early 1930s was interpreted as confirmation of a static folded or butterfly conformation for this system. At that time the very high barriers to pyramidal inversion for arsenic were not recognized. An alternative view that would explain the optical stability of the arsanthracene derivatives has been proposed. Accordingly, it is assumed that arsenic is configurationally stable but that the enantiomers exist in solution as rapidly interconverting, diastereomeric folded conformations (*260*):

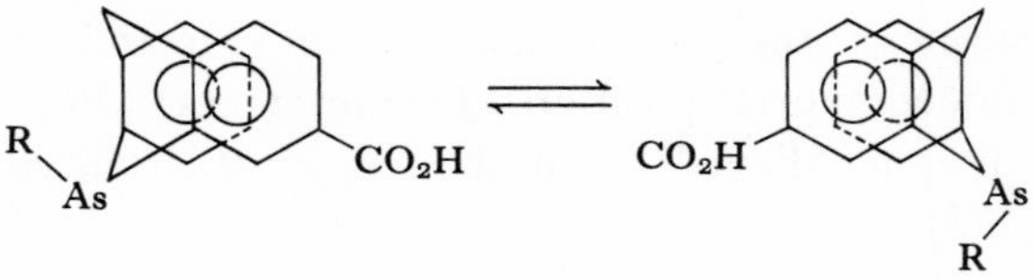

R–As … CO_2H ⇌ CO_2H … As–R (124)

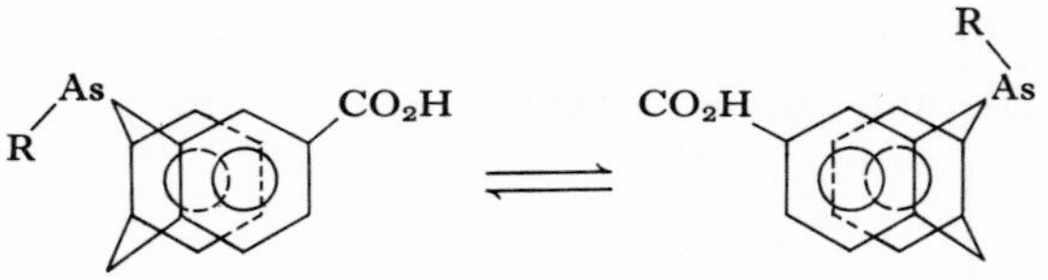

R–As … CO_2H ⇌ CO_2H … As–R (125)

Although the pyramidal inversion barrier for As in the arsanthracene system has not yet been reported, the $\Delta G^{\ddagger}$ for pyramidal inversion for a related sulfur derivative, 9,9-dimethyl-10-phenylthioxanthylium perchlorate, was calculated to be 25.4 kcal/mole at 200 ± 5° (*5*). Since considerable evidence demonstrating ring flexibility in 9,10-dihydroanthracene derivatives has accumulated, the isolation of optically active arsanthracene derivatives most probably is due to configurational stability of arsenic.

Preliminary studies of pyramidal inversion in arsindoles have also appeared (*45*). The inversion barriers for a number of related phosphorus and arsenic metallole derivatives are as follows:

	$Ph(CH_3)MR$	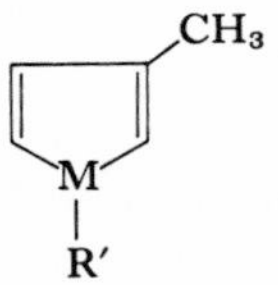 (3-CH_3 metallole, M–R′)	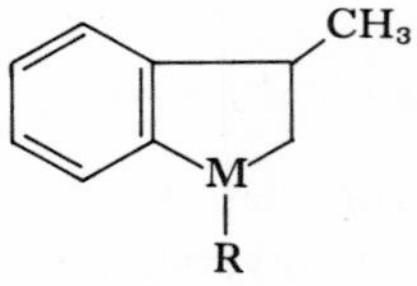 (3-CH_3 indole analog, M–R)
M = P:	33.1 kcal/mole	15.3 kcal/mole	35.3 kcal/mole
M = As:	43.1 kcal/mole		46–48 kcal/mole
	(XXVa)	(XXVb)	(XXVc)

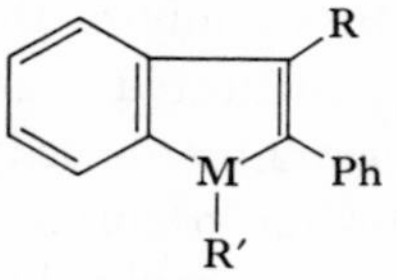

M = P: 23.7 kcal/mole
M = As: 35.2 kcal/mole
(XXVd)

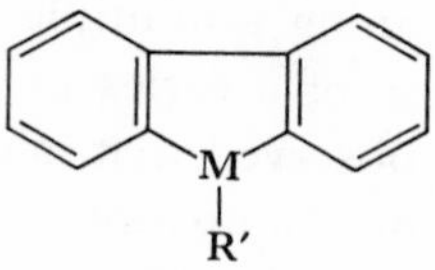

26.3 kcal/mole
(XXVe)

The lower barrier in phospholes relative to the acyclic phosphines is attributed to stabilization in the planar transition state. This stabilization is assigned to cyclic (3p-2p)π delocalization in P and (4p-2p)π delocalization for arsenic. The increased barrier upon annelation in the phosphorus system may be attributed to a disruption of the "phosphole" aromaticity. Similar effects have been noted in indoles.

Although (np-2p)π delocalization is a plausible explanation for the decreased barriers in the phospholes and arsoles, the argument would be more convincing if the absence of significant barrier decrease in arsacyclohexa-2,5-diene vs. arsacyclohexane could be demonstrated.

E. Mass Spectral Data

The majority of published mass spectral data for organometallic heterocycles consists of m/e values for parent ions. It is doubtful that any direct comparison of mass spectral data produced from organometallic heterocycles and the corresponding carbocycles can be made. There are at least two factors that preclude such a comparison. First, the electronegativity of the metals is less than that of carbon or common nonmetal heteroatoms. The greater electropositive character of the metal favors charge retention on the metal fragment in a process that results upon electron impact. Second, the metal–carbon bonds are weaker than the carbon–carbon or carbon–nonmetal bonds, thus metal–carbon cleavage processes may dominate in the fragmentation of the molecular ion. A third problem in a possible comparison of carbocycles and organometallic heterocycles involves the types of intermediates that are possible in a fragmentation. Frequently a fragmentation pattern for an organic derivative can be rationalized in terms of the formation (or elimination) of species containing a multiple bond. In a study of the mass spectral cracking patterns of a series of phenyl-substituted organosilicon compounds, it was observed that, although certain similarities existed with analogous carbon compounds, no cracking occurred in the silicon derivatives that would require a carbon–silicon double bond in either the ions or the neutrals (*120*). It is likely that this will prove the case for heavier metal atom systems as well.

The problems of analysis of fragmentation patterns generated from organometallic heterocycles is complicated by the fact that the metal heteroatoms frequently contain several isotopes of greater than 1% natural abundance, e.g., Bi and As have 1, B and Sb have 2, Si has 3, Pb has 4, Ge and Se have 5, and Sn and Te have 6 isotopes ($>1\%$). Frequently, data generated from organometallic derivatives contains clusters of fragments that differ by one hydrogen atom, and sorting these out manually is extremely tedious. Fortunately, a program is now available that reduces polyisotopic data to monoisotopic form, thus removing the more tedious aspects of the analysis and may simplify future work (*224*).

It is probably more reasonable to analyze data produced from organometallic heterocycles in terms of what is known or postulated for acyclic organometallic derivatives. The following is a brief summary of mass spectral characteristics of organometallic compounds of the general formula R_4M (M = Group IV atom) at 70 eV:

1. The majority of the ion current is carried by the metal fragments.
2. Parent ions are weak and generally decompose by elimination of an odd-electron neutral fragment.
3. R_3M^+ [M(IV)] ions are the most abundant metal fragments and their main decomposition is the result of elimination of even electron fragments.
4. Ions of the type RM^+ [M(II)] are usually more abundant than R_2M^+ except for molecules of the type Me_4M and Ar_4M.
5. Both odd- and even-electron ions appear to undergo transitions in which a neutral molecule is eliminated by cleavage of two bonds to the metal.

The general order of abundance, $R_3M^+ > RM^+ > R_2M^+$ probably reflects the greater stability of the M(IV) and M(II) oxidation states relative to M(III) and M(I).

Preliminary data on Group V derivatives suggest that certain generalizations from Group IV will be applicable in the analysis of data for As, Sb, and Bi. Parent ions are generally weak, and the majority of the ion current is carried by the metal fragments. Transitions in which a neutral molecule is eliminated by simultaneous cleavage of two M—C bonds from the molecular ion are common for aryl-substituted derivatives. For organobismuth and organolead derivatives the M^+ is either the base peak in the spectrum or about equal in intensity to the most abundant metal-containing fragment (*44*).

Much of the data that has accumulated for organometallic heterocycles is related to the dibenzcycloalkene series, i.e., metalloanthracene and dibenzo[*b*,*f*]metallepins. The majority of the fragments produced from the molecular ions of dibenzo[*b*,*f*]metallepins result from one of three processes (*78*): (1) cleavage of an exocyclic bond and elimination of an odd-electron radical; (2) elimination of a neutral molecule generally by cleavage of an exocyclic bond and abstraction of a hydrogen atom from the ethylene bridge (hereafter called the 1,4-elimination process); and (3) elimination of a neutral molecule by cleavage of two M—C ring bonds with charge retention either on the metal fragment or on the organic fragment.

The R_3M^+ ions generally fragment by processes described in 1 and 3 above as well as by elimination of several neutral molecules (generally not involving the exocyclic substituent). Although the published data on related systems are incomplete, it is obvious that these three processes dominate in derivatives of anthracene that contain a metal heteroatom as well as in the extensively documented 5,10-dihydrophenoxastannins (*212*), 5,10-dihydrophenoxasilins (*210*), 5,10-dihydrophenazastannins (*211*). Rearrangement processes also occur especially when the heterocycles contain an electronegative atom bonded to carbon; for example, the $SnBr^+$ ion appears in the spectrum of 2,8-dibromo-5-methyl-10,10-dimethyl-5,10-dihydrophenazastannin, and SnO and SiO species have been postulated in the fragmentation of phenoxastannins and phenoxasilins (*210–212*). Rearrangement processes of an electronegative atom from carbon to the metal atom are not restricted to heterocycles as $SiMe_2F^+$ and $SiCl^+$ are reported in the spectrum of $XC_6H_4CH_2SiMe_3$ derivatives (*120*).

1. *1,4-Elimination in Dibenzheterocycles and Aromaticity*

An interesting problem of interpretation is associated with the presence of 1,4-elimination in the spectral data generated from 5,10-dihydrometalloanthracene derivatives. In early work on 5,10-dihydrophenazastannins and phenoxastannins, the driving force for loss of an exocyclic M—R group was interpreted in terms of formation of the resonance-stabilized cation (XXVI) (*211*, *212*):

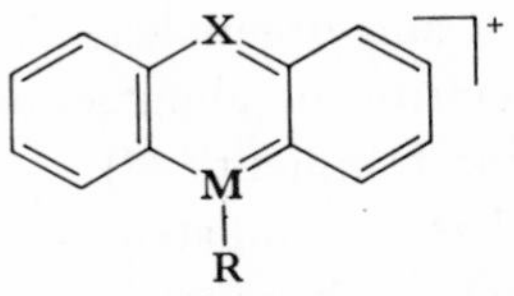

(XXVI)

A similar driving force was invoked for the elimination of RH from metalloanthracene derivatives of boron and silicon (*181*, *335*, *337*):

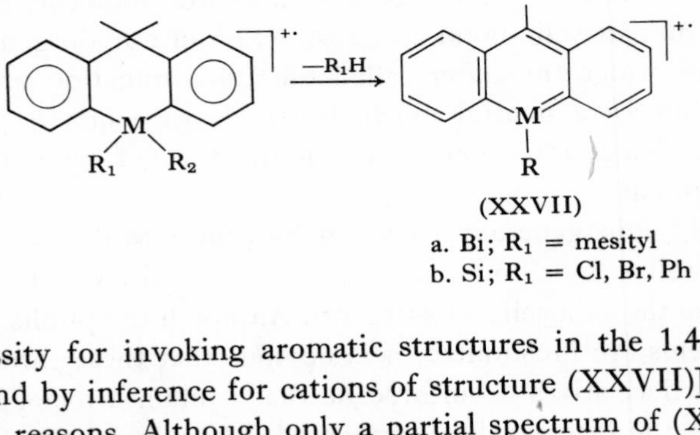

$$\xrightarrow{-R_1H} \quad (126)$$

(XXVII)

a. Bi; R_1 = mesityl
b. Si; R_1 = Cl, Br, Ph

The necessity for invoking aromatic structures in the 1,4-elimination process [and by inference for cations of structure (XXVII)] is doubtful for several reasons. Although only a partial spectrum of (XXVIIa) has been reported, the intensity data for the parent region is consistent only if the presence of both P^+ and $[P{-}H]^+$ are assumed. Likewise, the $P^+{-}RH$ region is consistent with a loss of $P^+{-}RH/P{-}R^{+\cdot}$ of ~4–5:1. It is not clear that loss of RH occurs in one step. The ratio of loss of $R^\cdot$ to loss of RH as a function of electronvolts could provide support for loss of RH. The problem is further complicated by the observation that RH and not RD is eliminated in 9-mesityl-10-deutero-10-phenyl-9,10-dihydro-9-boraanthracene (*339*). 10,11-Dihydrodibenz[*b*,*f*]metallepins also undergo the 1,4-elimination process, and aromaticity in the resultant $[P{-}RH]^+$ species is not necessary (*81*).

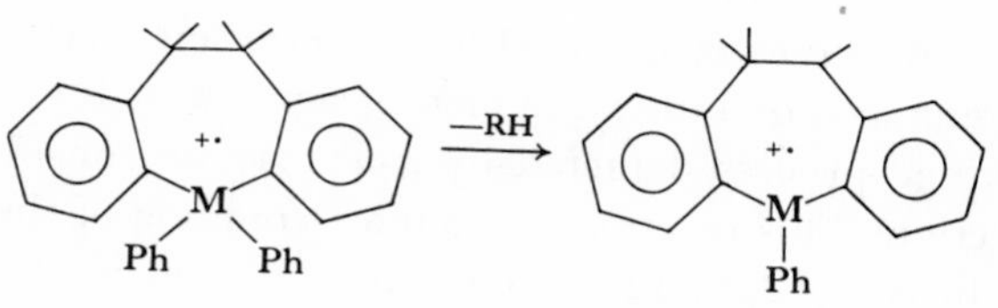

M	RH/R
Si	100
Ge	1
Sn	0.3

(127)

(XXVIII)

In the metallepin case support for the direct elimination of benzene is provided (rather than $P^+—H—C_6H_5$) by the observation that loss of benzene increases relative to loss of phenyl at lower electronvolts (*78*). Support for loss of hydrogen from the ethylene bridge is provided by the fact that 5-phenyl-5-methyl-10,11-dihydro-5*H*-dibenzo-[*b,f*]silepin-10,10,11,11-d_4 loses C_6H_5D exclusively upon electron impact (*76*).

It is probable that 1,4-elimination is a characteristic of compounds that exhibit the folded-boat conformation and not to any intrinsic aromaticity of the R_3M^+ cation. The decrease in 1,4-elimination for (XXVIII) in the series Si to Sn may be due to either a flattening of the tricyclic ring system or lengthening of the M—C bond, either of which increases the distance between the heteroatom substituent and the ethylene bridge.

2. *Cleavage of Two M—X Bonds*

Although processes involving a cleavage of a single M—X bond appear to involve only the exocyclic substituents, simultaneous cleavage of two M—X bonds occurs predominantly at the ring C—M bonds. The possible routes for simultaneous cleavages of two M—X bonds are illustrated in Eq. (128) for the molecular ions generated from 10,11-dihydrodibenzo[*b,f*]metallepins (*78*). A similar scheme can be postulated

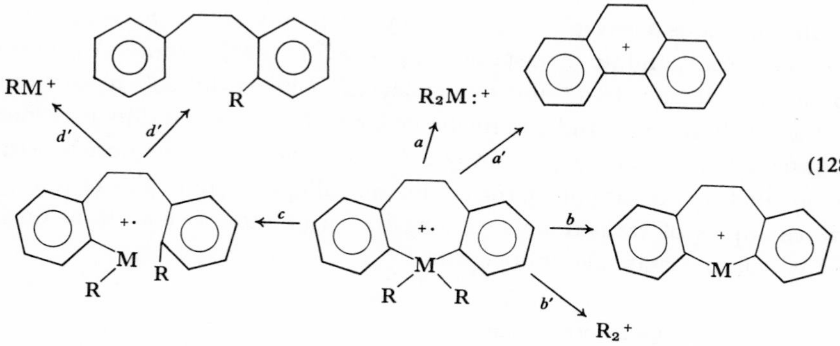

(128)

for the R_3M^+ species. As indicated in Eq. (128), the cleavage reaction may occur with charge retention on either the metal fragment or the organic fragment. The cleavage process depicted by *a-a′* may be considered a metal extrusion reaction. This reaction is the most common of the three processes depicted in Eq. (128) and has been reported (or is observed) in spectral data generated from metalloanthracenes, phenoxastannins, phenoxasilins, and phenazastannines. Metastable support is generally observed for both processes *a* and *a′* except for metallofluorenes (where biphenylene would be generated). The types of systems that undergo the metal extrusion reaction are

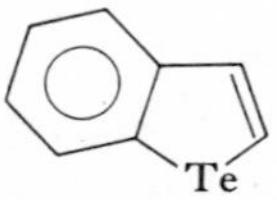

Te (*61*) (XXIX) Te (*158*) (XXX) Sb (*12*) (XXXI)

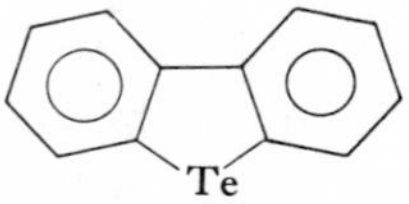

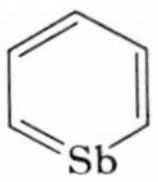

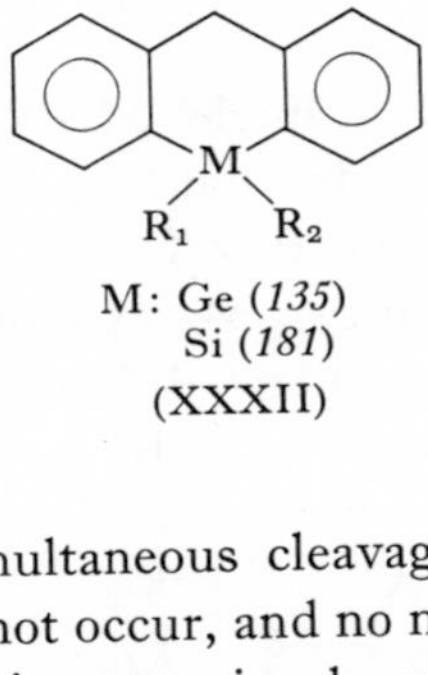

M R_1 R_2 M: Ge (*135*) Si (*181*) (XXXII)

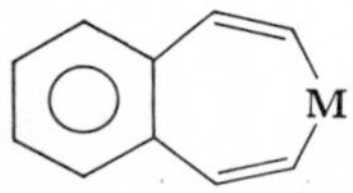

M M: BOH (*18*) $SnMe_2$ (*75*) (XXXIII)

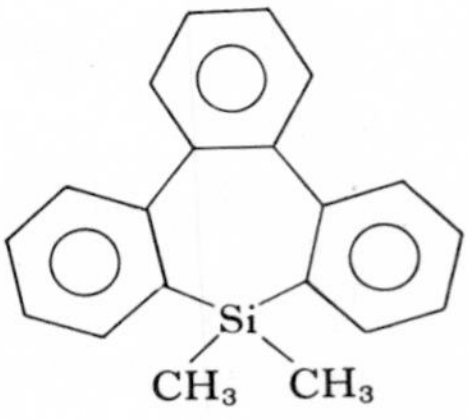

Si CH_3 CH_3 (*75*) (XXXIV)

Simultaneous cleavage of two exocyclic bonds (route *b*) apparently does not occur, and no metastable support exists for this route. A process involving one ring bond and one exocyclic bond is difficult to sort out as the resultant ion is indistinguishable from the parent, and its presence is inferred from its decomposition. A *c-d′* type transition occurs with strong metastable support in the dibenzometallepin system. The following are examples of transitions that may be formulated as *c-d′* transitions for which metastable support exists:[11]

[11] Asterisk indicates metastable support.

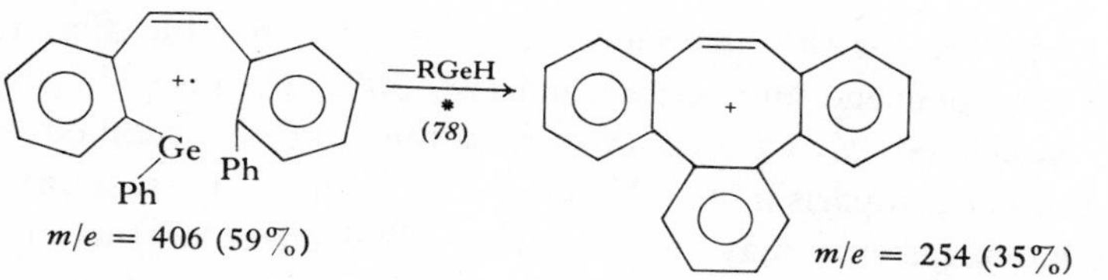

Ge, Ph, Ph; $m/e = 406$ (59%) $\xrightarrow[* (78)]{-\text{RGeH}}$ $m/e = 254$ (35%) (129)

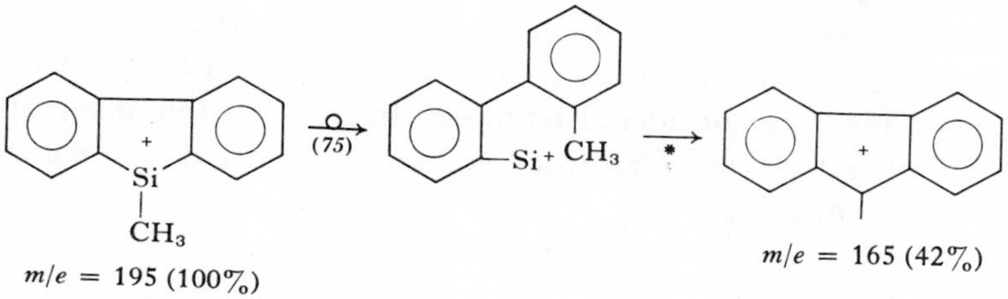

Si, CH_3; $m/e = 195$ (100%) $\xrightarrow{(75)}$ Si^+ CH_3 $\xrightarrow{*}$ $m/e = 165$ (42%) (130)

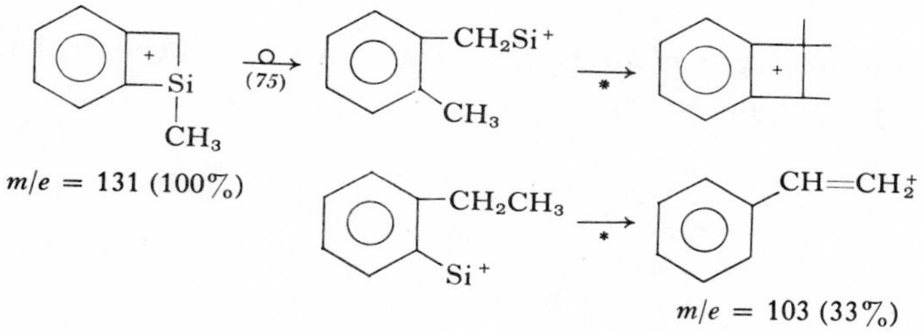

Si, CH_3; $m/e = 131$ (100%) $\xrightarrow{(75)}$ CH_2Si^+, CH_3 $\xrightarrow{*}$ (131)

CH_2CH_3, Si^+ $\xrightarrow{*}$ $CH{=}CH_2^+$; $m/e = 103$ (33%)

Mass spectral data generated by germacyclopentanes (*105*) exhibit the metal extrusion reaction [Eq. (128),*a*], but silacyclopentadienes show several rearrangement reactions (*247*). The data generated from metalloanthracenes and dibenzometallepins are similar to that reported for open-chain analogs, Ar_2MMe_2 and $Ar_2MAr'_2$.

F. Heterocyclic Systems with 6π, 10π, and 14π Electrons

Several heterocyclic derivatives that might exhibit aromatic properties have now been generated. Whether these systems that are formally analogous to benzene, cyclopentadienide ion, naphthalene, or anthracene but contain boron or an atom with a 3p or a 4p valence shell, are aromatic is uncertain. Ultraviolet and NMR spectral data that have been obtained for heteroaromatic species have been compared to analogous organic molecules. However, preliminary photoelectron spectral data obtained for arsabenzene (XXXVII_c) and arsaanthracene (XLIV) show that the orbital sequence for the heteroatom derivatives is not identical to that observed for benzene and anthracene (*28*, *307*). Because UV data are usually associated with transitions from the highest occupied energy levels, a direct comparison of UV data for pyridine and arsabenzene and other related systems may not be valid. The effect of the magnetic anisotropy of the heteroatoms on the chemical shifts of the ring protons cannot yet be assessed with accuracy. In view of these difficulties, conclusions concerning aromaticity of heterocycles containing a main group metal must be considered tentative. In the next few paragraphs the results that have been recorded for 6π-, 10π-, and 14π-electron systems will be discussed.

1. *6π-Electron Systems*

The following are types of derivatives that have been generated as possible 6π-heteroatomic species:

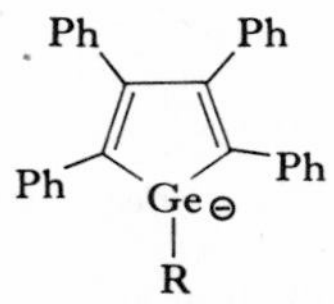

Ph Ph Ph Ph Ge⊖ R

(XXXV)

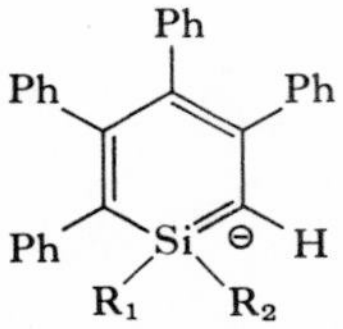

Ph Ph Ph Ph Si ⊖ H R_1 R_2

(XXXVI)

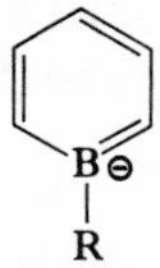

B⊖ R

(XXXVIIa)

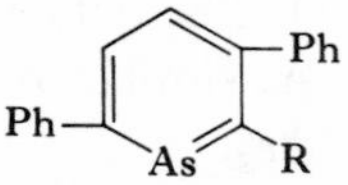

Ph Ph As R

(XXXVIIb)

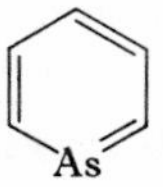

As

(XXXVIIc)

Sb

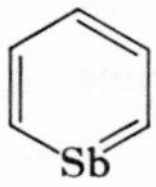

(XXXVIId)

Of the derivatives above, only (XXXVI) requires electron delocalization into empty *d* orbitals of the heteroatom, and of the remainder, all but (XXXVIIa) involve *d*-orbital participation. Heavy ring substitution precludes the use of NMR data in a structural analysis of (XXXV) and (XXXVI), but data obtained for the remaining derivatives have been interpreted as support for the presence of ring current in (XXXVIIa)–(XXXVIId). The PMR data for the trienes are given in Table XV. Because the chemical shift of the protons in $C_5H_5^-$ appear at $\tau = 4.6$, it is argued that the peak positions at lower field in $C_5H_5BPh^-$ imply substantial ring current and/or diminished negative charge at carbon-4 (*15*). Support for increased electron density at boron (a consequence of electron delocalization) is provided from ^{11}B nuclear resonance measurements. The ^{11}B chemical shift is +37 ppm relative to external Me_3B (−49 ppm relative to $BF_3 \cdot OEt_2$) in C_5H_6BPh and +59 ppm (−27 ppm) in the anion, $C_5H_5BPh^-$. This shift is consistent with an increase in negative charge on the boron atom. When the borabenzene anion is complexed to a transition metal the protons shift upfield and the ^{11}B chemical shift is −16 ppm (relative to $BF_3 \cdot Et_2O$) in $(C_5H_5BPh)_2Fe$ and −23.3 ppm in $C_5H_5CoC_5H_5BPh$ (*15, 160*).

The NMR spectra of pyridine, phosphabenzene, arsabenzene, and stibabenzene are quite similar in appearance (*11, 12*). The γ protons appear at lowest field partially as a result of remoteness from the heteroatom. The effect of the magnetic anisotropy of the heteroatom should decrease for the β and γ protons. The chemical shift differences for the γ protons in pyridine and arsabenzene are almost negligible. The low-field chemical shift value was interpreted in terms of the presence of ring current in the heteroatomic derivatives.

Ultraviolet spectral data for arsabenzene and stibabenzene have been reported (*11, 12*):

As (XXXVIIc)

λ_{max}	ϵ
219	1.5×10^4
268	1.1×10^4
305 (sh)	

Sb (XXXVIId)

λ_{max}	ϵ
236	$\sim 10^4$
312	$\sim 10^4$

The high-intensity bands in arsabenzene and stibabenzene were tentatively assigned to π-π^* transitions, although this may be a premature assignment in view of the published photoelectron spectral data.

The most interesting measurements produced from the heterocyclic analogs of benzene have been the (vertical) ionization potentials (*28*). The photoelectron spectra, supported by calculations, show inversion of the $a_2(\pi)$ and $b_1(\pi)$ orbitals in going from pyridine to the remaining Group V heterocyclic analogs. A portion of the orbital correlation diagram is shown here (Scheme 9) for the highest occupied molecular orbitals (HOMO). The crossover in the HOMO levels may preclude a direct comparison of the spectral data for pyridine and arsabenzene. Knowledge of the ordering of the antibonding levels will be necessary before analysis of the UV data may be completed.

TABLE XV

PROTON MAGNETIC RESONANCE DATA FOR 6π-ELECTRON SYSTEMS AND RELATED COMPOUNDS

Element		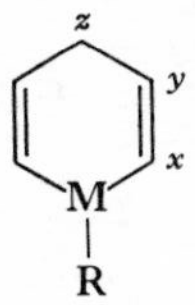 τ[a]	Reference	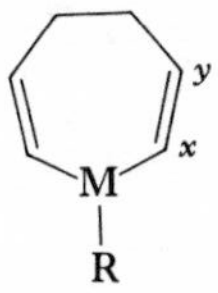 τ[a]	Reference	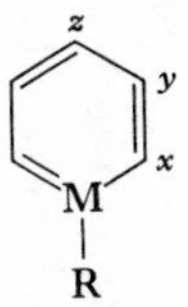 τ[a]	Reference	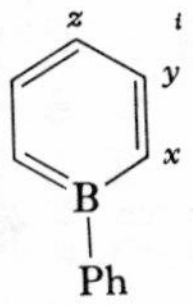 τ[a]	Reference
B	x	2.56[c]	*15*	3.25(4.47)[e]	*217*	3.05[c]	*15*	3.83(d)	*160*
	y	2.2–2.45[c,d]		2.82(3.67)[e]		2.4–2.8[d,c]		3.1(m)	
	z	6.52(m)		—		3.66(t)		—	
As	x			3.2–4.1(m)[f,c]	*238*	0.7	*11*	—	
	y			1.74–3.34(m)[g,c]		—		—	
	z					2.3–3.1(m)[h]		—	
Sb	x	3.5[c]	*12*	—	—	−0.7	*12*	—	
	y	3.1[c]		—		1.5–2.5(m)		—	
	z	6.6(m)		—		—			

[a] τ Values relative to TMS.
[b] Anion.
[c] Part of AB pattern.
[d] Overlap with exocyclic phenyl substituent.
[e] Figure in parentheses is for $R_1 = OCH_2CH_2NMe_2$ (quaternary boron).
[f] For both H_a and H_b, for $R_1R_2 = Ph$.
[g] For both H_a and H_b for quaternary salt, $R_1 = Ph$, $R_2 = Me$.
[h] Estimated from the published data.
[i] In $C_5H_5CoC_5H_5BPh$.

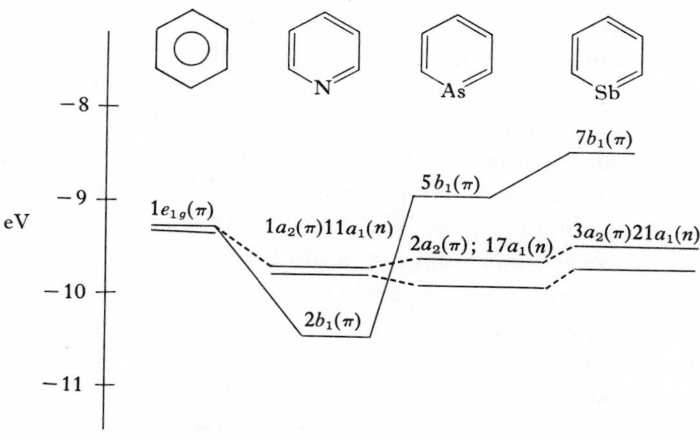

Scheme 9

The classic aromatic systems are planar molecules. Solid-state structural data have been obtained only for heavily substituted arsenic derivatives and for the borabenzene anion in a transition metal complex. The following data were obtained for these systems (*1*, *171*, *303*):

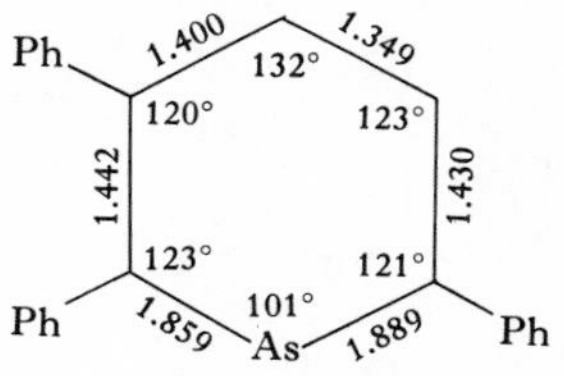

(XXXVIII)

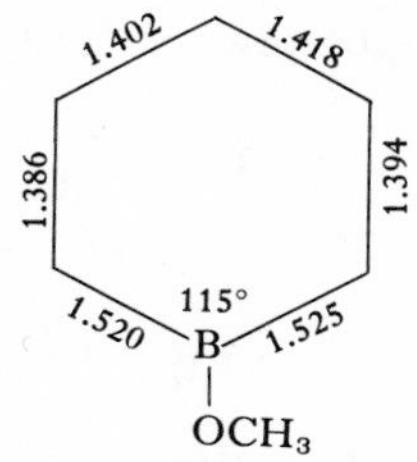

(XXXIX)
[in $Co(C_5H_5B{-}OCH_3)_2$]

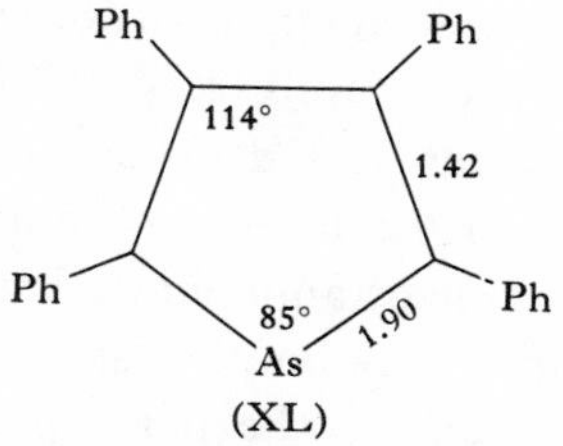

(XL)
[in $(C_6H_5)_4C_4AsMn(CO)_3$]

The rings in both 2,3,6-triphenylarsenin and the arsole ring in the complex $Ph_4C_4AsMn(CO)_3$ are planar with an As—C (ring) distance shorter than the sum of covalent radii for As and C_{sp^2}. The shortened As—C bond suggests electron delocalization through the heteroatom. The B—C (ring) distance is only slightly shorter than that expected for B_{sp^2}—C_{sp^2} bond but the ring is not strictly planar. The ring exhibits a slight chair conformation with a mean torsion angle of 5.3°. This non-planarity may be a result of the exocyclic methoxy substituent or a consequence of complexation. The cobalt metal atom is also displaced toward the C_3 atom away from the boron heteroatom.

The data obtained for arsenin seem to support the premise of aromatic character; however, the evidence is more tenuous for borabenzene anion. Future synthetic studies may produce a neutral, uncomplexed borabenzene derivative useful for structural analysis. This will be necessary before the presence or absence of aromaticity can be established.

2. *10π-Electron Systems*

Of the possible 10π-electron systems, only the following have been suggested to exhibit properties consistent with electron delocalization through the heteroatom (*190*, *217*):

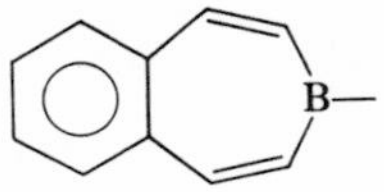

(XLI) 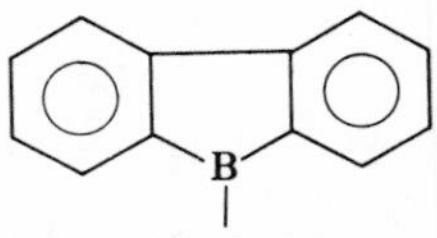 (XLII)

The carbon–boron heterocycle, 3-phenyl-3-benzoborepin, exhibits oxidative stability upon exposure to air, an unusual feature for a trivalent boron compound. In Table XVI are recorded the chemical shift data for the vinyl protons for the benzometallepins of B, Sn, and Si. The PMR spectrum of 3-phenyl-3-benzoborepin exhibits vinyl proton resonances at lower fields than would be expected for an olefinic boron compound (compared to trivinylboron or 4,5-dihydroborepin; see Table XV), and also at lower field than the benzostannepin derivative (*217*). The shift to lower field of 0.4 to 0.8 ppm may be consistent with the presence of a ring current, which would require the participation of the B_{p_z} orbital in the π-electron system. Support for increased electron density at boron might be provided from ^{11}B NMR measurements, but such data have not yet been reported. Complexation of boron, which converts the

TABLE XVI

NUCLEAR MAGNETIC RESONANCE AND ULTRAVIOLET DATA FOR BENZMETALLEPINS

Compound	NMR		UV		Reference
	$H_x(\tau)$	$H_y(\tau)$	λ_{max}	log ϵ	
$(\overset{y}{C}H_2{=}\overset{x}{C}H)_3B$	~3.45		234 nm	4.28	*217*
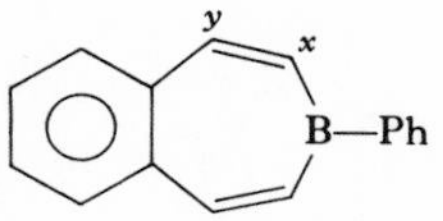 3-Phenyl-3-benzoborepin (B—Ph)	2.28	1.78	352[a]	4.21	*217*
3-Phenyl-3-benzoborepin dimethylamine complex (B(Ph)←NMe₂H)	3.64	2.95	237	4.66	*217*
$(\overset{y}{C}H_2{=}\overset{x}{C}H)_4Sn$	3.7	3.9	—	—	*217*
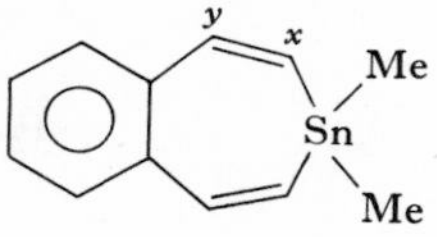 3,3-Dimethyl-3-benzostannepin (Sn(Me)₂)	3.68	2.50	227	4.51	*217*
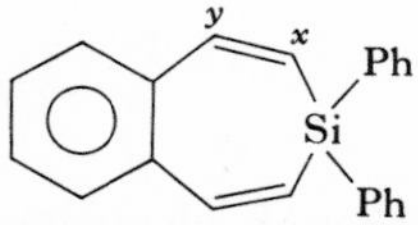 3,3-Diphenyl-3-benzosilepin (Si(Ph)₂)	3.72	2.53	—	—	*36*

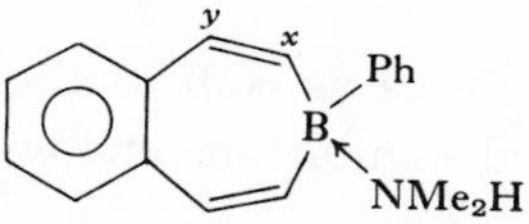

[a] Highest wavelength only. Other bands λ_{max}nm/log ϵ: 232/4.50, 285/4.68, 301/4.45, 322/3.98, 336/4.22.

boron center from trigonal to tetrahedral, would remove the B_{p_z} orbital from such a π-electron system, and should result in an upfield shift of the vinylic protons. This is indeed observed and the shift is greater than that observed in the model compound, 4,5-dihydrobenzoborepin (Table XV).

The UV spectral data (Table XV) also appear to support electron delocalization in 3-benzoborepin. A similarity between the spectra of 3-phenyl-3-benzoborepin and the isoelectronic ring system, benzotropylium ion, was reported. The spectrum of the dimethylamine complex of benzoborepin exhibits only one λ_{max} as compared to the uncomplexed system which contains six bands from 232 to 552 nm.

A second boron heterocycle for which electron delocalization is suggested is 9-borafluorene (*194*). The evidence is somewhat tenuous but follows the trends observed for benzoborepin. The 9-borafluorenes are yellow with a reported λ_{max} at $\sim$390 nm. Contribution of the resonance form,

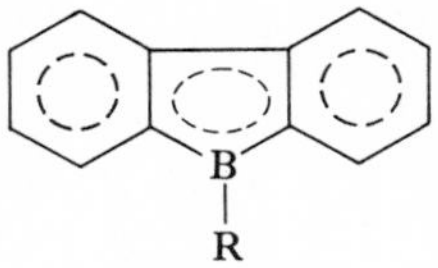

(XLIII)

was invoked to explain the color. The ether complexes as well as the borane dimer are colorless. This latter compound exhibits reversible dissociation at 80°C to give the colored monomer (*194*):

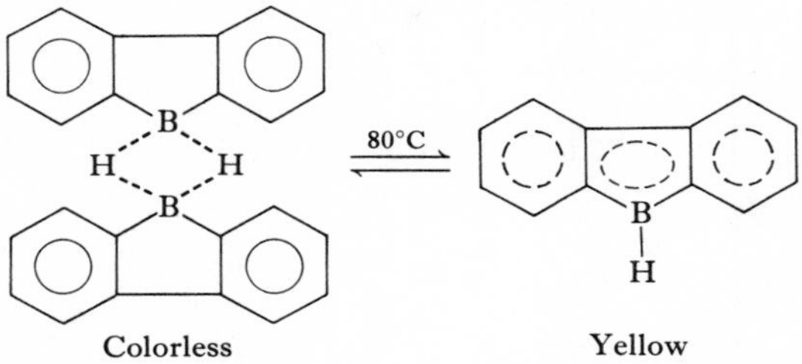

Colorless ⇌ (80°C) Yellow (132)

9-Alkyl- or arylaluminafluorenes, which are dimeric through electron-deficient bonding, are also colorless. Although electron delocalization through the boron heteroatom is a reasonable assumption, no ^{11}B NMR data were reported in support of this contention; however, the chemical shift differences for ^{11}B in Ph_3B and in 9-phenylborafluorene might be quite small.

Attempts to reduce 1,2-dihydro-2-boranaphthalene to the boranaphthalene anion failed (*285*).

3. *14π-Electron Systems*

Of the possible 14π-electron systems those receiving the greatest share of attention are the derivatives of anthracene. Both anionic and neutral derivatives are reported:

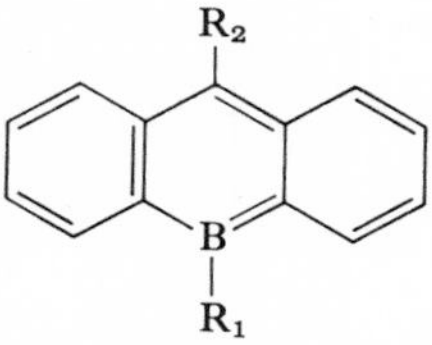

a. R_1 = Ph; R_2 = H (*179*)
b. R_1 = Mes; R_2 = H (*338*)
c. R_1 = Mes; R_2 = Ph (*339*)

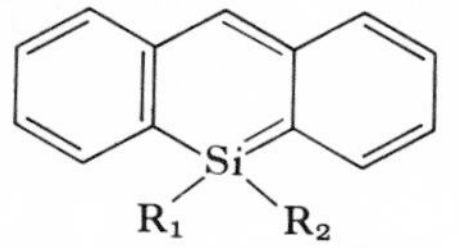

d. R_1 = R_2 = Me (*183*)
e. R_1 = Ph; R_2 = H (*183*)

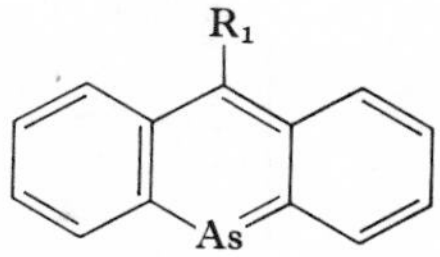

f. R_1 = H (*185, 342*)
g. R_1 = Ph (*343*)

(XLIV)

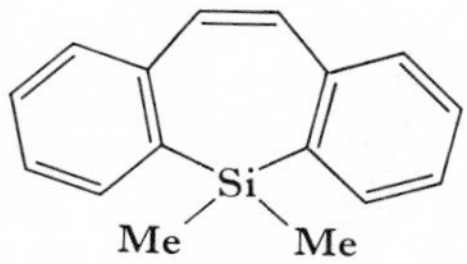

(XLV) (*26, 63, 80*)

Isolation of both solvated and unsolvated 9-boraanthracene derivatives has been reported and both are highly reactive toward oxygen or moisture. Only 10-phenyl-9-arsaanthracene has been isolated, and it is also an air-sensitive solid. Proton magnetic resonance spectroscopy and particularly UV spectroscopy have been used in the structural analysis of anthracene derivatives. The observed data are given in Table XVII. The small upfield shift of the C-10 proton in the boraanthracene anion may be a result of weak ring current effects which compensate for the charge at C-10. The increased negative charge on boron is clear from the ^{11}B NMR measurements: the ^{11}B resonance signal in the 9-phenylboraanthracene anion shifts upfield by 68 ppm relative to the neutral precursor (*179*).

Data for the silaanthracene derivative (XLIVe) are harder to evaluate. The ν(SiH) shifts from 2110 cm^{-1} in 9,9-dimethyl-9-silaanthracene to 2010 cm^{-1} in the anion, and the τ(SiH) shifts from 4.43 to <3.78 ppm in the PMR spectrum (*183*). These observations may be consistent with buildup of negative charge at the heteroatom which could result from electron delocalization into the *d* orbitals on silicon.

The UV spectra of all the metalloanthracene derivatives have been published. The longwave absorption maxima of 9-boraanthracene anion and 9-arsaanthracene are similar to anthracene, and 9-silaanthracene exhibits a UV spectrum similar to that of 9-fluorenyllithium (*183*). The spectrum of 9-mesityl-9-boraanthracene was qualitatively explained by

TABLE XVII

SPECTRAL DATA FOR 14π-ELECTRON SYSTEMS

Structure	MR_1R_2	τ—C_{10}—H	Solvent	λ_{max}	Solvent	Reference
	B—Ph	5.86	C_6D_6[a]	< 320 mm[b]	—	*179*
	B—Mes	5.53	C_6D_6	—	—	*338*
	$SiMe_2$	6.02	C_6D_6	—	—	*183*
	B—$Ph^{\ominus}Li^{+}\cdot 3.5$ THF	5.95	C_6D_6	~390, 490[c,d]	THF	*179*
	B—$Mes^{-}Li^{+}$			365, 435	C_6H_6—C_5H_{12}	*338*
	$\overset{\ominus}{Si}Me_2Li^{+}\cdot B$[e]	5.24	C_6D_6	470	—	*183*
	$Ph^{\ominus}$ Si H $Li^{+}\cdot B$[e]	4.98	C_6D_6	—	—	*183*
	As	—	—	454	THF	*185* *342*
	B—Mes	—	—	371, 445	C_6H_6—C_5H_{12}	*339*
	As	—	—	459	Toluene	*343*

[a] With added tetrahydrofuran (THF).
[b] Only end absorption above 320 nm.
[c] Estimated from published spectrum.
[d] Other bands also present.
[e] B is THF.

assuming overlap of a diphenylmethyl anion and an anthracene-type chromophoric system (*338*). However, a preliminary report of the PE. spectra for 9-phenylanthracene, 10-phenyl-9-phosphaanthracene, and 10-phenyl-9-arsaanthracene shows a similar crossover in the HOMO in going from the carbon derivative to the heterocyclic derivatives (*307*). The π—MO with a nonzero coefficient on the heteroatom is higher in energy than the π—MO with a node on the heteroatom. A similar crossover was exhibited in the benzene–metallobenzene series (*28*).

Analysis of the UV data obtained for the saturated and unsaturated dibenzosilepins shows no unusual features when compared to the carbon analogs (*63*). Thus, little to no electron delocalization through the heteroatom occurs for a seven-membered central ring.

References

1. Abel, E. W., Nowell, I. W., Modinos, A. G. J., and Towers, C., *Chem. Commun.* 258 (1973).
2. Abraham, R. J., Kricka, L. J., and Ledwith, A., *Chem. Commun.* 282 (1973).
3. Aeschlimann, J. A., Lees, N. D., McCleland, N. P., and Nicklin, G. N., *J. Chem. Soc.* 66 (1925).
4. Albert, A., "Heterocyclic Chemistry." Oxford Univ. Press, London and New York, 1968.
5. Andersen, K. A., Cinquini, M., and Papanikolaou, N. E., *J. Org. Chem.* **35**, 706 (1970).
6. Andrianov, K. A., Volkova, L. M., Delazari, N. V., and Chumaevskii, N. A., *Khim. Geterotskil. Soedin.* 435 (1967); *Chem. Abstr.* **67**, 108695c (1967).
7. Andrianov, K. A., Zhdanov, A. A., Karpova, I. V., and Odinets, V. A., *Izv. Akad. Nauk. SSSR, Ser. Khim.* 2733 (1969); *Chem. Abstr.* **72**, 79147w (1970).
8. Andrianov, K. A., Zhdanov, A. A., Odinets, V. A., and Karpova, I. V., *Kremniiorg. Soedin., Tr. Soveshch.* 5 (1967); *Chem. Abstr.* **69**, 106791y (1968).
9. Anet, F. A. L., and Kozerski, L., *J. Amer. Chem. Soc.* **95**, 3407 (1973).
10. Araki, T., Terunuma, D., Sato, T., Nagai, N., Furuichi, M., and Nakamura, S., *Bull. Chem. Soc. Jap.* **44**, 2725 (1971).
11. Ashe, A. J., III, *J. Amer. Chem. Soc.* **93**, 3293 (1971).
12. Ashe, A. J., III, *J. Amer. Chem. Soc.* **93**, 6690 (1971).
13. Ashe, A. J., III and Gordon, M. D., *J. Amer. Chem. Soc.* **94**, 7596 (1972).
14. Ashe, A. J., III and Shu, P., *Abstr. 165th Nat. Meeting Amer. Chem. Soc., Dallas, Tex., April 1973, No.* **052.**
15. Ashe, A. J., III and Shu, P., *J. Amer. Chem. Soc.* **93**, 1804 (1971).
16. Atwell, W. H., Ger. Offen, 1,921,833 (1970); *Chem. Abstr.* **72**, 31611r (1970).
17. Atwell, W. H., Weyenberg, D., and Gilman, H., *J. Org. Chem.* **32**, 885 (1967).
18. Axelrad, G., and Halpern, D., *Chem. Commun.* 291 (1971).
19. Baker, W., Banks, R., Lyon, D. R., and Mann, F. G., *J. Chem. Soc.* 27 (1945).
20. Barton, T. J., and Gottsman, E. E., *Syn. Inorg. Metalorg. Chem.* **3**, 201 (1973).
21. Barton, T. J., and Kline, E., *J. Organometal. Chem.* **42**, C21 (1972).

22. Barton, T. J., Kline, E. A., and Garvey, P. M., *J. Amer. Chem. Soc.* **95**, 3078 (1973).
23. Barton, T. J., and Nelson, A. J., *Tetrahedron Lett.* 5037 (1969).
24. Barton, T. J., Nelson, A. J., and Clardy, J., *J. Org. Chem.* **37**, 895 (1972).
25. Barton, T. J., and Roth, R. W., *J. Organometal. Chem.* **39**, C66, (1972).
26. Barton, T. J., Volz, W. E., and Johnson, J. L., *J. Org. Chem.* **36**, 3365 (1971).
27. Barton, T. J., Witiak, J. L., and McIntosh, C. L., *J. Amer. Chem. Soc.* **94**, 6229 (1972).
28. Batich, C., Heilbronner, E., Hornung, V., Ashe, A. J., III, Clark, D. T., Cobley, U. T., Kilcast, D., and Scanlan, I., *J. Amer. Chem. Soc.* **95**, 928 (1973).
29. Beeby, M. H., Cookson, G. H., and Mann, F. G., *J. Chem. Soc.* 1917 (1950).
30. Beeby, M. H., Mann, F. G., and Turner, E. E., *J. Chem. Soc.* 1923 (1950).
31. Benkeser, R. A., and Cunico, R. F., *J. Organometal. Chem.* **4**, 284 (1965).
32. Benkeser, R. A., Nagai, Y., Noe, J. L., Cunico, R. F., and Gund, P. H., *J. Amer. Chem. Soc.* **86**, 2446 (1964).
33. Benkeser, R. A., Noe, J. L., and Nagai, Y., *J. Org. Chem.* **30**, 378 (1965).
34. Benkeser, R. A., Smith, S. D., and Noe, J. L., *J. Org. Chem.* **33**, 597 (1967).
35. Bergmann, E. D., Shahak, I., and Aizenshat, Z., *Tetrahedron Lett.* 3469 (1968); *ibid.*, 2007 (1969).
36. Birkofer, L., and Haddad, H., *Chem. Ber.* **102**, 432 (1969).
37. Birkofer, L., Haddad, H., and Zamarlik, H., *J. Organometal. Chem.* **25**, C57 (1970).
38. Birkofer, L., and Krämer, E., *Chem. Ber.* **102**, 427 (1969).
39. Blackmore, W. R., and Abrahams, S. C., *Acta Crystallogr.* **8**, 317 (1955).
40. Blackmore W. R., and Abrahams, S. C., *Acta Crystallogr.* **8**, 323 (1955).
41. Blicke, F. F., Weinkauff, O. J., and Hargreaves, G. W., *J. Amer. Chem. Soc.* **52**, 780 (1930).
42. Boudjouk, P., and Sommer, L. H., *Chem. Commun.* 54 (1973).
43. Bowen, H. J. M., *Trans. Faraday Soc.* **50**, 463 (1954).
44. Bowie, J. H., and Nussey, B., *Org. Mass Spectrom.* **3**, 933 (1970).
45. Bowman, R. H., and Mislow, K., *J. Amer. Chem. Soc.* **94**, 2861 (1972).
46. Braye, E. H., Caplier, I., and Saussez, R., *Tetrahedron* **27**, 5523 (1971).
47. Braye, E. H., Hübel, W., and Caplier, I., *J. Amer. Chem. Soc.* **83**, 4406 (1961).
48. Brendel, G. J., and Shepherd, L. H., Jr., Ger. Offen. 1,953,713 (1970); *Chem. Abstr.* **73**, 77382r (1970).
49. Brieger, G., *Diss. Abstr.* **22**, 1824 (1961).
50. Brinkman, A. W., Gordon, M., Harvey, R. G., Rabideau, P. W., Strothers, J. B., and Ternay, A. L., Jr., *J. Amer. Chem. Soc.* **92**, 5912 (1970).
51. Brockway, L. O., and Jenkins, H. O., *J. Amer. Chem. Soc.* **58**, 2036 (1936).
52. Brook, A. G., and Pierce, J. B., *J. Org. Chem.* **30**, 2566 (1965).
53. Brown, R. D., Burden, F. R., and Godfrey, P. D., *J. Mol. Spectrosc.* **25**, 415 (1968).
54. Brunet, J. C., and Lemahieu, C., *J. Organometal. Chem.* **49**, 381 (1973).
55. Brunet, J. C., Resibois, B., and Bertrand, J., *Bull. Soc. Chim. Fr.* 3424 (1969).
56. Buchanen, G. W., *Tetrahedron Lett.* 665 (1972).
57. Burrows, G. J., and Turner, E. E., *J. Chem. Soc.* **119**, 426 (1921).
58. Bushweller, C. H., O'Neil, J. W., and Bilofsky, H. S., *Tetrahedron* **27**, 3065 (1971).
59. Butcher, S. S., *J. Chem. Phys.* **42**, 1830 (1965).
60. Butcher, S. S., *J. Chem. Phys.* **42**, 1833 (1965).
61. Buu-Hoi, N. P., Mangane, M., Renson, M., and Piette, J. L., *J. Heterocycl. Chem.* **7**, 219 (1970).

62. Campbell, I. G. M., *J. Chem. Soc.* 3109 (1950).
63. Cartledge, F. K., and Mollère, P. D., *J. Organometal. Chem.* **26**, 175 (1971).
64. Chao, T. H., Moore, S. L., and Laane, J., *J. Organometal. Chem.* **33**, 157 (1971).
65. Chakravartly, P. S., *Indian J. Chem.* **10**, 385 (1972).
66. Chernyshev, E. A., Komalenkova, N. G., and Shamshin, L. N., USSR Patent 295,765 (1971); *Chem. Abstr.* **75**, 49320s (1971).
67. Chernyshev, E. A., Komalenkova, N. G., Shamshin, L. N., and Bochkarev, V. N., *Zh. Obshch. Khim.* **42**, 1373 (1972); *Chem. Abstr.* **77**, 126745e (1972).
68. Chernyshev, E. A., Komalenkova, N. G., Shamshin, L. N. and Shchepinov. S. A., *Zh. Obshch. Khim.* **41**, 843 (1971); *Chem. Abstr.* **75**, 49195e (1971).
69. Chernyshev, E. A., Shchepinov, S. A., Krasnova, T. L., Filimonova, N. P., and Petrova, E. I., USSR Patent 217395 (1968); *Chem. Abstr.* **69**, P77464 (1968).
70. Christiaens, L., and Renson, M., *Bull. Soc. Chim. Belges* **77**, 153 (1968).
71. Clardy, J., and Barton, T. J., *Chem. Commun.* 690 (1972).
72. Cookson, G. H., and Mann, F. G., *J. Chem. Soc.* 2888 (1949).
73. Cookson, G. H., and Mann, F. G., *J. Chem. Soc.* 2895 (1949).
74. Corey, J. Y., *Syn. Inorg. Metalorg. Chem.* **2**, 85 (1972).
75. Corey, J. Y., unpublished work.
76. Corey, J. Y., and Armbruster, D., unpublished work.
77. Corey, J. Y., and Corey, E. R., *Tetrahedron Lett.* 4667 (1972).
78. Corey, J. Y., Corey, E. R., Glick, M. D., and Dueber, J. S., *J. Heterocycl. Chem.* **9**, 1379 (1972).
79. Corey, J. Y., and Dueber, M., unpublished work.
80. Corey, J. Y., Dueber, M. J., and Bichlmeir, B., *J. Organometal. Chem.* **26**, 167 (1971).
81. Corey, J. Y., Dueber, M. J., and Malaidza, M., *J. Organometal. Chem.* **36**, 49 (1972).
82. Corey, J. Y., and Francis, E. A., *Abstr. 8th Midwest Reg. Meeting Amer. Chem. Soc., Columbia, Missouri, Nov. 1972.*
83. Corey, J. Y., and Francis, E. A., *J. Organometal. Chem.* **61**, C20 (1973).
84. Corriu, R., Henner, B., and Massé, J., *Bull. Soc. Chim. Fr.* 3013 (1968).
85. Corriu, R. Leard, M., and Massé, J., *Bull. Soc. Chim. Fr.* 2555 (1968).
86. Corriu, R., and Massé, J., *C. R. Acad. Sci., Ser. C* **266**, 1709 (1968).
87. Corriu, R., and Massé, J., *Bull. Soc. Chim. Fr.* 3491 (1969).
88. Corriu, R. J. P., and Massé, J. P., *J. Organometal. Chem.* **22**, 321 (1970).
89. Corriu, R., Massé, J. P., and Royo, G., *Chem. Commun.* 252 (1971).
90. Coutant, R. W., and Levy, A., *J. Organometal. Chem.* **10**, 175 (1967).
91. Crossley, R., Downing, A. P., Nógradi, M., de Oliveira, A. B., Ollis, W. D., and Sutherland, I. O., *J. Chem. Soc., Perkin Trans. I* 205 (1973).
92. Curtis, M. D., *J. Amer. Chem. Soc.* **89**, 4241 (1967).
93. Curtis, M. D., *J. Amer. Chem. Soc.* **91**, 6011 (1969).
94. Curtis, M. D., and Wolber, P., *Inorg. Chem.* **11**, 431 (1972).
95. Damrauer, R., *Organometal. Chem. Rev., A* **8**, 67 (1972).
96. Das-Gupta, H. N., *J. Indian Chem. Soc.* **12**, 627 (1935).
97. Das-Gupta, H. N., *J. Indian Chem. Soc.* **14**, 231, 349 (1937).
98. Das-Gupta, H. N., *J. Indian Chem. Soc.* **14**, 397, 400 (1937).
99. Das-Gupta, H. N., *J. Indian Chem. Soc.* **15**, 495 (1938).
100. Davidson, J. M., and French, C. M., *J. Chem. Soc.* 191 (1960).
101. Davies, R. E., Openshaw, H. T., Spring, F. S., Stanley, R. H., and Todd, A. R., *J. Chem. Soc.* 295 (1948).

102. Debarge, A. E. J. J., French Patent 1,336,364 (1964); *Chem. Abstr.* **60**, P1793g (1964).
103. de Koe, P., and Bickelhaupt, F., *Angew. Chem. Int. Ed. Engl.* **6**, 567 (1967).
104. de Koe, P., and Bickelhaupt, F., *Angew. Chim., Int. Ed. Engl.* **7**, 889 (1968).
105. Duffield, A. M., Djerassi, C., Mazerolles, P., Dubac, J., and Manuel, G., *J. Organometal. Chem.* **12**, 123 (1968).
106. Dunogues, J., Calas, R., Dedier, J., and Pisciotti, F., *J. Organometal. Chem.* **25**, 51 (1970).
107. Eaborn, C., and Bott, R. W., *in* "Organometallic Compounds of the Group IV Elements," Part 1 (A. G. MacDiarmid, ed.), Dekker, New York, 1968.
108. Eaborn, C., Walton, D. M. R., and Chan, M., *J. Organometal. Chem.* **9**, 251 (1967).
109. Eisch, J. J., and Gonsior, L. J., *J. Organometal. Chem.* **8**, 53 (1967).
110. Eisch, J. J., and Harrell, R. L., Jr., *J. Organometal. Chem.* **20**, 257 (1969).
111. Eisch, J. J., Hota, N. K., and Kozima, S., *J. Amer. Chem. Soc.* **91**, 4575 (1969).
112. Eisch, J. J., and Kaska, W. C., *J. Amer. Chem. Soc.* **84**, 1501 (1962).
113. Eisch, J. J., and Kaska, W. C., *J. Amer. Chem. Soc.* **88**, 2976 (1966).
114. Featherman, S. I., and Quin, L. D., *J. Amer. Chem. Soc.* **95**, 1699 (1973).
115. Feitelson, B. N., and Petrow, F., *J. Chem. Soc.* 2279 (1951).
116. Fessenden, R., and Coon, M. D., *J. Org. Chem.* **26**, 2530 (1961).
117. Fessenden, R., and Freenor, F. J., *J. Org. Chem.* **26**, 2003 (1961).
118. Finkelshtein, E. S., Nametkin, N. S., Vdovin, V. M., Yatsenko, M. S., and Ushakov, N. V., *Symp. Int. Chim. Composés Org. Silicium, Resumes Commun., 2nd 1968*, p. 68.
119. Forbes, M. H., Heinekey, D. M., Mann, F. G., and Millar, I. T., *J. Chem. Soc.* 2762 (1961).
120. Freeburger, M. E., Hughes, B. M., Buell, G. R., Tiernan, T. O., and Spialter, L., *J. Org. Chem.* **36**, 933 (1971).
121. Fringuelli, F., Marino, G., Savelli, G., and Taticchi, A., *Chem. Commun.* 1441 (1971).
122. Frolov, S. I., Bubnov, Yu. N., and Mikhailov, B. M., *Izv. Akad. Nauk SSSR, Ser. Khim.* 1996 (1969); *Chem. Abstr.* **72**, 21731r (1970).
123. Gaidis, J. M., *J. Org. Chem.* **35**, 2811 (1970).
124. Gar, T. K., and Mironov, V. F., *Izv. Akad. Nauk SSSR, Ser. Khim.* 855 (1965); *Chem. Abstr.* **63**, 5666h (1965).
125. Garascia, R. J., and Mattai, I. V., *J. Amer. Chem. Soc.* **75**, 4589 (1953).
126. Gelius, R., *Angew. Chem.* **72**, 322 (1960).
127. Gelius, R., *Chem. Ber.* **93**, 1759 (1960).
128. Gerding, H., and Haak, F. A., *Rec. Trav. Chim. Pays-Bas* **68**, 293 (1949).
129. Gilman, H., and Atwell, W. H., *J. Amer. Chem. Soc.* **86**, 5589 (1964).
130. Gilman, H., and Atwell, W. H., *J. Amer. Chem. Soc.* **87**, 2678 (1965).
131. Gilman, H., and Atwell, W. H., *J. Org. Chem.* **28**, 2906 (1963).
132. Gilman, H., Cottis, S. G., and Atwell, W. H., *J. Amer. Chem. Soc.* **86**, 1596 (1964).
133. Gilman, H., Cottis, S. G., and Atwell, W. H., *J. Amer. Chem. Soc.* **86**, 5584 (1964).
134. Gilman, H., and Gorsich, R. D., *J. Amer. Chem. Soc.* **80**, 1883 (1958).
135. Gilman, H., and Gorsich, R. D., *J. Amer. Chem. Soc.* **80**, 3243 (1958).
136. Gilman, H., and Marrs, O. L., *Chem. Ind.* (*London*) 208 (1961).
137. Gilman, H., and Marrs, O. L., *J. Org. Chem.* **29**, 3175 (1964).
138. Gilman, H., and Marrs, O. L., *J. Org. Chem.* **30**, 325 (1965).
139. Gilman, H., and Marrs, O. L., *J. Org. Chem.* **30**, 1942 (1965).
140. Goubeau, J., Kalmar, T., and Hofmann, H., *Justus Liebigs Ann. Chem.* **659**, 39 (1962).

141. Grimme, W., Reinert, K., and Köster, R., *Tetrahedron Lett.* 624 (1961).
142. Gump, W., and Stoltzenberg, H., *J. Amer. Chem. Soc.* **53**, 1428 (1931).
143. Gusev, A. I., Shklover, V. E., Chernyshev, E. A., Krasnova, T. L., and Struchkov, Yu T., *Zh. Strukt. Khim.* **12**, 282 (1971); *Chem. Abstr.* **75**, 41743m (1971).
144. Gverdtsiteli, I. M., Doksopulo, T. P., Menteshashvili, M. M., and Abkhazava, I. I., *Soobshch. Akad. Nauk Gruz. SSR* **40**, 333 (1965); *Chem. Abstr.* **64**, 11239a (1966).
145. Hagen, V., and Ruehlmann, K., *Z. Chem.* **8**, 262 (1968).
146. Hagen, V., and Ruehlmann, K., *Z. Chem.* **9**, 309 (1969).
147. Hall, D. M., Lesslie, M. S., and Turner, E. E., *J. Chem. Soc.* 711 (1950).
148. Hanstein, W., Berwin, H. J., and Traylor, T. G., *J. Amer. Chem. Soc.* **92**, 7476 (1970).
149. Hawley, D. M., and Ferguson, G., *J. Chem. Soc. A* 2059 (1968).
150. Heaney, H., Heineky, D. M., Mann, F. G., and Millar, I. T., *J. Chem. Soc.* 3838 (1958).
151. Hedberg, K., reported in Dunitz, J. D., and Waser, J., *J. Amer. Chem. Soc.* **94**, 5645 (1972).
152. Heinekey, D. M., and Millar, I. T., *J. Chem. Soc.* 4414 (1963).
153. Heinekey, D. M., and Millar, I. T., *J. Chem. Soc.* 3101 (1959).
154. Hellwinkel, D., *Ann. N.Y. Acad. Sci.* **192**, 158 (1972).
155. Hellwinkel, D., and Bach. M., *J. Organometal. Chem.* **20**, 273 (1969).
156. Hellwinkel, D., and Bach, M., *J. Organometal. Chem.* **17**, 389 (1969).
157. Hellwinkel, D., and Bach, M., *Justus Liebigs Ann. Chem.* **720**, 198 (1969).
158. Hellwinkel, D., and Fahrbach, G., *Justus Liebigs Ann. Chem.* **712**, 1 (1968).
159. Hellwinkel, D., and Fahrbach, G., *Tetrahedron Lett.* 1823 (1965).
160. Herberich, G. E., and Greiss, G., *Chem. Ber.* **105**, 3413 (1972).
161. Herberich, G. E., Greiss, G., and Heil, H. F., *Angew. Chem., Int. Ed. Engl.* **9**, 805 (1970).
162. Hewett, C. L., Lermit, L. J., Openshaw, H. T., Todd, A. R., Williams, A. H., and Woodward, F. N., *J. Chem. Soc.* 292 (1948).
163. Holliman, F. G., and Mann, F. G., *J. Chem. Soc.* 547 (1943).
164. Holliman, F. G., and Thornton, D. A., unpublished work quoted on page 401 in Mann (*230*).
165. Hope, H., *Acta. Crystallogr.* **20**, 610 (1966).
166. Hope, H., Knobler, C., and McCullough, J. D., *Acta Crystallogr., Sect. B* **26**, 628 (1970).
167. Hosoya, S., *Acta Crystallogr.* **16**, 310 (1963).
168. Hota, N. K., and Willis, C., *J. Organometal. Chem.* **15**, 89 (1968).
169. Hubel, K. W., and Braye, E. H., U.S. Patent 3,280,017 (1966), *Chem. Abstr.* **66**, 2462t (1967); U.S. Patent 3,426,052 (1969), *Chem. Abstr.* **70**, 106663d (1969).
170. Hubel, K. W., Braye, E. H., and Caplier, I. H., U.S. Patent 3,151,140 (1964); *Chem. Abstr.* **61**, 16097c (1964).
171. Huttner, G., Krieg, B., and Gartzke, W., *Chem. Ber.* **105**, 3424 (1972).
172. Jackobs, J., and Sundaralingam, M., *Acta Crystallogr., Sect. B* **25**, 2487 (1969).
173. Johnson, F., U.S. Patent 3,234,239 (1966); *Chem. Abstr.* **64**, PC11251d (1966).
174. Johnson, F., and Gohlke, R. S., *Tetrahedron Lett.* 1291 (1962).
175. Johnson, F., Gohlke, R. S., and Nasutivicus, W. A., *J. Organometal. Chem.* **3**, 233 (1965).
176. Jones, E. R. H., and Mann, F. G., *J. Chem. Soc.* 294 (1958).
177. Jones, E. R. H., and Mann, F. G., *J. Chem. Soc.* 1719 (1958).

178. Jongsma, C., and van der Meer, H., *Tetrahedron Lett.* 1323 (1970).
179. Jutzi, P., *Angew. Chem., Int. Ed. Engl.* **11**, 53 (1972).
180. Jutzi, P., *Chem. Ber.* **104**, 1455 (1971).
181. Jutzi, P., *J. Organometal. Chem.* **16**, P71 (1969).
182. Jutzi, P., *J. Organometal. Chem.* **19**, P1 (1969).
183. Jutzi, P., *J. Organometal. Chem.* **22**, 297 (1970).
184. Jutzi, P., *Z. Naturforsch. B* **24**, 354 (1969).
185. Jutzi, P., and Deuchert, K., *Angew. Chem., Int. Ed. Engl.* **8**, 991 (1969).
186. Kabuss, S., Friebolin, H., and Schmid, H., *Tetrahedron Lett.* 469 (1965).
187. Kametani, T., Shibuya, S., and Ollis, W. D., *J. Chem. Soc. C* 2877 (1968).
188. Kolesnikov, S. P., Shiryaev, V. I., and Nefedov, O. M., *Izv. Akad. Nauk SSSR, Ser. Khim.* 584 (1966); *Chem. Abstr.* **65**, 6705e (1966).
189. Koos, E. W., Vanderkooi, J. P., Green, E. E., and Stille, J. K., *Chem. Commun.* 1085 (1972).
190. Köster, R., *Advan. Organometal. Chem.* **2**, 257 (1964).
191. Köster, R., Fr. 1,360,431 (1964); *Chem. Abstr.* **61**, 9526d (1964).
192. Köster, R., Ger. Patent 1,089,384 (1960); *Chem. Abstr.* **56**, 1478c (1962).
193. Köster, R., and Benedikt, G., *Angew. Chem. Int. Ed. Engl.* **1**, 507 (1962).
194. Köster, R., and Benedikt, G., *Angew. Chem., Int. Ed. Engl.* **2**, 323 (1963).
195. Köster, R., and Reinert, K., *Angew. Chem.* **71**, 521 (1959).
196. Kuivila, H. G., and Beumel, O. F., Jr., *J. Amer. Chem. Soc.* **80**, 3250 (1958).
197. Kursanov, D. M., and Volpin, M. E., *Zh. Vses. Khim. Obshchest.* **7**, 282 (1962); *Chem. Abstr.* **58**, 4389e (1963).
198. Laane, J., *J. Chem. Phys.* **50**, 776 (1969).
199. Laane, J., *J. Chem. Phys.* **52**, 358 (1970).
200. Laane, J., and Lord, R. C., *J. Chem. Phys.* **47**, 4941 (1967).
201. Lambert, J. B., Oliver, W. L., Jr., and Jackson, G. F., III, *Tetrahedron Lett.* 2027 (1969).
201a. Lambert, J. B., and Keske, R. G., *J. Org. Chem.* **31**, 3429 (1966).
201b. Lambert, J. B., Oliver, W. L., Jr., and Packard, B. S., *J. Amer. Chem. Soc.* **93**, 933 (1971).
201c. Lambert, J. B., Mixan, C. E., and Johnson, D. H., *J. Amer. Chem. Soc.* **95**, 4634 (1973).
202. Lambert, R. L., Jr., and Seyferth, D., *J. Amer. Chem. Soc.* **94**, 9246 (1972).
203. Laporterie, A., Dubac, J., Mazerolles, P., and Lesbre, M., *Tetrahedron Lett.* 4653 (1971).
204. Leavitt, F. C., Manuel, T. A., and Johnson, F., *J. Amer. Chem. Soc.* **81**, 3163 (1959).
205. Leavitt, F. C., Manuel, T. A., Johnson, F., Matternas, L. U., and Lehman, D. S., *J. Amer. Chem. Soc.* **82**, 5099 (1960).
206. Lehmkuhl, H., *Angew. Chem., Int. Ed. Engl.* **5**, 663 (1966).
207. Leites, L. A., *Izv. Akad. Nauk SSSR, Ser. Khim.* 1525 (1963); *Chem. Abstr.* **59**, 14767h (1963).
208. Leites, L. A., Dulova, V. G., and Volpin, M. E., *Izv. Akad. Nauk SSSR, Otd. Khim. Nauk* 731 (1963); *Chem. Abstr.* **59**, 10104h (1963).
209. Leites, L. A., Gar, T. K., and Mironov, V. F., *Dokl. Akad. Nauk SSSR* **158**, 400 (1964); *Chem. Abstr.* **62**, 14064g (1964).
210. Lengyel, I., and Aaronson, M. J., *J. Chem. Soc. B* 177 (1971).

211. Lengyel, I., and Aaronson, M. J., *J. Organometal. Chem.* **42**, 95 (1972).
212. Lengyel, I., Aaronson, M. J., and Dillon, J. P., *J. Organometal. Chem.* **25**, 403 (1970).
213. Lesbre, M., Manuel, G., Mazerolles, P., and Cauquy, G., *J. Organometal. Chem.* **40**, C14 (1972).
214. Lesbre, M., Mazerolles, P., and Satge, J., "The Organic Compounds of Germanium," Chapter 3. Wiley, New York, 1971.
215. Letzinger, R. L., and Skoog, I. H., *J. Amer. Chem. Soc.* **77**, 5176 (1955).
216. Leusink, A. J., Budding, H. A., and Noltes, J. G., *J. Organometal. Chem.* **24**, 375 (1970).
217. Leusink, A. J., Drenth, W., Noltes, J. G., and van der Kerk, G. J. M., *Tetrahedron Lett.* 1263 (1967).
218. Leusink, A. J., Noltes, J. G., Budding, H. A., and van der Kerk, G. J. M., *Rec. Trav. Chim. Pays-Bas* **83**, 1036 (1964).
219. Levy, H. A., and Brockway, L. O., *J. Amer. Chem. Soc.* **59**, 2085 (1937).
220. Lide, D. R., Jr., *Spectrochim. Acta* **15**, 473 (1959).
221. Lynch, E. R., *J. Chem. Soc.* 3729 (1962); P. J. Wheatley, *ibid.* 3733 (1962).
222. Lyon, D. R., and Mann, F. G., *J. Chem. Soc.* 30 (1945).
223. Lyon, D. R., Mann, F. G., and Cookson, G. H., *J. Chem. Soc.* 662 (1947).
224. McLaughlin, E., and Rozett, R. W., *J. Organometal. Chem.* **52**, 261 (1973).
225. McPhail, A. T., Breen, J. J., and Quin, L. D., *J. Amer. Chem. Soc.* **93**, 2574 (1973).
226. McPhail, A. T., Breen, J. J., Somers, J. H., Steele, J. C. H., Jr., and Quin, L. D., *Chem. Commun.* 1020 (1971).
227. McPhail, A. T., Luhan, P. A., Featherman, S. I., and Quin, L. D., *J. Amer. Chem. Soc.* **94**, 2126 (1972).
228. Mack, W., *Angew. Chem. Int. Ed. Engl.* **4**, 245 (1965).
229. Mack, W., *Angew. Chem., Int. Ed. Engl.* **5**, 896 (1966).
230. Mann, F. G., "The Heterocyclic Derivatives of Phosphorus, Arsenic, Antimony and Bismuth." Wiley (Interscience), New York, 1970.
231. Mann, F. G., see p. 439 in Mann (*230*).
232. Mann, F. G., Millar, I. T., and Baker, F. C., *J. Chem. Soc.* 6342 (1965).
233. Mann F. G., Millar, I. T., and Smith, B. B., *J. Chem. Soc.* 1130 (1953).
234. Mann, F. G., and Wilkinson, A. J., *J. Chem. Soc.* 3336 (1957).
235. Manuel, G., Mazerolles, P., and Florence, J. C., *C. R. Acad. Sci., Ser. C* **269**, 1553 (1969).
236. Manuel, G., Mazerolles, P., and Florence, J. C., *J. Organometal. Chem.* **30**, 5 (1971).
237. Märkl, G., Advena, J., and Hauptmann, H., *Tetrahedron Lett.* 3961 (1972).
238. Märkl, G., and Dannhardt, G., *Tetrahedron Lett.* 1455 (1973).
239. Märkl. G., and Hauptmann, H., *Angew. Chem., Int. Ed. Engl.* **11**, 439, 441 (1972).
240. Märkl, G., and Hauptmann, H. and Advena, J., *Angew. Chem., Int. Ed. Engl.* **11**, 441 (1972).
241. Märkl, G., and Hauptmann, H., *Tetrahedron Lett.* 3257 (1968).
242. Märkl, G., and Heier, K. H., *Angew. Chem., Int. Ed. Engl.* **11**, 1017 (1972).
243. Märkl, G., and Merz, P. L., *Tetrahedron Lett.* 1303 (1971).
244. Marsh, R. E., and McCullough, J. D., *J. Amer. Chem. Soc.* **73**, 1107 (1951).
245. Maruca, R. E., *J. Org. Chem.* **36**, 1626 (1971).
246. Maruca, R., Fischer, R., Roseman, L., and Gehring, A., *J. Organometal. Chem.* **49**, 139 (1973).

247. Maruca, R., Oertel, M., and Roseman, L., *J. Organometal. Chem.* **35**, 253 (1972).
248. Massol, M., Rivière, P., Barrau, J., and Satgé, J., *C. R. Acad. Sci., Ser. C.* **270**, 237 (1970).
248a. Mazerolles, P., and Faucher, A., *Bull. Soc. Chim. Fr.* 2134 (1967).
249. Mazerolles, P., and Manuel, G., *Bull. Soc. Chim. Fr.* 327 (1966).
250. Mazerolles, P., Manuel, G., and Thoumas, F., *C. R. Acad. Sci., Ser. C* **267**, 619 (1968).
251. Minh, T. Q., Christiaens, L., and Renson, M., *Tetrahedron* **28**, 5397 (1972).
252. Minh, T. Q., Thibaut, P., Christiaens, L., and Renson, M., *Tetrahedron* **28**, 5393 (1972).
253. Mironov, V. F., and Gar. T. K., *Izv. Akad. Nauk SSSR, Otd. Khim. Nauk* 578 (1963); *Chem. Abstr.* **59**, 3941h (1963).
254. Mironov, V. F., and Gar, T. K., *Dokl. Akad. Nauk SSSR* **152**, 1111 (1963); *Chem. Abstr.* **60**, 1786f (1964).
255. Mironov, V. F., and Gar, T. K., *Izv. Akad. Nauk SSSR, Ser. Khim.*, 291 (1965); *Chem. Abstr.* **62**, 14715a (1965).
256. Mironov, V. F., and Gar, T. K., *Izv. Akad. Nauk SSSR, Ser. Khim.* 482 (1966); *Chem. Abstr.* **65**, 5481g (1966).
257. Mironov, V. F., and Nepomnina, V. V., *Izv. Akad. Nauk SSSR, Otd. Khim. Nauk* 1231 (1959); *Chem. Abstr.* **54**, 1268f (1960).
258. Mislow, K., *Trans. N. Y. Acad. Sci.* **35**, 227 (1973).
259. Mislow, K., Glass, M. A. W., Hopps, H. B., Simon, E. and Wahl, G. H., Jr., *J. Amer. Chem. Soc.* **86**, 1710 (1964).
260. Mislow, K., Zimmerman, A., and Melillo, J. T., *J. Amer. Chem. Soc.* **85**, 594 (1963).
261. Montecalvo, D., St. Jacques, M., and Wasylishen, R., *J. Amer. Chem. Soc.* **95**, 2023 (1973).
262. Morgan, G. T., and Davies, G. R., *Proc. Roy. Soc., Ser. A* **127**, 1 (1930).
263. Morgan, G. T., and Davies, G. R., *Proc. Roy. Soc., Ser. A* **143**, 38 (1933).
264. Moritani, I., Murahashi, S. I., Yoshinaga, I., and Ashitaka, H., *Bull. Chem. Soc. Jap.* **40**, 1506 (1967).
265. Nagai, Y., Kono, H., Matsumoto, H., and Yamazaki, K., *J. Org. Chem.* **33**, 1966 1968).
266. Nakadaira, Y., and Sakurai, H., *J. Organometal. Chem.* **47**, 61 (1973).
267. Nakadaira, Y., and Sakurai, H., *Tetrahedron Lett.* 1183 (1971).
268. Nametkin, N. S., Vdovin, V. M., Finkel'shtein, E. Sh., Arkhipov, T. N., and Oppengeim, V. D., *Dokl. Akad. Nauk SSSR* **154**, 383 (1964); *Chem. Abstr.* **60**, 9304c (1964).
269. Nametkin, N. S., Vdovin, V. M., Finkel'shtein, E. H., Konobeevskii, K. S., and Oppengeim, V. D., *Dokl. Akad. Nauk SSSR* **162**, 585 (1965); *Chem. Abstr.* **63**, 7119 (1965).
270. Nefedov, O. M., and Kolesnikov, S. P., *Izv. Akad. Nauk SSSR, Ser. Khim.* 2615 (1971); *Chem. Abstr.* **76**, 99783n (1972).
271. Nefedov, O. M., Kolesnikov, S. P., Khachaturov, A. S., and Petrov, A. D., *Dokl. Akad. Nauk SSSR* **154**, 1389 (1964); *Chem. Abstr.* **60**, 12039f (1964).
272. Nefedov, O. M., and Manakov, M. N., *Izv. Akad. Nauk SSSR, Otd. Khim. Nauk* 769 (1963); *Chem. Abstr.* **59**, 8780d (1963).
273. Negishi, E., Katz, J-J., and Brown, H. C., *J. Amer. Chem. Soc.* **94**, 4025 (1972).
274. Neumann, W. P., "The Organic Chemistry of Tin," Chapter 20. Wiley, New York, 1970.

275. Nikishin, G. I., Petrov, A. D., and Sadykh-Zade, S. I., *Khim. Prakt. Primen. Kremneorg. Soed. Tr. Konf., 2nd, 1958, Leningrad*, 68 (1958); *Chem. Abstr.* **53**, 17097f,g (1959).
276. Nöth, H., and Schmid, G., *Z. Annorg. Allg. Chem.* **345**, 69 (1966).
277. Noltes, J. G., and van der Kerk, G. J. M., *Chimia* **16**, 122 (1962).
278. Noltes, J. G., and van der Kerk, G. J. M., *Rec. Trav. Chim. Pays-Bas* **81**, 41 (1962).
279. Nogradi, M., Ollis, W. D., and Sutherland, I. O., *Chem. Commun.* 158 (1970).
280. Oberhammer, H., and Bauer, S. H., *J. Amer. Chem. Soc.* **91**, 10 (1969).
281. Oki, M., Iwamura, H., and Hayakawa, N., *Bull. Chem. Soc. Jap.* **37**, 1865 (1964).
282. Okinoshima, H., Yamamoto, K., and Kumada, M., *J. Amer. Chem. Soc.* **94**, 9263 (1972).
283. Orlaff, H. D., Kolka, A. J., Calingaert, G., Griffing, M. E., and Kerr, E. R., *J. Amer. Chem. Soc.* **75**, 4243 (1953).
284. Ouellette, R. J., Baron, D., Stolfo, J., Rosenblum, A., and Weber, P., *Tetrahedron* **28**, 2163 (1973).
285. Paetzold, P. I., and Smolka, H. G., *Chem. Ber.* **103**, 289 (1970).
286. Passerini, R., and Purrello G., *Ann. Chim. (Rome)* **48**, 738 (1958).
287. Patterson, A. M., Capell, L. T., and Walker, D. F., "Ring Index," 2nd ed., Appendix. Amer. Chem. Soc., Washington, D.C., 1960.
288. Petrov, A. D., Nikishin, G. I., and Smetana, N. P., *Zh. Obshch. Khim.* **28**, 2085 (1958); *Chem. Abstr.* **53**, 3038 (1959).
289. Pitt, C. G., *J. Organometal. Chem.* **23**, C35 (1970).
290. Pulman, D. A., and Whiting, D. A., *Chem. Commun.* 831 (1971).
291. Quin, L. D., and Somers, J. H., *J. Org. Chem.* **37**, 1217 (1973).
292. Rabideau, P. W., and Paschal, J. W., *J. Amer. Chem. Soc.* **94**, 5801 (1972).
293. Rakita, R. E., and Rothschild, J., *Chem. Commun.* 953 (1971).
294. Rauch, M. D., and Klemann, L. P., *J. Amer. Chem. Soc.* **89**, 5732 (1967).
295. Resibois, B., and Brunet, J. C., *Ann. Chim. (Paris)* **5**, 199 (1970).
296. Resibois, B., Hode, C., Picart, B., and Brunet, J. C., *Ann. Chim. (Paris)* **4**, 203 (1969).
297. Roberts, E., Turner, E. E., and Bury, F. W., *J. Chem. Soc.* 1443 (1926).
298. Roberts, J. D., *Chem. Brit.* 529 (1966).
299. Rosenburg, E., and Zuckerman, J. J., *J. Organometal. Chem.* **33**, 321 (1971).
300. Ruehlmann, K., *Z. Chem.* **5**, 354 (1965).
301. Ruehlmann, K., Hagen, V., and Schiller, K., *Z. Chem.* **7**, 353 (1967).
302. Sakurai, H., and Murakami, M., *J. Amer. Chem. Soc.* **94**, 5080 (1972).
303. Sanz, F., and Daly, J. J., *J. Chem. Soc., Dalton Trans.* 511 (1973).
304. Sartain, D., and Truter, M. R., *J. Chem. Soc.* 4414 (1963).
305. Sato, T., and Moritani, *Tetrahedron Lett.* 3181 (1969).
306. Sato, T., Moritani, I., and Matsuyama, *Tetrahedron Lett.* 5113 (1969).
307. Schäfer, W. C., Schweig, A., Bickelhaupt, F., and Vermeer, H., *Angew. Chem., Int. Ed. Engl.* **11**, 924 (1972).
308. Scharpen, L. H., and Laurie, V. W., *J. Chem. Phys.* **43**, 2765 (1965).
309. Schmid, G., and Nöth, H., *Chem. Ber.* **100**, 2899 (1967).
310. Scibelli, J. V., and Curtis, M. D., *J. Organometal. Chem.* **40**, 317 (1972).
311. Senkler, G. H., Gust, D., Riccobono, P. X., and Mislow, K., *J. Amer. Chem. Soc.* **94**, 8626 (1972).
312. Severson, R. G., and Corey, J. Y., unpublished work.
313. Seyferth, D., Jula, T. F., Mueller, D. C., Mazerolles, P., Manuel, G., and Thoumas, F., *J. Amer. Chem. Soc.* **92**, 657 (1970).

314. Shapiro, H., and Frey, F. W., "The Organic Compounds of Lead," Chapter 13. Wiley, New York, 1968.
315. Sheehan, D., *Diss. Abstr.* **25**, 4417 (1965).
316. Sheehan, W. F., Jr., and Shomaker, V., *J. Amer. Chem. Soc.* **74**, 3956 (1952).
317. Shook, E., Jr., and Quin, L. D., *J. Amer. Chem. Soc.* **89**, 1841 (1967).
318. Sisido, K., Takeda, Y., and Kinugawa, Z., *J. Amer. Chem. Soc.* **83**, 538 (1961).
319. Smith, W. B., and Shoulders, B. A., *J. Phys. Chem.* **69**, 2022 (1965).
320. Springall, H. D., and Brockway, L. O., *J. Amer. Chem. Soc.* **60**, 996 (1938).
321. Stidham, H. D., *Spectrochim. Acta* **21**, 23 (1965).
322. St. Jacques, M., and Vaziri, C., *Can. J. Chem.* **51**, 1192 (1973).
323. Sutherland, I. O., and Ramsay, M. V. J., *Tetrahedron* **21**, 3401 (1965).
324. Ternay, A. L., Jr., Ens, L., Hermann, J., and Evans, S., *J. Org. Chem.* **34**, 941 (1969).
325. Thornton, D. A., *J. S. Afr. Chem. Inst.* **17**, 61 (1964).
326. Tochtermann, W., *Fortschr. Chem. Forsch.* **15**, 378 (1970).
327. Tochtermann, W., Schnabel, G., and Mannschreck, A., *Justus Liebigs Ann. Chem.* **705**, 169 (1967).
328. Tochtermann, W., Walter, U., and Mannschreck, A., *Tetrahedron Lett.* 2981 (1964).
329. Traetteberg, M., *J. Amer. Chem. Soc.* **86**, 4265 (1964).
330. Tribble, M. T., and Allinger, N. L., *Tetrahedron* **28**, 2147 (1972).
331. Trombelti, A., and Zauli, C., *Corsi Semin. Chim.* 60 (1968).
332. Truce, W. E., and Emrick, D. D., *J. Amer. Chem. Soc.* **78**, 6130 (1956).
333. Turner, E. E., and Bury, F. W., *J. Chem. Soc.* **123**, 2489 (1923).
334. Vafai, M., and Renson, M., *Bull. Soc. Chim. Belges* **75**, 145 (1966).
335. Van Mourik, G. L., and Bickelhaupt, F., *Rec. Trav. Chim., Pays-Bas* **88**, 868 (1967).
336. Van Tamelen, E. E., Brieger, G., and Untch, K. G., *Tetrahedron Lett.* **8**, 14 (1960).
337. Van Veen, R., and Bickelhaupt, F., *J. Organometal. Chem.* **24**, 589 (1970).
338. Van Veen, R., and Bickelhaupt, F., *J. Organometal. Chem.* **30**, C51 (1971).
339. Van Veen, R., and Bickelhaupt, F., *J. Organometal. Chem.* **43**, 241 (1972).
340. Van Veen, R., and Bickelhaupt, F., *J. Organometal. Chem.* **47**, 33 (1973).
341. Vdovin, V. M., Nametkin, N. S., Finkel'shtein, E. Sh., and Oppengeim, V. D., *Izv. Akad. Nauk SSSR, Ser. Khim.* 458 (1964); *Chem. Abstr.* **60**, 15902b (1964).
342. Vermeer, H., and Bickelhaupt, F., *Angew. Chem., Int. Ed. Engl.* **8**, 992 (1969).
343. Vermeer, H., and Bickelhaupt, F., *Tetrahedron Lett.* 3255 (1970).
344. Vilkov, L. V., and Mastryukov, V. S., *Zh. Strukt. Khim.* **6**, 811 (1965); *Chem. Abstr.* **64**, 12512e (1966).
345. Vilkov, L. V., Martryukov, V. S., Shcherbik, L. K., and Dulova, V. G., *Zh. Strukt. Khim.* **11**, 3 (1970); *Chem. Abstr.* **72**, 126065e (1970).
346. Vol'pin, M. E., Dulova, V. G., and Kursanov, D. N., *Izv Akad. Nauk SSSR, Otd. Khim. Nauk* 727 (1963); *Chem. Abstr.* **59**, 10104e, (1963).
347. Vol'pin, M. E., Dulova, V. G., Struchkov, Yu. T., Bokiy, N. K., and Kursanov, D. N., *J. Organometal. Chem.* **8**, 87 (1967).
348. Vol'pin, M. E., Koreshkov, Yu. D., Dulova, V. G., and Kursanov, D. N., *Tetrahedron* **18**, 107 (1962).
349. Vol'pin, M. E., Koreshkov, Yu. D., and Kursanov, D. N., *Izv. Akad. Nauk SSSR, Otd. Khim. Nauk* 1355 (1961); *Chem. Abstr.* **56**, 1470f (1962).
350. Vol'pin, M. E., and Kursanov, D. N., *Izv. Akad. Nauk SSSR, Otd. Khim. Nauk* 1903 (1960); *Chem. Abstr.* **55**, 14419e (1961).

351. Vol'pin, M. E., and Kursanov, D. N., *Zh. Obshch. Khim.* **32**, 1137 (1962); *Chem. Abstr.* **58**, 1332c (1963).
352. Vol'pin, M. E., and Kursanov, D. N., *Zh. Obshch. Khim.* **32**, 1142 (1962); *Chem. Abstr.* **58**, 1332e (1963).
353. Vol'pin, M. E., and Kursanov, D. N., *Zh. Obshch. Khim.* **32**, 1455 (1962); *Chem. Abstr.* **58**, 9111b (1963).
354. Vol'pin, M. E., Struchkov, Yu. T., Vilkov, L. V., Mastryukov, V. S., Dulova, V. G., and Kursanov, D. N., *Izv. Akad. Nauk SSSR, Ser. Khim.* 2067 (1963); *Chem. Abstr.* **60**, 5532h (1964).
355. Vranka, R. G., and Amma, E. L., *J. Amer. Chem. Soc.* **89**, 3121 (1967).
356. Washburne, S. S., and Chawla, R. R., *J. Organometal. Chem.* **31**, C20 (1971).
357. Weber, W., and Laine, R., *Tetrahedron Lett.* 4169 (1970).
358. West, R., and Bailey, R. E., *J. Amer. Chem. Soc.* **85**, 2871 (1963).
359. Weyenberg, D. R., Torporcer, L. H., and Nelson, L. E., *J. Org. Chem.* **33**, 1975 (1968).
360. Wittenberg, D., and Gilman, H., *J. Amer. Chem. Soc.* **80**, 2677 (1958).
361. Wittenberg, D., Talukar, P. B., and Gilman, H., *J. Amer. Chem. Soc.* **82**, 3608 (1960).
362. Wittig, G., and Benz, E., *Chem. Ber.* **91**, 873 (1958).
363. Wittig, G., and Hellwinkel, D., *Chem. Ber.* **97**, 769 (1964).
364. Wittig, G., and Hellwinkel, D., *Chem. Ber.* **97**, 789 (1964).
365. Wittig, G., and Herwig, W., *Chem. Ber.* **88**, 962 (1955).
366. Wong, C. H., and Shomaker, V., *J. Chem. Phys.* **28**, 1007 (1958).
367. Yale, H. L., Sowinksi, F., and Spitzmiller, E. R., *J. Heterocycl. Chem.* **9**, 899 (1972).
368. Zavistoski, J. G., and Zuckerman, J. J., *J. Org. Chem.* **34**, 4197 (1969).
369. Zhdanov, A. A., Andrianov, K. A., Odinets, V. A., and Karpova, I. V., *Zh. Obshch. Khim.* **36**, 521 (1966); *Chem. Abstr.* **65**, 735c (1966).

REFERENCES ADDED IN PROOF

370. Allinger, N. L., Greengard, R. A., and Finder, C. J., *Tetrahedron Lett.* 3095 (1973).
371. Balasubramanian, R., and George, M. V., *Tetrahedron* **29**, 2395 (1973).
372. Beagley, B., and McAloon, K. T., *J. Mol. Struct.* **17**, 429 (1973).
373. Brunet, J. C., Bertrand, J., and Lesenne, C., *J. Organometal Chem.* **71**, C8 (1974).
374. Chernyshev, E. A., Komalenkova, N. G., and Bashkirova, S. A., *Dokl. Akad. Nauk SSSR* **205**, 868 (1972).
375. Ferrier, W. G., and Ball, J. I., *Chem. Ind.* (*London*) 1296 (1954).
376. Gilman, H., and Gaj, B. J., *J. Org. Chem.* **22**, 447 (1957).
377. Jutzi, P., and Fetz, H., *Chem. Ber.* **106**, 3495 (1973).
378. Lasheen, M. J., *Acta Cryst.* **5**, 593 (1952).
379. Märkl, G., and Kneidl, F., *Angew. Chem., Int. Ed. Engl.* **12**, 931 (1973).
380. Meinema, H. A., Crispim Romao, C. J. R., and Noltes, J. G., *J. Organometal Chem.* **55**, 139 (1973).
381. Mueller, E., and Zountas, G., *Chem. Ztg.* **97**, 271 (1973).
382. Mueller, E., and Zountas, G., *Chem. Ztg.* **97**, 447 (1973).
383. Pasternik, R. A., *Acta Cryst.* **4**, 316 (1951).
384. Sakurai, H., and Hayashi, J., *J. Organometal. Chem.* **63**, C10 (1973).

Organometallic Reactions Involving Hydro–Nickel, –Palladium, and –Platinum Complexes

D. MAX ROUNDHILL

Department of Chemistry
Washington State University
Pullman, Washington

I
INTRODUCTION

Hydro[1] complexes of the transition metals have been known for many years. The subject has been reviewed previously, although in most of the earlier work the coverage included all the transition metals (*228*, *164*, *158*, *67*, *195*, *224*). A review of a less general nature has recently been published which covers the subject of hydro and alkyl complexes of nickel (*148*). In this article the synthesis and chemistry of the hydro complexes of the nickel triad are discussed in the early sections, and then later an attempt is made to show the involvement of compounds of this kind in reactions such as the isomerization, oligomerization, hydrosilylation, and hydrogenation of unsaturated hydrocarbons. Although hydro complexes of other transition metals have been shown to be important with respect to these reactions, the nickel triad of elements is particularly appropriate for separate consideration because these three elements have been widely used as *heterogeneous* catalysts in a variety of reactions. This article discusses the hydro complexes of this triad in some depth, although it is not entirely comprehensive since the spectroscopic section is brief and the table of hydro complexes at the end (see Table XIII) has some omissions. Nevertheless, it is hoped that most of the more important studies in this area are included and that the article will be a useful introductory guide into the involvement of compounds of this type in homogeneous catalysis.

[1] The term *hydro* has been used throughout in this article in preference to the more commonly used terms *hydride* or *hydrido*. These latter terms imply that the hydrogen atom bonded to a transition metal has a high electron density comparable to the saline hydrides of Groups IA and IIA. A number of hydro-transition metal compounds do, indeed, show chemical behavior characteristic of a hydridic hydrogen. However, this is in no way general, and, since it is a particularly poor assumption for the compounds of platinum, the use of the term hydro should avoid any misconception on the part of the reader not familiar with this area of chemistry.

There are numerous examples in the literature where a hydro complex of the nickel triad has been proposed as a reactive intermediate. It is beyond the scope of this article to discuss work of this type in detail, however, a number of examples can be presented in order to give the reader some idea of the range of reactions in which these hydro complexes may be involved. For example, a hydronickel intermediate was postulated in the nickel-catalyzed hydroalumination of alkynes where small amounts of $Ni(acac)_2$ were found to accelerate markedly the rate of *cis*-hydroalumination (*146*). Other reactions of hydro complexes involve the reduction of carbonium ions by $PtHCl(PEt_3)_2$ (*131*), the catalytic decomposition of formic acid by platinum complexes (*110, 135*), the preparation of α,β-unsaturated aldehydes and ketones from saturated analogs with $PdCl_2(PPh_3)_2$ (*278*), and the preparation of $Pt(CNO)_2(PPh_3)_2$ from $Pt(PPh_3)_2$ and CH_3NO_2 (*33*).

A reaction that has been studied in some detail is the homogeneous exchange reaction with isotopic hydrogen in benzene which occurs in the presence of catalytic amounts of Pt(II) and Pt(IV) salts (*154*). The exchange rate was found to follow the kinetics: k (exchange) α $[PtCl_4^{2-}][C_6H_6][Cl^-]^{-1}$. The proposed mechanism is shown in Fig. 1 and

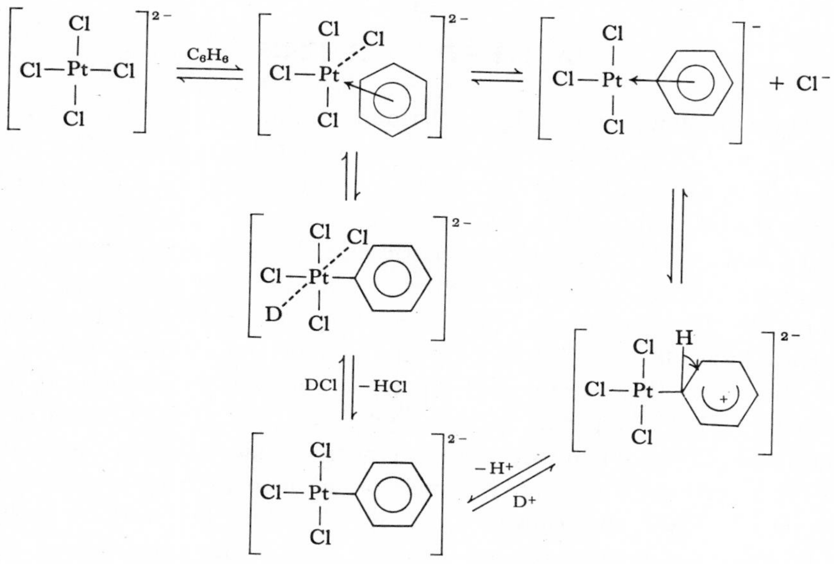

FIG. 1. Proposed mechanism for hydrogen–deuterium exchange on aromatics.

involves a hydroplatinum intermediate (*181*). The study has been extended to polycyclics which showed that the rate of exchange in an acetic acid medium increased with decreasing double-bond localization energy for the condensed polycyclics (*182*). When this reaction was carried out with ethylbenzene, it was found that isotopic hydrogen exchange occurred in the alkyl group, and a mechanism involving a π-allylic process was suggested (*155*). This reaction occurred when the reaction was catalyzed either homogeneously or heterogeneously. Although it is always possible to carry out these exchange reactions under homogeneous or heterogeneous conditions, the homogeneous method was found to be preferable since no competing dimerization was observed in this exchange work (*59*). This study is closely paralleled by the H—D exchange reaction in saturated alkanes which was found to be catalyzed by platinum complexes (*183–185*).

In the next sections, the synthesis, reactions, and relevance to catalysis of hydro complexes of the nickel triad are discussed. A short section on spectroscopy of these compounds is included, along with an extensive table of hydro complexes that have been characterized or isolated.

II

SYNTHETIC METHODS

The first synthesis of a hydroplatinum complex was that of $PtHCl(PEt_3)_2$ in 1957 (*70*). This compound was considered remarkable since it could be purified by distillation under reduced pressure. The complex showed a band in the IR at 2230 cm^{-1} characteristic of a Pt—H stretch, and a band at 824 cm^{-1} characteristic of a Pt—H bending mode. The corresponding deutero complex showed a band at 1601 cm^{-1} assignable to the Pt—D stretch mode. The ^1H NMR spectrum showed a resonance for the hydro ligand +22 ppm upfield of the water standard and small satellite peaks that were considered to be due to ^{195}Pt coupling. Even though $PtHCl(PEt_3)_2$ was the first hydro complex to be isolated in a pure state, Wilkinson in 1955 had been the first to show that the resonance of the hydro ligand occurs to high field of tetramethylsilane, as evidenced by the ^1H NMR spectrum of $ReH(cp)_2$ (*298*). Further work on hydro complexes has proven the value of IR and NMR spectroscopy for verification and structure determination. However, the latter provides the most diagnostic evidence, and any result claiming the existence of a hydro complex based on IR evidence alone should be treated with some caution. This is important since there have been several errors made in the claims for the existence of hydroplatinum complexes, and in nearly every case the compound was proposed on the sole evidence of IR spectroscopy.

Hydro complexes of nickel, palladium, and platinum have been synthesized by several methods employing acidic, basic, and neutral media. The preparative procedures discussed in the following sections do not represent the complete range of methods available for the synthesis of these complexes; nevertheless, they do include many of the more widely applicable methods and show the diversity of the chemistry involved.

A. Reduction with Hydrazine

This method is significant since it was the first one used in the preparation of $PtHCl(PEt_3)_2$ from $PtCl_2(PEt_3)_2$ (*70*). The yield is good. However, when $PtCl_2(PPh_3)_2$ is reduced with hydrazine in the presence of excess triphenylphosphine, the product is $Pt(PPh_3)_n$ ($n = 3$ or 4) (*216*). This difference may be due to triphenylphosphine being better able to act as an acceptor, thereby stabilizing the zero-valent platinum complex. Hydrazine has been used in the preparation of hydrotriphenylphosphine platinum complexes since the reaction of $PtCl_2(PPh_3)_2$ with hydrazine in the presence of 5-chloro-, 5-bromo-, or 5-phenyltetrazole leads to the formation of the two isomeric hydrotetrazolatoplatinum(II) complexes, structures (I) and (II) (*194*):

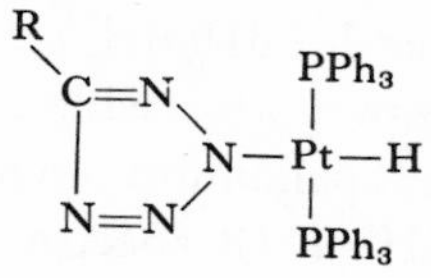

```
R                              R
 \                            /
  C=N          PPh3        N=C          PPh3
  |   \         |          |   \         |
  |    N—Pt—H              |    N—Pt—H
  |   /         |          |   /         |
  N=N          PPh3        N=N          PPh3

       (I)                       (II)
```

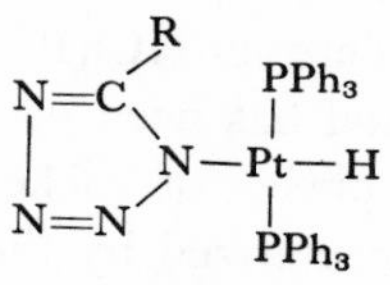

B. Reduction with Metallohydrides Such as Borohydride and Aluminum Hydride

The complex $PtHCl(PEt_3)_2$ has been prepared by reacting $PtCl_2(PEt_3)_2$ with $LiAlH_4$ (*78*). Sodium borohydride can be used to prepare the complexes $PtHCl(PMe_2Ph)_2$, $PtHCl(PMePh_2)_2$, and $PtHBr(PMe_2Ph)_2$ by reaction with the respective dihaloplatinum(II) compounds (*100*). Sodium borohydride has also been used to synthesize $NiHCl(PCy_3)_2$ from $NiCl_2(PCy_3)_2$ in tetrahydrofuran–ethanol solvent (*167*). Later these authors extended the reaction to the synthesis of further complexes of this type (*168*). Reaction of sodium borohydride in mixed solvents with $NiX_2(PR_3)_2$ (R = Cy and X = Cl, Br, SCN; or R = Pr^i and X = Cl) gave the monohydro complexes $NiHX(PR_3)_2$.

By metathetical replacement reactions, this range of complexes can be extended (R = Cy; X = I, CN and R = Pr^i; X = Br, I] (*168*):

$$NiHCl(PR_3)_2 + MX \longrightarrow NiHX(PR_3)_2 + MCl \quad (1)$$

Treatment of *trans*-hydrochlorobis(tricyclohexylphosphine)nickel(II) with sodium borohydride in a mixed solvent (acetone and ethanol) under argon yields *trans*-hydroborohydrobis(tricyclohexylphosphine)nickel(II), which has been shown to have the pentacoordinate structure (III) (*169*).

PR₃ / H—Ni(···H···)₂B(H)₂ / PR₃ (R = cyclohexyl)

(III)

This complex itself acts as a hydridic reducing agent and, in benzene, reacts with $MX_2(PR_3)_2$ (M = Ni or Pd) to form new hydro complexes:

$$NiH(BH_4)(PCy_3)_2 + NiX_2(PR_3)_2 \longrightarrow NiHX(PCy_3)_2 + cis/trans\text{-}NiHX(PR_3)_2 + B_2H_6 \quad (2)$$

In later work, Green *et al.* (*230*) used the reducing ability of $NiH(BH_4)(PCy_3)_2$ to prepare $PdHCl(PR_3)_2$ and $PdH(BH_4)(PR_3)_2$ (R = Cy or Pr^i). This method has been used to prepare a wide range of hydro and borohydro complexes of nickel and palladium, many of which have been solely characterized by their ^1H NMR spectra. It was considered (*166*) that in all cases the first step in the reaction was reduction, Eq. (3), which was followed by phosphine exchange reactions to form a mixture of six hydro complexes (M = Ni, Pd; R = Et, Pr^i, Bu^n; all complexes have *trans* stereochemistry):

$$MX_2(PR_3)_2 + NiH(BH_4)(PCy_3)_2 \longrightarrow MHX(PR_3)_2 + NiHX(PCy_3)_2 + \tfrac{1}{2}B_2H_6 \quad (3)$$

$$MHX(PR_3)_2 + NiHX(PCy_3)_2 \longrightarrow MHX(PR_3)_2 + MHX(PR_3)(PCy_3) + MHX(PCy_3)_2 + NiHX(PR_3)_2 + NiHX(PR_3)(PCy_3) + NiHX(PCy_3)_2 \quad (3a)$$

Complexes NiH(π-allyl)(PR_3) (R = Ph, F) have been prepared by low-temperature reaction of the corresponding bromo complexes with Na[$HBMe_3$] (*42*). The low-temperature reaction of bis(2,4-pentanedionato)nickel(II) with triisobutylaluminum in the presence of triethylphosphine or triisobutylphosphine gives the hydro complexes NiH(N_2)(PR_3)$_2$ (*268*). Trialkylaluminum compounds have also been used in the preparation of dihydro complexes of palladium and platinum. The complexes PtH_2L_2 (L = PCy_3, PCy_2Pr^i, PCy_2Et) have been obtained from Pt(acac)$_2$ and AlR_3 (R = Et, Bu^i) with the appropriate phosphine in ether (*205*). A similar reaction has been used to prepare $PdH_2(PCy_3)_2$. The compounds have been characterized by the formation of equimolar quantities of $CHCl_3$ from excess CCl_4, although it is surprising that both $PdH_2(PCy_3)_2$ and $PtH_2(PCy_3)_2$ are insoluble in organic solvents, making it doubtful whether the compounds are monomeric in structure.

C. Protonation Reactions

1. *Oxidative-Addition Processes*

a. *Platinum*. Protonation of $Pt(PPh_3)_4$ with a variety of acids has been shown to give hydroplatinum(II) complexes. This reaction was reported in 1964 when an equilibrium was proposed, involving both ionic and covalent hydro complexes (*60*):

$$Pt(PPh_3)_3 \underset{KOH}{\overset{HX}{\rightleftharpoons}} [PtH(PPh_3)_3]X \underset{PPh_3}{\overset{-PPh_3}{\rightleftharpoons}} PtHX(PPh_3)_2 \tag{4}$$

Subsequently (*61*) these authors suggested that with HCl the following equilibria were involved (L = PPh_3):

$$PtL_4 \underset{L}{\overset{-L}{\rightleftharpoons}} PtL_3 \underset{KOH}{\overset{HCl}{\rightleftharpoons}} [PtHL_3]Cl \underset{L}{\overset{-L}{\rightleftharpoons}} PtHClL_2 \underset{-HCl}{\overset{HCl}{\rightleftharpoons}} PtH_2Cl_2L_2 \tag{5}$$

When the conjugate base X of HX is a poor ligand for platinum, such as NO_3, ClO_4, BF_4, CH_3OSO_3, HSO_4, BPh_4, the product is the ionic hydro complex [$PtH(PPh_3)_3$]X. If, however, the conjugate base is an anion such as Cl, CN, or $SCOCH_3$ (*253*), which will coordinate strongly to platinum(II), the product is the covalent hydro complex $PtHX(PPh_3)_2$. The proposed equilibria in Eq. (5) suggest that $PtHCl(PPh_3)_2$ will oxidatively add HCl to form $PtH_2Cl_2(PPh_3)_2$, and indeed the isolation of this dihydroplatinum(IV) complex has been claimed (*61*). Later work has shown that this reaction does not occur to any appreciable extent, and that $PtH_2Cl_2(PPh_3)_2$ cannot be isolated or identified under these reaction conditions (*141*). Complex $Pt(PPh_3)_4$ will, however, oxidatively add 2 molecules of 1-ethynylcyclohexanol (*252*, *232*) to form the complex $PtH_2(C{\equiv}CC_6H_{10}OH)_2(PPh_3)_2$ (IV). Preliminary X-ray crystallographic and neutron diffraction data indicate that this complex crystallizes as monoclinic crystals of space group $P2_{1/c}$ with the platinum centrosymmetrically surrounded by a coplanar arrangement of two phosphorus and two carbon atoms. The hydrogen atoms have not yet

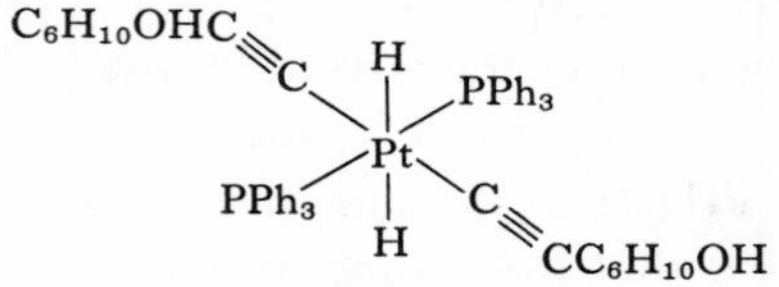

$C_6H_{10}OHC{\equiv}C$, H, PPh_3, Pt, PPh_3, H, $C{\equiv}CC_6H_{10}OH$

(IV)

been definitively located (*233*, *245*). This double oxidative addition of a terminal acetylene is apparently the only example of the formation of a dihydroplatinum(IV) complex from $Pt(PPh_3)_4$. The formation of such a compound cannot be related to the fact that a terminal acetylene is a weak acid since imides, which are also weak acids, react with $Pt(PPh_3)_4$ to form hydroplatinum(II) complexes of the type $PtHX(PPh_3)_2$ (*250*, *251*). A possible explanation for this reaction with 1-ethynylcyclohexanol could be the stabilizing influence of π-bonding which is possible for a ligand such as acetylide. Such an explanation, however, fails to agree with the concept that π-bonding is generally considered to favor the stabilization of low oxidation states. Nevertheless, if this explanation has validity, the complexes $PtHCN(PPh_3)_2$ and $PtH(NCS)(PPh_3)_2$ should perhaps oxidatively add a second molecule of a protonic acid such as HCN or 1-ethynylcyclohexanol.

When the phosphine coordinated to platinum is more basic than triphenylphosphine, the oxidative addition of a protonic acid to hydroplatinum(II) compounds to form a dihydroplatinum(IV) complex is more favored. In 1962, Chatt and Shaw (*78*) reported that $PtHCl(PEt_3)_2$ would oxidatively add HCl to form $PtH_2Cl_2(PEt_3)_2$:

$$PtHCl(PEt_3)_2 + HCl \longrightarrow PtH_2Cl_2(PEt_3)_2 \quad (6)$$

This and similar reactions have been recently studied by Ebsworth *et al.* (*14*) who have prepared a series of mono and dihydro complexes of six-coordinated platinum by the reaction of hydrogen halides with the compounds *trans*-$PtY_2(PEt_3)_2$ or *trans*-$PtHY(PEt_3)_2$, where Y = Cl, Br, I. The products have been characterized by ^1H NMR spectroscopy, although the complexes *cis-cis-trans*-$PtH_2X_2(PEt_3)_2$ (X = Cl, Br) have also been verified by microanalysis. The reaction between HX and *trans*-$PtHY(PEt_3)_2$ (X, Y = Cl, Br, I and X $\neq$ Y) gave a compound that showed two Pt—H resonances. On warming this solution, resonances due to $PtH_2X_2(PEt_3)_2$ and $PtH_2Y_2(PEt_3)_2$ increased in size. These results have been interpreted as involving the initial formation of $PtH_2XY(PEt_3)_2$ which is followed by the "scrambling" of the halogens on a roughly statistical basis,

$$2PtH_2XY(PEt_3)_2 \rightleftharpoons PtH_2X_2(PEt_3)_2 + PtH_2Y_2(PEt_3)_2 \tag{7}$$

The reaction of HX with *trans*-$PtX_2(PEt_3)_2$ appears to be a rapid equilibrium, the rates and instability constants decreasing in the order X = Cl > Br > I for the equilibrium

$$PtHX_3(PEt_3)_2 \rightleftharpoons PtX_2(PEt_3)_2 + HX \tag{8}$$

The reaction between HX and *trans*-$PtY_2(PEt_3)_2$ leads ultimately to halogen exchange, but the scrambling is less facile than occurs with $PtH_2XY(PEt_3)_2$. The rate of scrambling in the monohydro system appears to decrease in the order Cl > Br > I, and the product mixture is predominant in the compounds where H is trans to the lightest halogen present.

Since the stability of a hydro complex formed by the oxidative addition of a protonic acid depends both on the strength of the acid and the nephelauxetic effect of the conjugate base, it follows that stable complexes are to be anticipated from the addition of acids that are strong and have conjugate bases which will coordinate strongly. Very few acids fulfill both these functions adequately; however, it appears that a stable hydro complex will be formed if either property is possessed by the acid. It follows, therefore, that sulfur acids will very readily undergo oxidative addition, and complexes of the type $PtHX(PPh_3)_2$ have been obtained from the oxidative addition of PhSH, H_2S [the product is considered to be a mixture of (V) and (VI)] (*227*), mercaptobenzothiazole, and thioacetic acid (*253*). Addition of para-substituted benzenethiols to $Pt(PPh_3)_3$ or $Pt(PPh_3)_4$, where the para substituents are NO_2, Br, Cl, F, H, Me, NH_2, has shown good correlations between ν (PtH) or J(Pt—H) when plotted against the Hammett substituent parameter σ_p because of the electron density changes at platinum due to the mesomeric and inductive effects of the para substituent (*197*) (see Section VII).

$$\begin{array}{ccc} Ph_3P & & H \\ & Pt \leftarrow S & \\ Ph_3P & & H \end{array} \qquad \begin{array}{ccc} H & & PPh_3 \\ & Pt & \\ Ph_3P & & SH \end{array}$$

(V) (VI)

Formation of a stable hydro complex with thioacetic acid is particularly interesting, since the reaction with formic or acetic acids, which have similar acidities, leads to an equilibrium in which the hydro complex reductively eliminates in the absence of excess acid (*279*).

$$Pt(PPh_3)_3 + CH_3COSH \longrightarrow PtH(SCOCH_3)(PPh_3)_2 + PPh_3 \quad (9)$$

$$Pt(PPh_3)_3 + HCO_2H \rightleftharpoons [PtH(PPh_3)_3]HCO_2 \quad (10)$$

When $Pt(PPh_3)_4$ is protonated with a strong carboxylic acid such as CF_3CO_2H, or CF_2ClCO_2H the product is $[PtH(PPh_3)_3](CF_3CO_2)_2H$ or the corresponding $(CF_2ClCO_2)_2H$ complex, respectively, the ^1H NMR spectra of which have been discussed in detail (*279*) (see Section VII). In order to prepare $PtH(OCOCF_3)(PPh_3)_2$, the ethylene complex $Pt(C_2H_4)(PPh_3)_2$ must be used (*279*):

$$Pt(C_2H_4)(PPh_3)_2 + CF_3CO_2H \longrightarrow PtH(OCOCF_3)(PPh_3)_2 + C_2H_4 \quad (11)$$

Addition of CF_3CO_2H to the *N*-methylhexafluoroisopropylideneamine platinum complex (VII) gives the hydro complex (VIII) (*266*):

$$\underset{(VII)}{Pt[(CF_3)_2CNMe](PPh_3)_2} + CF_3CO_2H \longrightarrow \underset{(VIII)}{PtH(OCOCF_3)[(CF_3)_2CNMe](PPh_3)_2} \quad (12)$$

When a more basic phosphine is used as a ligand, protonation is facilitated. Thus $Pt(PMe_2Ph)_4$ can be protonated with phenol (*201*), nitromethane, and 2,4-pentanedione (*202*). A somewhat surprising result is the report of the protonation of $Pt(PPh_2OBu^n)_4$ with phenol (*203*). Since PPh_2OBu^n is anticipated to have a slightly higher "π-acceptor strength" than PPh_3 (*280*), and $Pt(PPh_3)_4$ does not react detectably with phenol, it can be surmised from this difference in reactivity that steric effects may also be important. It is possible that the decrease in bond angle on going from ML_3 to the product $[MHL_3]X$ or $MHXL_2$ is sufficiently unfavorable for steric effects to be significant in these reactions. However, a much more detailed study of such effects must be made before any definitive statement can be offered.

b. *Addition Reactions of Silanes*. The preparation of hydroplatinum compounds by the oxidative addition of silanes is placed in a separate subsection because of the current interest in the uses of low-valent transition metal complexes as hydrosilylation catalysts (see Section V). A convenient route for the preparation of a broad range of hydrosilyl platinum(II) complexes is the reaction of the silyl with $Pt(C_2H_4)(PPh_3)_2$ in the absence of solvent. The complexes obtained are of general formula $PtHX(PPh_3)_2$ (*143*), where X = $SiPh_3$, $SiPh_2Me$, $SiPh_2H$, $SiEt_3$, $Si(OEt)_3$, $SiMe_2OSiMe_2H$, $SiMe(OSiMe_3)_2$. When more reactive silanes such as $SiMe_2HCl$ or $SiMeHCl_2$ are used, the product is $PtX_2(PPh_3)_2$, where X is the respective silyl. When, however, $Pt(C_2H_4)(PPh_3)_2$ is treated with $SiMeHCl_2$ in the presence of hexene-1, the complex $PtH(SiMeCl_2)(PPh_3)_2$ (*301*) is obtained. When $SiHCl_3$ is allowed to react with Pt-$(diphos)_2$, the hydro complex (IX) is produced; however, if the reaction is carried out under reflux conditions the bis(silyl) complex (X) is formed (*73, 74*). When $SiHCl_3$ reacts with $Pt(PPh_3)_4$ the product is $Pt(SiCl_3)_2$-$(PPh_3)_2$. Triarylsilanes $Si(XC_6H_4)_3H$, in which X is an electron-with-

$$Pt(diphos)_2 + SiHCl_3 \xrightarrow{25^\circ} \underset{(IX)}{PtH(SiCl_3)(diphos)} + diphos$$

$$Pt(diphos)_2 + SiHCl_3 \xrightarrow{reflux} \underset{(X)}{Pt(SiCl_3)_2(diphos)} + diphos \qquad (13)$$

$$Pt(PPh_3)_4 + 2SiHCl_3 \longrightarrow Pt(SiCl_3)_2(PPh_3)_2 + 2PPh_3 + H_2 \qquad (14)$$

drawing substituent such as *m*-F, *m*-CF_3, and *p*-CF_3, react with $Pt(PPh_3)_4$ in refluxing benzene to give $PtH[Si(C_6H_4X)_3](PPh_3)_2$ (XI). Triphenylsilane and methyldiphenylsilane did not undergo this reaction. Complex (XI) did not react further with an excess of the triarylsilane, steric hindrance being suggested as the reason why the bis(triarylsilane) complex did not form (*74*). The oxidative addition of $SiMeCl_2H$ to $Pt(PPh_3)_4$ has been independently studied, and three types of complex have been obtained which contain a Pt—Si bond (*151*).

$$Pt(PPh_3)_4 + SiMeCl_2H \longrightarrow PtH(SiMeCl_2)(PPh_3)_2 \xrightarrow[-H_2]{SiMeCl_2H} Pt(SiMeCl_2)_2(PPh_3)_2$$

$$Pt(SiMeCl_2)_2(PPh_3)_2 \underset{}{\overset{SiMeCl_2H}{\rightleftharpoons}} [Pt(SiMeCl_2)(PPh_3)_2]_2 \qquad (15)$$

The addition of the silylacetylenes, $SiH_3C{\equiv}CH$ and $SiH_3C{\equiv}CCF_3$, to $PtI_2(PEt_3)_2$ gives the hydrosilylplatinum(IV) complexes (XII) and (XIII) (*13*):

$$PtI_2(PEt_3)_2 + SiH_3C{\equiv}CX \longrightarrow PtHI_2(SiH_2C{\equiv}CX)(PEt_3)_2 \quad (16)$$
$$[X = H\,(XII),\ CF_3(XIII)]$$

The high field line for Pt—H has been observed in the PMR spectrum although the compounds tend to decompose readily. The complex $PtHI_2(SiH_2I)(PEt_3)_2$ has been prepared both by the reaction of $PtI(SiH_2I)(PEt_3)_2$ with HI and by the reaction of $PtI_2(PEt_3)_2$ with SiH_3I (*38*, *145*).

$$PtI(SiH_2I)(PEt_3)_2 + HI \longrightarrow PtHI_2(SiH_2I)(PEt_3)_2 \longleftarrow SiH_3I + PtI_2(PEt_3)_2 \quad (17)$$

In 1968, Glockling and Hooton proposed that the oxidative addition of HCl to $[Pt(GeMe_3)(diphos)(PEt_3)]Cl$ gave the hydroplatinum(IV) complex (XIV) (*161*):

$$[Pt(GeMe_3)(diphos)(PEt_3)]Cl + HCl \longrightarrow \underset{(XIV)}{[PtHCl(GeMe_3)(diphos)(PEt_3)]Cl} \quad (18)$$

However, more recent work (*186*) has shown this complex to be the platinum(II) compound, i.e. $[Pt(GeMe_3)(diphos)(PEt_3)]HCl_2$ (XV).

Hydrogermylplatinum complexes have, however, been obtained by oxidative addition reactions. When $PtHCl(PEt_3)_2$ and an excess of GeH_3Cl were mixed together, the ^1H NMR spectrum showed four sets of high field lines (*39*). These compounds are considered to have structures (XVI)–(XIX), and to be formed by the scheme shown in Fig. 2. Compound (XX) of Fig. 2 would give species such as (XVIII) or (XIX) by addition of GeH_3Cl or GeH_2Cl_2; further addition to (XXI), followed by elimination of H_2, would form similar products. There was no

(XVI): Pt bearing H, PEt_3, ClH_2Ge, GeH_2Cl, Et_3P, GeH_2Cl

(XVII): Pt bearing H, PEt_3, Cl_2HGe, GeH_2Cl, Et_3P, $GeHCl_2$

(XVIII): Pt bearing H, PEt_3, ClH_2Ge, GeH_2Cl, Et_3P, GeH_2Cl

(XIX): Pt bearing H, PEt_3, ClH_2Ge, $GeHCl_2$, Et_3P, GeH_2Cl

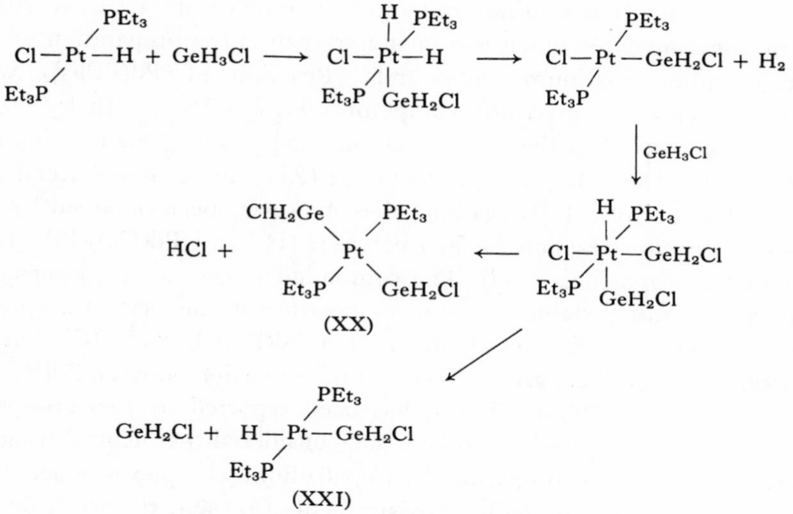

$$\text{Cl—Pt(PEt}_3\text{)(Et}_3\text{P)—H} + \text{GeH}_3\text{Cl} \longrightarrow \text{Cl—Pt(H)(PEt}_3\text{)(Et}_3\text{P)(GeH}_2\text{Cl)—H} \longrightarrow \text{Cl—Pt(PEt}_3\text{)(Et}_3\text{P)—GeH}_2\text{Cl} + \text{H}_2$$

$$\downarrow \text{GeH}_3\text{Cl}$$

$$\text{HCl} + \text{Pt(ClH}_2\text{Ge)(PEt}_3\text{)(Et}_3\text{P)(GeH}_2\text{Cl)} \ \text{(XX)} \longleftarrow \text{Cl—Pt(H)(PEt}_3\text{)(Et}_3\text{P)(GeH}_2\text{Cl)—GeH}_2\text{Cl}$$

$$\swarrow$$

$$\text{GeH}_2\text{Cl} + \text{H—Pt(PEt}_3\text{)(Et}_3\text{P)—GeH}_2\text{Cl} \ \text{(XXI)}$$

FIG. 2. Reaction of $PtHCl(PEt_3)_2$ with GeH_3Cl.

evidence for the presence of significant concentrations of four-coordinated species containing Pt—H bonds.

Compound $Pt(diphos)_2$ does not react with $SiMe_3H$ or $GeMe_3H$, but it reacts rapidly with $SnMe_3H$ (Fig. 3) to give the stable complex $PtH(SnMe_3)_3(diphos)$ (XXII) (*109*). The reaction scheme involves a series of oxidative-addition and reductive-elimination steps. The complex $PtHCl(SnMe_3)_2(diphos)$ (XXIII) has been prepared by treating $PtCl_2(diphos)$ with excess $SnMe_3H$; an oxidative step and an acid exchange step are involved (*108*, *109*):

$$\text{PtCl}_2\text{(diphos)} + 2\text{SnMe}_3\text{H} \longrightarrow \underset{\text{(XXIII)}}{\text{PtHCl(SnMe}_3)_2\text{(diphos)}} + \text{HCl} \underset{\text{SnMe}_3\text{H}}{\overset{\text{C}_6\text{H}_6}{\rightleftharpoons}} \text{PtCl(SnMe}_3\text{)(diphos)} \quad (19)$$

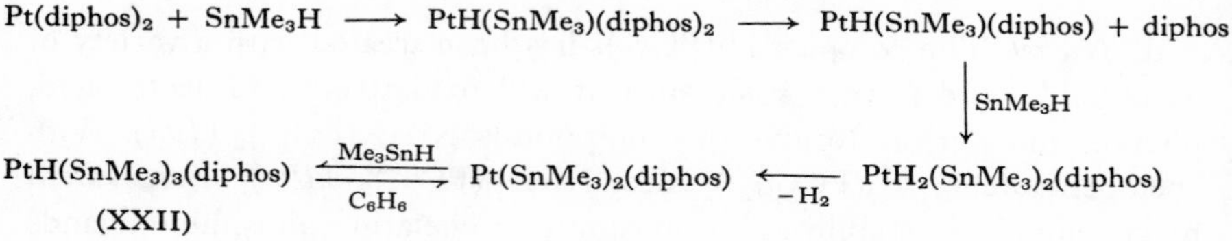

$$\text{Pt(diphos)}_2 + \text{SnMe}_3\text{H} \longrightarrow \text{PtH(SnMe}_3\text{)(diphos)}_2 \longrightarrow \text{PtH(SnMe}_3\text{)(diphos)} + \text{diphos}$$

$$\downarrow \text{SnMe}_3\text{H}$$

$$\underset{\text{(XXII)}}{\text{PtH(SnMe}_3)_3\text{(diphos)}} \underset{\text{C}_6\text{H}_6}{\overset{\text{Me}_3\text{SnH}}{\rightleftharpoons}} \text{Pt(SnMe}_3)_2\text{(diphos)} \xleftarrow[-\text{H}_2]{} \text{PtH}_2\text{(SnMe}_3)_2\text{(diphos)}$$

FIG. 3. Reaction of $Pt(diphos)_2$ with $SnMe_3H$.

c. *Palladium.* The synthesis of hydropalladium complexes by reactions in protonic media is much less favorable than is the preparation of the corresponding platinum compounds. Reaction of $Pd(PPh_3)_4$ with protonic acids HX gives the compounds $PdX_2(PPh_3)_2$ with hydrogen evolution rather than the hydropalladium complexes. This reaction has been reported with HCl (*61*), succinimide (*251*), and terminal acetylenes (*234*). The complex $PdHCl(PPh_3)_2$ has, however, been obtained by the low-temperature reaction (−50°) between HCl and $Pd(CO)(PPh_3)_3$ in equimolar proportions (*204*). The compound is extremely air-sensitive but is moderately stable at room temperature in an inert atmosphere. The complex can also be obtained from $Pd(PPh_3)_4$ and HCl, but no reaction conditions are given (*204*). A similar reaction between $Pd(PCy_3)_2$ and HCl to give $PdHCl(PCy_3)_2$ has been reported by two groups of workers (*204*, *212*). Earlier workers had considered that neutral reagents were necessary for preparing hydropalladium complexes since they decompose rapidly in acidic or basic media (*52*) and the work on the protonation of $Pd(PPh_3)_4$ appeared to support this assumption. Formation of $PdCl_2(PPh_3)_2$ from the reaction in Eq. (20) could result from the intermediate $PdHCl(PPh_3)_2$ oxidatively adding a second molecule of HCl, followed by a rapid reductive-elimination of H_2. An alternative

$$Pd(PPh_3)_4 + HCl \longrightarrow [PdHCl(PPh_3)_2 + 2PPh_3] \xrightarrow{HCl} PdCl_2(PPh_3)_2 + H_2 + 2PPh_3 \quad (20)$$

explanation is that the complex $PdHCl(PPh_3)_2$ is hydridic, and rapidly gives hydrogen with excess acid. It would appear, therefore, that from electronic considerations, tricyclohexylphosphine is a particularly inappropriate ligand for palladium in this reaction since its high basicity (*281*) would facilitate both mechanisms for the decomposition of the hydro complex. In Uchida's work (*204*) this problem may possibly have been averted by the use of only 1 mole of HCl. However, de Jongh (*212*) does not mention that this is a necessary precaution, and it is to be inferred that excess HCl can be used.

d. *Nickel.* The complex $Ni(PCy_3)_2$ has been treated with a variety of weak acids, and is very basic since it will oxidatively add acetic acid, phenol, and pyrrole to give the compounds $NiHX(PCy_3)_2$ (*193*). With cyclopentadiene, $Ni(PCy_3)_2$ yields $NiHcp(PCy_3)_2$ (*193*). Zero-valent nickel complexes stabilized by phosphite or chelating phosphine ligands are protonated by strong acids to give pentacoordinate compounds

(*137*, *258*, *285*) [L = $P(OEt)_3$, $P(O\text{-}p\text{-}C_6H_4OMe)_3$, $P(O\text{-}p\text{-}tolyl)_3$, $P(OCH_2CH_2Cl)_3$, $PPh(OEt)_2$, diphos, PMe_3; HX = HCl, H_2SO_4, CF_3CO_2H]:

$$NiL_4 + HX \longrightarrow [NiHL_4]X \tag{21}$$

The pentacoordinate geometry is confirmed by the observation of a quintet at τ 24.5 (J_{PH} = 26Hz) in the ^1H NMR spectrum and a doublet 135 ppm downfield of H_3PO_4(J_{PH} = 26 Hz) in the ^{31}P NMR spectrum (*137*). These complexes were not isolated except for $[NiH(diphos)_2]X$ (X = $AlCl_4$, BF_4, HCl_2) (*258*). These results contrast with the previously reported reactions of $Ni(diphos)_2$ and $Ni(PPh_3)_4$ in which nickel(II) salts and hydrogen were obtained (*61*).

2. *Replacement Reactions*

A recent example of this type of reaction is the preparation of PtH$(OCOCF_3)(PPh_3)_2$ from $PtHCl(PPh_3)_2$ and CF_3CO_2H (*247*) (a reaction that fails with weaker carboxylic acids):

$$PtHCl(PPh_3)_2 + CF_3CO_2H \longrightarrow PtH(OCOCF_3)(PPh_3)_2 + HCl \tag{22}$$

A similar reaction in Eq. (23) differs in that hydrogen is now eliminated (*13*):

$$PtHCl(PEt_3)_2 + SiH_3C{\equiv}CH \longrightarrow PtCl(SiH_2C{\equiv}CH)(PEt_3)_2 + H_2 \tag{23}$$

Before processes of the type shown in Eq. (22) become general for the synthesis of hydro complexes of the platinum metals, it will be necessary to study this reaction more fully in order to be able to predict the direction of the possible equilibrium. This depends on the relative proton affinities of the conjugate bases, the strength of the metal–ligand bond, and possibly on the relative volatilities or solubilities of the acids. A further complication arises from the type of reaction shown in Eq. (23) which must also be considered when predicting the final result.

A number of other reactions of this type have been reported. Treating $PtCl(SiMe_3)(PEt_3)_2$ with PhC≡CH gave a mixture of the hydro and

$$PtCl(SiMe_3)(PEt_3)_2 + PhC{\equiv}CH \longrightarrow \underset{\text{(XXIV)}}{PtH(C{\equiv}CPh)(PEt_3)_2} + SiMe_3H$$

$$\searrow \quad \underset{\text{(XXV)}}{PtCl(C{\equiv}CPh)(PEt_3)_2} + H_2 \tag{24}$$

chloro complexes (XXIV) and (XXV) (*160*). Among other reactions that lead to hydro complexes are reactions between $PdCl_2(PEt_3)_2$ and $GeMe_3H$ (*50*) or $GePh_3H$ (*52*) [Eq. (25)], $PtH(GePh_3)(PEt_3)_2$ and HCl (*124*) [Eq. (26)], and $PtCl(SiPh_2Me)(PMe_2Ph)_2$ and HCl (*72*) [Eq. (27)]:

$$PdCl_2(PEt_3)_2 + GeR_3H \longrightarrow PdHCl(PEt_3)_2 + GeR_3Cl \ (R = Me,Ph) \quad (25)$$

$$PtH(GePh_3)(PEt_3)_2 + HCl \longrightarrow PtHCl(PEt_3)_2 + GePh_3H \quad (26)$$

$$PtCl(SiPh_2Me)(PMe_2Ph)_2 + HCl \longrightarrow PtHCl(PMe_2Ph)_2 + SiPh_2MeCl \quad (27)$$

D. Reactions Involving Hydrogen

The oxidative addition of H_2 to low-valent d^8 complexes such as $IrCl(CO)(PPh_3)_2$ is a common synthetic method for dihydro complexes. Vaska considered the thermodynamic aspects of d^8 complexes of general formula $IrX(CO)L_2$ and interpreted the results in terms of an acid–base reaction where H_2 behaves as the acid or acceptor (*293*). This implies that low-valent compounds that are strongly basic should add hydrogen to form a dihydro complex. Although $Pt(PPh_3)_4$ is at least as strong a base as $IrCl(CO)(PPh_3)_2$, the Pt(0) compound will not add hydrogen to form a dihydro complex. Such a complex has been suggested, and, indeed, compounds $Pt(PPh_3)_4$ and $Pt(PPh_3)_3$ (*216*) have been incorrectly formulated as $PtH_2(PPh_3)_4$ and $PtH_2(PPh_3)_3$ (*84*). This work was refuted by Malatesta and Ugo who showed that these compounds were correctly formulated as the Pt(0) complexes (*217*). However, they claimed that when these complexes were dissolved in benzene under a nitrogen atmosphere, after 1 week conversion into $PtH_2(PPh_3)_2$ (XXVI) did occur. The hydrogen source was supposedly traces of water in the benzene. This controversy was finally solved when it was shown that these hydro complexes did not exist and that the alleged dihydro compounds were the carbonatoplatinum(II) complex (XXVII) formed by the reaction of O_2 and CO_2 with the Pt(O) complexes (*236*, *179*, *180*).

$$Pt(PPh_3)_2 + O_2 + CO_2 \longrightarrow \underset{\text{(XXVII)}}{Pt(O_2OCO)(PPh_3)_2} \longrightarrow Pt(O_2CO)(PPh_3)_2 + OPPh_3 \quad (28)$$

The failure of $Pt(PPh_3)_4$, which can readily achieve coordinate unsaturation, to add hydrogen has been explained in general terms for d^{10} complexes. These authors considered that the reaction between transition metals and hydrogen takes place by the interaction of the bonding molecular orbital of the hydrogen molecule and an empty (antibonding) *d* orbital on the metal (*172, 113*).

Recently, however, the addition of H_2 to the triethylphosphine-stabilized platinum(0) complex has been reported, the product being $PtH_2(PEt_3)_3$ (*157*):

$$Pt(PEt_3)_3 + H_2 \longrightarrow \underset{(XXVIII)}{PtH_2(PEt_3)_3} \tag{29}$$

The complex is thermally unstable. A similar reaction of H_2 occurs with the complexes $Ni(PEt_3)_3$ and $Pd(PEt_3)_3$ (*259*). ^1H Nuclear magnetic resonance spectroscopy confirms formation of hydro complexes; however, when the hydrogen is removed the resonances are lost. The hydro complexes could not be isolated, and the suggestion of a dimeric structure for the palladium complex is *very* tentative (*259*).

Hydro complexes of metals of this triad have been prepared by the hydrogenolysis of σ-bonded ligands. These ligands are commonly those of Group IV, and examples of the hydrogenolysis of these and other substituents are shown in Eqs. (30)–(37), giving $PtH(GePh_3)(PEt_3)_2$ (*124, 123*), $PtH(GePh_3)$(diphos) (*53*), $PdH(GePh_3)(PEt_3)_2$ (*51*), $PtH(SnMe_3)(PPh_3)_2$ (*10*), $PtH(PbPh_3)(PEt_3)_2$ (*130*), $PtHCl(PEt_3)_2$ (*78*), $[PtH(PPh_3)_3]X$ (*62*):

$$Pt(GePh_3)_2(PEt_3)_2 + H_2 \longrightarrow PtH(GePh_3)(PEt_3)_2 + GePh_3H \tag{30}$$

$$Pt(GePh_3)_2(\text{diphos}) + H_2 \longrightarrow PtH(GePh_3)(\text{diphos}) + GePh_3H \tag{31}$$

$$Pd(GePh_3)_2(PEt_3)_2 + H_2 \longrightarrow PdH(GePh_3)(PEt_3)_2 + GePh_3H \tag{32}$$

$$Pt(SnMe_3)_2(PPh_3)_2 + H_2 \longrightarrow PtH(SnMe_3)(PPh_3)_2 + SnMe_3H \tag{33}$$

$$Pt(PbPh_3)_2(PEt_3)_2 + H_2 \longrightarrow PtH(PbPh_3)(PEt_3)_2 + PbPh_3H \tag{34}$$

$$PtCl_2(PEt_3)_2 + H_2 \longrightarrow PtHCl(PEt_3)_2 + HCl \tag{35}$$

$$PtCl(Ph)(PEt_3)_2 + H_2 \longrightarrow PtHCl(PEt_3)_2 + C_6H_6 \tag{36}$$

$$[Pt(N{=}NC_6H_4R)(PPh_3)_3]X + H_2 \longrightarrow [PtH(PPh_3)_3]X + C_6H_5R + N_2 \tag{37}$$

E. Synthesis from Water and Alcohols

The isolation of hydro complexes where the hydrogen has been transferred from water or ethanol to the metal are important because of their relevance to the possible intermediacy of hydro complexes in reactions in these media. The complex *trans*-$PtCl(SiMe_3)(PEt_3)_2$ is rapidly hydrolyzed to $PtHCl(PEt_3)_2$,

$$2PtCl(SiMe_3)(PEt_3)_2 + H_2O \longrightarrow 2PtHCl(PEt_3)_2 + (Me_3Si)_2O \tag{38}$$

but the corresponding triphenylgermyl complex is stable in aqueous diglyme (*159*). Water also reacts with $Pt(MgI)_2(PPr_3^n)_2$ to give $PtH_2(PPr_3^n)_2$ (XXIX) which shows an IR stretch at 1731 cm^{-1} for Pt—H (*124*), a frequency similar to that for Pt—H in $PtH_2(PEt_3)_3$ (*157*).

$$Pt(MgI)_2(PPr_3^n)_2 + 2H_2O \longrightarrow \underset{\text{(XXIX)}}{PtH_2(PPr_3^n)_2} + 2Mg(OH)I \tag{39}$$

Clark and his co-workers have shown that water will provide a convenient source of the hydro ligand in reactions with cationic platinum(II) complexes (Eq. 40). Reaction occurs when L is triethylphosphine or triphenylphosphine (*93*, *81*), when X is a halogen, and also when L is an alkylarsine (*80*). The mechanism has been studied and is considered to proceed via an intermediate carboxylate (*95*, *96*).

$$[PtX(CO)L_2]BF_4 + H_2O \longrightarrow PtHXL_2 + CO_2 + H^+ + BF_4^- \tag{40}$$

The use of ethanol as a source of hydrogen has been reported in the synthesis of $PtHCl(PEt_3)_2$ from $PtCl_2(PEt_3)_2$ (*77*). In a study of the methanolysis reaction of $Ni(PCl_3)_4$, a hydro complex has been suggested as an intermediate (*20*). This complex has not been isolated but its existence has been inferred from the observation of a band in the IR spectrum at 1610 cm^{-1} which shifts to 1135 cm^{-1} when CH_3OD is used. These bands have been tentatively assigned as the Ni—H and Ni—D stretch, respectively. A reported preparation of $AsPh_4[PdHCl_2(CO)]$ (XXX) from palladium(II) chloride and carbon monoxide in 2-methoxyethanol probably involves the incorporation of a labile hydrogen atom from the solvent into the complex (*198*). The cation can be replaced by pyridinium or quinolinium. However, the characterization of these complexes is based mainly on the observation of a band in the IR spectrum in the hydropalladium region. Formation of the hydro complexes $PtH(OPPh_2)(HOPPh_2)(Bu^nOPPh_2)$, $PtH(OPPh_2)(HOPPh_2)(Pr^iOPPh_2)$, and $PtH(OPPh_2)(HOPPh_2)_2$ has been reported from the controlled solvolysis of the zero-valent platinum complexes in aqueous alcohol (*202*).

Although the hydrogen may come from the base, rather than from the alcohol, it is appropriate to include in this section the synthesis of $PtHCl(PEt_3)_2$ in excellent yield by treating $PtCl_2(PEt_3)_2$ with potassium hydroxide in ethanol (*78*, *76*):

$$PtCl_2(PEt_3)_2 + KOH + C_2H_5OH \longrightarrow PtHCl(PEt_3)_2 + CH_3CHO + KCl + H_2 \tag{41}$$

F. Alkyl Elimination

Reaction of $PtCl_2(PEt_3)_2$ or $PtBr_2(PEt_3)_2$ with cyclohexylmagnesium bromide gives $PtHBr(PEt_3)_2$ (*78*, *75*). This product was presumed to arise from elimination of cyclohexene from the intermediate cyclohexylplatinum(II) complex (*78*, *75*). Later workers conclude that this olefin elimination pathway is only responsible for 30% of the product. They conclude that the role of the solvent is important, and that 70% of the hydro complex is formed by the hydrolysis of the intermediate $PtBr(MgBr)(PEt_3)_2$. When this reaction is carried out in D_2O rather than H_2O, the complex $PtDBr(PEt_3)_2$ is formed (*125*).

The thermolysis of $PtCl(COC_8H_{17})(PPh_3)_2$ gives $PtHCl(PPh_3)_2$ and octene (*300*). A mechanism involving an intermediate alkyl complex is proposed.

G. Ligand Replacement

The general reaction shown in Eq. (42) can be used to prepare many hydro complexes if a source for the initial compound is available. This method has been used to prepare $PtHCN(PPh_3)_2$, $PtH(SCN)(PPh_3)_2$,

$$MHXL_2 + M'Y \longrightarrow MHYL_2 + M'X \qquad (42)$$

$PtH(NO_2)(PPh_3)_2$, and $PtH(OCN)(PPh_3)_2$ from $PtHCl(PPh_3)_2$ and an appropriate salt (*23*). A series of complexes $PtHL(PEt_3)_2$, where L is a substituted carboxylate, have been prepared by reacting the silver salt of the carboxylic acid with $PtHCl(PEt_3)_2$ (*19*). These reactions are favored by the high trans influence of the hydro ligand (*29*). There appears to be relatively few examples of the synthesis of covalent hydro complexes by replacement of neutral ligands. One example is the preparation of $Pt_2H_2(PPh_2)_2(PEt_3)_2$ (XXXI) from $PtHCl(PEt_3)_2$ and diphenylphosphine (*69*):

$$2PtHCl(PEt_3)_2 + 2PPh_2H \longrightarrow$$

$$2[PtH(PPh_2)(PEt_3)_2]Cl \longrightarrow \qquad (43)$$

Et₃P, PPh₂, H / Pt, Pt / H, Ph₂P, PEt₃

(XXXI)

H. Displacement of Anionic Ligands

Cationic hydro complexes have been obtained by the replacement of a coordinated anionic ligand (commonly a halide) with a neutral ligand in the presence of a noncoordinating anion. A representative example is the synthesis of $[PtH(CO)(PEt_3)_2]ClO_4$ from $PtHCl(PEt_3)_2$ in the presence of CO and sodium perchlorate (*86*):

$$PtHCl(PEt_3)_2 + CO + NaClO_4 \longrightarrow [PtH(CO)(PEt_3)_2]ClO_4 + NaCl \quad (44)$$

Neutral ligands that can be used in this reaction include Me_3CNC, *p*-$MeOC_6H_4NC$, pyridine, PPh_3, PEt_3, $P(OMe)_3$, and $P(OPh_3)$ (*87, 92*). A similar method has been used to prepare the corresponding triethylarsine compounds (*88*). Reaction of ethylene (1 atm) with $PtH(NO_3)(PEt_3)_2$ gives the cationic complex (XXXII) (*129*):

$$PtH(NO_3)(PEt_3)_2 + C_2H_4 \xrightarrow{NaBPh_4} \underset{(XXXII)}{[PtH(C_2H_4)(PEt_3)_2]BPh_4} + NaNO_3 \quad (45)$$

The analogous triphenylphosphine compound has been prepared as well as the analogs with propene, ammonia, and substituted amines coordinated to Pt(II) (*156, 286*). The reaction of $PtHCl(PEt_3)_2$ with diphos gives $[PtH(diphos)(PEt_3)]Cl$ (*170*). The tendency to form cationic hydro complexes follows the order $PEt_3 \simeq PBu_3 > PPh_2Et > PPh_3$ (*170*) for phosphine complexes. A more extensive generalization of the tendency to form hydro complexes in ethanol has shown that the order for the donor atoms is $P > S > As \geq Sb \sim N > Bi$ (*286*).

The first example of a cationic hydropalladium complex has been reported by Green *et al.* (*165*) from the reaction of $PdHCl(PR_3)_2$ with diphos and ammonium hexafluorophosphate ($R = Cy$ or Pr^i):

$$PdHCl(PR_3)_2 + diphos + NH_4PF_6 \longrightarrow [PdH(diphos)(PR_3)]PF_6 + PR_3 + NH_4Cl \quad (46)$$

When the analogous reaction was carried out with the nickel compound, the product was $[NiH(diphos)_2]PF_6$ (*165*).

III

STRUCTURE AND REACTIONS

A. Geometry and Structure

The geometry of hydro complexes has been determined primarily by IR and ^1H NMR spectroscopy. Using these techniques, the stereochemistry of a large number of complexes of the type $MHXL_2$ (M = Ni, Pd, Pt) has been determined, and in each case the monodentate L groups are mutually trans to each other. Indeed, the actual existence of *cis*-hydrohalide complexes of platinum having this formulation, where L is a monodentate phosphine ligand, has been questioned. These authors (*107*) proposed that instability of such complexes would occur because both H and X would be strongly labilized by a *trans*-phosphorus atom. Although such an explanation may be valid in the above case, it does, however, fail to account for the ready isolation of complexes such as $PtHCN(PEt_3)_2$ (*78*), $PtHCN(PPh_3)_2$ (*61*), and $PtH(SCN)(PEt_3)_2$ (*78*), where both the hydro and X ligands are opposite a group of high trans influence. A simple explanation, which may be valid, is that electronic effects are relatively unimportant in determining stereochemistry and that the complex with the substituted phosphine ligands mutually *trans* is the sterically favored isomer. The first *cis*-hydrohalide complex to be isolated, PtHCl(diphos) (*107*), was obtained from the hydrogenolysis of $Pt(SiMe_3)Cl$(diphos), although it was necessary to use a chelating phosphine ligand in order to obtain a mutually cis stereochemistry for the hydro and chloro ligands.

The structures of $PtHBr(PEt_3)_2$ (*238*), $PtHCl(PPh_2Et)_2$ (*147*), and $PdHCl(PEt_3)_2$ (*257*) have been obtained by single-crystal diffraction techniques. The geometries are close to square planar with the P—M—X angles falling in the range of 92° to 96°. The Pt–Cl distance of 2.422 Å in $PtHCl(PPh_2Et)_2$ is lengthened because of the high trans influence of the hydro ligand. The Pd–Cl distance in $PdHCl(PEt_3)_2$ is closely similar (2.427 Å) because of the identity of the ionic radii of the two metals. A single-crystal structure determination of $PtH(SB_9H_{10})(PEt_3)_2$ has shown the Pt–H bond distance to be 1.66 Å and that the platinum is bonded to one sulfur and three boron atoms (*196*).

B. Reactions

1. *Insertion Reactions*

Although the term *insertion reaction* in no way implies anything about the mechanism, it is used to describe the reaction between an unsaturated species and a metal–ligand bond (*299*), e.g.,

$$M{-}H + X{-}Y \longrightarrow M{-}X{-}Y{-}H \tag{47}$$

a. *Olefins*. Insertion of ethylene into the Pt—H bond of $PtHCl(PEt_3)_2$ was reported by Chatt and Shaw (*78*, *75*) who obtained a 25% yield of the ethyl derivative (XXXIII) after 18 hours when the reaction was carried out in cyclohexane at 95°/40 atm:

$$PtHCl(PEt_3)_2 + C_2H_4 \rightleftharpoons \underset{\text{(XXXIII)}}{PtClEt(PEt_3)_2} \quad (48)$$

The reaction is reversible. If a fluorinated olefin is used, it would be expected that the resulting fluoroalkyl compound would be readily obtained, since the greater strength of the metal–fluorocarbon bond as compared to the metal–hydrocarbon bond favors the insertion reaction (*267*). The reaction of hydroplatinum(II) complexes with fluoroolefins has been extensively studied. A surprising product from a number of these reactions is a fluorovinyl compound rather than the anticipated fluoroalkyl complex. In Fig. 4 are shown the reactions of $PtHCl(PEt_3)_2$ with a series of fluoroolefins to give $PtCl(CF{=}CF_2)(PEt_3)_2$ (XXXIV) (*104*, *105*), $PtCl(\overline{C{=}CF{-}CF_2CF_2})(PEt_3)_2$ (XXXV) (*105*, *82*), *cis*- and *trans*-$PtCl(CF{=}CFH)(PEt_3)_2$ (XXXVI) (*105*, *103*), $PtCl(CF{=}CFCF_3)(PEt_3)_2$ (XXXVII) (*105*), and $PtCl[C(CF_2H){=}CF_2](PEt_3)_2$ (XXXVIII) (*105*). The reaction of $PtHCl(PPh_3)_2$ with C_2F_4 is very similar to that of its triethylphosphine analog giving the fluorovinyl compounds $PtCl(CF{=}CF_2)(PPh_3)_2$ and $PtCl[C(CF_2H){=}CF_2](PPh_3)_2$ (*94*). The reaction of $PtHCl(PEt_3)_2$ with C_2F_4 in a silica tube gave a small yield of a complex which was initially considered to be the hydro complex $PtHCl(C_2F_4)(PEt_3)_2$ (*104*, *105*). Further work showed that the compound is $[PtCl(CO)(PEt_3)_2]SiF_5$ (XXXIX) (*94*), the anion having been formed by interaction

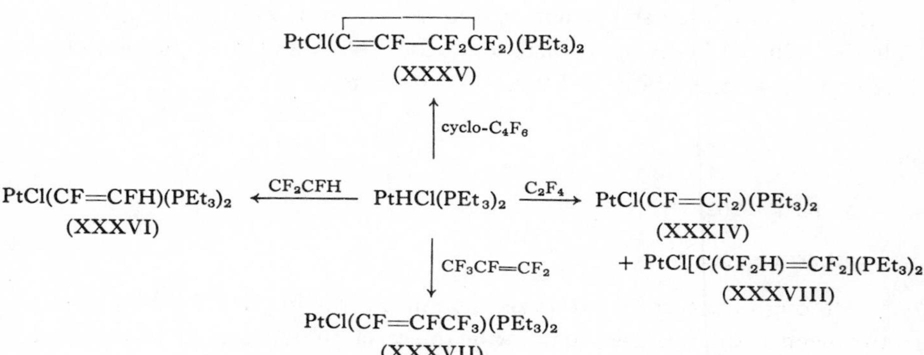

FIG. 4. Reactions between $PtHCl(PEt_3)_2$ and fluoroolefins.

of the fluoroolefin with glass. The main reactions leading to the formation of complex (XXXIX) are

$$PtHCl(PEt_3)_2 + C_2F_4 \longrightarrow \underset{\text{(XL)}}{PtCl(CF_2CF_2H)(PEt_3)_2} \longrightarrow PtCl(CF{=}CF_2)(PEt_3)_2 + HF \quad (49)$$

$$HF + \text{glass} \longrightarrow H_2O,\ BF_3,\ SiF_4,\ BF_4^-,\ SiF_6^{2-} \quad (50)$$

$$2PtCl(CF{=}CF_2)(PEt_3)_2 + H_2O + SiF_4 \longrightarrow \underset{\text{(XXXIX)}}{[PtCl(CO)(PEt_3)_2]SiF_5} + PtCl[C(CF_2H){=}CF_2](PEt_3)_2 + HF \quad (51)$$

The intermediate compound $PtCl(CF_2CF_2H)(PEt_3)_2$ (XL), which is the anticipated product from this insertion reaction, can be prepared by addition of C_2F_4 to $PtHCl(PEt_3)_2$ in a steel autoclave (*94*). When (XL) is heated in benzene solution in a glass tube, $PtCl(CF{=}CF_2)(PEt_3)_2$ and $PtCl[C(CF_2H){=}CF_2](PEt_3)_2$ are formed, along with traces ($\sim 5\%$) of $[PtCl(CO)(PEt_3)_2]^+$. A detailed spectroscopic study of many of these compounds has been made giving useful IR and NMR data on fluorovinyl complexes (*106*).

The reaction of tetracyanoethylene with $PtHX(PPh_3)_2$ (X = Cl, Br) gave $Pt(TCNE)(PPh_3)_2$; however, a similar reaction with $PtHCN(PEt_3)_2$ gave the complex $PtHCN(TCNE)(PEt_3)_2$ (*292*). A reaction that can be described as an insertion–elimination reaction has also been reported by Lappert *et al.*, namely, the reaction of $Me_3SnN{=}C(CF_3)_2$ with $PtHCl(PPh_3)_2$ to give $Pt[NH(CF_3)_2](PPh_3)_2$ (XLI) (*210*):

$$PtHCl(PPh_3)_2 + Me_3SnN{=}C(CF_3)_2 \longrightarrow (PPh_3)_2Pt\begin{smallmatrix}\diagup NH \\ \diagdown C(CF_3)_2\end{smallmatrix} \quad \text{(XLI)} \quad (52)$$

Formation of isolable complexes in good yield from the insertion reaction of olefins which are not substituted with electron-withdrawing groups has been reported for the cases where cyclic compounds are formed in the reaction. A reaction of this type occurs in the interaction of (*o*-vinylphenyl)diphenylphosphine with $PtHCl(PPh_3)_2$ as shown in Fig. 5 (*48, 47*). The cationic hydroplatinum(II) complex $[PtH(CO)(PEt_3)_2]^+$ reacts with (*o*-vinylphenyl)diphenylphosphine to give the complex (XLII) containing a five-membered ring:

$$[PtH(CO)(PEt_3)_2]^+ + VP \longrightarrow \underset{\text{(XLII)}}{[Ph_2PC_6H_4(CHMe)Pt(PEt_3)_2]^+} \quad (53)$$

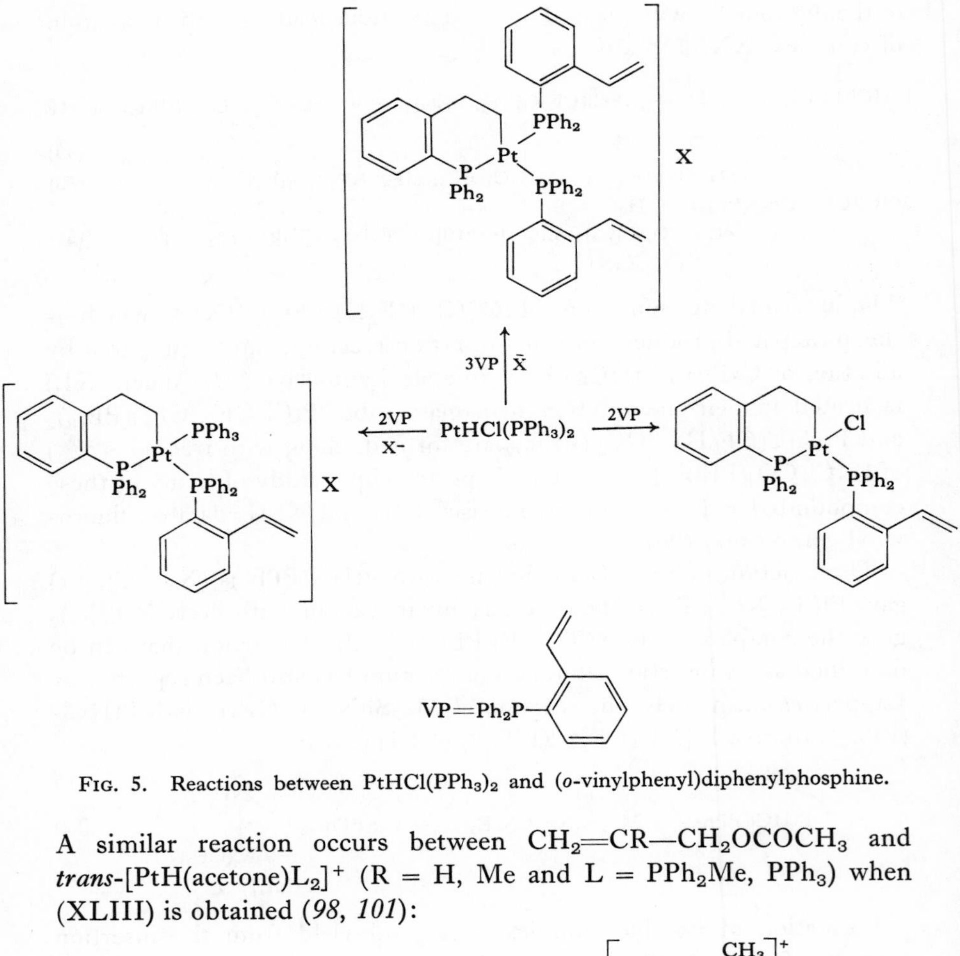

Fig. 5. Reactions between $PtHCl(PPh_3)_2$ and (*o*-vinylphenyl)diphenylphosphine.

A similar reaction occurs between $CH_2{=}CR{-}CH_2OCOCH_3$ and *trans*-$[PtH(acetone)L_2]^+$ (R = H, Me and L = PPh_2Me, PPh_3) when (XLIII) is obtained (*98*, *101*):

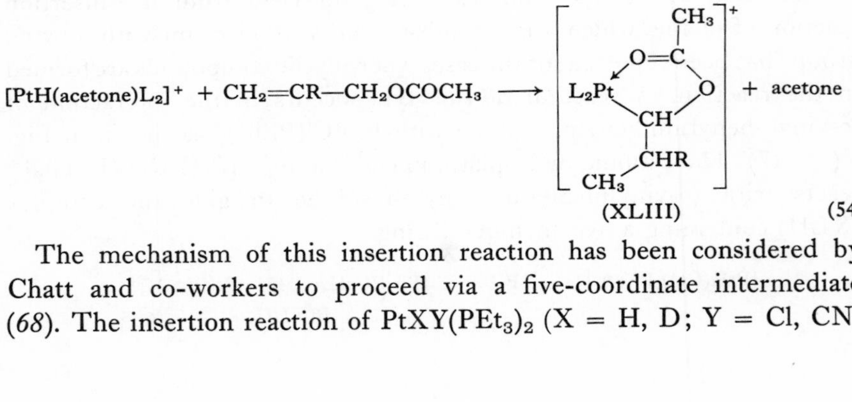

$$[PtH(acetone)L_2]^+ + CH_2{=}CR{-}CH_2OCOCH_3 \longrightarrow \left[L_2Pt(O{=}C(CH_3){-}O{-}CH{-}CHR{-}CH_3)\right]^+ + \text{acetone} \quad (54)$$

(XLIII)

The mechanism of this insertion reaction has been considered by Chatt and co-workers to proceed via a five-coordinate intermediate (*68*). The insertion reaction of $PtXY(PEt_3)_2$ (X = H, D; Y = Cl, CN) with olefins has been studied, along with the reverse reaction, the decomposition of $PtBr(CD_2CH_3)(PEt_3)_2$. The reaction of $PtHCN(PEt_3)_2$ with C_2H_4 (100° 80 atm) to form $PtCNEt(PEt_3)_2$,

$$PtHCN(PEt_3)_2 + C_2H_4 \longrightarrow PtCNEt(PEt_3)_2 \tag{55}$$

gives a side product $Pt(CN)_2(PEt_3)_2$ which is presumed to be formed by disproportionation of the hydro or ethyl complex. When a similar reaction is carried out with $PtHCl(PEt_3)_2$ and propene, a low yield of the *n*-propyl compound is obtained; however, it was impossible to obtain the isopropyl complex by this method. Reaction of octene-1 with $PtDCl(PEt_3)_2$ gives $PtHCl(PEt_3)_2$ in 100% yield; because of the small equilibrium constant for the insertion, the hydro complex is formed by the elimination reaction (*68*).

An alternative mechanism has been proposed for insertion of ethylene into $PtHCl(PEt_3)_2$ involving the intermediacy of a four-coordinate ionic species rather than a five-coordinate covalent one (*97*). The insertion reaction, which is represented by the forward reaction in Eq. (48), can be accelerated by addition of 1 mole % of tin(II) chloride, since under these conditions equilibrium is established within 30 minutes at 25° and 1 atm (*122*). The basis of the suggestion of a four-coordinate cationic intermediate arose from the observation of the similar reaction shown in Eq. (56). In this study it was found that the insertion reaction proceeded rapidly when $PtHBr(PPh_2Me)_2$ is treated with ethylene and silver fluorophosphate in acetone as solvent, provided an added base is present.

$$\begin{array}{c} PtHBr(PPh_2Me)_2 + AgPF_6 \xrightarrow{Me_2CO} [PtH(PPh_2Me)_2(Me_2CO)]PF_6 + AgBr \\ \Big\downarrow \text{i. } C_2H_4 \quad \text{ii. Base} \\ [PtEt(PPh_2Me)_2(\text{base})]PF_6 + Me_2CO \end{array} \tag{56}$$

Bases that can be used include carbon monoxide and 2,4,6-trimethylpyridine (*97*). This reaction has been further studied (*99*) with the report that *trans*-$[PtH(PPh_2Me)_2(Me_2CO)]PF_6$ reacts with C_2H_4 at −50° to give *trans*-$[PtH(PPh_2Me)_2(C_2H_4)]PF_6$ (XLIV). The proposal that compounds similar to (XLIV) are the intermediates in the insertion reaction is further strengthened by the observation that when this reaction is carried out at room temperature rather than at −50° the product is the ethyl complex *trans*-$[PtEt(PPh_2Me)_2(C_2H_4)]PF_6$ (XLV):

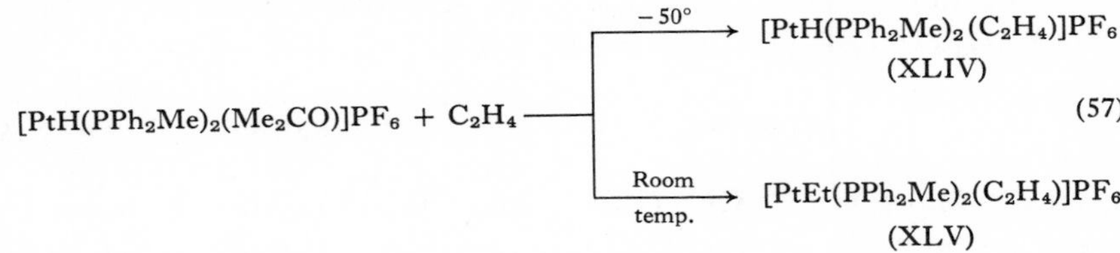

$$[PtH(PPh_2Me)_2(Me_2CO)]PF_6 + C_2H_4 \xrightarrow{-50^\circ} [PtH(PPh_2Me)_2(C_2H_4)]PF_6 \text{ (XLIV)}$$
$$[PtH(PPh_2Me)_2(Me_2CO)]PF_6 + C_2H_4 \xrightarrow{\text{Room temp.}} [PtEt(PPh_2Me)_2(C_2H_4)]PF_6 \text{ (XLV)} \quad (57)$$

Complexes $PtHCl(PR_3)_2$ (R = PMe_2Ph, PPh_3) were also found to undergo similar ethylene insertion reactions in acetone in the presence of an equimolar amount of $AgPF_6$ at room temperature. The triethylphosphine complex, however, is somewhat less reactive to insertion since *trans*-$[PtH(PEt_3)_2(C_2H_4)]PF_6$ [formed from $PtHCl(PEt_3)_2$ and ethylene in an acetone–methylene chloride solution in the presence of $AgPF_6$] was found to be stable toward further reaction with ethylene at room temperature. When the complex was allowed to react further with ethylene in boiling acetone for 40 minutes, insertion occurred to give *trans*-$[PtEt(PEt_3)_2L]PF_6$ (L = C_2H_4 or acetone).

This insertion reaction is sensitive to the nature of the X group in $PtHXL_2$. Thus, whereas $PtH(NO_3)(PPh_2Me)_2$ reacted very rapidly with ethylene under ambient conditions to give $PtEt(NO_3)(PPh_2Me)_2$ in good yield, $PtHBr(PPh_2Me)_2$ failed to insert C_2H_4 readily at room temperature, and to obtain $PtEtCl(PEt_3)_2$ required very vigorous conditions. Also it was found that several cationic hydroplatinum(II) complexes, such as *trans*-$[PtH(PPh_2Me)_2L]^+$ (L = 2,4,6-trimethylpyridine, PPh_2Me, *p*-$CH_3C_6H_4NC$), did not give any insertion products under mild conditions. These results are interpreted as indicating that the ease with which ethylene insertion into the Pt—H bond occurs under very mild conditions is closely related to the ease of displacement of the ligand trans to the hydro ligand (*122*). Acetone and nitrate can be replaced by ethylene as shown by the formation of hydroethylene complexes (*97*, *129*), whereas other ligands such as halides, pyridines, phosphines, and isocyanides cannot be readily substituted by ethylene. Although the intermediacy of a pentacoordinate compound is not discounted, it is concluded that the second intermediate in the insertion reaction, which is the one of lower energy, is a four-coordinate cationic species. When the scope of this insertion reaction with *trans*-$[PtH(PPh_2Me)_2(Me_2CO)]^+$ was studied by extending it to higher olefins, it was found that both propene and butene-1 would insert to form the *n*-propyl and *n*-butyl complexes, respectively [Eq.(58). However, the reaction failed with *cis*-butene-2, vinyl fluoride, and methyl vinyl ether. The reaction has been extended to dienes with insertion of allene and buta-

$$[PtH(PPh_2Me)_2(Me_2CO)]^+ \xrightarrow[C_4H_8-1]{C_3H_6} [Pt(n\text{-}C_nH_{2n+1})(PPh_2Me)_2L]^+ \quad (58)$$
$$(L = \text{acetone, olefin};\ n = 3,\ 4)$$

diene into the Pt—H bond of $[PtH(PR_3)_2(acetone)]^+$ to give the corresponding π-allyl and π-crotylplatinum(II) complexes, respectively.

$$[PtH(PR_3)_2(Me_2CO)]PF_6 + C_nH_m \longrightarrow [Pt(\pi\text{-}C_nH_{m+1})(PR_3)_2]PF_6 + Me_2CO \quad (59)$$
$$(PR_3 = PPh_3,\ PPh_2Me,\ PMe_2Ph;\ n = 3,\ m = 4;\ n = 4,\ m = 6)$$

b. *Acetylenes.* Insertion of acetylenes into hydroplatinum bonds has also received a considerable amount of study. An example is the reaction between 3,3,3-trifluoropropyne and $PtHCl(PEt_3)_2$ to give $PtCl(CH{=}CHCF_3)(PEt_3)_2$ (XLVI), which has the trans stereochemistry about the vinyl group (*174*). The reaction of $PtClMe(PEt_3)_2$ with 3,3,3-trifluoropropyne proceeds in a different manner to give $PtCl(C{\equiv}CCF_3)(PEt_3)_2$

$$PtHCl(PEt_3)_2 + CF_3C{\equiv}CH \longrightarrow \underset{\text{(XLVI)}}{PtCl(CH{=}CHCF_3)(PEt_3)_2} \quad (60)$$

(XLVII) (*16*). The mechanism proposed involves the intermediacy of a hydroplatinum complex, which leads to the final product via the elimination of methane (Eq. 61). The reaction between dicyanoacetylene and $PtHCl(PEt_3)_2$ or $PtHCl(PPh_3)_2$ has been studied and the product

$$PtClMe(PEt_3)_2 + CF_3C{\equiv}CH \rightleftharpoons PtClMe(\pi\text{-}CF_3C{\equiv}CH)(PEt_3)_2 \longrightarrow$$
$$PtHClMe(C{\equiv}CCF_3)(PEt_3)_2 \xrightarrow{-CH_4} \underset{\text{(XLVII)}}{PtCl(C{\equiv}CCF_3)(PEt_3)_2} \quad (61)$$

obtained depends on the solvent used (*223*). Insertion to form the vinyl compound (XLVIII) has only been reported for the reaction with $PtHCl(PEt_3)_2$ when benzene is used as solvent. When tetrahydrofuran is used as solvent the π-bonded acetylene complex is formed (XLIX).

$$\underset{\text{(XLIX)}}{Pt(\pi\text{-}C_4N_2)(PEt_3)_2} \xleftarrow[\text{THF}]{C_4N_2} PtHCl(PEt_3)_2 \xrightarrow[\text{benzene}]{C_4N_2} \underset{\text{(XLVIII)}}{PtCl[C(CN){=}CH(CN)](PEt_3)_2} \quad (62)$$

The protonation of zero-valent platinum–acetylene complexes has been discussed by several workers, and a mechanism proposed involving insertion of the acetylene into a hydroplatinum bond (*287*). The reaction of $Pt(CF_3C{\equiv}CCF_3)(PPh_3)_2$ with CF_3CO_2H gives the vinyl "insertion product" $Pt(OCOCF_3)[C(CF_3){=}C(CF_3)H](PPh_3)_2$ (L). A similar reaction between $Pt(CF_2{=}CF_2)(PPh_3)_2$ and CF_3CO_2H gives the fluoroalkyl complex $Pt(OCOCF_3)C_2F_5(PPh_3)_2$ (*28*). Protonation of $Pt(PhC{\equiv}CPh)(PPh_3)_2$ with HX proceeds beyond the vinyl intermediate to give $PtX_2(PPh_3)_2$ and *trans*-stilbene. If the platinum–acetylene complex and the protonic acid are mixed in equimolar proportions, the vinyl complex, with cis stereochemistry about the double bond, is obtained (*287*, *288*, *218*, *219*). This vinyl compound is an intermediate in the reaction of $Pt(PhC{\equiv}CPh)(PPh_3)_2$ with excess acid, since PtX_2-$(PPh_3)_2$ and *trans*-stilbene are obtained when $PtX(CPh{=}CHPh)(PPh_3)_2$ is treated with a further molecule of HX. This series of reactions is as follows (R = Ph, Me; X = Cl, Br, CO_2H, CO_2CF_3, $COSCH_3$, $COSC_6H_5$, picrate):

$$Pt(RC{\equiv}CR)(PPh_3)_2 + HX \longrightarrow PtX(CR{=}CHR)(PPh_3)_2 \xrightarrow{HX} PtX_2(PPh_3)_2 + RCH{=}CHR \tag{63}$$

This reaction is very facile, and the use of 1 mole of acid provides a convenient method for the synthesis of vinyl complexes. The proposed mechanism (*287*) is shown in Fig. 6 which suggests an oxidative addition of HX followed by insertion into the Pt—H bond to form the intermediate vinyl compound. If excess acid is present, a second oxidative addition of HX occurs to give a hydrovinyl platinum(IV) intermediate. If a second insertion reaction occurs, the olefin is formed along with $PtX_2(PPh_3)_2$.

Although *trans*-stilbene formed quantitatively in the reaction between $Pt(PhC{\equiv}CPh)(PPh_3)_2$ and excess acid, the analogous reaction with the

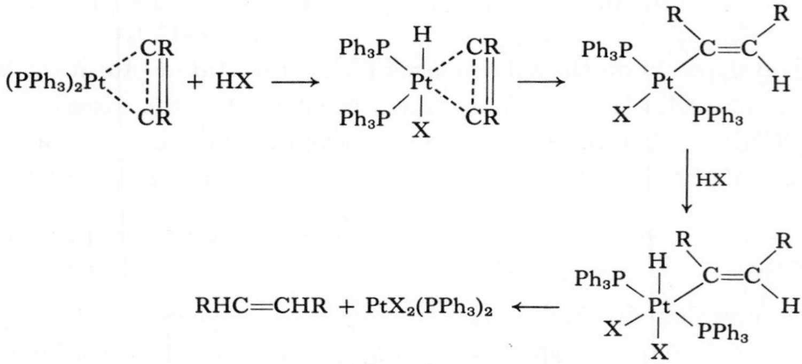

$$(PPh_3)_2Pt(RC{\equiv}CR) + HX \longrightarrow PtH(X)(RC{\equiv}CR)(PPh_3)_2 \longrightarrow PtX(CR{=}CHR)(PPh_3)_2 \xrightarrow{HX} PtH(X)_2(CR{=}CHR)(PPh_3)_2 \longrightarrow RHC{=}CHR + PtX_2(PPh_3)_2$$

FIG. 6. Reaction of $Pt(alkyne)(PPh_3)_2$ with protonic acids.

butyne-2 complex $Pt(MeC{\equiv}CMe)(PPh_3)_2$ gives mainly *cis*-butene-2, with the trans isomer being formed only to the extent of 20%. The intermediate vinyl complex has a cis stereochemistry about the double bond, and if the solvolysis of the vinyl compound occurred with retention of configuration, a quantitative yield of the *cis*-olefin would be obtained. Since this reaction gives significant quantities of the trans isomer, it has been interpreted to mean that solvolysis of vinyl complexes may not necessarily occur with retention of configuration about the double bond. An alternative explanation is that olefin isomerization occurs after completion of the reaction, although no evidence could be found to support this view (*288*).

The corresponding palladium complexes $Pd(CF_3C{\equiv}CCF_3)(PPh_3)_2$ and $Pd(CF_3C{\equiv}CCF_3)(diphos)$ have been prepared. These complexes react with halocarboxylic acids to form the corresponding vinyl compounds as follows:

$$Pd(CF_3C{\equiv}CCF_3)(PPh_3)_2 + CF_3CO_2H \longrightarrow Pd(OCOCF_3)[C(CF_3){=}CHCF_3](PPh_3)_2 \quad (64)$$

$$Pd(CF_3C{\equiv}CCF_3)(diphos) + CF_3CO_2H \longrightarrow Pd(OCOCF_3)[C(CF_3){=}CHCF_3](diphos) \quad (65)$$

$$Pd(CF_3C{\equiv}CCF_3)(PPh_3)_2 + CCl_3CO_2H \longrightarrow Pd(OCOCCl_3)[C(CF_3){=}CHCF_3](PPh_3)_2 \quad (66)$$

The reaction between $Pd(CF_3{\equiv}CCF_3)(diphos)$ and CF_3CO_2H, CF_2-HCO_2H, CCl_3CO_2H in chloroform as solvent follows a second-order rate law, the activation parameters for the reaction with CF_3CO_2H being $\Delta H^{\ddagger} = 11.2$ kcal mole^{-1} and $\Delta S^{\ddagger} = -35$ cal deg^{-1} mole^{-1} (*57*). The reaction of $Pd(CF_3C{\equiv}CCF_3)(PPh_3)_2$ with halogenated carboxylic acids follows a more complicated kinetic pattern which has been attributed to a further reaction involving the loss of triphenylphosphine. These authors propose the mechanism shown in Eq. (67), involving direct protonation on the metal. An alternative explanation proposed by

$$Pd(CF_3C{\equiv}CCF_3)(diphos) + H^+ \xrightarrow[\text{Slow}]{} [PdH(CF_3C{\equiv}CCF_3)(diphos)]^+ \xrightarrow[\text{Fast}]{CO_2CX_3^-} Pd(OCOCX_3)[C(CF_3){=}CHCF_3](diphos) \quad (67)$$

these authors, as well as by other workers (*223*, *163*), is that protonation occurs on the coordinated acetylene, and choice between these two mechanisms remains an open question. A further mechanism proposed to explain the products obtained in this series of reactions has also not been proven conclusively. This proposal is that in the solvolysis of the vinylplatinum(II) complex to form the free olefin, the protonation occurs on the metal rather than on the α-carbon atom of the coordinated vinyl group (*287*). This reaction may, however, be worthy of further study since if a hexacoordinate vinylplatinum(IV) complex were to readily give the free olefin on reaction with protonic acids the mechanism involving protonation on the metal could be eliminated, providing, of course, that ligand dissociation did not occur to any significant extent. Reaction of the cyclohexyne complex $Pt(\pi\text{-}C_6H_8)(diphos)$ (*36*) with *weak* acids has also been reported to form the cyclohexenyl compound (LI) (*37*) ($X = p\text{-}MeC_6H_4O$, CH_2NO_2, CH_2COMe, CH_2COPh, $CH(CN)Ph$, OH, OMe):

$$Pt(\pi\text{-}C_6H_8)(diphos) + HX \longrightarrow \underset{\text{(LI)}}{PtX(C_6H_9)(diphos)} \tag{68}$$

This reaction differs from that of the analogous triphenylphosphine complex $Pt(\pi\text{-}C_6H_8)(PPh_3)_2$ which reacts in refluxing ethanol to give the σ-cyclohexenylhydro complex $PtH(C_6H_9)(PPh_3)_2$, presumably via intermediate alkoxy complexes.

When diphenylacetylene and $Pt(OCOCF_3)_2(PPh_3)_2$ are refluxed together in methanol, an almost quantitative yield of the alkenyl complex $Pt(OCOCF_3)(CPh{=}CHPh)(PPh_3)_2$ (LII) (*27*) was formed. This compound has a slightly different IR spectrum and melting point than was found for *trans*-$Pt(OCOCF_3)$(*cis*-$CPh{=}CHPh)(PPh_3)_2$ (*218*), and it is suggested that there may be a different geometry about the platinum atom, the double bond, or both of them. In this reaction there is also some oxidation of methanol to formaldehyde. These authors (*27*) consider that the first intermediate $Pt(OCOCF_3)(OMe)(PPh_3)_2$ generates a hydroplatinum species, possible $PtH(OCOCF_3)(PPh_3)_2$, via a hydride shift mechanism. This hydro complex can then react with diphenyl acetylene to form the vinyl complex (LII). Further evidence for hydride shift is provided by the observation that in CD_3OD as solvent, the deuterium-substituted derivative $Pt(OCOCF_3)(CPh{=}CDPh)(PPh_3)_2$, can be isolated; in CH_3OD, only the hydrogen-substituted vinyl complex (LII) is obtained. The analogous reaction between hexafluorobutyne-2 and $Pt(OCOCF_3)_2(PPh_3)_2$ in methanol at 70° gives both the alkenyl complex $Pt(OCOCF_3)[C(CF_3){=}CHCF_3](PPh_3)_2$ (LIII) and the zerovalent complex $Pt(\pi\text{-}CF_3C{\equiv}CCF_3)(PPh_3)_2$. This reaction can again be interpreted in terms of a hydro intermediate such as $PtH(OCOCF_3)$-(π-$CF_3C{\equiv}CCF_3$)(PPh_3)$_2$, which can either undergo insertion to form the alkenyl complex (LIII) or eliminate trifluoroacetic acid to form Pt(π-$CF_3C{\equiv}CCF_3$)(PPh_3)$_2$. The vinyl complex was not obtained from reaction of diphenylacetylene with $Pt(OCOCH_3)_2(PPh_3)_2$ in methanol since apparently the elimination of acetic acid from $PtH(OCOCH_3)$-(π-PhC≡CPh)(PPh_3)$_2$ is the preferred pathway.

A similar mechanism involving an initial oxidative addition of HX to form a hydroplatinum intermediate may occur in the reaction of *trans*-Pt(C≡CPh)$_2$(PPh_3)$_2$ and *cis*-Pt(C≡CPh)$_2$(PPh_3)$_2$ with HCl to form PtCl(C≡CPh)(PPh_3)$_2$ and $PtCl_2(PPh_3)_2$, respectively (*111*):

$$trans\text{-Pt(C}{\equiv}\text{CPh)}_2\text{(PPh}_3\text{)}_2 + \text{HCl} \longrightarrow \text{PtCl(C}{\equiv}\text{CPh)(PPh}_3\text{)}_2 + \text{HC}{\equiv}\text{CPh} \quad (69)$$

$$cis\text{-Pt(C}{\equiv}\text{CPh)}_2\text{(PPh}_3\text{)}_2 + 2\text{HCl} \longrightarrow \text{PtCl}_2\text{(PPh}_3\text{)}_2 + 2\text{HC}{\equiv}\text{CPh} \quad (70)$$

The difference in nature of the final products from these two isomers has been related to the trans influence. In the cis isomer, both acetylide groups are equally labilized by the two phosphine ligands and are easily replaced by chloride. In the trans isomer, however, PtCl(C≡CPh)-(PPh_3)$_2$ is obtained since the acetylide trans to the chloride is not replaced as a consequence of the lower trans influence of the chloride as against that of triphenylphosphine. The catalysis of the linear polymerization of phenylacetylene by $PtHCl(PPh_3)_2$ may also involve the insertion of an acetylene into the hydroplatinum bond (*153*, *115*).

Recently, Jonassen *et al.* have carried out calculations on the reactions and reaction mechanisms of platinum–olefin and acetylene complexes using noniterative semiempirical one-electron molecular-orbital approximations (*295*). These workers have carried out calculations on the various isomers of PtHCl(π-C_2Me_2)(PH_3)$_2$ and of the vinyl compounds PtCl-(CMe=CHMe)(PH_3)$_2$ and have found that the difference in overlap potential of the six-coordinate π-complexes (6.93 e) is significantly different from that of the vinyl compound (7.8 e). These differences are substantial and clearly favor the vinyl form. The overlap population also indicates a preference for the cis configuration around the unsaturation in the vinyl complex which is in agreement with the experimental results (*218*). A further conclusion is that the six-coordinate hydro-π-acetylene intermediate would be only very slightly stable in the absence of a polyatomic conjugate base. Indeed, these workers consider a mechanism involving a six-coordinate hydro-π-acetylene intermediate improbable and suggest other mechanisms should be considered. One mechanism proposed [Eqs. (71)–(74)] involves initial protonation of the free acetylene which is formed by dissociation of the complex. In the presence of the conjugate base (Cl^-) addition occurs to form the intermediate vinyl. A further mechanism proposed by these authors is that

$$Pt(RC{\equiv}CR')(PPh_3)_2 + n(\text{solvent}) \longrightarrow Pt(\text{solvent})_n(PPh_3)_2 + RC{\equiv}CR' \tag{71}$$

$$R{-}C{\equiv}C{-}R' + H^+ \longrightarrow [R{-}C{=}C{-}R']^+ \text{ (H bridging the C=C)} \tag{72}$$

$$Pt(\text{solvent})_n(PPh_3)_2 + Cl^- + R{-}C{=}C{-}R'^+ \text{ (H bridging)} \longrightarrow \textit{cis/trans-}PtCl(CR{=}CR'H)(PPh_3)_2 \tag{73}$$

$$PtCl(CR{=}CR'H)(PPh_3)_2 + HCl \longrightarrow \textit{cis-}PtCl_2(PPh_3)_2 + CHR{=}CHR' \tag{74}$$

the vinyl is obtained by the intermediacy of $PtHCl(PPh_3)_2$, which then inserts the acetylene to form the vinylplatinum(II) complex. However, this mechanism has been considered previously and was thought to be improbable because of the failure to insert diphenylacetylene into $PtHCl(PPh_3)_2$ under the conditions of the reaction (*287*).

A calculation has also been made on the relative energies of $Pt(\pi\text{-}MeC{\equiv}CH)(PH_3)_2$ and $PtH(C{\equiv}CMe)(PH_3)_2$ (*295*). Consideration of total overlap population showed that the π-complex is definitely less stable than either the cis or trans oxidative adduct. The acetylenic hydrogen on the π-complex is considered to be almost completely hydridic, and it is proposed that the monosubstituted π-acetylene complex rearranges to the hydroacetylide complex via an S_N1 (lim) mechanism involving loss of hydride, rearrangement to a new cationic complex, and recombination. The rearrangement reaction is visualized as follows:

$$[(H_3P)_2\overset{+}{Pt}(\pi\text{-}HC{\equiv}CMe)] \longrightarrow (H_3P)_2\overset{+}{Pt}{-}C{\equiv}C{-}Me \tag{75}$$

Chisholm and Clark have compared the reactivities of $[PtHL_2]^+$ and $[PtMeL_2]^+$ (*83*), where L is PMe_2Ph. The hydro cations $[PtHL_2]^+$ are more reactive toward insertion reactions. This result has been partly attributed to the thermodynamic properties of the Pt—H bond compared with those of the Pt—C bond either in the $[PtMeL_2]^+$ cation or in the products. This difference in reactivity may also be related to the availability of alternative reaction mechanisms. Thus reactions involving

$[PtMeL_2]^+$ proceed in a Markownikov manner by electrophilic attack of Pt^+; thus $[Pt(h^3\text{-2-methallyl})L_2]^+$ is formed from allene and $[PtMe(acetone)L_2]^+$, whereas the analogous 1,3-butadiene cation does not lead to a π-allylic derivative by Pt—Me insertion. The hydro cation, however, can react by either a Markownikov or an anti-Markownikov mechanism with either Pt^+ or H^+ attack on the unsaturated ligand. This apparent versatility leads to the formation of π-allylic complexes from both allenes and 1,3-dienes with $[PtHL_2]^+$.

c. *Azides.* Although these reactions may not be strictly classified as insertions, there is sufficient similarity to include them in this section. The reaction of the azides RN_3(R = Ph, PhCO, $PhSO_2$, p-O_2NPh, CO_2Et) with the hydro complexes $PtHX(PR'_3)_2$ (R′ = Et, Ph; X = Cl, NCO, N_3, CN) gave amido compounds $PtX(NHR)(PR'_3)_2$ (*32, 31*). The mechanism has been considered to involve an insertion of one of the nitrogen atoms of the azide into the Pt—H bond along with elimination of dinitrogen.

$$PtHX(PR'_3)_2 + RN_3 \longrightarrow PtHX(\overline{N{=}N{-}N}R)(PR'_3)_2 \longrightarrow$$
$$PtX(NHR)N_2(PR'_3)_2 \text{ or } [Pt(NHR)N_2(PR'_3)_2]X \longrightarrow PtX(NHR)(PR'_3)_2 \quad (76)$$

A further reaction reported is that of benzenediazonium tetrafluoroborate with $PtHCl(PEt_3)_2$ which yields the σ-bonded azo complex $PtCl(N{=}NPh)(PEt_3)_2$ (*242*).

d. *Carbon Disulfide.* The hydroplatinum(II) complex $PtHCl(PPh_3)_2$ reacts readily with CS_2 to form $PtCl(S_2CH)(PPh_3)_2$ (LIV), considered to contain a Pt—S bond. The kinetics of the reaction have been interpreted to imply initial coordination of CS_2 followed by insertion (*239*).

$$PtHCl(PPh_3)_2 + CS_2 \longrightarrow \underset{\text{(LIV)}}{PtCl[SCH(S)](PPh_3)_2} \quad (77)$$

2. *Reactions with Electrophiles*

When the hydro ligand is opposite a group of high trans influence, it shows a chemical behavior toward electrophiles characteristic of a hydridic hydrogen. Two examples are the interaction of HCl with

[PtH(diphos)(PEt_3)]Cl and PtHCN(PPh_3)$_2$ (*202*) to form [PtCl(diphos)-(PEt_3)]Cl and PtClCN(PPh_3)$_2$, respectively:

$$[PtH(diphos)(PEt_3)]Cl + HCl \longrightarrow [PtCl(diphos)(PEt_3)]Cl + H_2 \quad (78)$$

$$PtHCN(PPh_3)_2 + HCl \longrightarrow PtClCN(PPh_3)_2 + H_2 \quad (79)$$

In a similar manner, PtH($OPPh_2$)($HOPPh_2$)$_2$ reacts with HCl and Ph_3CCl to give PtCl($OPPh_2$)($HOPPh_2$)$_2$. These reactions of hydro complexes with protonic acids and electrophiles are discussed in more detail in Section V.

The hydro ligand is replaced by Cl in the reaction of PtHCl(PEt_3)$_2$ and PdHCl(PEt_3)$_2$ with CCl_4 (*70*) (a process that is characteristic of a transition metal hydride), and in the reaction of PtHCl(PMe_2Ph)$_2$ with GeHPhEt(1—$C_{10}H_7$) (*142*). This latter reaction [Eq. (80)] was carried out with an optically active ($[\alpha]_D^{25} = +15.0°$) 1-naphthyl germane which gave an optically active ($[\alpha]_D^{25} = +12.9°$) germylplatinum complex (LV). It has been interpreted to mean that the reaction proceeds with retention of configuration about the substrate.

$$PtHCl(PMe_2Ph)_2 + GeHPhEt(1\text{-}C_{10}H_7) \longrightarrow \underset{\text{(LV)}}{PtCl[GePhEt(1\text{-}C_{10}H_7)](PMe_2Ph)_2} + H_2 \quad (80)$$

In 1965, it was mentioned that reaction of mercuric halides with PtHCl(PPh_3)$_2$ gave metallic mercury (*235*). A further study on mercury–platinum compounds showed that the reaction of $HgCl_2$ with PtHCl-(PPh_3)$_2$ at 0° gave *trans*-$PtCl_2$(PPh_3)$_2$ and elemental mercury (*12*). Recently, however, the mercuriplatinum complex, $PtCl_2$($HgCl_2$)(PEt_3)$_2$ (LVI) has been isolated from the reaction of mercuric chloride with PtHCl(PEt_3)$_2$ (*49*).

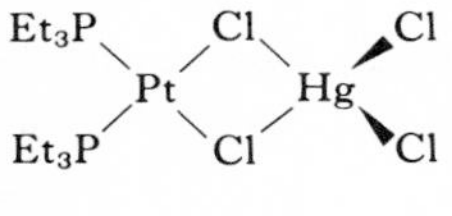

Et_3P, Et_3P — Pt(μ-Cl)$_2$Hg — Cl, Cl

(LVI)

A reaction that may involve an electrophilic attack is the exchange of hydroplatinum complexes with water. This reaction has been studied for the isotopic exchange reaction of PtHCl(PEt_3)$_2$ with D_2O, which is very slow in the absence of added acid (*149*).

3. *Reactions with Bases*

Reactions between $PtHX(PPh_3)_2$ and KOH, alkoxides, or lithium alkyls give the proposed 14-electron molecule $Pt(PPh_3)_2$ (*290, 291*). If triphenylphosphine is also present in the reaction mixture, or if $[PtH(PPh_3)_3]HSO_4$ is treated with KOH, the product is $Pt(PPh_3)_3$. A similar reaction involving elimination of HX from a hydrohaloplatinum(II) complex has been reported as a convenient preparation for carbonylphosphine cluster compounds [the bases used were NH_3 and N_2H_4 (*45*); R = Et, Ph]:

$$3PtHCl(PR_3)_2 + 3CO + \text{base} \longrightarrow Pt_3(CO)_3(PR_3)_4 + 2PR_3 + \text{salt} \qquad (81)$$

IV
HEXADIENE SYNTHESIS

The synthesis of hexa-1,4-diene has been achieved by the nickel-catalyzed homogeneous addition of ethylene to butadiene. The nickel is introduced in the form of the complex $Ni[P(OEt)_3]_4$. The reaction is carried out in acid media, and the active catalyst is the cationic complex $NiH[P(OEt)_3]_3{}^+$ which is a 16-electron molecule. In Fig. 7 the sequence of reactions that leads to the catalytic formation of isomeric hexadienes is

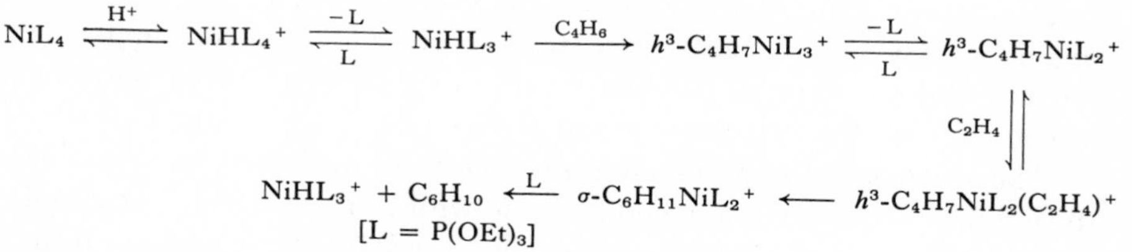

$$NiL_4 \underset{}{\overset{H^+}{\rightleftharpoons}} NiHL_4{}^+ \underset{L}{\overset{-L}{\rightleftharpoons}} NiHL_3{}^+ \xrightarrow{C_4H_6} h^3\text{-}C_4H_7NiL_3{}^+ \underset{L}{\overset{-L}{\rightleftharpoons}} h^3\text{-}C_4H_7NiL_2{}^+$$

$$\downarrow\uparrow C_2H_4$$

$$NiHL_3{}^+ + C_6H_{10} \xleftarrow{L} \sigma\text{-}C_6H_{11}NiL_2{}^+ \longleftarrow h^3\text{-}C_4H_7NiL_2(C_2H_4)^+$$

$$[L = P(OEt)_3]$$

FIG. 7. Catalytic synthesis of hexa-1,4-diene from ethylene and butadiene.

shown. An elegant cycle for this mechanism has been given by Tolman (*238, 119*). The various steps of this catalytic cycle have been studied in detail, and this section of the review article includes the portion of the work having direct relevance to hydronickel complexes.

A. *Protonation Step*

The protonation of $Ni[P(OEt)_3]_4$ to form a hydronickel complex was first reported by Drinkard *et al.* who showed that the pentacoordinate cation $NiH[P(OEt)_3]_4^+$ was formed in strong acid (*137*). The ^1H NMR spectrum in the high field region was found to be a doublet ($J_{P-H} = 26$ Hz), and the ^{31}P spectrum a quintet ($J_{P-H} = 26$ Hz). These data have been interpreted as being diagnostic of a square pyramidal structure (*137*). These authors do, however, concede that other possibilities cannot be eliminated, since the chemical shifts and coupling constants of a less symmetrical structure may be fortuitously very similar. A second possibility is that a fast exchange between the ligands may be occurring, although there was no observable broadening of the NMR resonance down to $-60°$. This complex ion $NiH[P(OEt)_3]_4^+$ is responsible for the yellow color (325 mμ) of $Ni[P(OEt)_3]_4$ in sulfuric acid. This protonation is reversible since the addition of NaOMe or Bu^nNH_2 causes reversion back to $Ni[P(OEt)_3]_4$ (*218*). In acetonitrile solution, the equilibrium shown in Eq. (82) was far to the right when strong acids were used. With CF_3CO_2H the equilibrium was not as far displaced, and when CH_3CO_2H was used in 1:1 ratio with the Ni(0) complex, there was no evidence for the formation of a hydro complex.

$$Ni[P(OEt)_3]_4 + H^+ \underset{}{\overset{K_i}{\rightleftharpoons}} NiH[P(OEt)_3]_4^+ \tag{82}$$

The equilibrium constant K_i using H_2SO_4 in methanol at 0° was determined spectrophotometrically as $33 \pm 3\ M^{-1}$. This value of K_i was obtained in a solution containing excess $P(OEt)_3$. An excess of ligand was introduced since this caused suppression of the decomposition of the hydro complex, without in any way affecting the absorption spectrum used for the measurement of K_i.

The kinetics of the protonation reaction were measured by stop-flow techniques. The reaction is endothermic with $\Delta H° = +5 \pm 2$ kcal and $\Delta S° = +24 \pm 6$ eu. The driving force was considered to come, not from the heat change, but from the increase in entropy associated with the freeing of solvent molecules bound to the proton. Addition of a large excess of $P(OEt)_3$ had no effect on the rate of hydro complex formation. This protonation reaction of $Ni[P(OEt)_3]_4$ occurs much more rapidly than does the dissociation reaction of this compound (*226*), clearly indicating that hydro formation occurs before dissociation of the phosphite can occur. In studying the effect of added $P(OEt)_3$ on the formation of the hydro complex, it was found that the initial absorbance at 325 mμ was unaffected. Since added ligand had no effect on the rate or extent of formation of $NiH[P(OEt)_3]_4$, but did have considerable inhibition on its rate of decay, it was proposed that decomposition of the hydronickel species occurred by attack of H^+ on a species such as $NiH[P(OEt)_3]_3{}^+$, which could be formed by dissociation from the pentacoordinate hydro complex.

$$NiH[P(OEt)_3]_3{}^+ + H^+ \rightleftharpoons Ni(II) + H_2 + 3P(OEt)_3 \quad (83)$$

The equilibrium constant for the reaction shown in Eq. (82) has been obtained for a range of different phosphorus ligands (*285*). The value of K_i was found to increase as the ligand L became more basic (*280*) (Table I). The large error for $Ni(diphos)_2{}^+$ (*258*) is due to the extremely low solubility of the complex in methanol.

TABLE I
FORMATION CONSTANTS FOR NiL_4 WITH H_2SO_4 IN CH_3OH AT 0°C

L	K_i (M^{-1})
Diphos	410 ± 120
$PPh(OEt)_2$	107 ± 13
$P(OEt)_3$	33 ± 3
$P(OMe)_3$	35 ± 2
$P(OCH_2CH_2Cl)_3$	1.2 ± 0.2
$P(OCH_2CCl_3)_3$	<0.1

B. Insertion Step

When $Ni[P(OEt)_3]_4$, sulfuric acid, and butadiene were mixed together in methanol at room temperature the initial yellow color of $NiH[P(OEt)_3]_4{}^+$ rapidly changed to deep red. This red color could be washed out from the crystals of the isolated compound, which was the yellow, π-crotyl complex, π-$C_4H_7Ni[P(OEt)_3]_3{}^+$. Formation of a π-crotyl complex did not occur in the absence of added acid. The isomer mixture resulting from this reaction consisted of 12% of the syn form and 88% of the anti form, and in none of this work was any evidence found for the occurrence of a σ-allynickel complex. This same ratio of isomers was formed irrespective of whether the reaction was carried out at 0° or at 25°. However, when the mixture was heated at 70° for 90 minutes the equilibrium distribution of isomers was found to be 95% syn and 5% anti (*283*). The rate of isomerization of the kinetically preferred anti isomer to the thermodynamically preferred syn was slowed by addition of excess phosphite. This effect is opposite to that expected if the isomerization occurred by the intermediate formation of a σ-crotyl complex. This result has been interpreted in terms of the necessity for a coordinately unsaturated intermediate for the syn–anti isomerization reaction to proceed. The proposed steps in this reaction are shown in Fig. 8.

$$\textit{anti-}\pi\text{-}C_4H_7NiL_3^+ \rightleftharpoons \textit{anti-}\pi\text{-}C_4H_7NiL_2^+ + L \quad \text{(fast)}$$
$$\textit{anti-}\pi\text{-}C_4H_7NiL_2^+ \rightleftharpoons \textit{syn-}\pi\text{-}C_4H_7NiL_2^+ \quad \text{(slow)}$$
$$\textit{syn-}\pi\text{-}C_4H_7NiL_2 + L \rightleftharpoons \textit{syn-}\pi\text{-}C_4H_7NiL_3^+ \quad \text{(fast)}$$
$$L = P(OMe)_3$$

FIG. 8. Proposed mechanism for syn–anti isomerization.

The insertion reaction of $NiH[P(OEt)_3]_4^+$ with a series of dienes has shown that the anti isomer is commonly the kinetically preferred one, and an interpretation has been advanced that the diene rotates into the cisoid configuration as the Ni—H bond is added across (*284*) [Eqs. (84), (85)]. This addition to the cisoid diene appeared to be the preferred mechanism; however, this did not occur when there was a cis substituent on one of the double bonds since cisoid *cis*-pentadiene-1,3 did not give *anti-anti*-dimethyl-1,3-π-allyl. The syn isomer appeared to be thermodynamically favored to a small extent for highly substituted π-allyls. The addition of the Ni—H bond across a double bond to form the π-allyl moiety was

$$+ \text{ NiH} \longrightarrow \text{Ni} \quad (100\%) \tag{84}$$

$$+ \text{ NiH} \longrightarrow \text{Ni} \quad (0\%) \tag{85}$$

found to be more favorable for terminal olefins. An exception seemed to be cyclopentadiene, which inserted to form a π-allyl complex with $NiH[P(OMe)_3]_4$ at a rate comparable with that of a terminal double bond.

This anomalously rapid reaction was considered to occur because the double bonds are held in a cisoid arrangement by the structure of the ring.

C. Olefin Isomerization

The isomerization of butene-1 has been reported in the presence of a mixture of $AlEt_3$ and $NiCl_2py_2$ (*79*), and Cramer has found that a very rapid isomerization of this olefin occurs in acidic solutions of $Ni[P(OEt)_3]_4$ (*122*). The initially formed butene-2 has a cis-to-trans ratio of 2.5:1, although as time progresses the ratio changes to 1:3 in favor of the trans isomer. The relative rate constants for the isomerization of linear butenes with $Ni[P(OEt)_3]_4$ and H_2SO_4 at 25° have been given as shown in Fig. 9 (*122*).

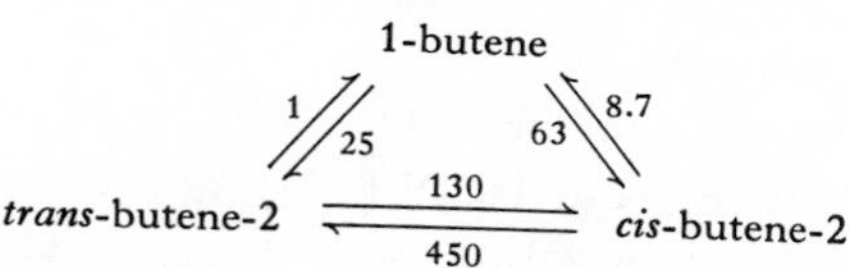

FIG. 9. Relative rate constants for the isomerization of linear butenes with Ni-$[P(OEt)_3]_4$.

This isomerization with $Ni[P(OEt)_3]_4$ in acid solution has been studied in some detail by Tolman (*282*). No evidence for the isomerization of butene-1 could be found in the absence of added acid, but, in the presence of acids, both butene-2 and a little butane were formed. The reaction was first order in $NiH[P(OEt)_3]_4$ and butene-1, and a rate order was found which was inversely proportional to the concentration of the phosphite. The rate expression is given by ($L = P(OEt)_3$ and K_{11} is a composite rate constant)

$$\frac{-d(\text{butene-1})}{dT} = \frac{K_2[\text{NiHL}_4{}^+]}{[\text{L}]} K_{11}[\text{butene-1}]$$

The isomerization reaction proceeded after an induction period, although the reaction was suppressed by oxygen, butadiene (π-allylic complexes were obtained), and triethylphosphite. Tolman has also studied the reaction of $NiH[P(OMe)_3]_4{}^+$ with pentadiene-1,4 since olefin isomerization and π-allyl formation can potentially occur in the same system.

These results showed that dimethyl-1,3-π-allylic products were formed, along with pentadiene-1,3, implying that double-bond migration occurred initially, to be followed by formation of π-allylic complexes. This preferential formation of π-allyls is particularly important since the favored route via π-allyls in the presence of butadiene explains why it is possible to obtain high yields of hexadienes from butadiene and ethylene so long as butadiene is still present.

The isomerization reaction of butene-1 has been carried out with D_2SO_4 in CH_3OD (*282*). The product contained both deuterated and nondeuterated olefins in a ratio consistent with a random scrambling model. The initial step in the reaction involves an insertion of the olefin into the Ni—H bond to form an alkyl. Elimination from the alkyl obtained by Markownikov addition to the olefin can lead to isomerization, whereas elimination from the anti-Markownikov addition product leads to butene-1 being re-formed. The rate of isomerization to deuteration of the olefin is of the order of 170.

The isomerization of pentene-1 has been reported with Ni[bis(diphenylphosphino)butane]$_2$ and HCN (*114*), and with $Ni[P(OEt)_3]_4$ and CF_3CO_2H (*40*). The isomerization with the triethylphosphite complex was carried out over a period of 20 hours, and the products were pentene-1 (3%), *cis*-pentene-2 (23%), and *trans*-pentene-2 (74%). Many of the conclusions of the study with pentene-1 were in agreement with those found for butene-1. The isomerization of pentene-1 which was deuterated at C-1 and C-2, $CHD{=}CD{-}CH_2CH_2CH_3$, gave products formed by movement of the double bond along the chain, but did not give products that contained the deuterium at C-3, C-4, or C-5. A further isomerization involved interchange of hydrogen and deuterium at C-1 and C-2. This 1,2-shift of deuterium cannot be explained on the basis of a π-allylic mechanism but can be accounted for by the proposal of an intermediate pentylnickel complex (*40*). One result from this study with pentene-1 which differs from that found with butene-1 is that with pentene-1 the rate of deuterium redistribution is greater than the rate of isomerization.

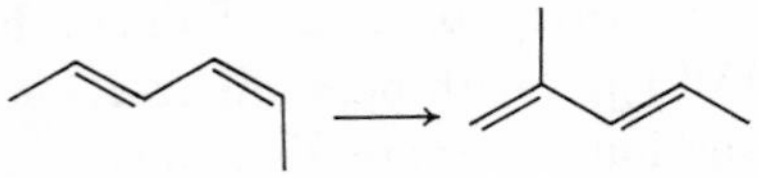

The hydronickel complex formed from tetrakis(tri-*o*-tolylphosphite)-nickel(0) and HCl in tetrahydrofuran is an active catalyst for the skeletal isomerization of hexa-2,4-diene as shown (*162*). The suggested mechanism involves protonation of the nickel–olefin complex to form a hydronickel-(II) intermediate. This compound undergoes insertion to form the alkyl. Skeletal isomerization proceeds via a cyclopropyl intermediate.

The isomerization of allyl methyl ether and allyl phenyl ether occurs readily with $PtH(ClO_4)(PPh_3)_2$, $PtH(NO_3)(PPh_2Me)_2$, $[PtH(PPh_3)_2(acetone)]BF_4$, and $PtH(SnCl_3)(PPh_3)_2$ (*102*). These authors also favor a mechanism involving addition of Pt—H across the terminal C=C bond before double-bond migration occurs, leading to catalytic formation of *cis*-propenyl alkyl ethers. A similar mechanism was considered for the reaction of butene-1, where again both Markownikov and anti-Markownikov addition occurred. The detailed mechanisms of these isomerizations are analogous in many respects to those studied for rhodium(I) complexes in acid solution (*120*).

D. Dimerization Reactions

A number of reactions have been reported that bear a strong similarity to the hexadiene synthesis from the addition of ethylene to butadiene. The dimerization of ethylene has been catalyzed by $R_3R'P[NiCl_3(PR_3)]$ ($R = Pr^i$, $R' =$ benzyl) and aluminum sesquichloride (*144*), and also by $NiBrL(PPh_3)_2$ (L = *o*-toyl, mesityl, 1-naphthyl) and boron trifluoride etherate (*220*). The former system is also effective for *n*-butenes, and propylene can be dimerized with a mixture containing bis(acetylacetone)nickel, ethylaluminum sesquichloride, and triphenylphosphine (*178*). A further reaction of this type where a hydro intermediate is suggested is the reported polymerization of acetylenes by Group VIII metal complexes in conjunction with alkali metal hydrides such as $NaBH_4$. Active complexes for this reaction are $NiX_2[P(CH_2CH_2CN)_3]_2$ and $PdX_2(PBu_3^n)_2$ (*215*).

Neutral nickel complexes that act as catalysts for the dimerization of ethylene have also been reported active in the catalysis of the isotope exchange between C_2H_4 and C_2D_4 (*220*, *15*). These authors suggest a hydro intermediate is involved and that it may arise from the oxidative addition of ethylene to nickel [Eq. (86)]. A similar catalytic effect has been found with $PdCl_2(PhCN)_2$ which also catalyzes the dimerization and isotope exchange in ethylene. The exchange reaction between

$$Ni + C_2H_4 \longrightarrow H—Ni—CH=CH_2 \qquad (86)$$

C_2H_4 and C_2D_4 occurs in dry benzene as solvent and was interpreted to support hydro formation by a reaction similar to that in Eq. (86) (*200*, *221*). This suggestion involves a very interesting reaction mechanism, and it would be valuable if compounds analogous to these intermediates could be isolated. A similar type of intermediate has been suggested in the H—D exchange in the alkenes $RCMe_2CH{=}CH_2$ by K_2PtCl_4. Initial olefin coordination is proposed followed by hydrogen abstraction from C-5 to give a hydroplatinum compound (*222*).

V

CATALYZED HX ADDITION TO MULTIPLE BONDS

A. Hydrosilylation

1. *Chloroplatinic Acid*

The hydrosilylation of olefins using compounds of platinum was described over 15 years ago by Speier *et al.* (*263*). These authors used chloroplatinic acid as a catalyst; but they also found that platinum black, platinum–carbon mixtures, and potassium tetrachloroplatinite effectively catalyzed the hydrosilylation of pentene-1 with methyldichlorosilane. The chloroplatinic acid-catalyzed addition of $SiCl_3H$ to heptene-3 has been used to prepare *n*-heptyltrichlorosilane in good yield (*255*). These workers also found that when the hydrosilylation reaction was carried out with $SiCl_3D$, deuteroalkylsilanes were obtained which were not usually monodeuterated. This effect is due to extensive exchange occurring between the Si—D and the olefinic C—H bonds, and was supported by the observation that at 100° H_2PtCl_6 would catalyze the exchange between $SiCl_3D$ and $SiMeCl_2H$ (*254*):

$$SiCl_3D + SiMeCl_2H \rightleftharpoons SiCl_3H + SiMeCl_2D \tag{87}$$

At 60°, chloroplatinic acid has been used to catalyze the hydrosilylation of butenes and the following activation energies were obtained: butene-1 (1.40 kcal/mole), butene-2 (16.0 kcal/mole), and isobutene (32.0 kcal/mole) (*274*).

The exchange reaction of silanes has also been effectively catalyzed using a palladium–carbon heterogeneous catalyst. A good example of this reaction is the trans-halogenation of silanes from halocarbons [Eq. (88)], although it is ineffective for the case where X = F because of the high energy of the C—F bond (*90*). A similar type of reaction was observed between silanes and compounds containing an exchangeable

$$R_3CX + R'_3SiH \xrightarrow{Pd/C} R'_3SiX + R_3CH \tag{88}$$

hydrogen. In the presence of chloroplatinic acid, methylphenylsilane reacts with propylamine to yield either the mono- or disubstituted silane depending on the ratio of the reactants (*231*). The exchange reaction of silanes with protonic acids, has also been found to be

$$MePhSiH_2 + PrNH_2 \longrightarrow MePhSiH(NHPr) + MePhSi(NHPr)_2 \tag{89}$$

effectively catalyzed by a palladium–charcoal heterogeneous catalyst (*262*).

2. *Substituted Phosphine Complexes*

The reaction between olefins or acetylenes with silicon hydrides in the presence of phosphinenickel complexes has been studied for a variety of substituents on the phosphine and on the silane. In many cases two products are obtained, one of them is the expected simple adduct and the other is an adduct, the formation of which has involved an interchange of hydrogen and chlorine on silicon in the course of hydrosilylation (*206*) Eq. (90). When methyldichlorosilane was added to octene-1 at temperatures in excess of 120°, the two silanes obtained were $C_8H_{17}SiMeClH$ and $C_8H_{17}SiMeCl_2$. The results outlined in Table II

$$RCH{=}CH_2 + SiR'_2ClH \longrightarrow RCH_2CH_2SiR'_2H + RCH_2CH_2SiR'_2Cl + SiR'_2Cl_2 \quad (R' = Cl_2,\ MeCl,\ Me_2) \tag{90}$$

show that only in the case of triphenylphosphine was there no product arising from the interchange of hydrogen and chlorine on the silicon. Hence the extent of the hydrogen–chlorine interchange is dependent on the basicity of the ligand; the more basic the phosphine, the higher the proportion of product formed by interchange on the silicon (*207*). When the silane was varied, it was found that the reactivity of the addition to octene-1 catalyzed by $NiCl_2(PEt_3)_2$ followed the order $SiCl_3H > SiMeCl_2H > SiMe_2ClH$, as shown in Table III.

TABLE II
HYDROSILYLATION AND INTERCHANGE OF HYDROGEN AND CHLORINE

Catalyst	Total yield (%)	$C_8H_{17}SiMeClH/C_8H_{17}SiMeCl_2$
$NiCl_2(PPh_3)_2$	3–5	0/100
$NiCl_2(PMe_2Ph)_2$	77	35/65
$NiCl_2(PBu_3^n)_2$	44	41/59
$NiCl_2(PEt_3)_2$	59	61/39
$NiCl_2(PMe_3)_2$	78	50/50
$NiCl_2$(diphos)	75	7/93
$NiCl_2(Ph_2PCH_2CH_2CH_2PPh_2)$	94	10/90
$NiCl_2(Ph_2PCH{=}CHPPh_2)$	84	6/94
$NiCl_2(Me_2PCH_2CH_2PMe_2)$	87	63/37
$NiCl_2$[1,1′-bis(dimethylphosphino)ferrocene]	95	83/17

When this hydrosilylation study using the complex from 1,1′-bis-(dimethylphosphine) ferrocene (dmpf) was extended further, it was found that olefins could be classified into two categories: those for which hydrosilylation was accompanied by hydrogen–chlorine change and those for which only the normal adduct was obtained. The first group in this classification include olefins with double bonds in the 1- and 2-positions: styrene, cyclohexene, and cyclooctadiene-1,5. The second group include acrylonitrile, methyl acrylate, vinyl acetate, cyclohexa-1,4-diene, and isoprene. Although some olefin isomerization occurred during the course of the hydrosilylation reaction, it was found to be slower than with the platinum catalysts. The complex $NiCl_2$(dmpf) also catalyzed the hydrosilylation of acetylenes. Addition of methyldichlorosilane to diphenylacetylene in the presence of $NiCl_2$(dmpf) proceeded without formation of the hydrogen–chlorine interchange product to give

TABLE III
HYDROSILYLATION OF OCTENE-1

Catalyst	SiX_2ClH	Total yield (%)	$C_8H_{17}SiX_2H/C_8H_{17}SiX_2Cl$
$NiCl_2(PEt_3)_2$	$SiCl_3H$	90	37/63
	$SiMeCl_2H$	61	49/51
	$SiMe_2ClH$	44	23/77

an isomeric mixture of *cis*- and *trans*-α-(methyldichlorosilyl)stilbene (in the ratio of 5:1):

$$PhC{\equiv}CPh + SiMeCl_2H \xrightarrow{NiCl_2(dmpf)} \underset{H}{\overset{Ph}{>}}C{=}C\overset{Ph}{\underset{SiMeCl_2}{<}} + \underset{H}{\overset{Ph}{>}}C{=}C\overset{SiMeCl_2}{\underset{Ph}{<}} \quad (91)$$

Phenylacetylene did not form an adduct under similar conditions. The complex between 1,2-bis(dimethylphosphino)-1,2-dicarbaclosododecaborane with nickel(II) chloride effectively catalyzed the hydrosilylation of olefins (*209*). Catalysis by this nickel complex differed, however, in that considerable amounts of internal adducts were formed:

$$RCH{=}CH_2 + SiMeCl_2H \longrightarrow RCH_2CH_2SiMeCl_2 + RCH(CH_3)SiMeCl_2 \quad (33\text{–}80\%) \quad (92)$$

Such products have not been previously encountered in the hydrosilylation reaction catalyzed by transition metal complexes.

Addition of methyldichlorosilane to octene-1 in the presence of either $NiCl_2$(diphos) or $Ni(C_2H_4)$(diphos) at 120° for 40 hours gave very similar results, both for the total yield of hydrosilylation addition product and for the extent of hydrogen–chlorine interchange. The shapes of the reaction against time curves were closely analogous, except that the divalent nickel complex showed a substantially longer induction period (*208*), implying that the same catalytic species may be formed from the two complexes. The divalent nickel complex must, therefore, undergo an initial 2-electron reduction to the zero-valent state (*208*, *35*). This suggestion of a 2-electron reduction has also been advanced to explain the similarity in the catalytic effect of chloroplatinic acid compared to that of divalent platinum complexes (*66*). A further difference was observed between nickel(II) and nickel(0) complexes when the reaction temperature was lowered. With the nickel(II) catalyst at 90°, no reaction occurred, however, with the nickel(0) catalyst hydrosilylation proceeded in a smooth manner. Hence a crucial step in the reaction could be the reduction of nickel(II) to nickel(0), which apparently does not occur below 100°. The suggested mechanism is shown in Fig. 10, which depicts the active catalyst as the zero-valent nickel–olefin complex [Ni(olefin)L_2]. The hydrosilyl–olefin complex was also considered to be an important species in the hydrogen–chlorine interchange reaction. An

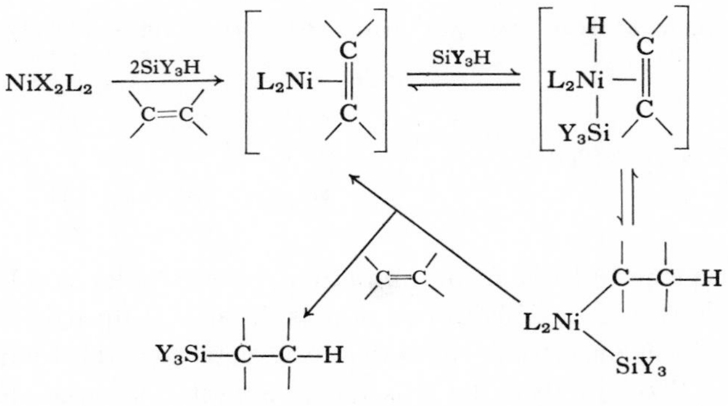

FIG. 10. Nickel-catalyzed hydrosilylation of olefins.

interesting feature is the suggestion that this exchange occurs between the incoming and coordinated silane, and not via an oxidative-addition and reductive-elimination mechanism (Fig. 11). The enhanced catalytic activity of $NiCl_2(PBu_3)_2$ as compared to that of $NiCl_2(PPh_3)_2$ may be related to the fact that the former has a square-planar structure and the latter a tetrahedral one (*35*).

Dichlorobis(triethylphosphine)nickel(II) is an effective catalyst for the reaction of symmetrical tetramethyldisilane with unsaturated hydrocarbons (*237*). A hydronickel complex was postulated as being involved in the reaction, although the mechanism may involve the intermediacy of a dimethylsilylene compound formed by Si—Si bond cleavage rather than the silyl-nickel bond being formed by Si—H bond cleavage. The complex was an effective catalyst for formation of the

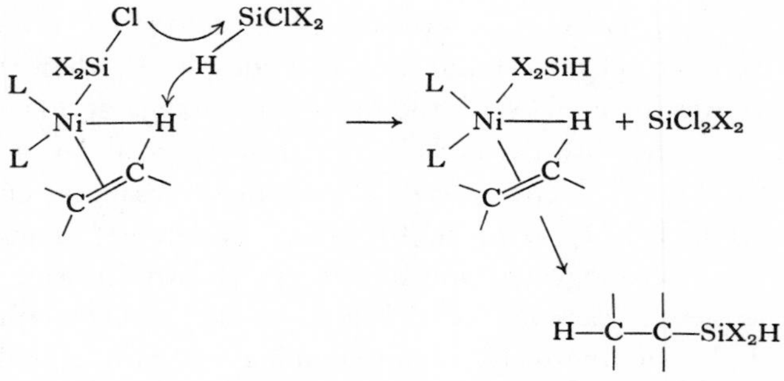

FIG. 11. Proposed mechanism for the interchange of hydrogen and chlorine on the silane.

cyclic silyl compound (LVII) from symmetrical tetramethyldisilane and an acetylene (*237*):

$$2RC{\equiv}CR + HMe_2SiSiMe_2H \xrightarrow[NiCl_2(PEt_3)_2]{90^\circ} \text{(LVII)} + SiMe_2H_2 \quad (93)$$

(LVII) [ring: $C_4R_4SiMe_2$ — R, R above; R–, –R at sides; Si with Me, Me below]

Several workers have reported that low-valent palladium complexes are effective catalysts for the hydrosilylation of double bonds. The reaction of hexene-1 with trichlorosilane to give 1-trichlorosilylhexane has been catalyzed by a range of compounds (Table IV). It is particularly

TABLE IV
HYDROSILYLATION OF HEXENE-1 WITH TRICHLOROSILANE

Catalyst	Temp. (°C)	Yield (%)
$Pd(PPh_3)_4$	100	90.0
$PdCl_2(PPh_3)_2$	120	91.0
$Pd(OAc)_2 + PBu_3$	80	75.6
$Pd + PPh_3$	120	92.3

interesting that the reaction is even catalyzed by a mixture of metallic palladium and triphenylphosphine (*173*). The reactivity of olefins to hydrosilylation with these catalysts follows the order: conjugated diene > olefin-1 > inner olefin. Ligands other than substituted phosphines can be used since the similar complexes $Pd(AsPh_3)_4$ and $Pd(SbPh_3)_4$ have been found to be effective catalysts (*289*).

The asymmetric hydrosilylation of α-methylstyrene with methyldichlorosilane has been catalyzed by (*R*)-benzylmethylphenylphosphine complexes of platinum(II) (*302*) or nickel(II) (*304*) to give a 5 or 17.6% excess of one enantiomer in the addition product, 2-phenylpropylmethyldichlorosilane. The corresponding palladium(II) complexes were, however, only slightly useful for asymmetric synthesis in hydrosilylation of olefins. Nevertheless, palladium(II) complexes of methyldiphenylphosphine or epimeric neomethyldiphenylphosphine, where the dissymmetry is remote from the phosphorus, are especially useful for the induction of asymmetry in the hydrosilylation of styrene and some cyclic conjugated dienes (*199*). A similar procedure has been used for asymmetric hydrosilylation of ketones. The dimeric platinum(II) complex of (*R*)-(+)-benzylmethylphenylphosphine was an effective catalyst for the synthesis of asymmetric alcohols (LVIII) from the reaction of methyldichlorosilane with aromatic ketones (*303*) (R = Me, Et, Pr^n, Pr^i, Bu^t):

$$RCOPh + SiMeCl_2H \xrightarrow{[PtCl_2{*}PR_3]_2} R(\overset{*}{C}HO)PhSiMeCl_2 \xrightarrow[H_3O^+]{MeLi} R(\overset{*}{C}HOH)Ph \quad (94)$$

(LVIII)

The hydrosilylation of butadiene with low-valent palladium complexes gives an alkylsilane formed by addition of 2 molecules of butadiene to 1 molecule of silane (*271*, *272*) [Eq. (95)]. Compounds Pd(maleic anhydride)$(PPh_3)_2$ and Pd(*p*-benzoquinone)$(PPh_3)_2$ were particularly effective catalysts for this reaction. Halopalladium(II) compounds and $Pd(PPh_3)_4$ were less effective as catalysts. The proposed mechanism is

$$2 \; CH_2{=}CH{-}CH{=}CH_2 + SiR_3H \xrightarrow{\text{Catalyst}} R_3Si{-}CH_2CH{=}CHCH_2CH_2CH{=}CHCH_3 \quad (95)$$

the anticipated one, involving formation of a hydropalladium intermediate that adds across the double bond to form the alkylpalladium. The 2:1 adduct was formed in high yield when PPh_3, $AsPh_3$, and $SbPh_3$ were used as ligands. When the ligand was a substituted phosphite, the 1:1 adduct was the major one, and, when trialkylphosphines were used, the 1:1 adduct was the sole product. The analogous platinum compounds are less effective as catalysts (*289*, *271*, *272*); nevertheless, $Pt(PPh_3)_4$ has been used to catalyze selectively the addition of the silanes $SiMeCl_2H$ and $SiMe_2PhH$ to terminal olefinic double bonds (*150*). With this catalyst, neither hydrosilylation of internal double bonds nor double-bond isomerization occurred. A further report on the hydrosilylation of olefins using platinum complexes as catalysts was the preparation of *n*-hexylmethyldichlorosilane from hexene-1 and methyldichlorosilane in the presence of $Pt(C_2H_4)(PPh_3)_2$ (*301*).

In Section II,C, the oxidative-addition reaction of protonic acids to low-valent complexes was discussed. This reaction can lead to formation of either the hydro complex or to the bis complex of the conjugate base. Both of these reactions have been reported for the addition of silanes to low-valent complexes of the nickel triad (*73*, *74*). It is to be expected that the nature of the product will depend on the strength of the acid HX and the trans influence of the conjugate base X. If the acid is strong and the conjugate base has a large trans influence, then it would be anticipated

$$
\begin{array}{ccc}
Pt(PPh_3)_3 + HCl & \longrightarrow & PtHCl(PPh_3)_2 + PPh_3 \\
\downarrow \text{HCN} & & \not\downarrow \text{HCN} \\
PPh_3 + PtHCN(PPh_3)_2 & \xrightarrow{HCl} & PtClCN(PPh_3)_2 + H_2
\end{array}
$$

FIG. 12. Reactions of $Pt(PPh_3)_3$ with strong and weak acids.

that the initially formed hydro complex $MHXL_2$ would react further with HX to form MX_2L_2 with liberation of hydrogen. A good example of this sequence of reactions has been reported for the protonation of $Pt(PPh_3)_3$. The addition of HCl and HCN to $Pt(PPh_3)_3$ gave *trans*-$PtHCl(PPh_3)_2$ and *trans*-$PtHCN(PPh_3)_2$, respectively. The former complex is resistant to further reaction because of the relatively low trans influence of Cl, and the latter because HCN is a weak acid. When, however, $PtHCN(PPh_3)_2$ is treated with HCl, hydrogen is liberated and $PtClCN(PPh_3)_2$ formed. As anticipated, there is no reaction between $PtHCl(PPh_3)_2$ and HCN (*44*) (Fig. 12). Considerations such as these must be used to determine the stability of a hydrosilylmetal complex as the intermediate in the hydrosilylation reaction; if the acid strength or trans influence is excessively high, the hydro intermediate may be converted into the bis-silyl complex before the olefin insertion reaction can take place. The trans influence for substituted silanes has been estimated to follow the order: $SiF_3 \approx SiCl_3 < SiMeCl_2 < Si(OEt)_3 \ll SiMe_2Cl$; $SiCl_3 < SiCl_2H < SiClH_2 < SiH_3 \ll SiMe_3$ (*176*). This order favors the stability of the hydro complex in the hydrosilylation reaction since the silanes that are anticipated to be the stronger protonic acids are those for which the trans influence of the conjugate base is lower.

3. *Olefin Complexes*

The olefin complexes of platinum(II) resemble those of rhodium(I) in being very effective catalysts for addition of silanes to olefinic compounds (*66*). The following results have been found for the hydrosilylation of hexene-1:

a. With $Si(OMe)_3H$ and $Si(OEt)_3H$, hydrosilylation proceeds rapidly to completion without any appreciable isomerization, either before or after completion of the addition reaction.

b. The most commonly observed behavior is with silanes such as $SiCl_3H$, $SiEtCl_2H$, and $SiPhCl_2H$, where rapid hydrosilylation is accompanied by extensive isomerization of excess olefin.

c. With $SiEt_3H$, $Si(CH_2Ph)_3H$, and $SiPh_3H$, hydrosilylation proceeds rapidly at first with concurrent isomerization, but the rates of both processes rapidly fall to zero.

The addition of methyldichlorosilane to styrene has been carried out homogeneously in the presence of $[PtCl_2(C_8H_8)]_2$ (*246*). This reaction is first-order with respect to the styrene and the silane, and first-order with respect to the platinum complex. In aromatic solvents, an induction period was observed which was attributed to coordination of the aromatic or a second molecule of styrene to the platinum. This necessity of a vacant site was further emphasized by the finding that addition did not occur in donor solvents such as tetrahydrofuran, ether, or pyridine.

B. *Hydrogermylation*

Hydrogermylation of phenylacetylene with triphenylgermane or α-naphthylphenylmethylgermane has been catalyzed by $RhCl(PPh_3)_3$, *cis*-$PtCl_2(PPh_3)_2$, or $H_2PtCl_6 \cdot 6H_2O$ (*116*). This addition gave three adducts, A, B, and C (Fig. 13) obtained from cis addition, trans addition, and anti-Markownikov addition. Isomerization in the presence of the hydrogermylation catalyst was very slow compared with the rate of addition (*118*). The addition of the asymmetric organogermanes, α-naphthylphenylmethylgermane $[\alpha]_D + 25.0°$ and α-naphthylphenylisopropylgermane $[\alpha]_D + 1.6°$, gave the pure trans adduct, and thus the hydrogermylation reaction is considered to proceed with retention of configuration.

$$\underset{\text{(C)}}{(Ph)(Ph_3Ge)C{=}CH_2} \xleftarrow{PhC\equiv CH} Ph_3GeH \xrightarrow{PhC\equiv CH} \underset{\text{(A)}}{(Ph_3Ge)(H)C{=}C(Ph)(H)}$$

$$Ph_3GeH \xrightarrow{PhC\equiv CH} \underset{\text{(B)}}{(Ph_3Ge)(H)C{=}C(H)(Ph)}$$

FIG. 13. Hydrogermylation of phenylacetylene with triphenylgermane.

The hydrogermylation of olefins has also been catalyzed by other complexes of the nickel triad (*117*). Complexes that have been used for addition of triphenylgermane to hexene-1 include $Ni(C_5H_8O_2)_2$, $NiCl_2(diphos)$, $PdCl_2(PPh_3)_2$, and $PtCl_2(PPh_3)_2$. Germanes have also been added to cyclohexene, styrene, and heptene-1. The addition of triphenylgermane, 1-naphthylphenylmethylgermane, or 1-naphthylphenylisopropylgermane to styrene gave either (LIX) or (LX). The

$$GeR_3H + CH_2{=}CH{-}Ph \xrightarrow{cis\text{-}PtCl_2(PPh_3)_2} \underset{\text{(LIX)}}{R_3GeCH_2CH_2Ph} + \underset{\text{(LX)}}{R_3GeCH(Me)Ph} \quad (96)$$

addition of asymmetric compounds was again considered to proceed with retention of configuration at the germanium atom, as was the isotopic exchange reaction of GeR_3H and GeR_3D compounds in the presence of hydrogermylation catalysts. The mechanism suggested for this hydrogermylation of olefins corresponds closely with those discussed for the catalyzed hydrosilylation of unsaturated hydrocarbons (*65*).

C. Acid Chloride Reactions

The conversion of acid chlorides to aldehydes has been achieved by platinum metal-catalyzed reaction with silanes. The general reaction,

$$SiR_3H + R'COCl \longrightarrow SiR_3Cl + R'CHO \quad (97)$$

has been catalyzed heterogeneously using a palladium–charcoal system (*89*). A number of platinum complexes have also been reported to be effective homogeneous catalysts for this reaction (*132*). For example, in the presence of *cis*-$PtCl_2(PPh_3)_2$ at 120°, benzaldehyde and *o*-methoxybenzaldehyde are obtained from the appropriate acyl chloride and triethylsilane. The suggested mechanism is shown in Fig. 14.

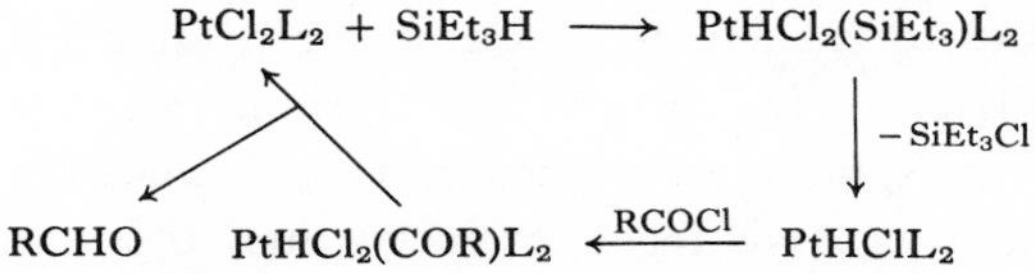

FIG. 14. Formation of aldehydes from acyl chlorides and triethylsilane in the presence of *cis*-$PtCl_2(PPh_3)_2$.

D. *Hydrocyanation*

Addition of HCN to ethylene to form propionitrile can be catalyzed heterogeneously with finely divided nickel (*127*). The hydrocyanation reaction can also be catalyzed by homogeneous catalysts, and nickel complexes have received considerable publicity in the patent literature. Adiponitrile has been obtained from HCN and $CH_2{=}CHCH_2CH_2CN$ using a mixture of tetrakis(triphenylphosphite)nickel and triphenylphosphite as catalyst. A similar synthesis of organic nitriles from 3-pentenenitrile and HCN has been reported by catalysis using a mixture of borohydride along with zero-valent nickel complexes. The effective compounds are $Ni(CO)_4$, $Ni(CO)_2(PPh_3)_2$, $Ni(CO)_2(AsPh_3)_2$, $Ni[P(OEt)_3]_4$ $Ni[P(OPh)_3]_4$, and $Ni(PF_3)_4$ (*140*, *139*). The compounds $Ni[Ph_2P(CH_2)_nPPh_2]_2$ ($n = 2, 3, 4$) have been used to catalyze the addition of HCN to butadiene to give pentenenitriles (*11*). The synthesis of allyl cyanide has been achieved from allyl acetate using $Ni[P(OPh)_3]_4$ and $ZnCl_2$ (*136*). Addition of HCN to acetylene has been catalyzed by $Ni[P(OPh)_3]_4$. The products from acetylene, propyne, and phenylacetylene were acrylonitrile, butenenitriles, and α-phenylacrylonitrile (*138*).

The reaction of HCN with bicyclo[2.2.1]heptene in the presence of $Pd[P(OPh)_3]_4$ gave only poor yields of 2-cyanobicyclo[2.2.1]heptane unless $P(OPh)_3$ was also present, in which case a product that was almost exclusively the exo isomer could be obtained (*38*). The suggested mechanism is outlined in Fig. 15. The palladium complex will also effectively catalyze the addition of HCN to ethylene.

$$PdL_4 + HCN \longrightarrow PdHCNL_2 + 2L$$
$$PdHCNL_2 + \text{olefin} \longrightarrow PdCN(\text{alkyl})L_2 \longrightarrow \text{alkyl cyanide} + PdL_2$$

FIG. 15. Suggested mechanism for the formation of 2-cyanobicyclo[2.2.1]heptane from bicyclo[2.2.1]heptene and HCN in the presence of $Pd[P(OPh)_3]_4$.

Hydrocyanation of hexene-1 to a mixture of heptanenitrile and 2-methylhexanenitrile, using Lewis acid promoted zero-valent nickel phosphite complexes, has been studied in some detail (*277*). The presence of a Lewis acid was advantageous since when the zero-valent nickel complex was used alone the rate of hydrocyanation was very slow. The role of excess ligand and the Lewis acid in the reaction appear to be

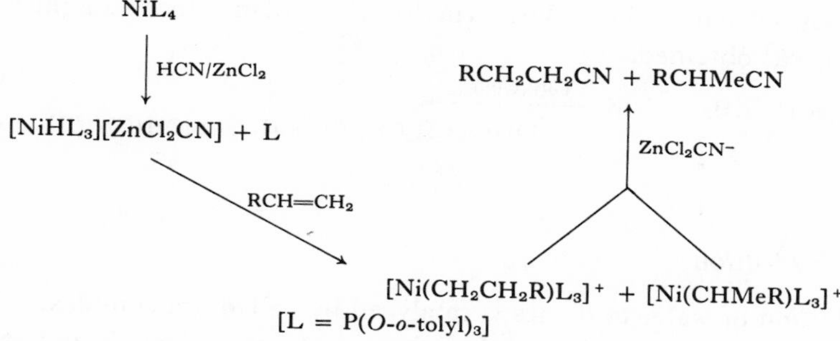

$[L = P(O\text{-}o\text{-tolyl})_3]$

FIG. 16. Proposed mechanism for the hydrocyanation of olefins.

associated with the activation of the Ni(0) complex, and stabilization of the catalytic intermediate. The excess ligand is considered to suppress formation of the catalytically inactive dicyano complex (*54*, *277*). The Lewis acid effectively acted as a cyanide ion acceptor in the reaction, and also affected the product distribution; in particular the ratio of normal to branched-chain isomers produced was dependent on the nature of the Lewis acid promoter. The proposed mechanism is shown in Fig. 16. Deactivation of the catalyst to the inactive dicyano complex $Ni(CN)_2L_2$ can occur by the series of reactions in Fig. 17. The presence of excess ligand would be involved in the final equilibrium step; the final product is coordinately saturated and would not readily react with HCN to form $[Ni(CN)L_3]^+$. The reaction mixture showed the anticipated catalytic activity for the olefin isomerization reaction. Since this reaction was considerably more facile than hydrocyanation, formation of the secondary alkyl could arise from insertion of either hexene-1 or hexene-2 into the Ni—H bond. Further study of this alternative showed that 85% of the branched nitrile product resulted from direct hydrocyanation of hexene-2 and only 15% from Markownikov addition of HCN to hexene-1. 2-Ethylpentanenitrile was not formed in the reaction.

Tetrakis(triphenylphosphite) palladium catalyzes addition of HCN to vinyltriethoxysilane, vinylmethyldiethoxysilane, vinyldimethylethoxysilane, and 1,3-divinyltetramethyldisiloxane to give the corresponding

$$[Ni(CH_2CH_2R)L_3]^+ + HCN \longrightarrow [Ni(CN)L_3]^+ + RCH_2CH_3$$
$$[Ni(CN)L_3]^+ + [ZnCl_2CN]^- \longrightarrow Ni(CN)_2L_2 + L + ZnCl_2$$
$$[Ni(CH_2CH_2R)L_3]^+ + L \rightleftharpoons [Ni(CH_2CH_2R)L_4]^+$$

FIG. 17. Mechanism for the deactivation of the hydrocyanation catalyst.

(cyanoethyl)silane (*55*). With vinyltriethoxysilane the straight-chain isomer was obtained.

$$(EtO)_3SiCH{=}CH_2 + HCN \xrightarrow{Pd[P(OPh)_3]_4} \underset{92\%}{(EtO)_3SiCH_2CH_2CN} + \underset{8\%}{(EtO)_3SiCH(CN)CH_3} \quad (98)$$

E. Hydration

Addition of water to dienes is catalyzed by palladium complexes. The reaction has been used for synthesizing unsaturated alcohols and ethers from aliphatic conjugated C_4 and C_6 olefins (*248*). In particular, the hydration of butadiene with water in the presence of bis(2,4-pentanedionato)palladium and triphenylphosphine gave 2,7-octadien-1-ol, 1,7-octadien-3-ol, and 1,3,5,7-octatetraene (*18*). The reaction was accelerated by carbon dioxide. Compounds $Pd(PPh_3)_4$ and $Pd(O_2CO)(PPh_3)_2$ were also effective.

F. Addition of C—H, N—H, and O—H Bonds to Olefins

Addition of active methylene compounds to butadiene has been catalyzed by complexes of the nickel triad. A catalyst mixture of $Ni(acac)_2$, $PPr(OR')_2$, and borohydride was effective for the addition of $R_1R_2CH_2$ (R_1 = Ph, R_2 = $COCH_3$; R_1 = Ph, R_2 = CN; R_1 = R_2 = CO_2Et; R_1 = $COCH_3$, R_2 = CO_2Et) to butadiene (*26*). Both Pd(0) and Pd(II) complexes are effective as catalysts for the addition of active methylene compounds such as 2,4-pentanedione and ethyl acetoacetate to butadiene (*177*, *269*). Several products were obtained from these reactions since both the 1:1 and 1:2 addition occurred.

Hydronickel complexes such as those obtained from $AlHBu_2$ and $PPh(OR)_2$ have been reported active for the addition of secondary amines to butadiene. The product obtained involved addition of 2 molecules of the amine to the dimer of butadiene (*249*):

$$2\ \text{(butadiene)} + 2NHR_2 \longrightarrow R_2N\text{-}CH_2CH{=}CHCH_2CH_2CH_2CH{=}CHCH_2\text{-}NR_2 \quad (99)$$

The reaction of allylic halides with diethylamine to give the allylamine is catalyzed by a mixture of $Pd(acac)_2$ and PPh_3 (*17*):

$$CH_3CH{=}CHCH_2X + NHEt_2 \xrightarrow{Pd(acac)_2 + PPh_3} CH_3CH{=}CHCH_2NEt_2 + HX \quad (100)$$

Reaction of hydroxylic compounds with butadiene has been studied with palladium compounds. A common feature is the dimerization of butadiene with incorporation of functional groups from alcohols (*270*). 1-Methoxyoctadiene-2,7 was obtained from butadiene and methanol in the presence of Pd(maleic anhydride)$(PPh_3)_2$ (*273*). Complex $Pt(PPh_3)_4$ has also been used, although platinum compounds were less effective. Octadienyl esters were obtained from butadiene and acetic acid in the presence of $Pd(acac)_2$ and either PPh_3 or $P(OPh)_3$ (*294*). Palladium complexes were effective for the synthesis of β,γ-unsaturated esters from butadiene, methanol, and CO. The favored mechanism involved addition of a hydropalladium complex to butadiene to give an allylpalladium intermediate (*46*).

VI

HOMOGENEOUS HYDROGENATION AND ISOMERIZATION

In 1963 a homogeneous catalyst was described that effected the facile homogeneous hydrogenation of ethylene and acetylene (*121*). The catalyst consisted of a mixture of chloroplatinic acid and stannous chloride dihydrate in methanol as solvent. It is suggested that stannous chloride both promotes coordination of ethylene to the metal and stabilizes platinum against reduction to the free metal by hydrogen. This catalyst system has also been used for hydrogenation of higher olefins, such as cyclohexene, and these authors observe that the rate of hydrogenation follows for olefins the order: terminal > 1,2-disubstituted > trisubstituted (*34*). This catalyst has also been used for isomerization of pentene-1 to a cis–trans mixture of pentene-2. The isomerization reaction occurs only when hydrogen is present, and a hydroplatinum(II) intermediate is suggested (*43*). Following isolation of the complexes $[Et_4N][PtH(SnCl_3)_2(PEt_3)_2]$ and $[Me_4N]_3[PtH(SnCl_3)_4]$ (*214*), Bond (*44*) suggested the more definitive intermediate in the hydrogenation reaction to be $[PtH(SnCl_3)Cl_2]^{2-}$ (Fig. 18). This author also found that in the isomerization of pentene-1 by this catalyst a high trans-to-cis

$$[Pt(SnCl_3)_2Cl_2]^{2-} + H_2 \longrightarrow [PtH(SnCl_3)Cl_2]^{2-} + H^+ + SnCl_3^-$$
$$[PtH(SnCl_3)Cl_2]^{2-} + C_5H_{10} \rightleftharpoons [PtC_5H_{11}(SnCl_3)Cl_2]^{2-}$$
$$[PtC_5H_{11}(SnCl_3)Cl_2]^{2-} + H_2 \longrightarrow [PtH(SnCl_3)Cl_2]^{2-} + C_5H_{12}$$

FIG. 18. Proposed intermediate formation in the hydrogenation reaction.

ratio of pentene-2 is obtained. The proposed mechanism involves insertion of olefin into the Pt—H bond, and the isomer ratio of pentene-2 is possibly determined by the preferential conformation of the 2-pentyl group σ-bonded to the platinum (*44*).

For the hydrogenation of ethylene and acetylene with chloroplatinic acid and stannous chloride as catalyst, a number of steps have been outlined which are characteristic of the mechanism (*305*). These steps are (1) competitive formation of a π-ethyleneplatinum and a hydroplatinum complex; (2) formation of the hydro-π-ethyleneplatinum complex; (3) rearrangement by insertion to form an ethylplatinum complex; and (4) attack of the protonic hydrogen on the metal–carbon bond to form ethane and the catalyst. The reduction of acetylene to ethane proceeds via the intermediate formation of ethylene.

A. *Complexes Used as Catalysts for the Hydrogenation of Double Bonds*

Bailar and co-workers have used several complexes as catalysts for reduction of linolenic ester to linoleic or oleic ester without any reduction to the saturated stearic ester. A considerable portion of this work was carried out using a mixed catalyst consisting of tin(II) chloride and a platinum(II) complex. However, catalytic work has also been carried out using palladium and nickel complexes, the former again being used along with tin(II) chloride (*191*). The experimental details have been recently reviewed (*190*) so that this article is concerned with the conclusions and mechanistic aspects rather than with the direct results.

Triphenylphosphine, triphenylarsine, and triphenylstibine (*23*) may be used as ligands in these complexes. Phenyldimethylphosphine is very effective as a ligand for these catalytically active complexes, but trialkylphosphines lead to inactive compounds. Diphenylsulfide and diphenylselenide have been successfully used as ligands in the preparation of these catalysts (*275*). However, if germanium(II) or lead(II) chloride are substituted for tin(II) chloride, the catalyst system becomes less effective (*23*). The homogeneous hydrogenation of methyl linoleate has also been carried out using $NiCl_2$ and sodium borohydride in DMF, although a mixture of $RhCl_3(py)_3$ and sodium borohydride was more effective (*2*). A similar system employed is a mixture of $AlBu_3{}^i$ along with either $NiCl_2(PBu_3)_2$ or $PdCl_2(PBu_3)_2$. The catalyst mixtures were effective for hydrogenation of hexene-1, and a hydrometal compound was suggested as an intermediate (*260*). Isomerization and hydrogenation of dimethyl maleate by cyanonickel complexes has been reported. The activity of the complexes increases in the order: $Ni(CN)_2(PPh_3)_2$ < $Ni(CN)_4^{2-}$ < $Ni(CN)_2(phen)$. The reaction was considered to involve $NiHCN(PPh_3)_2$ although conclusive proof was not presented (*256*).

B. Studies on Soybean Oil Methyl Ester and Methyl Linolenate Using Mixtures of $PtCl_2(PPh_3)_2$ and $SnCl_2.2H_2O$

This catalyst system is effective for hydrogenation of polyolefinic esters where it is desirable to terminate at the monoene stage (*22*). By using this homogeneous catalyst mixture, soybean oil methyl ester and methyl linoleate can be hydrogenated more specifically than with nickel or palladium as a heterogeneous catalyst. The use of a mixed catalyst is necessary since neither $PtCl_2(PPh_3)_2$ nor $SnCl_2 \cdot 2H_2O$ alone is effective. The homogeneous hydrogenation of methyl linolenate catalyzed by this system is similar to the others using various platinum–tin chloride complexes (*152*, *24*) since hydrogenation is accompanied by extensive conjugation, cis–trans isomerization, and double-bond migration. Reduction of soybean oil methyl ester at 90° and 575 psi hydrogen gives a mixture consisting of 11.8% palmitate, 4.4% stearate, 78.4% monoene, and 5.4% conjugated diene (*22*). Linoleate and linolenate were almost completely converted to monoene, and oleate was not reduced at all. The reduced esters are largely in the trans configuration. Cis–trans isomerization was observed for oleate, but the complex mixture would not effectively catalyze hydrogenation of isolated double bonds. With oleate, conversion to the trans configuration occurs to the extent of 81.7%, and to the extent to 82.7% for the soybean oil. Thus, when soybean oil methyl ester and methyl linoleate are hydrogenated the product is stereospecifically trans monoene.

Dependence on the nature of the ligands of the ability of the complex to function as a catalyst is shown by the extent of conversion of methyl linoleate by the complexes $PtCl_2L_2$ and tin(II) chloride in Table V. Hydrogenation with $PtBr_2(PPh_3)_2$ and tin(II) bromide is more effective than with the corresponding chloro analogs, but neither $Pt(CN)_2(PPh_3)_2$

TABLE V
HYDROGENATION STUDIES ON METHYL LINOLEATE

Complex	Conversion of methyl linoleate (%)
$PtCl_2[P(OPh)_3]_2$	94.1
$PtCl_2(AsPh_3)_2$	80.4
$PtCl_2(PPh_3)_2$	57.8
$PtCl_2(SbPh_3)_2$	25.4
$PtCl_2(PBu_3)_2$	21.6
$PtCl_2(en)$	2.3

nor $Pt(CN)_2(AsPh_3)_2$ is effective for hydrogenation. A mixture of $PtHCl(PPh_3)_2$ and tin(II) chloride has similar activity to the mixture with $PtCl_2(PPh_3)_2$, although the single complexes $PtHX(PPh_3)_2$ (X = CN, $SnCl_3$, NO_2, Cl, SCN, OCN) are ineffective.

These authors have summarized their work as follows (*23*):

1. A monoene such as oleate is not hydrogenated but is isomerized from the cis to the trans configuration.
2. Double-bond migration in methyl linolenate occurs to form conjugated dienes which are selectively hydrogenated to the monoenes.
3. Double-bond migration is accompanied by cis–trans isomerization.
4. In butanol solution, ester exchange reactions occur.
5. Active hydrogen can be supplied by hydrogen gas or by a solvent such as methanol.

C. Studies on Soybean Oil Methyl Ester and Methyl Linolenate Using Nickel and Palladium Catalysts

Dihalobis(triphenylphosphine)nickel(II) complexes are effective in both hydrogenation and isomerization of methyl linolenate to give the isomerized *trans*-monoene without forming the saturated stearate. The catalytic effects of $NiX_2(PPh_3)_2$ parallels the order of decreasing electronegativity of X: Cl > Br > I. With these complexes, hydrogenation occurs more rapidly than isomerization. However, although oleate undergoes cis–trans isomerization under hydrogen pressure, it does not occur under nitrogen pressure (*188*). Palladium complexes will catalyze hydrogenation and isomerization reactions that are catalyzed by the platinum complexes (*189*). The activity of $PdCl_2(PPh_3)_2$ is enhanced by addition of tin(II) chloride, but the activity of $Pd(CN)_2(PPh_3)_2$, $Pd(SCN)_2(PPh_3)_2$, and $PdCl_2(SbPh_3)_2$ cannot be enhanced in this manner. *n*-Butanol is an effective solvent for these reactions since it readily deprotonates, and there is a concomitant oxidation of the alcohol.

D. Hydrogenation of Olefins by Platinum and Palladium Complexes

In order to understand more completely the factors involved in the catalytic hydrogenation of oils, Bailar and co-workers extended their study to a variety of polyolefins. It was found that for C_8 series the order of ease of hydrogenation is octadiene-1,7 > cyclooctadiene-1,5 > cyclooctatetraene (*275*). The orders for C_6 and C_7 polyolefins are similar. Overall the sequence for the rate of hydrogenation decreases in the order: heptadiene-1,5 ~ heptadiene-1,6 ≫ cycloheptatriene, hexadiene-1,5 > cyclohexadiene-1,4 ≫ benzene. This has led to the assumption that hydrogenation of open-chain olefins is more facile than of cyclic olefins because of their greater ease in assuming the proper arrangement to the metal. The negligible difference in reactivity between conjugated and nonconjugated polyolefins is attributed to the much more rapid isomerization reactions that accompany the hydrogenation. The order of catalytic ability in the presence of tin(II) chloride is

$$PtCl_2(AsPh_3)_2 > PtCl_2(PPh_3)_2 > PtCl_2(SePh_2)_2 > PtCl_2(SPh)_2$$

with the reaction being most effective in a solvent that will not coordinate too strongly.

The hydrogenation of olefins using $PtCl_2(PPh_3)_2$ mixed with $SnCl_2$ is most effective when the olefin has terminal double bonds and removal of the double bond from a terminal position significantly reduces the amount of hydrogenation (*5*, *6*). This means that any compound that contains two terminal double bonds will give a small amount of saturated material. This was verified using hexadienes since hexadiene-1,5 gives 2.5% *n*-hexane, both hexadiene-1,4 and –1,3 yield only monoenes, and the hydrogenation of hexadiene-2,4 is difficult to achieve. In contradiction to some of the earlier results, it was found that hydrogenation can occur without prior conjugation of the diene since 2,3,3-trimethylpentadiene-1,4 can be reduced to a mixture of 2,3,3-trimethylpentene-1 and 3,3,4-trimethylpentene-1 (*6*). The results with this latter diene are important since it cannot isomerize to the conjugated diene in the presence of the catalyst.

In considering the mechanism of these reactions, it is significant that certain low molecular weight dienes, such as butadiene, are not reduced but can inhibit the reduction of other olefins. The double bond in $CH_2{=}CHCN$ can be effectively hydrogenated without any reduction of the nitrile group. The reduction of $CH_2{=}CCl(CN)$, $CHCl{=}CHCN$, and $CH_3CH{=}CHCN$ fails with the $PtCl_2(AsPh_3)_2 \cdot SnCl_2 2H_2O$ catalyst system—steric reasons are given as an explanation (*6*).

The scope of activity and selectivity in hydrogenations of unsaturated compounds catalyzed by a number of palladium complexes has been studied by Stern (*264*, *265*). For monoolefins the rate of hydrogenation declined in the order: $Pd(Ph_2PCH_2PPh_2)_3 > Pd(Ph_2PCH_2CH_2PPh_2)_2 > Pd(Ph_2PCH_2CH_2CH_2PPh_2)_2$. Selectivity was poor. The complex $Pd_2(Ph_2PCH_2PPh_2)_3$ catalyzed the facile hydrogenation of alkynes to alkenes. The hydrogenation of butadiene was markedly affected by solvents, and addition of CO or PPh_3 to the solution showed a strong poisoning effect (*264*). The rates of hydrogenation of pentadienes were found to follow the order: *trans*-pentadiene-1,3 > isoprene > *cis*-pentadiene-1,3 > pentadiene-1,4. The rate of reaction was increased by two orders of magnitude when the complex was exposed to oxygen prior to hydrogenation. This study was extended to the deuterium addition to ethylene, 3-methylbutene-1, and isoprene (*265*). With the monoolefins, in the presence of $Pd_2(Ph_2PCH_2PPh_2)_3$, it was found that monoexchange was faster than isomerization, diexchange, and deuteriogenation; however, with the oxygen-pretreated catalyst, deuteriogenation was faster than exchange and isomerization.

The heterogeneously catalyzed hydrogenation of olefins has been reported using dichloropalladium and platinum(II) complexes of a polymeric diphenylbenzylphosphine ligand (*56*).

E. *Mechanistic Aspects of the Hydrogenation and Isomerization Reactions*

It is widely assumed that hydro complex formation is an essential step in hydrogenation, and several hydro complexes have been isolated from hydrogenation reaction media. The formation of a hydro complex, however, is not the rate-determining step in the system, since the rate of hydrogenation is the same with $PtCl_2(PPh_3)_2$ or $PtHCl(PPh_3)_2$ as catalysts. Tin(II) chloride plays a significant role as a cocatalyst, and its presence is essential to hydrogenation with all these catalysts except $Pd(CN)_2(PPh_3)_2$, which has intrinsic catalytic activity. Hydrometal–olefin complex formation is also considered likely as evidenced by the isolation of several such complexes from the hydrogenation or isomerization reaction (*275*, *276*). Isomerization or double-bond migration occurs both before hydrogenation, as evidenced by formation of the conjugated dienes from nonconjugated ones, and after hydrogenation, as evidenced by the isomerization of monoenes under hydrogenation conditions. The importance of hydro complex formation is also shown by the fact that a source of hydride ions such as H_2 or CH_3OH must be present; no reaction was detected when CH_2Cl_2 was used as solvent in an atmosphere of argon. The importance of CH_3OH is shown when hexadiene-1,5 is catalytically isomerized in $CH_3OD—D_2$ and 2,4-hexadiene, deuterated on both the methyl and vinyl groups, is obtained. No evidence of deuteration was found when the reaction was run in nondeuterated methanol even in the presence of D_2; thus, transfer of the acidic proton of CH_3OH to the olefin accompanies isomerization even in the presence of H_2.

In a study of the hydrogenation and isomerization of octene-1 by $MX_2(PPh_3)_2$ and $SnCl_2$, it was found that there was no isomerization of octene-1 in the presence of CH_3OH if H_2 was absent (*1*).

It has been suggested that the hydroplatinum(II) intermediate is formed [Eq. (101)], with insertion–elimination leading to isomerization (*175*, *126*). The halogen acid is the source of reversibility and can be

$$MX_2L_2 + H_2 \rightleftharpoons MHXL_2 + HX \tag{101}$$

removed by use of a potassium acetate buffer (*1*). Complex PtHCl-$(PPh_3)_2$ was initially considered, on IR evidence, to be isolable as both the cis and trans isomers (*21*); however, more recently, the supposed "cis isomer" has been found to be a crystalline modification of the trans (*107*, *112*). This compound reacts with $SnCl_2$ to form $PtH(SnCl_3)$-$(PPh_3)_2$ (*21*) which is similar to the complex $PtH(SnCl_3)(PEt_3)_2$ (*213*). Coordination of the olefin followed by insertion to form the alkyl can then occur. It has been assumed (*275*) that the hydroolefin intermediate is a pentacoordinate platinum(II) complex (*276*), and complexes of this

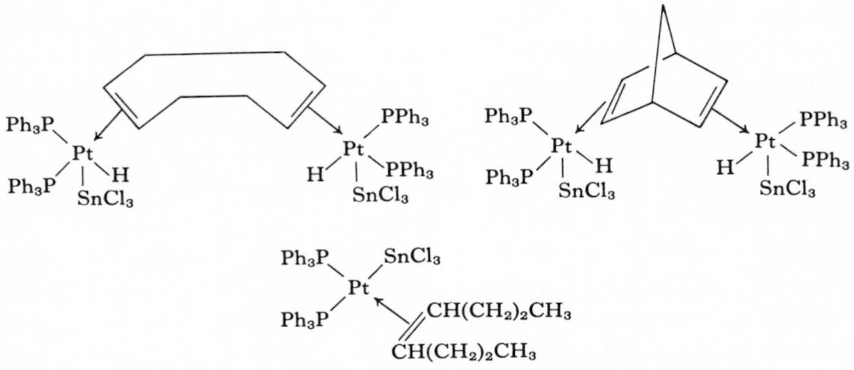

FIG. 19. Proposed structures of some platinum–olefin complexes.

type have been obtained from the hydrogenation or isomerization reactions of cyclooctadiene-1,5, bicycloheptadiene, and octene-4 (Fig. 19).

In Section III the insertion of an olefin into a Pt—H bond was discussed, and it was concluded from recent work that a four-coordinate cationic intermediate is preferable to a five-coordinate adduct. In view of this, it appears likely that an ionic compound such as $[PtH(C_6H_8)(PPh_3)_2][SnCl_3]\cdot CH_2Cl_2$ (*275*) is more closely analogous to the intermediate hydroolefin complex. The isomerization of olefins occurs by a rapid insertion–elimination reaction between the hydroolefin and alkyl complexes. In the decomposition of di-*n*-butyl bis(triphenylphosphine)-platinum(II) a hydro intermediate is proposed (*296*). A speculative mechanism involving a σ-π complex has been suggested to account for the hydrogenation of a conjugated diene to a monoene (Fig. 20). This σ-π complex is favored over the π-allylic complex as intermediate for the platinum-catalyzed reactions when the metal is bonded to a carbon atom in an α-position to the double bond (*275*). Migration of double bonds to form conjugated dienes takes place more rapidly with tin(II) chloride mixed with palladium and platinum complexes than it does with iodonickel and cyanopalladium complexes (*190*). This difference may be due to π-allylic intermediates being formed in the latter cases leading to 1,4-addition rather than the σ-π intermediate that gives 1,2-addition. Evidence for π-allylic intermediates in the iodonickel case comes from the catalyzed conversion of bicyclo[2.2.1]heptadiene to nortricyclene (*190*).

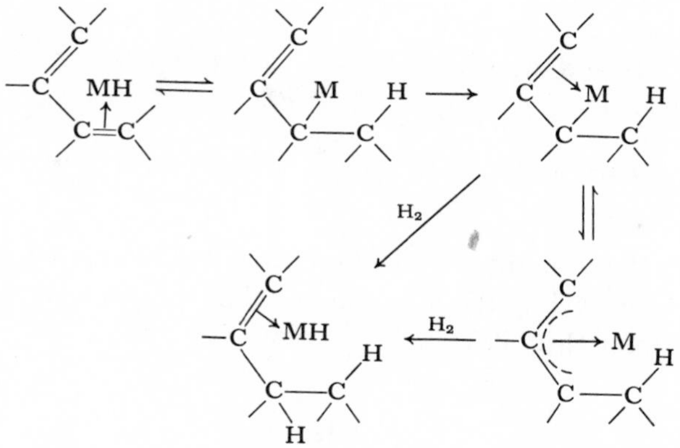

FIG. 20. Proposed intermediates in the hydrogenation of conjugated dienes to monoenes.

A similar type of mechanism (Fig. 21) was suggested for the hydrogenation of unsaturated hydrocarbons with $PtHCl(PEt_3)_2$ and HCl or $HClO_4$. Hexene-1, cyclohexene, octene-1, and 2-methylbutene-2 were effectively converted into the saturated hydrocarbon along with formation of $PtCl_2(PEt_3)_2$ (*171*). Acrylonitrile and *trans*-1,2-dichloroethylene could not be hydrogenated.

The hydro complex has been suggested as an intermediate in the ester exchange reaction of soybean oil methyl ester catalyzed by $PdCl_2(PPh_3)_2SnCl_2 \cdot 2H_2O$. The mechanism involves coordination of the ester

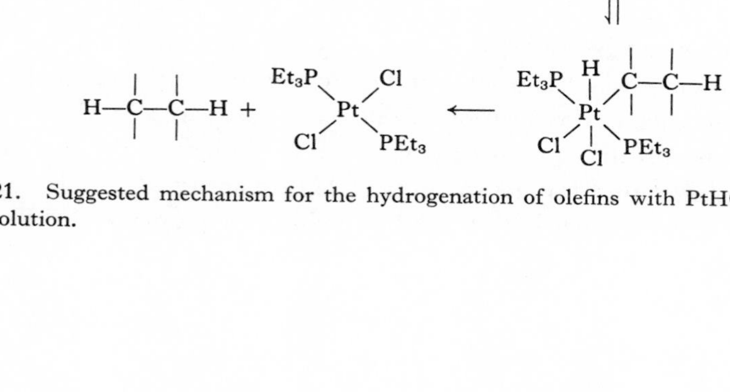

FIG. 21. Suggested mechanism for the hydrogenation of olefins with $PtHCl(PEt_3)_2$ in acid solution.

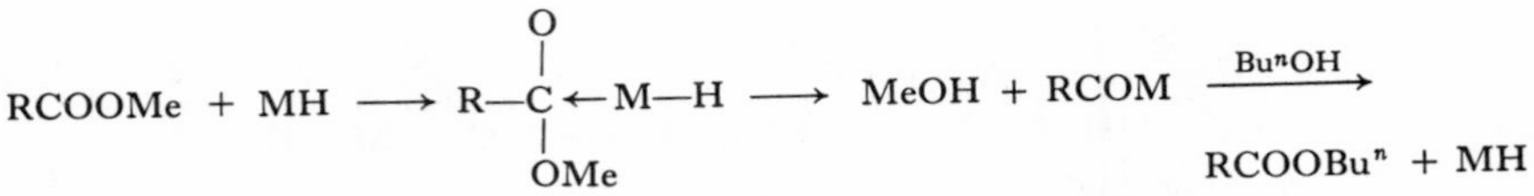

$$\mathrm{RCOOMe + MH \longrightarrow R{-}\overset{\displaystyle O}{\overset{|}{\underset{\underset{\displaystyle OMe}{|}}{C}}}{\leftarrow}M{-}H \longrightarrow MeOH + RCOM \xrightarrow{Bu^nOH} RCOOBu^n + MH}$$

FIG. 22. Suggested mechanism for the ester exchange reaction catalyzed by palladium complexes.

to the hydropalladium intermediate followed by elimination of methanol. Solvolysis of the acyl by the *n*-butanol used as solvent leads to formation of the butyl ester (Fig. 22).

VII
SPECTROSCOPIC STUDIES

This section is not intended as a comprehensive survey of the spectroscopic comparisons that have been made on hydro complexes of nickel, palladium, and platinum. A number of the more important studies are reviewed, however, and if this section is taken with Table XIII the reader should be able to develop a good framework of reference for interpretation of his results.

A. Infrared and ^1H Nuclear Magnetic Resonance Spectroscopy

The diagnostic value of IR and NMR spectroscopy for structural studies of hydro-transition metal complexes has led to wide use of these techniques for hydro complexes of the nickel triad of elements. The most comprehensive studies have been made on platinum complexes because of their ease of isolation and their good solution stability. Early IR studies showed that ν_{PtH} was very sensitive to the nature of the other ligands about the metal and that a good correlation existed between this stretching frequency and the trans effect of the ligand opposite to the hydro group (*71*). Before a valid comparison of a series of stretching frequencies can be made, it is necessary to find a common solvent since the position of ν_{PtH} has been found to be considerably solvent-dependent. The difference in the position of this band between, say, chloroform and hexane as solvent can vary by as much as 30 cm^{-1}, the higher-energy band being found for the solvents of high polarity (*3*). Chatt and Shaw (*78*) have shown that both the IR stretching frequency and the chemical

TABLE VI
SPECTRAL PARAMETERS FOR *trans*-$PtHX(PEt_3)_2$

X	ν_{PtH} (cm^{-1})	τ_{Pt-H}	J_{P-H} (Hz)
NO_3	2242	33.8	15.5
Cl	2183	26.9	14
Br	2178	25.6	13.8
I	2156	22.7	13
CN	2041	17.8	15.5

shift of the hydro ligand coordinated to platinum (*78*, *41*) decreased in the complexes *trans*-$PtHX(PEt_3)_2$ as the trans influence of ligand X increased (Table VI). The high trans influence of the hydro ligand is reflected in the low value of ν_{PtX} for complexes $PtHX(PEt_3)_2$ (X = Cl, (Br) (*4*). Further work has confirmed these spectral results. However, the values obtained for J_{Pt-H} in this study were found to have little correlation with either τ_{Pt-H} or J_{P-H} (*243*).

Green (*19*) has made a detailed study of the high field NMR spectra of a large number of acetato and benzoato complexes of general formula $PtHY(PEt_3)_2$. The results showed that there was a linear correlation between τ_{Pt-H}, J_{Pt-H}, ν_{PtH}, and the pK_a of the parent carboxylic acid. The variation of these parameters is small (Table VII). The chemical shift difference was considered to be a consequence of the variation in the metal–hydrogen distance, and the change in the Pt—H coupling

TABLE VII
SPECTRAL PARAMETERS FOR *trans*-$PtHY(PEt_3)_2$

Y	pK_a	τ_{Pt-H}	J_{Pt-H} (Hz)	ν_{PtH} (cm^{-1})	J_{P-H} (Hz)
p-$MeC_6H_4CO_2$	4.373	31.897	1176.2	2226	15.55
$PhCO_2$	4.212	31.918	1179.1	—	15.65
p-$ClC_6H_4CO_2$	3.977	32.036	1189.1	2231	15.6
m-$ClC_6H_4CO_2$	3.830	32.095	1195.5	2232	15.5
p-$NO_2C_6H_4CO_2$	3.425	32.200	1209.6	2235	15.8
$PhOCH_2CO_2$	3.171	32.342	1202.9	2234	15.55
o-$BrC_6H_4CO_2$	2.824	32.257	1205.2	2234	15.9
o-$NO_2C_6H_4CO_2$	2.173	32.515	1226.2	2241	15.55
$CHCl_2CO_2$	1.25	32.765	1255.8	2245	15.3
CF_3CO_2	0.23	33.013	1285.5	2258	15.35

constant to be a consequence of the variation in the *s*-character of the Pt—H bond. A similar study has been made on a series of four-substituted benzenethiol complexes of platinum(II). The substituents used were NO_2, Br, Cl, F, H, Me, OMe, NH_2, and a linear correlation was found by plotting ν_{PtH} or J_{Pt-H} against the Hammett substituent parameter σ_p (*197*) (Table VIII). There is a fair correlation between

TABLE VIII
SPECTRAL PARAMETERS AND HAMMETT SUBSTITUENT PARAMETERS FOR FOUR-SUBSTITUTED BENZENETHIOL COMPLEXES $PtHX(PPh_3)_2$

X	σ_p	τ_{Pt-H}	J_{Pt-H} (Hz)	ν_{PtH} (cm^{-1})
NO_2	0.78	19.86	1006	2130
Br	0.22	19.93	981	2120
Cl	0.22	19.96	980	2120
F	0.06	19.99	973	2118
H	0	19.87	969	2117
Me	−0.17	19.92	962	2113
OMe	−0.28	19.99	959	2117
NH_2	−0.67	19.98	954	2112

J_{Pt-H} and ν_{PtH}. This linear correlation has also been discussed for the complexes $PtHX(PPh_3)_2$ (X = CN, $SCOCH_3$, $(CH_2)_2(CO)_2N$, Cl, Br, NCS, $OCOCF_3$) and $PtHX(PPh_2Me)_2$ (X = CN, SCN, I, Br, Cl, NCS, NO_3) (*253*, *100*).

Several series of cationic hydroplatinum(II) complexes $[PtH(A)L_2]X$ have been prepared, and in each case detailed comparison has been made of the variation of spectral parameters between members of the series. Compounds with L = PEt_3 (*87*, *170*), $AsEt_3$ (*88*), and PPh_3 (*156*) have been studied, where A is a wide range of ligands (Tables IX, X, and XI). In general the ligand A is trans to the hydro group, although in one study the comparison was made when the ligand A was in the cis position (*88*). The Pt—H coupling constants are strongly dependent on the trans ligand and decrease with increasing trans influence of the ligand. For complexes $[PtHX(PPh_3)_2]ClO_4$ (*156*) the value of ν_{PtH} decreases with increasing trans influence of X. The anomalous position for the carbonyl compound likely is due to coupling between the closely spaced frequencies for ν_{PtH} and ν_{CO}. Church and Mays (*87*) concluded that changes in J_{Pt-H} could be attributed to changes in $(\alpha_{Pt})^2$, $|\psi Pt(6s)(0)|^2$, and the covalency of the Pt—H bond. An increase

TABLE IX
SPECTRAL PARAMETERS FOR $[PtHA(PEt_3)_2]^+$

A	τ	J_{Pt-H} (Hz)
Pyridine	29.32	1106
CO	14.76	967
$CNCMe_3$	17.13	895
PPh_3	16.51	890
$P(OPh)_3$	15.21	872
$P(OMe)_3$	14.54	846
PEt_3	16.24	790

TABLE X
SPECTRAL PARAMETERS FOR $[PtHA(AsEt_3)_2]^+$

A	τ	J_{Pt-H} (Hz)
Trans		
$P(OMe)_3$	15.28	699
$P(OPh)_3$	15.92	716
$CNCMe_3$	18.27	721
PPh_3	17.34	739
CO	15.65	768
$AsEt_3$	19.73	846
Cis		
PPh_3	18.51	881
$P(OPh)_3$	18.13	886
$P(OMe)_3$	17.81	936
PEt_3	19.05	945

TABLE XI
SPECTRAL PARAMETERS FOR
$[PtHX(PPh_3)_2]^+$

X	ν_{PtH} (cm^{-1})[a]
Pyridine	2203
NH_3	2202, 2186
$MeNH_2$	2195
CO	2188, 2087
PPh_3	2124
C_2H_4	2089
C_3H_6	2082

[a] Measured in hexachlorobutadiene.

in the σ-donor power of ligand A trans to H decreases both $(\alpha_{Pt})^2$ and the covalency of the Pt—H bond. It was pointed out (*88*) that the *s*-character of the Pt—H bond would be little altered by a change in the cis ligand, and, indeed, the effect on J_{Pt-H} of changing a cis ligand was found to be the reverse of the effect when the trans ligand was changed (Table X).

1. *Second-Order Effects in the NMR Spectra*

When the high field ^1H NMR spectrum of complexes $[PtH(PPh_3)_3]^+$ was measured using a 100-MHz spectrometer under high resolution, the spectrum could not be analyzed using simple first-order techniques. For a square-planar cation, the center-band spectrum for the hydro ligand should appear as a pair of triplets arising from a large coupling of the hydrogen with the phosphorus of a *trans*-triphenylphosphine and from a small coupling with the phosphorus atoms in the two equivalent cis positions. Because of its high solubility, a detailed study was made of the spectrum of $[PtH(PPh_3)_3](CF_3CO_2)_2H$ (*279*). The center-line band consisted of a pair of double doublets that could only arise, apparently, if the three triphenylphosphine ligands were in a non-equivalent environment. The ^{195}Pt satellite portion of the spectrum showed the anticipated double-triplet multiplicity. The other significant differences between the two portions of the spectrum were that the width of the center groups of four lines was 30 Hz compared to 26 Hz for the sattelites and that the apparent trans P—H coupling constant was 164 Hz for the center-line band and 160 Hz for the satellites. The likely explanation of these facts is that the cation $[PtH(PPh_3)_3]^+$ is square planar and that the observed splitting pattern is due to second-order effects. Computer-simulated spectra were generated which showed a double doublet for the centerline portion of the spectrum of the X portion of an AB_2X spectrum. It was found that inclusion of a coupling between X and the ^{195}Pt nucleus in the Laocoon III program caused the simulated spectrum to be observed as a double triplet.

A further study of these second-order effects has been made by Dingle and Dixon (*133*) who had earlier found that the ^{19}F NMR spectrum of $[PtF(PPh_3)_3]^+$ could not be interpreted using a simple first-order analysis procedure (*134*). Their conclusions on the hydro complexes (*133*) were essentially the same as those proposed earlier (*279*); however, they extended the study to cations $[PtH(PEt_3)_3]^+$ and *trans*-$[PtH(PPh_3)(PEt_3)_2]^+$ in addition to $[PtH(PPh_3)_3]^+$. These complex cations also showed differences between the center-band and ^{195}Pt satellite portions of the spectra. It was concluded that similar center-band and satellite spectra are expected to be observed only if $\delta_{AB}| \gg \frac{1}{2}|J_{AM} - J_{BM}|$. In addition, it may not always be possible to neglect the effect on the X spectrum of the remote, magnetically active nuclei that are coupled to A or B. This effect could arise in the case of tertiary phosphine complexes where it may not be possible to neglect the interactions of remote hydrogens in the alkyl or aryl groups.

2. *Ambidentate Ligands*

The ^1H NMR spectra have been studied of series of complexes $PtHXL_2$, where X is an ambidentate ligand such as NCO, NCS, CN, and NCSe, and L is P(*p*-tol)$_3$, PPh_3, PPh_2Et, PBu_3, PEt_3, and $AsEt_3$ (*7*, *9*). Linkage isomerism was only exhibited by the thiocyanate complexes. An accurate isomer ratio between these complexes was determined from the intensities of the hydro resonances. The value obtained showed a pronounced solvent dependence ranging from the minimum value of 1.3 for the —NCS to —SCN isomer ratio in chloroform solution to a maximum value of 2.4 in benzene solution. In chloroform, the —NCS to —SCN ratio was always greater than unity and increased in the order: L = $PEt_3 \sim PBu_3 < AsEt_3 < PPh_2Et < PPh_3 <$ P(*p*-tol)$_3$. Line broadening in complexes $PtHXL_2$ (X = —NCO, —NCS) has been ascribed to exchange rather than to the proposed quadrupole broadening of the hydro ligand with the ^{14}N nucleus (*8*). When free tertiary phosphine was added to solutions of hydro complexes of platinum(II) containing that phosphine, there was a considerable decrease in line width of the upfield resonance. Furthermore, these authors considered that, whereas the spectrum of *trans*-$PtHCN(PPh_3)_2$ at 35° exhibited three broad resonances with no P—H coupling, at $-60°$ a triplet was observed with $J_{P-H} = 13.7$ Hz and $\omega = 2$ Hz. This interpretation has been challenged by Pidcock (*244*) who has reaffirmed that the interaction of the hydro group with the quadrupolar ^{14}N nucleus is the cause of the broadening of lines in complexes containing Pt—N linkages. Moreover, broad lines were also found for *trans*-$PtHX(AsEt_3)_2$ (X = −NCS, —NCO), where the highly quadrupolar ^{75}As nuclei have no spin-spin interaction with the hydro ligand. Exchange of $AsEt_3$ would not, therefore, be expected to broaden the hydro resonances. Pidcock has discounted the connection between broadening of the high field resonance in the absence of phosphine and that caused by addition of phosphine.

B. ^{31}P, ^{195}Pt and Nuclear Magnetic Resonance Involving Other Nuclei

The ^{31}P NMR spectra of a series of complexes *trans*-$PtHX(PEt_3)_2$ has been obtained for a wide range of anions X (*128*, *261*) (Table XII). The order of ^{195}Pt shifts of the hydroplatinum complexes studied was $RCO_2 < NO_3 < NO_2 < Cl < SCN < Br < CN < I$, which resembles the nephelauxetic series for these ligands with the exception of cyanide. This implies that changes in covalency of the metal–ligand bond cause variation in platinum chemical shift. Socrates (*261*) has stated that the greater the electronegativity of the ligand X, the greater the observed shift to low field in the ^{31}P NMR spectra. A comparative study of ^{195}Pt NMR spectra by INDOR showed that chemical shifts ranged over 1600 ppm and that *trans*-$PtHCl(PEt_3)_2$ had one of the highest, resonating 1500 ppm upfield of *trans*-$PtCl_2(SEt_3)_2$ (*225*).

For *trans*-$PtHCN(PEt_3)_2$, it was observed that $J_{^{13}C-H} = 41.9$ Hz and that the chemical shift of the hydro group trans to ^{13}CN was about 0.006 ppm downfield from that which was trans to ^{12}CN (*297*).

A recent report on the chlorine ($2p_{3/2}$) binding energy value of 197.0 eV in *trans*-$PtHCl(PPh_3)_2$, and the ^{35}Cl NQR shift of 14.43 MHz in *trans*-$PtHCl(PMe_2Ph)_2$, indicates that each value was very low and could be related to the trans influence of the opposite ligand (*91*).

A collection of spectral data is shown in Table XIII.

TABLE XII

^{31}P NMR SPECTRA FOR *trans*-$PtHX(PEt_3)_2$

X	δ_{Pt} (ppm)	J_{Pt-H} (Hz)	J_{P-H} (Hz)	δ_P (ppm)	J_{Pt-P} (Hz)
NO_3	0	1330	15.2	0	2780
NO_2	60.3	1070	16.1	4.7	2800
Cl	137.4	1270	14.3	3.2	2720
—SCN	248.9	1220	14.1	6.3	2650
Br	249.3	1340	14.0	5.7	2660
CN	408.3	790	15.5	6.6	2560
I	442.9	1390	13.1	6.8	2660
—NCS	—	—	14.5	4.0	—

TABLE XIII
HYDRO COMPLEXES OF NICKEL, PALLADIUM, AND PLATINUM

Compound	ν_{MH} (cm^{-1})	^1H NMR (τ)	J_{P-H} (Hz)	J_{Pt-H} (Hz)	References
Nickel complexes					
$NiH(\pi\text{-allyl})(PF_3)_2$	—	—	—	—	*42*
$NiH(\pi\text{-allyl})(PPh_3)_2$	—	—	—	—	*42*
$NiH(C{\equiv}CPh)(PPh_3)_2$	—	—	—	—	*85*
$NiD(C{\equiv}CPh)(PPh_3)_2$	—	—	—	—	*85*
$NiH(C{\equiv}CPh)(PCy_3)_2$	—	—	—	—	*85*
$NiH[C{\equiv}C{-}C(Me){=}CH_2](PPh_3)_2$	—	—	—	—	*85*
$NiH(C{\equiv}C{-}CH{=}CH_2)(PPh_3)_2$	—	—	—	—	*85*
$NiH(BH_4)(PCy_3)_2$	1920	30.1	72	—	*166, 169*
$NiH(BH_4)(PPr_3^i)_2$	1986	29.7	70.5	—	*166, 230*
$NiH(OCOCH_3)_2(PCy_3)_2$	1920	—	—	—	*193*
$NiHCl(PCy_3)_2$	1916	34.6	73.5	—	*193, 168, 167*
$NiHBr(PCy_3)_2$	1917	33.0	77.9	—	*168*
$NiHBr(PPr_3^i)_2$	1979	32.7	73.5	—	*168*
$NiHCN(PCy_3)_2$	1870	25.3	64.5	—	*168*
$NiH(C_5H_5)(PCy_3)_2$	1920	—	—	—	*193*
$NiHCl(PBu_3^n)(PCy_3)$	—	34.0	75, 81	—	*166*
$NiHCl(PCy_3)(PEt_3)$	—	34.0	75, 81	—	*166*
$NiHCl(PCy_3)(PPr_3^n)$	—	34.0	75, 81	—	*166*
$NiHCl(PPr_3^i)_2$	1937	34.3	77.8	—	*166, 168*
$NiHBr(PBu_3^n)(PCy_3)$	—	32.4	75, 81	—	*166*
$NiHCl(PBu_3^n)(PPr_3^i)$	—	33.8	75, 81	—	*166*
$NiHI(PPr_3^i)_2$	1990	29.8	71.5	—	*168*
$NiHI(PCy_3)_2$	1976	30.3	72.4	—	*168*
$NiHCH_3(PCy_3)_2$	1800	—	—	—	*193*
$NiHN_2(PBu_3^n)_2$	1911	—	—	—	*268*
$NiHN_2(PEt_3)_2$	1914	—	—	—	*268*

Continued

TABLE XIII—*Continued*

Compound	ν_{MH} (cm^{-1})	^1H NMR (τ)	J_{P-H} (Hz)	J_{Pt-H} (Hz)	References
$NiHSCN(PCy_3)_2$	1928	33.5	72.9	—	*168*
$NiHOC_6H_5(PCy_3)_2$	1940	—	—	—	*193*
$NiHC_6H_5(PCy_3)_2$	1805	—	—	—	*193*
$NiHNC_4H_4(PCy_3)_2$	1910	—	—	—	*193*
$[NiH(diphos)_2]AlCl_4$	1950	23.0	6	—	*258*
$[NiH(diphos)_2]HCl_2$	1934	23.02	—	—	*258*
$[NiH(diphos)_2]BF_4$	1950	23.01	—	—	*258*
$[NiH(diphos)_2]HSO_4$	—	22.9	5.5	—	*285*
$[NiH\{P(O\text{-}p\text{-}C_6H_4OMe)_3\}_4]HSO_4$	—	23.1	35 ± 2	—	*285*
$[NiH\{P(OCH_2CH_2Cl)_3\}_4]HSO_4$	—	23.9	28.0	—	*285*
$[NiH\{P(OEt)_3\}_4]HSO_4$	—	24.3	26.5	—	*285*
$[NiH(PMe_3)_4]HSO_4$	—	27.2	4.5	—	*285*
$[NiH\{PPh(OEt)_2\}_4]HSO_4$	—	23.4	22.5	—	*285*
$[NiH\{P(OMe_3)\}_4]HSO_4$	—	24.3	26.5	—	*285*
$[NiH\{P(O\text{-}p\text{-tolyl})_3\}_4]HSO_4$	—	23.2	33 ± 2	—	*285*
$[NiH\{P(OEt)_3\}_4]CF_3CO_2$	—	24.5	26	—	*137*
$[NiH(diphos)_2]PF_6$	1930	23.0	5.0	—	*165*
$NiBr_2(PEt_3)_2 + Me_3GeH$	1937	—	—	—	*52*
Palladium complexes					
$PdHBH_4(PCy_3)_2$	2002	—	—	—	*166, 230*
$PdHBH_4(PPr_3{}^i)_2$	2013	23.2	9	—	*166, 230*
$PdHBr(PBu_3{}^n)_2$	—	22.3	—	—	*166*
$PdHBr(PBu_3{}^n)(PCy_3)$	—	22.8	~3	—	*166*
$PdHBr(PCy_3)_2$	1991	23.2	6.0	—	*166, 230*
$PdHBr(PEt_3)_2$	2029	22.5	—	—	*52*
$PdHCl(PBu_3{}^n)_2$	—	23.5	—	—	*166*
$PdHCl(PBu_3{}^n)(PCy_3)$	—	24.0	~3	—	*166*

$PdHCl(PBu_3^n)(PPr_3^i)$	—	24.0	~3	—	*166*
$PdHCl(PCy_3)_2$	2002	24.4 (27.4)	4.1 (4.5)	—	*166, 212, 230, 204*
$PdHCl(PCy_3)(PEt_3)$	—	24.0	~3	—	*166*
$PdHCl(PEt_3)_2$	2035	23.4	—	—	*50, 52, 51*
$PdHCl(PPh_3)_2$	2055	24.8	—	—	*204, 187*
$PdHCl(PPr_3^i)_2$	2010	24.5	4.6	—	*166, 230*
$PdHI(PCy_3)_2$	1966	21.3	8.2	—	*166*
$PdHI(PEt_3)_2$	2004	—	—	—	*52*
$PdHNCS(PCy_3)_2$	2022	25.4	—	—	*166*
$[PdH(diphos)(PCy_3)]PF_6$	1895	15.52	196, 5.6, 14.8	—	*165*
$[PdH(diphos)(PPr_3)]PF_6$	1959	15.31	19.8, 196	—	*165*
$AsPh_4[PdH(CO)Cl_2]$	1960	—	—	—	*198*
$pyH[PdH(CO)Cl_2]$	2010	—	—	—	*198*
$qnH[PdH(CO)Cl_2]$	2000	—	—	—	*198*
$PdHGePh_3(PEt_3)_2$	1890	—	—	—	*51*
$PdH_2(PCy_3)_2$	1740	—	—	—	*205*
Platinum complexes					
$PtHCH_2ClCO_2(PEt_3)_2$	2237	32.483	15.8	1216.7	*19*
$PtHCHCl_2CO_2(PEt_3)_2$	2245	32.765	15.3	1255.8	*19*
$PtHPhOCH_2CO_2(PEt_3)_2$	2234	32.342	15.55	1202.9	*19*
$PtHCF_3CO_2(PEt_3)_2$	2258	33.013	15.35	1285.5	*19*
$PtHPhCO_2(PEt_3)_2$	—	31.918	15.65	1179.1	*19*
PtH *m*-$BrC_6H_4CO_2(PEt_3)_2$	—	32.105	15.95	1196.2	*19*
PtH *o*-$BrC_6H_4CO_2(PEt_3)_2$	2234	32.257	15.9	1205.2	*19*
PtH *p*-$BrC_6H_4CO_2(PEt_3)_2$	2233	32.043	15.8	1191.3	*19*
PtH *m*-$ClC_6H_4CO_2(PEt_3)_2$	2232	32.095	15.5	1195.5	*19, 128*
PtH *o*-$ClC_6H_4CO_2(PEt_3)_2$	2234	32.226	15.65	1204.0	*19, 128*
PtH *p*-$ClC_6H_4CO_2(PEt_3)_2$	2231	32.036	15.6	1189.1	*19, 128*

Continued

TABLE XIII—*Continued*

Compound	ν_{MH} (cm^{-1})	^{1}H NMR (τ)	$J_{P—H}$ (Hz)	$J_{Pt—H}$ (Hz)	References
PtH *m*-$Me_2NC_6H_4CO_2(PEt_3)_2$	2224	31.857	15.7	1172.3	*19, 128*
PtH *p*-$Me_2NC_6H_4CO_2(PEt_3)_2$	2222	31.745	15.75	1160.9	*19, 128*
PtH *p*-$NCC_6H_4CO_2(PEt_3)_2$	—	32.170	15.65	1203.9	*19*
PtH-2,5 $(NO_2)_2C_6H_3CO_2(PEt_3)_2$	2247	32.715	15.35	1255.5	*19, 128*
PtH-3,5 $(NO_2)_2C_6H_3CO_2(PEt_3)_2$	2243	32.446	15.6	1236.8	*19, 128*
PtH *m*-$FC_6H_4CO_2(PEt_3)_2$	2232	32.077	15.55	1193.1	*19*
PtH *m*-$IC_6H_4CO_2(PEt_3)_2$	2232	32.123	15.7	1197.1	*19*
PtH *p*-$IC_6H_4CO_2(PEt_3)_2$	2231	32.061	15.45	1193.2	*19*
PtH *p*-$MeOC_6H_4CO_2(PEt_3)_2$	2226	31.869	15.7	1172.0	*19*
PtH *m*-$MeC_6H_4CO_2(PEt_3)_2$	2225	31.899	15.4	1176.5	*19, 128*
PtH *p*-$MeC_6H_4CO_2(PEt_3)_2$	2226	31.897	15.55	1176.2	*19, 128*
PtH *m*-$O_2NC_6H_4CO_2(PEt_3)_2$	2235	32.226	15.75	1210.6	*19, 128*
PtH *o*-$O_2NC_6H_4CO_2(PEt_3)_2$	2241	32.515	15.55	1226.2	*19, 128*
PtH *p*-$O_2NC_6H_4CO_2(PEt_3)_2$	2235	32.200	15.8	1209.6	*19, 128*
PtH-2,4,6 $(NO_2)_3C_6H_2CO_2(PEt_3)_2$	2263	33.671	15.7	1298.3	*19*
$PtH(C{\equiv}CPh)(PEt_3)_2$	2020	16.12	18.0	—	*160*
$PtH_2(C{\equiv}C—C_6H_{10}OH)_2(PPh_3)_2$	2120	22.88	—	—	*232*
$PtHBr(AsEt_3)_2$	2167	27.75	—	1185	*78, 243*
$PtHBr(PEt_3)_2$	2178	25.6	15	1346	*78, 286, 70, 243, 261*
$PtH_2Br_2(PEt_3)_2$	2243, 2252	26.88	−6.3	1204	*14*
$PtHBr(PMePh_2)_2$	2208	24.90	13.0	1302	*100*
$PtHBr(PPh_3)_2$	2215, 2280	—	—	—	*23*
$PtHCl(AsEt_3)_2$	2174	29.3	—	1117	*78, 243*
$PtHCl(PEt_3)_2$	2183 (2227)	26.9	14.5	1276	*78, 286, 70, 261, 241, 30*
$PtH_2Cl_2(PEt_3)_2$	2251, 2263	28.17	−6.5	1176	*14, 78*
$Pt_2H_2(PPh_2)_2(PEt_3)_2$	2005	—	—	—	*69*

$PtHCl(PEtPh_2)_2$	2206	—	—	—	*78*
$PtHCl(PEt_2Ph)_2$	2199	—	—	—	*71, 29*
$PtHCl(PMe_3)_2$	2182	—	—	—	*78*
$PtHCl(PMePh_2)_2$	2220	26.40	14.5	1260	*100*
$PtHCl(PMe_2Ph)_2$	2205	—	—	—	*72*
$PtHCl(AsPh_3)_2$	2170	—	—	—	*23, 61, 60*
$PtHCl(PPh_3)_2$	2220	26.1	13.3	1194	*78, 21, 112, 156, 61*
$PtHCl(PPr_3{}^n)_2$	2183	—	—	—	*78*
PtHCl(diphos)	2002	—	—	—	*107*
$PtHCl(SnMe_3)_2$(diphos)	1960	—	—	—	*109, 108*
$PtHOCN(PEt_3)_2$	2234	27.0	14.5	—	*78, 243*
$PtHOCN(PPh_3)_2$	2260	—	—	—	*23*
$PtHCN(PEt_3)_2$	2072	17.8	15.5 ($J_{H-C} = 41.9$)	778	*78, 243, 261, 297*
$PtHCN(TCNE)(PEt_3)_2$	2198	—	—	—	*292*
$PtHCN(PMePh_2)_2$	2059	17.37	14.0	1204	*100*
$PtHCN(PPh_3)_2$	2075	—	—	—	*23, 61, 60*
$PtHCl(CNMeCH_2CH_2CH_2NMe)(PEt_3)$	2210	27.4	18.0	1420	*63*
$PtH(C_6H_9)(PPh_3)_2$	1925	14.64	19.5	608	*37*
$PtH(OPPh_2)(HOPPh_2)_2$	1985	—	—	—	*202*
$PtH(OPPh_2)(HOPPh_2)(Pr^iOPPh_2)$	2000	—	—	—	*202*
$PtH(OPPh_2)(HOPPh_2)(n\text{-}Bu^nOPPh_2)$	2000	—	—	—	*202*
$PtHI_2(GeH_2I)(PEt_3)_2$	—	24.8	5.0	1111.2	*145*
$PtHGeMe_2(PEt_3)_2$	1923, 2037	—	—	—	*53*
$PtHGePh_3(PEt_3)_2$	2041	—	—	—	*53, 123, 124, 160*
$PtHGeMe_3(PPh_3)_2$	2055	—	—	—	*53*
$PtHGePh_3$(diphos)	1998	—	—	—	*53*
$PtHSeH(PPh_3)_2$	2140	14.2, 18.8	—	44.6, 993	*227*

Continued

TABLE XIII—*Continued*

Compound	ν_{MH} (cm^{-1})	^{1}H NMR (τ)	J_{P-H} (Hz)	J_{Pt-H} (Hz)	References
$PtHSH(PPh_3)_2$	2116	11.4, 19.2	10, 11	43.8, 932	*227*
$PtH(GeHCl_2)_2(GeH_2Cl)(PEt_3)_2$	—	21.16	7.5	752	*39*
$PtH(GeHCl_2)(GeH_2Cl)_2(PEt_3)_2$	—	21.55, 20.15	7.7, 7.0	751, 734	*39*
$PtH(GeH_2Cl)_3(PEt_3)_2$	—	20.70	8.0	733	*39*
$PtH_2(PEt_3)_3$	1766	23.37	18	635	*157*
$PtHGePh_3(PPr_3^n)_2$	1957	—	—	—	*124*
$PtH_2(PPr_3^n)_2$	1731	—	—	—	*124*
$PtH_2(PCy_3)_2$	1910	—	—	—	*205*
$PtH_2(PCy_2Pr^i)_2$	1920	16.9	20	610	*205*
$PtH[N(CO)_2(CH_2)_2](PPh_3)_2$	2190	24.1	—	—	*251*
$PtH[N(CO)_2C_6H_4](PPh_3)_2$	2200	24.0	12.3	—	*251*
$PtH[NCOSO_2C_6H_4](PPh_3)_2$	2230	25.7	12.8	—	*251*
$PtH[NCOSO_2C_6H_4](AsPh_3)_2$	2185	28.0	—	—	*251*
$PtH[N_2H(CO)_3](PPh_3)_2$	2210	24.6	—	—	*251*
$PtHN_3(PEt_3)_2$	2199	15.1	—	—	*261*
$PtH[C(CF_3)_2NMe](COOCF_3)(PPh_3)_2$	—	—	—	—	*266*
$PtH[C(NHPh)NMe](PEt_3)_2$	1953	—	—	—	*58*
$PtH[C(N\textit{p}\text{-}NO_2C_6H_4)NMe](PEt_3)_2$	1972	—	—	—	*58*
$PtHI(AsEt_3)_2$	2139	24.6	—	1220	*78, 243*
$PtHI(PEt_3)_2$	2156	22.7	13	1369	*78, 286, 70, 243, 261*
$PtHI(PEt_2Ph)_2$	2179	—	—	—	*71*
$PtHI(PEtPh_2)_2$	2189	—	—	—	*78*
$PtHI_2(SiH_2Cl)(PEt_3)_2$	—	23.4	5.5	1185	*145*
$PtHI_2(SiH_2I)(PEt_3)_2$	—	24.6	6.0	1160	*145, 38*
$PtHI(PMePh_2)_2$	2180	22.09	12.5	1332	*100*
$PtHI(PPh_3)_2$	2186	—	—	—	*78*
$PtHNCO(PEt_3)_2$	—	27.7	—	1080	*243, 261*

$PtHNCS(PEt_3)_2$	2195	27.6	14	1086	*78, 243, 261*
$PtHNCS(AsEt_3)_2$	—	29.9	—	—	*243*
$PtHNCS(PMePh_2)_2$	2177	22.18	14.0	1204	*100*
$PtHNO_3(PMePh_2)_2$	2275	32.90	15.0	1316	*100*
$PtHNO_3(PEt_3)_2$	2242	33.8	15.5	1322	*78, 243, 261*
$PtHNO_2(PEt_3)_2$	2150	29.7	16.5	1003	*78, 243, 261*
$PtHNO_2(PPh_3)_2$	2180	—	—	—	*23*
$PtHPbPh_3(PEt_3)_2$	1940	—	—	—	*130*
$PtHClO_4(PPh_3)_2$	2312	—	—	—	*156*
$PtHSiCl_3(PPh_3)_2$	2076	—	—	—	*74*
$PtHSiCl_2Me(PPh_3)_2$	2109	—	—	—	*151, 301*
$PtHSiMe_3$(diphos)	2000	—	—	—	*109, 108*
$PtHSiPh_3(PEt_3)_2$	2056	—	—	—	*25*
$PtH[Si(C_6H_4F\text{-}m)_3](PPh_3)_2$	2091	—	—	—	*73, 74*
$PtH[Si(C_6H_4F\text{-}p)_3](PPh_3)_2$	2096	—	—	—	*73, 74*
$PtH[Si(C_6H_4CF_3\text{-}m)_3](PPh_3)_2$	2048	19.38	139	—	*73, 74*
$PtH[Si(C_6H_4CF_3\text{-}p)_3](PPh_3)_2$	2105	—	—	—	*73, 74*
$PtHI_2(SiH_2C{\equiv}CH)(PEt_3)_2$	—	23.46	—	—	*13*
$PtHI_2(SiH_2C{\equiv}CCF_3)(PEt_3)_2$	—	23.26	—	—	*13*
$PtH(SiEt_3)(PPh_3)_2$	2105	—	—	—	*143*
$PtH[Si(OEt)_3](PPh_3)_2$	2090	—	—	—	*143*
$PtH(SiMe_2OSiMe_2H)(PPh_3)_2$	2090	—	—	—	*143*
$PtH[SiMe(OSiMe_3)_2](PPh_3)_2$	2080	—	—	—	*143*
$PtH(SiPh_3)(PPh_3)_2$	2090	—	—	—	*143*
$PtH(SiPh_2Me)(PPh_3)_2$	2095	—	—	—	*143*
$PtH(SiPh_2H)(PPh_3)_2$	2075	—	—	—	*143*
$PtHSPh(PPh_3)_2$	2130	19.93	12	960	*227, 197*
$PtHSC_6H_4Br(PPh_3)_2$	2132	19.93	13.4	981	*197*
$PtHSC_6H_4Cl(PPh_3)_2$	2133	19.96	13.4	980	*197*
$PtHSC_6H_4F(PPh_3)_2$	2132	19.99	13.6	973	*197*
$PtHSC_6H_4NH_2(PPh_3)_2$	2126	19.98	13.8	954	*197*

Continued

TABLE XIII—*Continued*

Compound	ν_{MH} (cm^{-1})	^1H NMR (τ)	J_{P-H} (Hz)	J_{Pt-H} (Hz)	References
$PtHSC_6H_4NO_2(PPh_3)_2$	2143	19.86	12.9	1006	*197*
$PtHSC_6H_4OMe(PPh_3)_2$	2122	19.99	13.8	959	*197*
$PtHSC_6H_4Me(PPh_3)_2$	2128	19.92	13.8	962	*197*
$PtHSCN(AsEt_3)_2$	2108	24.75	—	1046	*78, 243*
$PtHSCN(PEt_3)_2$	2112	22.95	14.5	1233	*71, 243, 261*
$PtHSCN(PMePh_2)_2$	2177	22.18	14.0	1204	*100*
$PtHSCN(PPh_3)_2$	2250	—	—	—	*23, 61, 60*
$PtH(B_9H_{10}S)(PEt_3)_2$	2214	—	—	—	*196*
$PtHSnCl_3(PPh_3)_2$	2100, 2056	—	—	—	*21, 25, 240*
$PtHSnCl_3(PEt_3)_2$	2105	19.2, 19.4	—	142	*213, 240*
$PtH(C_8H_{14})SnCl_3(PPh_3)_2$	2025	—	—	—	*275*
$PtHSnCl_3(C_8H_{16})(PPh_3)_2$	2220	—	—	—	*275*
$[PtHSnCl_3(PPh_3)_2]COD$	2100, 2200	—	—	—	*275*
$[PtHSnCl_3(PPh_3)_2]C_7H_8$	2220	—	—	—	*275*
$[PtH(C_6H_8)(PPh_3)_2]SnCl_3 \cdot CH_2Cl_2$	2030	—	—	—	*275*
$PtHSnMe_3(PPh_3)_2$	2042	—	—	—	*10*
$PtH(SnMe_3)_3$(diphos)	1960	15.25	—	—	*109, 108*
PtH(5-phenyltetrazole)$(PPh_3)_2$	2212, 2222	25.62, 26.13	13.2	964	*194*
PtH(5-bromotetrazole)$(PPh_3)_2$	2212, 2220	25.62, 26.43	12.5	~916	*194*
PtH(5-chlorotetrazole)$(PPh_3)_2$	2223, 2253	25.66, 26.38	10.2	—	*194*
$PtHCOOCF_3(PPh_3)_2$	2280	31.3	—	—	*253*
$PtHSCOCH_3$	2140	19.8	—	—	*253*
$PtH(N{=}C(CF_3)_2)(PPh_3)_2$	2146	22.8	16	748	*64*
$PtH(N{=}C(CF_3)_2)(PMe_2Ph)_2$	2115	24.3	17	—	*64*
Cationic platinum complexes					
$[PtH(CO)(AsEt_3)_2]ClO_4$	2149	15.65	—	768	*88*
$[PtH(AsEt_3)_3]ClO_4$	2099	19.73	—	846	*88*

$[PtH(Me_3CNC)(AsEt_3)_2]ClO_4$	—	18.27	—	721	*88*
cis-$[PtH(PEt_3)(AsEt_3)_2]ClO_4$	—	19.05	11.5	945	*88*
$[PtH(PPh_3)(AsEt_3)_2]ClO_4$	2069	17.34	168	739	*88*
cis-$[PtH(PPh_3)(AsEt_3)_2]ClO_4$	—	18.51	9.8	881	*88*
$[PtH\{P(OMe)_3\}(AsEt_3)_2]ClO_4$	2044	15.28	270	699	*88*
$[PtH\{P(OPh)_3\}(AsEt_3)_2]ClO_4$	2066	15.95	290	716	*88*
cis-$[PtH\{P(OMe)_3\}(AsEt_3)_2]ClO_4$	2089	17.81	2	936	*88*
cis-$[PtH\{P(OPh)_3\}(AsEt_3)_2]ClO_4$	2082	18.13	4.0	886	*88*
$[PtH(PEt_3)_3]ClO_4$	2090	16.24	15.0	790	*87, 170, 286, 157*
$[PtH(PEt_3)_3]BPh_4$	2126	15.9	15, 157	788	*92*
$[PtH(CO)(PEt_3)_2]ClO_4$	2167	14.76	13.5	967	*87, 286, 86*
$[PtH(Me_3CNC)(PEt_3)_2]ClO_4$	2104	17.13	14.4	895	*87, 286, 86*
$[PtH(CO)(PEt_3)_2]BPh_4$	2162	—	—	—	*92*
$[PtH(p\text{-}MeOC_6H_4NC)(PEt_3)_2]ClO_4$	2096	16.56	14.0	890	*87*
$[PtH(C_5H_5N)(PEt_3)_2]ClO_4$	2216	29.32	14.4	1106	*87, 286*
$[PtH(3\text{-}ClC_5H_4N)(PEt_3)_2]ClO_4$	2210	—	—	—	*286*
$[PtH(4\text{-}CNC_5H_4N)(PEt_3)_2]ClO_4$	2200	—	—	—	*286*
$[PtH(3\text{-}MeC_5H_4N)(PEt_3)_2]ClO_4$	2210	—	—	—	*286*
$[PtH(2,6\text{-}Me_2C_5H_3N)(PEt_3)_2]ClO_4$	2208	—	—	—	*286*
$[PtH(PCy_3)(PEt_3)_2]ClO_4$	2085	—	—	—	*286*
$[PtH(AsPh_3)(PEt_3)_2]ClO_4$	2136	—	—	—	*286*
$[PtH(PPh_3)(PEt_3)_2]ClO_4$	2100	16.51	14.4	890	*87, 286*
$[PtH(PPh_3)(PEt_3)_2]BF_4$	2072	—	—	—	*170*
$[PtH(PBu_3)(PEt_3)_2]Cl$	2090	—	—	—	*170*
$[PtH(SbPh_3)(PEt_3)_2]ClO_4$	2123	—	—	—	*286*
$[PtH\{P(OMe)_3\}(PEt_3)_2]ClO_4$	2067	14.54	15.2	846	*87*
$[PtH\{P(OPh)_3\}(PEt_3)_2]ClO_4$	2090	15.21	14.4	872	*87*
$[PtH(S{=}C(NH_2)_2)(PEt_3)_2]ClO_4$	2150	—	—	—	*286*
$[PtH(PMe_2Ph)_3]PF_6$	2079	17.10	20.0, 170	880	*100*
$[PtH\{C(NHp\text{-}NO_2C_6H_4)NMe_2\}(PEt_3)_2]ClO_4$	2060	18.3	15–17	—	*58*

Continued

TABLE XIII—*Continued*

Compound	ν_{MH} (cm^{-1})	^1H NMR (τ)	J_{P-H} (Hz)	J_{Pt-H} (Hz)	References
$[PtH\{C(NHPh)NHMe\}(PEt_3)_2]ClO_4$	2050	18.5	15–17	—	*58*
$[PtH\{C(NHPh)NMe_2\}(PEt_3)_2]ClO_4$	2050	—	—	—	*58*
$[PtH(\overline{CNMeCH_2CH_2N}Me)(PEt_3)_2]Cl$	2025	17.2	15.0	706	*63*
$[PtH(\overline{CNMeCH_2CH_2N}Me)(PEt_3)_2]BF_4$	2040	16.9	17.0	702	*63*
$[PtH(C_2H_4)(PEt_3)_2]BPh_4$	—	17.2	12.0	908	*129*
$[PtH(C_2H_4)(PMePh_2)_2]PF_6$	—	16.32	11.5	931	*99*
$[PtH(PMePh_2)_3]PF_6$	2082	15.39	18.0, 167	840	*100*
$[PtH(2\text{-}MeC_5H_4N)(PMePh_2)_2]PF_6$	2178	27.20	14.5	1080	*100*
$[PtH(2,4,6\text{-}Me_3C_6H_2N)(PMePh_2)_2]PF_6$	2188	26.53	14.0	1073	*100*
$[PtH(p\text{-}MeC_6H_4NC)(PMePh_2)_2]PF_6$	~2100	16.10	13.5	872	*100*
$[PtH(O{=}C(CH_3)_2)(PMePh_2)_2]PF_6$	2275	33.60	14.5	1458	*100*
$[PtH(PPh_3)_3]Cl$	2102	—	—	—	*61, 60*
$[PtH(PPh_3)_3]BF_4$	2112	—	—	—	*61, 60*
$[PtH(PPh_3)_3]BPh_4$	2122	—	—	—	*61, 92*
$[PtH(PPh_3)_3]NO_3$	2107	—	—	—	*61, 60*
$[PtH(PPh_3)_3]ClO_4$	2112	—	—	—	*61, 60, 156*
$[PtH(PPh_3)_3]HSO_4$	2112	—	—	—	*61*
$[PtH(PPh_3)_3]CH_3OSO_3$	2102	—	—	—	*61, 60*
$[PtH(PPh_3)_3]$picrate	2145	—	—	—	*279*
$[PtH(PPh_3)_3](CF_3COO)_2H$	2145	15.83	17, 160	774	*279*

$[PtH(PPh_3)_3](CF_2ClCOO)_2H$	2145	15.83	17, 160	774	*279*
$[PtH(PPh_3)_3](C_2F_5COO)_2H$	2145	15.83	17, 160	774	*279*
$[PtH(PPh_3)_3](CF_2COO)_2H$	2145	15.83	17, 160	774	*279*
$[PtH(PPh_3)_3](COO)_4H_3$	2145	15.83	17, 160	774	*279*
$[PtH(PPh_3)_3]F$	—	15.83	17, 160	774	*279*
$[PtH(PPh_3)_3](CF_3COCHCOCF_3)$	—	—	—	—	*202*
$[PtH(CO)(PPh_3)_2]ClO_4$	2087, 2188	—	—	—	*156*
$[PtH(C_2H_4)(PPh_3)_2]ClO_4$	2089	—	—	—	*156*
$[PtH(C_3H_6)(PPh_3)_2]ClO_4$	2082	—	—	—	*156*
$[PtH(NH_3)(PPh_3)_2]ClO_4$	2186, 2202	—	—	—	*156*
$[PtH(NH_2Me)(PPh_3)_2]ClO_4$	2195	—	—	—	*156*
$[PtH(NH_2Et)(PPh_3)_2]ClO_4$	2193	—	—	—	*156*
$[PtH(NHMe_2)(PPh_3)_2]ClO_4$	2194	—	—	—	*156*
$[PtH(C_5H_5N)(PPh_3)_2]ClO_4$	2203	—	—	—	*156*
$[PtH(SbPh_3)(PPh_3)_2]ClO_4$	2130	—	—	—	*156*
$[PtH(S{=}C(NH_2)_2)(PPh_3)_2]ClO_4$	2150	—	—	—	*156*
$[PtH(PEt_3)(diphos)]Cl$	2043	12.91	7.5, 19.4, 168.3	988	*161, 170, 186*
$[PtH(PEt_3)(diphos)]BPh_4$	2046	—	—	—	*161*
$[PtH(PEt_3)(diphos)]SnCl_3$	2050	—	—	—	*170*
$[PtH(C_6H_4PMePh)C_2F_3(PMePh_2)]Br$	2195	24.77	13.5	—	*229*
$[NEt_4][PtH(SnCl_3)(PEt_3)_2]$	2108	—	—	—	*214*
$[NMe_4][PtH(SnCl_4)_4]$	2072, 2052 sh	—	—	—	*214*

C. Mass Spectroscopy

The mass spectra of the compounds *trans*-$PtHX(PEt_3)_2$ (X = Cl, Br, CN, CNO) were found to be dependent on the group X. In all cases parent ions were observed, but, when X was CNO or Cl, no ion $[M—1]^+$ was found. When X was CN or Br, ions corresponding to $[M—H]^+$, $[M—X]^+$, and $[M—HX]^+$ were produced before fragmentation of the phosphine occurred (*192*, *211*).

ACKNOWLEDGMENTS

The author thanks the Cities Service Oil Company, Cranbury, New Jersey, the Research Corporation, and the Graduate School at Washington State University for support of his research work in this area. Thanks are due Dr. J. H. Nelson and Dr. K. R. Dixon for communication of results prior to publication. I wish also to express my gratitude to Mr. T. B. Rauchfuss for many hours of valuable discussion.

REFERENCES

1. Abley, P., and McQuillin, F. J., *Disc. Faraday Soc.* **46**, 31 (1968).
2. Abley, P., and McQuillin, F. J., *J. Catal.* **24**, 536 (1972).
3. Adams, D. M., *Proc. Chem. Soc.* 431 (1961).
4. Adams, D. M., Chatt, J., Gerratt, J., and Westland, A. D., *J. Chem. Soc.* 734 (1964).
5. Adams, R. W., Batley, G. E., and Bailar, J. C., *Inorg. Nucl. Chem. Lett.* **4**, 455 (1968).
6. Adams, R. W., Batley, G. E., and Bailar, J. C., *J. Amer. Chem. Soc.* **90**, 6051 (1968).
7. Adlard, M. W., and Socrates, G., *J. Inorg. Nucl. Chem.* **34**, 2339 (1972).
8. Adlard, M. W., and Socrates, G., *J. Chem. Soc., Chem. Commun.* 17 (1972).
9. Adlard, M. W., and Socrates, G., *J. Chem. Soc., Dalton Trans.* 797 (1972).
10. Akhtar, M., and Clark, H. C., *J. Organometal. Chem.* **22**, 233 (1970).
11. Albanese, P., Benzoni, L., Corain, B., and Turo, A., German Patent, 2,009,470 (1971), *Chem. Abstr.* **74**, 124890 (1971).
12. Allen, A. D., and Baird, M. C., *Chem. Ind.* (*London*) 139 (1965).
13. Anderson, D. W. W., Ebsworth, E. A. V., MacDougal, J. K., and Rankin, D. W. H., *J. Inorg. Nucl. Chem.* **35**, 2259 (1973).
14. Anderson, D. W. W., Ebsworth, E. A. V., and Rankin, D. W. H., *J. Chem. Soc., Dalton Trans.* 854 (1973).
15. Ando, N., Maruya, K., Mizoroki, T., and Ozaki, A., *J. Catal.* **20**, 299 (1971).
16. Appleton, T. G., Clark, H. C., and Puddephatt, R. J., *Inorg. Chem.* **11**, 2074 (1972).
17. Atkins, K. E., Walker, W. E., and Manyik, R. M. *Tetrahedron Lett.* 3821 (1970).
18. Atkins, K. E., Walker, W. E., and Manyik, R. M., *J. Chem. Soc. D* 330 (1971).
19. Atkins, P. W., Green, J. C., and Green, M. L. H., *J. Chem. Soc. A* 2275 (1968).
20. Bachman, D. F., Stevens, E. D., Lane, T. A., and Yoke, J. T., *Inorg. Chem.* **11**, 109 (1972).
21. Bailar, J. C., and Itatani, H., *Inorg. Chem.* **4**, 1618 (1965).

22. Bailar, J. C., and Itatani, H., *J. Amer. Oil Chemists Soc.* **43**, 337 (1966).
23. Bailar, J. C., and Itatani, H, *J. Amer. Chem. Soc.* **89**, 1592 (1967).
24. Bailar, J. C., Itatani, H., and Jayim, H., *Kayaku No Ryoika* **22**, 337 (1968).
25. Baird, M. C., *J. Inorg. Nucl. Chem.* **29**, 367 (1967).
26. Baker, R., Halliday, D. E., and Smith, T. N., *J. Organometal Chem.* **35**, C61 (1972).
27. Barlex, D. M., and Kemmitt, R. D. W., *J. Chem. Soc., Dalton Trans.* 1436 (1972).
28. Barlex, D. M., Kemmitt, R. D. W., and Littlecott, G. W., *J. Chem. Soc. D* 613 (1969).
29. Basolo, F., Chatt, J., Gray, H. B., Pearson, R. G., and Shaw, B. L., *J. Chem. Soc.* 2207 (1961).
30. Beattie, I. R., and Livingston, K. M. S., *J. Chem. Soc. A* 2201 (1969).
31. Beck, W., and Bauder, M., *Chem. Ber.* **103**, 583 (1970).
32. Beck, W., Bauder, M., Fehlhammer, W. P., Pollmann, P., and Schachl, H., *Inorg. Nucl. Chem. Lett.* **4**, 143 (1968).
33. Beck, W., Schorpp, K., and Kern, F., *Angew. Chem.* **83**, 43 (1971).
34. Bekkum, H. van, Gogh, J. van, and Minnen Pathuis, G. van, *J. Catal.* **7**, 294 (1967).
35. Bennett, E. W., and Orenski, P. J., *J. Organometal. Chem.* **28**, 137 (1971).
36. Bennett, M. A., Robertson, G. B., Whimp, P. O., and Yoshida, T., *J. Amer. Chem. Soc.* **93**, 3797 (1971).
37. Bennett, M. A., Robertson, G. B., Whimp, P. O., and Yoshida, T., *J. Amer. Chem. Soc.* **95**, 3028 (1973).
38. Bentham, J. E., and Ebsworth, E. A. V., *Inorg. Nucl. Chem. Lett.* **6**, 145 (1970).
39. Bentham, J. E., and Ebsworth, E. A. V., *Inorg. Nucl. Chem. Lett.* **6**, 671 (1970).
40. Bingham, D., Webster, D. E., and Wells, P. B., *J. Chem. Soc., Dalton Trans.* 1928 (1972).
41. Birnbaum, E. R., *Inorg. Nucl. Chem. Lett.* **7**, 233 (1971).
42. Boenneman, H., *Angew. Chem., Int. Ed. Engl.* **9**, 736 (1970).
43. Bond, G. C., and Hellier, M., *Chem. Ind.* (*London*) 35 (1965).
44. Bond, G. C., and Hellier, M., *J. Catal.* **7**, 217 (1967).
45. Booth, G., and Chatt, J., *J. Chem. Soc. A* 2131 (1969).
46. Brewis, S., and Hughes, P. R., *Chem. Commun.* 157 (1965).
47. Brookes, P. R., *J. Organometal. Chem.* **47**, 179 (1973).
48. Brookes, P. R., and Nyholm, R. S., *J. Chem. Soc. D* 169 (1970).
49. Brookes, P. R., and Shaw, B. L., *J. Chem. Soc., Dalton Trans.* 783 (1973).
50. Brooks, E. H., and Glockling, F., *Chem. Commun.* 510 (1965).
51. Brooks, E. H., and Glockling, F., *J. Chem. Soc. A* 1241 (1966).
52. Brooks, E. H., and Glockling, F., *J. Chem. Soc. A* 1030 (1967).
53. Brooks, E. H., Cross, R. J., and Glockling, F., *Inorg. Chim. Acta* **2**, 17 (1968).
54. Brown, E. S., and Rick, E. A., *J. Chem. Soc. D* 112 (1969).
55. Brown, E. S., Rick, E. A., and Mendicino, F. D., *J. Organometal. Chem.* **38**, 37 (1972).
56. Brunner, H., and Bailar, J. C., *Inorg. Chem.* **12**, 1465 (1973).
57. Burgess, J., Kemmitt, R. D. W., and Littlecott, G. W., *J. Organometal. Chem.* **56**, 405 (1973).
58. Busetto, L., Palazzi, A., Crociani, B., Belluco, U., Badley, E. M., Kilby, B. J. L., and Richards, R. L., *J. Chem. Soc., Dalton Trans.* 1800 (1972).
59. Calf, G. E., and Garnett, J. L., *J. Chem. Soc. D* 373 (1969).
60. Cariati, F., Ugo, R., and Bonati, F., *Chem. Ind.* (*London*) 1714 (1964).

61. Cariati, F., Ugo, R., and Bonati, F., *Inorg. Chem.* **5**, 1128 (1966).
62. Cenini, S., Ugo, R., and La Monica, G., *J. Chem. Soc. A* 3441 (1971).
63. Cetinkaya, B., Cetinkaya, E., and Lappert, M. F., *J. Chem. Soc., Dalton Trans.* 906 (1973).
64. Cetinkaya, B., Lappert, M. F., and McMeeking, J., *J. Chem. Soc. D* 215 (1971).
65. Chalk, A. J., *Ann. N.Y. Acad. Sci.* **172**, 533 (1971).
66. Chalk, A. J., and Harrod, J. F., *J. Amer. Chem. Soc.* **87**, 16 (1965).
67. Chatt, J., *Science* **160**, 723 (1968).
68. Chatt, J., and Coffey, R. S., *J. Chem. Soc. A* 190 (1968).
69. Chatt, J., and Davidson, J. M., *J. Chem. Soc.* 2433 (1964).
70. Chatt, J., Duncanson, L. A., and Shaw, B. L., *Proc. Chem. Soc.* 343 (1957).
71. Chatt, J., Duncanson, L. A., and Shaw, B. L., *Chem. Ind.* (*London*) 859 (1958).
72. Chatt, J., Eaborn, C., and Ibekwe, S., *Chem. Commun.* 700 (1966).
73. Chatt, J., Eaborn, C., and Kapoor, P. N., *J. Organometal. Chem.* **13**, P21 (1968).
74. Chatt, J., Eaborn, C., and Kapoor, P. N., *J. Chem. Soc. A* 881 (1970).
75. Chatt, J., and Shaw, B. L., *J. Chem. Soc.* 4020 (1959).
76. Chatt, J., and Shaw, B. L., *Chem. Ind.* (*London*) 931 (1960).
77. Chatt, J., and Shaw, B. L., *Chem. Ind.* (*London*) 290 (1961).
78. Chatt, J., and Shaw, B. L., *J. Chem. Soc.* 5075 (1962).
79. Chauvin, Y., and Lefevre, G., *C.R. Acad. Sci.* **259**, 2105 (1964).
80. Cherwinski, W. J., and Clark, H. C., *Can. J. Chem.* **47**, 2665 (1969).
81. Cherwinski, W. J., and Clark, H. C., *Inorg. Chem.* **10**, 2263 (1971).
82. Cherwinski, W. J., and Clark, H. C., *J. Organometal. Chem.* **29**, 451 (1971).
83. Chisholm, M. H., and Clark, H. C., *Inorg. Chem.* **12**, 991 (1973).
84. Chopoorian, J. A., Lewis, J., and Nyholm, R. S., *Nature* (*London*) **190**, 528 (1961).
85. Chukhadzhyan, G. A., and Evoyan, Z. K., *Arm, Khim. Zh.* **24**, 530 (1971).
86. Church, M. J., and Mays, M. J., *Chem. Commun.* 435 (1968).
87. Church, M. J., and Mays, M. J., *J. Chem. Soc. A* 3074 (1968).
88. Church, M. J., and Mays, M. J., *J. Chem. Soc. A* 1938 (1970).
89. Citron, J. D., *J. Org. Chem.* **34**, 1977 (1969).
90. Citron, J. D., Lyons, J. E., and Sommer, L. H., *J. Org. Chem.* **34**, 638 (1969).
91. Clark, D. T., Briggs, D., and Adams, D. B., *J. Chem. Soc., Dalton Trans.* 169 (1973).
92. Clark, H. C., and Dixon, K. R., *J. Amer. Chem. Soc.* **91**, 596 (1969).
93. Clark, H. C., Dixon, K. R., and Jacobs, W. J., *Chem. Commun.* 548 (1968).
94. Clark, H. C., Dixon, K. R., and Jacobs, W. J., *J. Amer. Chem. Soc.* **90**, 2259 (1968).
95. Clark, H. C., Dixon, K. R., and Jacobs, W. J., *J. Amer. Chem. Soc.* **91**, 1346 (1969).
96. Clark, H. C., and Jacobs, W. J., *Inorg. Chem.* **9**, 1229 (1970).
97. Clark, H. C., and Kurosawa, H., *J. Chem. Soc. D* 957 (1971).
98. Clark, H. C., and Kurosawa, H., *J. Chem. Soc., Chem. Commun.* 150 (1972).
99. Clark, H. C., and Kurosawa, H., *Inorg. Chem.* **11**, 1275 (1972).
100. Clark, H. C., and Kurosawa, H., *J. Organometal. Chem.* **36**, 399 (1972).
101. Clark, H. C., and Kurosawa, H., *Inorg. Chem.* **12**, 357 (1973).
102. Clark, H. C., and Kurosawa, H., *Inorg. Chem.* **12**, 1566 (1973).
103. Clark, H. C., Tsai, J. H., and Tsang, W. S., *Chem. Commun.* 171 (1965).
104. Clark, H. C., and Tsang, W. S., *Chem. Commun.* 123 (1966).
105. Clark, H. C., and Tsang, W. S., *J. Amer. Chem. Soc.* **89**, 529 (1967).
106. Clark, H. C., and Tsang, W. S., *J. Amer. Chem. Soc.* **89**, 533 (1967).
107. Clemmitt, A. F., and Glockling, F., *J. Chem. Soc. A* 2163 (1969).
108. Clemmitt, A. F., and Glockling, F., *J. Chem. Soc. D* 705 (1970).

109. Clemmitt, A. F., and Glockling, F., *J. Chem. Soc. A* 1164 (1971).
110. Coffey, R. S., *Chem. Commun.* 923 (1971).
111. Collamati, I., and Furlani, A., *J. Organometal. Chem.* **17**, 457 (1969).
112. Collamati, I., Furlani, A., and Attioli, G., *J. Chem. Soc. A* 1694 (1970).
113. Collman, J. P., *Accounts Chem. Res.* **1**, 136 (1968).
114. Corain, B., *Chem. Ind.* (*London*) 1465 (1971).
115. Corischi, D., Furlani, A., Bicev, P., and Russo, M. V., *Gazz. Chim. Ital.* **101**, 526 (1971).
116. Corriu, R. P., and Moreau, J. J. E., *J. Chem. Soc. D* 812 (1971).
117. Corriu, R. P., and Moreau, J. J. E., *J. Organometal. Chem.* **40**, 55 (1972).
118. Corriu, R. P., and Moreau, J. J. E., *J. Organometal. Chem.* **40**, 73 (1972).
119. Cotton, F., and Wilkinson, G., "Advanced Inorganic Chemistry," 3rd ed., p. 797. Wiley (Interscience), New York, 1972.
120. Cramer, R., *Trans. N.Y. Acad. Sci.* **33**, 97 (1971).
121. Cramer, R. D., Jenner, E. L., Lindsey, R. V., and Stolberg, U. G., *J. Amer. Chem. Soc.* **85**, 1691 (1963).
122. Cramer, R. D., and Lindsey, R. V., *J. Amer. Chem. Soc.* **88**, 3534 (1966).
123. Cross, R. J., and Glockling, F., *Proc. Chem. Soc.* **143** (1964).
124. Cross, R. J., and Glockling, F., *J. Chem. Soc.* 5422 (1965).
125. Cross, R. J., and Glockling, F., *J. Organometal. Chem.* **3**, 253 (1965).
126. Davies, N. R., *Nature* (*London*) **205**, 281 (1965).
127. Davis, D., and Scott, L., U.S. Patent, 3,278,575 (1966), *Chem. Abstr.* **66**, 55068w (1966).
128. Dean, R. R., and Green, J. C., *J. Chem. Soc. A* 3047 (1968).
129. Deeming, A. J., Johnson, B. F. G., and Lewis, J., *J. Chem. Soc. D* 598 (1970).
130. Deganello, G., Carturan, G., and Belluco, U., *J. Chem. Soc. A* 2873 (1968).
131. Deno, N. C., Friedman, N., Hodge, J. D., Mackay, F. P., and Saines, G., *J. Amer. Chem. Soc.* **84**, 4713 (1962).
132. Dent, S. P., Eaborn, C., and Pidcock, A., *J. Chem. Soc. D* 1703 (1970).
133. Dingle, T. W., and Dixon, K. R., *Inorg. Chem*, **13**, 846 (1974).
134. Dixon, K. R., and McFarland, J. J., *J. Chem. Soc., Chem. Commun.* 1274 (1972).
135. Dixon, K. R., and Hawke, D. J., *Can. J. Chem.* **49**, 3252 (1971).
136. Drinkard, W. C., U.S. Patent, 3,558,688 (1971), *Chem. Abstr.* **74**, 87445 (1971).
137. Drinkard, W. C., Eaton, D. R., Jesson, J. P., and Lindsey, R. V., *Inorg. Chem.* **9**, 392 (1970).
138. Drinkard, W. C., and Lindsey, R. V., Patent, Fr. 1,533,557 (1968), *Chem. Abstr.* **71**, 101353 (1969).
139. Du Pont de Nemours, British Patent, 1,112,539 (1968), *Chem. Abstr.* **69**, 26810p (1968).
140. Du Pont de Nemours, British Patent, 1,178,950 (1970), *Chem. Abstr.* **72**, 89831d (1970).
141. Dumler, J. T., and Roundhill, D. M., *J. Organometal. Chem.* **30**, C35 (1971).
142. Eaborn, C., Kapoor, P. N., Tune, D. J., Turpin, C. L., and Walton D. R. M., *J. Organometal. Chem.* **34**, 153 (1972).
143. Eaborn, C., Pidcock, A., and Ratcliff, B., *J. Organometal. Chem.* **43**, C5 (1972).
144. Eberhardt, G. G., and Myers, H. K., *J. Catal.* **26**, 459 (1972).
145. Ebsworth, E. A. V., Bentham, J. E., and Craddock, S., *J. Chem. Soc. A* 587 (1971).
146. Eisch, J. J., and Foxton, M. W., *J. Organometal. Chem.* **12**, P33 (1968).
147. Eisenberg, R., and Ibers, J. A., *Inorg. Chem.* **4**, 773 (1965).

148. Fahey, D. R., *Organometal. Chem. Rev.* **7**, 245 (1972).
149. Falk, C. D., and Halpern, J., *J. Amer. Chem. Soc.* **87**, 3523 (1965).
150. Fink, W., *Helv. Chim. Acta* **54**, 1304 (1971).
151. Fink, W., and Wenger, A., *Helv. Chim. Acta* **54**, 2186 (1971).
152. Frankel, E. N., Emken, E. A., Itatani, H., and Bailar, J. C., *J. Org. Chem.* **35**, 1447 (1967).
153. Furlani, A., Collamati, I., and Sartori, G., *J. Organometal. Chem.* **17**, 463 (1969).
154. Garnett, J. L., and Hodges, R. J., *J. Amer. Chem. Soc.* **87**, 4546 (1967).
155. Garnett, J. L., and Kenyon, R. S., *J. Chem. Soc. D* 698 (1970).
156. Gavrilova, V. I., Gel'fman, M. I., Ivannikova, N. V., and Razumovskii, V. V., *Russ. J. Inorg. Chem.* **16**, 596 (1971).
157. Gerlach, D. H., Kane, A. R., Parshall, G. W., Jesson, J. P., and Muetterties, E. L., *J. Amer. Chem. Soc.* **93**, 3544 (1971).
158. Ginsberg, A. P., *Transition Metal Chem.* **1**, 111 (1965).
159. Glockling, F., and Hooton, K. A., *Chem. Commun.* 218 (1966).
160. Glockling, F., and Hooton, K. A., *J. Chem. Soc. A* 1066 (1967).
161. Glockling, F., and Hooton, K. A., *J. Chem. Soc. A* 826 (1968).
162. Gosser, L. W., and Parshall, G. W., *Tetrahedron Lett.* 2555 (1971).
163. Greaves, E. O., Lock, C. J. L., and Maitlis, P. N., *Can. J. Chem.* **46**, 3879 (1968).
164. Green, M. L. H., and Jones, D. J., *Advan. Inorg. Chem. Radiochem.* **7**, 115 (1965).
165. Green, M. L. H., and Munakata, H., *J. Chem. Soc. D* 549 (1971).
166. Green, M. L. H., Munakata, H., and Saito, T., *J. Chem. Soc. A* 469 (1971).
167. Green, M. L. H., and Saito, T., *J. Chem. Soc. D* 208 (1969).
168. Green, M. L. H., Saito, T., and Tanfield, P. J., *J. Chem. Soc. A* 152 (1971).
169. Green, M. L. H., Munakata, H., and Saito, T., *J. Chem. Soc. D* 1287 (1969).
170. Guistiniani, M., Dolcetti, G., and Belluco, U., *J. Chem. Soc. A* 2047 (1969).
171. Guistiniani, M., Dolcetti, G., Pietropaolo, R., and Belluco, U., *Inorg. Chem.* **8**, 1048 (1969).
172. Halpern, J., and Nyholm, R. S., *Proc. Int. Congr. Catal. 3rd* **1** (1965).
173. Hara, M., Ohno, K., and Tsuji, J., *J. Chem. Soc. D* 247 (1971).
174. Harbourne, D. A., Rosevear, D. T., and Stone, F. G. A., *Inorg. Nucl. Chem. Lett.* **2**, 247 (1966).
175. Harrod, J. F., and Chalk, A. J., *Nature (London)* **205**, 280 (1965).
176. Haszeldine, R. N., Parish, R. V., and Setchfield, J. H., *J. Organometal. Chem.* **57**, 279 (1973).
177. Hata, G., Takahashi, K., and Miyake, A., *J. Org. Chem.* **36**, 2116 (1971).
178. Hayashi, E., Narui, S., Ota, M., Sakai, M., Sakakibara, Y., and Uchino, N., *Kogyo Kagaku Zasshi* **74**, 1834 (1971).
179. Hayward, P. J., Blake, D. M., Nyman, C. J., and Wilkinson G., *J. Chem. Soc. D* 987 (1969).
180. Hayward, P. J., Blake, D. M., Wilkinson, G., and Nyman, C. J., *J. Amer. Chem. Soc.* **92**, 5873 (1970).
181. Hodges, R. J., and Garnett, J. L., *J. Phys. Chem.* **72**, 1673 (1968).
182. Hodges, R. J., and Garnett, J. L., *J. Phys. Chem.* **73**, 1525 (1969).
183. Hodges, R. J., Webster, D. E., and Wells, P. B., *J. Chem. Soc. A* 3230 (1971).
184. Hodges, R. J., Webster, D. E., and Wells, P. B., *J. Chem. Soc., Dalton* 2571 (1972).
185. Hodges, R. J., Webster, D. E., and Wells, P. B., *J. Chem. Soc., Dalton Trans.* 2577 (1972).
186. Hooton, K. A., *J. Chem. Soc. A* 680 (1969).

187. Hosokawa, T., and Maitlis, P. M., *J. Amer. Chem. Soc.* **94**, 3238 (1972).
188. Itatani, H., and Bailar, J. C., *J. Amer. Chem. Soc.* **89**, 1600 (1967).
189. Itatani, H., and Bailar, J. C., *J. Amer. Oil Chemists Soc.* **44**, 147 (1967).
190. Itatani, H., and Bailar, J. C., *Ind. Eng. Chem., Prod. Res. Develop.* **11**, 146 (1972).
191. Jardine, I., Howsan, R. W., and McQuillin, F. J., *J. Chem. Soc. C* 260 (1969).
192. Johnson, B. F. G., Lewis, J., and Robinson, P. W., *J. Chem. Soc. A* 1684 (1970).
193. Jonas, K., and Wilke, G., *Angew. Chem., Int. Ed. Engl.* 519 (1969).
194. Jonassen, H. B., Nelson, J. H., Schmitt, D. L., Henry, R. A., and Moore, D. W., *Inorg. Chem.* **9**, 2678 (1970).
195. Kaesz, H. D., and Saillant, R. B., *Chem. Rev.* **72**, 231 (1972).
196. Kane, A. R., Guggenberger, L. J., and Muetterties, E. L., *J. Amer. Chem. Soc.* **92**, 2571 (1970).
197. Keskinen, A. E., and Senoff, C. V., *J. Organometal. Chem.* **37**, 201 (1972).
198. Kingston, J. V., and Scollary, G. R., *J. Chem. Soc. D* 455 (1969).
199. Kiso, Y., Yamamoto, K., Tamao, K., and Kumada, M., *J. Amer. Chem. Soc.* **94**, 4373 (1972).
200. Kitamura, T., Maruya, K., Moro-oka, Y., and Ozaki, A., *Bull. Chem. Soc. Jap.* **45**, 1457 (1972).
201. Kong, P-C., and Roundhill, D. M., unpublished results.
202. Kong, P-C., and Roundhill, D. M., *J. Chem. Soc., Dalton Trans.* 187 (1974).
203. Kong, P-C., and Roundhill, D. M., *Inorg. Chem.* **11**, 749 (1972).
204. Kudo, K., Hidai, M., Murayama, T., and Uchida, Y., *J. Chem. Soc. D* 1701 (1970).
205. Kudo, K., Hidai, M., and Uchida, Y., *J. Organometal. Chem.* **56**, 413 (1973).
206. Kumada, M., Kiso, Y., and Umeno, M., *J. Chem. Soc. D* 611 (1970).
207. Kumada, M., Sumitani, K., Kiso, Y., and Tamao, K., *J. Organometal. Chem.* **50**, 297 (1973).
208. Kumada, M., Sumitani, K., Kiso, Y., and Tamao, K., *J. Organometal. Chem.* **50**, 311 (1973).
209. Kumada, M., Sumitani, K., Kiso, Y., and Tamao, K., *J. Organometal. Chem.* **50**, 219 (1973).
210. Lappert, M. F., McMeeking, J., and Palmer, D. E., *J. Chem. Soc., Dalton Trans.* 151 (1973).
211. Lewis, J., and Johnson, B. F. G., *Accounts Chem. Res.* **1**, 245 (1968).
212. Linde R. van der, and de Jongh, R. O., *J. Chem. Soc. D* 563 (1971).
213. Lindsey, R. V., Parshall, G. W., and Stolberg, U. C., *J. Amer. Chem. Soc.* **87**, 658 (1965).
214. Lindsey, R. V., Cramer, R. D., Prewitt, C. J., and Stolberg, U. C., *J. Amer. Chem. Soc.* **87**, 658 (1965).
215. Luttinger, L. B., and Colthup, E. C., *J. Org. Chem.* **27**, 3752 (1962).
216. Malatesta, L., and Cariello, C., *J. Chem. Soc.* 2323 (1958).
217. Malatesta, L., and Ugo, R., *J. Chem. Soc.* 2080 (1963).
218. Mann, B. E., Shaw, B. L., and Tucker, N. I., *J. Chem. Soc. D* 1333 (1970).
219. Mann, B. E., Shaw, B. L., and Tucker, N. I., *J. Chem. Soc. A* 2667 (1971).
220. Maruya, K., Mizoroki, T., and Ozaki, A., *Bull. Chem. Soc. Jap.* **45**, 2255 (1972).
221. Maruya, K., Mizoroki, T., and Ozaki, A., *Bull. Chem. Soc. Jap.* **46**, 993 (1973).
222. Masters, C., *J. Chem. Soc., Chem. Commun.* 1258 (1972).
223. McLure, G. L., and Baddley, W. H., *J. Organometal. Chem.* **25**, 261 (1970).
224. McCue, J. P., *Coord. Chem. Rev.* **10**, 265 (1973).
225. McFarlane, W., *Chem. Commun.* 393 (1968).

226. Meier, M., Basolo, F., and Pearson, R. G., *Inorg. Chem.* **8**, 795 (1969).
227. Morelli, D., Segre, A., Ugo, R., La Monica, G., Cenini, S., Conti, F., and Bonati, F., *Chem. Commun.* 524 (1967).
228. Muetterties, E. L., "Transition Metal Hydrides." Dekker, New York, 1971.
229. Mukhedhar, A. J., Green, M., and Stone, F. G. A., *J. Chem. Soc. A* 947 (1970).
230. Munakata, H., and Green, M. L. H., *J. Chem. Soc. D* 881 (1970).
231. Nametkin, N. S., Chernysheva, T. I., Gevenyan, M. I., Lashenko, I. N., and Pritula, N. A., *Izv. Akad. Nauk. SSSR, Ser. Khim.* 2330 (1972).
232. Nelson, J. H., Roundhill, D. M., and Jonassen, H. B., *Inorg. Chem.* **8**, 2591 (1969).
233. Nelson, J. H., Personal communication.
234. Nelson, J. H., Verstuyft, A. W., Kelly, J. D., and Jonassen, H. B., *Inorg. Chem.* **13**, 27 (1974).
235. Nyholm, R. S., and Vrieze, K., *J. Chem. Soc.* 5331 (1965).
236. Nyman, C. J., Wymore, C. E., and Wilkinson, G., *Chem. Commun.* 407 (1967).
237. Okinoshima, H., Yamamoto, K., and Kumada, M., *J. Amer. Chem. Soc.* **94**, 9263 (1972).
238. Owston, P. G., Partridge, J. M., and Rowe, J. M., *Acta Cryst.* **13**, 246 (1960).
239. Palazzi, A., Busetto, L., and Graziani, M., *J. Organometal. Chem.* **30**, 273 (1973).
240. Parish, R. V., and Rowbotham, P. J., *J. Chem. Soc., Dalton Trans.* 37 (1973).
241. Parshall, G. W., *Inorg. Syn.* **12**, 26 (1970).
242. Parshall, G. W., *J. Amer. Chem. Soc.* **87**, 2133 (1965).
243. Powell, J., and Shaw, B. L., *J. Chem. Soc.*, 3879 (1965).
244. Pidcock, A., *J. Chem. Soc., Chem. Commun.* 249 (1973).
245. Rasmussen, S. E., and Mariezcurrena, R. A., *Acta Chem. Scand.* **27**, 2678 (1973).
246. Reikhsfel'd, V. O., and Astrakhanov, M. I., *Kremniiorg. Mater.* 50 (1971).
247. Robinson, S. D., and Uttley, M. F., *J. Chem. Soc., Chem. Commun.* 1047 (1972).
248. Romanelli, M. G., and Kelly, R. J., German Patent, 2,011,163 (1970), *Chem. Abstr.* **74**, 53040x (1971).
249. Rose, D., *Tetrahedron Lett.* **41**, 4197 (1972).
250. Roundhill, D. M., *J. Chem. Soc. D* 567 (1969).
251. Roundhill, D. M., *Inorg. Chem.* **9**, 254 (1970).
252. Roundhill, D. M., and Jonassen, H. B., *Chem. Commun.* 1233 (1968).
253. Roundhill, D. M., Tripathy, P. B., and Renoe, B. W., *Inorg. Chem.* **10**, 727 (1971).
254. Ryan, J. W., and Speier, J. L., *J. Amer. Chem. Soc.* **86**, 895 (1964).
255. Saam, J. C., and Speier, J. L., *J. Amer. Chem. Soc.* **80**, 4104 (1958).
256. Samejima, H., Mizuta, T., Yamamoto, H., and Kivan, T., *Bull. Chem. Soc. Jap.* **42**, 2722 (1969).
257. Schneider, M. L., and Shearer, H. M. M., *J. Chem. Soc., Dalton Trans.* 354 (1973).
258. Schunn, R. A., *Inorg. Chem.* **9**, 394 (1970).
259. Schunn, R. A., *6th Int. Conf. Organometal. Chem., Amherst, 1973.*
260. Sloan, M. F., Matlack, A. S., and Breslow, D. S., *J. Amer. Chem. Soc.* **85**, 4014 (1963).
261. Socrates, G., *J. Inorg. Nucl. Chem.* **31**, 1667 (1969).
262. Sommer, L. H., and Citron, J. D., *J. Org. Chem.* **32**, 2470 (1967).
263. Speier, J. L., Webster, J. A., and Barnes, G. H., *J. Amer. Chem. Soc.* **79**, 974 (1957).
264. Stern, E. W., and Maples, P. K., *J. Catal.* **27**, 120 (1972).
265. Stern, E. W., and Maples, P. K., *J. Catal.* **27**, 134 (1972).
266. Stone, F. G. A., Ashley-Smith, J., and Green, M., *J. Chem. Soc. A* 3161 (1970).
267. Stone, F. G. A., and Treichel, P. M., *Advan. Organometal. Chem.* **1**, 143 (1964).

268. Srivastava, S. C., and Bigorgne, M., *J. Organometal. Chem.* **18**, P30 (1969).
269. Takahashi, K., Miyake, A., and Hata, G., *Bull. Chem. Soc. Jap.* **45**, 1183 (1972).
270. Takahashi, S., Shibano. T., and Hagihara, N., *Bull. Chem. Soc. Jap.* **41**, 454 (1968).
271. Takahashi, S., Shibano, T., and Hagihara, N., *J. Chem. Soc. D* 161 (1969).
272. Takahashi, S., Shibano. T., Kojima, H., and Hagihara, N., *Organometal. Chem. Syn.* **1**, 193 (1971).
273. Takahashi, S., Yamazaki, H., and Hagihara, N., *Bull. Chem. Soc. Jap.* **41**, 254 (1968).
274. Tarasenko, V. N., Odabashyan, G. V., and Lakhtikov, A. I., *Trans. Mosk. Khim.—Technol. Inst.* **70**, 138 (1972).
275. Tayim, H. A., and Bailar, J. C., *J. Amer. Chem. Soc.* **89**, 4330 (1967).
276. Tayim, H. A., and Bailar, J. C., *J. Amer. Chem. Soc.* **89**, 3420 (1967).
277. Taylor, B. W., and Swift, H. E., *J. Catal.* **26**, 254 (1972).
278. Theissen, R. J., *J. Org. Chem.* **36**, 752 (1971).
279. Thomas, K., Dumler, J. T., Renoe, B. W., Nyman, C. J., and Roundhill, D. M. *Inorg. Chem.* **11**, 1795 (1972).
280. Tolman, C. A., *J. Amer. Chem. Soc.* **92**, 2953 (1970).
281. Tolman, C. A., *J. Amer. Chem. Soc.* **92**, 4217(1970).
282. Tolman, C. A., *J. Amer. Chem. Soc.* **94**, 2994 (1972).
283. Tolman, C. A., *J. Amer. Chem. Soc.* **92**, 6777 (1970).
284. Tolman, C. A., *J. Amer. Chem. Soc.* **92**, 6785 (1970).
285. Tolman, C. A., *Inorg. Chem.* **11**, 3128 (1972).
286. Toniolo, L., Giustiniani, M., and Belluco, U., *J. Chem. Soc. A* 2666 (1969).
287. Tripathy, P. B., and Roundhill, D. M., *J. Amer. Chem. Soc.* **92**, 3825 (1970).
288. Tripathy, P. B., Renoe, B. W., Adzamli, K., and Roundhill, D. M., *J. Amer. Chem. Soc.* **93**, 4406 (1971).
289. Tsuji, J., Hara, M., Ohno, K., German Patent, 1,942,798 (1970), *Chem. Abstr.* **73**, 15497y (1970).
290. Ugo, R., Cariati, F., and La Monica, G., *Chem. Commun.* 868 (1966).
291. Ugo, R., La Monica, G., Cariati, F., Cenini, S., and Conti, F., *Inorg. Chim. Acta* **4**, 390 (1970).
292. Uguagliati, P., and Baddley, W. H., *J. Amer. Chem. Soc.* **90**, 5446 (1968).
293. Vaska, L., and Werneke, M. F., *Trans. N.Y. Acad. Sci.* **33**, 70 (1971).
294. Walker, W. E., Manyik, R. M., Atkins, K. E., and Farmer, M. L., *Tetrahedron Lett.* 3817 (1970).
295. Wheelock, K. S., Nelson, J. H., Kelly, J. D., Jonassen, H. B., and Cusachs, L. C., *J. Chem. Soc., Dalton Trans.* 1457 (1973).
296. Whitesides, G. M., Gaasch, J. F., and Stedronsky, E. R., *J. Amer. Chem. Soc.* **94**, 5258 (1972).
297. Whitesides, G. M., and Maglio, G., *J. Amer. Chem. Soc.* **91**, 4980 (1969).
298. Wilkinson, G., and Birmingham, J. M., *J. Amer. Chem. Soc.* **77**, 3421 (1955).
299. Wojcicki, A., *Advan. Organometal. Chem.* **11**, 87 (1973).
300. Wright, D., *Chem. Commun.* 197 (1966).
301. Yamamoto, K., Hayashi, T., and Kumada, M., *J. Organometal. Chem.* **28**, C37 (1971).
302. Yamamoto, K., Hayashi, T., and Kumada, M., *J. Amer. Chem. Soc.* **93**, 5301 (1971)
303. Yamamoto, K., Hayashi, T., and Kunada, M., *J. Organometal. Chem.* **46**, C65 (1972).
304. Yamamoto, K., Uramoto, Y., and Kumada, M., *J. Organometal. Chem.* **31**, C9 (1971).
305. Yasumori, I., and Hirabayashi, K., *Trans. Faraday Soc.* **67**, 3283 (1971).

Palladium-Catalyzed Organic Reactions

PATRICK M. HENRY

Department of Chemistry
University of Guelph
Guelph, Ontario, Canada

I

INTRODUCTION AND SCOPE

Homogeneous catalysis by transition metals is one of the fastest growing fields of chemistry. No doubt much of the interest stems from the potential commercial importance of these reactions and for that reason most of the earlier work was concerned with the preparative aspects of homogeneous catalysis. However more recently a considerable amount of mechanistic work has been carried out on systems which are amenable to this type of study and we now have a fair understanding of two of the basic reactions of homogeneous catalysis, the oxidative addition (*31*, *38*, *39*, *80*, *276*, *277*) and insertion reactions (*168*, *293*, *294*).

Because of their commercial interest, homogeneous catalysis by Pd(II) is probably the most studied of all homogeneous catalytic systems.

Because of industrial implications most of the research in this field has been in the synthetic area and the mechanistic work has lagged behind many other fields of homogeneous catalysis. However, recently work on the mechanistic aspects have increased considerably and this review will emphasize these recent advances.

There have been a number of reviews of various aspects of catalytic palladium chemistry (*2*, *47*, *50*, *82*, *83*, *102–104*, *140*, *149*, *161*, *162*, *173*, *247*, *248*, *266–268*) the most comprehensive of which is the two-volume treatise by Maitlis (*173*). The complex chemistry of palladium is well covered by the recent book of Hartley (*83*).

Since the two volumes by Maitlis covers the field very well up to the end of 1970, the original aim was to cover in some depth the new advances in the field from the beginning of 1971 to the end of 1973. However, even this limited objective proved to be impossible within the scope of an *Advances in Organometallic Chemistry* article due to the large amount of work published in these 3 years. Thus, rather than just mention all the work done in these 3 years the writer chose to treat several topics in more detail. The choice of topics does not reflect their importance to the field but rather the writer's familiarity with the subject matter. Thus the very interesting work on dynamic π-allyl complexes as well as the large number of new catalytic reactions involving 1,2- and 1,3-dienes will not be discussed. Some of this work has recently been reviewed (*267*). Other topics given little or no mention is the coordination chemistry of Pd(II) and valence isomerization. There are also a large number of miscellaneous new reactions catalyzed by Pd(II) which will not be included. Examples are carbene formation from isocyanides, SO_2 insertions, formation of carboxamido complexes, additions of nucleophiles to isocyanides, carbonylation of alcohols, and aromatic substitution.

II

COMPLEXES OF PALLADIUM

A. *Inorganic Complexes*

There is, of course, an extensive literature available on the inorganic chemistry of palladium(II), and Hartley's book is recommended as a recent review (*83*). In this survey, we shall confine ourselves to the inorganic chemistry which is required for an understanding of the catalytic processes. A knowledge of the equilibria between Pd(II) and various ligands is especially important in interpreting kinetic data and unfortunately disregard of these equilibria has often led to incorrect assumptions as to mechanism.

1. *Chloride Complexes*

The equilibrium between Pd(II) and chloride,

$$\begin{array}{lll} Pd^{2+} + Cl^- \rightleftharpoons PdCl^+ & & K_1 \\ PdCl^+ + Cl^- \rightleftharpoons PdCl_2 & & K_2 \\ PdCl_2 + Cl^- \rightleftharpoons PdCl_3^- & & K_3 \\ PdCl_3^- + Cl^- \rightleftharpoons PdCl_4^{2-} & & K_4 \end{array}$$

has been studied by several workers who find values of log β_4 between 11 and 12 at 25°C (*17*, *27*, *58*, *78*). A recent study gives values of the various K's in essential agreement with the earlier workers (*59*).

The equilibrium in other solvents has not been studied to any great extent. In lower dielectric solvents, Pd(II) may tend to exist as chloride-bridged dimers that are isolable in the solid state (*52*). The cationic complexes $[Pd_2X_2(PR_3)_4]^{2+}$ (X = Br, I) (*57*) have recently been reported, and the 1:1 adducts of osazones (L) with $PdCl_2$, of composition Pd(L)Cl_2, were found to eliminate HCl to give the dimeric species $Pd_2(L—H)_2Cl_2$ (*29*).

$$2\ \text{MeC(=N(NHPh))–C(Me)=N(NHPh)}\rightarrow PdCl_2 \longrightarrow [\text{MeC(=N–NPh)–C(Me)=N(NHPh)}]Pd(\mu\text{-}Cl)_2Pd[\text{N(NHPh)=C(Me)–C(Me)=N–NPh}] \quad (1)$$

Mixtures of LiCl and $PdCl_2$ when heated in the presence of CH_3CN dissolve in a 1:1 ratio to give a species $LiPdCl_3$ (*88*) which has been interpreted to be a monomer $LiPdCl_3(CH_3CN)$ (*238*). However the results are just as consistent with the dimeric species, $Li_2Pd_2Cl_6$.

Mixtures of NaCl and $PdCl_2$ in acetic acid also dissolve in a 1:1 ratio to give a species with the stoichiometry, $NaPdCl_3$ (*51*). In this case, molecular weight measurements indicated the species as the dimer $Na_2Pd_2Cl_6$ (*105*) which almost certainly contains chloride bridges between the palladiums (*52*). In LiCl solutions at low [LiCl], spectral evidence indicates that the predominant species is also the dimer. Furthermore, it was shown that at higher [LiCl] the following equilibrium is operable:

$$Li_2Pd_2Cl_6 + 2LiCl \overset{K}{\rightleftharpoons} 2Li_2PdCl_4 \tag{2}$$

where K has a value of 0.1 M^{-1} at 25°C. In these studies the self-dimerization of LiCl had to be considered:

$$2LiCl \overset{K_D}{\rightleftharpoons} Li_2Cl_2 \tag{3}$$

K_D has a value of 2.4 M^{-1} at 37.5°C.

2. *Acetate Complexes*

Compound $Pd(OAc)_2$ has been reported to be monomeric in benzene at 80°C, but trimeric at 37°C (*246*). An X-ray study of $[Pd(OAc)_2]_3 \cdot H_2O$ shows that all six acetates are bridging (*240*). Compounds of the type $[X(Me_2PhE)Pd(O_2CR)]_2$ (X = Co, Br, I; E = P or As) have also been found to contain bridging carboxylates (*223*), whereas compound $[Pd(OAc)(ONCMe_2)]_3$ contains both bridging acetates and bridging acetoxinato groups (*188*).

Despite a report to the contrary (*258*), $Pd(OAc)_2$ is trimeric in acetic acid at 25°C (*213*). Furthermore, it reacts with sodium or lithium acetate according to the following two equilibria (M = Na or Li):

$$2Pd_3(OAc)_6 + 6MOAc \overset{K'}{\rightleftharpoons} 3M_2Pd_2(OAc)_6 \tag{4}$$

$$M_2Pd_2(OAc)_6 + 2MOAc \overset{K''}{\rightleftharpoons} 2M_2Pd(OAc)_4 \tag{5}$$

A particularly interesting feature of these equilibria is that they are not instantaneous but take hours to reach equilibrium. The rate of conversion of trimer to dimer is given by the expression:

$$\frac{-d[Pd_3(OAc)_6]}{dt} = k[Pd_3(OAc)_6][MOAc] \tag{6}$$

This expression suggests that the rate-determining step is the attack of acetate on one of the Pd(II)'s in the trimer followed by rapid decomposition. This S_N2 type of attack is general for Pd(II) complexes (*14*). These results indicate that the bridging acetates are nonlabile, a result not expected on the basis of known chemistry of Pd(II) acetates. This nonlability complicates the study of catalysis by Pd(II) in acetate systems and was not previously taken into account in kinetic studies (Section III,A,2a).

3. *Carbonyl Complexes*

Carbonyl complexes of Pd(II) are almost certainly involved in Pd(II)-catalyzed reactions in the presence of carbon monoxide. Thus $PdCl_2$ takes up CO in the presence of methanol vapor to give $Pd(CO)Cl_2$ (*184*). This carbonyl is decomposed in the presence of H_2O to give CO_2 and Pd(0), a reaction analogous to the oxidation of CO by other metal ions. Another Pd(II) carbonyl containing the trichlorotin group of formula $[Pd(CO)(SnCl_3)_2Cl]^-$ has recently been reported (*158*).

Carbonyls that apparently contain the Pd(I) oxidation state have also been reported. These include $Pd_2(CO)_2Cl$ (*64*), Pd(CO)Cl (*263*), and $Pd(CO)_2Cl_4^{2-}$ (*192*). However the formulations are not certain, and it is possible they may be hydrido carbonyls.

Carbonyls of zero-valent palladium have recently been prepared by reduction of Pd(II) compounds by triethylaluminum or sodium borohydride in the presence of triphenylphosphine (*165, 192*). More recently these same workers reported that $(PPh_3)_2PdCl_2$ reacts with CO in methanol–amine systems to give the carbonyls, $Pd(CO)(PPh_3)_3$, $Pd_3(CO)_3(PPh_3)_3$, and $Pd_3(CO)_3(PPh_3)_4$ when primary or secondary amines such as diethylamine or cyclohexylamine are used (*129*). These carbonyl complexes are interconvertible under the certain conditions. When tertiary amines such as triethylamine and tri-*n*-butylamine are used the product is a carbomethoxy complex $(PPh_3)_2PdCl(COOCH_3)$. This complex is analogous to the stable carbomethoxy Hg(II) compound, $ClHg(COOCH_3)$, and is of particular interest in Pd(II) catalytic chemistry because some oxidations involving CO almost certainly proceed through this type of intermediate (Section III, A, 3, b). The carbomethoxy complex is probably formed by attack of methoxide on Pd(II) carbonyl species:

$$(PPh_3)_3Pd(CO)Cl^+ + CH_3O^- \longrightarrow (PPh_3)_2Pd(COOCH_3)Cl \qquad (7)$$

Another reaction of considerable mechanistic interest reported by these workers (*165*) is the oxidative addition of methyl iodide, allyl chloride, or vinyl chloride to $Pd(CO)(PPh_3)_3$ to give the corresponding acyl complexes *trans*-$PdX(COR)(PPh_3)_2$. These reactions almost certainly occur by oxidative addition, followed by CO insertion:

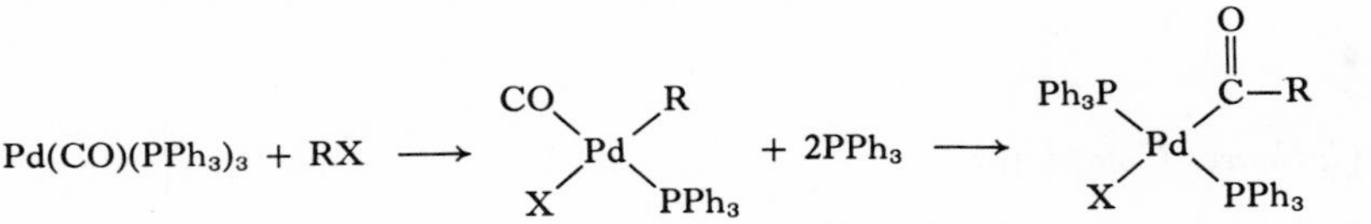

$$Pd(CO)(PPh_3)_3 + RX \longrightarrow \begin{matrix} CO & & R \\ & Pd & \\ X & & PPh_3 \end{matrix} + 2PPh_3 \longrightarrow \begin{matrix} Ph_3P & & C(=O)-R \\ & Pd & \\ X & & PPh_3 \end{matrix} \quad (8)$$

A similar reaction has been reported by Stille *et al.* (*251*) using chiral alkyl halide. This reaction proceeds with inversion of configuration at carbon. The configuration was established by the rotation of the carbonylated complex which was also prepared by oxidative addition of acyl halide.

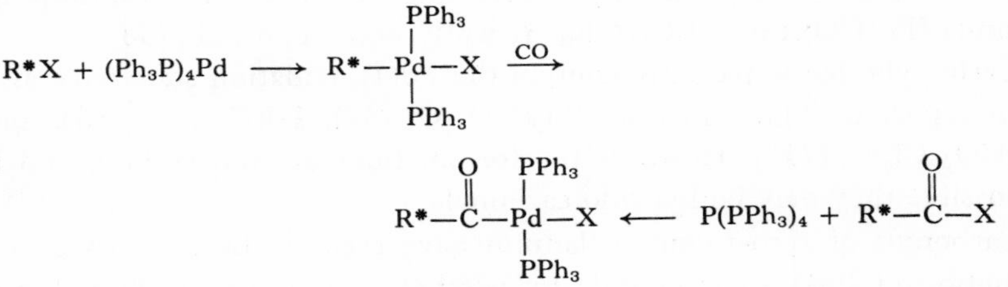

$$R^*X + (Ph_3P)_4Pd \longrightarrow R^*-\underset{PPh_3}{\overset{PPh_3}{Pd}}-X \xrightarrow{CO}$$

$$R^*-\overset{O}{\overset{\|}{C}}-\underset{PPh_3}{\overset{PPh_3}{Pd}}-X \longleftarrow P(PPh_3)_4 + R^*-\overset{O}{\overset{\|}{C}}-X \quad (9)$$

Zero-valent palladium carbonyls without phosphine ligands have been prepared by two groups of workers by the technique of matrix isolation at low temperatures. Carbonyls of the type $Pd(CO)_n$ (n = 1 to 4) have been characterized by IR spectra. Diffusion studies indicate that the lower carbonyls react readily with CO to give $Pd(CO)_4$ (*46, 167*).

4. *Hydride Complexes*

Palladium hydride complexes are of considerable interest in the catalytic chemistry of palladium because of their postulated occurrence as intermediates in a number of reactions such as hydrogenations, isomerizations, and oxidation reactions. In contrast to Pt(II), which forms stable hydrides, most Pd(II) hydrides are unstable. Although there are earlier reports of unstable Pd(II) hydride complexes being formed, the first stable hydride was prepared by Brooks and Glocking by the following reaction (*20, 21*):

$$\textit{trans}\text{-}(Et_3P)PdCl_2 + Me_3GeH \xrightarrow{40°C} \textit{trans}\text{-}(Et_3P)_2Pd(H)Cl + HCl + Me_3GeCl + Me_6Ge_6 \quad (10)$$

The hydride is stable up to about 55°C and has the trans structure in both solution and solid state. More recently, a Pd(II) hydride has been prepared by addition of HCl to $(Ph_3P)_3PdCO$ and $(Ph_3P)_4Pd$ at -50°C in ether. A later report describes the preparation of $(R_3P)_2PdH(BH_4)$ and $(R_3P)_2Pd(H)X$ (R = isopropyl and cyclohexyl; X = Cl, Br, I, or NCS). The latter was prepared by reduction of *trans*-$(R_3P)_2PdX_2$ with *trans*-$(R_3P)_2NiH(BH_4)$, and the former compound was obtained by reaction of *trans*-$(R_3P)_2Pd(H)Cl$ with sodium borohydride (*77*). These hydride complexes exchange phosphine groups fairly rapidly.

$$(R'_3P)_2PdHX + (R_3P)_2PdHX \rightleftharpoons (R'_3P)(R_3P)PdHX \qquad (11)$$

A new dihydride complex has also been reported (*166*).

5. *Pd(I) Complexes*

Univalent palladium complexes may very well be the initial products of decomposition of organopalladium intermediates to give the final products. Until recently the only Pd(I) complexes known were the arene complex $[PdAl_2Cl_7(C_6H_6)]_2$ (*3, 4*), and a compound postulated to be $[Pd(C_6H_6)(H_2O)(ClO_4)]_n$ (*49*). Otsuka and co-workers (*209*) have now reported the preparation of the Pd(I) isonitrile complexes $[PdClL_2]_2$-(C_6H_5Cl), $[PdB_2L_2]_2$, and $[PdIL_2]_2$ (L = *tert*-Bu, NC). These dimers are diamagnetic, suggesting strong magnetic exchange between the two d^9 nuclei. The complexes react with NO to give a nitro complex that must involve disproportionation of the NO. Palladium(II) does not undergo this type of reaction with NO.

$$[PdBrL_2]_2 + 4NO \rightleftharpoons [PdBr(NO_2)L_2] + N_2 \qquad (12)$$

Another new Pd(I) compound is reported under reactions of acetylenes (Section V, A).

A new cyclohexa-1,4-diene complex of Pd(I) has also been recently found in the autoxidation of this diolefin (*48*).

B. *Olefin and Acetylene π Complexes*

π-Olefin and acetylene complexes are important intermediates in homogeneous catalysis since complexing to a transition metal apparently activates the olefin or acetylene for further reaction. Thus free olefins generally undergo electrophilic type of attack. However, when bonded to a transition metal they become susceptible to nucleophilic attack. According to the Dewar–Chatt–Duncanson model (*33*, *54*) the bonding consists of (*a*) a σ-type bond formed by overlap of an empty metal hybrid orbital with the filled π orbital of the olefin and (*b*) a π-type bond formed by overlap of filled *d* orbitals of the metal with the unfilled π^* orbitals of the olefin, namely,

$$\text{Pd} \longleftarrow \begin{matrix} \backslash\,/ \\ \text{C} \\ \| \\ \text{C} \\ /\,\backslash \end{matrix}$$

The bonding in this type of complex has been extensively discussed and it would be beyond the scope of this review to consider it in any further detail.

1. *Pd(II) Complexes*

a. *Olefin Complexes.* A large number of olefin π complexes of mono- and diolefins have been prepared and characterized by IR and X-ray studies. The reader is referred to the works of Maitlis (*173*), Hartley (*82*), or Nelson and Jonassen (*204*) for background discussion.

The monoolefin complexes are often prepared by displacement of nitriles from Pd(II) (*156*):

$$2\,(\text{RCN})_2\text{PdCl}_2 + \text{olefin} \longrightarrow (\text{olefin PdCl}_2)_2 + 4\text{RCN} \tag{13}$$

Olefin complexes can often be prepared in solution by adding the olefin to a soluble Pd(II) salt. Thus Na_2PdCl_4, in HOAc, will absorb ethylene reversibly to give solutions of a different color than the original solution of the Pd(II) salt (*106*). In some cases the intermediate π complex can be detected in catalytic systems. Thus, in the oxidation of ethylene to acetaldehyde, formation of the intermediate π complex, according to the following equilibrium

$$\text{C}_2\text{H}_4 + \text{PdCl}_4{}^{2-} \rightleftharpoons (\text{C}_2\text{H}_4)\text{PdCl}_3{}^- + \text{Cl}^- \tag{14}$$

can be detected either by using fast reaction techniques (*107*) or studying the equilibria under conditions where further reaction of the π complex is slow (*217*, *218*).

The diolefin complexes are much more stable than monoolefin complexes and can be prepared by the direct reaction of olefin with Na_2PdCl_4 in various solvents. The 1,5-cyclooctadiene (COD) complex is prepared by treating COD with Na_2PdCl_4 in acetone, acetic acid, water, or ethanol (*34*, *87*):

$$COD + Na_2PdCl_4 \longrightarrow (COD)PdCl_2 + 2NaCl \quad (15)$$

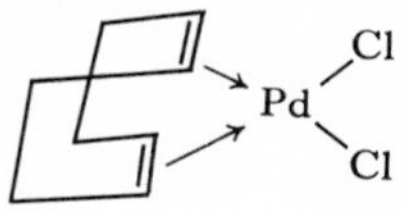

A new diolefin complex reported recently is the Pd(II) complex with cis, namely, *cis*-1,4-cyclononadiene (*16*). As with other diolefin complexes, it is quite stable, decomposing at 150°C. The compound is insoluble and very likely has a polymeric structure,

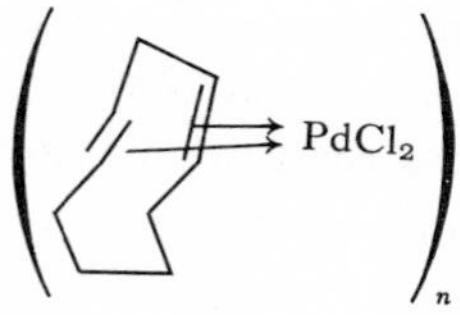

$$\left(C_9H_{14} \rightrightarrows PdCl_2 \right)_n$$

A 1,4-cyclooctadiene complex has also been recently reported (*260*).

Hartley and Wagner (*85*) recently reported some novel complexes of palladium(II) with the allylammonium ions in which coordination is through the olefin rather than the amine. These complexes, which have the advantage of higher stability over ordinary π complexes, have formulas $[Pd(CH_2{=}CHCH_2NH_3)Cl_3]$ and $Na^+[Cl_2PdCl_2PdCl(CH_2{=}CHCH_2NH_3Cl)]^-$. A third salt, $(CH_2{=}CH_2NH_3)_2{}^+[PdCl_4]^{2-}$, in which there is no covalent bonding, was also formed.

Wakatsuki and co-workers (*287*) have prepared vinyl ether complexes of Pd(II) and Pt(II) of composition $[Cl_2ML]$ (M = Pd or Pt; L = vinyl ethers, propenyl ethers, 1,2-dimethoxyethylene) by displacement of PhCN from $Pd(PhCN)_2Cl_2$. They are fairly stable at 25°C. Since these ethers would be good σ donors but poor π acceptors, the stability of these complexes suggest that σ donation is most important with π complexes of Pd(II) and Pt(II). This conclusion is in agreement with earlier IR studies, as well as a recent ^{13}C NMR study of Pt olefin and acetylene complexes (*35*). Furthermore, a study of complex formation of para-substituted styrenes with (6-acetoxynorbornenyl)Pd(benzoyltrifluoroacetonate) in $CDCl_3$,

$$\text{(6-acetoxynorbornenyl)Pd(bta)} + \text{Y}-C_6H_4-CH=CH_2 \underset{CDCl_3}{\overset{K}{\rightleftharpoons}} \text{Y}-C_6H_4-CH=CH_2 \rightarrow \text{Pd(bta)(acetoxynorbornenyl)} \quad (16)$$

(bta = benzoyltrifluoroacetonate)

also support the contention that the σ donation is the most important factor in the bonding (*13*). $\text{Log}_{10}K$ displayed a linear dependence on the σ_p^+ parameter of Y, which indicates the resonance electron donation of Y is of major importance in determining the value of K. The olefin $\pi \rightarrow \text{metal} - d$ σ-bonding component must be the dominant factor since, if metal $d \rightarrow \text{olefin} - \pi^*$ bonding was of prime importance, the opposite trend would have been observed. An NMR study of styrene complexes of Pt(II) and Pd(II) indicates rotation is faster in Pd(II) complexes, in keeping with the lower π-bonding strength and greater lability of Pd(II) complexes (*147*). Another interesting point is that bonding is weaker in bis(styrene) complexes than in monostyrene complexes.

Pregaglia and co-workers (*224*) have studied complexes of the type $[PdCl_2(\text{olefin})]_2$ (olefin = ethylene, 1-pentene, *cis*-2-pentene, *trans*-2-pentene). These compounds can exist in three possible geometries,

(I) Cl, olefin on first Pd; bridging Cl, Cl; olefin, Cl on second Pd (trans)

(II) olefin, Cl on first Pd; bridging Cl, Cl; olefin, Cl on second Pd (cis)

(III) olefin, olefin on first Pd; bridging Cl, Cl; Cl, Cl on second Pd

whereas the Kharasch complex (olefin = C_2H_4), prepared from $[PdCl_2(PhCN)_2]_2$, is known from X-ray crystallography to have structure I. However, a different isomer, which is assigned structure III, was obtained by displacing *cis*-2-pentene from $[PdCl_2(\textit{cis}\text{-2-pentene})_2]_2$. The *cis*-2-pentene complex can apparently exist in all three forms, depending on the method of preparation, whereas the 1-pentene and *trans*-2-pentene complexes could only be isolated either as structure I or III. A study of the displacement of one pentene by another in these complexes indicated the following order of stability in $CHCl_3$: *cis*-2-

pentene > 1-pentene > *trans*-2-pentene, which is the same order as for the silver complexes. However, for butenes in aqueous solution, the relative stability for the equilibria

$$\text{olefin} + PdCl_4^{2-} \xrightleftharpoons{K_1} (\text{olefin})\, PdCl_3^- + Cl^- \quad (17)$$

was in the order: 1-butene > *cis*-2-butene > *trans*-2-butene. This is the order expected on the basis of steric hindrance.

The equilibrium between Pd(II) salts and olefins in acetic acid has been studied by several workers. Cruikshank and Davies (*44*) studied the interaction of $Na_2Pd_2Cl_6$ with allylbenzene using spectral measurements. They postulated the presence of the species $Na[Pd_2Cl_5(\text{olefin})]$-$[Pd_2Cl_4(\text{olefin})_2]$, and $[PdCl_2(\text{olefin})_2]$. They proposed that the latter complex is the catalytically active species in olefin isomerization. Hartley and Wagner (*86*), using spectral measurements and 1-hexene and 1-octene as the olefins, also postulated that the former two species were formed in the following equilibria:

$$Na_2Pd_2Cl_6 + \text{olefin} \rightleftharpoons Na[Pd_2Cl_5(\text{olefin})] + NaCl \quad (18)$$

$$Na[Pd_2Cl_5(\text{olefin})] \rightleftharpoons [Pd_2Cl_4(\text{olefin})_2] + NaCl \quad (19)$$

Hartley and Wagner found no evidence for $PdCl_2(\text{olefin})_2$ and suggested that Cruikshank and Davies needed to postulate this complex because they did not take into account the formation of $Na[Pd_2Cl_5$-$(CH_3COOH)]$. The results of a study of ethylene uptake by Na_2PdCl_4 were most consistent with the following equilibrium (*106*):

$$Na_2Pd_2Cl_6 + 2\ \text{olefin} \rightleftharpoons 2[NaPdCl_3(\text{olefin})] \quad (20)$$

Hartley and Wagner postulated the formation of $[PdCl_4(\text{olefin})_2]$ because their results required a complex with a Pd-to-olefin ratio of 1, and the latter complex is formed when Na_2PdCl_4 is dissolved in pure olefin. However, this is not necessarily the case in acetic acid, and it is not clear whether they ruled out the formation of $Na[PdCl_3(\text{olefin})]$.

In the lithium chloride system containing excess LiCl, spectral and kinetic studies on allylic ester exchange (see Section III, B, 1) (*108*) indicated the main equilibrium between allylic esters and $Li_2Pd_2Cl_6$ was formation of a monomeric species,

$$Li_2Pd_2Cl_6 + 2\ \text{allyl ester} \rightleftharpoons 2Li[PdCl_3(\text{allyl ester})] \quad (21)$$

whereas the kinetics require that a dimeric π complex be formed to some extent.

$$Li_2Pd_2Cl_6 + \text{allyl ester} \rightleftharpoons Li[Pd_2Cl_5(\text{allyl ester})] + LiCl \quad (22)$$

Of course, these results do not rule out formation of $[Pd_2Cl_4(olefin)_2]$-type complexes in the NaCl case. Formation of this type of species would be strongly inhibited by excess LiCl, not present in the $Na_2Pd_2Cl_6$ system because of the low solubility of NaCl in acetic acid. Clearly the chloride-containing system in acetic acid needs further study to define conclusively the equilibria present.

The equilibria present between olefins such as styrene, allylic esters, or 3,3-dimethyl butene-1 and palladium(II) acetate in acetic acid, have recently been studied (*212*). In the rate of acetate concentrations in which the acetate-bridged dimer $[Na_2Pd_2(OAc)_6]$ is the main species (Section II, A, 2), some peculiar results were observed. When olefin was added to the solution, a rapid change in spectra occurred followed by a slower change. The fast equilibrium is

$$\begin{array}{c}{}^{2-}\text{OAc}\text{---}\text{OAc}\text{---}\text{OAc}\\ /\ \ \text{Pd}\ \ /\ \ \text{Pd}\ \ /\\ \text{OAc}\text{---}\text{OAc}\text{---}\text{OAc}\end{array} + \text{olefin} \rightleftharpoons \begin{array}{c}{}^{-}\text{OAc}\text{---}\text{OAc}\text{---}\text{olefin}\\ /\ \ \text{Pd}\ \ /\ \ \text{Pd}\ \ /\\ \text{OAc}\text{---}\text{OAc}\text{---}\text{OAc}\end{array} + \text{OAc}^- \quad (23)$$

whereas the slow reaction is the formation of monomeric species.

$$Na_2Pd_2(OAc)_6 + 2\ \text{olefin} \rightleftharpoons 2[NaPd(OAc)_3(\text{olefin})] \quad (24)$$

Thus, the terminal acetates are labile and the bridging acetates are not, as might be expected from the reaction of $Pd_3(OAc)_6$ or $Na_2Pd_2(OAc)_6$ with NaOAc (Section II, A, 2). The equilibria are analogous to those in the $Li_2Pd_2Cl_6$ system. In the chloride-containing system, however, the attainment of equilibrium is instantaneous.

b. *Acetylene.* Because of their high reactivity, few stable acetylene complexes of Pd(II) are known. A complex from di-*t*-butyl acetylene has been reported (*135*) and a complex of 2-butyne has been detected by NMR at $-50°C$ (*56*).

$$2Bu^tC{\equiv}CBu^t + [C_2H_4PdCl_2]_2 \longrightarrow \left[\begin{array}{c}Bu^t\\ C\\ |||\\ C\\ Bu^t\end{array} \longrightarrow PdCl_2\right]_2 + 2C_2H_4 \quad (25)$$

2. *Pd(0) Complexes*

a. *Olefins.* Olefin complexes are usually prepared by reaction of $(Ph_3P)_4Pd$ with the olefin. Until recently no complexes of simple olefins such as ethylene were known, but, in 1971, Van der Linde and De

Jongh reported the preparation of $(C_2H_4)Pd(PPh_3)_2$ (*273*). Tolman and co-workers studied the equilibria at 25°C,

$$M(PPh_3)_3 + C_2H_4 \underset{K}{\overset{C_6H_4}{\rightleftharpoons}} (C_2H_4)M(PPh_3)_2 + PPh_3 \tag{26}$$

and found values of 300, 0.12, and 0.013 for M = Ni, Pt, and Pd, respectively (*261*). The value of K is higher with $P(p\text{-}C_6H_4CH_3)_3$ than with PPh_3. This effect is explained by the former being a better electron donor, thus strengthening the π-olefin $\rightarrow$ metal σ contribution to the bonding. The relative values of K for the various metals is explained by the availability of electron density on the zero-valent metal for metal $\rightarrow$ olefin π^* bonding. This is in the order Ni $\gg$ Pt > Pd.

Another route to Pd(0) complexes reported recently is by reaction of $[Pd(CNPh)_2]$ with activated monoolefins (*18*, *220*). With fumaronitrile maleic anhydride and p-quinones, 1:1 adducts are formed, whereas, with tetrachloro-p-benzoquinone, an adduct is formed in which the Pd-to-ligand ratio is 2:1. Another new procedure is the use of bis-(dibenzlideneacetone) palladium(0) which reacts directly with olefins (*145*).

b. *Acetylenes.* Acetylene complexes have been prepared from acetylenes having electron-withdrawing substituents. They are prepared either by replacement of PPh_3 from $(Ph_3P)_4Pd$ or by reduction of $(R_3P)_2PdCl_2$ with N_2H_4 in the presence of the acetylene (*74*, *75*). Dimethylacetylene-dicarboxylate was recently reported to react directly with bis(dibenzyl-ideneacetone) palladium(0) to give a complex (*146*). The reaction, however, was very sensitive to the nature of the ligands.

C. π-Allylic Complexes

In the past 3 years there has been considerable work on various aspects of π-allylic Pd(II) complexes. Because to treat these studies in detail would itself require an entire review, only a few of the recent advances will be discussed.

1. *Preparation*

One popular method of syntheses of π-allyl complexes is reaction of allylic chlorides with palladium chloride in the presence of a reducing agent such as CO/H_2O or $SnCl_2$ (*53, 206, 235, 269*), which apparently reduce Pd(II) to Pd(0) followed by oxidative addition of the allyl chloride. The list of reducing reagents that will promote the reaction has been extended to $TiCl_3$, Fe, Zn, and Cu (*170*).

Another popular method of synthesis of π-allyls is by reaction of an olefin containing an allylic hydrogen with $PdCl_2$ directly (*142, 143*):

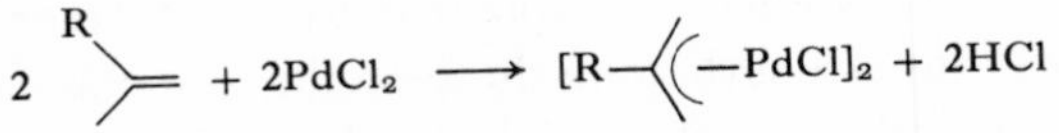

$$2\ \text{R-olefin} + 2PdCl_2 \longrightarrow [R\text{-}\pi\text{-allyl}\text{—}PdCl]_2 + 2HCl \quad (27)$$

Often a base such as NaOAc is necessary to make the reaction go smoothly (*281*). Basic solvents such as *N,N*-dimethylformamide (DMF) also aid in the formation of π-allyl compounds from 1-olefins (*200*); in this case $[(DMF)_2H][Pd_2Cl_6]$ is also formed. In a recent study of the formation of π-allylic complexes from 1-olefins in DMF under mild conditions, it was suggested that bases may not only shift the equilibrium to the right by neutralizing the acid but may also aid in removal of the proton from the allylic position (*40*). One possible mode of promotion might be the stabilization of intermediate Pd(IV) hydrido species. However, more work is required before the role of basic agents in Pd(II) catalysis is finally understood.

Several new types of allylic complexes have recently been prepared. Atkinson and Smith prepared π-allylic complexes from 2,2,4-trimethy-pent-3-en-1-ol of the type $\{[(CH_2\text{—}CHMeCH)CMe_2CH_2OH]PdCl\}_2$, in which the alcohol function is not coordinated to the Pd(II) (*10*). In contrast, the Rh(III) and Pt(II) complexes are olefin complexes, with the alcohol function complexed to the metal.

π-Allyl complexes with new coordinating groups in the other coordination positions of Pd(II) have also been reported. These include shift base complexes (structure IV) (*225*) and monothiodibenzoyl-methanatopalladium(II) complexes (structure V) (*169*). π-Allyl complexes containing the novel pyrazolide ligands have also been reported

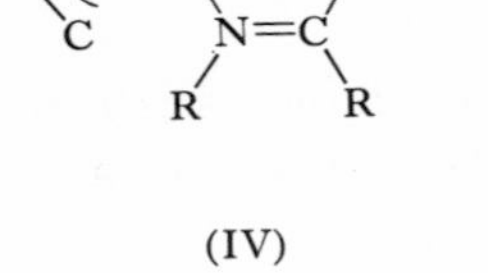

C C C Pd O N=C R R

(IV)

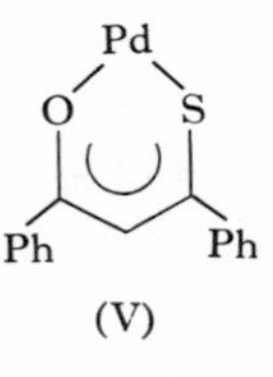

Pd O S Ph Ph

(V)

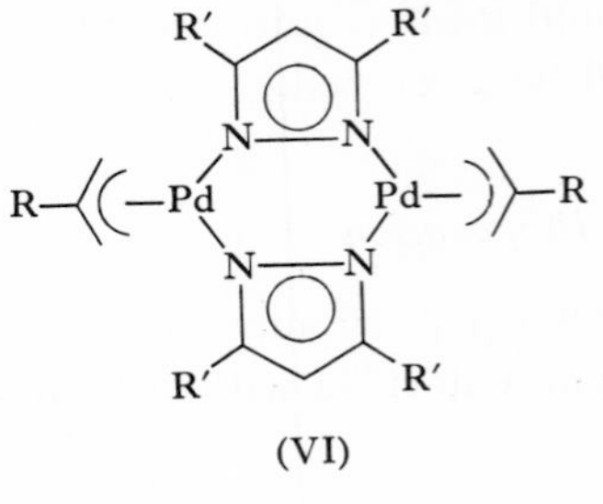

R′ R′ N N R Pd Pd R N N R′ R′

(VI)

(structure VI) (*264*). Finally, π-allyl complexes have been reported with azide or cyanide ligands (*241*). These can be mononuclear or contain cyanide or azide bridges.

A novel synthesis of oxo-π-allyl and π-allyl palladium complexes using diazo compounds has been reported by Yoshimura and co-workers (*298*). These reactions proceed by carbene insertion in the Pd(II)—Cl bond:

$$R{-}\underset{\underset{O}{\|}}{C}{-}CHN_2 + (PhCN)_2PdCl_2 \longrightarrow \left[R{-}C(O){=}CHCl\ \pi\text{-}PdCl\right]_2 \qquad (28)$$

$$CH_2{=}CH{-}CHN_2 + (PhCN)_2PdCl_2 \longrightarrow \left[(\pi\text{-}C_3H_4CH_2Cl)PdCl\right]_2 \qquad (29)$$

Equation (28) is of particular interest because a π-oxopropenylmanganese complex is the only such complex previously isolated.

Another new synthesis of π-allyl complexes is the reaction of divinyl carbinols with palladium(II) salts in methanol or ethylene glycols (*270*). In methanol, either 1,5- or 1,2-addition of methoxide occurred, whereas ethylene glycol gave a dioxanyl π-allylic palladium complex.

$$H_2C{=}CH{-}\underset{HO}{CH}{-}CH_2{=}CH_2 + Na_2PdCl_4 \longrightarrow \left[H_3COH_2C{-}C{\cdots}\underset{H}{C}{\cdots}C{-}CH_2OCH_3\ (\pi\text{-}PdCl)\right]_2 \qquad (30)$$

$$C_6H_5{-}CH{=}CH{-}\underset{OH}{CH}{-}CH_2{=}CH_2 + Na_2PdCl_4 \longrightarrow \left[C_6H_5{-}\underset{H}{C}{\cdots}\underset{H}{C}{\cdots}\underset{H}{C}{-}\underset{OCH_3}{CH}CH_2OCH_3\ (\pi\text{-}PdCl)\right]_2 \qquad (31)$$

$$H_2C{=}\underset{CH_3}{C}{-}\underset{OH}{CH}{-}CH{=}CH_2 + (CH_2OH)_2 + 2Na_2PdCl_4 \longrightarrow \left[H{-}\underset{H}{C}{\cdots}\underset{CH_3}{C}{\cdots}CH{-}CH(OCH_2CH_2CH_2O)\ (\pi\text{-}PdCl)\right]_2 \qquad (32)$$

New cationic Pd(II) complexes have also been reported. Abstraction of chloride or hydride from γ positions gave a novel tetrahapto structure, rather than the previously reported π-allylcarbonium ion structure (*171*).

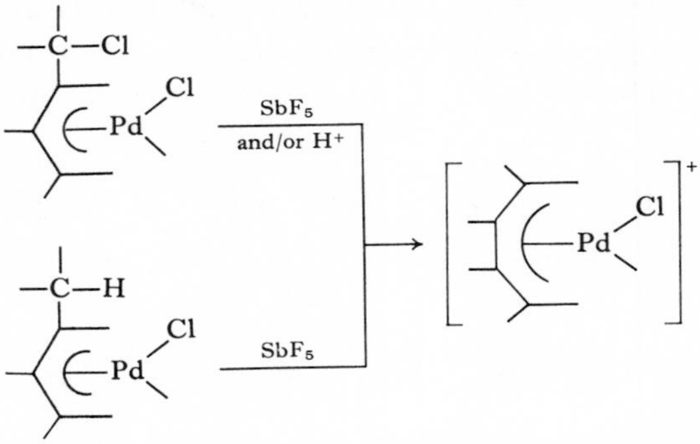

$$\text{—C—Cl (}\pi\text{-allyl)PdCl} \xrightarrow[\text{and/or H}^+]{\text{SbF}_5} ;\quad \text{—C—H (}\pi\text{-allyl)PdCl} \xrightarrow{\text{SbF}_5} \left[\text{(tetrahapto)PdCl}\right]^+ \tag{33}$$

Chloride extraction from [2-(dichloromethyl)-π-allyl] palladium chloride gave a trimethylenemethane palladium complex (*171*):

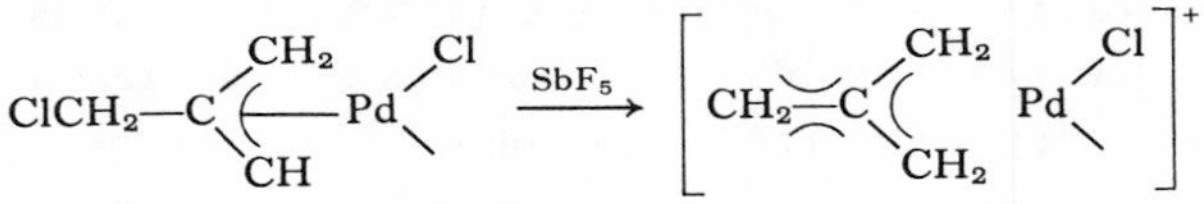

$$\text{ClCH}_2\text{—C(CH}_2\text{)(CH)—PdCl} \xrightarrow{\text{SbF}_5} \left[\text{CH}_2\text{=C(CH}_2\text{)(CH}_2\text{) PdCl}\right]^+ \tag{34}$$

Cationic π-allyl compounds have also been prepared by removal of chloride using $AgPF_6$ (*236*):

$$\left[(\pi\text{-allyl})\text{PdCl}\right]_2 + 2\text{AgPF}_6 \longrightarrow 2[\text{Pd}(\pi\text{-allyl})]^+ + 2\text{AgCl} \tag{35}$$

The coordinately unsaturated species react rapidly with ligands such as dipyridyl, diolefins, and phosphines to give the coordinately saturated species.

An interesting series of chiral π-allyl complexes have recently been prepared and resolved (*60*, *172*). Studies of the epimerization and syn- and anticonfigurations of these complexes have elucidated the mechanisms of their geometric and optical isomerization.

III
REACTIONS OF MONOOLEFINS

A. Oxidation Reactions

1. *Reaction in Aqueous Solution*

No doubt the most studied reaction in Pd(II) chemistry is the basic Wacker process—the oxidation of ethylene to acetaldehyde in H_2O.

This reaction was known for a number of years (*219*), but it was the discovery by Smidt and co-workers that addition of $CuCl_2$ prevented the precipitation of Pd(0) which made the reaction commercially interesting (*244*). Since CuCl is readily regenerated by O_2, the net reaction is the oxidation of ethylene by air to acetaldehyde:

$$PdCl_4^{2-} + C_2H_4 + H_2O \longrightarrow CH_3CHO + Pd(0) + 2HCl + 2Cl^- \quad (36)$$

$$2Cl^- + Pd(0) + 2CuCl_2 \longrightarrow PdCl_4^{2-} + 2CuCl \quad (37)$$

$$2CuCl + \tfrac{1}{2}O_2 + 2HCl \longrightarrow 2CuCl_2 + H_2O \quad (38)$$

$$C_2H_4 + \tfrac{1}{2}O_2 \longrightarrow CH_3CHO \quad (39)$$

There is general agreement that at low [Pd(II)] the rate expression is (*107, 152, 199*)

$$\frac{-d[C_2H_4]}{dt} = \frac{k[PdCl_4^{2-}][C_2H_4]}{[Cl^-]^2[H^+]} \quad (40)$$

and the generally accepted mechanism is as follows:

$$PdCl_4^{2-} + C_2H_4 \overset{K}{\rightleftharpoons} PdCl_3(C_2H_4)^- + Cl^- \quad (41)$$

$$PdCl_3(C_2H_4)^- + H_2O \rightleftharpoons PdCl_2(H_2O)(C_2H_4) + Cl^- \quad (42)$$

$$H_2O + PdCl_2(H_2O)(C_2H_4) \rightleftharpoons PdCl_2(OH)(C_2H_4) + H_3O^+ \quad (43)$$

$$Cl_2Pd(OH)(CH_2{=}CH_2) \xrightarrow{H_2O} Cl_2Pd(H_2O)(CH_2CH_2OH) \quad (44)$$

$$Cl_2Pd(H_2O)(CH_2CH_2OH) \longrightarrow CH_3CHO + Pd(0) + 2Cl^- \quad (45)$$

Although there is now agreement about the general scheme there is still discussion as to some details of the mechanism. First there is the mode of hydroxypalladation. The original suggestion was that the insertion occurred by attack of the coordinated OH on the coordinated ethylene in the slow step of the reaction.

$$[Cl_2Pd(OH)(CH_2{=}CH_2)]^- \xrightarrow[H_2O]{slow} [Cl_2Pd(H_2O)(CH_2CH_2OH)]^- \quad (46)$$

This step was designated the slow step on the basis of isotope effects. Deuterium-labeling experiments indicate that the decomposition step requires a 1,2-shift of halogen to form the acetaldehyde product (*244*). Thus, if ethylene is oxidized in D_2O, nondeuterated acetaldehyde is produced.

$$\mathrm{Cl_2(D_2O)Pd{-}CH_2CH_2{-}OD} \xrightarrow[\mathrm{D_2O}]{} \mathrm{CH_3CHO + DCl + Cl^- + Pd(0)} \tag{47}$$

Now, if a 1,2-hydride shift occurs, an isotope effect greater than 1 would be expected if C_2D_2 was used in place of C_2H_4. The actual value of k_H/k_D is about 1, suggesting that the hydroxypalladation step, in which C—H bonds are not broken, is the rate-determining step (*107*).

It has recently been suggested that hydroxypalladation occurs by attack of OH^- on the aquo π complex, a route that is also consistent with the kinetics (*84*).

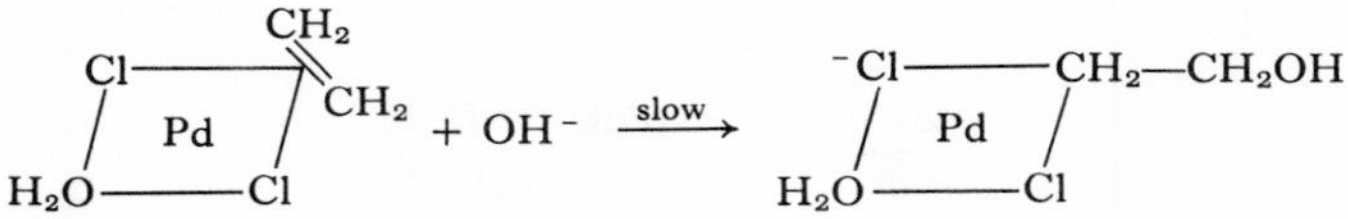

$$\begin{array}{ccc} \mathrm{Cl} & \text{———} & \mathrm{CH_2{=}CH_2} \\ & \mathrm{Pd} & \\ \mathrm{H_2O} & \text{———} & \mathrm{Cl} \end{array} + \mathrm{OH^-} \xrightarrow{\text{slow}} \begin{array}{ccc} {}^{-}\mathrm{Cl} & \text{———} & \mathrm{CH_2{-}CH_2OH} \\ & \mathrm{Pd} & \\ \mathrm{H_2O} & \text{———} & \mathrm{Cl} \end{array} \tag{48}$$

However, as has been pointed out previously, this route can be eliminated because it would require a rate constant that is faster than that for a diffusion-controlled process (*107*). This, of course, is impossible.

Another route that cannot be eliminated is attack of water on the hydroxy π complex. However, it is not apparent why the anionic hydroxy complex should be attacked any more readily than the trichloro π complex formed by Eq. (41) (see Section VI).

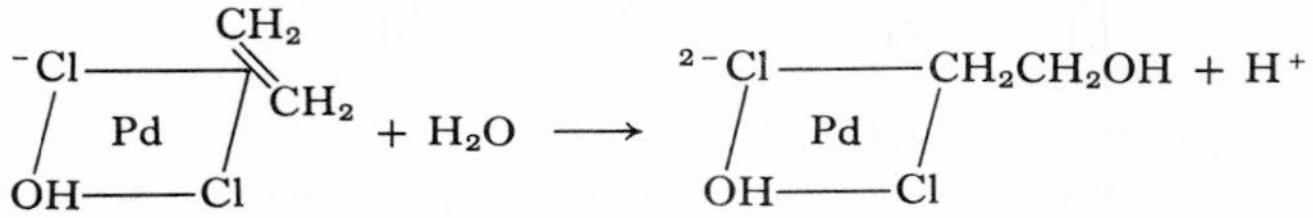

$$\begin{array}{ccc} {}^{-}\mathrm{Cl} & \text{———} & \mathrm{CH_2{=}CH_2} \\ & \mathrm{Pd} & \\ \mathrm{OH} & \text{———} & \mathrm{Cl} \end{array} + \mathrm{H_2O} \longrightarrow \begin{array}{ccc} {}^{2-}\mathrm{Cl} & \text{———} & \mathrm{CH_2CH_2OH} \\ & \mathrm{Pd} & \\ \mathrm{OH} & \text{———} & \mathrm{Cl} \end{array} + \mathrm{H^+} \tag{49}$$

Another mechanistic proposal is the following (*152*):

$$\mathrm{Cl_2(H_2O)Pd(C_2H_4)} \overset{\mathrm{H_2O}}{\rightleftharpoons} {}^{-}\mathrm{Cl_2(H_2O)PdCH_2CH_2OH} \tag{50}$$

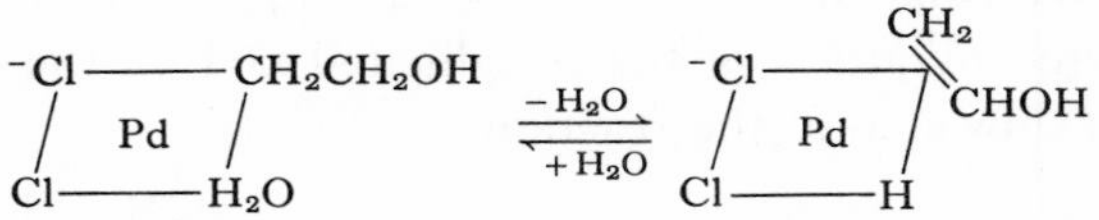

$$\begin{array}{ccc} {}^{-}\mathrm{Cl} & \text{———} & \mathrm{CH_2CH_2OH} \\ & \mathrm{Pd} & \\ \mathrm{Cl} & \text{———} & \mathrm{H_2O} \end{array} \underset{+\mathrm{H_2O}}{\overset{-\mathrm{H_2O}}{\rightleftharpoons}} \begin{array}{ccc} {}^{-}\mathrm{Cl} & \text{———} & \mathrm{CH_2{=}CHOH} \\ & \mathrm{Pd} & \\ \mathrm{Cl} & \text{———} & \mathrm{H} \end{array} \tag{51}$$

$$\begin{array}{ccc} {}^{-}\mathrm{Cl} & \text{———} & \mathrm{CH_2{=}CHOH} \\ & \mathrm{Pd} & \\ \mathrm{Cl} & \text{———} & \mathrm{H} \end{array} \underset{-\mathrm{H_2O}}{\overset{\mathrm{H_2O}}{\rightleftharpoons}} \begin{array}{ccc} {}^{-}\mathrm{Cl} & \text{———} & \mathrm{CH(OH)CH_3} \\ & \mathrm{Pd} & \\ \mathrm{Cl} & \text{———} & \mathrm{H_2O} \end{array} \tag{52}$$

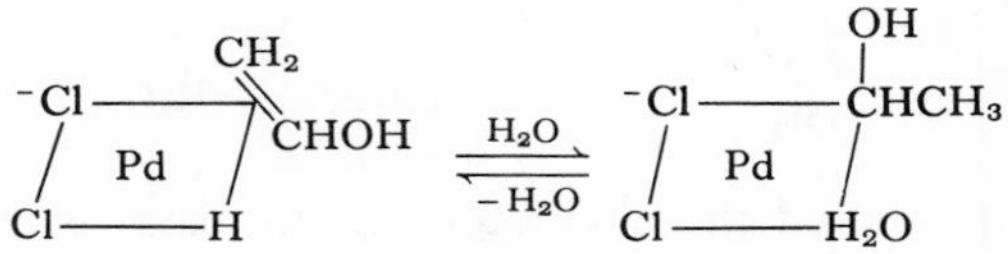

$$\begin{array}{l} {}^{-}\text{Cl}-\!-\!-\text{CH(OH)}-\text{CH}_3 \\ \;/\quad \text{Pd} \quad / \\ \text{Cl}-\!-\!-\text{H}_2\text{O} \end{array} \xrightarrow[\text{H}_2\text{O}]{\text{slow}} \text{CH}_3-\text{CH(OH)}_2 + \text{PdCl}_2^{2-} \qquad (53)$$

$$\text{CH}_3\text{CH(OH)}_2 \longrightarrow \text{CH}_3\text{CHO} + \text{H}_3\text{O}^+$$

In this sequence the hydroxypalladation is no longer the slow step but an equilibrium. It is argued that the low isotope effects occur because the deuterium isotope effects in Eqs. (51) and (52) would tend to cancel giving a low net isotope effect. This is a reasonable argument since the actual isotope effect for decomposition was not known. However, this isotope effect was recently measured using CHDCHD in a competitive experiment (*109*).

$$^{-}\text{Cl}_2(\text{OH})\text{Pd}(\text{CHDCHD}) \xrightarrow{\text{H}_2\text{O}} {}^{-}\text{Cl}_2(\text{H}_2\text{O})\text{PdCHDCHDOH} \begin{cases} \xrightarrow{\text{H shift}} \text{CH}_2\text{DCDO} \\ \xrightarrow{\text{D shift}} \text{CHD}_2\text{CHO} \end{cases} \qquad (54)$$

The ratios of the two deuterated acetaldehydes, which is a measure of the isotope effect, is easily measured by mass spectrometry. The value of k_H/k_D was found to be 1.7 which is higher than the experimentally determined value of k_H/k_D. This is strong evidence that hydroxypalladation is the slow step and, thus, at least part of this scheme is incorrect.

The proposed mode of decomposition, however, deserves serious consideration. This suggestion was made as an alternative to a mode of decomposition originally suggested by the writer in 1964. In this scheme, Pd(II) is viewed as promoting an incipient hydride shift.

$$\left[\text{H}\cdots\text{O}\!=\!\text{C(H)(H)}-\text{C(H)(H)}\cdots\text{Pd(Cl)}_2\right]^{-} \longrightarrow \text{CH}_3\text{CHO} + \text{Pd(0)} + 2\text{Cl}^{-} \qquad (55)$$

However, before that shift can be completed, a rapid rearrangement of electrons occurs to give a carbonyl that puts a positive charge on the carbon not containing the hydroxyl. This is neutralized by the hydride to give, in effect, a Pd(II)-*assisted* hydride shift. Now this mode of decomposition was proposed to emphasize two aspects of Pd(II) chemistry not generally recognized at that time. First, in contrast to oxidations by other metal ions, which proceed via oxymetallation adducts, the decomposition does not occur by way of carbonium-type mechanisms. For instance, in the case of Tl^{3+} the decomposition of the intermediate gives both glycol and acetaldehyde by a type of decomposition in which thallium acts as an electron sink (*102*).

$$Tl^{3+} + C_2H_4 + H_2O \longrightarrow {}^{2+}Tl\text{—}CH_2CH_2OH \longrightarrow$$

$$\begin{array}{l} \xrightarrow{\sim H} CH_3CHO \\ {}^{+}Tl\cdots{}^{\oplus}CH_2\text{—}CH_2OH \\ \xrightarrow{H_2O} (CH_2OH)_2 + H^{+} \end{array} \tag{56}$$

The intermediate hydroxypalladation adduct does not decompose in this manner because Pd(0) would be the leaving group as opposed to Tl^{+} in the Tl^{3+} oxidations. Monomeric Pd(0) is unstable in aqueous solution and its formation would add considerably to the activation energy.

Second, the proposed decomposition emphasized that decomposition occurred by some type of Pd(II)–β-hydrogen interaction. This fact was also not generally recognized at the time the proposal was made.

A mechanism such as that given in Eqs. (50)–(53) was not considered because it was felt that the vinyl π complex would decompose to acetaldehyde which would contain a deuterium if the reaction was run in D_2O. Moreover, the final decomposition does involve a solvolysis reaction, and, for the reason stated above, Pd(II) alkyls do not tend to decompose in this fashion in the absence of added oxidants. Since the original proposal was made, stable vinyl alcohol complexes of Pt(II) have been prepared (*271*), therefore, kinetically stable Pd(II) complexes are possible as well. Also, the presence of an α-OH group would be expected greatly to increase rates of solvolysis, and there is recent evidence that electron-releasing groups facilitate depalladation (Section V, A). Furthermore, this mechanism is more satisfying in that it pictures definite hydride eliminations and readditions rather than vague dotted lines. In addition, the mechanism can be slightly altered to be more in keeping with known Pd(II) chemistry if Pd(II) is pictured as leaving as a hydride.

$$CH_3CH(OH)(Pd\lessdot) \longrightarrow CH_3CHO + HPd\lessdot \tag{57}$$

Another possibility is that another Pd(II) aids in the removal to make the leaving group a Pd(I) dimer which would eventually disproportionate to Pd(0) and Pd(II):

$$\underset{\underset{H_2O}{\uparrow}}{CH_3\overset{OH}{CH}}\cdots Pd(II)\cdots Pd(II) \longrightarrow CH_3\underset{OH}{\overset{OH}{|}}CH + Pd(I)—Pd(I) + H^+ \longrightarrow Pd(0) + Pd(II) \quad HgCl_2 + CH_3CH \tag{58}$$

Palladium(I) dimers are reported to be the Pd(II) reduction product in the coupling of aromatics.

In practice the two are very similar, and it would be difficult to distinguish between them. One possibility is reaction of $CH_3\overset{OH}{\overset{|}{C}}HHgCl$ with $PdCl_2$ in D_2O to see if the acetaldehyde product contains deuterium. If the Pd(II)–carbon bond does undergo solvolysis the acetaldehyde would contain no deuterium.

$$CH_3\overset{OH}{\overset{|}{C}}H—HgCl + PdCl_4^- \longrightarrow HgCl_2 + CH_3\overset{OH}{CH}\diagdown PdCl_3^{2-} \tag{59}$$

$$CH_3\overset{OH}{\overset{|}{C}}HPdCl_3^{2-} \begin{cases} \xrightarrow{H_2O} CH_3\overset{OH}{C}HOH_2 + PdCl_2 \xrightarrow{D_2O} CH_3CHO \\ \longrightarrow CH_2{=}\overset{OH}{C}H + HPdCl_3^- \xrightarrow{D_2O} CH_2DCHO \end{cases} \tag{60}$$

Actually this type of proposal is more important in nonaqueous solvents where different types of products would be formed (see Sections III, A, 2 and 3).

Another serious discussion concerning mechanism relates to the form of the rate expression at high [Pd(II)]. Thus, Moiseev and co-workers have reported that, at high [Pd(II)] concentrations, a second term that is second order in $[PdCl_4^{2-}]$ becomes important (*194*, *198*). The total rate expression thus becomes

$$\text{rate} = k_I \frac{[PdCl_4^{2-}][C_2H_4]}{[Cl^-]^2[H_3O^+]} + k_{II} \frac{[PdCl_4^{2-}]^2[C_2H_4]}{[Cl^-]^3[H_3O^+]} \tag{61}$$

They suggest the second term arises from the interaction $PdCl_4^{2-}$ with the hydroxy π complex to give a dimeric species:

$$Cl_2Pd(OH)(C_2H_4) + PdCl_4^{2-} \rightleftharpoons Cl_2PdCl_2Pd(OH)Cl(C_2H_4)^{2-} + Cl^- \tag{62}$$

This dimeric species then undergoes the hydroxypalladium step faster than does the monomeric species in Eq. (45). This is an interesting suggestion and would indicate pathways in Pd(II) chemistry not previously suspected.

However, a serious doubt to this mechanism has been raised. Moiseev and co-workers used a reaction system without a gas phase to measure their kinetics. In a recent study in which a gas phase was present no evidence for the second term was found (*110*). It was suggested that the second term in Moiseev's work may have resulted from a systematic error arising from his method of measurement. Moiseev believes the failure to observe the second term results from mass transfer limitations of gas-to-liquid mixing in the reactor using a gas phase (*195*). This does not seem to be a complete explanation because a wide variety of reaction conditions and rates were studied and all gave values of k' in Eq. (40) which were in experimental error of that found at low [Pd(II)]. Furthermore, these reactors had been used previously to measure rates of this magnitude. However, it must be stated that the reactor used was not the highest-efficiency-type mixing apparatus such as was used in the original studies of the Wacker reaction (*107*). Thus the question will not be conclusively settled until gas phase reactors with very high gas–liquid mixing are used to study the Wacker reaction at high [Pd(II)].

Another point to be settled is the effect of $CuCl_2$ or reaction rate in the catalytic system containing both Pd(II) and $CuCl_2$. Matveev and co-workers suggested two paths because addition of $CuCl_2$ increased the rate (*187*). However, as Aquiló has pointed out (*2*), $CuCl_2$ could merely be decreasing the Cl^- concentration by forming $CuCl_3^-$. The kinetic studies of François, however, do suggest a path catalytic in $CuCl_2$ (*67*, *68*). This is reasonable since such a path is definitely operative in acetic acid (Section III, A, 2, b). However, as found in acetic acid, one would expect different products (i.e., $HOCH_2CH_2Cl$) for the $CuCl_2$-catalyzed path. Actually, 2-chloroethanol is the main product at high $[CuCl_2]$ and $[Cl^-]$ concentrations (*245*). Thus the $CuCl_2$-catalyzed path appears to be operative, but careful studies of the equilibria, kinetics, and product distributions in the system will be necessary before the exact mechanism will be known.

A study of the reaction on low $[H^+]$ and $[Cl^-]$ gave a kinetic expression of the form (*149*, *152*)

$$\text{rate} = \frac{a[H^+][Cl^-]}{b + [H^+]^2[Cl^-]^3} \tag{63}$$

which was interpreted in terms of the following equilibria:

$$PdCl_4^{2-} + C_2H_4 + 2H_2O \overset{K_1}{\rightleftharpoons} [C_2H_4PdCl(OH)_2]^- + 3Cl^- + 2H^+ \quad (64)$$

$$[C_2H_4PdCl(OH)_2] + H^+ + Cl^- \overset{K_2}{\rightleftharpoons} cis\text{-}[C_2H_4PdCl_2(OH)]^- + H_2O \quad (65)$$

$$cis\text{-}[C_2H_4PdCl_2(OH)]^- \overset{k}{\longrightarrow} CH_3CHO + Pd + 2Cl^- + H^+ \quad (66)$$

If K_2 is assumed small, the rate expression

$$\text{rate} = \frac{kK_1K_2[Pd][C_2H_4][H^+][Cl^-]}{K_1[C_2H_4] + [H^+]^2[Cl^-]^3} \quad (67)$$

is derived (Pd = total palladium). The equation is of the same form as Eq. (63). Now this scheme cannot be completely eliminated, but, as has been pointed out (*111*), at low $[H^+]$ and $[Cl^-]$ the equilibrium changes and the observed kinetics in this region most likely reflect these changes rather than any scheme such as given above. Thus, until equilibrium studies are carried out in this region of $[H^+]$ and $[Cl^-]$, the above scheme should be viewed as tentative.

An interesting variation on the oxidation of olefins to carbonyl compounds has been described by Rodeheaver and Hunt (*227*). These workers prepared the hydroxy mercurials and exchanged these mercurials with $PdCl_2$ to give the hydroxypalladation adduct which then decomposed in the expected fashion to give ketones:

$$R'CH(OH){-}CH(HgCl)R + PdCl_4^{2-} \longrightarrow R'CH(OH){-}CH(PdCl_3^{2-})R + HgCl_2 \quad (68)$$

$$\downarrow$$

$$R'C(=O){-}CH_2R$$

Apparently this procedure gives higher yields of carbonyl products than obtained with Pd(II) and the olefin alone. This procedure overcomes the low solubility of the olefins in aqueous solution. There is the possibility that the reaction proceeds via deoxymercuration, followed by direct reaction of Pd(II) with the olefin. However this seems unlikely, and the transmetallation reaction for oxymercurials have been demonstrated previously and the yields using mercurials are much higher than in the case of direct reaction with olefins.

There have been several recent studies of the oxidation of allylic alcohols and chlorides with Pd(II) salts in aqueous solution. Allyl alcohols have been known to be oxidized to acrolein, and both crotyl and α-methallyl alcohol give mixture of crotonaldehyde and methyl vinyl ketone. Jira proposes the following scheme for this oxidation (*148*):

$$H_3C—CH{=}CHCH_2OH \xrightarrow[H_2O]{PdCl_2} H_3C—CO—CH_2—CH_2OH \longrightarrow H_3C—COCH{=}CH_2 \quad (69)$$

$$H_3C—CH(OH)—CH{=}CH_2 \xrightarrow[H_2O]{PdCl_4^{2-}} H_3CCH(OH)—CH_2—CHO \longrightarrow H_3CCH{=}CH—CHO \quad (70)$$

$$CH_2{=}CHCH_2OH \xrightarrow[H_2O]{PdCl_4^{2-}} OCH—CH_2—CH_2OH \longrightarrow OCHCH{=}CH_2 \quad (71)$$

This scheme would suggest that acrolein is formed by dehydration of β-hydroxypropionaldehyde. This aldehyde was detected in the reaction mixture, providing evidence for the proposed route.

Another reaction that has received continuing attention is the formation of π-allyl Pd(II) complexes from allylic alcohols, chlorides, and ethers. In pure allyl alcohol, $PdCl_2$ catalyzes the disproportionation of allyl alcohol to propylene and a product $[C_6H_{10}O_2]$ which results from oxidative dimerization of allyl alcohol (*79*):

$$3C_3H_5OH \xrightarrow{PdCl_2} C_3H_6 + \begin{matrix} H_2C{=}C—CH_2 \\ | \quad\; | \\ H_2C \quad CH \\ \diagdown O \diagup \diagdown CH_2OH \end{matrix} + H_2O \quad (72)$$

As this reaction proceeds, $[\pi\text{-allyl PdCl}]_2$ is gradually formed. Since the π-allyl group represents a reduction of allyl alcohol, it has been suggested that allyl alcohol reacts with $PdCl_2$ to give the π-allyl complex and $C_6H_{10}O_2$ (*150*):

$$6C_3H_5OH + 2PdCl_2 \longrightarrow (\pi\text{-}C_3H_5PdCl)_2 + 2C_6H_{10}O_2 + 2HCl + H_2O \quad (73)$$

A study of π-allyl formation from allyl alcohol in aqueous solution indicated the oxidation product was acrolein (*222*):

$$4C_3H_5OH + 2PdCl_4^{2-} \longrightarrow (\pi\text{-}C_3H_5PdCl)_2 + 2CH_2{=}CHCHO \quad (74)$$

The kinetics were identical to that found for the oxidation of simple olefins in aqueous solution which led to the following mechanism:

$$\begin{matrix} Cl \\ \quad\diagdown \\ Pd \\ \quad\diagup \\ Cl \end{matrix}\!\!\begin{matrix} CH_2 \\ \| \\ CHCH_2OH \\ \diagdown OH \end{matrix} \xrightarrow[H_2O]{slow} \begin{matrix} Cl \quad\; CH_2CH(OH)CH_2OH \\ \diagdown Pd \diagup \\ Cl \diagup \quad \diagdown H_2O \end{matrix} \xrightarrow[C_3H_5OH]{fast} \begin{matrix} H_2C \\ CH \leftarrow PdCl_2 \\ CH_2 \end{matrix} + CH_2{=}CHCHO \quad (75)$$

The exact way in which the fast conversion was made is unclear. Kinetic studies on the reaction with diallyl ether in aqueous methanol led to the suggestion of a similar mechanism in which the last step was again not clearly defined (*221*).

$$2(CH_2{=}CHCH_2)_2O + 2PdCl_4^{2-} \longrightarrow [(\pi\text{-}C_3H_5)PdCl]_2 + 3Cl^- + CH_2{=}CHCHO \tag{76}$$

Crotyl chloride or 3-chloro-1-butene reacts in aqueous media to give π-allylic complexes and a hydroxyl carbonyl compound, which reacts again to give a dicarbonyl (*151*):

$$2C_4H_7Cl + PdCl_2 + H_2O \longrightarrow HC\begin{smallmatrix}\diagup CH_2 \\ \diagdown CH \\ | \\ CH_3\end{smallmatrix}\!\!\leftarrow Pd\begin{smallmatrix}\diagup Cl \\ \ddots\end{smallmatrix} + \underset{(C_4H_8O_2)}{H_3CCHOHCOCH_3} \tag{77}$$

$$C_4H_7Cl + C_4H_8O_2 + PdCl_2 \longrightarrow HC\begin{smallmatrix}\diagup CH_2 \\ \diagdown CH \\ | \\ CH_3\end{smallmatrix}\!\!\leftarrow Pd\begin{smallmatrix}\diagup Cl \\ \ddots\end{smallmatrix} + H_3CCOCOCH_3 \tag{78}$$

It would seem that the mechanisms for these reactions are exactly analogous to that for formation of π-allyls from allylic chlorides plus reducing agents (Section II, C, 1). The mechanism, in its simplest form, amounts to reduction of Pd(II) to Pd(0) which oxidatively adds allyl chloride, alcohol, or ether to give the π-allyl. Thus, with H_2O/CO as reductant in the presence of allyl chloride the scheme would be

$$PdCl_4^{2-} + CO + H_2O \longrightarrow {}^{-2}Cl_3Pd\overset{\overset{O}{\|}}{C}{-}OH + Cl^- + H^+ \longrightarrow Pd(0) + CO_2 + HCl + 2Cl^- \tag{79}$$

$$Pd(0) + C_3H_5Cl \longrightarrow CH\begin{smallmatrix}\diagup CH_2 \\ \diagdown CH_2\end{smallmatrix}\!\!\leftarrow Pd\begin{smallmatrix}\diagup Cl \\ \ddots\end{smallmatrix} \tag{80}$$

When no other reductant is present, the allylic group itself is oxidized to give Pd(0) which can then add oxidatively to give the π-allyl complex.

A reaction closely related to olefin oxidation in aqueous solution is the oxidation of cyclopropanes by Na_2PdCl_4 in 2:1 glyme–water solvents (*210*). The direct oxidation product of phenyl cyclopropane is propiophenone. If the reaction is carried out in glyme–D_2O, no deuterium is incorporated in the product. A deuterium-labeling experiment suggested isomerization of the original oxypalladation adduct by Pd(II)–hydride intermediates:

$$PdCl_4^{2-} + C_6H_5\text{-}(\text{D-cyclopropane}) \xrightarrow{H_2O} C_6H_5C(D)(OH)\text{-}CH(H)\text{-}CH_2Pd \rightleftharpoons C_6H_5C(D)(OH)\text{-}CH{=}CH_2 \;(\leftarrow Pd\text{—}H) \rightleftharpoons C_6H_5C(D)(OH)\text{-}CH(Pd)\text{-}CH_2H \qquad (81)$$

As in the Wacker reaction, the 1,2-oxypalladation adduct then decomposes by a deuterium shift.

$$C_6H_5C(D)(OH)\text{-}CH(Pd)\text{-}CH_2H \longrightarrow C_6H_5C(=O)\text{-}CH(D)\text{-}CH_2H \qquad (82)$$

There is a precedent for the isomerization of oxypalladation adducts (Section III, A, 2a and b), but it is surprising that no allylic alcohol is formed simply by Pd(II)–hydride elimination from the original oxypalladation adduct.

2. *Reactions in Acetic Acid*

a. *In the absence of Added Oxidant.* Oxidation of ethylene in acetic acid solvent gives a variety of products including vinyl acetate, ethylidene diacetate, acetaldehyde, acetic anhydride, ethylene glycol, mono- and diacetates, and β-chloroethylacetate (*174*). Although not recognized until recently (*112*), the last three products are obtained only in the presence of other oxidants, such as $CuCl_2$, as discussed in Section III, A, 2, b. The acetaldehyde and acetic anhydride are formed from a secondary Pd(II)-catalyzed decomposition of vinyl acetate (see Section III, B, 1). The primary products then, in the absence of oxidants, are vinyl acetate and ethylidene diacetate. Actually at 25°C, with palladium acetate alone, little if any ethylidene diacetate is formed. It is reported to be a major product at higher temperatures (*275*), but because the importance of added oxidants was not realized, it is not clear whether or not it is an important product in the absence of oxidants.

The vinyl acetate is almost certainly formed by acetoxypalladation followed by Pd(II)–hydride elimination:

$$\gtrless Pd(OAc) + C_2H_4 \longrightarrow \gtrless PdCH_2CH_2OAc \xrightarrow{-(\gtrless PdH)} CH_2{=}CHOAc \quad (83)$$

Since Pd(II)–hydride is generally not stable to decomposition to Pd(0) and H^+, the intermediate hydride has not been isolated. Stable Pd(II) hydrides have been reported, but this is not direct evidence they are intermediates in the oxidation. However, Maitlis, in his model system for Pd(II) catalysis, has been able to demonstrate Pd(II)–hydride elimination to give a stable hydride (*133*, *182*) (see Section V, A).

A study of the products resulting from oxidation of straight-chain olefins has indicated that these olefins are oxidized by the same mechanism (*163*). In this case products that indicate Pd(II) movement down the chain are formed. This apparently proceeds by way of Pd(II)–hydride eliminations and readdition. For instance, in the oxidation of 1-butene, some 3-buten-1-yl acetate is formed:

$$CH_3CH_2CH{=}CH_2 + \gtrless Pd{-}OAc \longrightarrow CH_3CH_2\underset{\underset{Pd\lesseqgtr}{|}}{CH}{-}CH_2OAc \longrightarrow CH_5CH\underset{\underset{Pd}{\downarrow}}{=}CH{-}CH_2OAc$$

$$\longrightarrow CH_3\underset{\underset{Pd\lesseqgtr}{|}}{CH}{-}CH_2CH_2OAc \longrightarrow CH_2\underset{\underset{HPd\lesseqgtr}{\downarrow}}{=}CHCH_2CH_2OAc \longrightarrow CH_2{=}CHCH_2CH_2OAc + H^+ + Pd(0) \quad (84)$$

The ethylidene diacetate could be formed by a mechanism similar to that proposed by Smidt and co-workers for the Pd(II) oxidation of ethylene in water (*244*) (Section III, A, 1).

$$\gtrless Pd{-}CH_2CH_2OAc \longrightarrow CH_2\underset{\underset{HPd\lesseqgtr}{\downarrow}}{=}CHOAc \longrightarrow CH_3{-}\underset{\underset{Pd\lesseqgtr}{|}}{CHOAc} \xrightarrow{OAc^-} CH_3CH(OAc)_2 + Pd(0) \quad (85)$$

In keeping with this mechanism, Moiseev and Vargaftik reported that when the reaction was carried out in CH_3COOD, no deuterium was incorporated in the ethylidene diacetate (*197*). However, no experimental details were given, and no complete report has appeared. Because of the complications of running the reaction in the presence of oxidants, this system needs further study.

The oxidation of cyclohexene has been the subject of considerable discussion, and it is now apparent that it behaves differently from the straight-chain olefins. Cyclohexene was originally reported to yield both cyclohex-2-en-1-yl acetate, structure (VII), and cyclohex-3-en-1-yl acetate, structure (VIII), in chloride-containing acetic acid (*76*) and only the allylic isomer with $Pd(OAc)_2$ in chloride-free acetic acid (*6*). However, it has now been demonstrated that if no oxidants are present to regenerate the Pd(0) to Pd(II) in neutral or basic HOAc, the Pd(0) formed will disproportionate the cyclohexene to give benzene (*22*, *295*). In acetic acid containing perchloric acid, cyclohexanone (structure VIII) and cyclohex-1-en-1-yl acetate are formed (*22*). If Pd(0) is prevented from precipitating by use of oxidants in neutral or basic acetic acid, the allylic and homoallylic acetates are formed.

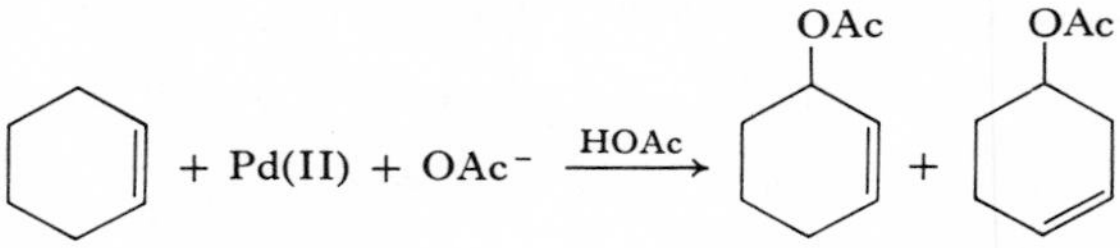

$$\text{cyclohexene} + Pd(II) + OAc^- \xrightarrow{HOAc} \text{(VII)} + \text{(VIII)} \qquad (86)$$

Studies of the oxidation of 3,3,6,6-d_4-cyclohexene (*127*, *295*) indicated the allylic product arose from a π-allyl intermediate, whereas the homoallylic acetate arose from the mechanism suggested for oxidation of straight-chain olefins. Thus the allylic product was a 50:50 mixture of the two deuterium-labeled isomers, structures (IX) and (X), which would be expected from a symmetrical intermediate. An acetoxypalladation mechanism would have predicted only X. It has also been demon-

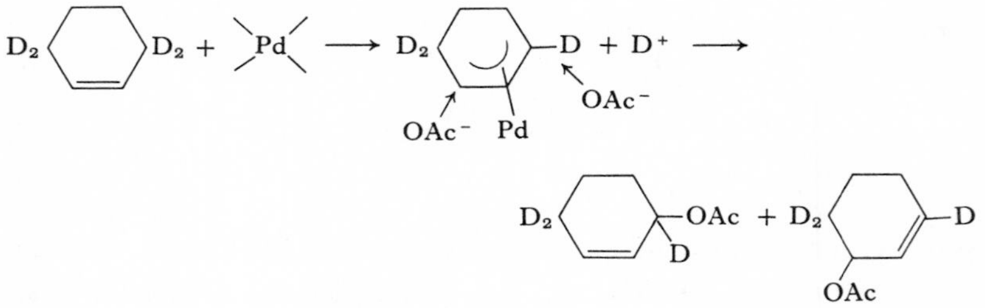

$$D_2\text{-cyclohexene-}D_2 + Pd \longrightarrow \pi\text{-allyl-Pd complex} + D^+ \longrightarrow \text{(IX)} + \text{(X)} \qquad (87)$$

strated that the π-allylic intermediate will decompose to allylic acetate (*296*).

The homoallylic acetate contained all the deuteriums initially present in the deuterated cyclohexene. One of the deuteriums had been stereospecifically transferred to an adjacent carbon. The stereochemistry of the product is most consistent with trans acetoxypalladation and cis Pd(II)–H(D) eliminations and additions (*127*, *295*):

(88)

Free-radical processes are also present in the cyclohexene oxidation, giving cyclohex-2-en-1-ol and cyclohex-2-en-1-one (*22*).

The oxidation of hex-1-ene, hex-*cis*-2-ene, and 3,3-dimethylbut-1-ene catalytically under O_2 proved to be complicated. Hex-1-ene gave mainly hex-1-en-2-yl acetate, the product expected by analogy with the acetoxypalladation Pd(II)–hydride mechanism. 3,3-Dimethylbut-1-ene also gave the expected product (*23*). Hex-*cis*-2-ene gave oxidation via a free-radical mechanism. Hex-1-ene also undergoes isomerization to hex-2-ene.

The regeneration step may be rate determining and proceed via a peroxide mechanism. π-Allyl complexes are also formed; one being a novel complex of formula $Pd_3(\pi\text{-allyl})_2(OAc)_4$ in which one $Pd(OAc)_2$ serves as a bridging group.

$$Pd_3(OAc)_6 + \text{hex-1-ene, hex-}cis\text{-2-ene} \longrightarrow (\pi\text{-allyl})Pd(\mu\text{-OAc})_2Pd(\mu\text{-OAc})_2Pd(\pi\text{-allyl}) \quad (89)$$

There has been several kinetic studies of olefin oxidation in acetic acid (*189*, *196*, *207*, *274*). However, all these studies were undertaken before a knowledge of the equilibria involving the Pd(II) species was known. In comparison with kinetic studies with organic systems, mechanistic studies of metal ion catalysis have the further complication that the metal can exist in several forms depending on reaction conditions. Since each species will most likely have different reactivities, the various equilibria in the system must be known before the kinetics can be interpreted properly.

Probably the most complete study to date is that of Moiseev *et al.* (*196*) who measured the kinetics of oxidation of ethylene by palladium(II) acetate. These workers found that initially the rate increased sharply with increasing [NaOAc] until a maximum rate was reached at [NaOAc] = 0.2–0.3 *M*, depending on temperature; further increase in [NaOAc] caused a sharp decrease in rate. Moiseev *et al.* attributed the initial increase in rate to the dissociation of polymeric palladium acetate species to give $Na_2Pd(OAc)_4$.

$$[Pd(OAc)_2]_n + 2n\text{NaOAc} \rightleftharpoons n\text{Na}_2\text{Pd(OAc)}_4 \tag{90}$$

Also the interaction of ethylene with the Pd(II) species was studied, and the important equilibrium was proposed to be

$$C_2H_4 + Na_2Pd(OAc)_4 \overset{K}{\rightleftharpoons} (C_2H_4)Pd(OAc)_2 + 2NaOAc \tag{91}$$

where K has the value 3.0 moles/liter at 35°C. The rate expression was proposed to be

$$\text{rate} = \frac{kK[Na_2Pd(OAc)_4][C_2H_4]}{[NaOAc]^2} \tag{92}$$

where k has a value of 1.3×10^{-3} sec^{-1} at 35°C.

Since the time of this study, the equilibria in this system were determined to be trimer → dimer → monomer [see Eqs. (4) and (5), Section II, A, 2].

Furthermore, a study of vinyl ester exchange, which proceeds by a mechanism analogous to the acetoxypalladation mechanism for olefin oxidation [see Section III, B, 1, Eqs. (174) and (175)], indicates that the dimer is the most catalytically active species, with the trimer next, and the monomer, ($Na_2Pd(OAc)_4$), unreactive. In the study of Moiseev *et al.*, a maximum rate is attained at the point at which the concentration of $Na_2Pd_2(OAc)_6$ reaches a maximum. Thus the dimer is the reactive species rather than $Na_2Pd(OAc)_4$. However, the decrease in rate with increasing [NaOAc] is greater than can be explained on the basis of conversion of dimer to unreactive monomer [Eq. (5), Section II, A, 2].

Analysis of the data of Moiseev *et al.*, using the known values of K'' suggests that the rate expression for ethylene oxidation is

$$\text{rate} = \frac{k[Na_2Pd_2(OAc)_6][C_2H_4]}{[NaOAc]} \tag{93}$$

It is interesting to compare this tentative rate expression with that found for vinyl ester exchange [Section III, B, 1, Eq. (173)] which does not contain the [NaOAc] inhibition term because of a cancellation of the inhibition term with a catalytic term in [NaOAc]. The rate-determining step for the vinyl ester exchange is the acetoxypalladation step. The added term in the oxidation rate expression suggests that acetoxypalladation is not the slow step but rather the decomposition of the acetoxypalladation adduct. Furthermore, the decomposition must be inhibited by acetate. The following scheme is consistent with other Pd(II) as well as Pt(II) chemistry:

$$[Pd_2(OAc)_6(CH_2CH_2OAc)]^{2-} \underset{}{\overset{K_D}{\rightleftharpoons}} [Pd_2(OAc)_4(CH_2CH_2OAc)]^{-} \text{ (XI)} + OAc^- \tag{94}$$

$$\text{(XI)} \xrightarrow{\text{slow}} [Pd_2(OAc)_4(H)(CH_2{=}CHOAc)]^{-} \tag{95}$$

Thus in exchange studies evidence was found for the need for a vacant coordination site on Pd(II) before elimination from an organometallic intermediate can occur. Recent studies on decomposition of Pt(II) alkyls by hydride elimination also indicate the need for an empty coordination site on Pt(II) before hydride elimination can occur (*290*). If the analysis of the data is correct, this is the first evidence for the need of vacant coordination sites for hydride elimination in Pd(II) chemistry. However, this scheme is tentative because of the complicated equilibria. Work in progress using liquid olefins will hopefully clarify the mechanism (*212*).

This scheme can also explain the changes in product distributions with changes in [NaOAc] for the oxidation of 1-olefins (*37*, *186*, *189*). For example, it was recently reported that the oxidation of propylene gave only isopropenyl acetate in the absence of [NaOAc], but at [NaOAc] = 0.9 *M* the products consisted of 90% allyl acetate and 10% isopropenyl acetate (*189*). The formation of isopropenyl acetate arises from Markownikoff acetoxypalladation, whereas the allyl acetate results from non-Markownikoff acetoxypalladation.

$$\begin{array}{c} \\ \diagdown \\ -\text{Pd(OAc)} + \text{CH}_3\text{CH}{=}\text{CH}_2 \\ \diagup \end{array} \begin{array}{c} \rightleftharpoons \text{CH}_3\underset{\text{OAc}}{\underset{|}{\text{C}}}\text{HCH}_2\text{Pd}\lessgtr \ \text{(XII)} \xrightarrow{-\text{HPd}\lessgtr} \text{CH}_3\overset{\text{OAc}}{\overset{|}{\text{C}}}{=}\text{CH} \ \text{(XIII)} \\ \rightleftharpoons \text{CH}_3\underset{\text{Pd}\lessgtr}{\underset{|}{\text{C}}}\text{H}{-}\text{CH}_2\text{OAc} \ \text{(XIV)} \xrightarrow{-\text{HPd}\lessgtr} \text{CH}_2{=}\text{CHCH}_2\text{OAc} \ \text{(XV)} \end{array} \quad (96)$$

In the low [NaOAc] range, decomposition of the acetoxypalladation adduct would be fast. The proportion of compounds (XIII) and (XV) would depend on the relative rates of formation of compounds (XII) and (XIV), and the product distribution would be kinetically controlled. At high [NaOAc], the rate of decomposition of compounds (XII) and (XIV) would be inhibited, and they would have the opportunity to come to an equilibrium distribution. In this case the product distribution would be equilibrium controlled.

b. *In the Presence of Oxidants.* As mentioned in the last section, one complicating factor in both product distribution as well as kinetic studies was the effect of oxidants present in the reaction mixture. Cupric chloride was commonly used, as in the Wacker reaction, to reoxidize the Pd(0) to Pd(II) and thereby make the reaction catalytic in Pd(II) [Eq. (97)]. Originally it was thought that $CuCl_2$ had no other effect on the reaction. However, it is now recognized that $CuCl_2$ plus Pd(II) salts result in the formation of saturated esters and very likely are responsible

$$C_2H_4 + PdCl_2 + CuCl_2 \xrightarrow{HOAc} ClCH_2CH_2OAc,\ HOCH_2CH_2OAc,\ AcOCH_2CH_2OAc \quad (97)$$

for at least some of the ethylidene diacetate formed in these systems (*112*, *237*).

Later it was observed that $LiNO_3$ gives saturated esters in oxidation of olefins in the presence of palladium(II) acetate (*259*). This result demonstrates one of the pitfalls in interpreting the results of earlier studies of oxidations by palladium(II) acetate, since, as usually prepared by the nitric acid oxidations of Pd sponge, it contains some nitrogeneous impurities (*24*).

It has now been shown that a variety of oxidants will produce saturated esters from ethylene in the presence of Pd(II). In some cases appreciable amounts of ethylidene diacetate are formed (*113*).

The most reasonable mechanism appears to be the interception of the acetoxypalladation adduct by oxidant:

$$\gtrdot Pd(OAc) + C_2H_4 \longrightarrow \gtrdot Pd{-}CH_2CH_2OAc \xrightarrow{-HPd\lessdot} CH_2{=}CHOAc \quad (98)$$

$$\downarrow \text{oxidant}$$

saturated products

The mode of interaction of the oxidant with the acetoxypalladation adduct is not certain. The oxidant could be removing electrons from Pd as the Pd(II)—C bond is broken and Pd(0) is never formed, or the Pd(II) could be oxidized to Pd(IV) which would leave much more easily than Pd(II). Another possibility is that the organic radical is transferred to the oxidant followed by decomposition. It would be difficult to distinguish between the various possibilities. Related reactions are the cleavage of σ-bonded palladium complexes with Collins reagent (*280*), decomposition of π-allyls with oxidants (*164*), and the decomposition of oxypalladation adducts of diolefins with oxidants (Section IV, B).

A recent study of the product distribution from cyclohexene in the presence of $CuCl_2$ sheds some light of the stereochemical aspects of the reaction (*114*, *127*). The products consisted of 1,2,- 1,3,- and 1,4-chloro- and diacetates plus the allylic and homoallylic acetates found in the absence of $CuCl_2$. No enol acetate was detected. Only certain isomers were found. The diacetates were always the cis isomer, whereas the 1,3- and 1,4-chloroacetates were almost exclusively trans. The 1,2-chloroacetates were about in equal mixtures of cis and trans isomers. Also, deuterium-labeling studies indicated that acetoxypalladation was a trans process. Thus, the chloroacetates must be formed mainly by cis displacement of Pd(II) and diacetates by trans attack on the Pd(II)—carbon bond. For example, the scheme for formation of 1,2 and 1,3 products is shown in Eq. (99).

Note that when Pd(II) is replaced by acetate to give diacetates, inversion of configuration occurs, whereas chloride can replace with either inversion or retention. Both processes occur for the 1,2-chloroacetates to give both cis and trans isomers. In the case of the 1,3-chloroacetate, only the *trans*-chloroacetate is formed. These results are in keeping with other Pd(II) chemistry such as the exchange studies

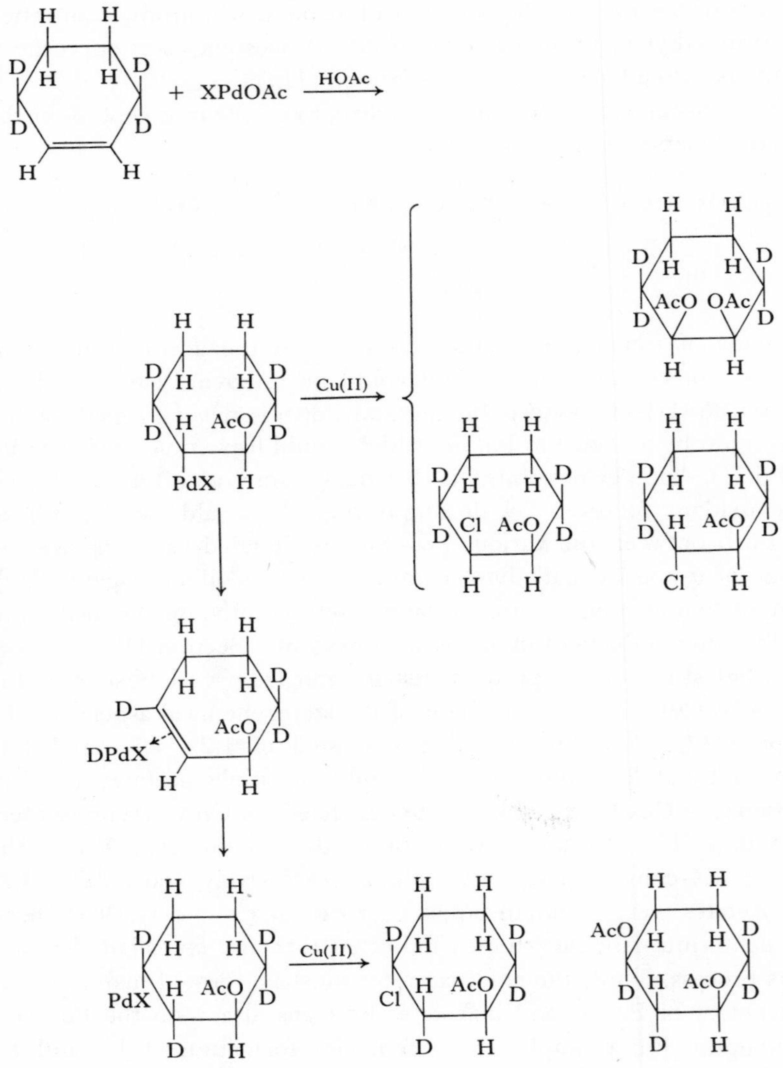

$$\text{(99)}$$

(Section III, B, 1). Thus, in chloride-containing acetic acid, acetate is not complexed to Pd(II) and acetoxypalladation is trans; displacement of Pd(II) occurs in an S_N2 fashion. Chloride is in the coordination sphere as well as in solution and can thus replace Pd(II) with either retention or inversion of configuration at carbon.

In this particular system it was felt that neither the Pd(IV) nor the transfer to $CuCl_2$ routes were likely. Oxidation of cyclohexene by Pd(IV) salts gave, as the only ester product, *trans*-2-chlorocyclohexan-1-yl acetate. Thus Pd(IV) does not give diacetates or the *cis*-chloroacetates. Furthermore, $CuCl_2$ is a weak oxidant and would not be expected to be capable of oxidizing Pd(II) to Pd(IV). The transfer of alkyl to $CuCl_2$ was also considered unlikely, since, if Pt(II) is used in place of Pd(II), different product distributions are obtained. This result would not be expected if both were transferring alkyl to a common Cu(II) alkyl intermediate (*106*).

3. *Reactions in Alcohol*

a. *In the Absence of CO.* The oxidation of olefins in alcohols gives, in the absence of CO, vinyl ethers, dimethyl acetals, and aldehydes or ketones from oxidation of the solvent (*175*):

$$R'\text{—}CH\text{=}CH_2 + Pd(II) \xrightarrow{RCH_2OH} R'C(OCH_2R)_2CH_3,\ R'\overset{\displaystyle OCH_2R}{\overset{|}{C}}H\text{=}CH_2,\ R'CHO \qquad (100)$$

There has been little mechanistic study of this system. The reaction almost certainly proceeds via an alkoxypalladation adduct. The vinyl ether would arise from PdH elimination from this intermediate:

$$\geqslant Pd(II) + C_2H_4 + CH_3O^- \longrightarrow \geqslant PdCH_2CH_2OR \xrightarrow{-HPd(II)} CH_2\text{=}CHOR \qquad (101)$$

The acetal product could arise from addition of alcohol to the vinyl ether catalyzed by Pd(II):

$$CH_2\text{=}CHOR + HOR \xrightarrow{Pd(II)} CH_3\text{—}CH(OR)_2 \qquad (102)$$

or by a mechanism analogous to that proposed for decomposition of the hydroxypalladation adduct in the Wacker reaction (Section III, A, 1):

$$\begin{array}{l} {}^-Cl_3PdCH_2CH_2OR \rightleftharpoons {}^-Cl_2(H)Pd(CH_2\text{=}CHOR) \longrightarrow Cl_2Pd\text{—}\overset{\displaystyle CH_3}{\overset{|}{C}}HOR \\ \qquad\qquad\qquad\qquad\qquad\qquad\qquad\qquad\qquad \downarrow ROH \\ \qquad\qquad\qquad\qquad\qquad\qquad CH_3CH(OR)_2 + PdCl_2^{2-} + H^+ \end{array} \qquad (103)$$

In support of the latter mechanism, it has been reported that oxidation of ethylene in CH_3OD gave the acetate $MeCH(OMe)_2$ in which no deuterium incorporation was observed (*197*). A kinetic study of the reaction in the presence of $CuCl_2$ indicated a mechanism similar to that found in water (*67, 68*). At higher $CuCl_2$ there was a term indicating acceleration by $CuCl_3^-$. However, the equilibria involved in this complicated system are not well understood, so that the catalytic term in $CuCl_2$ should be considered tentative. Thus, at present the deuterium-labeling results favor a mechanism similar to that in water, with decomposition occurring by a mechanism involving a α-methoxy Pd(II) alkyl. This result is in agreement with the studies of Maitlis *et al.* (*182*) on lightly stabilized, Pd(II) model systems (Section V, A) which suggest that Pd(0) will leave a carbon containing strongly electron-releasing groups.

Rodeheaver and Hunt also studied the use of oxymercurials to give oxidation products with Pd(II) in alcohols (*227*). Interestingly, in methanol, ketones were formed rather than the expected vinyl ethers or dimethyl acetate:

$$\mathrm{ClHg{-}CH(R){-}\underset{}{\overset{\displaystyle OCH_3}{\overset{|}{C}}}H(R') + PdCl_2 \longrightarrow HgCl_2 + CH_2(R){-}\overset{\displaystyle O}{\overset{\|}{C}}{-}R' + Pd(0) + Cl^-} \qquad (104)$$

In a 1:1 ethylene glycol–tetrahydrofuran solvent, ethylene ketals would be expected, since these would be the products if the olefins were directly oxidized by Pd(II) salts in ethylene glycol (*139*):

$$\mathrm{Hg(OAc)_2 + RCH{=}CH_2 + HOCH_2CH_2OH \longrightarrow \underset{\displaystyle OCH_2CH_2OH}{\underset{|}{RCHOH_2HgOAc}}} \qquad (105)$$

$$\mathrm{PdCl_2 + \underset{\displaystyle OCH_2CH_2OH}{\underset{|}{RCHCH_2HgOAc}} \longrightarrow \underset{\displaystyle OCH_2CH_2OH}{\underset{|}{RCH{-}CH_2PdCl}} + Hg(OAc)Cl} \qquad (106)$$

$$\mathrm{\underset{\displaystyle OCH_2CH_2OH}{\underset{|}{RCHCH_2PdCl}} \longrightarrow \text{2-R-2-methyl-1,3-dioxolane (O, O ring; R, } CH_3\text{)} + Pd + HCl} \qquad (107)$$

Evidence that deoxymercuration does not occur to give free olefin, followed by oxidation by Pd(II) to give product, is provided by the fact that undec-1-ene was not oxidized by $PdCl_2$ under the reaction conditions in 2 hours, but a 63% yield was obtained in 30 minutes when the mercurial was employed. A very significant deuterium-labeling experiment was also carried out. 2-*d*;-3,3-Dimethylbut-1-ene was prepared, treated with mercuric acetate, and oxidized by $PdCl_2$. The ketal product retained 86% of the deuterium label.

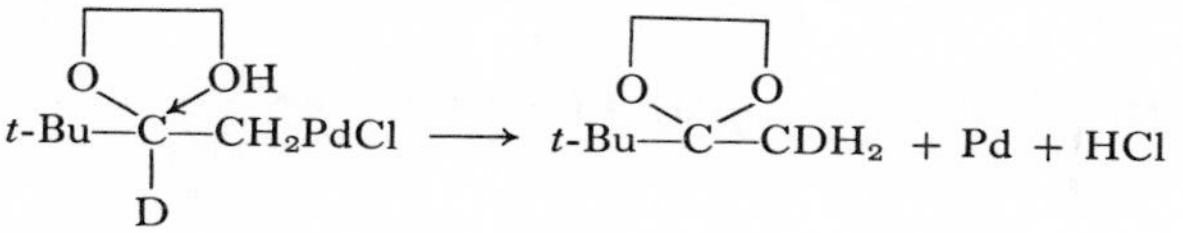

$$t\text{-Bu}-\underset{\text{D}}{\overset{\text{O}\quad\text{OH}}{\text{C}}}-CH_2PdCl \longrightarrow t\text{-Bu}-\overset{\text{O}\quad\text{O}}{\text{C}}-CDH_2 + Pd + HCl \tag{108}$$

This result agrees with that reported for formation of acetals in CH_3OH and suggests that the mechanism does not proceed via a vinyl ether intermediate.

A similar reaction has been observed with the oxythallation adduct of styrene (*272*). In the presence of NaOAc the product is acetophenone dimethylacetal. When the reaction was run in CH_3OD, again there was no deuterium incorporation in the product:

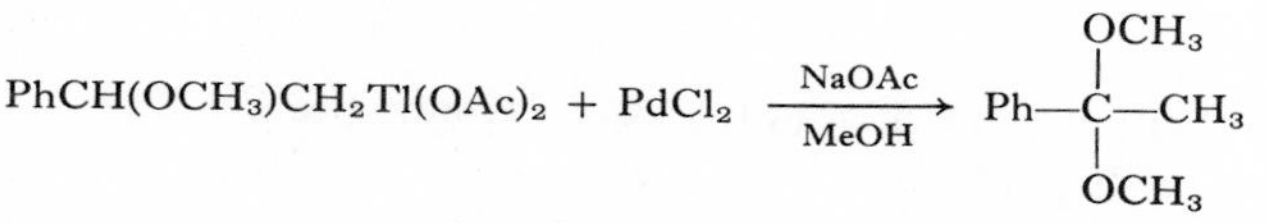

$$PhCH(OCH_3)CH_2Tl(OAc)_2 + PdCl_2 \xrightarrow[\text{MeOH}]{\text{NaOAc}} Ph-\underset{OCH_3}{\overset{OCH_3}{C}}-CH_3 \tag{109}$$

b. *In the Presence of CO.* Olefins will react with nucleophiles in alcohols to give acrylic acids, β-substituted acids, and diacids, depending on reaction conditions. For instance, with ethylene the reaction (*63, 89–91, 100, 252*) is

$$C_2H_4 + ROH + CO \xrightarrow{PdCl_2} CH_2{=}CHCOOR,\ ROCH_2{-}CH_2COOR,\ (CH_2COOR)_2 \tag{110}$$

The diacid is formed in the presence of excess CO and must arise from the following scheme:

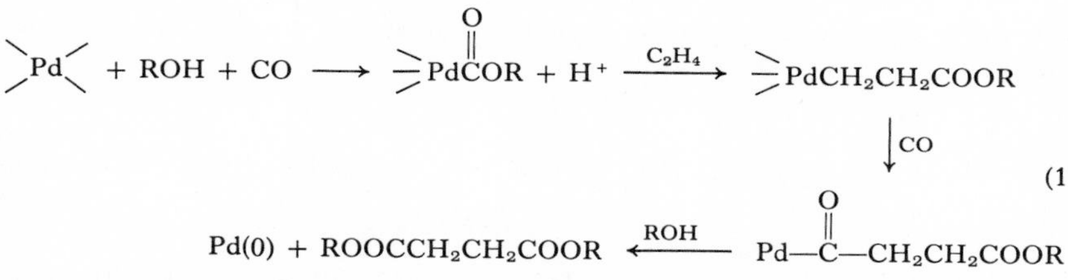

$$\begin{array}{l} \rangle Pd\langle + ROH + CO \longrightarrow \ \geq Pd\overset{O}{\overset{\|}{C}}OR + H^+ \xrightarrow{C_2H_4} \ \geq PdCH_2CH_2COOR \\ \qquad\qquad\qquad\qquad\qquad\qquad\qquad\qquad\qquad\qquad \downarrow CO \\ Pd(0) + ROOCCH_2CH_2COOR \xleftarrow{ROH} Pd-\overset{O}{\overset{\|}{C}}-CH_2CH_2COOR \end{array} \tag{111}$$

The carbalkoxy Pd(II) is analogous to the known carbalkoxy Hg(II), and very recently a stable carbomethoxy Pd(II) has been prepared (Section II, A, 3). Evidence for this scheme is that PdCOOR, formed *in situ* by exchange with ClHgCOOR, gives the same products (*89, 90*).

The monoacids may be prepared either by methoxypalladation followed by CO insertion,

$$\gt Pd \lt + C_2H_4 + ROH \longrightarrow \gt PdCH_2CH_2OR \xrightarrow{CO} \gt Pd\overset{O}{\overset{\|}{C}}CH_2CH_2OR \xrightarrow{ROH} RO\overset{O}{\overset{\|}{C}}CH_2CH_2OR + Pd(0) \xrightarrow{-ROH} ROOCCH{=}CH_2 \quad (112)$$

or by insertion of olefin into the carbalkoxypalladium(II) followed by decomposition,

$$\gt PdCOOR + C_2H_4 \longrightarrow \gt Pd{-}CH_2CH_2COOR \begin{cases} \xrightarrow{-PdH} CH_2{=}CHCOOR \\ \xrightarrow{CH_3O^-} CH_3OCH_2CH_2COOR \end{cases} \quad (113)$$

It is, in fact, likely that Eq. (112) may explain the β-alkoxy ester formation, whereas Eq. (113) may be the route to methyl acrylate. Thus, preparation of carboalkoxypalladium(II) *in situ* by exchange with the stable ROOCHgCl in the presence of ethylene gives only acrylate esters (*90*):

$$ROOCHgCl + PdCl_2 \longrightarrow [\gt PdCOOR] \xrightarrow{\gt C_2H_4} PdCH_2CH_2COOR \xrightarrow{\gt PdH} CH_2{=}CHCOOR \quad (114)$$

If this reaction is carried out in the presence of excess CO, the product is mainly diester, which confirms the route given above for diester formation (*63*, *91*, *252*). In regard to decomposition of the Pd(II) acyl to ester, a recent study on "lightly stabilized," Pd(II) model systems has shown that decomposition occurs by attack of methoxide (Section V, A).

$$\gt Pd{-}\overset{O}{\overset{\|}{C}}{-}R + CH_3O^- \longrightarrow \gt Pd{-}\underset{OCH_3}{\underset{|}{\overset{O^-}{\overset{|}{C}}}}{-}R \longrightarrow Pd(0) + CH_3OOCR \quad (115)$$

This could be another piece of evidence that Pd will leave as Pd(0) when there are strongly electron-releasing groups on the carbon to which it is attached.

Stille and co-workers (*251*) have studied the stereochemistry of the carbonylation of *cis*- and *trans*-2-butene. They found that *cis*-2-butene gave the *threo*-methyl-3-methoxy-2-butane-carboxylate, whereas *trans*-2-butene gave the erythro isomer. On the other hand, *cis*-2-butene gave *meso*-dimethyl-2,3-butanedicarboxylate, whereas *trans*-2-butene gave the *d,l*-isomer. This result (Section IV, B), plus earlier evidence that CO insertion in Pd–C bonds is cis (*130*), is consistent with the following scheme for the *cis*-2-butene, involving trans methoxypalladation and *cis*-carbon monoxide insertion as well as cis insertion of —PdCOOR:

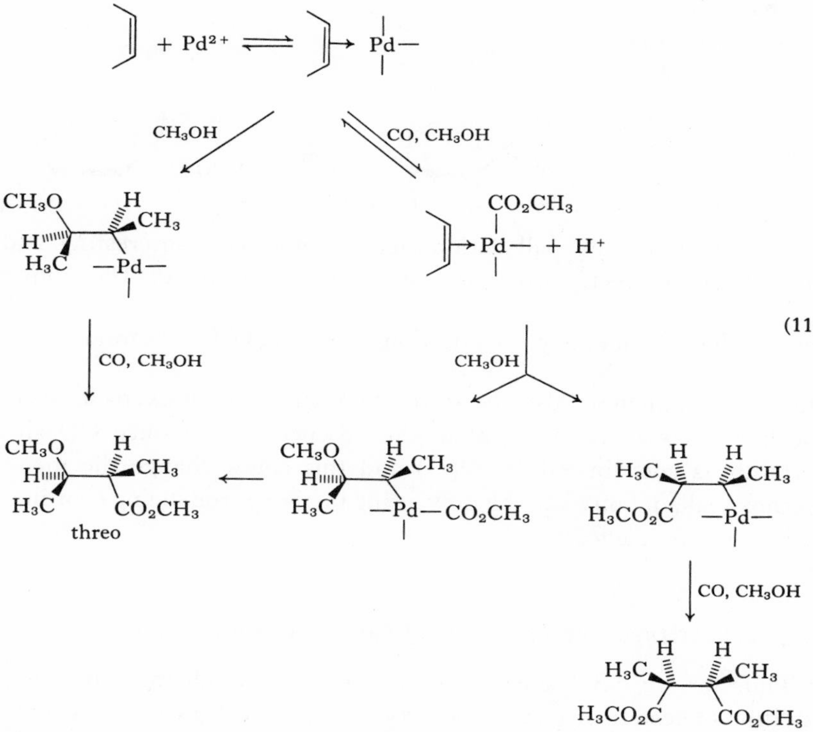

$$\text{olefin} + Pd^{2+} \rightleftharpoons \text{olefin} \rightarrow Pd—$$

CH_3OH → CH_3O, H, CH_3, H, H_3C, —Pd—

CO, CH_3OH ⇌ olefin → Pd(CO_2CH_3)— + H^+

CO, CH_3OH → CH_3O, H, CH_3, H, H_3C, CO_2CH_3 (threo)

CH_3OH → CH_3O, H, CH_3, H, H_3C, —Pd—CO_2CH_3 ; and H_3C, H, H, CH_3, H_3CO_2C, —Pd—

CO, CH_3OH → H_3C, H, H, CH_3, H_3CO_2C, CO_2CH_3

(116)

Also consistent with this scheme is the formation of *cis*-diesters with cyclic olefins. In this case 1,3- as well as 1,2-diesters are formed. The scheme for cyclopentene is shown in Eq. (117).

As might be expected, higher CO pressures favor the 1,2-isomer. This ability of Pd(II) to move around a ring has been noted previously by Heck (*89*) in his carbonylation of cyclic olefins to give unsaturated cyclic esters as well as by a number of workers in other reactions (see Sections III, A, 2 and 4).

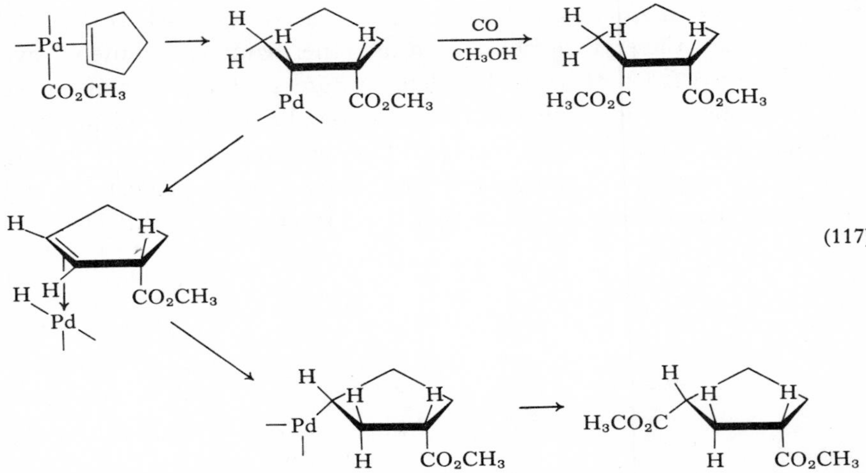

(117)

The trans methoxypalladation suggestion is very interesting and is most likely correct, although trans elimination of Pd(II) from the intermediate formed by cis addition of $\rangle PdCO_2CH_3$ cannot be conclusively eliminated. Also the stereochemistry of methoxypalladation of the butenes may be different in the absence of CO since CO almost certainly is coordinated to Pd(II) and this could change the mode of methoxypalladation either by preventing methoxy from being coordinated or by electronic effects.

4. *Reactions Involving Formation of Carbon–Carbon Bonds*

Thus far we have considered oxidations in which the initial step is addition of acetate or methoxide (oxypalladation). Now we shall consider oxidations in which the first step is addition of Pd(II)–carbon bonds.

a. *Aromatics.* The most studied reaction of this type is the addition of "arylpalladium" to olefins. This generally unstable reagent is prepared *in situ* by the exchange of aryl mercuric chloride with palladium(II) chloride. Thus, styrene can be prepared as follows:

$$PhHgCl + \rangle PdCl \longrightarrow HgCl_2 + PhPd\langle + C_2H_4 \longrightarrow PhCH_2CH_2Pd\langle \xrightarrow{-HPd\langle} PhCH{=}CH_2 \qquad (118)$$

This reaction has been studied extensively by Heck, and a large number of new reactions have been discovered (*90*, *92*, *176*).

One reaction of particular interest, in relation to the oxidation discussed in Section III, A, 2b, is the formation of saturated chlorides in the presence of $CuCl_2$ (*93*):

$$\begin{array}{c} PhPd\lessgtr + C_2H_4 \longrightarrow PhCH_2CH_2Pd\lessgtr \xrightarrow{\text{no } CuCl_2} PhCH{=}CH_2 + Pd(0) + H^+ \\ \downarrow 2CuCl_2 \\ 2PhCH_2CH_2Cl + 2CuCl + ClPd\lessgtr \end{array} \tag{119}$$

Another source of phenylpalladium involves aromatic sulfinic acid which evolves SO_2 in the presence of Pd(II) to give the desired intermediates (*73*).

The reaction can also proceed, although somewhat slower, if the aromatic itself is used:

$$PhH + Pd(OAc)_2 + C_2H_4 \longrightarrow PhCH{=}CH_2 + Pd(0) + OAc^- + HOAc \tag{120}$$

It has been claimed, however, that this reaction proceeds by a different mechanism than the insertion of phenylpalladium across an olefinic bond (*45*, *70*, *201*, *202*) (see discussion under mechanism below).

A novel reaction, apparently also involving direct palladation, has been reported (*234*). It is the production of cinnamic acid derivatives from saturated acids such as propionic and *n*- and isobutyric acids.

$$R{-}C_6H_4 + CH_3CH_2COOH \xrightarrow{PdCl_2} R{-}C_6H_4{-}CH{=}CHCOOH \tag{121}$$

Somehow the Pd(II) must be dehydrogenating the saturated acid to acrylic acid which then is arylated by the usual route.

There have been a number of studies recently aimed at both improving the reaction and determining the details of its mechanism.

The reaction can be made catalytic in Pd(II) by using aryl halides as the source of aromatics. The first report of this reaction used phenyl iodide in the presence of potassium acetate (*193*):

$$C_6H_5I + CH{=}CHX + CH_3COOK \longrightarrow C_6H_5CH{=}CHX + CH_3COOH + KI \tag{122}$$

($X = H$, Ph, CH_3, or $COOCH_3$). Palladium black could also be used as a catalyst, and triethylamine or pyridine as bases. Heck also studied this reaction, and proposed that phenylpalladium was formed by oxidative addition to Pd(0), which was formed by reduction of the Pd(II) (*101*).

$$RX + Pd \rightleftharpoons [RPdX] \tag{123}$$

$$[RPdX] + \overset{H}{\diagdown} C{=}C \diagup \quad [R{-}\underset{|}{\overset{H}{\overset{|}{C}}}{-}\underset{|}{\overset{|}{C}}{-}PdX] \longrightarrow \overset{R}{\diagdown} C{=}C \diagup \tag{124}$$

The compound $Pd[P(C_6H_5)_3]_4$ is known to add oxidatively organic halides to give $[P(C_6H_5)_3Pd(X)R$ (*42*, *65*, *66*). A significant improvement in this reaction was made when triphenylphosphine was added to the reaction mixture (*94*). Aryl bromides could now be used in place of the more difficult to obtain aryl iodides. Also, the reaction becomes partially stereospecific. Thus, *cis*- or *trans*-1-phenyl-1-propene gave 88% or more of the corresponding 1,2-diphenyl-1-propene of the same stereochemistry. This is the same result found previously for the reaction using mercurated aromatics (*90*).

The reaction was also made catalytic by operating under on O_2 pressure (*242*). Apparently O_2 can oxidize the hydrido-palladium intermediate proposed for oxidations of olefins under O_2 (*23*).

The arylation of enol esters has also been improved (*95*). Previously a wide range of products were produced including β-aryl carbonyls, arylated enol esters, styrene, and stilbene derivatives (*96*). It has also been found that arylated enol esters can be obtained as major products if the reactions are carried out with stoichiometric amounts of aryl mercuric acetate and palladium acetate in anhydrous acetonitrile or in excess enol ester solution. The products are those arising from addition of the phenyl group to the carbons not containing the ester. Thus, with vinyl acetate and phenyl mercuric acetate, the product is the enol acetate of phenyl acetaldehyde:

$$C_6H_5HgOAc + Pd(OAc)_2 + CH_2{=}CHOAc \longrightarrow C_6H_5CH{=}CHOAc \tag{125}$$

In other variations, mercurated ferrocene was treated with olefins in the presence of Pd(II) salts to give vinyl ferrocene derivatives (*9*, *154*). Aryl cobalt derivatives have also been used to produce the phenyl-palladium species (*284–286*). In another study it was found that even aryl phosphine ligands will transfer the aromatic to palladium (*297*).

An oxyphenylation reaction of certain olefins such as indene in alcohol solvents has also been reported (*131*). The reaction apparently does not proceed by addition of ROH to phenylated indene but the exact mechanism is uncertain.

$$\text{Indene} + C_6H_5HgCl \xrightarrow[\text{ROH}]{Li_2PdCl_4} \text{2-phenylindene} + \text{1-OR-2-phenylindane} \quad (126)$$

The two mechanisms suggested for the olefin arylation reaction are Eq. (118) and the rate-determining formation of a vinyl Pd(II) complex.

$$\text{Pd(II)} + \text{H}-\text{C}{=}\text{C} \longrightarrow \text{Pd(II)}-\text{C}{=}\text{C} + \text{C}_6\text{H}_5\text{R} \longrightarrow \text{RC}_6\text{H}_4-\text{C}{=}\text{C} + \text{Pd(0)} \quad (127)$$

In the case when the aromatic is supplied as arylmercury, there seems little doubt that the first mechanism is operative. First, the reaction is much faster than when the nonmercurated aromatic is used. This would not be expected if the formation of vinylmercurial is rate determining. Second, if olefins are not present, coupled products are formed by reaction of phenylpalladiums with each other (*275*). Third, the phenylpalladium can be trapped by reaction with CO to give acids (*115*). Finally, products are formed that can only be explained by movement of Pd(II) around the ring *after* addition of the aryl group (89). This movement almost certainly occurs by a series of Pd(II)–hydride additions and eliminations.

As mentioned earlier, Heck found that reaction occurred with retention of configuration, which is consistent with phenylpalladation and hydride elimination having the same stereochemistry. Since free hydride or phenyl radical would not exist free in hydroxylic solvents such as acetic acid, it seems reasonable to assume that both are cis. Support for this view comes from studies with 3,3,6,6,-tetradeuterated cyclohexene (*128*). The main product was the homoallylic acetate which contained all four deuteriums originally present. Moreover, one of the deuteriums had undergone a stereospecific shift to the adjoining carbon. The stereochemistry requires that phenylpalladation and Pd(II)–hydride elimination–readdition have the same stereochemistry. In a previous study with this same olefin, it was determined that acetoxypalladation and Pd(II)–hydride elimination–readdition have opposite stereochemistries (*127*). If the choice is between acetoxypalladation and phenylpalladation, the acetoxypalladation must certainly be trans, confirming that phenylpalladation is cis. The scheme is thus as follows:

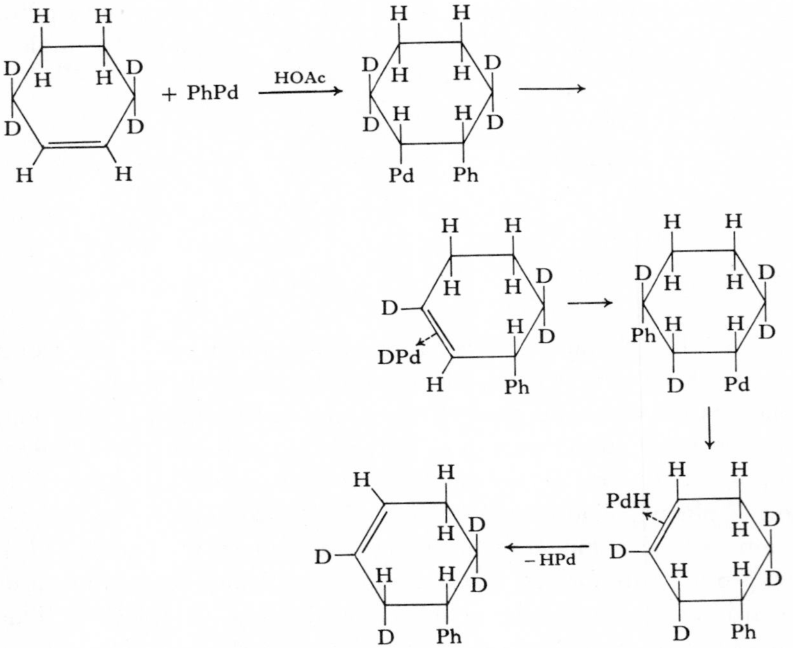

+ PhPd $\xrightarrow{\text{HOAc}}$ … $\xleftarrow{-\text{HPd}}$ (128)

In a study of electronic and steric factors in the olefin and arylation reaction, Heck determined that the addition is generally sterically controlled with the phenyl group mainly adding to the least-substituted carbon of the double bond (*89*). The reaction scheme with propylene is depicted by Eq. (129).

In regard to electronic effects, the aryl group prefers to add to the more positive carbon of the double bond. A chelation effect may be operative in the arylation of allylic alcohols in which the phenyl adds to the more substituted carbon of the double bond.

$$[\mathrm{ArPdOAc}] + \mathrm{CH_2{=}CHCH_3} \longrightarrow [\mathrm{Ar{-}CH_2{-}CH(CH_3){-}PdOAc}] + [\mathrm{AcOPd{-}CH_2{-}CH(CH_3){-}Ar}]$$

$$\downarrow \qquad\qquad\qquad \downarrow \tag{129}$$

$$(E)\text{-}\mathrm{ArCH{=}CHCH_3} + (Z)\text{-}\mathrm{ArCH{=}CHCH_3} + \mathrm{ArCH_2CH{=}CH_2} + \mathrm{CH_2{=}C(CH_3){-}Ar}$$

$$\mathrm{(CH_3)_2C{=}CH{-}CH(OH){-}CH_3} + [\mathrm{C_6H_5PdOAc}] \longrightarrow$$

$$[\mathrm{CH_3{-}C(C_6H_5)(CH_3){-}CH(Pd(OAc)){-}CH(OH){-}CH_3}] \xrightarrow{-\mathrm{HPdOAc}} [\mathrm{CH_3{-}C(C_6H_5)(CH_3){-}CH{=}C(OH){-}CH_3}] \tag{130}$$

$$\downarrow$$

$$\mathrm{(CH_3)_2{-}C(C_6H_5){-}CH_2COCH_3}$$

Note that, since there are no hydrogens on the vinylic carbon on which the phenyl group adds, the product could not have arisen from Eq. (127).

The mechanism given by Eq. (127) has been put forward in the case where the aromatic itself is used. It is consistent with the observed chemistry, i.e., retention of configuration of the olefin. Vinyl Pd(II) intermediates are reasonable and are probably present in olefinic coupling reactions (Section III, A, 4b), therefore this mechanism must be seriously considered. However, it has been found that when benzene d_6 is used, there is an isotope effect, $k_H/k_D = 5.0$ (*243*). This cannot be an equilibrium isotope effect because, when a mixture of deuterated and nondeuterated benzenes were reacted, no scrambling of deuteriums occurred. This indicated that palladium–aryl σ-bond formation is the slow irreversible step in olefin arylation. Furthermore, styrene β,β-d_2 gave only a secondary isotope effect, $k_H/k_D = 1.25$. This is also in keeping with the rate-determining step *not* being palladium–olefin σ-bond formation. These results do not eliminate the mechanism given by

Eq. (127) but indicate that the scheme must be modified so formation of phenylpalladium occurs first. Such a reaction scheme has recently been put forward (*45*). Evidence to support this reaction is that phenylation of styrene, or substituted styrene β,β-d_2, gives stilbenes with only one deuterium. However, if phenylpalladation is followed by palladium deuteride elimination, this is exactly the result which would have been predicted.

$$\left[\begin{array}{c} \text{AcO, HOAc} \\ \text{D}_2\text{C}{=}\text{CHAr} \rightarrow \text{Pd–Ph} \end{array}\right] \longrightarrow \left[\begin{array}{c} (\text{HOAc})_2\text{Pd–Ph} \\ \text{D}_2\text{C}{=}\text{CHAr} \end{array}\right] \longrightarrow \text{Ar(H)C}{=}\text{C(D)Ph} \qquad (131)$$

$$\text{Ph—Pd(II)} + \text{D}_2\text{C}{=}\text{CHAr} \longrightarrow \text{PhCD}_2\text{—CHAr(Pd(II))} \xrightarrow{-\text{DPd(II)}} \text{PhCD}{=}\text{CHAr} \qquad (132)$$

Further evidence for this mechanism is the fact that retention of configuration occurs. However, as pointed out earlier, this is consistent with phenylpalladation and Pd(II)–hydride elimination having the same stereochemistry. One interesting result was the phenylation of *cis*- and *trans*-1,2-dichloroethylenes to give phenyl-substituted dichloroethylene of the same configuration; *cis*-dichloroethylene gave a 42% yield (*201*).

$$\text{Cl(H)C}{=}\text{C(H)Cl} + \text{PhH} \xrightarrow[\text{reflux, 8 hr}]{\text{Pd(OAc)}_2\text{AcOH}} \text{Cl(H)C}{=}\text{C(Cl)Ph} \qquad (133)$$

No monochlorophenylated ethylenes were reported. The monochlorinated product would be expected from both mechanisms since Pd—Cl bonds would be easily substituted rather than C—H in the vinylic mechanism, whereas in the phenylpalladation mechanism Pd—Cl elimination would be much more facile than Pd—H elimination.

Further evidence to support Eq. (127) is that stable vinyl Pd(II) derivatives react with aromatics to give phenylated olefins (*202*):

$$\text{Cl}_2\text{C}{=}\text{C(H)–Pd(PPh}_3)_2\text{Cl} + \text{CH}_3\text{C}_6\text{H}_5 \xrightarrow[\text{reflux, 1.5 hr}]{\text{AcOH, AgOAc}} p\text{-CH}_3\text{C}_6\text{H}_4\text{–C(H)}{=}\text{CCl}_2 \qquad (134)$$

However, it should be pointed out that because this reaction occurs does not mean that the reaction without preformation of vinyl Pd(II) is by the same route. Certainly more work is required before the mechanism of the reaction is known for certain, but at present there is no reason to believe the reaction with free aromatic is by a different route from that involving formation of phenylpalladium by exchange.

b. *Aliphatic.* The oxidative coupling of α-substituted styrenes was first reported by Hüttel *et al.* (*141*, *144*) who found that α-substituted styrenes were coupled by $PdCl_2$ in acetic acid. A number of other workers have reported variations on this reaction (*177*).

$$2\ (Ph)(R)C{=}CH_2 \longrightarrow (Ph)(R)C{=}CHCH{=}C(Ph)(R) \qquad (135)$$

In a reaction analogous to the catalytic coupling of aryl halides, Heck (*94*, *101*) found that benzyl and vinyl halides would also undergo the reaction:

$$(CH_3)_2C{=}CHBr + CH_2{=}CHCOOCH_3 + Et_3N \xrightarrow[2P(Ph)_3]{Pd(OAc)_2} (CH_3)_2C{=}CH{-}CH{=}CH{-}COOCH_3 + Et_3NH^+Br^- \qquad (136)$$

The first step almost certainly involves oxidative addition to Pd(0) to give a vinyl Pd(II) derivative, as this is a known reaction. The vinyl group is then added across the double bond followed by a Pd(II)–hydride elimination to give the observed product. An interesting recent variation on this scheme is the use of vinyl silanes to give 1,3-butadiene derivatives (*288*). The proposed route is as follows:

$$(Ph)(H)C{=}C(H)(Si(CH_3)_3) + PdCl_2 \rightleftharpoons \left[PhCHCl{-}CH(PdCl)(Si(CH_3)_3)\right] \longrightarrow \left[(Ph)(H)C{=}C(H)(PdCl)\right] + (CH_3)_3SiCl \qquad (137)$$

$$\text{PhCH=CHSi(CH}_3)_3 + \text{PhCH=CHPdCl} \longrightarrow [(\text{PhCH=CH})_2\text{Pd}] \longrightarrow (\text{PhCH=CH})_2 + \text{Pd(0)} \quad (138)$$

If the reaction is run in the presence of other olefins, unsymmetrically substituted 1,3-butadienes are formed.

$$[\text{PhCH=CHPdCl}] + \text{CH}_2\text{=CHCO}_2\text{CH}_3 \longrightarrow \text{PhCH=CHCH=CHCO}_2\text{CH}_3 \quad (139)$$

Alkylpalladium(II) intermediates that add to olefinic bonds can also be prepared by exchange with alkylmercurials without β-hydrogens. An example is the neopentyl Pd(II) acetate prepared from the corresponding mercurial (*97*):

$$[(\text{CH}_3)_3\text{CCH}_2\text{PdOAc}] + \text{CH}_2\text{=CHCOOCH}_3 \longrightarrow (\text{CH}_3)_3\text{CCH}_2\text{CH=CHCOOCH}_3 + [\text{HPdOAc}] \quad (140)$$

With the neophyl group an interesting rearrangement takes place.

$$[\text{C}_6\text{H}_5\text{C(CH}_3)_2\text{CH}_2\text{PdOAc}] \rightleftharpoons [o\text{-C}_6\text{H}_4(\text{C(CH}_3)_2\text{CH}_2)\text{Pd(H)(OAc)}] \rightleftharpoons [o\text{-(CH}_3)_3\text{CC}_6\text{H}_4\text{PdOAc}] \xrightarrow{\text{CH}_2\text{=CHCOOCH}_3} [o\text{-(CH}_3)_3\text{CC}_6\text{H}_4\text{CH}_2\text{CH(PdOAc)COOCH}_3] \xrightarrow{-[\text{HPdOAc}]} o\text{-(CH}_3)_3\text{CC}_6\text{H}_4\text{CH=CHCOOCH}_3 \quad (141)$$

This rearrangement is analogous to the *o*-metallation of aryl phosphine ligands (*216*).

A novel reaction that probably involves π-allyl intermediates is the reaction of allylic compounds with cycloalkanone enamines (*208*):

$$\text{1-piperidinocyclohexene} + CH_2{=}CHCH_2OPh \longrightarrow$$

$$\text{2-}(CH_2CH{=}CH)\text{cyclohexanone} + CH_2{=}CHCH_2\text{-cyclohexanone-}CH_2CH{=}CH_2 \quad (142)$$

A similar reaction in the allylic alkylation of olefins,

$$\text{isobutylene} \longrightarrow (\pi\text{-allyl-}Pd\text{-}Cl)_2 \xrightarrow{N} CH_2{=}C(CH_3)CH_2\text{-}N \quad (143)$$

where N is (*265*)

$^-CH(CO_2C_2H_5)_2$, $^-C(SOCH_3)(CO_2CH_3)$, or $^-CH(SO_2CH_3)(CO_2CH_3)$

The mechanism originally proposed for the vinylic coupling reactions involved decomposition of a diolefin π-complex (*281*, *282*). However, the information developed since then on reactions of vinylic Pd(II) compounds suggests that these compounds are intermediates in the coupling reaction. The vinyl Pd(II) compounds could either oxidatively couple, since this is known to occur, or the vinyl Pd(II) intermediate could add across the olefinic double bond:

$$2\ R'RC{=}CHPd(II) \longrightarrow (R'RC{=}CH)_2 + Pd(0) + Pd(II) \quad (144)$$

$$R'RC{=}CHPd(II) + R'RC{=}CH_2 \longrightarrow Pd(II){-}CR'R{-}CH_2{-}CH{=}CR'R \xrightarrow{-HPd(II)} (R'RC{=}CH)_2 \quad (145)$$

In light of the reactivity of vinyl Pd(II) compounds toward olefins discussed above, the latter would appear more likely. Thus the vinyl compound, prepared *in situ* from β-bromostyrene, preferentially reacted with methyl acrylate rather than self-couple according to Eq. (144) (*94*).

The mode of formation of the vinyl compound in the olefin coupling reaction is unknown. It could be prepared by substitution of a vinyl hydrogen by oxidative addition, followed by reductive elimination:

$$\text{>C=C<}_{\text{H}} + Pd(OAc)_2 \longrightarrow \text{>C=C<}^{\text{H}}_{\text{Pd(IV)(OAc)(OAc)}} \longrightarrow \text{>C=C<}_{\text{Pd(II)}} + HOAc \quad (146)$$

$$\text{>C=C<}_{\text{H}} + Pd(OAc)_2 \longrightarrow {}^{+}\text{>C—C<—}Pd(OAc)_2^{-} \longrightarrow \text{>C=C<}_{\text{PdOAc}} + HOAc \quad (147)$$

The route could also be analogous aromatic substitution. One factor that favors the latter mechanism is that the coupling reaction goes most readily with those olefins that have two aliphatic substituents on the carbon next to the carbon engaged in the coupling. These substituents would tend to stabilize the positive charge.

B. Nonoxidative Reactions

1. *Exchange Reactions*

One potentially very useful synthetic reaction in Pd(II) catalysis is the vinyl and allylic exchange reactions,

$$\left.\begin{matrix} CH_2{=}CHX \\ CH_2{=}CHCH_2X \end{matrix}\right\} + Y^- \underset{}{\overset{Pd(II)}{\rightleftharpoons}} \left.\begin{matrix} CH_2{=}CHY \\ CH_2{=}CHCH_2Y \end{matrix}\right\} + X^- \quad (148)$$

where X and Y are nucleophiles such as —OH, —OR, —OOCR, halide, $—NR_2$, and —Ph (*178*). The reaction, first reported by Smidt *et al.* (*244*) for X and Y = carboxylates, is nonoxidative and thus truly catalytic in Pd(II), since oxidants are not required to regenerate Pd(II) from Pd(0).

The evidence, to be discussed below, very strongly favors an addition–elimination type of mechanism. For vinyl exchange, the scheme is

$$\mathrm{Pd(II)} + \mathrm{Y}^- + \mathrm{CH_2{=}CHX} \rightleftharpoons \mathrm{Pd{-}CH_2{-}CH}\begin{matrix}\diagup \mathrm{X} \\ \diagdown \mathrm{Y}\end{matrix} \rightleftharpoons \mathrm{Pd(II)} + \mathrm{X}^- + \mathrm{CH_2{=}CHX} \qquad (149)$$

This type of reaction has received considerable study recently. McKeon and co-workers (*191*) have developed a useful synthetic procedure for the preparation of vinyl ethers by the exchange reaction. At room

$$\mathrm{CH_2{=}CHOR} + \mathrm{R'OH} \rightleftharpoons \mathrm{CH_2{=}CHOR'} + \mathrm{ROH} \qquad (150)$$

temperature, acetals are the main product and Pd(0) is formed, but, at temperatures below $-25°C$, palladium(II) salts are efficient catalysts for the interchange and Pd(II) was not reduced. They also studied the stereochemistry of the reaction and found that it involved inversion of configuration. Thus *trans*-ethyl propenyl ether exchanged with propanol at $-30°C$ to give *cis* propyl propenyl ether [Eq. (151)]. This agrees with the results reported earlier by Sabel *et al.* (*232*), who indicated that addition of Pd(II)–OC_3H_7 and elimination of C_2H_5O–Pd(II) have the same stereochemistry [Eq. (149); X = OC_2H_5 and Y = OC_3H_7]. They favored cis addition and cis elimination.

$$\begin{matrix}\mathrm{C_2H_5O} & & \mathrm{H} \\ & \mathrm{C} & \\ & \| & \\ & \mathrm{C} & \\ \mathrm{CH_3} & & \mathrm{H}\end{matrix} + \mathrm{C_3H_7OH} \xrightleftharpoons{\mathrm{Pd(II)}} \begin{matrix}\mathrm{H} & & \mathrm{OC_3H_7} \\ & \mathrm{C} & \\ & \| & \\ & \mathrm{C} & \\ \mathrm{CH_3} & & \mathrm{H}\end{matrix} \qquad (151)$$

In another study, these workers found that cis complexes of palladium acetate with bidentate ligands such as *o*-phenanthroline will catalyze the exchange at higher temperatures (*190*). As trans complexes were inactive, it is likely that two adjacent cis coordination positions are required for catalysis. These new catalysts are extremely interesting and deserve further mechanistic study.

In an elaborate study, Takahashi and co-workers investigated the exchange of allylic ethers and esters with a variety of active hydrogen compounds (*255*), including esters, ethers, carbanoid species, and amines.

Phenylpalladium had previously been shown to exchange with allyl chloride (*98*):

$$[PhPdCl] + CH_2{=}CHCH_2Cl \longrightarrow [PhCH_2{-}\underset{}{\overset{\displaystyle PdCl}{\overset{|}{C}}}H{-}CH_2Cl] \longrightarrow PdCl_2 + PhCH_2CH{=}CH_2 \quad (152)$$

An interesting variation on this reaction was carried out with β-unsaturated acid chlorides to give ketenes (*99*):

$$[PhPdCl] + CH_2{=}CHCOCl \longrightarrow [PhCH_2{-}\overset{\displaystyle PdCl}{\overset{|}{C}}HCOCl] \longrightarrow PdCl_2 + PhCH_2CH{=}C{=}O \quad (153)$$

A possible commercial route to vinyl acetate is the exchange of vinyl chloride with acetate. A mechanistic study of this reaction using *trans*-2-deuterovinyl chloride gave an equal mixture of *cis*- and *trans*-2-deutero-vinyl acetate (*283*):

$$(D)(H)C{=}C(H)(Cl) \longrightarrow (D)(H)C{=}C(H)(OAc) + (H)(D)C{=}C(H)(OAc) \quad (154)$$

This must have resulted from the Pd(II)-catalyzed isomerization of both the vinyl chloride and vinylic acetate during the course of the reaction. Another study with *cis*- and *trans*-propenyl chlorides indicated that this reaction proceeds mainly with retention of configuration (*232*),

$$(H)(Cl)C{=}C(H)(CH_3) \xrightarrow[Pd(II)]{OAc^-} (H)(OAc)C{=}C(H)(CH_3) \quad (155)$$

suggesting that oxypalladation and dechloropalladation have opposite stereochemistries. A related reaction is the exchange of allylic chloride with acetate:

$$CH_2{=}CHCH_2Cl + OAc^- \xrightarrow{PdCl_2/DMF} CH_2{=}CHCH_2OAc + Cl^- \quad (156)$$

The author also suggested an acetoxypalladation-dechloropalladation mechanism (*19*). Since these reactions do not result in the reduction of Pd(II) and give readily stereochemical information, they are well suited for mechanistic studies. A series of studies on various exchanges has recently been carried out to determine the factors involved in Pd(II)-catalyzed additions and eliminations (*104*).

The first series of exchanges studied were carried out in acetic acid containing chloride. In this system, Pd(II) exists mainly as the chloride-bridged dimer, $Li_2Pd_2Cl_6$ (Section II, A, 1). The first system studied was vinyl ester exchange

$$CH_2{=}CHOCOR + HOAc \underset{}{\overset{Pd(II)}{\rightleftharpoons}} CH_2{=}CHOAc + HOOCR \quad (157)$$

which was found to have the following rate expression (*116*, *117*):

$$\text{rate} = \frac{[Li_2Pd_2Cl_6][C_2H_3OOCR]}{[LiCl]}(k_1' + k_1''[LiOAc]) \quad (158)$$

This expression is consistent with attack of acetate on a dimeric Pd(II) π complex from outside the coordination sphere of Pd(II):

$$\left[\begin{array}{c} \text{Cl—Cl—}CH_2{=}CHOOCR \\ \text{Pd} \quad \text{Pd} \\ \text{Cl—Cl—Cl} \end{array}\right]^{-} OAc^- \longrightarrow \left[\begin{array}{c} \text{Cl—Cl—}CH_2CH(OOCR)OAc \\ \text{Pd} \quad \text{Pd} \\ \text{Cl—Cl—Cl} \end{array}\right]^{2-} \quad (159)$$

Note that this oxypalladation is different from that suggested for the Wacker reaction because in the latter case the kinetics suggested the coordination of both OH and ethylene. Trans acetoxypalladation in chloride-containing media was confirmed by stereochemical studies in the olefin oxidation reaction (Section III, A, 2a). Acetoxypalladation could be shown to be stereospecific since propenyl acetates exchanged with inversion and 1-acetoxy-1-cyclopentene did not exchange. The former result is analogous to that obtained with propenyl ethers and indicates that addition and elimination have the same stereochmeistry. The scheme for exchange of *cis*-propenyl acetate with deuterated acetic acid ($OAc_D = CD_3COO$) is

$$\underset{H\quad H}{\overset{H_3C\quad OAc}{C{=}C}} + -PdOAc_D \underset{\text{trans E}}{\overset{\text{trans A}}{\rightleftharpoons}} [\text{Newman: front } -Pd, H, OAc; \text{ back } H, CH_3, OAc_D] \rightleftharpoons$$

$$[\text{Newman: front } -Pd, Ac_DO, H; \text{ back } H, CH_3, OAc] \underset{\text{trans A}}{\overset{\text{trans E}}{\rightleftharpoons}} -PdOAc + \underset{H\quad OAc_D}{\overset{H_3C\quad H}{C{=}C}} \quad (160)$$

The second result confirms that acetoxypalladation is very stereospecific since only very pure, trans acetoxypalladation–deacetoxypalladation would prevent exchange in this system:

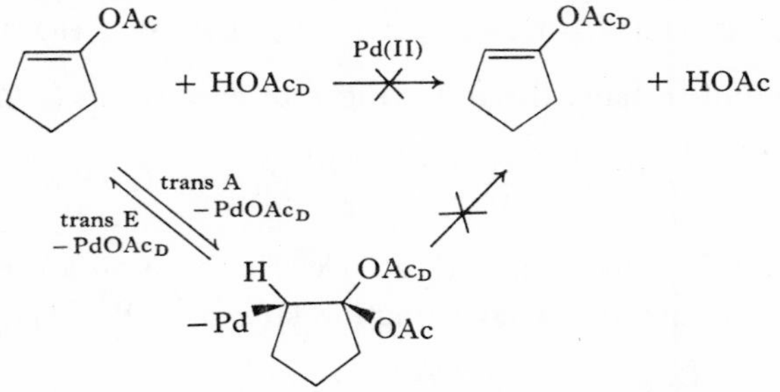

(161)

Allylic ester exchange gave the same rate expression indicating that a dimeric π complex is the reactive species (*108*, *118*). However, in this case, allyl ester actually inhibited the exchange. This could be due to formation of unreactive monomeric π complexes.

$$Li_2Pd_2Cl_6 + 2 \text{ allyl ester} \underset{}{\overset{K}{\rightleftharpoons}} 2LiPdCl_3 \text{ (allyl ester)} \qquad (162)$$

This π complex is believed to be much less reactive than the dimeric π complex because of the higher negative charge on the Pd(II) involved in π-complexing in the monomer.

The exchange of vinyl chloride with radioactive lithium chloride was also studied (*119*).

$$CH_2{=}CHCl + LiCl^* \longrightarrow CH_2{=}CHCl^* + LiCl \qquad (163)$$

In this case the rate expression and the stereochemical results were consistent with a nonstereospecific chloropalladation–dechloropalladation sequence. This can be explained by chloride being both inside and outside the coordination sphere of Pd(II), and it can thus attack the olefin in either a cis or trans manner.

The exchange of vinyl chloride with acetate [Eq. (155)] had a rate expression with a $[LiCl]^2$ inhibition term (*120*):

$$\text{rate} = \frac{[Li_2Pd_2Cl_6][C_2H_3Cl]}{[LiCl]^2}(k' + k''\,[LiOAc]) \qquad (164)$$

A careful study of the exchange of *cis*- and *trans*-propenyl chlorides indicated that the reaction did not proceed with complete retention of configuration. This *cis*-1-chloropropene gave 85% *cis*-1-acetoxypropene and 15% of the trans isomer. If acetoxypalladation is trans, as the earlier evidence indicates, then dechloropalladation must not be stereospecific. This is consistent with the finding that chloropalladation was not stereospecific in the vinyl chloride–radioactive chloride exchange. Furthermore, the rate expression indicates that a vacant coordination site is required before dechloropalladation can occur.

$$\left[\begin{array}{c} \quad\quad\quad\quad\quad CH_2 \\ Cl — Cl — \| \\ /\ Pd\ /\ Pd\ /\ CHCl \\ Cl — Cl — Cl \end{array}\right]^- + OAc^- \rightleftharpoons \left[\begin{array}{c} \quad\quad\quad\quad\quad\quad\quad /Cl \\ Cl — Cl — CH_2CH \\ /\ Pd\ /\ Pd\ /\quad \backslash \\ Cl — Cl — Cl \quad OAc \end{array}\right]^{2-} \quad (165)$$

$$\rightleftharpoons \left[\begin{array}{c} \quad\quad\quad\quad\quad\quad\quad /OAc \\ Cl — Cl — CH_2CH \\ /\ Pd\ /\ Pd\ /\quad \backslash \\ Cl — Cl \quad\quad\quad Cl \end{array}\right]^- + Cl^- \xrightarrow{\text{slow}} \left[\begin{array}{c} \quad\quad\quad\quad\quad CH_2 \\ Cl — Cl — \| \\ /\ Pd\ /\ Pd\ /\ CHOAc \\ Cl — Cl — Cl \end{array}\right]^- \quad (166)$$

The hydration of vinyl acetate to give acetaldehyde in wet acetic acid is of special interest because it almost certainly involves a hydroxypalladation step analogous to that for the Wacker reaction.

$$CH_2{=}CHOAc + H_2O + Pd(II) \xrightarrow{HOAc} Pd(II){—}CH_2{—}CH\begin{array}{l} /OAc \\ \backslash OH \end{array} \longrightarrow Pd(II){—}CH_2{—}CHO + HOAc + H^+ \xrightarrow{H^+} Pd(II) + CH_3CHO \quad (167)$$

The kinetics indicated that, in fact, the hydroxypalladation occurred by trans attack of water on the dimeric π complex (*121*):

$$^-\begin{array}{c} \quad\quad\quad\quad\quad CH_2 \\ Cl — Cl — \| \\ /\ Pd\ /\ Pd\ /\ CHOAc \\ Cl — Cl — Cl \end{array} + H_2O \longrightarrow {}^{2-}\begin{array}{c} \quad\quad\quad\quad\quad\quad\quad /OH \\ Cl — Cl — CH_2CH \\ /\ Pd\ /\ Pd\ /\quad \backslash \\ Cl — Cl — Cl \quad OAc \end{array} + H^+ \quad (168)$$

Thus, the mechanism is quite different from that found in water for ethylene (Section III, A, 1), suggesting that hydroxypalladation, as well as other oxymetallations, are very sensitive to reaction conditions.

A reaction that is apparently closely related to the hydration of vinyl acetate in wet acetic acid is the Pd(II)-catalyzed decomposition of vinyl acetate in dry acetic acid. The kinetics are very complicated (*122*).

$$CH_2{=}CHOAc + \xrightarrow{Pd(II)} CH_3CHO + Ac_2O \qquad (169)$$

At high [Pd(II)] concentrations, the kinetics are similar to those for hydration of vinyl acetate in wet acetic acid. At low [Pd(II)] and [LiCl] concentrations, the rate is essentially independent of Pd(II). One possibility is that trace amounts of water are being formed by the following equilibrium which may be catalyzed by metal salts in the system:

$$2HOAc \rightleftharpoons Ac_2O + H_2O \qquad (170)$$

The equilibrium is continuously forced to the right as the water is consumed to saponify the vinyl acetate.

One exchange that did not fit in the general scheme is the acid-catalyzed exchange of 1-cyclo-hexen-1-yl propionate (*205*). In contrast with the vinyl ester exchange of straight-chain enol esters, addition of sodium acetate actually inhibited the reaction and acids, such as CF_3COOH, strongly accelerated the rate. Deuterium-labeling experiments indicated that a symmetrical intermediate was involved. One possible intermediate involves a Pd(IV) π-allylic species.

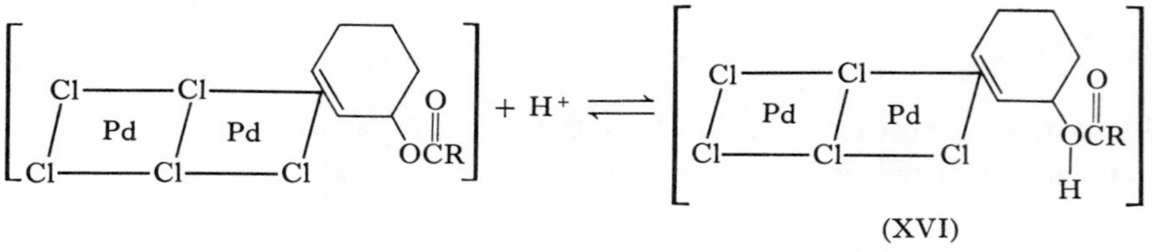

[Cl—Cl—; Pd; Pd; Cl—Cl—Cl; cyclohexenyl; O; OCR] + H⁺ ⇌ [Cl—Cl—; Pd; Pd; Cl—Cl—Cl; cyclohexenyl; O; OCR; H]

(XVI)

(171)

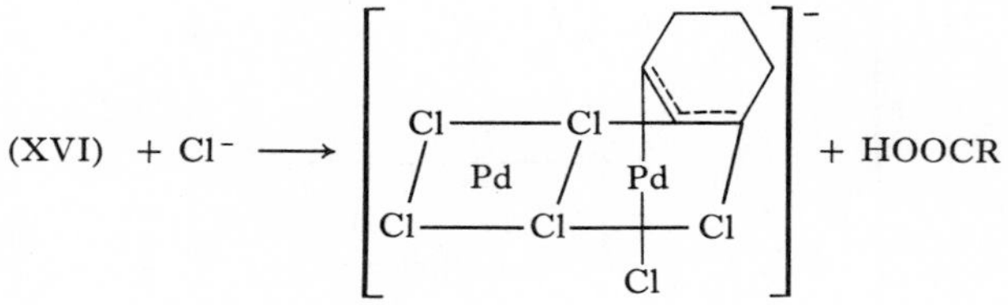

(XVI) + Cl⁻ ⟶ [Cl—Cl—; Pd; Pd; Cl—Cl—Cl; Cl; π-allyl cyclohexenyl]⁻ + HOOCR (172)

It is interesting that cyclohexene reacts so differently from straight-chain olefins in most Pd(II) oxidations. Thus, it apparently also proceeds via π-allylic intermediates during oxidation in acetic acid (Section III, A, 2a), and even in water does not give the same kinetics as straight-chain olefins and so may very well be reacting via a π-allyl intermediate. The formation of the π-allyl could occur by a mechanism analogous to the allylic ester exchange where hydrogen would replace the ester group. Elimination of HX (X = Cl or OAc) would give a Pd(II) π-allyl.

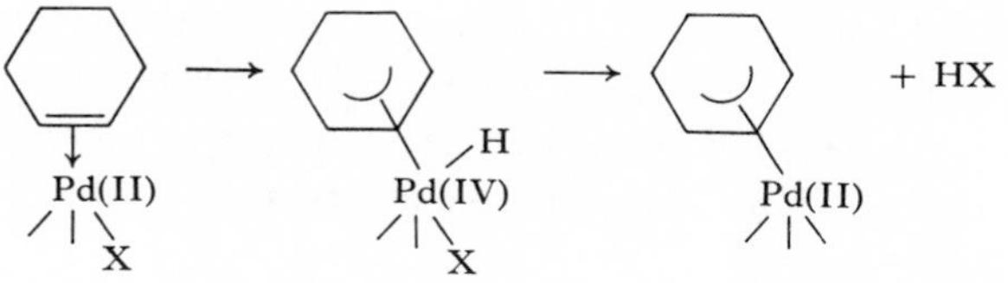

Pd(II) (X) → Pd(IV) (H, X) → Pd(II) + HX (173)

The reason cyclohexene does not give the normal type of oxidation may be related to steric inhibition to oxypalladation (*205*).

The vinyl ester exchange has also been studied in the chloride-free palladium(II) acetate system, using vinyl propionate as substrate. Of the three Pd(II) species present in this system (Section II, A, 2), the dimer is most reactive, with the trimer $Pd_3(OAc)_6$ next, and the monomer unreactive. The rate expression for exchange catalyzed by the dimer is (*214*).

$$\text{rate} = k[Na_2Pd_2(OAc)_6][C_2H_3OAc] \tag{174}$$

Stereochemical studies indicate that in this system acetoxypalladation is not stereospecific, as contrasted with the chloride-containing system. However, in this system acetate is a ligand and is, thus, available in the coordination sphere. The lack of an [NaOAc] term in the rate expression is believed to result in a cancellation in the two equilibria given by Eqs. (175) and (176):

$$Na_2Pd_2(OAc)_6 + C_2H_3O_2CC_2H_5 \rightleftharpoons NaPd_2(OAc)_5(C_2H_3O_2CC_2H_5) + NaOAc \tag{175}$$

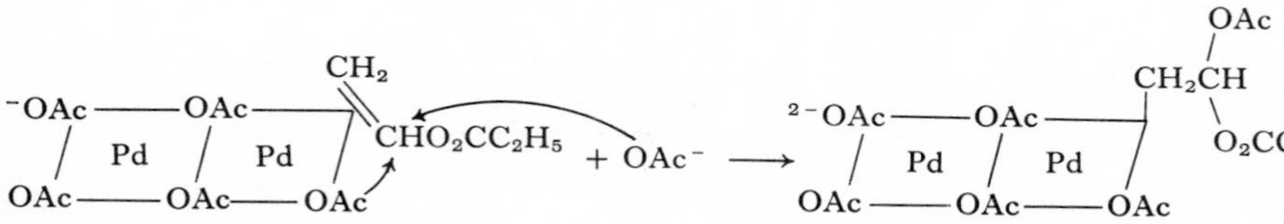

$^-$OAc—OAc—(CH$_2$=CHO$_2$CC$_2$H$_5$), Pd, Pd, OAc—OAc—OAc + OAc$^-$ → $^{2-}$OAc—OAc—(CH$_2$CH(OAc)O$_2$CC$_2$H$_5$), Pd, Pd, OAc—OAc—OAc (176)

2. *Isomerization Reactions Not Involved Pd(II)–Hydrides*

Palladium(II) salts are known to isomerize double bonds along a hydrocarbon chain as well as to cause cis–trans isomerizations (*50*, *82*, *179*). These isomerizations are believed to occur either by Pd(II)–hydride or via π-allyl complexes. Recently, Pd(II)-catalyzed cis–trans isomerizations that apparently do not proceed by either mechanism have been reported. For instance, *cis*- and *trans*-1,2-dideuteroethylenes are isomerized without deuterium scrambling—a result that is inconsistent with hydride mechanisms (*136*). Likewise vinyl chlorides and esters are isomerized, and again it could be shown that Pd(II)–hydrides were not involved (*117*, *123*, *124*). The isomerization of vinyl ethers has also been reported recently (*257*). Plausible mechanisms for this isomerization include the following:

a. *π-σ Rearrangement to a diradical:*

$$>\!Pd\leftarrow\Vert\,(\text{HRC}{=}\text{CHX}) \rightleftharpoons >\!Pd{-}\underset{\text{H}}{\overset{\text{R}}{\text{C}}}{-}\underset{\text{H}}{\overset{\text{X}}{\text{C}}}\cdot \longrightarrow >\!Pd\leftarrow\Vert\,(\text{HRC}{=}\text{CXH}) \qquad (177)$$

(X = H, OOCR, Cl, OR). A similar mechanism has been suggested for metal carbonyl-catalyzed thermal isomerization of olefins (*185*).

b. *π-σ Rearrangement to a carbonium ion:*

$$>\!Pd\leftarrow\Vert\,(\text{HXC}{=}\text{CH(CH}_3)) \longrightarrow Pd{-}\underset{\text{CH}_3}{\overset{\text{H}}{\text{C}}}{-}\text{C}^{+}(\text{X})(\text{H}) \longrightarrow >\!Pd\leftarrow\Vert\,(\text{XHC}{=}\text{CH(CH}_3)) \qquad (178)$$

A similar rearrangement has been postulated for Pt(II)–olefin π complexes (*153*).

c. *Nonstereospecific chloropalladation–dechloropalladation:*

$$\underset{\text{Cl}}{>\!Pd}\leftarrow\Vert\,(\text{HXC}{=}\text{CH(CH}_3)) \xrightarrow{\text{cis A}} Pd{-}\underset{\text{X}}{\overset{\text{H}}{\text{C}}}{-}\underset{\text{Cl}}{\overset{\text{CH}_3}{\text{C}}}{-}\text{H} \xrightarrow{\text{trans E}} >\!Pd\leftarrow\Vert\,(\text{XHC}{=}\text{CH(CH}_3)) + \text{Cl}^- \qquad (179)$$

Certainly more work is required before a definite mechanism can be proposed and it is very likely that more than one mechanism is operative.

Another new isomerization involves allylic esters (*118*, *125*).

$$\mathrm{RCH{=}CH{-}CH_2OOCC_2H_5} \xrightleftharpoons{\mathrm{Pd(II)}} \mathrm{R\overset{\overset{\displaystyle OOCC_2H_5}{|}}{C}H{-}CH{=}CH_2} \tag{180}$$

When C_2H_5 is replaced with the more electron withdrawing CF_3, the rate is slower, indicating that reaction does not proceed via allylic carbanions, the path for thermal isomerization of allylic esters. Oxygen-18-labeling experiments and the electronic effects are consistent with an *internal* oxypalladation mechanism:

$$\mathrm{CH_3CH{=}CHCH_2(O^*{-}C(C_2H_5){=}O)}\;(\downarrow \mathrm{Pd}) \rightleftharpoons \mathrm{CH_3\overset{|}{C}H{-}\overset{|}{C}H{-}\overset{|}{C}H_2}\ (\text{cyclic } \mathrm{O{-}\overset{+}{C}(C_2H_5){-}O^*},\ \mathrm{Pd}) \rightleftharpoons \mathrm{CH_3\overset{|}{C}HCH{=}CH_2}\ (\mathrm{O{-}C(C_2H_5){=}O^*})\;(\downarrow \mathrm{Pd}) \tag{181}$$

3. *Preparation of Stable Pd(II) Adducts*

Most Pd(II)-catalyzed reactions of monoolefins proceed via addition of Pd(II) across the olefin bond to give Pd(II)–σ-bonded carbon intermediates. Evidence for this scheme has been provided by isolation of stable adducts of this type. Usually these adducts are prepared from nonconjugated diolefins because of the special stability of this type of complex, but in some cases they have been prepared from monoolefins.

a. *Oxypalladation.* Methoxypalladation adducts have been prepared from chelating monoolefins such as allyl amines (*41*):

$$\mathrm{2Me_2NCH_2{-}CH{=}CH_2 + 2PdCl_2} \xrightarrow{\mathrm{MeOH}} \left[\mathrm{MeO{-}CH(CH_2{-}Pd(Cl)(\mu){\leftarrow}N(Me)_2{-}CH_2)}\right]_2 + \mathrm{2HCl} \tag{182}$$

This reaction has now been extended to allylic sulfides (*257*):

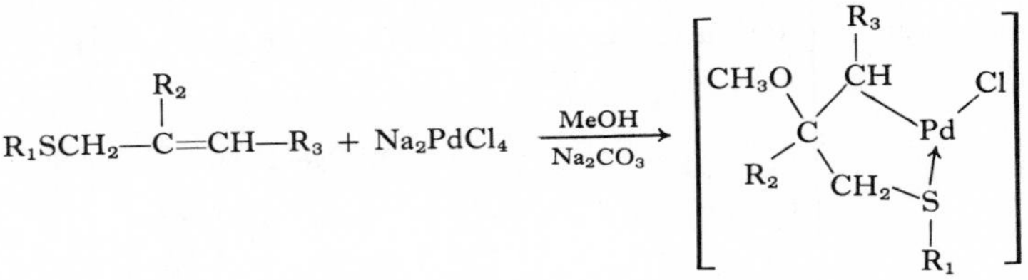

$$R_1SCH_2{-}\overset{R_2}{\overset{|}{C}}{=}CH{-}R_3 + Na_2PdCl_4 \xrightarrow[Na_2CO_3]{MeOH} \left[\begin{array}{c} CH_3O \quad \overset{R_3}{\overset{|}{CH}} \quad Cl \\ C \qquad\quad Pd \\ R_2 \quad CH_2{-}\underset{R_1}{\underset{|}{S}} \end{array} \right] \tag{183}$$

b. *Addition of Pd(II) across Carbon Bonds.* There have been several reports recently of addition of allyl Pd(II) into norbornene-type bonds. Most have been carried out with dienes and are discussed in Section IV, A. However, there have been two recent reports of the insertion of π-methallyl Pd(II) across norbornene itself (*71, 137*). The cis–exo structure has been confirmed by X-ray studies (*299, 300*).

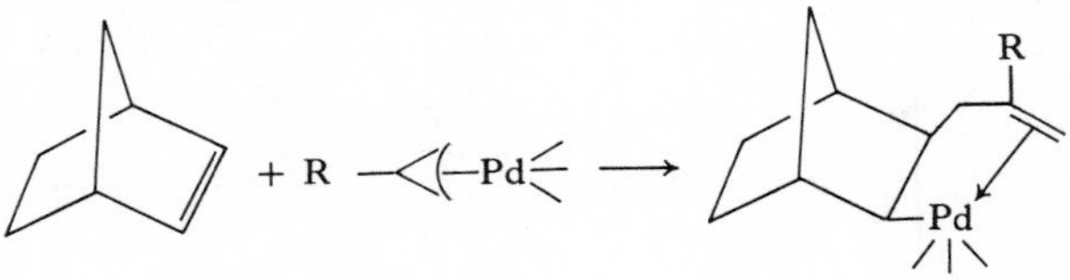

$$+ \text{R} \longrightarrow \tag{184}$$

4. *Hydrogenation and Related Reactions*

a. *Hydrogenation.* Although metallic palladium has long been used as a hydrogenation catalyst, there has been relatively little study of homogeneous hydrogenation by Pd(II) salts. It has been reported that triphenylphosphine complexes of $PdCl_2$ will catalyze hydrogenation of long-chain polysaturated esters. During hydrogenation, considerable double-bond isomerization occurred (*11*). Recent studies using Pd(0)–phosphine complexes also indicate considerable isomerization (*249*). The results are explained by both isomerization and hydrogenation proceeding via Pd(II)–hydrides. Pretreatment of catalyst with O_2 increased the rate of hydrogenation. Studies on the deuteration of olefins provided more insight into the mechanism (250). Monoexchange of H for D was faster than isomerization, diexchange, or deuterogenation with untreated catalyst. With oyxgen-treated catalyst, deuterogenation became the predominant reaction. These results support completely reversible addition of palladium–hydride species to olefins.

b. *Hydrosilation.* The hydrosilation of olefins by platinum has been intensively studied. Until recently, however, there has been little study of Pd(II) as a catalyst for this purpose. The only previous report was the use of a zero-valent palladium–phosphine complex (*256*). Recently, it has been found that Pd(0) and Pd(II) will catalyze addition of trimethyl- or trichlorosilane to olefins, with the chloro compound the more reactive (*81*, *159*):

$$Cl_3SiH + CH_2{=}CHC_4H_9 \longrightarrow Cl_3SiCH_2{-}CH_2C_4H_9 \qquad (185)$$

Another recent report described the asymmetric hydrosilation of styrene using chiral phosphine ligands such as methyldiphenylphosphine (*160*). Enantiomeric excesses of 5% of S isomer of 2-phenylethyltrichlorosilane were obtained. The mechanism almost certainly involved insertion of Pd(II) into a silicon–hydride bond, followed by addition of Pd–hydride across the olefinic bond and elimination of Pd(II) to give the silicon–carbon bond.

c. *Addition of HX.* In this class would be the addition of HCN to olefins to give nitrile, catalyzed by Pd(0)–phosphine complexes (*25*). Unfortunately the reaction has large catalyst requirements and only went well with ethylene and with norbornene-type olefins. Recently, the reaction was found to proceed quite well with olefinic silanes (*26*). Thus vinyltriethoxysilane gave predominantly 1-(cyanoethy)-triethoxysilane when tetrakis(triphenylphosphite)palladium was used. Catalyst turnover was 278.

$$\underset{}{(C_2H_5O)_3SiCH{=}CH_2} + HCN \longrightarrow \underset{92\%}{(C_2H_5O)_3SiCH_2CHCN} + \underset{8\%}{(C_2H_5O)_3Si\overset{\overset{\displaystyle CN}{|}}{C}H_2{-}CH_3} \qquad (186)$$

Another more complicated reaction of this type is the hydrocarboxylation of olefins using palladium phosphines as catalysts:

$$RCH{=}CH_2 + CO + ROH \longrightarrow R{-}\overset{\overset{\displaystyle COOR}{|}}{C}H{-}CH_3 + R{-}CH_2{-}CH_2COOR \qquad (187)$$

This reaction as originally carried out surprisingly gives mainly the branched ester. More recent work by Fenton (*61*, *62*) has shown that under certain reaction conditions the straight-chain isomer can be made the predominant product, although considerable amounts of the branched-chain isomer is still found. The catalyst was found to undergo a complicated series of changes. Thus, triphenylphosphine gave PPh_2Cl, *o*-metallated ligand, and *o*-chlorinated phenyl rings on the PPh_3.

The two most reasonable mechanisms involve addition of palladium–hydride followed by CO insertion or addition of $Pd{-}\overset{\overset{\displaystyle O}{\|}}{C}{-}X$ (X = OH or Cl) to the olefin:

$$
\begin{array}{l}
{>}Pd{-}H + RCH{=}CH_2 \longrightarrow {>}Pd{-}\overset{\overset{\displaystyle R}{|}}{C}H{-}CH_3 \quad + \quad PdCH_2{-}CH_2R \\
\qquad\qquad\qquad\qquad\qquad \downarrow CO \qquad\qquad\qquad\qquad \downarrow CO \\
\qquad\qquad\qquad\qquad Pd{-}\overset{\overset{\displaystyle O}{\|}}{C}{-}\overset{\overset{\displaystyle R}{|}}{C}H{-}CH_3 \qquad Pd\overset{\overset{\displaystyle O}{\|}}{C}CH_2{-}CH_2{-}R \\
\qquad\qquad\qquad\qquad\qquad \downarrow ROH \qquad\qquad\qquad\qquad \downarrow ROH \\
\qquad\qquad\qquad Pd(0) + RO\overset{\overset{\displaystyle O}{\|}}{C}\overset{\overset{\displaystyle R}{|}}{C}H{-}CH_3 \quad Pd(0) + RO\overset{\overset{\displaystyle O}{\|}}{C}CH_2CH_2R
\end{array}
\tag{188}
$$

$$
\begin{array}{l}
{-}\overset{\diagdown}{\underset{H\diagup}{Pd}}{-}\overset{\overset{\displaystyle O}{\|}}{C}{-}X + RCH{=}CH_2 \longrightarrow R{-}\underset{}{\overset{\overset{\displaystyle X}{|}}{\overset{C{=}O}{|}}}\!\!CH{-}CH_2Pd\overset{\diagup}{\underset{\diagdown H}{-}} + R{-}\underset{\underset{\displaystyle H\diagup\,|\,\diagdown}{Pd-}}{\underset{|}{C}}H{-}CH_2COX \\
\qquad\qquad\qquad\qquad\qquad\qquad\qquad \downarrow \qquad\qquad\qquad\qquad\qquad \downarrow \\
\qquad\qquad\qquad\qquad R{-}\overset{\overset{\displaystyle X}{|}}{\overset{C{=}O}{|}}\!\!CH{-}CH_3 + Pd(0) \qquad R{-}CH_2CH_2COX + Pd(0)
\end{array}
\tag{189}
$$

The first mechanism was thought unlikely, since alcohols or esters were not produced by reaction of the intermediate Pd(II) alkyl with water or carboxylic acid. These products, it was argued, would be expected from analogy with the formation of acetaldehyde and vinyl ester by the oxidation of ethylene in water and carboxylic acids, respectively (Sections III, A, 1 and 2). However, this is a poor analogy since Pd(II)–C bonds do not ordinarily tend to decompose in this fashion, namely,

$$RCH_2{-}CH_2Pd\overset{\diagup}{\underset{\diagdown}{-}} + ROH \longrightarrow RCH_2CH_2OR \tag{190}$$

(R = H or alkyl). Furthermore, additions of preformed carboalkoxyl

Pd(II) groups ($PdC(=O)OR$) to olefins have been investigated, and the results are quite different from those proposed in this study (*90*). First, the products are unsaturated rather than saturated esters. Second, addition of the $—C(=O)OR$ group is always to the terminal carbon. Third, under CO pressure, saturated diesters are produced by the insertion of CO into the Pd–C intermediate (*91*). These are not found in the present study. Now the fact that phosphine is present and Pd(II)–hydrides may be involved, as shown in Eq. (189) could very well change the mode of decomposition. However, more work is required before a definite mechanism can be proposed.

Another reaction of this type is the nonoxidative dimerization of olefins. They very likely proceed via palladium–hydride species which may be formed by a small amount of oxidation of the olefin. Often Pd(II)–olefin π complexes are used. Kawamoto *et al.* (*155*) have recently reported the dimerization of styrene and vinyl compounds using the styrene Pd(II) π complex. Also, it has been reported (*254*) that phosphine complexes, $Pd(PPh_3)_4$ or $(PPh_3)_2PdX_2$ ($X = N_3$ or NCO), have been employed to give a novel dimerization of malonotrile:

$$2(NC)_2CH_2 \longrightarrow NCCH_2—C(NH_2){=}C(CN)_2 \qquad (191)$$

This reaction probably proceeds via a Pd(II)-stabilized imine derivative, NC—CH═C═NH.

IV
REACTIONS OF DIOLEFINS

A. Formation of Stable Adducts with Pd(II)–Carbon σ Bonds

The chemistry of diolefin π complexes of Pd(II) has a special place in the catalytic chemistry of Pd(II) because their reactions with nucleophiles gave adducts that served as stable models for the types of unstable intermediates postulated in many reactions of Pd(II) with olefins.

The first oxypalladation adducts from dienes were obtained by the reaction of dienes in basic alcohol media

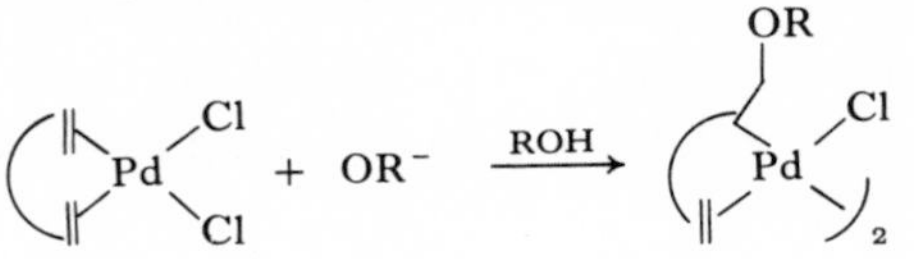

$$\text{(diolefin)PdCl}_2 + \text{OR}^- \xrightarrow{\text{ROH}} [\text{(ROC-diolefin)PdCl}]_2 \quad (192)$$

(diolefin = diolefin such as 1,5-cyclooctadiene norbornadiene and dicyclopentadiene).

The gross structural features of this type of complex were first postulated by Chatt *et al.* (*34*) and later confirmed by X-ray studies for the platinum(II) analogs (*211*, *291*). In this type of complex, the OR group is trans to the Pd(II), with the alkoxy group exo and the Pd(II) endo. Adducts have also been prepared in which the nucleophile is an acetate or a carbanoid species such as malonate or acetoacetate (*180*).

A number of new adducts have been reported recently with a variety of nucleophiles. Benzylamine gives an adduct with the 1,5-cyclooctadiene $PdCl_2$–π complex which is a tetramer because of intermolecular N–Pd bonds (*1*),

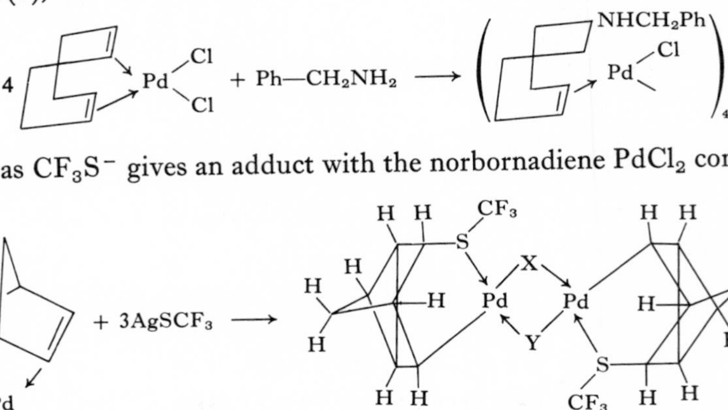

$$4\ (\text{C}_8\text{H}_{12})\text{PdCl}_2 + \text{Ph—CH}_2\text{NH}_2 \longrightarrow (\text{C}_8\text{H}_{12}\text{NHCH}_2\text{Ph})\text{PdCl})_4 \quad (193)$$

whereas CF_3S^- gives an adduct with the norbornadiene $PdCl_2$ complex:

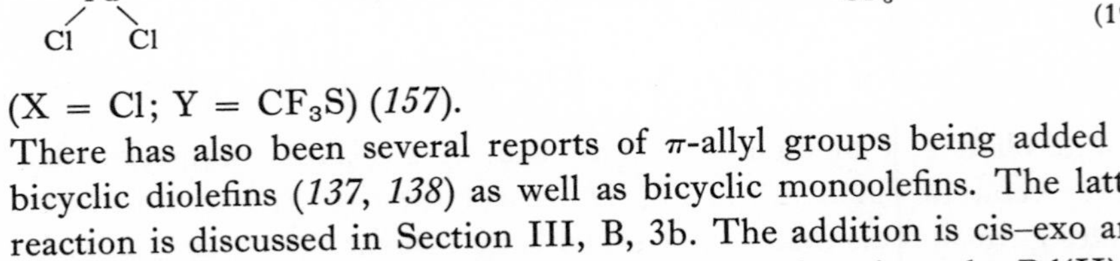

$$2\ (\text{C}_7\text{H}_8)\text{PdCl}_2 + 3\text{AgSCF}_3 \longrightarrow \quad (194)$$

(X = Cl; Y = CF_3S) (*157*).

There has also been several reports of π-allyl groups being added to bicyclic diolefins (*137*, *138*) as well as bicyclic monoolefins. The latter reaction is discussed in Section III, B, 3b. The addition is cis–exo and the diolefins are, in effect, behaving as monoolefins since the Pd(II) is not complexed to the second double bonds of the diene. Thus, they will not be considered further in this section. The structure of the norbornadiene adduct has been confirmed by X-ray diffraction (*233*).

A reaction of considerable interest is the addition of phenyl to the norbornadiene $PdCl_2$ π complex (*239*):

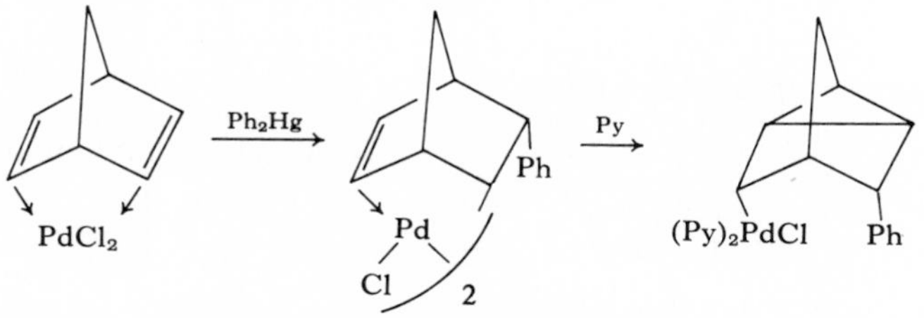

$PdCl_2$ $\xrightarrow{Ph_2Hg}$ Ph, Pd, Cl, 2 $\xrightarrow{Py}$ $(Py)_2PdCl$ Ph (195)

In this case the Pd(II) is bonded to both double bonds and thus held in the endo position. The addition occurs in the unfavorable endo position to give a diendo product—the first report of this process. This stereochemical result is consistent with that found for addition of phenyl groups to monoolefins (Section III, A, 4).

Another novel addition is that of Pd–Cl to vinyl norbornene (*292*), which occurs in the trans–exo mode found for noncarbanoid nucleophiles.

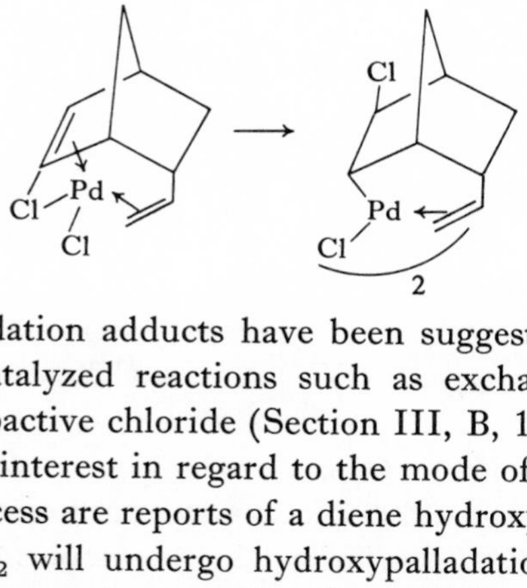

Cl, Pd, Cl $\longrightarrow$ Cl, Pd, Cl, 2 (196)

These chloropalladation adducts have been suggested as intermediates in other Pd(II)-catalyzed reactions such as exchange of Cl in vinyl chloride with radioactive chloride (Section III, B, 1).

Of considerable interest in regard to the mode of hydroxypalladation in the Wacker process are reports of a diene hydroxypalladation adduct. Thus $(COD)PdBr_2$ will undergo hydroxypalladation in basic H_2O to give an adduct of uncertain stereochemistry (*289*):

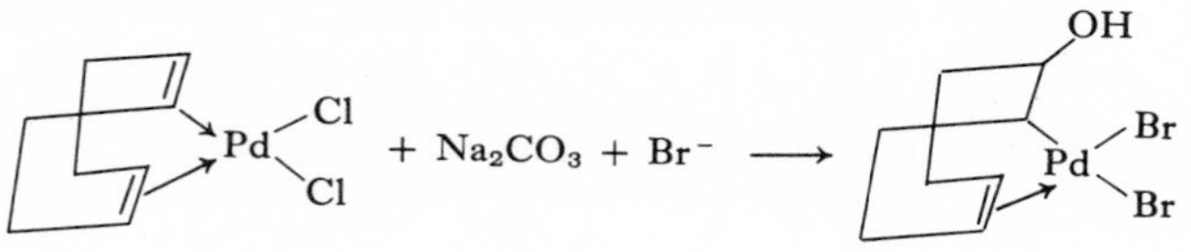

$$\text{Pd(Cl)}_2 + Na_2CO_3 + Br^- \longrightarrow \text{OH, Pd(Br)}_2 \qquad (197)$$

An analogous adduct can be obtained from 5-vinylnorbornene (*292*). In this case the stereochemistry can be shown to be trans–exo in keeping with the addition of other noncarbanoid nucleophiles.

In the Wacker reaction, it may be noted, the kinetics strongly suggest cis addition (Section III, A, 1).

B. *Reactions of Diene σ Adducts*

There has now been a number of reports of the rearrangement of the norbornadiene σ adduct to the nortricyclyl system first reported by Coulsen (*43*). Thus Maitlis *et al.* (*239*) found that their phenylpalladation adduct in the presence of pyridine rearranged to this adduct [Eq. (195)].

Along this same line the reaction of the 3-acetoxy *endo*-5-norticyclyl bis(pyridine) palladium with diphenylmercury gave the *endo*-3-phenyl-5-mercury derivative. This was postulated to proceed via attack of coordinated phenyl on the carbon–acetate bond (*278*):

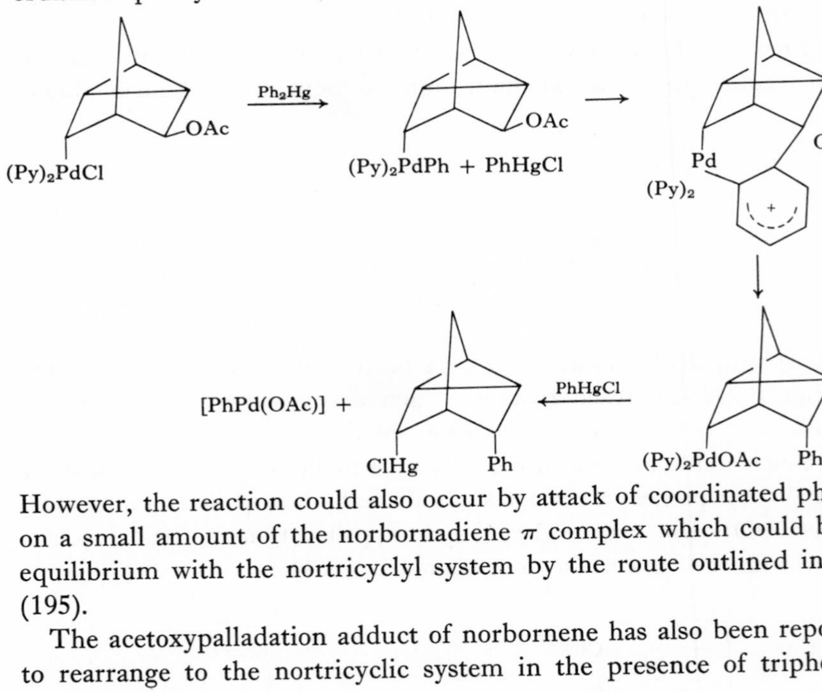

OAc; $(Py)_2PdCl$ $\xrightarrow{Ph_2Hg}$ OAc; $(Py)_2PdPh + PhHgCl$ ⟶ Pd; $(Py)_2$; OAc^-; +

(198)

$[PhPd(OAc)] +$ ClHg; Ph $\xleftarrow{PhHgCl}$ $(Py)_2PdOAc$; Ph

However, the reaction could also occur by attack of coordinated phenyl on a small amount of the norbornadiene π complex which could be in equilibrium with the nortricyclyl system by the route outlined in Eq. (195).

The acetoxypalladation adduct of norbornene has also been reported to rearrange to the nortricyclic system in the presence of triphenylphosphine (*15*).

The carbonylation of the adducts has been studied as well. A careful analysis of the stereochemistry of the esters obtained from carbonylation of the methoxypalladation adducts has shown that CO insertion occurred with retention of configuration in all cases (*130*). Thus, carbonylation of the norbornadiene adduct in alcohol gave the nortricyclic ester ether.

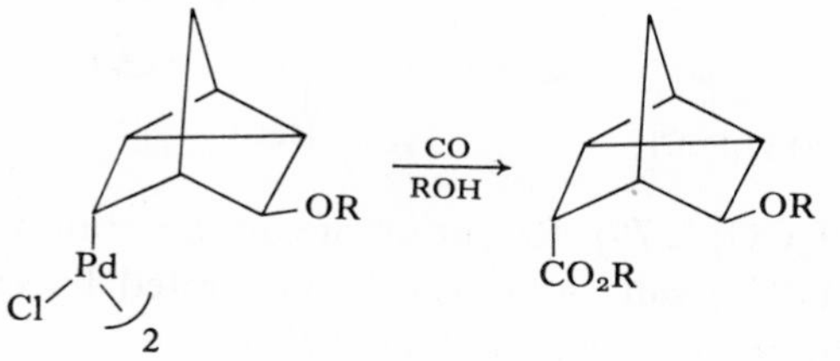

$$\xrightarrow[\text{ROH}]{\text{CO}} \tag{199}$$

The 1,5-cyclooctadiene methoxypalladation adduct gave the *trans*-β-methoxy ester.

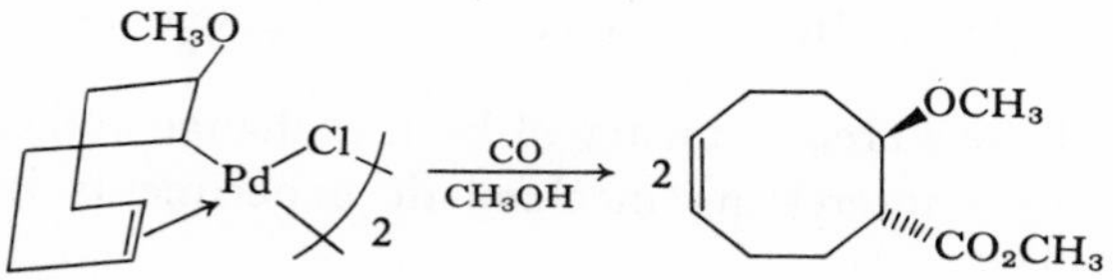

$$\xrightarrow[\text{CH}_3\text{OH}]{\text{CO}} 2 \tag{200}$$

This result is of some mechanistic importance because it permits the determination of stereochemistry of Pd(II) oxymetallation by stereospecific replacement of Pd(II) by CO. This technique was used for determining the mode of methoxypalladation of the butenes (Section III, A, 3, b).

If the Pd(II) in the adduct has phosphine ligands, an unstable Pd(II) acyl is formed. This intermediate can be decomposed with HCl to give aldehydes, or with MeI or LiR to give ketones (*31*).

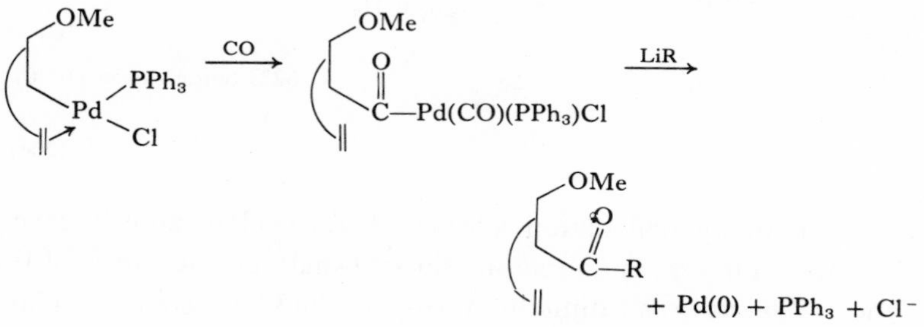

$$\xrightarrow{\text{CO}} \quad \text{C—Pd(CO)(PPh}_3\text{)Cl} \quad \xrightarrow{\text{LiR}} \quad \text{C—R} + \text{Pd(0)} + \text{PPh}_3 + \text{Cl}^- \tag{201}$$

Exchange of the norbornadiene or nortricyclyl Pd(II) adducts with Hg(II) salts gives the mercurial of the same configuration indicating that exchange proceeds with retention of configuration

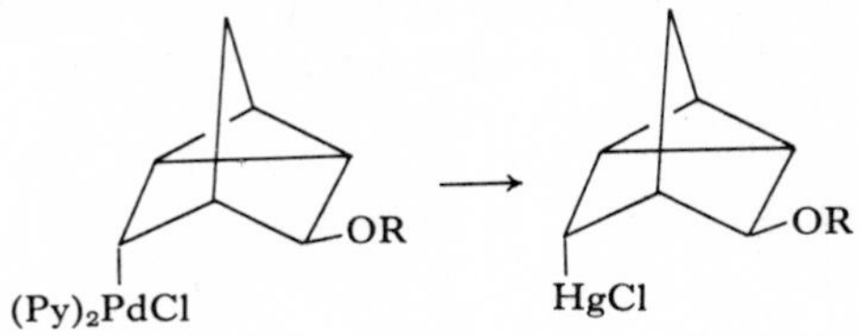

(202)

(R = Me or CH_3CO) (*279*). Retention of configuration when mercurials exchange with Pd(II) salts was also demonstrated by carbonylation of the Pd(II) products as discussed above (*251*).

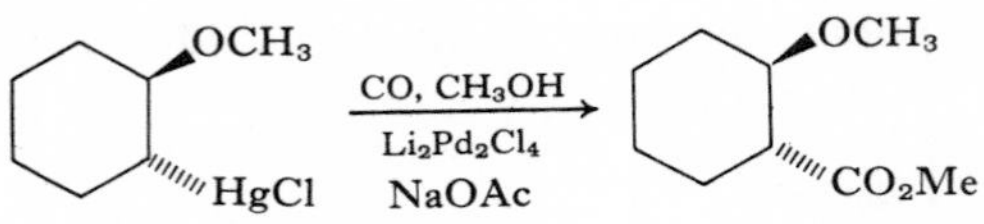

(203)

A knowledge of the stereochemistry of both exchange and carbonylation should allow some interesting mechanistic experiments to be carried out.

The reaction of diene adducts with oxidants has also been studied. In both cases cis insertion of the Pd(II)–carbon bond into the second olefinic bond occurred. Thus Stille *et al.* (*126*) found that reaction of the norbornadiene(methoxy) adduct or the norticyclyl(methoxy) adduct with Cl_2 or with Br_2 in CH_3OH gave 3-*exo*-methoxy-5-halometricyclenes plus the dimethoxy adduct:

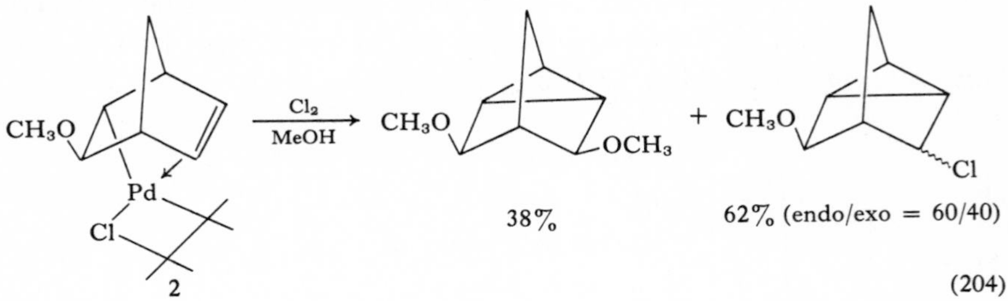

38%

62% (endo/exo = 60/40)

(204)

However, the methoxy palladation adduct of 1,5-cyclooctadiene gave one of the two epimers of 2-*endo*-methoxy-6-halo-*cis*-bicyclo(3.3.0)-octane and 2,6-*endo*,*endo*-dimethoxy-*cis*-bicyclo(3.3.0)octane. The halide epimer depends to a large extent on solvent. The proposed scheme is given in Eq. (205).

CH_3O ... Pd ... X ... 2 — X_2, cis insertion into Pd–C σ bond → [CH_3O ... X–Pd(–X)–X ≡ OCH_3 ... X–Pd–X, X]; CH_2Cl_2 → OCH_3, X; CH_3OH, X^-, S_N2 attack → OCH_3, X + OCH_3, OCH_3 (205)

In an analogous reaction the acetoxypalladation adduct of 1,5-cyclooctadiene reacted in acetic acid, containing traces of chloride in the presence of $Pb(OAc)_4$, to give 2,6-*endo*,*endo*-diacetoxy-*cis*-bicyclo-(3.3.0)octane (*126*). These results are consistent with a scheme involving trans oxypalladation, cis insertion into the second double bond, and trans elimination of Pd(II) with OAc^-:

H, H, Pd, H, H — trans addition → OAc, H, H, Pd, H, H — cis addition → OAc, H, H, H, Pd, Pb(IV), OAc^-, H — S_N2 → OAc, H, H, H, H, OAc (206)

The two schemes given by Eqs. (205) and (206) are consistent with those of cyclohexene oxidation by $PdCl_2$–$CuCl_2$ in acetic acid. In chloride-containing media, the halide which is strongly complexed to the Pd(II) may displace palladium in either a cis or trans fashion, whereas acetate which is not complexed can only displace palladium by trans attack (Section III, A, 2, b). In a nonpolar solvent (i.e., CH_2Cl_2) in which chloride salts have little solubility, the cis attack would be most important, whereas in polar solvents, the trans attack becomes more important. Once again the question of the mode of action of the oxidant arises. Stille (*251*) suggests that the Cl_2 or Br_2 is oxidizing Pd(II) to Pd(IV), whereas in Eq. (206) the Pd(II) species are represented. As discussed previously, the present results do not permit a unique mechanism for this general type of reaction, and it is possible that different routes may be operative with different oxidants.

The decomposition of adducts of 1,5-cyclooctadiene to give 3.3.0-octanes has precedent. Thus, Anderson and Burreson (*5*) decomposed the oxymetallation adduct by light to give an unsaturated bicyclooctane, whereas Takahashi and Tsuji (*253*) used strong base to produce a disubstituted bicyclooctane. Both reactions must also occur by addition of Pd(II)–C bonds across the second olefinic bond as in Eqs. (205) and (206).

Another reaction that must occur by a similar route is the oxidation of 1,5-hexadiene to give cyclopentene derivatives (*186*):

$$(\text{1,5-hexadiene})Pd(OAc)_2 \longrightarrow AcO\text{-}CH_2\ldots Pd(OAc) \longrightarrow AcO\text{-cyclopentyl-}CH_2PdOAc \xrightarrow{-HPdOAc} AcO\text{-methylenecyclopentane} \qquad (207)$$

C. Catalytic Reactions of Unconjugated Dienes

In this category would fall the novel reaction of quinone with various XY groups catalyzed by Pd(0)–phosphine complexes (*230*):

$$O{=}C_6H_4{=}O \quad M(PPh_3)_2 + XY \longrightarrow XO\text{-}C_6H_4\text{-}OX + M(PPh_3)_2Y_2 \qquad (208)$$

(XY = CH_3I, CH_3COCl, NOCl, CH_3COCl, C_2H_5COCl and $CH_2{=}CHCH_2Cl$). The halide always ends up on the palladium, whereas the other group is attached to the hydroquinone.

The hydrocarbonylation of unsaturated fatty acids has recently been reported to give acid or esters (*69*). A mechanism involving π allylic Pd(II) intermediates has been proposed.

Another novel reaction of quinone is the formation of anthraquinone derivatives catalyzed by triphenylphosphine Pd(0) complexes (*231*):

$$2RCH_2CH_2CH{=}CH_2 + 7\,[\text{quinone}] \longrightarrow R\text{-}[\text{anthraquinone}]\text{-}R + 6\,[\text{hydroquinone}] \qquad (209)$$

It is proposed that the olefin is first dehydrogenated to give a conjugated diene which then undergoes a Diels–Alder reaction with the quinone, followed by dehydrogenation:

$$RCH_2CH_2CH{=}CH_2 + [\text{quinone}] \longrightarrow RCH{=}CHCH{=}CH_2 + [\text{hydroquinone}] \qquad (210)$$

$$2RCH{=}CHCH{=}CH_2 + [\text{quinone}] \longrightarrow R\text{-}[\text{octahydroanthraquinone}]\text{-}R \xrightarrow{O=C_6H_4=O} R\text{-}[\text{anthraquinone}]\text{-}R \qquad (211)$$

V

REACTIONS OF ACETYLENES

Acetylenes have a special place in transition metal chemistry because of their high reactivity and the large variety of products which they afford. The field of palladium-catalyzed reactions of acetylenes has made considerable progress in the past 3 years, mainly due to the efforts of Maitlis and co-workers. Also much interesting work has been carried out on Pt(II) catalysis of acetylenes. This work has recently been reviewed (*36*).

A. *Oligomerization Reactions*

Probably the first well-characterized reaction of acetylenes is the reaction of diphenylacetylene with palladium(II) chloride to give a crystalline compound (*181*). It was later shown to be a π-allyl complex which could be converted to a tetraphenyl cyclobutadienyl complex by acid:

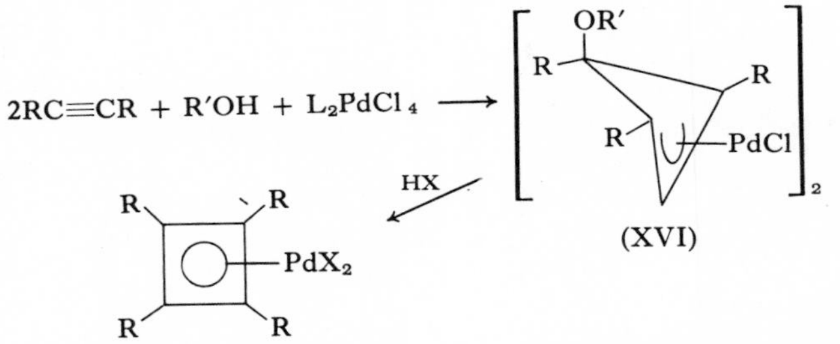

$$2RC{\equiv}CR + R'OH + L_2PdCl_4 \longrightarrow [\text{(XVI)}]_2 \xrightarrow{HX} R_4C_4PdX_2 \quad (212)$$

(XVI)

In aprotic solvents, hexaphenyl benzene and the tetraphenyl cyclobutadiene palladium(II) chloride are obtained directly. The brilliant work of Maitlis and his co-workers has clarified the pathways for this reaction.

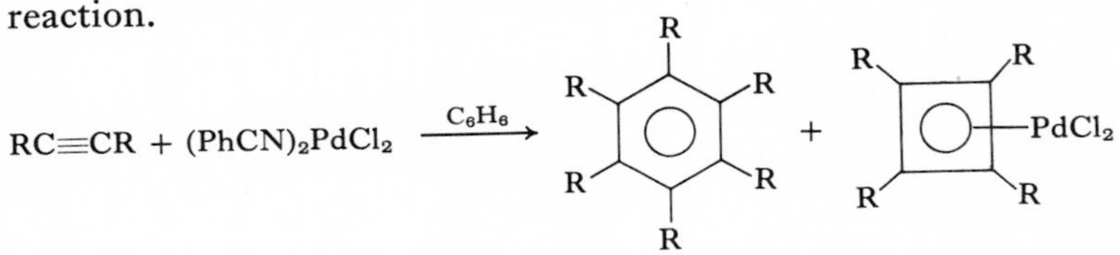

$$RC{\equiv}CR + (PhCN)_2PdCl_2 \xrightarrow{C_6H_6} C_6R_6 + R_4C_4PdCl_2 \quad (213)$$

A general scheme [Eqs. (214) and (215)] is given below (X = Cl) (*55*, *56*, *226*).

One fact that this route must explain is the formation of structure (XVII) from cyclization of $CD_3C{\equiv}CCH_3$. This was explained by the

(XVIII)

tetramers

(214)

$$CD_3C{\equiv}CCH_3 \longrightarrow \text{(XVII)}\ 10\%$$ (215)

demonstration that structure (XVIII) is a fluxional intermediate, thus scrambling the CD_3 and CH_3 groups (*182*).

The formation of the *endo*-π-allyl structure (XIX), can be explained by trans attack of alcohol followed by a thermally allowed conrotatory cyclization step (X = OR) (*183*):

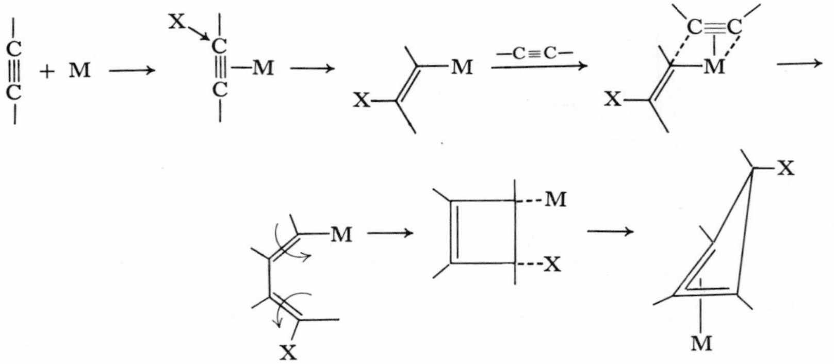

(216)

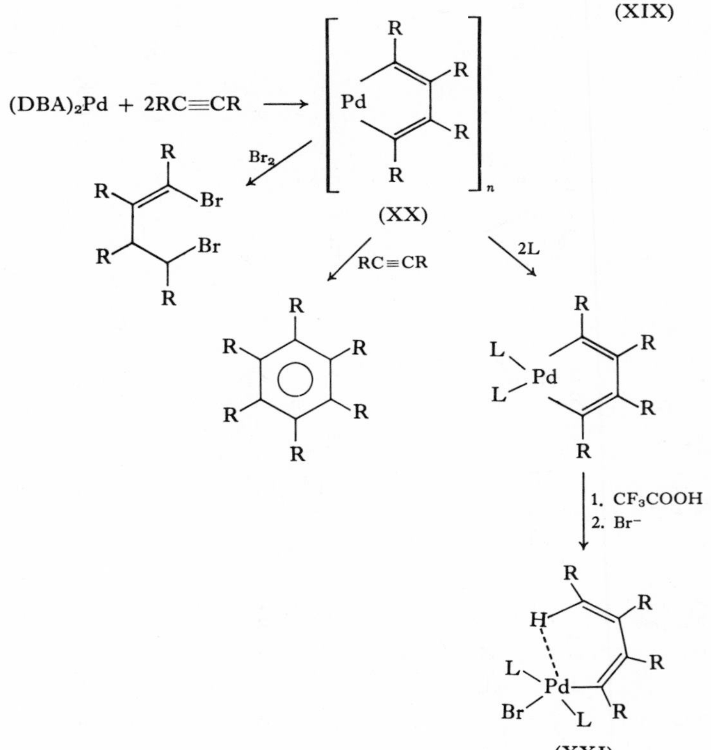

(XIX)

$(DBA)_2Pd + 2RC{\equiv}CR \longrightarrow$ (XX)

Br_2 ; $RC{\equiv}CR$; 2L

(217)

1. CF_3COOH
2. Br^-

(XXI)

Mosely and Maitlis have recently shown that (dibenzylideneacetone)-palladium(0) [$(DBA)_2Pd$] reacts with dimethylacetylenedicarboxylate to give a palladiacyclopentadiene species, structure (XX), which will undergo a series of reactions (*203*).

In the structure (XXI), there is evidence for the Pd(II)–hydrogen interaction indicated by the dotted line (*228*).

When dimethylacetylenedicarboxylate is treated with $(PhCN)_2$-$PdCl_2$ another type of intermediate is obtained in which there is an oxygen–Pd(II) interaction (*229*):

$$3RC{\equiv}CR + (PhCN)_2PdCl_2 \longrightarrow [\ldots]_2 \qquad (218)$$

The initial step in Eq. (214) involves chloropalladation. The work of Heck indicates phenylpalladation occurs more readily than chloro-palladation (Section III, A, 4). The phenylpalladium is prepared *in situ* from the reaction of Ph_2Hg with $PdCl_2$:

$$Ph_2Hg + PdCl_2 \longrightarrow \text{'PhPdCl'} + PhHgCl \qquad (219)$$

This reaction in the presence of 2-butyne gave a relatively stable intermediate analogous to structure (XVIII) in Eq. (214).

$$Ph_2Hg + (PhCN)_2PdCl_2 + MeC{\equiv}CMe \xrightarrow{C_6H_6} [\ldots] \qquad (220)$$

(XXII)

The structure of the *p*-tolyl derivative has been determined by X-ray (*30*, *132*). Complex (XXII) is of particular interest because it serves as a "lightly stabilized" model for catalytic Pd(II) chemistry (*133*, *134*). For instance, on standing it undergoes internal cyclization analogous to insertion of Pd–C bonds into olefins (Section III, A, 4).

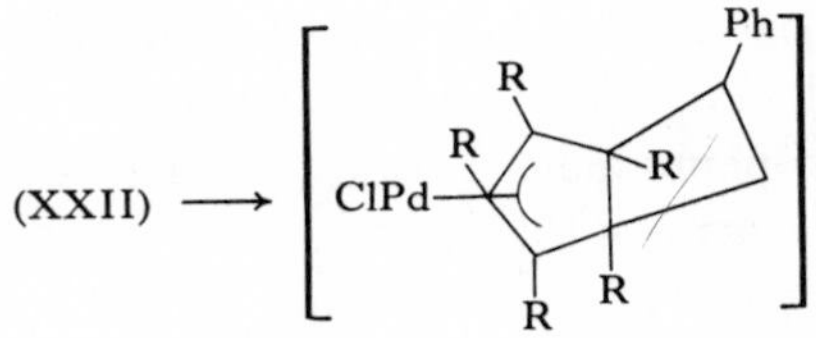

(XXII) ⟶ [ClPd– R R R R R Ph] (221)

Upon reaction with hydrazine, hydrogen, or CH_3O^-, the Pd–C bond is reductively cleaved:

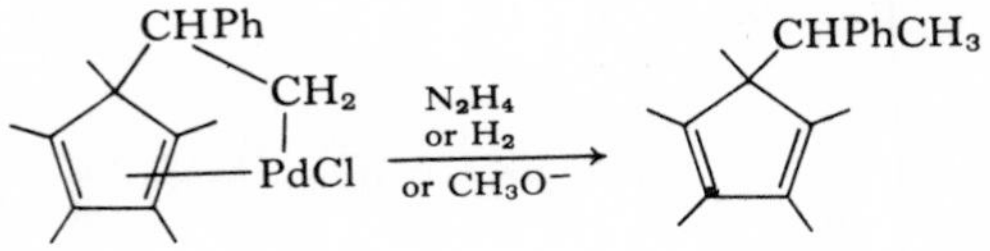

$$\text{(cyclopentadiene–CHPh–CH}_2\text{–PdCl)} \xrightarrow[\text{or } CH_3O^-]{N_2H_4 \text{ or } H_2} \text{(cyclopentadiene–CHPhCH}_3\text{)} \qquad (222)$$

Presumably in all three cases the reaction proceeds through Pd(II)–hydride species.

Carbonylation of (XXII) depends on the basicity of the medium. In neutral or acid methanol a cyclization occurs, whereas in the presence of base, ester is formed.

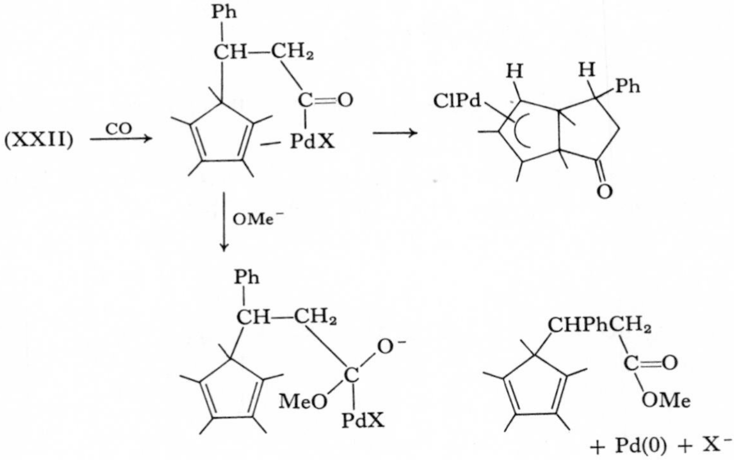

(XXII) $\xrightarrow{CO}$ (cyclopentadiene–CH(Ph)–CH$_2$–C(=O)–PdX) ⟶ ClPd–(bicyclic ketone with H, H, Ph) (223)

↓ OMe⁻

(cyclopentadiene–CH(Ph)–CH$_2$–C(O$^-$)(MeO)–PdX) (cyclopentadiene–CHPhCH$_2$–C(=O)OMe) + Pd(0) + X^-

This latter route is of considerable interest in Pd(II) catalytic chemistry since it suggests that Pd(II) will be induced to leave as Pd(0) when strongly electron-releasing groups are on the α-carbon atom (Sections III, A, 1, 2, and 3). When triphenylphosphine is added to compound

(XXII). The Pd(II) is induced to leave as a stable Pd(II) hydride:

$$\text{(XXII)} + 4Ph_3P \xrightarrow[\text{Ar}]{CD_3Cl_3} \text{[1-(1-phenylvinyl)-1,2,3,4-tetramethylcyclopentadiene]} + 2(Ph_3P)PdHCl \xrightarrow{[O]/CDCl_3} 2(Ph_3P)PdCl_2 \quad (224)$$

This is the first example of a Pd(II)–hydride elimination producing a stable hydride, although this type of elimination is often postulated in Pd(II) catalytic chemistry (Sections III, A, 2 and 4).

A novel Pd(I) complex has been obtained from the cyclization of diphenylacetylene with palladium acetate in methanol (*12*):

$$3Ph_2C_2 + Pd(OAc)_2 + MeOH \longrightarrow \text{[hexaphenyl palladacycle, Pd–OMe]} \longrightarrow \text{[pentaphenylcyclopentadiene–C(Ph)(OMe)–Pd(II)–OMe]} \longrightarrow \text{[pentaphenylcyclopentadiene–C(Ph)(Me)(OMe), Pd(0)]} \xrightarrow[-PhC(OMe)_3,\ -H^+]{HOMe} (C_5Ph_5)Pd(I) \longrightarrow \underset{\text{(XXIII)}}{Pd_2C_{12}Ph_{12}} \quad (225)$$

An X-ray study of compound (XXIII) indicates that it contains 2-cyclopentadienyl Pd(I) groups with a bridging diphenylacetylene ligand. The general scheme given above accounts for this product, although the exact mode of formation of the pentaphenylcyclopentadienyl Pd(I) monomer is uncertain. Apparently, this is another example of Pd(0) being induced to leave a carbon with strongly electron-donating groups on it.

B. *Addition to Acetylenes*

The kinetics of the addition of trifluoro-, difluorochloro-, and trichloroacetic acid to Pd(0) complexes of $CF_3C{\equiv}CCF_3$ has recently been studied (*28*). Compound $Pd(CF_3C{\equiv}CCF_3)(Ph_2PCH_2CH_2PPh_2)$ gave second-order kinetics with $\Delta H^{\ddagger} = 11.2$ kcal/mole, whereas $\Delta S^{\ddagger} = -35$ cal/deg mole. The product is *cis*-$Pd(OCOCR)(CF_3C{=}CHCF_3)$-$(Ph_2PCH_2CH_2PPh_2)$ in which the addition of Pd–H has occurred cis. The $Pd(CF_3C{\equiv}CCF_3)(PPh_3)_2$ gave more complicated kinetics which can be attributed to reversible loss of PPh_3.

Three reports of the addition of carbanionoid species have recently been reported. The Grignard CH_3MgBr reacts with diphenylacetylene in the presence of L_2PdCl_2 (L = benzonitrile or norbornadiene) to give α,α'-dimethylstilbene as major product (*72*).

$$PhC{\equiv}CPh + 2CH_3MgBr + L_2PdCl_2 \longrightarrow PhC(CH_3){=}C(CH_3)Ph + Pd(0) \quad (226)$$

The dimethylstilbene is mainly trans. Stilbene and α-methylstilbene are side products. When strong nucleophiles such as Ph_3P, Ph_2CH_2-CH_2PPh_2, or $(PhO)_3P$ are added to the reaction mixture at $-70°C$, $L'_2Pd(CH_3)_2$ (L′ = phosphine or phosphite) is formed. This compound is believed to be an intermediate in the reaction.

The insertion of hexafluoro-2-butyne into an allyl palladium bond is the third insertion (*8*). The addition is, as expected, cis.

$$CH_3\text{-}(\pi\text{-allyl})Pd L_2 \underset{}{\overset{C_4F_6}{\rightleftharpoons}} \left[H_2C{=}C(CH_3)CH_2\text{-}PdLCl(CF_3C{\equiv}CCF_3) \right] \xrightarrow{L} H_2C{=}C(CH_3)CH_2(CF_3)C{=}C(CF_3)PdL_2Cl \quad (227)$$

(L = PMe_2Ph)

C. Oxidation

Heck (*91*) reported that acetylenes can be carbonylated to give substituted maleate esters. Propiolate esters are formed as side products. The reaction scheme is as follows:

$$\mathrm{HC{\equiv}C{-}R} + \mathrm{Pd\overset{O}{\overset{\|}{C}}OR} \xrightarrow{\mathrm{ROH}} \mathrm{ROCO\overset{H}{\overset{|}{C}}{=}\overset{R}{\overset{|}{C}}{-}Pd}\lessgtr \tag{228}$$

$$\mathrm{ROCO\overset{H}{\overset{|}{C}}{=}\overset{R}{\overset{|}{C}}{-}Pd}\lessgtr \xrightarrow{-\mathrm{HPd}\lessgtr} \mathrm{ROCOC{\equiv}CR}$$

$$\mathrm{ROCO\overset{H}{\overset{|}{C}}{=}\overset{R}{\overset{|}{C}}{-}Pd}\lessgtr \xrightarrow{\mathrm{CO}} \mathrm{ROCO\overset{H}{\overset{|}{C}}{=}\overset{R}{\overset{|}{C}}{-}COPd}\lessgtr \xrightarrow{-\mathrm{ROH}} \mathrm{ROCO{-}\overset{H}{\overset{|}{C}}{=}\overset{R}{\overset{|}{C}}{-}COOR} + \mathrm{Pd(0)} + \mathrm{H^+}$$

Yields are greater than 100% based on Pd(II) probably because of regeneration of the hydridopalladium species by reduction of acetylene.

$$[\mathrm{HPdCl}] + \mathrm{RC{\equiv}CR'} \longrightarrow \left[\begin{matrix}\mathrm{R} & & \mathrm{R} \\ & \mathrm{C{=}C} & \\ \mathrm{H} & & \mathrm{PdCl}\end{matrix}\right] \xrightarrow{\mathrm{HCl}} \begin{matrix}\mathrm{R} & & \mathrm{R} \\ & \mathrm{C{=}C} & \\ \mathrm{H} & & \mathrm{H}\end{matrix} + \mathrm{PdCl_2} \tag{229}$$

VI

STEREOCHEMISTRY OF ADDITION OF Pd(II) AND NUCLEOPHILES TO UNSATURATED SUBSTRATES

The stereochemistry of oxypalladation and other additions of nucleophiles with palladium(II) has received considerable attention, and in the last 3 years a number of stereochemical studies have been carried out. These have been mentioned in the appropriate portion of the text. It seems fitting to conclude with a brief discussion of this subject.

The kinetics of the Wacker reaction (Section III, A, 1) suggested that the hydroxypalladation step involves cis addition of Pd(II) and coordinated OH [Eq. (46)]. Before stereochemical work on monoolefins had been carried out, this result caused many workers to believe that other additions of nucleophiles would proceed in the same fashion.

The stereochemistry of addition of nucleophiles to Pd(II) cyclic diolefin adducts was next studied, and in all cases it was found to be trans (Section IV, A). However, strained diolefins such as norbornadiene tend to add nucleophiles exo, whereas the metal involved is held in the endo position. Thus, mercury(II) does not form strong π bonds and hydroxymercuration of norbornene is cis–exo (*7*, *262*). However, the trans addition to unstrained olefins could not be explained in this way.

More recently, however, results with monooelfins indicate that trans addition can occur to both Pd(II) and Pt(II) π complexes. The first demonstration was the trans addition of amines to Pt(II) complexes (*215*), and more recently the trans attack of acetate on cyclohexene has been demonstrated (Section III, A, 2, a). However, cis attack can also occur such as the addition of phenylpalladium to cyclohexene (Section III, A, 4) or addition of PdCOOR to cyclic olefins (Section III, A, 3). Also, in the exchange studies (Section III, B, 1) the stereochemistry indicates that some nucleophiles can attack cis or trans. Thus chloride ion containing acetic acid can attack the Pd(II) olefin π complex from either inside or outside the coordination sphere. What are the factors involved? The most important appears to be the ability of the nucleophile to coordinate to Pd(II). Thus, phenyl is covalently bonded to Pd(II) and is therefore always in the coordination sphere. Chloride is both inside the coordination sphere as well as outside the coordination sphere and can thus attack both cis and trans. Acetate is not complexed to Pd(II) in chloride-containing media, and thus can only attack trans. On the other hand, in chloride-free acetic acid, acetate is both inside and outside the coordination. Stereochemical results indicate that in this system acetate can attack in both a cis and trans fashion.

Another factor may be electronic. In the acetate allylic ester exchange studies, it was shown that the dimeric π complex [Eqs. (159), (165), (168); Section III, B, 1)] is reactive, whereas the monomeric π complex [Eq. (162)] is unreactive. This is believed due to the higher negative charge on the Pd(II) π-bonded to the olefin in the monomeric π complex.

Thus the factors involved in stereochemistry of addition can be fairly subtle. For instance, the methoxypalladation of *cis*- and *trans*-2-butenes in methanol by palladium(II) chloride in the presence of CO was shown to be trans (Section III, A, 3, b). This is not surprising in the light of the above discussion, but in this system it should be realized the CO could very well be affecting the mode of addition by complexing to Pd(II).

One possibility is

$$\mathrm{PdCl_3(olefin)^- + CO \longrightarrow PdCl_2(olefin)(CO) + Cl^-} \tag{230}$$

Thus, the anionic species is converted to a neutral species that may be more susceptible to nucleophilic attack. In the absence of CO the route may well be different. Certainly kinetic studies are necessary to define completely this system.

In this light the results with diolefins do not appear surprising. In fact with the Pd(II)–carbon systems, also addition has recently been shown to be cis (Section IV, A). When Pd(II) is not π-complexed to the double bonds, the addition is cis–exo, and when Pd(II) is complexed in the endo position the addition is cis–endo. Another factor in the diene complexes is the fact two cis coordination positions are taken up. Since chloride may not be as readily displaced to give the nucleophile, attack may tend to be trans. Moreover these are neutral complexes and thus may be more susceptible to nucleophilic attack.

Stereochemistry of addition for acetylenes (Section V, A) has not been studied to as large an extent, but additions can also be cis or trans depending on conditions. With Pt(II)–acetylene complexes, addition can also be either cis or trans depending on the nucleophile (*36*).

Finally, let us return to the Wacker process. It has recently been stated that it is unfortunate that the suggestion of cis attack of hydroxyl has been generally accepted and assumed to apply to other reactions, since it was based primarily on kinetic results and the kinetics could also be interpreted in terms of trans addition (*252*). Actually, well before this publication appeared, it had been demonstrated that monoolefins could be attacked in both a cis and trans fashion. It had also been previously pointed out by the writer (*102*) that the kinetics do not absolutely require cis addition. Now the only reasonable alternative to cis addition [Eq. (46)] is attack of water on the hydroxyl π complex. If this is the mode, why does water not attack the trichloro π complex?

$$[\mathrm{PdCl_3(CH_2{=}CH_2)}]^- + \mathrm{H_2O} \longrightarrow [\mathrm{PdCl_3(CH_2CH_2OH)}]^{2-} + \mathrm{H^+} \tag{231}$$

Also the allylic exchange work indicated that nucleophiles do not like to attack the negatively charged monomeric π complex (Section III, B, 1). Furthermore, the kinetics of hydration of vinyl esters in wet acetic acid indicates that attack of H_2O is trans in this case [Eq. (168)]. The dimeric π complex is the reactive species, as it is with other nucleophiles, so that water also does not like to attack the monomeric π complex. Since palladium(II) chloride does not form dimeric π complexes in water, the route in wet acetic acid is not available.

$$\begin{array}{c} \text{Cl}—\text{Pd}(\text{H}_2\text{O})(\text{Cl})(\text{CH}_2{=}\text{CH}_2) + \text{H}_2\text{O} \longrightarrow {}^-\text{Cl}—\text{Pd}(\text{H}_2\text{O})(\text{Cl})—\text{CH}_2\text{CH}_2\text{OH} + \text{H}^+ \end{array} \quad (232)$$

Why, then, does water not attack the neutral aquo π complex, a route that is not consistent with the kinetics? The answer could simply be that once the aquo π complex is formed, the lower-energy route is to lose a proton to give the better nucleophile, hydroxyl, which then attacks the olefin from the coordination sphere. However, the important point is that in the light of recent studies, trans attack still does not appear likely.

Further studies will hopefully advance our knowledge of the details of addition of Pd(II) and nucleophiles across unsaturated systems. I believe the most useful work will be those studies in which both kinetic and stereochemical investigations are carried out simultaneously.

References

1. Agami, C., Levisalles, J., and Rose-Munch, F., *J. Organometal. Chem.* **39**, C17 (1972).
2. Aguiló, A., *Advan. Organometal. Chem.* **5**, 321 (1967).
3. Allegra, G., Immirzi, A., and Porri, L., *J. Amer. Chem. Soc.* **87**, 1394 (1965).
4. Allegra, G., Cassagrande, G. T., Immirzi, A., Porri, L., and Vitulfi, G., *J. Amer. Chem. Soc.* **92**, 289 (1970).
5. Anderson, C. B., and Burreson, B. J., *Chem. Ind.* (*London*) 620 (1967).
6. Anderson, C. B., and Winstein, S., *J. Org. Chem.* **28**, 605 (1963).
7. Anderson, M. M., and Henry, P. M., *Chem. Ind.* (*London*) 2053 (1961).
8. Appleton, T. G., Clark, H. C., Poller, R. C., and Puddephatt, R. J., *J. Organometal. Chem.* **39**, C13 (1972).
9. Asano, R., Moritani, I., Sonoda, A., Fujiwara, Y., and Teranishi, S., *J. Chem. Soc., C* 3691 (1971).
10. Atkinson, L. K., and Smith, D. C., *J. Chem. Soc. A* 3592 (1971).
11. Bailar, J. C., and Itatani, H., *J. Amer. Chem. Soc.* **89**, 1592 (1967).
12. Ban, E., Cheng, P.-T., Jack, T., Nyburg, S. C., and Powell, J., *Chem. Commun.* 368 (1973).
13. Ban, G. M., Hughes, R. P., and Powell, J., *Chem. Commun.* 591 (1973).
14. Basolo, F., and Pearson, R. G., "Mechanisms of Inorganic Reactions," Chapter 5. Wiley, New York, 1967.

15. Betts, S. J., Harris, A., Haszeldine, R. N., and Parish, R. V., *J. Chem. Soc. A* 3699 (1971).
16. Bhagwat, M. M., and Devaprabhakara, D., *J. Organometal. Chem.* **52**, 425 (1973).
17. Biryukov, A. A., and Schlenskaya, V. A., *Russ. J. Inorg. Chem.* **9**, 450 (1964).
18. Boschi, T., Vguagliati, P., and Crociani, B., *J. Organometal. Chem.* **30**, 283 (1971).
19. Brady, D. G., *Chem. Commun.* 434 (1970).
20. Brooks, E. H., and Glockling, J., *J. Chem. Soc. A* 1030 (1967).
21. Brooks, E. H., and Glockling, J., *J. Chem. Soc. A* 1241 (1966).
22. Brown, R. G., and Davidson, J. M., *J. Chem. Soc. A* 1321 (1971).
23. Brown, R. G., and Davidson, J. M., *Advan. Chem. Ser.* **132,** 49 (1974).
24. Brown, R. G., Davidson, J. M., and Triggs, C., *Amer. Chem. Soc., Div. Petrol. Chem. Prepr.* **14**, B23 (1969).
25. Brown, E. S., and Rick, E. A., *Amer. Chem. Soc., Div. Petrol. Chem. Prepr.* **14**, B29 (1969).
26. Brown, E. S., Rick, E. A., and Mendicino, F. D., *J. Organometal. Chem.* **38**, 37 (1972).
27. Burger, K., and Dyrssen, D., *Acta. Chem. Scand.* **17**, 1489 (1963).
28. Burgess, J., Kemmitt, R. D. W., and Littlecott, G. W., *J. Organometal. Chem.* **56**, 405 (1973).
29. Caglioti, L., Cattalini, L., Ghedini, M., Gasparrini, F., and Vigato, P. A., *J. Chem. Soc., Dalton Trans.* 514 (1972).
30. Calvo, C., Hosokawa, T., Reinheimer, H., and Maitlis, P. M., *J. Amer. Chem. Soc.* **94**, 3237 (1972).
31. Carro, S., and Ugo, R., *Inorg. Chim. Acta.* **1**, 49 (1967).
32. Carturan, G., Graziani, M., Ros, R., and Belluco, V., *J. Chem. Soc., Dalton Trans.* 262 (1972).
33. Chatt, J., and Duncanson, L. A., *J. Chem. Soc.* 2939 (1953).
34. Chatt, J., Vallarino, L. M., and Venanzi, L. M., *J. Chem. Soc.* 3413 (1957).
35. Chisholm, M. H., Clark, H. C., Manzer, L. E., and Stothers, J. B., *J. Amer. Chem. Soc.* **94**, 5087 (1972).
36. Chisholm, M. H., and Clark, H. C., *Accounts Chem. Res.* **6**, 202 (1973).
37. Clark, D., Hayden, P., and Smith, R. D., *Disc. Faraday Soc.* **46**, 98 (1968).
38. Collman, J. P., *Accounts Chem. Res.* **1**, 136 (1968).
39. Collman, J. P., and Roper, W. R., *Advan. Organometallic Chem.* **7**, 54 (1968).
40. Conti, F., Donati, M., Pregaglia, G. F., and Ugo, R., *J. Organometal. Chem.* **30**, 421 (1971).
41. Cope, A. C., Kliegman, J. M., and Friedrich, E. C., *J. Amer. Chem. Soc.* **89**, 287 (1967).
42. Coulson, D. R., *Chem. Commun.* 1530 (1968).
43. Coulson, D. R., *J. Amer. Chem. Soc.* **91**, 200 (1969).
44. Cruikshank, B. I., and Davies, N. R., *Aust. J. Chem.* **25**, 919 (1972).
45. Danno, S., Moritani, I., Fujiwara, Y., and Teranishi, S., *J. Chem. Soc. B* 196 (1971).
46. Darling, J. H., and Ogden, J. S., *J. Chem. Soc., Dalton Trans.* 1079 (1973).
47. Davidson, J. M., *MTP* (*Med. Tech. Publ. Co.*) *Int. Rev. Sci. Inorg. Chem.* **6**, 349–391 (1972).
48. Davidson, J. M., *Chem. Commun.* 1019 (1971).
49. Davidson, J. M., and Triggs, C., *J. Chem. Soc. A* 1324 (1968).
50. Davies, N. R., *Rev. Pure Appl. Chem.* **17**, 83 (1967).
51. Davies, N. R., *Aust. J. Chem.* **17**, 212 (1964).

52. Dempsey, J. N., and Baenziger, N. C., *J. Amer. Chem. Soc.* **77**, 4984 (1955).
53. Dent, W. T., Long, R., and Wilkinson, A. J., *J. Chem. Soc. A* 1839 (1967).
54. Dewar, M. J. S., *Bull. Soc. Chim. Fr.* **18**, C79 (1951).
55. Dietl, H., and Maitlis, P. M., *Chem. Commun.* 481 (1968).
56. Dietl, H., Reinheimer, H., Moffat, J., and Maitlis, P. M., *J. Amer. Chem. Soc.* **92,** 2276 (1970).
57. Dixon, K. R., and Hawke, D. J., *Can. J. Chem.* **49**, 3252 (1971).
58. Droll, H. A., Block, B. P., and Fernelius, W. C., *J. Phys. Chem.* **61**, 1000 (1957).
59. Elding, L. I., *Inorg. Chim. Acta* **6**, 647 (1972).
60. Faller, J. W., and Tully, M. T., *J. Amer. Chem. Soc.* **94**, 2676 (1972) and preceeding papers.
61. Fenton, D. M., *J. Org. Chem.* 3192 (1973).
62. Fenton, D. M., and Oliver, K. L., *Chemtech*, 220 (1972).
63. Fenton, D. M., and Steinwand, P. J., *J. Org. Chem.* **37**, 2034 (1972).
64. Fischer, E. O., and Vogler, A., *J. Organometal. Chem.* **3**, 161 (1965).
65. Fitton, P., Johnson, M. P., and McKeon, J. E., *Chem. Commun.* 6, (1968).
66. Fitton, P., and Rick, E. A., *J. Organometal. Chem.* **28**, 287 (1971).
67. François, P., *Ann. Chim.* (*Paris*) [*14*] **4**, 371 (1969).
68. François, P., and Trambouze, Y., *Bull. Soc. Chim. Fr.* 51 (1969).
69. Frankel, E. N., Thomas, F. L., and Rohwedder, W. K., *Advan. Chem. Ser.* **132,** 145 (1974).
70. Fujiwara, Y., Moritani, I., Asano, R., Tanaka, H., and Teranishi, S., *Tetrahedron* **25**, 4815 (1969).
71. Gallazzi, M. C., Hanlon, T. L., Vitulli, G., and Porri, L., *J. Organometal. Chem.* **33**, C45 (1971).
72. Garty, N., and Michman, M., *J. Organometal. Chem.* **36**, 391 (1972).
73. Garves, K., *J. Org. Chem.* **35**, 3273 (1970).
74. Greaves, E. O., Lock, C. J. L., and Maitlis, P. M., *Can. J. Chem.* **46**, 3879 (1968).
75. Greaves, E. O., and Maitlis, P. M., *J. Organometal. Chem.* **6**, 104 (1966).
76. Green, M., Haszeldine, R. N., and Lindley, J., *J. Organometal. Chem.* **6**, 107 (1966).
77. Green, M. L. H., and Munakata, H., *J. Chem. Soc. A* 469 (1971).
78. Grinberg, A. A., Kiseleva, N. V., and Gel'fman, M. I., *Dokl. Akad. Nauk., SSSR* **153**, 1327 (1963).
79. Hafner, W., Prigge, H., and Smidt, J., *Liebigs Ann. Chem.* **693**, 109 (1966).
80. Halpern, J., *Accounts Chem. Res.* **3**, 386 (1970).
81. Hara, M., Okno, K., and Tsuji, J., *Chem. Commun.* 247 (1971).
82. Hartley, F. R., *Chem. Rev.* **69**, 799 (1969).
83. Hartley, F. R., "The Chemistry of Platinum and Palladium." Appl. Sci. Publ., London, 1973.
84. Hartley, F. R. (*83*), p. 389.
85. Hartley, F. R., and Wagner, J. L., *J. Chem. Soc., Dalton Trans.* 2282 (1972).
86. Hartley, F. R., and Wagner, J. L., *J. Organometal. Chem.* **55**, 395 (1973).
87. Haszeldine, R. N., Parish, R. V., and Robbins, D. W., *J. Organometal. Chem.* **23**, C33 (1970).
88. Heck, R. F., *J. Amer. Chem. Soc.* **90**, 317 (1968).
89. Heck, R. F., *J. Amer. Chem. Soc.* **93**, 6896 (1971).
90. Heck, R. F., *J. Amer. Chem. Soc.* **91**, 6707 (1969).
91. Heck, R. F., *J. Amer. Chem. Soc.* **94**, 2712 (1972).
92. Heck, R. F., *J. Amer. Chem. Soc.* **90**, 5518 (1968) and following papers.

93. Heck, R. F., *J. Amer. Chem. Soc.* **90**, 5538 (1968).
94. Heck, R. F., Private communication.
95. Heck, R. F., *Organometal. Chem. Syn.* **1**, 455 (1972).
96. Heck, R. F., *J. Amer. Chem. Soc.* **90**, 5535 (1968).
97. Heck, R. F., *J. Organometal. Chem.* **37**, 389 (1972).
98. Heck, R. F., *J. Amer. Chem. Soc.* **90**, 5531 (1968).
99. Heck, R. F., *J. Organometal. Chem.* **33**, 399 (1971).
100. Heck, R. F., and Henry, P. M., U.S. Patent 3,579,568 (1971).
101. Heck, R. F., and Nolley, J. P., *J. Org. Chem.* **37**, 2320 (1972).
102. Henry, P. M., *Advan. Chem. Ser.* **70**, 126 (1968).
103. Henry, P. M., *Trans. N.Y. Acad. Sci.* **33**, 41 (1971).
104. Henry, P. M., *Accounts Chem. Res.* **6**, 16 (1973).
105. Henry, P. M., *Inorg. Chem.* **10**, 373 (1971).
106. Henry, P. M., unreported data.
107. Henry, P. M., *J. Amer. Chem. Soc.* **86**, 3246 (1964).
108. Henry, P. M., *J. Amer. Chem. Soc.* **94**, 1527 (1972).
109. Henry, P. M., *J. Org. Chem.* **38**, 2415 (1973).
110. Henry, P. M., *J. Amer. Chem. Soc.* **94**, 4437 (1972).
111. Henry, P. M., *J. Amer. Chem. Soc.* **88**, 1595 (1966).
112. Henry, P. M., *J. Org. Chem.* **32**, 2575 (1967).
113. Henry, P. M., *J. Org. Chem.* **38**, 1681 (1973).
114. Henry, P. M., *J. Amer. Chem. Soc.* **94**, 7305 (1972).
115. Henry, P. M., *Tetrahedron Lett.* 2285 (1968).
116. Henry, P. M., *J. Amer. Chem. Soc.* **93**, 3853 (1971).
117. Henry, P. M., *J. Amer. Chem. Soc.* **94**, 7316 (1972).
118. Henry, P. M., *J. Amer. Chem. Soc.* **94**, 5200 (1972).
119. Henry, P. M., *J. Org. Chem.* **37**, 2443 (1972).
120. Henry, P. M., *J. Amer. Chem. Soc.* **94**, 7311 (1972).
121. Henry, P. M., *J. Org. Chem.* **38**, 2766 (1973).
122. Henry, P. M., *J. Org. Chem.* **38**, 3596 (1973).
123. Henry, P. M., *J. Amer. Chem. Soc.* **93**, 3547 (1971).
124. Henry, P. M., *J. Org. Chem.* **38**, 1140 (1973).
125. Henry, P. M., *Chem. Commun.* 328 (1971).
126. Henry, P. M., Davies, M., Ferguson, G., Phillips, S., and Restivo, R., *Chem. Commun.*, 112 (1974).
127. Henry, P. M., and Ward, G. A., *J. Amer. Chem. Soc.* **93**, 1494 (1971).
128. Henry, P. M., and Ward, G. A., *J. Amer. Chem. Soc.* **94**, 673 (1972).
129. Hidai, M., Kokura, M., and Uchida, Y., *J. Organometal. Chem.* **52**, 431 (1973).
130. Hines, L. F., and Stille, J. K., *J. Amer. Chem. Soc.* **94**, 485 (1972).
131. Horino, H., and Inoue, N., *Bull. Chem. Soc. Jap.* **44**, 3210 (1971).
132. Hosokawa, T., Calvo, C., Lee, H. B., and Maitlis, P. M., *J. Amer. Chem. Soc.* **95**, 4914 (1973).
133. Hosokawa, T., and Maitlis, P. M., *J. Amer. Chem. Soc.* **95**, 4924 (1973).
134. Hosokawa, T., and Maitlis, P. M., *J. Amer. Chem. Soc.* **94**, 3238 (1972).
135. Hosokawa, T., Moritani, I., and Nishioka, S., *Tetrahedron Lett.* 3833 (1969).
136. Hudson, B., Webster, D. E., and Wells, P. B., *J. Chem. Soc., Dalton Trans.* 1204 (1972).
137. Hughes, R. P., and Powell, J., *J. Organometal. Chem.*, **60**, 387 (1973).
138. Hughes, R. P., and Powell, J., *J. Organometal. Chem.* **30**, C45 (1971).

139. Hunt, D. F., and Rodeheaver, G. T., *Tetrahedron Lett.* 3595 (1972).
140. Hüttel, R., *Synthesis* **2**, 228 (1970).
141. Hüttel, R., and Bechter, M., *Angew. Chem.* **71**, 456 (1959).
142. Hüttel, R., and Christ, H., *Chem. Ber.* **96**, 3101 (1963).
143. Hüttel, R., and Christ, H., *Chem. Ber.* **97**, 1439 (1964).
144. Hüttel, R., Kratzer, J., and Bechter, M., *Chem. Ber.* **94**, 766 (1961).
145. Ito, T., Haseyawa, S., Takahashi, Y., and Ishii, Y., *Chem. Commun.* 629 (1972).
146. Ito, T., Yakahashi, Y., and Ishii, Y., *Chem. Commun.* 629 (1972).
147. Iwao, T., Saika, A., and Kinugasa, T., *Inorg. Chem.* **11** 3106 (1972).
148. Jira, R., *Tetrahedron Lett.* 1225 (1971).
149. Jira, R., and Freiseleben, W., *Organometal. React.* **3**, 5 (1972).
150. Jira, R., and Freiesleben, W., see (*149*), p. 82.
151. Jira, R., and Sedlmeier, J., *Tetrahedron Lett.* 1227 (1971).
152. Jira, R., Sedlmeier, J., and Smidt, J., *Ann. Chem.* **693**, 99 (1966).
153. Kaplan, P. D., Schmidt, S., and Orchin, M., *J. Amer. Chem. Soc.* **90**, 4175 (1968).
154. Kasahara, A., Izumi, T., Saito, G., Yodono, M., Saito, R., and Goto, Y., *Bull. Chem. Soc. Jap.* **45**, 895 (1972).
155. Kawamoto, K., Tatani, A., Toshinobu, T., and Teranishi, S., *Bull. Chem. Soc., Jap.* **44**, 1239 (1971).
156. Kharasch, M. S., Seyler, R. C., and Mayo, F. R., *J. Amer. Chem. Soc.* **60**, 882 (1938).
157. King, R. B., and Efraty, A., *Inorg. Chem.* **10**, 1376 (1971).
158. Kingston, J. V., and Scollary, G. R., *J. Chem. Soc. A* 3765 (1971).
159. Kiso, Y., Kumada, M., Tamao, K., and Umeno, M., *J. Organometal. Chem.* **50**, 297 (1973).
160. Kiso, Y., Yamamoto, K., Tamao, K., and Kumada, M., *J. Amer. Chem. Soc.* **94**, 4373 (1972).
161. Kitching, W., *Organometal. Chem. Rev.* **3**, 61 (1968).
162. Kitching, W., *Organometal. React.* **3**, 319 (1972).
163. Kitching, W., Rappoport, Z., Winstein, S., and Young, W. G., *J. Amer. Chem. Soc.* **88**, 2054 (1966).
164. Kitching, W., Sakakiyama, T., Rappoport, Z., Sleezer, P. D., Winstein, S., and Young, W. G., *J. Amer. Chem. Soc.* **94**, 2329 (1972).
165. Kudo, K., Hidai, M., and Uchida, Y., *J. Organometal. Chem.* **33**, 393 (1971).
166. Kudo, K., Hidai, M., and Uchida, Y., *J. Organometal. Chem.* **56**, 413 (1973).
167. Kündig, E. P., Moshovits, M., and Ozin, G. A., *Can. J. Chem.* **50**, 3587 (1972).
168. Lappert, M. F., and Prokai, B., *Advan. Organometal. Chem.* **5**, 225 (1967).
169. Lippard, S. J., and Morehourse, S. M., *J. Amer. Chem. Soc.* **94**, 6949 (1972).
170. Lukas, J., and Blom, J. E., *J. Organometal. Chem.* C25 (1971).
171. Lukas, J., and Kramer, P. A., *J. Organometal. Chem.* **31**, 111 (1971).
172. Maglio, G., Musco, A., and Palumbo, R., *Inorg. Chim. Acta* **4**, 153 (1970).
173. Maitlis, P. M., "Organic Chemistry of Palladium", Vols. I and II. Academic Press, New York, 1971.
174. Maitlis, P. M. (*173*), Vol. II, pp. 93–101.
175. Maitlis, P. M. (*173*), Vol. II, pp. 105–108.
176. Maitlis, P. M. (*173*), Vol. II, pp. 9–16.
177. Maitlis, P. M. (*173*), Vol. II, pp. 60–70.
178. Maitlis, P. M. (*173*), Vol. II, pp. 108–115.
179. Maitlis, P. M. (*173*), Vol. II, pp. 128–142.

180. Maitlis, P. M. (*173*), Vol. I, pp. 74–79.
181. Maitlis, P. M. (*173*), Vol. II, pp. 47–58.
182. Maitlis, P. M., *Pure Appl. Chem.* **33**, 489 (1973).
183. Maitlis, P. M., *Pure Appl. Chem.* **30**, 427 (1972).
184. Manchot, W., and König, J., *Chem. Ber.* **59**, Part II, 883 (1926).
185. Manuel, T. A., *J. Org. Chem.* **27**, 3941 (1962).
186. Matsuda, T., Mitanyasu, T., and Nakamura, Y., *Kogyo Kagaku Zasshi* **72**, 1751 (1969).
187. Matveev, K. I., Bukhtoyarov, I. F., Shul'ts, N. N., and Emel'yanova, O. A., *Kinet. Katal.* (*USSR*) **5**, 572 (1964).
188. Mawby, A., and Pringle, G. E., *J. Inorg. Nucl. Chem.* **33**, 1989 (1971).
189. McCaskie, J. E., Ph.D. Thesis, University of California, Los Angeles, 1971.
190. McKeon, J. E., and Fitton, P., *Tetrahedron* **28**, 233 (1972).
191. McKeon, J. E., Fitton, P., and Griswold, A. A., *Tetrahedron* **28**, 227 (1972).
192. Misono, A., Uchida, Y., Hidai, M., and Kudo, J., *Organometal. Chem.* **20**, P7 (1969).
193. Mizoroki, T., Mori, K., and Ozaki, A., *Bull. Chem. Soc. Jap.* **44**, 581 (1971).
194. Moiseev, I. I., *Amer. Chem. Soc., Div. Petrol. Chem., Prepr.* **14**, (2) B49 (1969).
195. Moiseev, I. I., personal communication.
196. Moiseev, I. I., Belov, A. P., Igoshin, V. A., Syrkin, Ya.K., *Dokl. Akad. Nauk SSSR* **173**, 863 (1967).
197. Moiseev, I. I., and Vargaftik, M. N., *Izv. Akad. Nauk. SSSR* 759 (1965) (Engl. Transl., p. 744).
198. Moiseev, I. I., Vargaftik, M. N., Pestnikov, S. V., Levanda, O. G., Romanova, T. N., and Syrkin, Ya.K., *Dokl. Akad. Nauk SSSR* **171**, 1365 (1966).
199. Moiseev, I. I., Vargaftik, M. N., and Syrkin, Ya.K., *Dokl. Akad. Nauk SSSR* **153**, 140 (1963).
200. Morelli, D., Ugo, R., Conti, F., and Donati, M., *Chem. Commun.* 801 (1967).
201. Moritani, I., Danno, S., Fujiwara, Y., and Teranishi, S., *Bull. Chem. Soc., Jap.* **44**, 578 (1971).
202. Moritani, I., Fujiwara, Y., and Danno, S., *J. Organometal. Chem.* **27**, 279 (1971).
203. Moseley, K., and Maitlis, P. M., *Chem. Commun.* 1604 (1971).
204. Nelson, J. H., and Jonassen, H. B., *Coord. Chem. Rev.* **6**, 27 (1971).
205. Ng, F. T. T., and Henry, P. M., *J. Org. Chem.* **38**, 3338 (1973).
206. Nicholson, J. K., Powell, J., and Shaw, B. L., *Chem. Commun.* 174 (1966).
207. Ninomiya, R., Sato, M., and Shiba, T., *Bull. Jap. Petrol. Inst.* **7**, 31 (1965).
208. Onoue, H., Moritani, I., and Murahashi, S.-I., *Tetrahedron Lett.*, 121 (1973).
209. Otsuka, S., Tatsuno, Y., and Ataka, K., *J. Amer. Chem. Soc.* **93**, 6705 (1971).
210. Ouellette, R. J., and Levin, C., *J. Amer. Chem. Soc.* **93**, 471 (1971).
211. Panattoni, C., Bombieri, G., Forsellini, E., Crociani, B., and Belluco, U., *Chem. Commun.* 187 (1969).
212. Pandey, R. N., unreported data.
213. Pandey, R. N., and Henry, P. M., *Can. J. Chem.*, **52**, 1241 (1974).
214. Pandey, R. N., and Henry, P. M., *Advan. Chem. Ser.* **132**, 33 (1974).
215. Panunzi, A., De Renzi, A., and Paiaro, G., *J. Amer. Chem. Soc.* **92**, 3488 (1970).
216. Parshall, G. W., *Accounts Chem. Res.* **3**, 139 (1970).
217. Pestrikov, S. V., Moiseev, I. I., and Romanova, T. M., *Russ. J. Inorg. Chem.* **10**, 1199 (1965).
218. Pestrikov, S. V., Moiseev, I. I., and Tsvilikhovskaya, B. A., *Russ. J. Inorg. Chem.* **11**, 930 (1966).

219. Philips, F. C., *Amer. Chem. J.* **16**, 255 (1894).
220. Pietropaolo, D., Boschi, T., Zanella, R., and Belluco, U., *J. Organometal. Chem.* **49**, C88 (1973).
221. Pietropaolo, R., Faraone, F., Sergi, S., and Pietropaolo, D., *J. Organometal. Chem.* **42**, 177 (1972).
222. Pietropaolo, R., Uguagliati, P., Boschi, T., Crociani, B., and Belluco, U., *J. Catal.* **18**, 338 (1970).
223. Powell, J., and Jack, T., *Inorg. Chem.* **11**, 1039 (1972).
224. Pregaglia, G. F., Conti, F., Minasso, B., and Ugo, R., *J. Organometal. Chem.* **47**, 165 (1973).
225. Reichert, B. E., and West, B. O., *J. Organometal. Chem.* **36**, C29 (1972).
226. Reinheimer, H., Moffat, J., and Maitlis, P. M., *J. Amer. Chem. Soc.* **92**, 2285 (1970).
227. Rodeheaver, G. T., and Hunt, D. F., *Chem. Commun.* 818 (1971).
228. Roe, D. M., Bailey, P. M., Moseley, K., and Maitlis, P. M., *Chem. Commun.* 1273 (1972).
229. Roe, D. M., Calvo, C., Krishnamachari, N., Moseley, K., and Maitlis, P. M., *Chem. Commun.* 436 (1973).
230. Roffia, P., Conti, F., Gregorio, G., and Pregaglia, G. F., *J. Organometal. Chem.* **54**, 357 (1973).
231. Roffia, P., Conti, F., Gregorio, G., Pregaglio, G. F., and Ugo, R., *J. Organometal. Chem.* **56**, 391 (1973).
232. Sabel, A., Smidt, J., Jira, R., and Prigge, H., *Chem. Ber.* **102**, 2939 (1969).
233. Sadownick, J. A., and Lippard, S. J., *Inorg. Chem.* **12**, 2659 (1973).
234. Sakakibara, T., Nishimura, S., Kimura, K., Minato, I., and Odaira, Y., *J. Org. Chem.* **35**, 3884 (1970).
235. Sakakibara, T., Rakahashi, Y., Sakai, S., and Ishii, Y., *Chem. Commun.* 396 (1969).
236. Schrock, R. R., and Osborn, J. A., *J. Amer. Chem. Soc.* **93**, 3089 (1971).
237. Schultz, R. G., and Gross, D. E., *Advan. Chem. Ser.* **70**, 97 (1968).
238. Schultz, R. G., and Rony, P. R., *J. Catal.* **16**, 133 (1970).
239. Segnitz, A., Bailey, P. M., and Maitlis, P. M., *Chem. Commun.* 698 (1973).
240. Shapski, A. C., and Smart, M. L., *Chem. Commun.* 658 (1970).
241. Shaw, B. L., and Shaw, G., *J. Chem. Soc. A* 3533 (1971).
242. Shue, R. S., *J. Catal.* **20**, 112 (1972).
243. Shue, R. S., *J. Amer. Chem. Soc.* **93**, 7116 (1971).
244. Smidt, J., Jira, R., Sedlmeier, J., Sieber, R., Rüttinger, R., and Kojer, H., *Angew Chem. Int. Ed. Engl.* **1**, 80 (1962).
245. Stangl, H., and Jira, R., *Tetrahedron Lett.* 3589 (1970).
246. Stephenson, T. A., Morehouse, S. M., Powell, A. R., Heffer, J. P., and Wilkinson, G., *J. Chem. Soc.* 3632 (1965).
247. Stern, E. W., *Catal. Rev.* **1**, 74 (1968).
248. Stern, E. W., *in* "Transition Metals in Homogeneous Catalysis" (G. M. Schrauzer, ed.), pp. 93–141. Dekker, New York, 1971.
249. Stern, E. W., and Maples, P. K., *J. Catal.* **27**, 120 (1972).
250. Stern, E. W., and Maples, P. K., *J. Catal.* **27**, 134 (1972).
251. Stille, J. K., Hines, L. F., Fries, R. W., Wong, P. K., and James, D., *Advan. Chem. Ser.* **132**, 90 (1974).
252. Stille, J. K., James, D. E., and Hines, L. F., *J. Amer. Chem. Soc.* **95**, 5062 (1973).
253. Takahashi, H., and Tsuji, J., *J. Amer. Chem. Soc.* **90**, 2387 (1968).
254. Takahashi, K., Miyake, A., and Hata, G., *Bull. Chem. Soc. Jap.* **44**, 3484 (1971).

255. Takahashi, K., Miyake, A., and Hata, G., *Bull. Chem. Soc. Jap.* **45**, 230 (1972).
256. Takahashi, S., Shibano, T., and Hagihara, N., *Chem. Commun.* 161 (1969).
257. Takahashi, Y., Tokuda, A., Sakai, S., and Ishii, Y., *J. Organometal. Chem.* **35**, 415 (1971).
258. Tamura, M., and Yasui, T., *J. Chem. Soc. Jap. Ind. Chem. Sect.* **71**, 1855 (1968) (Engl. summary, p. A116); *Chem. Abstr.* **70**, 61532 (1969). Quoted in (*83*), p. 185.
259. Tamura, M., and Yasui, T., *Kogyo Kagaku Zasshi* **72**, 575, 578, 581 (1969).
260. Tayim, H. A., and Kharboush, M., *Inorg. Chem.* **10**, 1827 (1971).
261. Tolman, C. A., Seidel, W. C., and Gerlach, D. H., *J. Amer. Chem. Soc.* **94**, 2669 (1972).
262. Traylor, T. G., *Accounts Chem. Res.* **2**, 152 (1969).
263. Trieber, A., *Tetrahedron Lett.* 2831 (1966).
264. Trofimenko, S., *Inorg. Chem.* **10**, 1372 (1971).
265. Trost, B. M., and Fullerton, T. J., *J. Amer. Chem. Soc.* **95**, 292 (1973).
266. Tsuji, J., *Fortschr. Chem. Forsch.* **28**, 41 (1972).
267. Tsuji, J., *Accounts Chem. Res.* **6**, 8 (1973).
268. Tsuji, J. *Advan. Org. Chem.* **6**, 109 (1968).
269. Tsuji, J., and Iwamoto, N., *Chem. Commun.* 828 (1966).
270. Tsukiyama, K., Takahashi, Y., Sakai, S., and Ishii, Y., *J. Chem. Soc. A* 3112 (1971).
271. Tsutsui, M., Ori, M., and Francis, J., *J. Amer. Chem. Soc.* **94**, 1414 (1972).
272. Uemura, S., Zuchi, K., and Okano, M., *Chem. Commun.* 234 (1972).
273. Van der Linde, R., and De Jongh, R. O., *Chem. Commun.* 563 (1971).
274. van Helden, R., Kohll, C. F., Medema, D., Verberg, G., and Jonkhoff, T., *Rec. Trav. Chim. Pays-Bas* **87**, 961 (1968).
275. van Helden, R., and Verberg, G., *Rec. Trav. Chim. Pays-Bas* **84**, 1263 (1965).
276. Vaska, L., *Accounts Chem. Res.* **1**, 335 (1968).
277. Vaska, L., *Trans. N.Y. Acad. Sci.* **33**, 70 (1971).
278. Vedejs, E., and Salomon, M. F., *Chem. Commun.* 1582 (1971).
279. Vedejs, E., and Salomon, M. F., *J. Org. Chem.* **37**, 2075 (1972).
280. Vedejs, E., Salomon, M. F., and Weeks, P. D., *J. Organometal. Chem.* **40**, 221 (1972).
281. Volger, H. C., *Rec. Trav. Chim. Pays-Bas* **87**, 225 (1969).
282. Volger, H. C., *Rec. Trav. Chim. Pays-Bas* **86**, 677 (1967).
283. Volger, H. C., *Rec. Trav. Chim. Pays.-Bas* **87**, 501 (1968).
284. Vol'pin, M. E., Levitin, I. Ya., and Volkova, L. G., *Izv. Akad. Nauk SSSR* 1124 (1971).
285. Vol'pin, M. E., Taube, R., Drevs, H., Volkova, L. G., Levitin, I. Ya., and Ushakova, T. M., *J. Organometal. Chem.* **39**, C79 (1972).
286. Vol'pin, M. E., Volkova, L. G., Levitin, I. Ya, Boronina, N. N., and Yurkevich, A. M., *Chem. Commun.* 849 (1971).
287. Wakatsuki, Y., Nozakura, S., and Murahashi, S., *Bull. Chem. Soc. Jap.* **45**, 3426 (1971).
288. Weber, W. P., Felix, R. A., Willard, A. K., and Koenig, K. E., *Tetrahedron Lett.* 4701 (1971).
289. White, D. A., *J. Chem. Soc. A* 145 (1971).
290. Whitesides, G. M., Gaasch, J. F., and Stedronsky, E. R., *J. Amer. Chem. Soc.* **94**, 5258 (1972).
291. Whitta, W. A., Powell, H. M., and Veranzi, L. M., *Chem. Commun.* 310 (1966).
292. Wipke, W. T., private communication.
293. Wojcicki, A., *Accounts Chem. Rev.* **4**, 344 (1971).

294. Wojcicki, A., *Advan. Organometal. Chem.* **11**, 87 (1973).
295. Wolfe, S., and Campbell, P. G. C., *J. Amer. Chem. Soc.* **93**, 1497 (1971).
296. Wolfe, S., and Campbell, P. G. C., *J. Amer. Chem. Soc.* **93**, 1499 (1971).
297. Yamane, T., Kikukawa, K., Takagi, M., and Matsuda, T., *Tetrahedron* **29**, 955 (1973).
298. Yoshimura, N., Murahashi, S.-I., and Moritani, I., *J. Organometal. Chem.* **52**, C58 (1973).
299. Zocchi, M., Tieghi, G., and Albinati, A., *J. Organometal Chem.* **33**, C47 (1971).
300. Zocchi, M., Tieghi, G., and Albinati, A., *J. Chem. Soc., Dalton Trans.* 883 (1973).

The Organometallic Chemistry of the Main Group Elements—A Guide to the Literature

J. D. SMITH and D. R. M. WALTON

School of Molecular Sciences
University of Sussex
Brighton, England

In Volume 10 of *Advances in Organometallic Chemistry*, M. I. Bruce contributed a guide (*1*) to the literature, on the organometallic chemistry of the transition metals, for the period 1950–1970. This guide was brought up-to-date in two supplements (*2*, *3*) which appeared in subsequent volumes. Our article on the literature concerning the organometallic chemistry of the Main Group elements is conceived as a companion to those of Bruce, and is set out in very similar style. Inevitably, there is some overlap—particularly in discussion of general books and of those areas of organometallic chemistry, e.g., Ziegler-Natta catalysis and compounds with metal–metal bonds, which involve both Main Group and Transition Metals. Our field has been defined as the organometallic chemistry of the alkali and alkaline earth elements, the elements Zn—Hg, B—Tl, Si—Pb, and As—Bi; the very extensive organic chemistry of phosphorus, which is discussed in numerous other publications (e.g., *4–6*) and the organic chemistry of the elements S—Te have been excluded. We have not attempted fully to document the use of Grignard and similar reagents in organic syntheses; only major reviews or those articles that deal specifically with structures or mechanisms have been listed. More details in this field may be found in references A32, A34, and G3.

Partly from lack of space, we have restricted our general discussion of the use of the literature and information retrieval services. The comments of Bruce and the references he cites are valuable here. Also we have not given statistical analyses of published papers. The relative importance of literature on various elements is, however, apparent from our book lists and the appendix of review articles: clearly silicon and boron compounds have been most thoroughly studied. We broadly endorse the comments of Bruce about the relative importance of various journals in organometallic chemistry—except to add that the organometallic chemistry of the Main Group elements has been particularly widely investigated in the USSR. Books and papers from Russian authors are therefore of considerable importance.

Organometallic compounds of Main Group elements have been studied since as long ago as 1760 when Cadet made his "fuming liquid," a mixture of cacodyl $(Me_2As)_2$ and cacodyl oxide $(Me_2As)_2O$. Some 12,000 organometallic compounds were documented by Krause and von Grosse (A22) in 1937; since then, and especially since 1950, there has been an enormous growth in the subject. Our guide gives details of a few of the more important books from before 1950, and of most of the books and reviews published between then and the end of 1973.

A. *Textbooks*

Most people meet the organometallic chemistry of the Main Group elements for the first time during their study of organic or inorganic chemistry. With few exceptions (e.g., A1), recent general inorganic textbooks used in the United Kingdom (A1–A4) do not draw together the organometallic chemistry of all the Main Group elements. Therefore, the subject is often not seen as a whole. Most of the organic textbooks in wide use mention organometallic compounds (mainly Grignard reagents) and reactions such as the Wurtz–Fittig reaction or hydroboration—but in paragraphs scattered throughout the text and easily traced only through the index. A few books (A5–A7) have specific chapters on organometallic compounds.

A1. C. S. G. Phillips and R. J. P. Williams, "Inorganic Chemistry," Oxford Univ. Press, London and New York, Vol. 1 (1956), Vol. 2 (1966). Volume 2, Chapter 33 (pp. 550–580) deals with the organometallic compounds of the non-transition metals.

A2. F. A. Cotton and G. Wilkinson, "Advanced Inorganic Chemistry," 3rd ed. Wiley (Interscience), New York, 1972. The main group elements are covered group by group and there are concise sections on organometallic compounds in most chapters.

A3. P. J. Durrant and B. Durrant, "Introduction to Advanced Inorganic Chemistry," 2nd ed. Longmans, London, 1972. Main Group organometallic chemistry is discussed element by element in several separate sections.

A4. H. J. Emeléus and A. G. Sharpe, "Modern Aspects of Inorganic Chemistry." Routledge and Kegan Paul, London, 1973. There are sections on carboranes, borazines, and silicones.

A5. R. C. Fuson, "Advanced Organic Chemistry." Wiley, New York, 1950. See Chapter VII (pp. 154–176) on the uses of organometallic compounds (mainly Grignard reagents) in synthesis.

A6. I. L. Finar, "Organic Chemistry," 4th ed. Longmans, London, 1963. Chapter XV (pp. 348–367) gives good general coverage.

A6a. J. D. Roberts and M. C. Caserio, "Basic Principles of Organic Chemistry." Benjamin, New York, 1964. Chapter 12 (25 pp.) gives a good introduction to organometallic compounds.

A7. L. O. Smith and S. J. Cristol, "Organic Chemistry." Reinhold, New York, 1966. Chapter 19 is a reasonably balanced account.

Several texts at school (A8) or undergraduate level (A9–A15) do attempt to bring together material on organometallic chemistry from a variety of sources. These books are often original and stimulating and reflect the particular standpoints of the individual authors.

A8. P. Simpson, "Organometallic Chemistry of the Main Group Elements." Longmans, London, 1970. 101 pp. A very simple account, mainly for use in schools

A9. G. E. Coates, M. L. H. Green, P. Powell, and K. Wade, "Principles of Organometallic Chemistry." Methuen, London, 1968. 259 pp. An undergraduate text, with emphasis on general features of organometallic chemistry.

A10. J. J. Eisch, "The Chemistry of Organometallic Compounds. The Main Group Elements." Macmillan, London, 1967. 178 pp. An undergraduate text; a good account of principles and reactions.

A11. P. L. Pauson, "Organometallic Chemistry." Arnold, London, 1967. 202 pp. An undergraduate text from the organic point of view.

A12. E. G. Rochow, "Organometallic Chemistry." Reinhold, New York, 1964. 104 pp. An excellent short introduction to organometallic chemistry.

A13. O. Yu. Okhlobystin, "Tret'ya Khimiya. Elementoorganicheskiya Soedineniya." Nauka, Moscow, 1965. 199 pp.

A14. Yu. A. Ol'dekop and N. A. Maier, "Elementoorganicheskaya Khimiya." Znanie, Moscow, 1971. No. 7 in the "New in Life, Science and Technology. Chemistry Series." 63 pp.

A15. K. Wade, "Electron Deficient Compounds." Nelson, London, 1971. An excellent introduction to the reactions of and bonding in metal alkyls and carboranes.

Of the more detailed books suitable for postgraduate reading, the best general account of organometallic chemistry, balanced and well documented, is the two-volume work by Coates, Green, and Wade (A16). At a similar level, there are accounts of Main Group organometallic chemistry in several major comprehensive treatises (A19–A20, A25). The person beginning research in organometallic chemistry will also find much of interest and value in some of the classic accounts of organometallic chemistry (A21–A24). In particular, the section by Gilman in the organic text, which he edited (A21), and the book by Krause and von Grosse (A22), both written more than thirty years ago, retain their importance. Many of the laboratory techniques, which are now commonplace in organometallic chemistry, were collected together for the first time in Krause and von Grosse. Their concise section on techniques is still a useful source of ideas.

A16. G. E. Coates, M. L. H. Green, and K. Wade, "Organometallic Compounds," 3rd ed. Methuen, London. Vol. 1: The main group elements (1967), 573 pp.; Vol. 2: The transition elements (1968), 348 pp. The first edition, by G. E. Coates, was published in 1956 and a much expanded second edition in 1960.

A17. E. G. Rochow, D. T. Hurd, and R. N. Lewis, "The Chemistry of Organometallic Compounds." Wiley, New York, 1957. 344 pp. A well-balanced book, but rather out of date now.

A18. I. A. Shikhiev, "Khimiya Elementoorganicheskikh Soedinenii." Maarif, Baku, 1965. 194 pp.

A19. S. Coffey, ed., "Rodd's Chemistry of Carbon Compounds." Elsevier, Amsterdam. See C. B. Milne and A. N. Wright, Vol. IB (1965), pp. 165–277, Aliphatic organometallic and organometalloidal compounds, and R. Livingston, Vol. IVA (1973), pp. 539–549, 5-Membered heterocyclic compounds with a single heteroatom from groups III, IV, V in the ring.

A20. J. C. Bailar, H. J. Emeléus, R. S. Nyholm, and A. F. Trotman-Dickenson, "Comprehensive Inorganic Chemistry." Pergamon, Oxford, 1973, 5 vols. Volume 1 has a short chapter by M. L. H. Green, An introduction to the organic chemistry of the metallic elements. Organometallic compounds are discussed in several other chapters in Vols. 1 and 2, which deal with Main Group elements.

A21. H. Gilman, ed., "Organic Chemistry. An Advanced Treatise," 2nd ed. Wiley, New York, 1943. See especially Vol. 1, Chapter 5, pp. 489–580, by Gilman and Vol. 2, Chapter 24, pp. 1806–20, by G. Calingaert and H. A. Beatty.

A22. E. Krause and A. von Grosse, "Die Chemie der metallorganischen Verbindungen." Bornträger, Berlin, 1937. Reprinted by Sänding, Wiesbaden, 1965. 926 pp.

A23. F. Runge and J. Schmidt, "Organometallverbindungen." Wissenschaftliche, Stuttgart, 1932, 1934. 2 Vols.

A24. J. Newton Friend, ed., "Textbook of Inorganic Chemistry." Griffin, London, See chapters by A. E. Goddard and D. Goddard, Vol. XI, No. 1 (1928), Groups 1–4, 379 pp.; No. 2 (1930), As, 547 pp.; No. 3 (1936), P, Sb, Bi, 293 pp.

A25. N. V. Sidgwick, "The Chemical Elements and Their Compounds." Oxford Univ. Press, London and New York, 1950. 2 Vols. Organometallic chemistry is discussed element by element.

The enormous increase in the amount of material in Main Group organometallic chemistry over the last ten or fifteen years has meant that no single book can now possibly be comprehensive. As a result, a series of monographs describing particular elements, or groups of elements, has appeared and these constitute the best introductions to the detailed literature. There are several series covering various groups in the Periodic Table with a common approach. For example, the series (A26) edited by Nesmeyanov and Kocheshkov and the Houben–Weyl volumes (A27) are both particularly useful for preparative chemists. Another series of monographs, e.g., A115, A117, A119, A124, is edited by D. Seyferth. In general, the element-by-element monographs deal better with the chemistry (preparations, reactions, types of derivative) than with structural aspects of organometallic compounds. Volumes on organosilicon compounds and boron-nitrogen compounds (A129) in the new revised eighth edition of Gmelin have appeared, but other organometallic compounds of the Main Group elements have not yet been covered.

A26. A. N. Nesmeyanov and K. A. Kocheshkov, eds., "Metody elementoorganicheskoi Khimii." Nauka, Moscow, 1965.
Also published in several volumes:
26.1 Vol. 1, Bor, alyuminii, gallii, indii, tallii (1964). 499 pp.
26.2 Vol. 2, Magnii, beryllii, kal'tsii, strontsii, barii (1963). 561 pp.
26.3 Vol. 3, Tsink, kadmii (1964). 235 pp.
26.4 Vol. 4, Rtut' (1965). 438 pp.
These have all been revised and translated into English (A57, A34, A35, A37).
An earlier series of books, "Sinteticheskie metody v oblasti metalloorganicheskikh soedinenii," by the same authors was published in 1945.

A27. O. Bayer, E. Müller, and K. Ziegler, eds., "Methoden der organischen Chemie" (Houben-Weyl), Vol. XIII: Metallorganische Verbindungen. Thieme, Stuttgart.
27.1 No. 1 (1970) G. Bähr, P. Burba, H. F. Ebel, A. Lüttringhaus, and U. Schöllkopf, Li, Na, K, Rb, Cs, Cu, Ag, Au. 940 pp.
27.2 No. 2a (1973) K. Nützel, G. Bähr, H. Gilman, H. O. Kalinowski, and G. F. Wright, Be, Mg, Ca, Sr, Ba, Zn, Cd. 949 pp.
27.3 No. 4 (1970) G. Bähr, P. Burba, H. Lehmkuhl, and K. Ziegler, Al, Ga, In, Tl. 430 pp.

A28. B. J. Wakefield, "The Chemistry of Organolithium Compounds." Pergamon, Oxford, 1974. 336 pp. More than 2000 references.

A29. A. A. Morton, "Solid Organoalkali Metal Reagents. A New Chemical Theory for Ionic Aggregates." Gordon & Breach, New York, 1964. 217 pp. The emphasis is on heterogeneous catalysis.

A30. M. Schlosser, "Struktur and Reaktivität polarer Organometalle; eine Einführung in die Chemie organischer Alkali– und Erdalkalimetall–Verbindungen." Springer, Berlin, 1973. 161 pp.

A31. C. Courtot, "Le magnésium en chimie organique." Lorraine-Rigot, Nancy, 1926. 250 pp.

A32. M. S. Kharasch and O. Reinmuth, "Grignard Reactions of nonmetallic Substances." Prentice-Hall, Englewood Cliffs, New Jersey, 1954. 1384 pp.

A33. D. A. Everest, "The Chemistry of Beryllium." Elsevier, Amsterdam, 1964. Chapter 7, 20 pp. (37)* deals with organoberyllium compounds.

A34. S. T. Ioffe and A. N. Nesmeyanov, "The Organic Compounds of Magnesium, Calcium, Strontium and Barium." North-Holland Publ., Amsterdam, 1967. 745 pp. For Russian edition, see A26.2.

A35. N. I. Sheverdina and K. A. Kocheshkov, "The Organic Compounds of Zinc and Cadmium." North-Holland Publ., Amsterdam, 1967. 252 pp. For Russian edition, see A26.3.

A36. J. Boersma and J. G. Noltes, "Organozinc Co-ordination Chemistry." International Lead Zinc Research Organization, New York, 1968. 119 pp.

A37. L. G. Makarova and A. N. Nesmeyanov, "The Organic Compounds of Mercury." North-Holland Pub., Amsterdam, 1967. 540 pp. For Russian edition, see A26.4.

A38. F. Jensen and B. Rickborn, "Electrophilic Substitution of Organomercurials." McGraw-Hill, New York, 1968. 203 pp.

A39. P. L. Bidstrup, "Toxicity of Mercury and Its Compounds." Elsevier, Amsterdam, 1964. One chapter (10 pp.) is on organomercury compounds.

A40. L. Friberg and J. Vostal, "Mercury in the Environment." C.R.C. Press, Cleveland, Ohio, 1972. 215 pp. This has a good account of epidemiology and toxicology.

A41. F. M. d'Itri, "The Environmental Mercury Problem." C.R.C. Press, Cleveland, Ohio, 1972. 124 pp. An introduction and guide to literature.

A42. W. Gerrard, "The Organic Compounds of Boron." Academic Press, New York, 1961. 231 pp.

A43. H. C. Brown, "Hydroboration," Benjamin, New York, 1962. 290 pp.

A44. H. Steinberg, "Organoboron Chemistry." Wiley (Interscience), New York. Volume 1 (1964), Boron-oxygen and boron-sulfur compounds. 950 pp.; Vol. 2 (1966), by H. Steinberg and R. J. Brotherton, Boron-nitrogen and boron-phosphorus compounds. 515 pp.

A45. R. M. Adams, ed., "Boron, Metallo-boron Compounds and Boranes." Wiley (Interscience), New York, 1964. 765 pp. Mainly inorganic boron chemistry, but organometallic compounds are described briefly in Chapters 6 and 7. Chapter 8 contains a section (7 pp.) on the toxicity of organoboron compounds.

A46. K. Niedenzu and J. W. Dawson, "Boron-Nitrogen Compounds." Springer, Berlin, 1965. 158 pp.

A47. E. L. Muetterties, ed., "The Chemistry of Boron and Its Compounds." Wiley, New York, 1967. See K. Niedenzu and J. W. Dawson, Chapter 7, B-N compounds, 56pp. (376); M. F. Lappert, Chapter 8, B-C compounds, 142 pp. (1213)—this chapter has a complete literature survey up to 1964; G. W. Parshall, Chapter 9, B-P compounds, 25 pp. (174); E. L. Muetterties, Chapter 10, S and Se compounds, 18 pp. (76).

A48. R. L. Hughes, I. C. Smith, and E. W. Lawless, "Production of the Boranes and Related Research." Academic Press, New York, 1967. Organometallic compounds are discussed in Chapters 8 (alkylboranes), 9 (carboranes), and 11 (B-N chemistry).

A49. B. M. Mikhailov, "Khimiya Borovodorodov." Nauka, Moscow, 1967. 530 pp. There is some discussion of organoboron chemistry in a major book about boron hydrides.

* The number in parentheses indicates the number of references given.

A50. A. L. Muetterties and W. H. Knoth, "Polyhedral Boranes." Dekker, New York, 1968. 197 pp.

A51. G. R. Eaton and W. N. Lipscomb, "NMR Studies of Boron Hydrides and Related Compounds." Benjamin, New York, 1969. There are extensive tabulations including data on carboranes (89 pp.) and organoboron compounds.

A52. R. N. Grimes, "Carboranes." Academic Press, New York, 1970. 272 pp. A full account.

A53. M. Grassberger, "Organische Borverbindungen." Chemie., Weinheim, 1971. Chemical pocketbooks, Vol. 15. 152 pp.

A54. H. C. Brown, "Boranes in Organic Chemistry." Cornell Univ. Press, Ithaca, New York, 1972. 462 pp. The second half (pp. 255–450) is a major review showing the versatility of organoboron compounds in organic syntheses.

A55. G. M. L. Cragg, "Organoboranes in Organic Synthesis." Dekker, New York, 1973. 432 pp.

A56. I. A. Sheka, I. S. Chaus, and T. T. Mityureva, "Chemistry of Gallium." Elsevier, Amsterdam, 1966. 298 pp. Organogallium compounds are discussed on pp. 158–173.

A57. A. N. Nesmeyanov and R. A. Sokolik, "The Organic Compounds of Boron, Aluminium, Gallium, Indium and Thallium." North-Holland Publ., Amsterdam, 1967. 628 pp. For Russian edition, see A26.1.

A58. G. Bruno, "The Use of Aluminium Alkyls in Organic Synthesis." Ethyl Corporation, Baton Rouge, Louisiana, 1970.

A59. A. G. Lee, "The Chemistry of Thallium." Elsevier, Amsterdam, 1971. There is one chapter, 72 pp. (232), on organothallium compounds.

A60. T. Mole and E. A. Jeffery, "Organoaluminium Compounds." Elsevier, Amsterdam, 1972. 465 pp. This is a well-presented and comprehensive account.

A61. B. N. Dolgov, "Khimiya kremniiorganicheskikh soedinenii." Goskhimtekhizdat, Leningrad, 1933.

A62. M. G. Voronkov, "Khimiya kremniiorganicheskikh soedinenii v rabotakh russkikh i sovetskikh uchenykh." Leningradskii gosudarstvennyi universitet. Leningrad, 1952. An account describing the chemistry of organosilicon compounds in the works of Russian scientists.

A63. A. P. Kreshkov *et al.*, "Analiz kremniiorganicheskikh soedinenii." Goskhimizdat, Moscow, 1954. 255 pp. On the analysis of organosilicon compounds.

A64. K. A. Andrianov, "Kremniiorganicheskie soedineniya." Goskhimizdat, Moscow, 1955. 520 pp.

A65. A. D. Petrov and V. F. Mironov, "Darstellung und Eigenschaften von Siliciumkohlenwasserstoffen." Akademie-Verlag, Berlin, 1955. 48 pp.

A66. M. Kumada and R. Okawara, "Organosilicon Chemistry," Maki, Tokyo, 1969. M. Kumada, M. Ishikawa, K. Yamamoto, and K. Tamao, "The Chemistry of Organosilicon Compounds," Kagaku Dojin, Kyoto, 1972. Both in Japanese.

A67. C. Eaborn, "Organosilicon Compounds." Butterworth, London, 1960. 530 pp.

A68. E. A. V. Ebsworth, "Volatile Silicon Compounds." Pergamon, Oxford, 1963. 250 pp.

A69. A. D. Petrov, B. F. Mironov, V. A. Ponomarenko, and E. A. Chernyshev, "Synthesis of Organosilicon Monomers." Heywood, London, 1964. A translation of "Sintez Kremniiorganicheskikh Monomerov." Academy of Sciences Press, Moscow, 1961. 551 pp.

A70. L. H. Sommer, "Stereochemistry, Mechanism and Silicon." McGraw-Hill, New York, 1965. 189 pp.

A71. E. Ya. Lukevits and M. G. Voronkov, "Organic Insertion Reactions of Group IV Elements." Consultants Bureau, New York, 1966. 372 pp. This is a translation of a revised version of a book on hydrometallation originally published with the title Gidrosililirovanie, gidrogermilirovanie i gidrostannilirovanie. Acad. Sci. Latvian SSR, Riga, 1964.

A72. K. A. Andrianov, "Kremnii (metody elementoorganicheskoi khimii)." Nauka Moscow, 1968. 699 pp.

A73. A. E. Pierce, "Silylation of Organic Compounds." Pierce Chemical Company, Rockford, Illinois, 1968. 487 pp.

A74. A. G. MacDiarmid, ed., "Organometallic Compounds of the Group IV Elements." Dekker, New York. This is an important, comprehensive review.

Vol. 1 "The Bond to Carbon," Part 1 (1968). E. A. V. Ebsworth, Physical basis of the chemistry of the group IV elements. 88 pp. (556). C. Eaborn and R. W. Bott, The synthesis and reactions of the silicon-carbon bond. 350 pp. (2414).

Part 2 (1968). F. Glockling and K. A. Hooton, Synthesis and properties of the germanium-carbon bond. 80 pp. (300). J. G. A. Luijten and G. J. M. van der Kerk, Synthesis and properties of the tin-carbon bond. 80 pp. (551). L. C. Willemsens and G. J. M. van der Kerk, Synthesis and properties of the lead-carbon bond. 30 pp. (264).

Vol. 2 "The Bond to Halogens and Halogenoids," Part 1 (1972). C. H. Van Dyke, Synthesis and properties of the silicon-halogen and silicon-halogenoid bond. 296 pp. (1248).

Part 2 (1972). J. J. Zuckerman, Synthesis and properties of the germanium-halogen and germanium-halogenoid bond. 57 pp. (538). H. C. Clark and R. J. Puddephatt, Synthesis and properties of the tin-halogen and tin-halogenoid bond. 62 pp. (493). S. E. Cook, F. W. Frey, and H. Shapiro, Synthesis and properties of the lead-halogen and lead-halogenoid bond. 40 pp. (287).

A75. S. N. Borisov, M. G. Voronkov, and E. Ya. Lukevits, "Organosilicon Derivatives of Phosphorus and Sulphur." Plenum, New York, 1971. 350 pp. An updated translation of Kremneorganicheskie Proizvodnye Fosfora i Sery." Khimiya, Leningrad, 1968. 292 pp. A comprehensive account.

A76. M. G. Pomerantseva, Z. V. Belyakova, S. A. Golubtsov, and N. S. Shvats, "Poluchenie karbofunktsional'nykh organosilanov po reaktsii prisoedineniya (obzor)." Nitekhim, Moscow, 1971. 89 pp. A review of the preparation of carbofunctional organosilanes by addition reactions.

A77. M. G. Voronkov, G. I. Zelchan, and E. Ya. Lukevits, "Kremnii i Zhizn'." Zinatne, Riga, 1971. 327 pp. This is a comprehensive book on "Silicon and Life." Part 1 (2571) covers silicon in nature and Part 2 (2564) the biological action of silicon compounds. Most of the references are to the biochemical and medical literature.

A78. R. Soder, "Die Silikone." Novelelectric A. G., Zurich, 1947. 59 pp.

A79. H. W. Post, "Silicones and Other Organic Silicon Compounds." Reinhold, New York, 1948. 230 pp.

A80. K. A. Andrianov and O. I. Gribanova, "Kremniiorganicheskie polimernye soedineniya." Gosenergoizdat, Moscow, 1946.

A81. K. A. Andrianov, "Vysokomolekularnye kremniiorganicheskie soedineniya." Moscow, 1949

A82. K. A. Andrianov and M. V. Sobolevskii, "Vysokomolekularnye kremniiorganicheskie soedineniya." Oborongiz, Moscow, 1950.

A83. E. G. Rochow, "An Introduction to the Chemistry of the Silicones," 2nd ed. Wiley, New York, 1951. 213 pp. A French edition was published by H. Dunod, Paris, 1952.

A84. V. Bažant, V. Chvalovský, and J. Rathouský, "Silikony, organokřemičité sloučeniny, jejich příprava, vlastnosti a použití." SNTL, Prague, 1954. 312 pp. A full account (2245) of silicones, organosilicon compounds, their preparation, properties, and use. Translated into English by Liaison Office Technical Information Centre, Wright Patterson Air Force Base, Dayton, Ohio, 1960. A second edition in Polish was published by Państwowe wydawnictwa techniczne, Warsaw, 1955, and a third edition by Goskhimizdat, Moscow, 1960.

A88. R. R. McGregor, "Silicones and Their Uses." McGraw-Hill, New York, 1954. 302 pp.

A89. A. P. Kreshkov, "Kremniiorganicheskie soedineniya v tekhnike." Promotroiizdat, Moscow, 1956. 289 pp.

A90. R. Levin, "The Pharmacy of Silicones and Their Uses in Medicine." Morgan Bros., London, 1958.

A91. V. Bažant, V. Chvalovský, and J. Rathouský, "Technické použití silikonů." SNTL, Prague, 1959.

A92. N. N. Sokolov, "Metody sinteza poliorganosiloksanov." Gosudarstvennoe Energeticheskoe, Moscow, 1959. 187 pp.

A93. A. Hunyar, "Chemie der Silikone," 2nd ed. VEB, Berlin, 1959. 343 pp. Many references are to patent literature.

A94. R. N. Meals and F. M. Lewis, "Silicones." Reinhold, New York, 1959. 304 pp.

A95. K. A. Andrianov and A. I. Petrashko, "Kremniiorganicheskie polimery v narodnom khozyaistve." Academy of Sciences, Moscow, 1959. 77 pp. A short introductory account of organosilicon polymers in the national economy.

A96. I. A. Shikhiev, M. F. Shostakovskii, and N. V. Komarov, "Novye kislorodsoderzhashchie kremneorganicheskie soedineniya." Aznefteizdat, Baku, 1960. 191 pp. On new oxygen-containing organosilicon compounds.

A97. S. Fordham, ed., "Silicones." Newnes, London, 1960. 252 pp.

A98. "Silicones for Use in Medicine," published jointly by Midland Silicones Ltd., and Hopkins and Williams Ltd., London, 1962. 7 pp. A brief survey (232).

A99. P. Rościszewski, "Zastosowanie silikonów." Wydawnictwa Naukowo-Techniczne, Warsaw, 1964. 286 pp. A discussion of the technological aspects of silicones.

A100. V. V. Severnyi and E. G. Novitskii, "Kremniiorganicheskie kompozitsii i ikh primenenie." Goskomitet po Delam Izobretenii i Otkrytii SSSR, 1965. 35 pp. Organosilicon compositions and their uses.

A101. R. J. M. Voorhoeve, "Organohalosilanes. Precursors to Silicones." Elsevier, Amsterdam, 1967. 400 pp.

A102. N. F. Orlov, M. V. Androsova, and N. V. Vvedenskii, "Kremniiorganicheskie soedineniya v tekstil'noi y legkoi promyshlennosti." Legkaya Industriya, Moscow, 1966. 238 pp. An account of organosilicon compounds in textiles and light industry.

A103. V. V. Arkharova, "Kremniiorganicheskie smazochnye materially." Tsent-Nauch-Issled. Inst. Patent Inform. Tekh-Ekon Issled, Moscow, 1967. 29 pp. Organosilicon lubricating materials.

A104. W. Noll, "Chemistry and Technology of Silicones." Academic Press, New York, 1968. 706 pp. Translated from second German edition, "Chemie und Technologie der Silikone." Chemie, Weinheim, 1965.

A105. K. C. Frisch, "Cyclic Monomers." Wiley, New York, 1972. See the chapter by O. K. Johannson and C. Lee on cyclic siloxanes and silazanes.

A106. V. I. Zhunko. N. P. Kharitonov, M. P. Sinitsyn *et al.*, "Svoistva i primery primeneniya organosilikatnykh materialov." Nauka Leningrad Otd., Leningrad, 1972. 22 pp. On properties and examples of the use of organosilicate materials.

A107. F. Rijkens, "Organogermanium Compounds." Germanium Research Committee, Schotanus and Jens, Utrecht, 1960. 97 pp. A survey of the literature from January 1950 to July 1960 (266).

A108. L. C. Willemsens, "Organolead Chemistry." International Lead Zinc Research Organization, New York, 1964. 82 pp. A concise review with special reference to the literature covering the period January 1953 to July 1963 (421).

A109. F. Rijkens and G. J. M. van der Kerk, "Investigations in the Field of Organogermanium Chemistry." Germanium Research Committee, Schotanus and Jens, Utrecht, 1964. 146 pp. (441).

A110. J. G. A. Luijten and G. J. M. van der Kerk, "Investigations in the Field of Organotin Chemistry." Tin Research Institute, Greenford, Middlesex, United Kingdom, 1955. 125 pp. (204).

A111. L. C. Willemsens and G. J. M. van der Kerk, "Investigations in the Field of Organolead Chemistry." International Lead Zinc Research Organization, New York, 1965. 118 pp. (197).

A112. V. F. Mironov and T. K. Gar, "Organicheskie Soedineniya Germaniya." Nauka, Moscow, 1967. 362 pp.

A113. H. M. J. C. Creemers, "Hydrostannolysis. A General Method for Establishing Tin-Metal Bonds." Schotanus and Jens, Utrecht, 1967. 251 pp.

A114. D. A. Kochkin and I. N. Azerbaev, "Olovo- i svinets-organicheskie monomery i polimery." Nauka, Alma Ata, 1968. 299 pp. Tin and lead organic monomers and polymers.

A115. H. Shapiro and F. W. Frey, "The Organic Compounds of Lead." Wiley (Interscience), New York, 1968. 468 pp.

A116. F. Glockling, "The Chemistry of Germanium." Academic Press, New York, 1969. 234 pp. This book contains 117 pp. on organogermanium compounds.

A117. M. Lesbre, P. Mazerolles, and J. Satgé, "The Organic Compounds of Germanium." Wiley, New York, 1971. 701 pp. This is an extremely thorough review with 200 pp. of tabulations and 2127 references up to 1968 and part of 1969.

A118. F. J. A. des Tombe, "Reactions of Organotin Halides with Zinc. The Occurrence of 1, 2-Intermetallic Shifts." Elinkwijk, Utrecht, 1970. 64 pp.

A119. W. P. Newmann, "The Organic Chemistry of Tin." Wiley (Interscience), New York, 1970. 282 pp. A translation of "Die organische Chemie des Zinns." Enke, Stuttgart, 1967.

A120. R. C. Poller, "The Chemistry of Organotin Compounds." Logos Press, London, 1970. 315 pp.

A121. A. K. Sawyer, ed., "Organotin Compounds." Dekker, New York, 1971. This is a comprehensive account in 3 volumes. The chapters are as follows:

Vol. 1 G. J. M. van der Kerk and J. G. A. Luijten, Introduction. 6 pp.
E. J. Kupchik, Organotin hydrides. 67 pp. (260).
G. P. van der Kelen, E. V. Van der Berghe, and L. Verdonck, Organotin halides. 55 pp. (554).
A. J. Bloodworth and A. G. Davies, Organotin compounds with Sn—O bonds. Alkoxides. 89 pp. (135).

Vol. 2 R. Okawara and M. O'Hara, Organotin compounds with Sn—O bonds. Carboxylates, salts. 40 pp. (135).

H. Schumann and M. Schmidt, Organotin compounds with Sn—S, Sn—Se, Sn—Te bonds. 211 pp. (396).
K. Jones and M. F. Lappert, Organotin compounds with Sn—N bonds. 64 pp. (230).
H. Schumann, Organotin compounds with Sn—As, Sn—Sb, Sn—Bi bonds. 42 pp. (115).

Vol. 3 M. Gielen and J. Nasielski, Organotin compounds with Sn—C without Sn—Sn bonds. 197 pp. (466).
A. K. Sawyer, Organotin compounds with Sn—Sn bonds. 56 pp. (124).
M. J. Newlands, Compounds with Sn—other metal bonds. 49 pp. (237).
J. G. A. Luijten, Applications and biological effects. 43 pp. (285).
M. C. Henry and W. E. Davidson, Organotin polymers. 21 pp. (86).
C. R. Dillard, Analysis. 9 pp. (87).

A122. I. N. Azerbaev and D. A. Kochkin, "Organicheskie soedineniya olova i svintsa. Monomery i polimery." Znanie, Moscow, 1972. 31 pp. Organic tin and lead compounds. Monomers and polymers. No. 8 in "Life, Science and Technology. Chemistry Series"

A123. A. W. Johnson, "Ylid Chemistry." Academic Press, New York, 1966. This book is mainly about organophosphorus compounds, but there is one chapter of 20 pp. (39) on arsenic and antimony ylids.

A124. G. O. Doak and L. D. Freedman, "Organometallic Compounds of Arsenic, Antimony and Bismuth." Wiley (Interscience), New York, 1970. 509 pp.

A125. F. G. Mann, "Chemistry of Heterocyclic Compounds," Vol. 1: Heterocyclic Derivatives of Phosphorus, Arsenic, Antimony and Bismuth, 2nd ed. Wiley (Interscience), New York, 1970. A much shorter first edition was published in 1950.

A126. C. A. McAuliffe, ed., "Transition Metal Complexes of Phosphorus, Arsenic and Antimony ligands." Macmillan, New York, 1973.
This contains the following articles:
A. Pidcock, Group Vb to transition metal bonds. 28 pp. (130).
J. C. Cloyd and C. A. McAuliffe, Transition metal complexes containing monotertiary arsines and stibines. 47 pp. (587).
B. Chiswell, The chemistry of multidentate ligands containing heavy group Vb donors. 33 pp. (132).
E. C. Alyea, Metal complexes with ditertiary arsines." 54 pp. (250).

A127. I. Lindqvist, "Inorganic Molecules of Oxo-Compounds." Springer-Verlag, Berlin and New York, 1963. 114 pp. This is marginal to organometallic chemistry, but there is some useful information about arsine oxide complexes.

A128. E. Wiberg and E. Amberger, "Hydrides of the Main Groups I–IV." Elsevier, Amsterdam, 1971. A very thorough survey with 3500 references. 760 pp.

A129. "Gmelins Handbuch der anorganischen Chemie," 8th ed. Verlag Chemie, Weinheim.
A129. 1 G. Hantke, U. Krüerke, and C. Huschke, Vol. 15 (1958). "Silicium Organische Siliciumverbindungen." 486 pp.
A129. 2 E. H. E. Pietsch, A. Kotowski, M. Becke-Goehring, and K-C. Buschbeck, Vol. 13, Part 1 (1974) "Borverbindungen" (on B–N compounds). 331 pp.

Besides the element-by-element monographs, there are a number of books which deal with particular features of organometallic chemistry.

An important group of these cover mechanisms (A130–A139, C18), some aspects of which are described in more general books not specifically devoted to organometallic chemistry. There are also a number of books dealing with thermochemical properties of organometallic compounds (A147–A150), with electronic (A151), vibrational (A152–A154, G9, G10), nuclear magnetic resonance spectra (A155, G9), and mass spectra (A156–A157). The series of monographs edited by M. Tsutsui on the characterization of organometallic compounds (C16) also contain important accounts of the use of spectroscopic methods. A few books deal specifically with analysis (A63, A158–A159, C15).

A130. O. A. Reutov and I. P. Beletskaya, "Reaction Mechanisms of Organometallic Compounds." North-Holland Publ., Amsterdam, 1968. 476 pp.

A131. T. G. Brilkina and V. A. Shushunov, "Reactions of Organometallic Compounds with Oxygen and Peroxides." Iliffe Books, London, 1969. 234 pp. A translation of "Reaktsii metalloorganicheskikh soedinenii s kislorodom i perekisyami," Nauka, Moscow, 1966.

A132. J. C. Lockhart, "Redistribution Reactions." Academic Press, New York, 1970. 173 pp.

A133. O. A. Reutov, I. P. Beletskaya, and V. I. Sokolov, "Mekhanizmy reaktsii metalloorganicheskikh soedinenii." Khimiya, Moscow, 1972. 367 pp.

A134. D. S. Matteson, "Organometallic Reaction Mechanisms of the Non-transition Elements." Academic Press, New York, 1973. 375 pp.

A135. C. K. Ingold, "Structure and Mechanism in Organic Chemistry," 2nd ed. Bell, London, 1969. 1266 pp. Chapter 8 covers mechanisms of electrophilic substitution at saturated carbon, especially with reference to the Hg—C bond.

A136. K. U. Ingold and B. P. Roberts, "Free Radical Substitution Reactions: Bimolecular Homolytic Substitutions (S_H2 Reactions) at Saturated Multivalent Atoms." Wiley (Interscience), New York, 1971. 245 pp. There is a section on S_H2 reactions in organo-group IV compounds.

A137. W. Kirmse, "Carbene Chemistry, 2nd ed. Academic Press, New York, 1971. 615 pp. Halomethyl derivatives (especially of Li, Zn, Hg) are described in Chapter 3. There are also sections on reactions of carbenes with Si—H, Ge—H, Si—Cl, Si—C bonds and on bivalent Si, Ge, Sn.

A138. L. Kaplan, "Bridged Free Radicals." Dekker, New York, 1972. 481 pp. There is only a short section on organometallic-substituted radicals, but reactions of radicals with organometallic compounds are covered.

A139. J. K. Kochi, ed., "Free Radicals." Wiley (Interscience), New York, 1973. Vol. 1, 683 pp.; Vol. 2, 852 pp. Volume 1, Chapter 6 contains a useful summary of organometallic-alkyl halide reactions, mainly involving RLi species. Chapter 10 by A. G. Davies and B. P. Roberts is entitled "Bimolecular Homolytic Substitution at Metal Atoms." Volume 2 contains a chapter of 67 pp. (266) by H. Sakurai on group IVB radicals.

A140. G. L. Jenkins, W. H. Hartung, K. E. Hamlin, and J. B. Data, "Chemistry of Organic Medicinal Products," 4th ed. Wiley, New York, 1957. 569 pp. Chapter 15 is on As and Hg compounds.

A141. N. G. Gaylord, "Reduction with Complex Metal Hydrides." Wiley (Interscience), New York, 1956. 1024 pp. There is a short section in Chapter 4 on reactions of hydrides with organometallics.

A142. M. Hudlický, "Chemistry of Organic Fluorine Compounds." Pergamon, Oxford, 1962. 536 pp. A short section (14 pp.) deals with organometallic compounds.

A143. H. Ulrich, "Cycloaddition Reactions of Heterocumulenes." Academic Press, New York, 1967. 364 pp. Thirty pages deal with organometallic compounds.

A144. T. F. Rutledge, "Acetylenic Compounds." Reinhold, New York.
Vol. 1 Preparation and Substitution Reactions (1968). 342 pp.
Vol. 2 Acetylenes and Allenes (1969). 432 pp.

A145. H. G. Viehe, ed., "Chemistry of Acetylenes." Dekker, New York, 1969. See the chapter by P. Cadiot and W. Chodkiewicz, Acetylenic derivatives of groups III, IVB, and VB. 50 pp. (368).

A146. R. D. Chambers, "Fluorine in Organic Chemistry." Wiley, New York, 1973. 416 pp. There is a chapter on organometallics.

A147. T. Cottrell, "The Strengths of Chemical Bonds," 2nd ed. Butterworth, London, 1958. 317 pp.

A148. C. T. Mortimer, "Reaction Heats and Bond Strengths." Pergamon, Oxford, 1962. 201 pp. Metal-carbon and metal-halide bonds are considered in Chapter 8, and bond strengths in silicon, phosphorus, and sulfur compounds in Chapter 10.

A149. J. D. Cox and G. Pilcher, "Thermochemistry of Organic and Organometallic Compounds." Academic Press, New York, 1970. 643 pp. The most comprehensive account.

A150. J. B. Pedley, ed., "Computer Analysis of Thermochemical Data (CATCH Tables)." Univ. of Sussex, Brighton, 1972. (i) Halogen compounds (1972), (ii) nitrogen compounds (1972), (iii) phosphorus compounds (1972), (iv) silicon compounds (1972), (v) chromium, molybdenum, and tungsten compounds (1974). Booklets (iv), (v) give data on organometallic compounds. The book by Cox and Pilcher (A149) is to be updated under the same system.

A151. B. G. Ramsay, "Electronic Transitions in Organometallics." Academic Press, New York, 1969. 297 pp.

A152. L. J. Bellamy, "Infrared Spectra of Complex Molecules." Methuen, London, 1958. There is a chapter (8 pp) on organosilicon compounds.

A153. D. M. Adams, "Metal-Ligand and Related Vibrations: A Critical Survey of the Infrared and Raman Spectra of Metallic and Organometallic Compounds." Arnold, London, 1967. 360 pp. There are tabulated data and references on metal-carbon and related frequencies (especially on Zn, Al, Sn groups) and on metal-ligand frequencies involving heavier group VB and VIB donors.

A154. J. R. Ferraro, "Low Frequency Vibrations of Inorganic and Coordination Compounds." Plenum, New York, 1971. 309 pp.

A155. J. B. Stothers, "^{13}C NMR Spectroscopy." Academic Press, New York, 1973. Only a very short section (8 pp.) on organometallic compounds.

A156. J. H. Benyon, R. A. Saunders, and A. E. Williams, "The Mass Spectra of Organic Molecules." Elsevier, Amsterdam, 1968. 510 pp. Chapters cover B and Si compounds.

A157. M. R. Litzow and T. R. Spalding, "Mass Spectrometry of Inorganic and Organometallic Compounds." Elsevier, Amsterdam, 1973. 616 pp.

A158. T. R. Crompton, "Analysis of Organoaluminium and Organozinc Compounds." Pergamon, Oxford, 1968. 384 pp.

A159. J. G. A. Luijten, "A Bibliography of Organotin Analysis," 2nd ed. Tin Res. Inst. Greenford, Middlesex, 1970. 48 pp.

Organometallic compounds are used in industry as initiators and catalyst components for polymerization, as anti-knock agents and various other additives, and in silicone manufacture. Several collections of articles and books deal with industrial applications (A160–A162). We have made no attempt fully to document the literature on polymer chemistry, but have listed some of the books, particularly those on anionic polymerizations and Ziegler–Natta catalysis, in which organometallic compounds figure prominently (A163–A170). The literature on silicones is very extensive (A78–A105). Books range from short introductions (e.g., A85, A95, A97, A100) to scholarly compilations (A84, A104). Several other books deal with "inorganic polymers." Some (A176, A177) are very diffuse and mention a number of topics that are only tenuously connected. There are, however, under this heading some good reviews of metallosiloxane polymers, i.e., those like silicones in which some of the silicon atoms have been replaced by other elements. Industrial applications have also been described in Conference Reports (E7, E12, E13, E16).

A160. G. J. H. Harwood, "Industrial Applications of Organometallic Compounds. A Literature Survey." Chapman & Hall, London, 1963. 451 pp. An extensive compilation of patent literature is included.

A161. "Kirk-Othmer Encyclopedia of Chemical Technology," 2nd ed. Wiley (Interscience), New York, 1963–1972. Twenty volumes plus supplement. The emphasis is on industrial aspects and many references are to patents. There are chapters on organolithium compounds, 10 pp. (45), Grignard reagents, 14 pp. (23), organoboron compounds, 20 pp. (185), organoaluminum compounds, 16 pp. (54), silylating agents, 9 pp. (47), silicones, 40 pp. (151), and organolead compounds, 18 pp. (43).

A162. G. A. Razuvaev, B. G. Gribov, G. A. Domrachev, and B. A. Salamatin, "Metalloorganicheskie soedineniya v elektronike." Nauka, Moscow, 1972. 476 pp. Sections deal with the possible use of organometallic compounds in electronics and in the production of coatings by decompositions in the gas phase or in solution.

A163. N. G. Gaylord and H. F. Mark, "Linear and Stereoregular Addition Polymers." Wiley (Interscience), New York, 1959. 533pp.

A164. J. Furukawa and T. Saegusa, "Polymerization of Aldehydes and Oxides." Wiley (Interscience), New York, 1963. 457 pp.

A165. D. J. Cram, "Fundamentals of Carbanion Chemistry." Academic Press, New York, 1965. 256 pp.

A166. L. Reich and A. Schindler. "Polymerization by Organometallic Compounds." Wiley (Interscience), New York, 1966. 705 pp.

A167. G. Natta and F. Danusso, "Stereoregular Polymers and Stereospecific Polymerisation." Pergamon, Oxford, 1967. 871 pp. This contains 170 papers by the Natta school (1954–1959), some unabridged and some as abstracts.

A168. A. D. Ketley, ed., "The Stereochemistry of Macromolecules," Vol. 1. Dekker, New York, 1967. 387 pp. This deals with Ziegler-Natta polymerization.

A169. M. Szwarc, "Carbanions, Living Polymers and Electron Transfer Processes." Wiley (Interscience), New York, 1968. 659 pp. A major work covering all aspects of anionic polymerization.

A170. E. C. Leonard, ed., "Vinyl Monomers" (in 3 parts). Wiley (Interscience), New York, 1971.

A171. G. E. Ham, ed., "Kinetics and Mechanisms of Polymerization." Dekker, New York.
- 171.1 Vol. 1, G. E. Ham, ed., Vinyl polymerization, Part 1 (1967), 396 pp.; Part 2 (1969) 509 pp.
- 171.2 Vol. 2, K. C. Frisch and S. L. Reegen, Ring opening polymerizations (1969) 517 pp.
- 171.3 Vol. 3, D. H. Solomon, ed., Step-growth polymerization (1972) 372 pp.

A172. C. E. Schildknecht, "Allyl Compounds and Their Polymers (Including Polyolefins)." Wiley, New York, 1973. There are chapters on allylsilicon compounds, 22 pp. (106) and allylboron compounds, 6 pp. (41).

A173. T. Keii, "Kinetics of Ziegler-Natta Polymerisation." Chapman & Hall, London, and Kodansha, Tokyo, 1972. 262 pp.

A174. F. G. A. Stone and W. A. G. Graham, eds., "Inorganic Polymers." Academic Press, New York, 1962. 631 pp. There are sections on P, S, B polymers, silicones, and other organometallic polymers.

A175. M. F. Lappert and G. J. Leigh, eds., "Developments in Inorganic Polymer Chemistry." Elsevier, Amsterdam, 1962. 285 pp. Contains articles on B—N, P—N compounds, silicones, and polymetallosiloxanes.

A176. F. G. R. Gimblett, "Inorganic Polymer Chemistry." Butterworth, London, 1963. 434 pp.

A177. D. N. Hunter, "Inorganic Polymers." Blackwell, Oxford, 1963. 107 pp.

A178. K. A. Andrianov, "Metalorganic Polymers." Wiley (Interscience), New York, 1965. 371 pp. A translation of an earlier book in Russian, "Polimery s neorganicheskimi glavnymi tsepyami molekul," Acad. Sci. USSR, Moscow, 1962. 327 pp.

A179. H. R. Allcock, "Heteroatom Ring Systems and Polymers." Academic Press, New York, 1967. 401 pp. See especially the sections on B and Si chemistry.

A180. S. N. Borisov, M. G. Voronkov, and E. Ya. Lukevits, "Organosilicon Heteropolymers and Heterocompounds." Plenum, New York, 1970. 633 pp. A translation of "Kremne elementoorganicheskie soedineniya," Khimiya, Leningrad, 1966. 542 pp. (1735). Many references are to the patent literature.

A181. I. Haiduc, "The Chemistry of Inorganic Ring Systems," Parts 1 and 2. Wiley (Interscience), New York, 1970. 1190 pp. Part 1 covers homocyclic and heterocyclic systems with atoms of groups III and IV. Part 2 covers heterocyclic systems with atoms of groups V and VI and metals. This is a much expanded version of an earlier book in Romanian, "Introducere in chimia ciclurilor anorganice," Academici Rupublicii Populure Romîne, Bucharest, 1964, which was later also revised and published in Polish.

A182. D. A. Armitage, "Inorganic Rings and Cages." Arnold, London, 1972. 387 pp.

Finally, we list a number of books that contain collections of articles by particular authors or collections of separate articles grouped round a particular theme. Several of these contain important reviews. For example, until the publication of Coates, Green, and Wade, the book (A184) edited by H. Zeiss was probably the best book in English on organometallic chemistry and as such it had considerable influence. A number of similar authoritative, up-to-date summaries of various aspects of organometallic chemistry appear as published lectures at major conferences (Section E).

A183. V. I. Kuznetsov, "Razvitie khimii metalloorganicheskikh soedinenii v SSSR." Acad. Sci. USSR, Moscow, 1956. 220 pp. A summary of the development of the chemistry of organometallic compounds in the USSR.

A184. H. Zeiss, ed., "Organometallic Chemistry," ACS Monograph 147. Reinhold, New York, 1960. This contains the following:
J. W. Richardson, Carbon-metal bonding, 35 pp. (41).
R. Huisgen, Benzyne chemistry, 52 pp. (125).
H. D. Kaesz and F. G. A. Stone, Vinylmetallics, 62 pp. (154).
H. C. Brown, Organoboranes, 44 pp. (103).
K. Ziegler, Organoaluminum compounds, 76 pp. (380).
H. Gilman and H. J. S. Winkler, Organosilylmetallic chemistry, 76 pp. (185).

A185. L. C. Povarov, ed., "Issledovaniya v Oblasti Kremniiorganicheskikh Soedinenii Sintez i Fizikokhimicheskie Svoistva." Acad. Sci. USSR, Moscow, 1962. A collection of 14 papers on studies in the area of organosilicon chemistry: synthesis and physicochemical properties.

A186. A. N. Nesmeyanov, "Selected Works in Organic Chemistry." Pergamon, Oxford, 1964. 1200 pp. There are sections on diazo-method of synthesis of organometallic compounds, synthesis of organometallic compounds; quasicomplex and unsaturated organometallic compounds—tautomerism and beta-conjugation; onium compounds; organic compounds of Si, Ti, F. Similar collections in Russian have been published under the title "Izbrannye Trudy." Acad. Sci. USSR, Moscow (see also A187).

A187. A. V. Topchiev, "Polimerizatsiya. Kremniiorganicheskie Soedineniya." Nauka, Moscow, 1966. 525 pp. This contains (i) 18 articles, 176 pp. on polymerization of hydrocarbons; (ii) 7 articles, 50 pp. on properties of polymers and transformations in molecular chains; (iii) 4 articles, 23 pp. on polymerization of heterocyclic compounds; (iv) 4 articles, 28 pp. on polymeric semiconductors; and (v) 15 articles, 246 pp. on organosilicon compounds.

A188. E. A. V. Ebsworth, ed., "New Pathways in Inorganic Chemistry." Cambridge Univ., London and New York, 1968. 392 pp. Three articles refer to organometallic chemistry:
J. J. Lagowski, Fluoroalkylmercurials, 10 pp. (17).
H. C. Clark, Organometallic cations, 13 pp. (36).
F. G. A. Stone, Transition metal derivatives of Si, Ge, Sn and Pb, 20 pp. (87).

A189. A. N. Nesmeyanov, "Elementoorganischeskaya Khimiya (Izbrannye Trudy) 1959–1969." Nauka, Moscow, 1970. 874 pp. Selected works.

A190. E. Wolski, "Chemistry of Organometallic Compounds," Part A: Organosilicon and Organogermanium Chemistry. CFSTI, Springfield, Virginia, 1966. A survey of communist world scientific and technical literature.

A191. G. Bähr, "Organometallic Compounds" (Ger.). Field intelligence agency technical (FIAT) review of German science (1939–1946). Inorg. Chem. Part II, pp. 155–179 (1948). A useful summary as much of the material has not been published elsewhere.

B. Review—Periodic Surveys

Organometallic chemistry is currently reviewed element by element in at least three independent publications. The first in-depth review appeared as a single volume in 1965 entitled *Annual Survey of Organometallic Chemistry* (B1). This covered the literature of the previous year. After three annual issues the publication became multi-authored and was incorporated into *Organometallic Chemistry Reviews* as *Series B* (as distinct from *Series A, subject reviews*). In January 1973, the surveys became part of the *Journal of Organometallic Chemistry*. At the time of writing, coverage is complete for all elements to the end of 1973, except Si (1970); Be, Ca, Sr, Ba, Si, Pb (1971); Si (1972) and Li, Na, K, Ge, Pb, As, Sb, Bi (1973).

B1. R. B. King and D. Seyferth, eds., *Annual Survey of Organometallic Chemistry*. Elsevier, Amsterdam, vols. 1 (1965), 2 (1966), 3 (1967). *Organometallic Chemistry Reviews, Series B, Annual Surveys*, vols. 4 (1968), 5 (1969), 6 (1970) all elements except Si.
Vol. 7 (1971), Si for 1968 and 1969.
Vol. 8 (1971), all elements for 1970 except Si, Ge, Sn, and Pb.
Vol. 9 (1971), Ge, Sn, and Pb for 1970.
Journal of Organometallic Chemistry (1973–1974) (Main Group surveys only cited)*
Vol. 45, Mg, Al, Ga, In, Tl for 1971.
Vol. 48, Li, Na, K, Zn, Cd, Ge, Pb, As, Bi for 1971.
Vol. 53, Sn for 1971, B (Pt. II) for 1972.
Vol. 58, B (Pt. I) for 1972, Sn for 1972.
Vol. 62, Na, K, Be, Ca, Sr, Ba, Hg, Al, Ga, In, Tl, Pb for 1972.
Vol. 68, Li, Mg, Ge, As, Sb, Bi for 1972.
Vol. 75, Be, Ca, Sr, Ba, B (Pts I and II), Al for 1973.
Vol. 79, Tl, Sn for 1973.
Vol. 83, Zn, Cd for 1972 and 1973; Ga, In, Si for 1973.

Annual Reports, published by the Chemical Society, London (B2) carried a separate chapter entitled Organometallic Chemistry (in practice devoted to Main Group elements) for the first time in 1965. A similar chapter appeared in 1966 and in Section B (Organic Chemistry) the

* Individual volumes may be purchased from Elsevier, Amsterdam.

following year. For the years 1968–1970 (inclusive) Chapter 9, Pt. (1), in Section B dealt with Main Group organometallics. These necessarily brief reviews were superseded in 1971 by the series of *Specialist Reports* (B3) which contain more detailed accounts.

B2. *Annual Reports*. The Chemical Society, London. Volume 62 (1965), 63 (1966), and 64 (1967) B: chapters entitled "Organometallic Compounds" by A. G. Davies. Volumes 65 (1968), B, Pt. (1) and 66 (1969), B, Pt. (1)—"The Main Group Elements" by M. F. Lappert, J. D. Smith, and D. R. M. Walton. Volume 67 (1970), B, Pt. (1)—"The Main Group Elements" by D. J. Cardin, M. F. Lappert, J. D. Smith, and D. R. M. Walton.

B3. *Specialist Periodical Reports*. The Chemical Society, London.

3.1 E. W. Abel and F. G. A. Stone, eds., "Organometallic Chemistry." Volumes 1 (1972) and 2 (1973) contain surveys for 1971 and 1972, respectively. The following chapters in Vol. 1 refer to Main Group organometallic compounds.

Ch. 1 Group I. The alkali and coinage metals by B. C. Crosse (includes Li, Na, K).

Ch. 2 Group II. The alkaline earths and Zn and its cogeners by B. C. Crosse (Be, Mg, Ca, Sr, Ba, Zn, Cd, Hg), pp. 17–39 (155).

Ch. 3 Group III. B, Al, Ga, In and Tl by J. P. Maher, pp. 40–103 (269).

Ch. 4 Group III. The carboranes by T. Onak, pp. 104–116 (38).

Ch. 5 Group IV. The Si group (Si, Ge, Sn, Pb) by D. A. Armitage, pp. 117–183 (357).

Ch. 7 Organometallic compounds containing metal-metal bonds by J. D. Cotton, pp. 194–227 (179).

Ch. 14 4. Carborane complexes, 5. Complexes with other related ligands by R. J. Mawby, pp. 384–387 (15).

Ch. 17 X-ray and electron diffraction studies by R. F. Bryan, pp. 468–501 (255).

Volume 2 is organized on similar lines.

3.2 N. N. Greenwood, ed., "Spectroscopic Properties of Inorganic and Organometallic Compounds," vols. 1 (1968), 2 (1969), 3 (1970), 4 (1971), 5 (1972), 6 (1973). Each volume surveys the previous year under the following main chapter headings: nuclear magnetic resonance spectroscopy; nuclear quadrupole resonance spectra; electron spin resonance spectroscopy; microwave spectroscopy; vibrational spectra; electronic spectra; and Mössbauer spectroscopy.

3.3 D. H. Williams, ed., "Mass Spectrometry of Organic and Organometallic Compounds," vol. 1 (June 1968–June 1970), vol. 2 (June 1970–June 1972).

3.4 D. H. Ridd, ed., "Organic Compounds of Sulphur, Selenium and Tellurium," vol. 1 (April 1969–March 1970), vol. 2 (April 1970–March 1972).

3.5 G. A. Sim and L. E. Sutton, eds., "Molecular Structure by Diffraction Methods," vol. 1 (January 1971–March 1972), vol. 2 (April 1972–March 1973).

The Inorganic Chemistry series (11 volumes plus index volume) of the MTP International Review of Chemistry (B4) is now in production. The first set surveys the period 1967–1971, and future series are planned biennially.

B4. MTP International Review of Science, Inorganic Chemistry Series 1, Butterworth, London. Volume 4 is devoted to Main Group organometallic chemistry and vols. 1 and 6 contain relevant articles.

4.1 M. F. Lappert, ed., vol. 1, "Main Group Elements, H and Groups I–IV," contains:

Ch. 4 Carboranes and metallocarboranes by R. Snaith and K. Wade, pp. 139–183 (203).

4.2 B. J. Aylett, ed., vol. 4 "Organometallic Derivatives of the Main Group Elements."

Ch. 1 Organoalkali metal compounds by P. West, pp. 1–40 (212).
Ch. 2 Organo derivatives of Zn, Cd and Hg by K. C. Bass, pp. 41–72 (140).
Ch. 3 B by K. Niedenzu, pp. 73–104 (305).
Ch. 4 Al, Ga, In, Tl by J. B. Farmer and K. Wade, pp. 105–140 (251).
Ch. 5 Si, Pt. 1 by V. Chvalovský, pp. 141–204 (580).
Ch. 6 Si, Pt. 2 by H. Bürger, pp. 205–246 (310).
Ch. 7 Ge, by E. J. Bulten, pp. 247–274 (166).
Ch. 8 Sn, Pb, by A. J. Bloodworth, pp. 275–354 (1358).
Ch. 9 Organic compounds of As, Sb, and Bi, by J. P. Crow and W. R. Cullen, pp. 355–412 (288).

4.3 M. J. Mays, ed., vol. 6 "Transition Metals," Part 2, contains:

Ch. 3 Organometallic complexes containing Group III (B to Tl) and Group IV (Si to Pb) ligands by F. Glockling and S. R. Stobart, pp. 63–120 (282).
Ch. 4 Transition metal complexes containing P, As, Sb and Bi donor ligands by S. D. Robinson, pp. 121–169 (474).

Other annual surveys, although primarily concerned with other themes, contain substantial sections on organometallic compounds:

B5. "Organic Reactions Mechanisms," Wiley (Interscience,) London. Literature reviews for 1966 and 1967 by B. Capon, M. J. Perkins, and C. W. Rees: Chapter 3 is entitled "Electrophilic Aliphatic Substitution." Subsequent volumes (eds. Capon and Rees) each contain a Chapter 3 by J. M. Brown (1968, 1969, and 1970) and by D. C. Ayres (1971 and 1972) entitled "Carbanions and Electrophilic Aliphatic Substitution" which is generally useful, as are Chapters 7 by A. R. Butler (1968–1972) entitled "Electrophilic Aromatic Substitution" which contains subsections on metal cleavage and on metallation.

B6. K. Jones and E. F. Mooney, eds., "Annual Reports on NMR Spectroscopy." Academic Press, New York, contains material on organometallic compounds.

C. *Reviews—Continuing Series*

Several review publications are concerned with organometallic compounds in general (C1–C3) or with specific Main Group elements (C4, C5; see also A27, A44, A74, A121).

C1. *Organometallic Chemistry Reviews*, Elsevier, Amsterdam, vols. 1, 2 (1966–1967); *Series A*, *Subject reviews*, vols. 3–8 (1968–1972). Beginning in 1973 subject reviews were incorporated into the *Journal of Organometallic Chemistry*.

C2. F. G. A. Stone and R. West, eds., *Advances in Organometallic Chemistry*, Academic Press, New York, vols. 1 (1964), 2 (1964), 3 (1965), 4 (1966), 5 (1967), 6 (1968), 7 (1968), 8 (1970), 9 (1971), 10 (1972), 11 (1973), 12 (1974).

C3. E. J. Becker and M. Tsutsui, eds., *Organometallic Reactions*, Wiley, New York, vols. 1 (1970), 2 (1970), 3 (1972), 4 (1973), 5 (1974).

C4. *International Quarterly Scientific Reviews Journal*, Reviews on Si, Ge, Sn, and Pb compounds, Freund, Tel Aviv, Israel, vol. 1 (1972).

C5. H. Steinberg and A. L. McCloskey, eds., *Progress in Boron Chemistry*, Pergamon, Oxford, vol. 1 (1964); R. J. Brotherton and H. Steinberg, eds., vols. 2 (1970); 3 (1971).

Articles of direct relevance to organometallic chemistry of the Main Group elements are also published in various series devoted to inorganic (C6, C7) or organic chemistry or polymer chemistry (C19, C20).

C6. H. J. Emeléus and A. G. Sharpe, eds., *Advances in Inorganic Chemistry and Radiochemistry*, Academic Press, New York, vols. 1 (1959), 2 (1960), 3 (1961), 4 (1962), 5 (1963), 6 (1964), 7 (1965), 8 (1966), 9 (1966), 10 (1967), 11 (1968), 12 (1969), 13 (1970), 14 (1972), 15 (1972), 16 (1974).

C7. F. A. Cotton, ed., *Progress in Inorganic Chemistry*, Wiley (Interscience), New York, vols. 1 (1959), 2 (1960), 3 (1962), 4 (1962), 5 (1963), 6 (1964), 7 (1966), 8 (1967), 9 (1968), 10 (1968); S. J. Lippard, ed., vols. 11 (1970), 12 (1970); J. O. Edwards, ed., 13 (1970); S. J. Lippard, ed., vols. 14 (1971), 15 (1972), 16 (1972); J. O. Edwards, ed., vols. 17 (1972), 18 (1973). Volumes 13 and 17 constitute parts (1) and (2), respectively, of *Inorganic Reaction Mechanisms*.

Most of the remaining entries in the appendix have come from the following journals (C8–C12).

C8. *Chemical Reviews*, The American Chemical Society, vols. 1–53 (1925–1953); annual volumes from vol. 54 (1954 ff.).

C9. *Quarterly Reviews*, The Chemical Society, London, vols. 1–25 (1947–1971). This has been succeeded by *Chemical Society Reviews* beginning with vol. 1 (1972).

C10. *Angewandte Chemie*, Gesellschaft Deutscher Chemiker, vols. 1–82 (1889–1970); English translation (*International Edition in English*), Verlag Chemie, Weinheim/ Academic Press, New York, vols. 1–12 (1962–1973) (with different pagination).

C11. *Accounts of Chemical Research*, The American Chemical Society, vols. 1–6 (1968–1973).

C12. *Uspekhi Khimii*, Nauka, Moscow, vols 1–42 (1932–1973). English translation (*Russian Chemical Reviews*) published since 1960 by the Chemical Society, London, vols. 29–42 (1960–1973) (with different pagination).

Two bibliographies of review articles (C13, C14) are a useful source of material on topics such as reactions of Grignard and allied reagents with particular functional groups.

C13. Bibliography of Chemical Reviews, Chemical Abstracts Service (1958–1962). Collected abstracts of reviews, arranged by the usual *Chemical Abstracts* sections.

C14. D. A. Lewis, "Index of Reviews—Organic Chemistry," I.C.I. Ltd., Plastics Division, Welwyn Garden City, Herts, England. The first issue (1968) and the supplements (1969 and 1970) have been published in a cumulative volume (1971—covers literature to December 1970), and annual supplements for 1972 and 1973 are now available. Triennial cumulative volumes are planned.

A number of publications planned as a whole but scheduled to appear as series spread over a number of years cover general aspects of organometallic chemistry (C15–C17). Other series with various themes have large sections devoted to organometallic chemistry (C18–C20).

C15. R. Belcher and S. M. W. Anderson, eds., *The Analysis of Organic Materials*, Academic Press, New York. No. 4 in this international series of monographs is entitled "Chemical Analysis of Organometallic Compounds." Vol. 1 (1973) by T. R. Crompton contains the following chapters:

Group 1A elements: Li, Na, and K—organolithium compounds.

Group 2A elements: Be, Mg, and Ca—organoberyllium and organocalcium compounds.

Group 2B elements: Cd and Hg—organozinc and organomercury compounds.

Group 3B elements: B, Al, Ga, In and Tl—organoboron, organoaluminium and, organolithium compounds.

Two further volumes are planned.

C16. M. Tsutsui, ed., *Characterisation of Organometallic Compounds*, Wiley (Interscience), New York. Vol. 26 in a series of monographs on analytical chemistry.

16.1 Vol. 26.1 (1969) contains:

F. K. Cartledge and H. Gilman: Introduction to organometallic chemistry, pp. 1–33 (183).

O. Schwarzkopf and F. Schwarzkopf: Chemical characterisation of organometallic compounds, pp. 35–72 (67).

K. Nakamoto: Characterisation of organometallic compounds by infrared spectroscopy, pp. 73–135 (186).

R. W. Kiser: Mass spectroscopy of organometallic compounds, pp. 137–211 (267).

N. C. Baenziger: The determination of organometallic structures by X-ray diffraction, pp. 213–276 (280).

R. Varma: Characterisation of metalloid and organometallic compounds by microwave spectroscopy, pp. 277–314 (73).

R. H. Herber: Characterisation of organometallic compounds by Mössbauer spectroscopy, pp. 315–340 (46).

16.2 Vol. 26.2 (1971) contains:

R. G. Kidd: Nuclear magnetic resonance spectroscopy of organometallic compounds, pp. 373–437 (164).

L. N. Mulay and J. T. Dehn: Magnetic susceptibility—characterisation and elucidation of bonding in organometallics, pp. 439–480 (43).

F. J. Smentowski: Characterisation of organometallic compounds by electron spin resonance, pp. 481–651 (1236).

W. T. Reichle: Preparation, physical properties, and reactions of sigma-bonded organometallic compounds, pp. 653–826 (696).

C17. D. B. Denny, ed., *Techniques and Methods of Organic and Organometallic Chemistry*, Dekker, New York, 1969. The first volume of the series contains one article of general interest, namely:

A. C. Bond: High vacuum techniques, pp. 32–50 (9).

C18. C. H. Bamford and C. F. H. Tipper, eds., *Comprehensive Chemical Kinetics*, Elsevier, Amsterdam.

18.1 Vol. 4 (1972), Decomposition of inorganic and organometallic compounds, contains:
K. H. Hermann and A. Haas: Kinetics of homogeneous decomposition of hydrides.
S. J. W. Price: The decomposition of metal alkyls, aryls, carbonyls and nitrosyls.

18.2 Vol. 7 (1972), Reactions of metallic salts and complexes, and organometallic compounds contains:
C. H. Langford and M. Parvis: Reactions in solution between various metal ions of the same element in different oxidation states.

18.3 Vol. 12 (1973), Electrophilic substitution at a saturated carbon atom. This volume (270 pp.), written entirely by M. H. Abraham, contains an authoritative and detailed discussion of mechanisms of reactions of carbon(sp^3)-metal bonds.

18.4 Vol. 13 (1973), Reaction of aromatic compounds, contains:
R. Taylor: Kinetics of electrophilic aromatic substitution. Sections 7 (Mercuration) and 9 (Protodemetallation) are noteworthy for their comprehensive coverage.

C19. H. F. Mark and E. H. Immergut, eds., *Polymer Reviews*, Wiley (Interscience), New York; see vols. 2 (A163), 3 (A164), 8 (A178), and 12 (A166).

C20. H. F. Mark, C. S. Marvel, and H. W. Melville, eds., *High Polymers*, Wiley, New York; see vols. 24 (A170), 26 (A105), and 28 (A172).

D. Abstracting and Primary Journals

Chemical Abstracts (D1) is undoubtedly the most convenient and comprehensive reference source, particularly since 1962 when a section devoted to organometallic and organometalloidal compounds was introduced. Information concerning books, review articles, and conferences is, however, extremely brief, and usually the subject matter of a particular entry must be judged solely from the title. Comparable publications in other languages include *Chemisches Informationsdienst* (D2) and *Referativnyi Zhurnal, Khimya.*

D1. *Chemical Abstracts*, vols. 1–43 (1907–1949); in vols. 44–52 (1950–1958), see sections 6 (Inorganic chemistry) and 10 (Organic chemistry); from vol. 53 (1959) section 10 was regrouped; from vol. 56 (1962), see sections 8 (Crystallization and crystal structure), 14 (Inorganic chemicals and reactions) and 33 (Organometallic and organometalloidal compounds); from vol. 58, section 33 was renumbered as 39; from vol. 66 (1967) the sections were again renumbered: (old numbering in parentheses) 29 (39), 70 (8), 78 (14). At present published weekly, 2 volumes per annum.

D2. *Chemisches Informationsdienst*, Bayer AG., Germany, weekly. Brief abstracts plus author and keyword index. The section entitled "Organoelementverbindungen" is particularly relevant.

One publication, *Organometallic Compounds* (D3), contains brief abstracts of papers under various section headings.

D3. *Organometallic Compounds*, R. H. Chandler Ltd., London. Vol. 1 (1962), then vols. 2–23 (1963–1973) at two volumes per annum.

An excellent series of abstracts covering the use of organolithium compounds in organic synthesis (D4) first appeared in 1949. Unfortunately publication ceased in 1963, but the eleven volumes available contain the most comprehensive list of organolithium reagents yet published.

D4. *Annotated Bibliography on the Use of Organolithium Compounds in Organic Synthesis*, Lithium Corporation of America, Rand Tower, Minneapolis, Minnesota. First issue 1949 (223 abstracts of papers from previous years), then nine supplements: 1 (1950), 2 (1952), 3 (1954), 4 (1956), 5 (1958), 6 (1959), 7 (1960), 8 (1961), 9 (1962) followed by vol. 11* (1963), each containing abstracts of articles from the previous year.

Organometallic chemistry has had its own journal since 1963 (D5), the frequency of issues reflecting the massive growth in subject matter. Another journal (D6) covering the use of organometallic compounds in synthesis, ceased publication after one volume.

D5. *Journal of Organometallic Chemistry*, Elsevier, Amsterdam. Vol. 1 (1963–1964), 2 (1964), 3, 4 (1965), 5, 6 (1966), 7–10 (1967), 11–15 (1968), 16–20 (1969), 21–25 (1970), 26–33 (1971), 34–46 (1972), 47–63 (1973), now published weekly. Volume 20 contains an index for vols. 1–20 and vol. 49, no. 2 a similar index for vols. 21–50.

D6. *Organometallics in Chemical Synthesis*, Elsevier, Amsterdam, vol. 1 (1970–1971).

Several journals published in Russian are translated into English from cover to cover. The most important for the organometallic chemistry of the Main Group elements are *Proceedings of the Academy of Sciences of the U.S.S.R.* (*Doklady Akademii Nauk SSSR*) published since 1956, *Bulletin of the Academy of Sciences of the USSR* (*Izvestiya Akademii Nauk SSSR, Seriya Khimicheskaya*) published since 1952 and *Journal of General Chemistry* (*Zhurnal obshchei khimii*) published since 1949, all by Consultants Bureau, New York, and *Russian Journal of Inorganic Chemistry* (*Zhurnal neorganicheskoi khimii*) published since 1959 and *Russian Chemical Reviews* (*Uspekhi khimii*) published since 1960 by The Chemical Society, London. Pagination is different in Russian and English versions. Before the complete cover-to-cover translations were begun, Consultants Bureau published translations of selected papers from Russian authors. There were several collections of articles on organosilicon and organoboron chemistry.

* Erroneous numbering; should be vol. 10.

E. Conference Reports

The meetings of the widest interest to Main Group organometallic chemists are the International Conferences on Organometallic Chemistry (ICOMC) (E1) held every two years since 1963. Apart from the first two, these Conferences have been held under the auspices of the International Union of Pure and Applied Chemistry (IUPAC). The main lectures have been published in the journal *Pure and Applied Chemistry*, and as separate volumes, produced by Butterworth (London). Details of lectures on Main Group organometallic compounds are given below.

E1. International Conferences on Organometallic Chemistry

- 1.1 Cincinnati, Ohio, 1963: Theme of the first conference "Current Trends in Organometallic Chemistry." For titles and authors of papers, see *J. Organometal. Chem.* **1**, 205–207 (1964).
- 1.2 Madison, Wisconsin, 1965: For titles and authors of papers, see *J. Organometal. Chem.* **4**, 421–425 (1965).
- 1.3 Munich, Germany, 1967: Plenary lectures: *Pure Appl. Chem.* **17**, 179–272 (1968) and Butterworth, London (1968). Of relevance to Main Group chemistry: M. F. Hawthorne: Recent advances in the chemistry of polyhedral complexes derived from transition metals and carboranes, pp. 195–210 (27).
- 1.4 Bristol, England, 1969: Plenary lectures, F. G. A. Stone and M. I. Bruce, eds., *Pure Appl. Chem.* **23**, 375–503 (1970) and Butterworth, London (1970).
  - 1.4.1 D. Seyferth: Divalent carbon insertions into Group IV hydrides and halides, pp. 391–412 (70).
  - 1.4.2 W. P. Neumann: Recent developments in the field of organic derivatives of Group IVB elements, pp. 433–446 (20).
  - 1.4.3 T. L. Brown: Structures and reactivities of organolithium compounds, pp. 447–462 (43).
- 1.5 Moscow, 1971: Plenary lectures, Z. N. Parnes, ed., *Pure Appl. Chem.* **30**, 335–365 (1972) and Butterworth, London (1972).
  - 1.5.1 G. J. M. van der Kerk: Organozinc coordination chemistry and catalytic effects of organozinc coordination compounds, pp. 389–408 (11).
  - 1.5.2 I. F. Lutsenko: O- and C-isomeric organoelement derivatives of keto-enol systems: rearrangements and elementotropism, pp. 409–425 (25).
  - 1.5.3 J. Nasielski: Two aspects of penta-coordination in organometallic chemistry, pp. 449–462 (60).
  - 1.5.4 H. Normant: Organomagnesium chemistry in France after Grignard (Fr.), pp. 463–498 (101).
  - 1.5.5 R. Okawara: Some recent advances in the organometallic chemistry of thallium, pp. 499–508 (40).
  - 1.5.6 S. Pasynkiewicz: Reactions of organoaluminium compounds with electron donors, pp. 509–521 (29).
  - 1.5.7 L. J. Todd: Recent developments in the study of carboranes, pp. 587–598 (27).
  - 1.5.8 T. G. Traylor *et al.*: σ–π Conjugation: occurrence and magnitude, pp. 599–606 (9).

Lectures 1, 3, 5, and 6 were also published in *Usp. Khim.* **41**, 1161–1241 (1972).*

1.6 Amherst, Massachusetts, 1973: Abstracts of papers, eds. M. D. Rausch and S. A. Gardner, University of Massachusetts (1975).

1.6.1 H. Schmidbaur: Organometallic anion salts and ylides.

1.6.2 P. Timms, The direct preparation of organometallic compounds from atoms or small molecules.

Symposia on organosilicon chemistry (E2) have also been organized at regular intervals. The plenary lectures from some conferences, sponsored by IUPAC, are published in *Pure Appl. Chem.* and are also available in separate Butterworth publications. Work by Russian scientists on organosilicon chemistry has been described at several conferences and detailed volumes of proceedings (E3) have appeared.

E2. International Symposia on Organosilicon Chemistry

2.1 Prague, Czechoslovakia, 1965: Special lectures, eds. V. Chvalovský, F. Mareš, and J. Hetflejš, *Pure Appl. Chem.* **13**, 1–327 (1966) and Butterworth, London (1966).

2.1.1 R. West: Recent advances on two classical problems of organosilicon chemistry, pp. 1–13 (23).

2.1.2 M. Schmidt: Analogies and differences between organometallic compounds of Si, Ge and Sn (Ger.), pp. 15–33 (31).

2.1.3 M. G. Voronkov: Silatranes: intra-complex heterocyclic compounds of pentacoordinated silicon, pp. 35–59 (85).

2.1.4 R. Calas: Aspects of silicon hydride reactivity in organic chemistry (Fr.), pp. 61–79 (36).

2.1.5 J. Goubeau: Heterocyclic ring-systems containing silicon (Ger.), pp. 81–91 (9).

2.1.6 W. Sundermeyer: Preparation of organosilicon halides in molten salts as reaction media, pp. 93–99 (16).

2.1.7 W. Noll: Spreading behaviour and acidolysis of the siloxane linkage as varying with the donor-acceptor properties of the organic substituents, pp. 101–110 (8).

2.1.8 R. C. Mehrotra: Synthesis and properties of alkoxy- and acyloxy-silanes, pp. 111–131 (90).

2.1.9 R. A. Benkeser: Silane addition reactions—their synthetic utility and mechanism, pp. 133–140 (20).

2.1.10 R. N. Meals: Hydrosilation in the synthesis of organosilanes, pp. 141–157 (84).

* In the remainder of this guide, we have given the titles of reviews and articles in English and have shown the language used in the original by the *Chemical Abstracts* nomenclature i.e., Cz., Czech; Dan., Danish; Engl., English; Fr., French; Ger., German; Hung., Hungarian; Ital., Italian; Japan., Japanese; Pol., Polish; Rom., Romanian; Russ., Russian; Span., Spanish. Where no language is given the article is in English. Where reviews have appeared in more than one journal in more than one language, the language is the same as that of the journal title. Otherwise the original language is English.

2.1.11 D. Seyferth, G. Singh, and R. Suzuki: Steric effects and π-bonding in organosilicon chemistry: their assessment by means of comparative organic chemistry of carbon and silicon, pp. 159–166 (16).

2.1.12 M. Kumada: Recent research on organopolysilanes, pp. 167–187 (45).

2.1.13 E. A. V. Ebsworth: Structural chemistry of Si–H compounds, pp. 189–202 (47).

2.1.14 H. Kriegsmann: New spectroscopic investigations of bonding in organosilicon compounds (Ger.), pp. 203–213 (4).

2.1.15 A. G. Brook: Reactions and rearrangements of carbon-functional organosilicon compounds, pp. 215–229 (15).

2.1.16 V. Chvalovský: Problems of bond utilization of silicon d-orbitals, pp. 231–245 (9).

2.1.17 E. G. Rochow: Polymeric methylsilazanes, pp. 247–262 (4).

2.1.18 U. Wannagat: Novel ways in the preparation of cylic silicon-nitrogen compounds, pp. 263–279 (39).

2.1.19 G. Fritz: Preparation and properties of carbosilanes (Ger.), pp. 281–295 (16).

2.1.20 R. A. Shaw: The radiation chemistry of silicon compounds, pp. 297–312 (57).

2.1.21 V. Bažant: Direct synthesis of organohalogenosilanes, pp. 313–327 (25).

2.2 Bordeaux, France, 1968: Plenary lectures, *Pure Appl. Chem.* **19**, 291–538 (1969) and Butterworth, London (1969).

2.2.1 R. West: Anionic rearrangements of organosilicon compounds, pp. 291–307 (34).

2.2.2 K. A. Andrianov: On polymerisation of organosilicon cyclic compounds, pp. 309–327 (9).

2.2.3 U. Wannagat: N-metallated silicon-nitrogen derivatives: Preparation, structure and reactions, pp. 329–342 (0).

2.2.4 D. R. Weyenberg and W. H. Atwell: Divalent silicon intermediates in the pyrolysis of alkoxypolysilanes, pp. 343–351 (16).

2.2.5 G. A. Razuvaev and N. S. Vyazankin: Organosilicon and organogermanium derivatives with Si–metal and Ge–metal bonds, pp. 353–374 (72).

2.2.6 C. Eaborn: Some recent studies of the cleavages of carbon-silicon and related bonds, pp. 375–388 (35).

2.2.7 R. A. Benkeser: The chemistry of trichlorosilane in the presence of tertiary amines, pp. 389–397 (15).

2.2.8 M. G. Voronkov: Biologically-active organosilicon compounds (Fr.), pp. 399–416 (203).

2.2.9 H. Kriegsmann *et al.*: Intra- and inter-molecular exchange effects in organosilicon compounds (Ger.), pp. 417–430 (15).

2.2.10 A. G. MacDiarmid *et al.*: Properties of silicon derivatives of Co, Mn and Fe carbonyls, pp. 431–448 (73).

2.2.11 H. Gilman *et al.*: Silylations of some polyhalogenated compounds, pp. 449–472 (65).

2.2.12 V. Bažant: On the mechanism of the direct synthesis of organohalogenosilanes, pp. 473–488 (0).

2.2.13 E. Frainnet: Recent work on the reactivity of organosilicon hydrides (Fr.), pp. 489–523 (74).

2.2.14 L. H. Sommer: Mechanistic pathways of the Si–H bond—stereochemical studies, pp. 525–538 (30).

2.3 Madison, Wisconsin, 1972: Abstracts were available only to conference participants. There are no plans to publish details of plenary lectures. However, a list of titles and lecturers is given below.

2.3.1 D. R. Bennett: Metabolism of organosilicon compounds in animal systems.
2.3.2 H. Bock: Photoelectron spectra of organosilicon compounds.
2.3.3 A. G. Brook: Photochemistry of acylsilanes.
2.3.4 V. Chvalovský: Carbofunctional alkylsilanes, their preparation and properties.
2.3.5 I. M. T. Davidson: Gaseous reaction intermediates in organosilicon chemistry.
2.3.6 J. Dunogues: Magnesium in organosilicon chemistry.
2.3.7 C. L. Frye: Ring-chain equilibria in organosilicon chemistry.
2.3.8 M. Ishikawa and M. Kumada: Skeletal rearrangement of methylpolysilanes.
2.3.9 K. Itoh: Some recent researches on the addition reactions of silicon-nitrogen bonds and related reactions.
2.3.10 R. A. Jackson: Radical and molecular reactions of organosilicon compounds.
2.3.11 M. Kumada: Some reactions involving silicon and metal complexes of the nickel triad.
2.3.12 M. F. Lappert: Silyl- and silylmethyl-derivatives of transition metals.
2.3.13 A. G. MacDiarmid: Even-electron paramagnetic silicon compounds.
2.3.14 J. L. Margrave: Chemistry of divalent silicon.
2.3.15 K. Mislow: Pyramidal inversion barriers of silicon compounds.
2.3.16 H. Sakurai: Reactions of silyl radicals and anions in solution.
2.3.17 H. Schmidbaur: The role of silicon in ylid chemistry.
2.3.18 P. S. Skell: Reactive silicon intermediates.
2.3.19 M. G. Voronkov: Intramolecular interaction between the silicon atom and geminal substituents in organosilicon compounds ("α-effect").
2.3.20 D. R. M. Walton: Organosilicon compounds in the service of acetylene chemistry.
2.3.21 N. Wiberg: Silyl derivatives of diimine.

E3. Russian Conferences on Organosilicon Chemistry

3.1 Conference entitled "Chemistry and Practical Applications of organosilicon compounds," Leningrad, 1958. The proceedings (Russ.) were published in 6 volumes by the Academy of Sciences USSR (1961). Almost all papers are by Russian authors.

3.2 The 38 papers (Russ.) presented at another conference (Leningrad, 1966) with the same title as 3.1 were published by Khimiya, Leningrad, 1968 (251 pp.). The three sections deal with monomers, polymers, and materials and applications.

3.3 Conference on Organosilicon Compounds, 1966. Published in 5 volumes by Niiteckhim, Moscow, 1966 (Russ.).

Other IUPAC-sponsored regular meetings occasionally contain lectures of interest to main-group chemists, notably the Macromolecular

Chemistry series (E4) and International Conferences on Coordination Chemistry (ICCC) (E5), and the International Conferences on Pure and Applied Chemistry (E6).

E4. Conferences on Macromolecular Chemistry
  4.1 Prague, Czechoslovakia, 1965. Plenary lectures published in *Pure Appl. Chem.* **12**, 1–643 (1965). Lectures dealing with anionic polymerization and Ziegler-Natta catalysts.
  4.2 Helsinki, Finland, 1972. Plenary lectures:
    3.2.1 J. P. Kennedy: Alkylaluminium compounds in carbonium ion polymerization.
    3.3.2 P. H. Plesch: Development in the theory of cation polymerisation, A new theory of initiation by aluminium halides.

E5. Fourteenth International Coordination Chemistry Conference, Toronto, 1972. Plenary lectures, ed. A. B. P. Lever, published in *Pure Appl. Chem.* **33**, 453–652 (1973).
  M. F. Hawthorne: New routes to, and reactions of, polyfunctional transition metal carborane species, pp. 475–488 (30).

E6. Seventeenth International Conference on Pure and Applied Chemistry.
  E. G. Rochow: From structure to synthesis of organometallic compounds. *Pure Appl. Chem.* **1**, 136 (1959).

Details of other international and smaller meetings, together with relevant lectures are given below (E7–E20).

E7. Symposium on Metal-Organic Compounds, Industrial Chemistry Division, 131st National ACS Meeting, Miami, Florida, 1957. Thirty-six articles mostly on Main Group organometallics published in *Advan. Chem. Ser.* No. 23 (371 pp.) ACS Washington, D.C., 1959.

E8. Symposium at Inorganic Chemistry Division, 133rd National ACS Meeting, San Francisco, California, 1958 and at ACS Meeting, Boston, Massachussetts, 1959. Published in *Advan. Chem. Ser.* No. 32. ACS, Washington, D.C., 1961.
  8.1 R. M. Washburn, F. A. Billig, M. Bloom, and C. F. Allright: Organoboron compounds.
  8.2 S. L. Clark, J. R. Jones, and H. Stange: Synthesis of B-C ring compounds.
  8.3 L. F. Hohnstedt and G. W. Schaeffer: Borazine chemistry.

E9. The thirty-fourth Priestly lectures (1960) Penn. State University, were given by E. G. Rochow on "Unnatural products: new and useful materials from silicon."

E10. Summer School of Organic Chemistry, Shillong, India, 1961.
  10.1 B. C. Subba Rao: Organoboranes and the hydroboration reaction, pp. 73–79 (23).
  10.2 S. V. Sunthankar: Some new reactions of organometallic compounds and their synthetic applications. [Review of benzyne intermediates and reactions of organometallics with conjugated enol ethers.] pp. 117–127 (27).

E11. International Organometallic Conference, Paris, 1962. Proceedings published in *Bull. Soc. Chim. Fr.* 1345–1508 (1962). Many of the papers are in French and deal with the chemistry of organolithium and organomagnesium compounds.

E12. International Symposium sponsored by the US Army Research Office, Duke University, Durham, North Carolina, 1963. Lecture by K. Niedenzu entitled

"Boron-nitrogen chemistry." Published in *Advan. Chem. Ser.* No. 42. ACS, Washington, D.C., 1964.

E13. Industrial syntheses and applications of organometallics. A conference held by the New York Academy of Sciences (1964). The lectures were published in *Ann. N.Y. Acad. Sci.* **125**, Art. 1, 1–248 (1965).

13.1 W. J. Considine: Alkylation (arylation) of metal halides with organometallics, pp. 4–11 (11).

13.2 E. M. Marlett: Electrochemical synthesis of organometallics, pp. 12–24 (121).

13.3 G. J. M. van der Kerk and J. G. Noltes: Hydride additions, pp. 25–42 (31).

13.4 R. E. Dessy, T. Psarras, and S. Green: Redistribution reactions, pp. 43–56 (48).

13.5 A. Ross: Industrial applications of organotin compounds, pp. 107–123 (39).

13.6 H. Lehmkuhl: Organoaluminium compounds, pp. 124–136 (94).

13.7 R. N. Meals: Silicones, pp. 137–146 (44).

13.8 J. Kollonitsch: Industrial use of organometallics in organic synthesis, pp. 161–171 (27).

13.9 M. C. Henry and W. E. Davidsohn: Organometallic polymers, pp. 172–182 (70).

13.10 J. J. Smith, W. L. Carrick and A. K. Ingherman: Organometallic compounds in olefin polymerisation, pp. 183–188 (19).

13.11 H. L. Yale: Therapeutic applications of organometallic compounds. pp. 189–197 (25).

13.12 E. F. Marshall and R. A. Wirth: Uses for organometallics in fuels and lubricants, pp. 198–217 (0).

13.13 H. E. Podall and M. M. Mitchell, Jr.: The use of organometallic compounds in chemical vapour deposition, pp. 218–228 (36).

13.14 R. J. Daum: Agricultural and biocidal applications of organometallics, pp. 229–241 (84).

13.15 C. Tamborski: High temperature additives. pp. 242–248 (12).

E14. Robert A. Welch Foundation Conferences on Chemical Research, held at Houston. The ninth meeting (1965) was devoted to organometallic compounds. The lectures were published in a single volume (ed. W. O. Milligan), together with verbatim reports of the subsequent formal discussions. Several dealt with Main Group chemistry.

14.1 H. Gilman: Organometallic chemistry, pp. 7–10 (0).

14.2 G. Wittig: Role of ate complexes as reaction determining intermediates, pp. 13–32 (35).

14.3 G. E. Coates: Studies on some second group elements, pp. 49–79 (84).

14.4 D. Seyferth: Halomethylmercury compounds, pp. 89–125 (87).

The seventeenth meeting (1973) included the following lectures:

14.5 H. C. Brown: Recent developments in hydroboration and organoboranes.

14.6 A. McKillop: Some applications of simple salts to organic syntheses, pp. 153–182 (0).

E15. Conference entitled "Modern Methods in Organic Synthesis" (Ital.), Frascati, 1967. Published by the Accademia Nazionale dei Lincei, Rome, 1968. Three lectures (Eng.) by H. C. Brown under the heading "Boron in Organic Synthesis" (pp. 31–103) were entitled:

15.1 Selective reductions.

15.2 Hydroboration.
15.3 Organoboranes in synthesis.

E16. International Symposium on the Decomposition of Organometallics to Refractory Ceramics, Metals, and Metal Alloys, ed. K. S. Masdiyasni, University of Dayton Press, Dayton, Ohio (1968).
16.1 W. H. Atwell and H. Gilman: Catenated organic compounds of Si, Ge, Sn, and Pb, pp. 1–28 (151).
16.2 W. Kroll and W. Naegle: Dialkyaluminium acetylacetonates, pp. 307–318 (14).
16.3 M. C. Henry: Synthesis and utilization of new organolead compounds, pp. 55–62 (14).

E17. Faraday Society Discussion No. 47 Cambridge, England, 1969, entitled "Bonding in Metal-Organic Compounds."
17.1 W. Zeil, J. Haase, and M. Dakkouri: Structures of molecules of type $Me_3XC{\equiv}CY$: X=Si, Ge; Y=H, Cl determined by electron-diffraction, pp. 149–156 (10).
17.2 D. B. Chambers, G. E. Coates, and F. Glockling: Electron-impact studies on organo-beryllium and -aluminium compounds, pp. 157–164 (3).
17.3 K. A. Levison and P. G. Perkins: Bonding in hexamethyldialuminium and related compounds, pp. 183–189 (8).

E18. W. O. George, ed., "Spectroscopic Methods in Organometallic Chemistry." Butterworth, London (1970). Proceedings of a conference, Kingston College of Technology, July, 1969.
18.1 A. J. Downs: Vibrational spectroscopy, pp. 1–32 (58).
18.2 T. C. Gibb: Mössbauer spectroscopy, pp. 33–60 (22).
18.3 W. McFarlane: Nuclear magnetic resonance, pp. 61–94 (51).
18.4 T. R. Spalding: Mass spectrometry, pp. 95–133 (86).
18.5 P. B. Ayscough: Electron spin resonance, pp. 134–177 (25).
18.6 P. G. Perkins: Electronic spectroscopy, pp. 178–193 (16).
18.7 Abstracts of 11 research papers, pp. 194–209.

E19. International Meeting on Boron Compounds, Castle Liblice, Prague, 1971. Published ed. B. Stíbr, *Pure Appl. Chem.* **29**, 493–595 (1972) and by Butterworth, London (1972).
19.1 L. I. Zakharkin: Recent advances in the chemistry of dicarba*closo*decaboranes (12), pp. 513–526 (41).
19.2 H. C. Brown and E. Negishi: Cyclic hydroboration of dienes, pp. 527–545 (49).
19.3 M. F. Hawthorne: Recent developments in the chemistry of polyhedral compounds from metals and carboranes, pp. 547–567 (32).
19.4 R. E. Williams: Carborane polymers, pp. 569–576 (0).

E20. G. J. M. van der Kerk, "Recent Trends in Organolead Chemistry" in "Lead '68." Proceedings of 3rd International Conference, Pergamon, Oxford (1969), 12 pp. (23).

F. Synthesis

Organic derivatives of zinc, first made about 1850 by Frankland, were shown to be powerful and versatile reagents in organic synthesis. After

1900, zinc compounds were almost superseded by the more easily handled Grignard reagents. Only in the last 25 years has there been wide exploitation in organic syntheses of organic derivatives of other Main Group elements, such as B, Tl, and Si. This historical development has led to a rather artificial division of subject matter in the literature. Information on Grignard, organolithium, and related reagents is best found in compendia of organic reactions (F1–F6). Syntheses involving derivatives of the less electropositive elements are often found most easily in the inorganic literature (F7–F10). Those wishing to make well-established organometallic compounds would do well to consult the indexes of compendia (F1, F8), in which the preparations have been checked by independent assessors. Articles by experts, e.g., F2, F3, and F7, in which a large number of similar preparations are compared, also contain much useful information about possible difficulties. Experimental details for many preparations of standard organometallic compounds may be found in publications such as A22, A26, A27, A32, A34, A35, A37, A57; C10; F11–F13. See also C17.

F1. "Organic Syntheses." Wiley, New York, vols. 1–53 (1921–1973). Collected vols. 1 (2nd ed., 1958), 2 (1943), 3 (1955), 4 (1963), and 5 (1973) contain syntheses (with appropriate revisions) for the previous decade. These are useful source books for operations with Grignard, organo-sodium and -lithium reagents.

F2. "Organic Reactions," Wiley, New York, vols. 1–20 (1942–73). Volume 20 contains cumulative subject and author indexes for all previous volumes.

F3. W. Foerst, ed., "Newer Methods of Preparative Organic Chemistry," Academic Press, New York, vols. 1 (1948), 2 (1963), 3 (1964), 4 (1968). These books give English translations of articles (sometimes updated) published in *Angew. Chem.* (*Angew. Chem., Int. Ed. Engl.*). C10.

F4. W. Theilheimer, ed., "Synthetic Methods of Organic Chemistry. A thesaurus." Wiley (Interscience), New York. English edition of original Ger. publication. Volume 1 (1948) covers literature for the period 1942–1944; vol. 2 (1949) covers literature for 1945–1946; vols. 3 (3rd ed., 1966) covering literature for 1946–1947 and 4 (2nd ed., 1966) covering literature for 1947–1948 are available only in Ger. All subsequent annual vols., 5 (2nd ed., 1966) covering literature 1948–1949, and 6–27 (1952–1973) are in Engl. Volumes 5, 10 (1956), 15 (1961), 20 (1966), and 25 (1971) contain cumulative indexes for the preceding five year periods.

F5. C. A. Buehler and D. E. Pearson, "Survey of Organic Synthesis." Wiley (Interscience), New York, 1971. The arrangement is based on functional groups and there are subsections on organometallics.

F6. J. McMurry and R. Bryan Miller, eds., "Annual Reports in Organic Synthesis," Academic Press, New York. Surveys for 1970, 1971, and 1972 are available. Each contains sections on organometallic reagents.

F7. W. L. Jolly, ed., "Preparative Inorganic Reactions." Wiley, New York, vols. 1–7 (1964–1971).

F8. "Inorganic Syntheses." McGraw-Hill, New York, vols. 1–13 (1939–1972). These volumes are extremely valuable for preparations of derivatives of Main Group elements, especially of B, Si, Ge, Sn, As, and Sb. Information on particular compounds is easily located in the cumulative indexes: vol. 10 (vols. 1–10) and vol. 13 (vols. 11–13).

F9. G. Marr and B. W. Rockett, "Practical Inorganic Chemistry." van Nostrand-Reinhold, Princeton, New Jersey, 1972, 444pp. Contains 36 pp. on organometallic compounds.

F10. W. L. Jo'ly, "The Synthesis and Characterization of In^rganic Compounds." Prentice-Hall, Englewood Cliffs, New Jersey, 1970.

Apart from the now defunct journal, *Organometallics in Chemical Synthesis* (D6), three other regular publications (F15–F17) contain a proportion of papers and review articles on organometallic compounds of Main Group elements.

F11. G. Brauer, "Handbook of Preparative Inorganic Chemistry," 2nd ed. Academic Press, New York, 1963. Translation from Ger. Volume 1 contains preparative details for several organometallic derivatives of main group elements.

F12. A. I. Vogel, "Practical Organic Chemistry," 3rd ed., Longmans, London, 1956.

F13. W. Caruthers, "Modern Methods of Organic Synthesis." Cambridge University Press, London and New York, 1971. Useful summary of synthetic uses of organoboranes.

F14. R. B. Wagner and H. D. Zook, "Synthetic Organic Chemistry," 3rd ed. Wiley, New York, 1961.

F15. *Synthesis*. Thieme, Stuttgart Academic Press, New York, 1969 ff, an international journal of methods in synthetic organic chemistry published monthly. Contains review articles, short communications, and a cumulative index of numbered abstracts of new synthetic methods.

F16. *Synthesis in Inorganic and Metalorganic Chemistry*, Dekker, New York, vols. 1 (1971), 2 (1972), 3 (1973).

F17. *Organic Preparations and Procedures*, Dekker, New York, vol. 1 (1969), 2 (1970), 3 (1971), 4 (1972), 5 (1973).

G. Compound and Formulae Registers

Two single-volume works (G1, G2) aim to encompass the whole of organometallic chemistry. Their coverage, however, partly owing to the method of data presentation, is somewhat uneven, noncritical, and far from comprehensive.

G1. H. C. Kaufmann: "Handbook of Organometallic Compounds." Van Nostrand, New York, 1961, 1546 pp.

G2. N. Hagihara, M. Kumada, and R. Okawara: "Handbook of Organometallic Compounds." Benjamin, New York (1968), 1043 pp. This is an English translation of a Japanese publication issued by Assakura, Tokyo.

Several more detailed indexes cover individual metals or groups of metals (G3–G6).

G3. S. T. Ioffe and A. N. Nesmeyanov: "Handbook of Magnesium-organic compounds." Pergamon, Oxford (1956), 3 vols. 2020 pp. (Russ.). Originally published by Academy of Sciences USSR Press, Moscow (1950), the work lists 13,395 reactions involving organomagnesium compounds, and products and literature.

G4. M. Dub and R. Weiss, eds.: "Organometallic Compounds: Methods of Synthesis, Physical Constants, and Chemical Reactions." Springer-Verlag, Berlin and New York. The second edition covers the literature for 1937–1964. Volume 2, Compounds of Ge, Sn and Pb (1967) 697 pp.; vol. 3, Compounds of As, Sb and Bi (1968) 925 pp.; formula index (1970) 343 pp. Literature from 1965–1968 is covered in vol. 2, first supplement (1973) 1116 pp. and vol. 3, first supplement (1972) 613 pp. This work is thorough, but uncritical.

G5. V. Bažant, V. Chvalovský, and J. Rathouský: "Chemistry of Organosilicon Compounds." Publishing house of the Czechoslovak Academy of Sciences, Prague, and Academic Press, New York. Volumes 1 and 2 (two parts) (1965); 3 and 4 (three parts) (1973). Volume 1 summarizes the chemistry of organosilicon compounds and documents the literature to September 1961 (616 pp.). Volume 2 is a formula register with the same cut-off date. Volume 3 reviews selected topics (NMR, IR and Raman spectroscopy; π-bonding; penta- and hexa-co-ordinate silicon; stereochemistry) up to December 1969 (761 pp.) and Volume 4 is a formula index for the period October 1961–December 1969 (2349 pp.).

G6. Sheng Lieh Liu: "Melting Point Table of Organosilicons." Natural Science Council, Taipai, 1968. Of little use. 520 pp.

Excellent compilations of compounds are given in several other books (e.g., A22, A27, A84, A129, A149, A150, A153) and reviews (see index, especially: acetylene compounds of groups III–V (4.9); compounds of Li (5.5), Be (6.7; 6.9); B (9.4), Al (13.5), Tl (14.4), Ge, In, Tl (14.11), Ge (22.9), Sn (23.9), As (25.2), Bi (25.6). Other compound registers deal with specific physical properties (G7–G9 and index).

G7. O. Kennard and D. G. Watson: "Molecular Structure and Dimensions." Oosthoek, Utrecht, Crystallographic Data Centre, Cambridge and International Union of Crystallography. Volume 1 (1970) on general organic crystal structures and Vol. 2 (1970), on complexes and organometallic structures, list compound names, formulae, and references for the years 1935–1969. Similar data for organic and organometallic structures up to 1972 are given in Volumes 3 (1972) and 4 (1973). Volume A1 (1974) gives molecular dimensions, including bond lengths, bond angles, and torsion angles, for about 1300 structures published during 1960–1965.

G8. L. E. Sutton, ed.: "Interatomic Distances," *Chem. Soc. Spec. Publ.* **11** (1958), supplement, *ibid.* **18** (1965). A number of organometallic compounds are listed in this critical compilation of bond lengths and angles. See also B3.5.

G9. K. Licht and P. Reich: "Literature Data for IR, Raman and NMR Spectroscopy of Si, Ge, Sn and Pb Organic Compounds." VEB Deutscher Verlag der Wissenschaften, Berlin, 1971, 623 pp. The data are comprehensive to 1966 and the introduction is in Ger., Eng., Fr., Russ., and Span.

G10. N. N. Greenwood, E. J. F. Ross, and B. P. Straughan:
"Index of Vibrational Spectra of Inorganic and Organometallic Compounds," Vol. 1 (1935–1960), Butterworth, London, 1972, 754 pp.

Acknowledgments

We thank Mrs. D. Petrak and Prof. Z. Lasocki for help with literature in Slavonic languages.

References

1. M. I. Bruce, *Advan. Organometal. Chem.* **10**, 273 (1972).
2. M. I. Bruce, *Advan. Organometal. Chem.*, **11**, 447 (1973).
3. M. I. Bruce, *Advan. Organometal. Chem.*, **12**, 380 (1974).
4. E. J. Griffith and M. Grayson, eds., "Topics in Phosphorus Chemistry," Wiley (Interscience), New York, Vols. 1–8 (1964–1973).
5. S. Trippett, Senior reporter, *Organophosphorus chemistry*, *Chem. Soc. Spec. Publ.*, Vols. 1–4 (1970–1973).
6. G. M. Kosolapoff and L. Maier, "Organic Phosphorus Compounds," 4 vols. Wiley (Interscience), New York, 1972.

Appendix

This table lists *ca.* 630 articles that provide further detailed references to the primary literature. We think we have included most of the important reviews on the organometallic chemistry of the Main Group elements published since 1950, but our coverage of articles written in languages other than English has no doubt been far from complete. We should be grateful if omissions could be drawn to our attention. Extensive cross references to earlier sections of this guide are given at the end of each subsection, which is organized according to the somewhat arbitrary classification as follows:

1. Historical and biographical
2. General reviews
3. General methods and techniques
4. Inter-group reviews
5. Lithium
6. Na, K, Rb, Cs, Be, Ca, Sr, Ba
7. Magnesium–Grignard reagents
8. Zinc, cadmium, and mercury
9. Boron—General reviews
10. Boron—boranes, hydroboration
11. Boron—carboranes
12. Boron—B–B, B–Hal, B–N, B–O, and B–S
13. Aluminum (excluding Ziegler–Natta catalysts)
14. Gallium, indium, and thallium
15. General reviews—Group IVB
16. Silicon—general, industrial synthesis, Si–Hal. Si–H bonds

17. Silicon—Si–O compounds
18. Silicon—Si–N compounds
19. Silicon—polysilanes
20. Silicon—carbosilanes, miscellaneous cyclic compounds
21. Silicon—uses in synthesis
22. Germanium
23. Tin and lead—general
24. Tin and lead—M–H, M–N, M–O, and M–S compounds
25. Arsenic, antimony, bismuth and tellurium
26. Compounds containing metal–metal bonds
27. Redistribution reactions
28. Fluorocarbon compounds
29. Stereochemistry
30. Kinetics and mechanism—heterolyses of C–M bonds
31. Kinetics and mechanism—homolytic processes, oxidation, autoxidation, organometallic peroxides
32. Kinetics and mechanism—carbenes, carbenoids
33. Kinetics and mechanism—benzynes, hetarynes
34. Catalysis and related topics—polymerization, organometallic polymers
35. Physical methods—infrared and Raman spectroscopy
36. Physical methods—nuclear magnetic resonance
37. Physical methods—mass spectrometry
38. Physical methods—Mössbauer spectroscopy
39. Physical methods—photoelectron spectroscopy
40. Physical methods—X-ray and electron diffraction studies
41. Energetics
42. Miscellaneous topics

Ref. No.	Authors	Reference	Title	No. of pages (No. of ref.)
1. *Historical and biographical*				
1.1	R. Kh. Freidlina, M. I. Kabachnik, and V. V. Korshak	*Usp. Khim.* **38**, 1539 (1969); *Russ. Chem. Rev.* **38**, 684 (1969)	New contributions to the development of organoelement and organic chemistry	15* (164)
1.2	H. Gilman	*Advan. Organometal. Chem.* **7**, 1 (1968)	Some personal notes on more than one half century of organometallic compounds	52 (46)
1.3	A. N. Nesmeyanov	*Advan. Organometal. Chem.* **10**, 1 (1972)	My way in organometallic chemistry	78 (455)
1.4	W. P. Neumann	*Naturwissenschaften* **55**, 553 (1968)	Increasing significance of organometallic chemistry (Ger.)	5 (26)
1.5	O. Yu. Okhlobystin	*Zh. Obshch. Khim.* **37**, 2376 (1967)	Development of organometallic chemistry in the USSR	17 (179)
1.6	O. Yu. Okhlobystin	*Priroda* (*Moscow*) *No.* **9**, 40 (1969)	Elements of the periodic system and organic chemistry (Russ.)	5
1.7	S. Pasynkiewicz	*Wiad. Chem.* **23**, 167 (1969)	Development of organometallic chemistry (Pol.)	16 (14)
1.8	E. G. Rochow	*Advan. Organometal. Chem.* **9**, 1 (1970)	Of time and carbon–metal bonds	19 (28)
1.9	T. Tsuruta	*Kagaku No Ryoiki* **26**, 283 (1972)	Organometallic Chemistry (Japan)	4
1.10	N. A. Vol'kenau, B. L. Dyatkin, S. T. Ioffe, V. I. Kuznetsov, L. G. Makarova, O. Yu. Okhlobystin, and R. A. Sokolik	*Razv. Org. Khim. SSSR, 1917–1967*; *Akad. Nauk SSSR, Inst. Istor. Estestvoza Tekh.* 100 (1967)	Chemistry of heteroorganic compounds (Russ.)	85 (957)
1.11	K. Ziegler	*Chem. Soc. Spec. Publ. No.* **13**, 1 (1959)	New aspects of some organometallic complex compounds	12 (7)

* Here and elsewhere the number of pages refers to the English translation.

1.12	K. Ziegler	*Advan. Organometal. Chem.* **6**, 1 (1967)	A forty year stroll through the realms of organometallic chemistry	17 (51)
	See also: A54, A62, A183, A186, A189			
2. *General Reviews*				
2.1	J. M. Barnes and L. Magos	*Organometal. Chem. Rev.* **3**, 137 (1968)	The toxicology of organometallic compounds	13 (57)
2.2	J. J. Eisch and H. Gilman	*Advan. Inorg. Chem. Radiochem.* **2**, 61 (1960)	Organometallic compounds	42 (145)
2.3	J. H. Harwood	*Chem. Process Eng.* **47**, 65 (1966)	Developments in organometallics	5 (103)
2.4	E. Hoggarth	*Advan. Sci.* **23**, 652 (1967)	Organometallic chemistry and its growing significance in industry	6 (13)
2.5	M. Kumada	*Yuki Gosei Kagaku Kyoki Shi* **20**, 177 (1962)	Chemistry of organometallic compounds (Japan)	13 (127)
2.6	M. F. Lappert	*Chem. Brit.* **5**, 342 (1969)	Organometallic chemistry	5 (46)
2.7	G. W. Marr and B. W. Rockett	*Educ. Chem.* **6**, 105 (1969)	Organometallic chemistry of main group elements	6 (14)
2.8	O. Yu. Okhlobystin	*Usp. Khim.* **36**, 34 (1967); *Russ. Chem. Rev.* **36**, 17 (1967)	The influence of coordination on the reactivity of organometallic compounds	8 (102)
2.9	A. K. Prokof'ev, V. I. Bregadze, O. Yu. Okhlobystin	*Usp. Khim.* **39**, 412 (1970); *Russ. Chem. Rev.* **39**, 196 (1970)	Intermolecular coordination in organic derivatives of the elements	17 (210)
2.10	F. G. A. Stone	*Nature (London)* **232**, 534 (1971)	Perspectives in organometallic chemistry	5 (62)
2.11	W. Wardlaw and D. C. Bradley	*Endeavour* **14**, 140 (1955)	Organic compounds of the metals	6 (19)
2.12	T. P. Whaley and A. P. Giraitis	*Chem. Eng. Progr.* **58** (8), 65 (1962)	Organometallics as vehicles for metals	3 (0)

(*Continued*)

Ref. No.	Authors	Reference	Title	No. of pages (No. of ref.)
2.13	K. Ziegler	*In* "Perspectives in Organic Chemistry" (Todd, ed.), pp. 185–213. Wiley (Interscience), New York, 1956.	New discoveries in organometallic synthesis	19 (54)
	See also: A8–A25, A183, A184, A191; B2, B4.1; C16.1, C16.2; E1.5, E1.6, E6, E13, E14			
3.	*General methods and techniques*			
3.1	K. Arakawa and K. Tanikawa	*Bunseki Kagaku* **15**, 398 (1966)	Gas chromatography of organometallic compounds (Japan)	15 (38)
3.2	G. B. Deacon	*Organometal. Chem. Rev., Sect. A* **5**, 355 (1970)	Synthesis of organometallic compounds by thermal decarboxylation	17 (67)
3.3	K. Itoh and Y. Ishii	*Kagaku (Kyoto)* 420 (1969)	Addition-elimination reactions of organometallic compounds I (Japan)	13 (179)
3.4	Y. Iwashita	*Kagaku Kogyo* **23**, 1596 (1972)	Organometallic compounds containing carbon dioxide (Japan)	7 (22)
3.5	S. Herzog and J. Dehnert	*Z. Chem.* **4**, 1 (1964)	A rational method for working in the absence of air (Ger.)	11 (14)
3.6	R. G. Jones and H. Gilman	*Chem. Rev.* **54**, 835 (1954)	Methods of preparation of organometallic compounds	56 (471)
3.7	W. Kitching	*Organometal. Chem. Rev., Sect. A* **3**, 35 (1968)	Recent aspects of oxymetallation	25 (72)
3.8	W. Kitching and C. W. Fong	*Organometal. Chem. Rev., Sect. A* **5**, 281 (1970)	Insertion of sulphur dioxide and sulphur trioxide into metal–carbon bonds	41 (83)
3.9	M. F. Lappert and J. S. Poland	*Advan. Organometal. Chem.* **9**, 397 (1970)	α-Heterodiazoalkanes and the reactions of diazoalkanes with derivatives of metals and metalloids	38 (213)
3.10	M. F. Lappert and B. Prokai	*Advan. Organometal. Chem.* **5**, 225 (1967)	Insertion reactions of compounds of metals and metalloids involving unsaturated substrates	94 (294)

3.11	A. Lattes	*Afinidad* **29**, 153 (1972)	Application of organometallic compounds to the synthesis of amines and nitrogenated heterocycles (Span.)	18 (0)
3.12	B. L. Laube and C. D. Schmulbach	*Progr. Inorg. Chem.* **14**, 65 (1971)	Inorganic electrosynthesis in nonaqueous solvents	54 (451)
3.13	H. Lehmkuhl	*Synthesis* 377 (1973)	Preparative scope of organometallic electrochemistry	20 (93)
3.14	M. Matzner, R. P. Kurkjy, and R. J. Cotter	*Chem. Rev.* **64**, 645 (1964)	The chemistry of chloroformates	23 (750)
3.15	V. D. Nefedov, M. A. Toropova, and E. N. Sinotova	*Usp. Khim.* **38**, 1913 (1969); *Russ. Chem. Rev.* **38**, 873 (1969)	Radiochemical methods for preparing organometal (metalloid) compounds	20 (147)
3.16	J. P. Oliver	*Advan. Organometal. Chem.* **8**, 167 (1970)	Fast exchange reactions of group I, II, and III organometallic compounds	42 (161)
3.17	O. A. Ptitsyna	*Probl. Org. Khim.* 73 (1970)	Production of organometallic compounds via ArN_2^+ and Ar_2I^+ salts (Russ.)	13 (65)
3.18	H. Schumann and M. Schmidt	*Angew. Chem.* **77**, 1049 (1965); *Angew. Chem., Int. Ed. Engl.* **4**, 1007 (1965)	Reactions of organometallic compounds with sulphur, selenium, tellurium, and phosphorus	7 (61)
3.19	W. Sundermeyer	*Angew. Chem.* **77**, 241 (1965); *Angew. Chem., Int. Ed. Engl.* **4**, 222 (1965)	Fused salts and their use as reaction media	16 (229)
3.20	W. Sundermeyer and W. Verbeek	*Angew. Chem.* **78**, 107 (1966); *Angew. Chem., Int. Ed. Engl.* **5**, 1 (1966)	Method for the preparation of methylmetal compounds	5 (32)
3.21	M. E. Vol'pin and I. S. Kolomnikov	*Organometal. React.* **5**	The reaction of organometallic compounds with carbon dixodide	
3.22	H. R. Ward	*Accounts Chem. Res.* **5**, 25 (1972)	Chemically induced dynamic nuclear polarization (CIDNP) I The phenomenon, examples and applications	6 (32)

(Continued)

Ref. No.	Authors	Reference	Title	No. of pages (No. of ref.)
3.23	M. Weidenbruch	*Chem. Ztg.* **97**, 355 (1973)	Synthesis of organometallic compounds by carbene insertion (Ger.)	6 (93)
3.24	D. R. Wiles	*Advan. Organometal. Chem.* **11**, 207 (1973)	The radiochemistry of organometallic compounds	46 (101)
	See also: A22, A130–A159; B3.2, B5, B6; C16, C17; E13.1			
4. *Inter-group reviews*				
4.1	H. G. Ang	*Chem. Ind.* (*London*) 863 (1969)	Aspects of bis(trifluoromethyl)amino chemistry	3 (17)
4.2	W. Beck	*Organometal. Chem. Rev., Sect. A* **7**, 159 (1971)	Metal fulminate complexes (Ger.)	41 (118)
4.3	F. Bonati	*Organometal. Chem. Rev.* **1**, 379 (1966)	Organometallic derivatives of β-diketones	10 (85)
4.4	F. Bonati	*Stereochim. Inorg. Accad. Naz. Lincei, Corso Estivo Chim., 9th, 1965*, pp. 441–453 (1967)	Organometallic derivatives of β-diketones (Ital.)	13 (46)
4.5	D. C. Bradley	*Progr. Inorg. Chem.* **2**, 303 (1960)	Metal alkoxides	59 (130)
4.6	D. C. Bradley	*Coordination Chem. Rev.* **2**, 299 (1967)	Metal oxide alkoxide (trialkylsiloxide) polymers	20 (25)
4.7	D. C. Bradley	*Advan. Inorg. Chem. Radiochem.* **15**, 259 (1972)	Metal alkoxides and dialkylamides	64 (235)
4.8	T. Chivers	*Organometal. Chem. Rev., Sect. A* **6**, 1 (1970)	Chlorocarbon and bromocarbon derivatives of metals and metalloids	64 (305)
4.9	W. E. Davidsohn and M. C. Henry	*Chem. Rev.* **67**, 74 (1967)	Organometallic acetylenes of the main groups III–V	44 (258)
4.10	M. L. H. Green and P. L. I. Nagy	*Advan. Organometal. Chem.* **2**, 325 (1964)	Allyl metal complexes	38 (106)

4.11	M. Hlevca and I. J. Itaru	*Rev. Fiz. Chim. Ser. A* **7**, 321 (1970)	Cyclopentadienyl metal compounds (Rom.)	8 (15)
4.12	M. F. Lappert and H. Pyszora	*Advan. Inorg. Chem. Radiochem.* **9**, 133 (1966)	Pseudohalides of group IIIB and IVB elements	51 (215)
4.13	L. Malatesta	*Progr. Inorg. Chem.* **1**, 283 (1959)	Isocyanide complexes of metals	98 (117)
4.14	T. A. Manuel	*Advan. Organometal. Chem.* **3**, 181 (1965)	Lewis base-metal carbonyl complexes	80 (373)
4.15	D. J. Peterson	*Organometal. Chem. Rev., Sect. A* **7**, 295 (1972)	α-Neutral heteroatom-substituted organometallic compounds	63 (172)
4.16	D. Seyferth	*Progr. Inorg. Chem.* **3**, 129 (1962)	Vinyl compounds of metals	15 (496)
4.17	O. J. Scherer	*Angew. Chem.* **81**, 871 (1969); *Angew. Chem., Int. Ed. Engl.* **8**, 861 (1969)	Elemento-organic amines and imines	16 (77)
4.18	H. Schmidbaur	*Fortschr. Chem. Forsch.* **13**, 167 (1970)	Isosteric organometallic compounds	59 (181)
4.19	H. Schmidbaur	*Advan. Organometal. Chem.*, **9**, 260 (1970)	Isoelectric species in the organophosphorus, organosilicon, and organoaluminium series	99 (461)
4.20	H. Schmidbaur	*Allg. Prakt. Chem.* **18**, 138 (1967)	Preparative and spectroscopic studies on isosteric organometallic compounds (Ger.)	7 (39)
4.21	J. S. Thayer	*Organometal. Chem. Rev.* **1**, 157 (1968)	Azide derivatives of organometallic compounds	21 (80)
4.22	J. S. Thayer and R. West	*Advan. Organometal. Chem.* **5**, 169 (1967)	Organometallic pseudohalides	55 (198)
4.23	W. Tochtermann	*Angew. Chem.* **78**, 355 (1966); *Angew. Chem., Int. Ed. Engl.* **5**, 351 (1966)	Structures and reactions of organic ate-complexes	20 (183)
	See also: A128, A143–A145, A186, A188			

(*Continued*)

Ref. No.	Authors	Reference	Title	No. of pages (No. of ref.)
5. *Lithium*				
5.1	C. Agami	*Bull. Soc. Chim. Fr.* 1619 (1970)	Aprotic solvents IV. Use of N,N,N′,N′-tetramethylethylenediamine as solvent in the chemistry of organometallic compounds (Fr.)	6 (73)
5.2	E. A. Braude	*Progr. Org. Chem.* **3**, 172 (1955)	Organic compounds of lithium	46 (201)
5.3	J. M. Brown	*Chem. Ind.* (*London*) 454 (1972)	Organolithium reagents in synthesis	3 (23)
5.4	T. L. Brown	*Advan. Organometal. Chem.* **3**, 365 (1965)	The structures of organolithium compounds	30 (77)
5.5	H. Gilman	*Org. React.* **8**, 258 (1954)	The metallation reaction with organolithium compounds	66 (170)
5.6	J. K. Hamblin	*Chem. Brit.* **5**, 354 (1969)	Reactions of alkanes in the presence of alkali metals	6 (11)
5.7	H. Heaney	*Organometal. Chem. Rev.* **1**, 27 (1965)	Grignard and organolithium reagents derived from di- and poly-halogen compounds	15 (104)
5.8	R. G. Jones and H. Gilman	*Org. React.* **6**, 339 (1951)	The halogen-metal interconversion reaction with organolithium compounds	27 (63)
5.9	M. E. Jorgenson	*Org. React.* **18**, 1 (1970)	Preparation of ketones from the reaction of organolithium reagents with carboxylic acids	97 (242)
5.10	V. Kalyanaraman and M. V. George	*J. Organometal. Chem.* **47**, 225 (1973)	Alkali metal additions to unsaturated systems	56 (612)
5.11	R. I. Katkevich and L. I. Vereschagin	*Usp. Khim.* **38**, 1964 (1969); *Russ. Chem. Rev.* **38**, 900 (1969)	Advances in the synthesis of α-ethynylcarbonyl compounds	13 (215)
5.12	R. B. King	*Advan. Organometal. Chem.* **2**, 157 (1964)	Reactions of alkali metal derivatives of metal carbonyls and related compounds	99 (251)

5.13	R. B. King	*Accounts Chem. Res.* **3**, 417 (1970)	Some applications of metal carbonyl anions in the synthesis of unusual organometallic compounds	11 (75)
5.14	G. Köbrich	*Chem. Ztg.* **97**, 349 (1973)	Chemistry of dichloromethyllithium (Ger.)	7 (31)
5.15	E. A. Kovrizhnykh and A. I. Shatenshtein	*Usp. Khim.* **38**, 1836 (1969); *Russ. Chem. Rev.* **38**, 840 (1969)	Effects of electron donor solvents on the reactivity of lithium alkyls	10 (115)
5.16	I. Kuwajima	*Yuki Gosei Kagaku Kyokai Shi.* **29**, 616 (1971)	New reagents. Dialkylcopper lithium	4 (23)
5.17	J. M. Mallan and R. L. Bebb	*Chem. Rev.* **69**, 693 (1969)	Metallations by organolithium compounds	62 (488)
5.18	H. Normant	*Angew. Chem.* **79**, 1029 (1967); *Angew. Chem., Int. Ed. Engl.* **6**, 1046 (1967)	Hexamethylphosphoramide	21 (184)
5.19	W. E. Parham and E. E. Schweizer	*Org. React.* **13**, 55 (1963)	Halocyclopropanes from halocarbenes	36 (140)
5.20	W. Reid	*Newer Prep. Methods Org. Chem.* **4**, 95 (1968)	Ethynylation reactions	44 (63)
5.21	D. Seebach	*Synthesis* 17 (1969)	Nucleophilic acylation with 2-lithium-1,3-dithianes and -1,3,5-trithianes	20 (77)
5.22	S. Tsutsumi and H. Ryang	*Kagaku Kogyo* **23**, 481 (1972)	Organic synthesis by anionic carbonyl metal complexes (Japan)	9 (27)
5.23	N. Ya. Turova and A. V. Novoselova	*Usp. Khim.* **34**, 385 (1965); *Russ. Chem. Rev.* **34**, 161 (1965)	Alcohol derivatives of the alkali and alkaline earth metals, magnesium and thallium (I)	15 (350)
5.24	G. Wittig	*Newer Prep. Methods Org. Chem.* **1**, 571 (1948)	Syntheses with organolithium compounds	21 (87)
	See also: A27.1, A28, A139, A161, A165, A166, A169, A184; B3.1, B4.2; C15; D4; E1.4, E11			
6.	*Sodium, potassium, rubidium, caesium, beryllium, calcium, strontium, and barium*			
6.1	G. A. Balueva and S. T. Ioffe	*Usp. Khim.* **31**, 940 (1962); *Russ. Chem. Rev.* **31**, 439 (1962)	Organic compounds of Be, Ca, Sr, and Ba	13 (93)

(Continued)

Ref. No.	Authors	Reference	Title	No. of pages (No. of. ref.)
6.2	R. A. Benkeser, D. J. Foster, D. M. Sauve, and J. F. Nobis	*Chem. Rev.* **57**, 867 (1957)	Metallations with organosodium compounds	28 (94)
6.3	F. Bertin and G. Thomas	*Bull. Soc. Chim. Fr.* 3951 (1971)	Coordination chemistry of Be(II). Organic compounds of Be and their complexes (Fr.)	8 (60)
6.4	E. de Boer	*Advan. Organometal. Chem.* **2**, 115 (1964)	Electronic structure of alkali metal adducts of aromatic hydrocarbons	42 (117)
6.5	H. Christensen	*Dan. Kemi* **46**, 177 (1965)	Some industrial applications for organosodium compounds (Dan.)	6 (31)
6.6	G. E. Coates	*Rec. Chem. Progr.* **28**, 3 (1967)	Some advances in the organic chemistry of Be, Mg, and Zn	21 (38)
6.7	G. E. Coates	*Quart. Rev.* **4**, 217 (1950)	Organometallic compounds of the first three periodic groups	19 (50)
6.8	G. E. Coates and G. L. Morgan	*Advan. Organometal. Chem.* **9**, 195 (1970)	Organoberyllium compounds	62 (165)
6.9	G. Eberhardt	*Organometal. Chem. Rev.* **1**, 491 (1966)	Recent developments in the catalytic applications of the organoalkali metal compounds	10 (23)
6.10	N. R. Fetter	*Organometal. Chem. Rev., Sect. A* **3**, 1 (1968)	Organoberyllium compounds	34 (92)
6.11	H. Ruschig, R. Fugmann, and W. Meixner	*Newer Prep. Methods Org. Chem.* **2**, 361 (1963)	Continuous preparation of phenylsodium	6 (4)
6.12	M. Schlosser	*Angew. Chem.* **76**, 124 (1964); *Angew. Chem., Int. Ed. Engl.* **3**, 287 (1964)	Organosodium and organopotassium compounds I	20 (218)

6.13	M. Schlosser	*Angew. Chem.* **76**, 258 (1964); *Angew. Chem., Int. Ed. Engl.* **3**, 362 (1964)	Organosodium and organopotassium compounds II	12 (162)
	See also: A26.2, A27.1, A27.2, A29, A30, A34; B1–B4; C15; E17.2			
7. *Magnesium—Grignard reagents*				
7.1	E. C. Ashby	*Bull. Soc. Chim. Fr.* 2131 (1972)	The composition of Grignard compounds in ether solvents as inferred from molecular association and nmr studies. The relevance of Grignard composition to reaction mechanisms and stereochemistry	10 (0)
7.2	R. A. Benkeser	*Synthesis* 347 (1971)	Chemistry of allyl and crotyl Grignard reagents	12 (93)
7.3	B. Blagoev and D. Ivanov	*Synthesis* 615 (1970); see also *Izv. Otd. Khim. Nauki, Bulg. Akad. Nauk* **1**, 629 (1970)	Syntheses with polyfunctional organomagnesium compounds	13 (86)
7.4	C. Blomberg	*Bull. Soc. Chim. Fr.* 2143 (1972)	Reaction mechanisms in the chemistry of organomagnesium compounds (Fr.)	6 (25)
7.5	W. A. Bonner	*Advan. Carbohydrate Chem.* **6**, 251 (1951)	Friedel Crafts and Grignard processes in the carbohydrate series	38 (68)
7.6	C. Courtot	"Traité de chimie organique" (V. Grignard, G. Dupont, and E. Locquin, eds.), Vol. 1, pp. 86–501. Masson, Paris, 1937	Combinaisons organomagnesiennes (Fr.)	416
7.7	T. Eicher	*Chem. Carbonyl Group* 621 (1966)	Reactions of carbonyl compounds with organometallic compounds	73 (192)
7.8	L. F. Elsom, J. D. Hunt, and A. McKillop	*Organometal. Chem. Rev., Sect. A* **8**, 135 (1972)	The catalytic effects of cobalt (II) chloride on the reactions of Grignard reagents	17 (65)
7.9	B. L. Erusalimskii, I. G. Krasnosel'skaya, and I. V. Kulevskaya	*Usp. Khim.* **37**, 2003 (1968); *Russ. Chem. Rev.* **37**, 874 (1968)	Organomagnesium compounds and their complexes as polymerisation initiators	12 (215)

(*Continued*)

Ref. No.	Authors	Reference	Title	No. of pages (No. of ref.)
7.10	R. C. Fuson	*Advan. Organometal. Chem.* **1**, 221 221 (1964)	Conjugate addition of Grignard reagents to aromatic systems	17 (59)
7.11	R. A. Heacock and S. Kašpárek	*Advan. Heterocycl. Chem.* **10**, 43 (1969)	The indole Grignard reagent	70 (202)
7.12	S. T. Ioffe	*Usp. Khim.* **27**, 1010 (1958)	Vinylation by means of Grignard reaction	14 (41)
7.13	J. Munch-Petersen	*Bull. Soc. Chim. Fr.* 471 (1966)	Reactions of organomagnesiums with unsaturated $\alpha\beta$ carbonyl compounds (Fr.)	10 (74)
7.14	J. G. Noltes	*Bull. Soc. Chim. Fr.* 2151 (1972)	Application of Grignard reactivity to the preparation of organometallics and organometalloids	10 (73)
7.15	H. Normant	*Ind. Chim. Belge.* **29**, 759 (1964)	The organic magnesium compounds. By an extension of the Grignard reactions (Fr.)	8
7.16	H. Normant	*Advan. Org. Chem.* **2**, 1 (1961)	Alkenylmagnesium halides	65 (88)
7.17	H. Normant	*Sci. Progr. Decouverte No.* **3434**, 10 (1971)	Organomagnesium compounds (Fr.)	9
7.18	H. Normant	*Afinidad* **29**, 141 (1972)	The preparation of organomagnesium compounds (Fr.)	13 (52)
7.19	I. Partchamazad	*Quart. Bull. Fac. Sci. Tehran Univ.* **2**, 5 (1970)	Rearrangements observed during reactions of benzylmagnesium chloride	6 (11)
7.20	R. M. Salinger	*Survey Progr. Chem.* **1**, 301 (1963)	The structure of the Grignard reagent and the mechanisms of the reactions	23 (70)
7.21	D. A. Shirley	*Org. React.* **8**, 28 (1954)	The synthesis of ketones from acid halides and organometallic compounds of magnesium, zinc and cadmium	30 (146)
7.22	K-H. Thiele and P. Z. Dunneck	*Organometal. Chem. Rev.* **1**, 331 (1966)	Electron donor-acceptor complexes of zinc, cadmium, mercury and beryllium alkyls	26 (85)

7.23	J. Villiéras	*Organometal. Chem. Rev., Sect. A* **7**, 81 (1971)	α-Haloalkyl and related Grignard reagents, preparation and synthetic utility	13 (47)
7.24	B. J. Wakefield	*Advan. Inorg. Chem. Radiochem.* **11**, 341 (1968)	Alkyl derivatives of the group II metals	83 (502)
7.25	B. J. Wakefield	*Organometal. Chem. Rev.* **1**, 131 (1966)	Recent advances in the chemistry of organomagnesium compounds (a) The nature of the Grignard reagents (b) New organomagnesium compounds	25 (162)
7.26	B. J. Wakefield	*Chem. Ind. (London)* 450 (1972)	Organomagnesium compounds in organic synthesis	3 (24)
	See also: A26.2, A27.2, A31–A34, A161; B1–B4; C15; E11; G3			
8. *Zinc, cadmium, and mercury*				
8.1	D. Breitinger and K. Brodersen	*Angew. Chem.* **82**, 379 (1970); *Angew. Chem., Int. Ed. Engl.* **9**, 357 (1970)	Development of and problems in the chemistry of mercury-nitrogen compounds	11 (111)
8.2	M. Gaudemar	*Organometal. Chem. Rev., Sect. A* **8**, 183 (1972)	Thirty years of the Reformatsky reaction (Fr.)	50 (543)
8.3	D. Gibson	*Coord. Chem. Rev.* **4**, 225 (1969)	Carbon-bonded β-diketone complexes	16 (30)
8.4	D. Grdenić	*Quart. Rev.* **19**, 303 (1965)	The structural chemistry of mercury	26 (137)
8.5	R. G. Henein	*Gyögyszereszet* **15**, 165, 201 (1971)	Organic mercury compounds (Hung.) (in two parts)	4 + 7 (0)
8.6	N. Inamoto	*Yuki Gosei Kagaku Kyokai Shi* **26**, 551 (1968)	Reactions of organomercury compounds (Japan)	12 (89)
8.7	L. G. Makarova	*Organometal. React.* **1**, 119 (1970)	Reactions of organomercury compounds 1	230 (886)
8.8	L. G. Makarova	*Organometal. React.* **2**, 335 (1971)	Reactions of organomercury compounds 2	89 (438)

(*Continued*)

Ref. No.	Authors	Reference	Title	No. of pages (No. of ref.)
8.9	F. G. Reynolds	*Ann. N.Y. Acad. Sci.* **159** (Pt. 1), 143 (1969)	Redistribution equilibriums on mercury	9 (25)
8.10	H. L. Roberts	*Advan. Inorg. Chem. Radiochem.* **11**, 309 (1968)	Some general aspects of mercury chemistry	30 (62)
8.11	E. Sebe, S. Kitamura, K. Hayakawa, and K. Sumino	*Nippon Yakurigaku Zasshi* **66**, 45 (1970)	Side reactions accompanying catalytic hydration of acetylene (Japan)	19 (21)
8.12	D. Seyferth	*Accounts Chem. Res.* **5**, 65 (1972)	Phenyl(trihalomethyl)mercury compounds: exceptionally versatile dihalocarbene precursors	9 (109)
8.13	R. L. Shriner	*Org. React.* **1**, 1 (1942)	The Reformatsky reaction	37 (109)
8.14	N. S. Zefirov	*Usp. Khim.* **34**, 1272 (1965); *Russ. Chem. Rev.* **34**, 527 (1965)	Stereochemistry of the hydroxymercuration of alkenes	10 (70)
	See also A26.3, A26.4, A35–A41, A135, A140; B1–B5; C15, C18.3; E1.5, E14.4			
9. *Boron—general reviews*				
9.1	C. E. H. Bawn and A. Ledwith	*Progr. Boron Chem.* **1**, 345 (1964)	Reactions of diazoalkanes with B compounds	24 (46)
9.2	T. D. Coyle and F. G. A. Stone	*Progr. Boron Chem.* **1**, 83 (1964)	Some aspects of the coordination chemistry of boron	84 (360)
9.3	T. L. Heying	*Progr. Boron Chem.* **2**, 119 (1970)	Polymers and clusters of B atoms	21 (38)
9.4	M. F. Lappert	*Chem. Rev.* **56**, 959 (1956)	Organic compunds of boron	106 (387)
9.5	D. S. Matteson	*Progr. Boron Chem.* **3**, 117 (1970)	Neighbouring-group effects of boron in organoboron compounds	60 (88)
9.6	Y. Proux	*Pitture Vernici* **39**, 377 (1963)	Chemistry of boron (Ital.)	7 (20)

9.7	R. Schaeffer	*Progr. Boron Chem.* **1**, 417 (1964)	NMR spectroscopy of B compounds	46 (70)
9.8	G. Schmid	*Angew. Chem.* **82**, 920 (1970); *Angew. Chem., Int. Ed. Engl.* **9**, 819 (1970)	Metal boron compounds—problems and perspectives	12 (89)
9.9	O. P. Shitov, S. L. Ioffe, V. A. Tartakovskii, and S. S. Novikov	*Usp. Khim.* **39**, 1913 (1970); *Russ. Chem. Rev.* **39**, 905 (1970)	Cationic boron complexes	18 (189)
9.10	A. H. Soloway	*Progr. Boron Chem.* **1**, 203 (1964)	Boron compounds in cancer therapy	32 (128)
9.11	A. V. Topchiev, A. A. Prokhorova, and M. V. Kurashev	*Usp. Khim.* **33**, 1033 (1964); *Russ. Chem. Rev.* **33**, 453 (1964)	Unsaturated organoboron compounds	10 (124)
9.12	V. A. Zamyatina and N. I. Bekasova	*Usp. Khim.* **30**, 48 (1961); *Russ. Chem. Rev.* **30**, 22 (1961)	Polymeric compounds of boron	6 (100)
9.13	V. A. Zamyatina and N. I. Bekasova	*Usp. Khim.* **33**, 1216 (1964); *Russ. Chem. Rev.* **33**, 524 (1964)	Polymeric boron compounds	9 (120)
	See also: A26.1, A42, A44, A47, A48, A53, A57; B1–B4; C5, C15, E8			
10. *Boron—boranes, hydroboration*				
10.1	H. C. Brown	*Accounts Chem. Res.* **2**, 65 (1969)	Organoborane-carbon monoxide reactions. A new versatile approach to the synthesis of carbon structures	8 (38)
10.2	H. C. Brown	*Chem. Brit.* **7**, 458 (1971)	Versatile organoboranes	8 (64)
10.3	H. C. Brown and M. M. Midland	*Angew. Chem.* **84**, 702 (1972); *Angew. Chem., Int. Ed. Engl.* **11**, 692 (1972)	Organic syntheses via free-radical displacement reactions of organoboranes	9 (55)
10.4	H. C. Brown and M. M. Rogic	*Organometal. Chem. Syn.* **1**, 305 (1972)	Organoboranes as alkylating and arylating agents	23 (75)

(Continued)

Ref. No.	Authors	Reference	Title	No. of pages (No. of ref.)
10.5	H. C. Brown	*Advan. Organometal. Chem.* **11**, 1 (1973)	Boranes in organic chemistry	20 (67)
10.6	G. W. Campbell	*Progr. Boron Chem.* **1**, 167 (1964)	Structures of the boron hydrides	35 (119)
10.7	J. Casanova, Jr.	*In* "Isonitrile Chemistry" (I. Ugi, ed.), p. 109. Academic Press, New York, 1971	Reaction of isonitriles with boranes	23 (71)
10.8	T. D. Coyle and J. J. Ritter	*Advan. Organometal. Chem.* **10**, 237 (1972)	Organometallic aspects of diboron chemistry	36 (120)
10.9	H. D. Johnson and S. G. Shore	*Fortschr. Chem. Forsch.* **15**, 87 (1970)	New results in boron chemistry. Lower boron hydrides	58 (226)
10.10	C. F. Lane	*Aldrichim. Acta* **6**, 21 (1973)	The versatile boranes	14 (115)
10.11	K. Kösswig	*Chem. Ztg.* **96**, 373 (1971)	1,5,9-Cyclododecatriene as a key substance in organic chemistry (Ger.)	11 (102)
10.12	R. Köster	*Angew. Chem.* **75**, 1079 (1963); *Angew. Chem., Int. Ed. Engl.* **3**, 174 (1964)	Transformation of organoboranes at elevated temperatures	11 (60)
10.13	L. H. Long	*Progr. Inorg. Chem.* **15**, 1 (1972)	Recent studies of diborane	99 (335)
10.14	B. M. Mikhailov	*Organometal. Chem. Rev. Sect. A* **8**, 1 (1972)	Allylboron compounds	65 (157)
10.15	B. M. Mikhailov	*Usp. Khim.* **31**, 417 (1962); *Russ. Chem. Rev.* **31**, 207 (1962)	The chemistry of diborane	17 (238)
10.16	T. Ogushi	*Kagaku to Kogyo (Osaka)* **42**, 47 (1968)	Chemistry of organoboron compounds II. Trialkylboranes (Japan)	7 (35)
10.17	E. Schenker	*Newer Methods Prep. Org. Chem.*, **4**, 197 (1968)	Use of complex borohydrides and of diborane in organic chemistry	139 (1890)

10.18	F. G. A. Stone	*Advan. Inorg. Chem. Radiochem.* **2**, 279 (1960)	Chemical reactivity of the boron hydrides and related compounds	34 (156)
10.19	A. Suzuki	*Kagaku No Ryoiki. Zokan* **89**, 213 (1970)	Reactions using organoboranes (Japan)	21 (123)
10.20	A. Suzuki	*Yuki Gosei Kagaku Kyokai Shi* **28**, 288 (1970)	Hydroboration and chemistry of organoboranes (Japan)	21 (212)
10.21	A. Suzuki	*Yuki Gosei Kagaku Kyokai Shi* **29**, 995 (1971)	New organic synthesis by the use of organoboranes. Mainly on radical reactions (Japan.)	13 (90)
10.22	L. J. Todd	*Progr. Boron Chem.* **2**, 1 (1970)	Chemistry of polyhedral borane ions	36 (81)
10.23	I. Uzarewicz, M. Zaidalewicz, and A. Uzarewicz	*Wiad. Chem.* **24**, 1 (1970)	Diborane and its alkyl derivatives in organic synthesis I. Methods of synthesis of diborane; the hydroboration mechanism (Pol.)	26
10.24	I. Uzarewicz, M. Zaidalewicz, and A. Uzarewicz	*Wiad. Chem.* **25**, 699 (1971)	Diborane and its alkyl derivatives in organic synthesis II. Hydroboration of acetylenes (Pol.)	33 (133)
10.25	I. Uzarewicz, M. Zaidalewicz, and A. Uzarewicz	*Wiad. Chem.* **26**, 33 (1972)	Diborane and its alkyl derivatives in organic synthesis III. Hydroboration of olefins (Pol.)	30 (75)
10.26	I. Uzarewicz, M. Zaidalewicz, A. Uzarewicz	*Wiad. Chem.* **26**, 193 (1972)	Diborane and its alkyl derivatives in organic synthesis IV. Hydroboration of dienes (Pol.)	11 (31)
10.27	G. Zweifel and H. C. Brown	*Org. React.* **13**, 1 (1963)	Hydration of olefins. Dienes and acetylenes via hydroboration	54 (70)
	See also: A15, A43, A45, A47–A51, A54, A55, A161, A172, A184; E10.1, E14.5, E15, E19.2			
11. *Boron—carboranes*				
11.1	H. Beall and C. Hackett Bushweller	*Chem. Rev.* **73**, 465 (1973)	Dynamical processes in boranes, borane complexes, carboranes, and related compounds	22 (119)
11.2	V. I. Bregadze and O. Yu. Okhlobystin	*Usp. Khim.* **37**, 353 (1968); *Russ. Chem. Rev.* **37**, 173 (1968)	Organoelement derivatives of barenes (carboranes)	12 (90)

(Continued)

Ref. No.	Authors	Reference	Title	No. of pages (No. of ref.)
11.3	V. I. Bregadze and O. Yu. Okhlobystin	*Organometal. Chem. Rev., Sect. A* **4**, 345 (1969)	Organoelement derivatives of barenes (carboranes)	32 (106)
11.4	G. B. Dunks and M. F. Hawthorne	*Accounts Chem. Res.* **6**, 124 (1973)	The non-icosahedral carboranes: synthesis and reactions	8 (69)
11.5	A. Finch and P. J. Gardener	*Progr. Boron Chem.* **3**, 177 (1970)	Thermochemistry of boron compounds	34 (135)
11.6	D. T. Haworth	*Endeavour* **31**, 16 (1972)	Chemistry of the carboranes	6 (29)
11.7	M. F. Hawthorne	*Endeavour* **25**, 146 (1966)	Polyhedral boranes, carboranes and carbametallic boron hydride derivatives	8 (40)
11.8	M. F. Hawthorne	*Accounts Chem. Res.* **1**, 281 (1968)	The chemistry of the polyhedral species derived from transition metals and carboranes	8 (40)
11.9	K. Issleib, R. Lindner, and A. Tzschach	*Z. Chem.* **6**, 1 (1966)	Carboranes (Ger.)	7 (108)
11.10	V. V. Korshak, I. G. Sarishvili, A. F. Zhigach, and M. V. Sobolevskii	*Usp. Khim.* **36**, 2068 (1967); *Russ. Chem. Rev.* **36**, 903 (1967)	Polycarboranes	10 (55)
11.11	R. Köster and M. A. Grassberger	*Angew. Chem.* **79**, 197 (1967); *Angew. Chem., Int. Ed. Engl.* **6**, 218 (1967)	Structures and syntheses of carboranes	22 (122)
11.12	K. Niedenzu	*Naturwissenschaften* **56**, 305 (1969)	Carboranes and the significance of polyhedral structures in boron chemistry (Ger.)	4 (39)
11.13	J. D. Odom and R. Schaeffer	*Progr. Boron Chem.* **2**, 141 (1970)	Use of isotopic labels in the study of carboranes and binary compounds of B and H	32 (137)
11.14	T. Onak	*Advan. Organometal. Chem.* **3**, 263 (1965)	Carboranes and organo-substituted boron hydrides	100 (355)

11.15	F. R. Scholar and L. J. Todd	*Prep. Inorg. React.* **7**, 1 (1971)	Polyhedral boranes and heteroatom boranes	93 (237)
11.16	V. I. Stanko, Yu. A. Chapovskii, V. A. Brattsev, and L. I. Zakharkin	*Usp. Khim.* **34**, 1011 (1965); *Russ. Chem. Rev.* **34**, 424 (1965)	The chemistry of decarborane and its derivatives	15 (129)
11.17	L. J. Todd	*Advan. Organometal. Chem.* **8**, 87 (1970)	Transition metal-carborane complexes	28 (49)
11.18	R. E. Williams	*Progr. Boron Chem.* **2**, 37 (1970)	Carboranes	82 (201)
	See also: A15, A50–A52; B3.1, B4.1, B4.2; E1.3, E1.5, E5, E19.1, E19.3			
12.	*Boron—B–B, B–Hal, B–N, B–O, and B–S compounds*			
12.1	I. Bally and A. T. Balaban	*Stud. Cercet. Chim.* **17**, 431 (1969)	Unsaturated organoboron heterocycles with boron-oxygen bonds (Rom.)	32 (164)
12.2	R. J. Brotherton	*Progr. Boron Chem.* **1**, 1 (1964)	Chemistry of compounds which contain B–B bonds	81 (189)
12.3	R. H. Cragg and M. F. Lappert	*Organometal. Chem. Rev.* **1**, 43 (1966)	Organic boron-sulphur compounds	22 (115)
12.4	R. H. Cragg	*Quart. Rep. Sulfur Chem.* 3 (1968)	Chemistry of boron-sulphur compounds for the period 1950–1967	80
12.5	A. G. Davies	*Progr. Boron Chem.* **1**, 265 (1964)	Organoperoxyboranes	24 (61)
12.6	M. J. S. Dewar	*Progr. Boron Chem.* **1**, 235 (1964)	Heteroaromatic boron compounds	29 (45)
12.7	A. Finch, J. B. Leach and J. H. Morris	*Organometal. Chem. Rev., Sect. A* **4**, 1 (1969)	Boron-nitrogen ring systems	45 (76)
12.8	R. A. Geanangel and S. G. Shore	*Prep. Inorg. React.* **3**, 123 (1966)	Boron-nitrogen compounds	116 (428)
12.9	W. Gerrard	*Chem. Ind.* (*London*) 832 (1966)	Boron chemistry: organic analogues of heterocycles	9 (71)

(*Continued*)

Ref. No.	Authors	Reference	Title	No. of pages (No. of ref.)
12.10	V. Gutmann and A. Meller	*Oesterr. Chem. Ztg.* **66**, 324 (1965)	Substitution of borazines (Ger.)	6 (42)
12.11	A. K. Holliday and A. G. Massey	*Chem. Rev.* **62**, 303 (1962)	Boron subhalides and related compounds with boron–boron bonds	16 (108)
12.12	W. Kliegel	*Organometal. Chem. Rev., Sect. A* **8**, 153 (1972)	Boron-nitrogen betaines (Ger.)	28 (144)
12.13	R. Köster	*Progr. Boron Chem.* **1**, 289 (1964)	Organoboron heterocycles	56 (71)
12.14	R. Köster	*Advan. Organometal. Chem.* **2**, 257 (1964)	Heterocyclic organoboranes	67 (90)
12.15	P. M. Maitlis	*Chem. Rev.* **62**, 223 (1962)	Heterocyclic organic boron compounds	23 (127)
12.16	J. A. Marshall	*Synthesis* 229 (1971)	Diene synthesis via boronate fragmentation	7 (31)
12.17	A. G. Massey	*Advan. Inorg. Chem. Radiochem.* **10**, 1 (1967)	The halides of boron	152 (879)
12.18	E. Matejcikova and A. Sopkova	*Chem. Listy* **65**, 486 (1971)	Dialkylamino compounds of boron of the $B_n(NR_2)_{n+2}$ type (Czech.)	13 (62)
12.19	D. S. Matteson	*Organometal. Chem. Rev.* **1**, 1 (1966)	Organofunctional boronic esters	25 (41)
12.20	D. S. Matteson	*Accounts Chem. Res.* **3**, 186 (1970)	Boronic ester neighbouring groups	8 (51)
12.21	A. Meller	*Fortschr. Chem. Forsch.* **15**, 146 (1970)	New results in boron chemistry. Boron-nitrogen ring compounds (Ger.)	43 (205)
12.22	A. Meller	*Fortschr. Chem. Forsch.* **26**, 37 (1972)	Chemistry of iminoboranes (Ger.)	40 (48)
12.23	A. Meller	*Organometal. Chem. Rev.* **2**, 1 (1967)	Infrared spectra of organoboron-nitrogen compounds (Ger.)	60 (200)

12.24	E. K. Mellon, Jr. and J. J. Lagowski	*Advan. Inorg. Chem. Radiochem.* **5**, 259 (1963)	The borazines	46 (194)
12.25	B. M. Mikhailov	*Usp. Khim.* **37**, 2121 (1968); *Russ. Chem. Rev.* **37**, 935 (1968)	Sulphur-containing organic compounds of boron	19 (124)
12.26	B. M. Mikhailov	*Usp. Khim.* **29**, 972 (1960); *Russ. Chem. Rev.* **29**, 459 (1960)	Borazole and its derivatives	11 (71)
12.27	B. M. Mikhailov	*Progr. Boron Chem.* **3**, 313 (1970)	Organic B–S compounds	58 (128)
12.28	K. Niedenzu	*Organometal. Chem. Rev.* **1**, 305 (1966)	Synthesis of organohaloboranes	24 (124)
12.29	K. Niedenzu and C. D. Miller	*Fortschr. Chem. Forsch.* **15**, 191 (1970)	New results in boron chemistry: 1,3,2-diazaboracycloalkanes	14 (37)
12.30	K. Niedenzu	*Allg. Prakt. Chem.* **17**, 596 (1966)	Azaboracycloalkanes (Ger.)	4 (36)
12.31	K. Niedenzu	*Angew. Chem.* **76**, 165 (1964); *Angew. Chem., Int. Ed. Engl.* **3**, 86 (1964)	The chemistry of aminoboranes	7 (90)
12.32	H. Nöth	*Progr. Boron Chem.* **3**, 211 (1970)	Some recent developments in B–N chemistry	101 (329)
12.33	Z. Polivka and M. Ferles	*Chem. Listy* **62**, 869 (1968)	Haloborylation of unsaturated compounds (Czech.)	26 (69)
12.34	B. Serafinowa	*Wiad. Chem.* **22**, 819 (1968)	Boronic acids and their derivatives (Pol.)	21 (180)
12.35	J. C. Sheldon and B. C. Smith	*Quart. Rev.*, **14**, 200 (1960)	The borazoles	20 (79)
12.36	P. L. Timms	*Accounts Chem. Res.* **6**, 118 (1973)	Chemistry of boron and silicon subhalides	6 (25)
12.37	K. Torssel	*Progr. Boron Chem.* **1**, 369 (1964)	Chemistry of boronic and borinic acids	47 (388)
12.38	S. Trofimenko	*Accounts Chem. Res.* **4**, 17 (1971)	Polypyrazolylborates, a new class of ligands	6 (37)

See also: A44, A46–A48, A129.2, A175–A179, A181–A182; E12

(*Continued*)

Ref. No.	Authors	Reference	Title	No. of pages (No. of ref.)
13. *Aluminum (excluding Zielger–Natta catalysts)*				
13.1	L. F. Albright	*Chem. Eng. (New York)* **74**, 179 (1967)	The properties, chemistry and synthesis of alkylaluminums	9 (84)
13.2	M. Bartholin	*Acta Chim. (Budapest)* **65**, 239 (1970)	Organoaluminium siloxane polycondensates (Fr.)	6 (6)
13.3	A. Bottini	*Nuova Chim.* **46**, 67 (1970)	Alkylaluminium compounds (Ital.)	8 (70)
13.4	D. C. Bradley	*Progr. Stereochem.* **3**, 1 (1962)	The stereochemistry of some elements of group III	52 (180)
13.5	R. Köster and P. Binger	*Advan. Inorg. Chem. Radiochem.* **7**, 263 (1965)	Organoaluminum compounds	85 (326)
13.6	H. Lehmkuhl	*Chimia* **24**, 182 (1970)	Organoaluminium complexes with hydrocarbons having electron affinity (Ger.)	13 (53)
13.7	H. Lehmkuhl	*Angew. Chem.* **75**, 1090 (1963); *Angew. Chem., Int. Ed. Engl.* **3**, 107 (1964)	Complex formation with organoaluminium compounds	8 (32)
13.8	H. Reinheckel, K. Haage, and D. Jahnke	*Organometal. Chem. Rev., Sect. A* **4**, 47 (1969)	Organic reactions of organoaluminium compounds (Ger.)	89 (225)
13.9	K. Wade	*Chem. Brit.* **4**, 503 (1968)	Some developments in aluminium chemistry	6 (58)
13.10	J. Weidlein	*J. Organometal. Chem.* **49**, 257 (1973)	Organometallic 8-membered ring compounds containing Al, Ga, In and Tl (Ger.)	26 (125)
	See also: A15, A26.1, A27.3, A57–A58, A60, A158, A161, A184; B1–B4; C15; E1.5, E13.6, E16.2, E17.2, E17.3; index 18.6			
14. *Gallium, indium, and thallium*				
14.1	N. N. Greenwood	*Advan. Inorg. Chem. Radiochem.* **5**, 91 (1963)	The chemistry of gallium	43 (310)

14.2	H. Kurosawa and R. Okawara	*Trans. N.Y. Acad. Sci.* **30**, 962 (1968)	Mono- and diorganothallium chemistry	6 (19)
14.3	H. Kurosawa and R. Okawara	*Organometal. Chem. Rev., Sect. A* **6**, 65 (1970)	Recent advances in organothallium chemistry	52 (225)
14.4	A. G. Lee	*Quart. Rev.* **24**, 310 (1970)	Organothallium chemistry	20 (101)
14.5	A. G. Lee	*Organometal. React.* **5**	Reactions of organothallium compounds	
14.6	A. McKillop and E. C. Taylor	*Chem. Brit.* **9**, 4 (1973)	Thallium in organic synthesis	8 (54)
14.7	A. McKillop and E. C. Taylor	*Advan. Organometal. Chem.* **11**, 147 (1973)	Recent advances in organothallium chemistry	69 (173)
14.8	C. T. Pedersen	*Dan. Kemi* **51**, 163 (1970)	Use of thallium and thallium-containing compounds in organic synthesis (Dan.)	3 (14)
14.9	E. C. Taylor and A. McKillop	*Aldrichim. Acta* **3**, 4 (1973)	Organothallium chemistry—new horizons in synthesis	6 (27)
14.10	E. C. Taylor and A. McKillop	*Accounts Chem. Res.* **3**, 338 (1970)	Thallium in organic synthesis	8 (56)
14.11	K. Yasuda and R. Okawara	*Organometal. Chem. Rev.* 2, 255 (1967)	Organo-gallium, -indium and -thallium compounds	22 (206)
	See also: A26.1, A27.3, A57–A59; B1–B4; C15; E1.5, E14.6			
15.	*General reviews—Group IVB*			
15.1	E. W. Abel and S. M. Illingworth	*Organometal. Chem. Rev., Sect. A* **5**, 143 (1970)	Phosphines, arsines, stibines and bismuthines containing Si, Ge, Sn and Pb	39 (144)
15.2	E. W. Abel, M. O. Dunster, and A. Waters	*J. Organometal. Chem.* **49**, 287 (1973)	Cyclopentadienyl compounds of Si, Ge, Sn and Pb	34 (114)
15.3	E. W. Abel and D. A. Armitage	*Advan. Organometal. Chem.* **5**, 2 (1967)	Organosulfur derivatives of Ge, Sn and Pb	90 (374)
15.4	Yu. A. Alexandrov	*Organometal. Chem. Rev., Sect. A* **6**, 209 (1970)	Oxidation of organic derivatives of non-transition elements of group IV (other than carbon) by ozone	17 (34)

(Continued)

Ref. No.	Authors	Reference	Title	No. of pages (No. of ref.)
15.5	C. J. Attridge	*Organometal. Chem. Rev., Sect. A* **5**, 323 (1970)	π-Bonding in group IVB	40 (159)
15.6	Yu. I. Baukov and I. F. Lutsenko	*Organometal. Chem. Rev., Sect A* **6**, 355 (1970)	Organo-element (Si, Ge, Sn and Pb) derivatives of ketoenols	90 (380)
15.7	A. G. Brook	*Advan. Organometal. Chem.* **7**, 96 (1968)	Keto-derivatives of group IV organometalloids	59 (102)
15.8	E. H. Brooks and R. J. Cross	*Organometal. Chem. Rev., Sect. A* **6**, 227 (1970)	Group IVB metal derivatives of the transition elements	56 (209)
15.9	V. Chvalovský and V. Bažant	*Helv. Chim. Acta* **52**, 2398 (1969)	Differences between the organic chemistry of silicon and some other group IV elements	20 (23)
15.10	D. D. Davis and C. E. Gray	*Organometal. Chem. Rev., Sect. A* **6**, 283 (1970)	Alkali metal and magnesium derivatives of organo-Si, -Ge, -Sn, -Pb, -P, -As, -Sb and -Bi compounds	35 (236)
15.11	J. B. Dence	*Chemistry* **46**, 6 (1973)	Covalent carbon-metal(loid) compounds	8 (0)
15.12	R. E. Dessy and L. A. Bares	*Accounts Chem. Res.* **5**, 415 (1972)	Organometallic electrochemistry	7 (36)
15.13	A. N. Egorochkin, N. S. Vyazankin, and S. Ya. Khorshev	*Usp. Khim.* **41**, 828 (1972); *Russ. Chem. Rev.* **41**, 425 (1972)	Effect of d_π-p_π interaction in organic compounds of group IVB elements	14 (222)
15.14	G. Fritz	*Angew. Chem.* **80**, 2 (1968); *Angew. Chem., Int. Ed. Engl.* **7**, 1 (1968)	Formation and properties of medium and high molecular weight compounds of elements of group III, IV and V	6 (20)
15.15	M. Gielen and N. Sprecher	*Organometal. Chem. Rev.* **1**, 455 (1966)	Influence of d-orbitals on coordination at group IVB metals (Fr.)	34 (388)
15.16	M. Gielen, C. Dehouck, H. Mokhtar-Jamai, and J. Topart	*Rev. Si, Ge, Sn, Pb Compounds* **1**, 9 (1972)	Stereochemistry of four-, five-, and six-coordinate complexes of group IV metals	25 (82)

15.17	H. Gilman, W. H. Atwell, and F. K. Cartledge	*Advan. Organometal. Chem.* **4**, 1 (1966)	Catenated organic compounds of Si, Ge, Sn and Pb	94 (368)
15.18	K. Itoh, S. Sakai, and Y. Ishii	*Yuki Gosei Kagaku Kyokai Shi* **24**, 729 (1966)	Some addition reactions of group IVA organometallics to unsaturated compounds (Japan)	12 (95)
15.19	N. E. Kolobova, A. B. Antonova, and K. N. Anisimov	*Usp. Khim.* **38**, 1802 (1969); *Russ. Chem. Rev.* **38**, 822 (1969)	Derivatives of metal carbonyls containing a bond between atoms of transition metals and group IVB elements	18 (176)
15.20	J. G. A. Luijten, F. Rijkens, and G. J. M. van der Kerk	*Advan. Organometal. Chem.* **3**, 397 (1965)	Organometallic nitrogen compounds of Ge, Sn and Pb	49 (132)
15.21	L. K. Luneva	*Usp. Khim.* **36**, 1140 (1967); *Russ. Chem. Rev.* **36**, 467 (1967)	Ethynyl derivatives of Si, Ge, Sn and Pb	10 (141)
15.22	K. M. Mackay and R. Watt	*Organometal. Chem. Rev., Sect A* **4**, 137 (1969)	Chain compounds of Si, Ge, Sn and Pb	76 (507)
15.23	I. Omae	*Rev. Si, Ge, Sn, Pb Compounds* **1**, 59 (1972)	Organometallic intramolecular-coordination compounds containing carbonyl groups	38 (100)
15.24	S. Sakai, K. Itoh, and Y. Ishii	*Yuki Gosei Kagaku Shi* **28**, 1109 (1970)	Addition reactions of the group IV organometallic compounds and their synthetic applications (Japan)	18 (195)
15.25	H. Sakurai and M. Kira	*Kagaku No Ryoiki* **22**, 897 (1968)	Aryl substituted group IVB compounds (Japan)	9 (39)
15.26	O. J. Scherer	*Organometal. Chem. Rev., Sect. A* **3**, 281 (1168)	Cleavage of organo-silicon, -germanium and -tin-nitrogen compounds by halides of the main group elements (Ger.)	28 (115)
15.27	H. Schumann	*Angew. Chem.* **81**, 970 (1969); *Angew. Chem., Int. Ed. Engl.* **8**, 937 (1969)	Organogermyl, organostannyl and organoplumbyl phosphines, arsines, stibines and bismuthines	14 (105)
15.28	C. F. Shaw III and A. L. Allred	*Organometal. Chem., Rev. Sect. A* **5**, 95 (1970)	Non-bonded interactions in organometallic compounds of group IVB	47 (376)

(Continued)

Ref. No.	Authors	Reference	Title	No. of pages (No. of ref.)
15.29	M. F. Shostakovskii, B. A. Trofimov, and A. S. Atavin	*Khim. Atsetilena, Tr. Vses. Konf., 3rd, 1968*, pp. 17–24, Nauka, Moscow, 1972	Acetylene chemistry (Russ.)	8 (71)
15.30	C. H. Yoder and J. J. Zuckerman	*Prep. Inorg. React.* **6**, 81 (1971)	Heterocyclic compounds of the group IV elements	75 (390)
15.31	S. F. Zhil'tsov and O. N. Druzhkov	*Usp. Khim.* **40**, 226 (1971); *Russ. Chem. Rev.* **40**, 126 (1971)	Reactions of organic derivatives of the elements with polyhalogenomethanes	16 (250)
	See also: A7, A74, A190; B3.1, B4.2; C4; E1.4, E16.1			
16.	*Silicon—general, Si–Hal, Si–H compounds*			
16.1	W. H. Atwell and D. R. Weyenberg	*Angew. Chem.* **81**, 485 (1969); *Angew. Chem., Int. Ed. Engl.* **8**, 469 (1969)	Divalent silicon intermediates	9 (94)
16.2	V. Bažant	*Rozpr. Cesk. Akad. Ved. Rada Mat. Prirod. Ved.* **71**, No. 11, (1961)	Synthesis and reactions of organosilicon compounds	57 (102)
16.3	H. Bürger	*Fortschr. Chem. Forsch.* **9**, 1 (1967/1968)	Bonding to silicon (Ger.)	59 (390)
16.4	H. Bürger	*Angew. Chem.* **85**, 519 (1973); *Angew. Chem., Int. Ed. Engl.* **12**, 474 (1973)	Anomalies in the structural chemistry of silicon	13 (124)
16.5	A. W. P. Jarvie	*Organometal. Chem., Rev. Sect. A* **6**, 153 (1970)	The anomalous properties of β-functional organosilicon compounds: the β-effect	54 (196)
16.6	J. F. Klebe	*Accounts Chem. Res.* **3**, 299 (1970)	Silyl-proton exchange reactions	7 (17)
16.7	A. G. MacDiarmid	*Prep. Inorg. React.* **1**, 165 (1964)	Halogen and halogenoid derivatives of the silanes	38 (156)

16.8	D. H. O'Brien	*Organometal. Chem. Rev., Sect. A* **7** 95 (1971)	The reactions of organosilanes with Lewis acids	62 (184)
16.9	C. A. Roth	*Ind. Eng. Chem. Prod. Res. Develop.* **11**, 134 (1972)	Silylation of organic chemicals	6 (33)
16.10	M. G. Voronkov and E. Ya. Lukevits	*Usp. Khim.* **38**, 2173 (1969); *Russ. Chem. Rev.* **38**, 975 (1969)	Biologically active compounds of silicon	12 (358)
16.11	M. G. Voronkov	*Chem. Brit.* **9**, 411 (1973)	Bio-organosilicon chemistry	5 (27)
16.12	K. A. Andrianov and A. I. Petrashko	*Usp. Khim.* **38**, 408 (1969); *Russ. Chem. Rev.* **38**, 211 (1969). See also: *Organometal. Chem. Rev.* **2**, 383 (1967)	Halogen derivatives of alkyl (or aryl) halogenosilanes and tetra-substituted silanes	36 (677)
16.13	B. J. Aylett	*Advan. Inorg. Chem. Radiochem.* **11**, 249 (1968)	Silicon hydrides and their derivatives	58 (280)
16.14	V. Bažant, J. Joklík, and J. Rathouský	*Angew. Chem.* **80**, 133 (1968); *Angew. Chem., Int. Ed. Engl.* **7**, 112 (1968)	Direct synthesis of organohalogenosilanes	9 (120)
16.15	R. A. Benkeser	*Accounts Chem. Res.* **4**, 94 (1971)	The chemistry of trichlorosilane—tertiary amine combinations	8 (52)
16.16	E. Bonitz	*Angew. Chem.* **78**, 475 (1966); *Angew. Chem., Int. Ed. Engl.* **5**, 462 (1966)	Reactions of elementary silicon	8 (85)
16.17	G. M. Cameron and J. G. Marsden	*Chem. Brit.* **8**, 381 (1972)	Silane coupling reagents	5 (15)
16.18	A. J. Chalk	*Ann. N.Y. Acad. Sci.* **72** (*Art.* 13), 533 (1971)	Olefin hydrosilation catalysed by group VIII metal complexes	8 (35)
16.19	A. J. Chalk and J. F. Harrod	*Advan. Organometal. Chem.* **6**, 119 (1967)	Catalysis by cobalt carbonyls	51 (161)

(*Continued*)

Ref. No.	Authors	Reference	Title	No. of pages (No. of ref.)
16.20	J. E. Drake and C. Riddle	*Quart. Rev.* **24**, 263 (1970)	Volatile compounds of the hydrides of silicon and germanium with elements of groups V and VI	15 (78)
16.21	D. Elad	*In* "Organic Photochemistry" (O. L. Chapman, ed.), Vol. 2, pp. 168–212. Dekker, New York, 1969	Photochemical additions to double bonds	45 (190)
16.22	I. Ito, M. Takamizawa, S. Tanaka, and J. Suzuki	*Yuki Gosei Kagaku Kyokai Shi* **20**, 213 (1962)	Manufacture of dimethyldichlorosilane (Japan)	4 (18)
16.23	J. L. Margrave and P. W. Wilson	*Accounts Chem. Res.* **4**, 145 (1971)	Silicon difluoride, a carbene analogue. Its reactions and properties	7 (32)
16.24	G. V. Motsarev, K. A. Andrianov, and V. I. Zetkin	*Usp. Khim.* **40**, 980 (1971); *Russ. Chem. Rev.* **40**, 485 (1971)	Progress in halogenation of organic silicon compounds	30 (335)
16.25	G. V. Odabashy, V. A. Ponamarenko, and A. D. Petrov	*Usp. Khim.* **30**, 947 (1961); *Russ. Chem. Rev.* **30**, 407 (1961)	Organosilicon compounds of fluorine	19 (254)
16.26	H. Schmidbaur	*20 Jahre Fonds Chem. Ind. Beitr. Wiss. Veranst.* p. 47 (1970); see also *20 Jahre Fonds Chem. Ind. Förder Forsch. Wiss. Lehre, Beitr. Wiss. Veranst.* p. 129 (1970)	Organosilicon chemistry of phosphorus ylides (Ger.)	7 (30) 7 (59)
16.27	M. E. Vol'pin and I. S. Kolomnikov	*Usp. Khim.* **38**, 561 (1969); *Russ. Chem. Rev.* **38**, 273 (1969)	Homogeneous hydrogenation	17 (282)
16.28	J. J. Zuckerman	*Advan. Inorg. Chem. Radiochem.* **6**, 383 (1964)	The direct synthesis of organosilicon compounds	49 (489)

See also: A61–A72, A74, A77, A84, A101, A129, A161, A172, A185–A187; B1–B4; E1.5, E1.6, E2, E3, E9, E13; G5, G6

17. *Silicon—Si–O compounds*				
17.1	T. N. Balykova and V. V. Rode	*Usp. Khim.* **38**, 662 (1969); *Russ. Chem. Rev.* **38**, 306 (1969)	Progress in the study of the degradation and stabilisation of siloxane polymers	12 (141)
17.2	P. D. George, M. Prober, and J. R. Elliot	*Chem. Rev.* **56**, 1065 (1956)	Carbon functional silicones	155 (555)
17.3	D. V. N. Hardy and N. J. L. Megson	*Quart. Rev.* **2**, 25 (1948)	The chemistry of silicon polymers	21 (77)
17.4	E. Hengge	*Fortschr. Chem. Forsch.* **9**, 145 (1967/1968)	Siloxanes and layer-structured silicon compounds (Ger.)	20 (32)
17.5	H. W. Kohlschütter	*Fortschr. Chem. Forsch.* **1**, 1 (1949)	Chemistry of silicones (Ger.)	60 (147)
17.6	R. R. McGregor	Dow Corning Corp. publ.	Silicones in medicine and surgery	
17.7	K. Nakano	*Yuki Gosei Kagaku Kyokai Shi* **24**, 598 (1966)	Silicone water repellants (Japan)	11 (64)
17.8	N. S. Nametkin, T. Kh. Islamov, L. E. Gusel'nikov, and V. M. Vdovin	*Usp. Khim.* **41**, 203 (1972); *Russ. Chem. Rev.* **41**, 111 (1972)	Cyclocarbosiloxanes	20 (135)
17.9	F. Schindler and H. Schmidbaur	*Angew. Chem.* **79**, 697 (1967); *Angew. Chem., Int. Ed. Engl.* **6**, 683 (1967)	Siloxane compounds of the transition metals	12 (149)
17.10	H. Schmidbaur	*Angew. Chem.* **77**, 206 (1965); *Angew. Chem., Int. Ed. Engl.* **4**, 201 (1965)	Recent developments in the chemistry of heterosiloxanes	10 (83)
17.11	M. G. Voronkov and N. F. Sviridova	*Usp. Khim.* **40**, 1761 (1971); *Russ. Chem. Rev.* **40**, 819 (1971)	Methods of synthesis of $\alpha\omega$-difunctional oligosiloxanes with two organic substituents per unit	16 (187)

(*Continued*)

Ref. No.	Authors	Reference	Title	No. of pages (No. of ref.)
17.12	J. A. C. Watt	*Chem. Brit.* **6**, 519 (1970)	Silicone liquid rubbers	6 (9)
	See also: A67–A72, A76–A106, A161, A174–A182; E2, E3, E13.7			
18.	*Silicon—Si–N, Si–S and Si–P compounds*			
18.1	B. J. Aylett	*Organometal. Chem. Rev., Sect. A* **3**, 151 (1968)	Silicon-nitrogen polymers	21 (122)
18.2	B. J. Aylett	*Prep. Inorg. React.* **2**, 93 (1965)	Silicon-nitrogen compounds	46 (202)
18.3	E. A. Chernyshev and E. F. Bugerenko	*Organometal. Chem. Rev., Sect. A* **3**, 469 (1968)	Organosilicon compounds containing phosphorus	33 (284)
18.4	R. Fessenden and J. S. Fessenden	*Chem. Rev.* **61**, 361 (1961)	Chemistry of silicon-nitrogen compounds	20 (221)
18.5	W. Fink	*Angew. Chem.* **78**, 803 (1966); *Angew. Chem., Int. Ed. Engl.* **5**, 760 (1966)	Silicon-nitrogen heterocycles	17 (186)
18.6	G. Fritz	*Angew. Chem.* **78**, 80 (1966); *Angew. Chem., Int. Ed. Engl.* **5**, 53 (1966)	Compounds of phosphorus with silicon and aluminium	5 (34)
18.7	A. Haas	*Angew. Chem.* **77**, 1066 (1965); *Angew. Chem., Int. Ed. Engl.* **4**, 1014 (1965)	The chemistry of silicon-sulphur compounds	10 (74)
18.8	H. Holtschmidt and G. Oertel	*Angew. Chem.* **74**, 795 (1962); *Angew. Chem., Int. Ed. Engl.* **1**, 617 (1962)	Isocyanates of esters of some acids of phosphorus and silicon	4 (23)
18.9	N. N. Motovola and E. I. Starovoitova	*Kauch. Rezina* **25**, 4 (1966)	Organosilicon compounds containing heteroatoms (P, B)	7 (38)

18.10	R. Müller	*Organometal. Chem. Rev.* **1**, 359 (1966)	Organofluorosilicates and their use in the preparation of organometallic compounds (Ger.)	18 (64)
18.11	E. Ya. Lukevits, L. Liberts, and M. G. Voronkov	*Usp. Khim.* **39**, 2005 (1970); *Russ. Chem. Rev.* **39**, 953 (1970	Organosilicon derivatives of aminoalcohols	10 (204)
18.12	E. Ya. Lukevits and A. E. Pestunovich	*Usp. Khim.* **41**, 1994 (1972); *Russ. Chem. Rev.* **41**, 929 (1972)	Organosilicon derivatives of monoazaheterocycles	17 (350)
18.13	J. Preizner	*Wiad. Chem.* **23**, 601 (1969)	Isocyanate silicon derivatives (Pol.)	18 (75)
18.14	U. Wannagat	*Fortschr. Chem. Forsch.* **9**, 102 (1967/1968)	Tertiary silyl-substituted amines (Ger.)	42 (126)
18.15	U. Wannagat	*Advan. Inorg. Chem. Radiochem.* **6**, 225 (1964)	The chemistry of silicon-nitrogen compounds	53 (186)
18.16	N. Wiberg	*Angew. Chem.* **83**, 379 (1971); *Angew. Chem., Int. Ed. Engl.* **10**, 374 (1971)	Bis(trimethylsilyl)diimine: preparation structure and reactivity	14 (59)
18.17	E. Wiberg, O. Stecher, H. J. Andrascheck, L. Kreuzbichler, and E. Staude	*Angew. Chem.* **75**, 516 (1963); *Angew. Chem., Int. Ed. Engl.* **2**, 507 (1963)	Recent developments in the chemistry of metal silyls of the type $M(SiR_3)_n$	8 (24)
18.18	D. Wittenberg and H. Gilman	*Quart Rev.* **13**, 116 (1959)	Organosilylmetallic compounds: their formation and reactions, and comparison with related types	30 (155)
18.19	M. Yamaguchi, H. Takahashi, and S. Isao	*Shinku Kagaku* **16**, 78 (1969)	Silicon phthalocyanines (Japan)	6 (15)
18.20	Yu. K. Yur'ev and Z. V. Belyakova	*Usp. Khim.* **29**, 809 (1960); *Russ. Chem. Rev.* **29**, 383 (1960)	Acyloxysilanes	12 (151)
	See also: A67, A72, A74, A75, A129, A181, A182; E2, E3			

(Continued)

Ref. No.	Authors	Reference	Title	No. of pages (No. of ref.)
19. *Silicon—polysilanes*				
19.1	M. Kumada	*Nippon Kagaku Zasshi* **90**, 425 (1969)	Organopolysilanes: preparation and reactions (Japan)	20
19.2	M. Kumada	*Yuki Gosei Kagaku Kyokai Shi* **28**, 915 (1970)	New reactions of organosilicon compounds (Japan)	15 (112)
19.3	M. Kumada	*Kagaku To Kogyo Tokyo* **21**, 1129 (1968)	Chemistry of organopolysilanes (Japan.)	7 (20)
19.4	M. Kumada and K. Tamao	*Advan. Organometal. Chem.* **6**, 19 (1967)	Aliphatic organopolysilanes	98 (214)
19.5	A. G. MacDiarmid	*Quart. Rev.* **10**, 208 (1956)	Silyl compounds	22 (67)
19.6	A. G. MacDiarmid	*Advan. Inorg. Chem. Radiochem.* **3**, 207 (1961)	Silanes and their derivatives	49 (208)
19.7	H. Sakurai	*Yuki Gosei Kagaku Kyokai Shi* **25**, 555 (1967)	Synthesis, structures and reactions of organopolysilanes. Part 1 (Japan)	12 (90)
19.8	H. Sakurai	*Yuki Gosei Kagaku Kyokai Shi* **25**, 642 (1967)	Synthesis, structures and reactions of organopolysilanes. Part 2 (Japan)	16 (117)
19.9	G. Schott	*Fortschr. Chem. Forsch.* **9**, 60 (1967/1968)	Oligo- and poly-silanes and their derivatives (Ger.)	41 (266)
19.10	G. Urry	*Accounts Chem. Res.* **3**, 306 (1970)	Systematic synthesis in the polysilane series	7 (32)
	See also: A66–A68, A72; E2, E.3			
20. *Silicon—Carbosilanes, miscellaneous cyclic compounds*				
20.1	K. A. Andrianov, I. Haiduc, and L. M. Khananashvili	*Usp. Khim.* **32**, 539 (1963); *Russ. Chem. Rev.* **32**, 243 (1963)	Inorganic cyclic silicon-containing compounds and their organic derivatives	26 (563)

20.2	K. A. Andrianov and L. M. Khananashvili	*Organometal. Chem. Rev.* **2**, 141 (1967)	Cyclic organosilicon compounds	88 (808)
20.3	T. Araki	*Kagaku No Ryoiki* **25**, 1123 (1971)	Chemistry of silicon-containing small-ring compounds (Japan)	11 (64)
20.4	R. Damrauer	*Organometal. Chem. Rev., Sect. A* **8**, 67 (1972)	Cyclobutanes containing heterocyclic silicon and germanium	85 (142)
20.5	G. Fritz	*Fortschr. Chem. Forsch.* **4**, 459 (1963/1965)	Chemistry of silicon compounds. Discoveries based upon pyrolysis (Ger.)	95 (121)
20.6	G. Fritz, J. Grobe, and D. Kummer	*Advan. Inorg. Chem. Radiochem.* **7**, 349 (1965)	Carbosilanes	69 (104)
20.7	G. Fritz	*Angew. Chem.* **79**, 657 (1967); *Angew. Chem., Int. Ed. Engl.* **6**, 677 (1967)	Formation and properties of carbosilanes	6 (29)
20.8	G. Fritz	*Chem. Ztg.* **97**, 111 (1973)	Chemistry of carbosilanes	6 (30)
20.9	H. Gilman and G. L. Schwebke	*Advan. Organometal. Chem.* **1**, 90 (1964)	Organic substituted cyclosilanes	51 (107)
20.10	R. G. Neville	*J. Chem. Educ.* **39**, 276 (1962)	Functionally substituted aromatic silanes	5 (103)
	See also: A67, A72, A74, A105, A129, A171, A172, A179, A181, A182; E2, E3			
21.	*Silicon—uses in synthesis*			
21.1	L. Birkofer and A. Ritter	*Angew. Chem.* **77**, 414 (1965); *Angew. Chem., Int. Ed. Engl.* **4**, 417 (1965)	The use of silylation in organic synthesis	12 (81)
21.2	V. Chvalovský	*Organometal. React.* **3**, 191 (1972)	Cleavage of the carbon-silicon bond	127 (520)
21.3	K. Dey	*J. Sci. Ind. Res.* **30**, 458 (1971)	Ketone synthesis through organosilicon compounds	7 (83)

(*Continued*)

Ref. No.	Authors	Reference	Title	No. of pages (No. of ref.)
21.4	J. F. Klebe	*Advan. Org. Chem.* **8**, 97 (1972)	Silylation in organic synthesis	82 (187)
21.5	N. V. Komarov and V. K. Roman	*Usp. Khim.* **39**, 1220 (1970); *Russ. Chem. Rev.* **39**, 578 (1970)	Organosilicon ketones	13 (180)
21.6	J. Lipowitz and S. A. Bowman	*Aldrichim. Acta* **6**, 1 (1973)	The use of polymethylhydrosiloxane (PMHS) as a reducing agent for organic compounds	6 (30)
21.7	H. W. Post	*J. Med. (Basel)* **1**, 242 (1970)	Organosilyl ethers of penicillin and hormones	4 (10)
21.8	K. Rühlmann	*Z. Chem.* **5**, 130 (1965)	Uses of organosilicon compounds in preparative organic chemistry	12 (230)
21.9	K. Rühlmann	*Synthesis* 236 (1971)	The decomposition of carboxylic acid esters in the presence of sodium and trimethylchlorosilane	18 (125)
21.10	V. M. Vdovin and A. D. Petrov	*Usp. Khim.* **31**, 793 (1962); *Russ. Chem. Rev.* **31**, 393 (1962)	Organosilicon compounds containing a cyano group	15 (157)
21.11	E. Zbiral	*Synthesis* 285 (1972)	Organic syntheses using lead (IV) acetate azides	18 (74)
	See also: A67, A68, A71–A74; E2, E3			
22. *Germanium*				
22.1	E. J. Bulten	Thesis, Organisch Chemisch Instituut TNO, Utrecht. Publ. Germanium Research Committee (1969)	Chemistry of alkylpolygermanes	128 (185)
22.2	E. Gastinger	*Fortschr. Chem. Forsch.* **3**, 603 (1955)	Research developments in germanium chemistry (Ger.)	54 (310)
22.3	F. Glockling	*Quart. Rev.* **20**, 45, (1966)	Organogermanium chemistry	21 (125)

22.4	K. A. Hooton	*Prep. Inorg. React.* **4**, 85 (1968)	Organogermanium compounds	92 (337)
22.5	A. A. Lavigne and J. M. Tancrede	*Coordination Chem. Rev.* **3**, 497 (1968)	Coordination compounds of germanium	12 (68)
22.6	M. Lesbré	*Bull. Union Physiciens* **65** (528), 1 (1970)	Germanium chemistry (Fr.)	11
22.7	M. Lesbré	*Afinidad* **29**, 171 (1972)	Reactivity of hydrogen-germanium compounds (Fr.)	6 (24)
22.8	V. F. Mironov and T. K. Gar	*Organometal. Chem. Rev. Sect. A* **3**, 311 (1968)	Trichlorogermane chemistry	10 (45)
22.9	D. Quane and R. S. Bottei	*Chem. Rev.* **62**, 403 (1962)	Organogermanium chemistry	40 (221)
22.10	I. Ruidisch, H. Schmidbaur, and H. Schumann	*In* "Halogen Chemistry" (V. Gutmann, ed.), p. 333. Academic Press, New York, 1967.	Organoelement halides of germanium	17 (711)
22.11	J. Satgé, M. Massol, and P. Rivière	*J. Organometal. Chem.* **56**, 1 (1973)	Divalent germanium species as starting materials and intermediates in organogermanium chemistry	40 (160)
22.12	R. Zablotna	*Wiad. Chem.* **22**, 861 (1968)	Alkylhalogermanes (Pol.)	11 (57)
	See also: A107, A109, A112, A116, A117; B3.1, B4.2; G4			
23.	*Tin and lead—General*			
23.1	A. J. Bloodworth and A. G. Davies	*Chem. Ind. (London)* 490 (1972)	Organotin compounds in organic synthesis	4 (41)
23.2	A. Bokranz and H. Plum	*Fortschr. Chem. Forsch.* **16**, 365 (1971)	Industrial preparation and applications of organotin compounds (Ger.)	39 (127)
23.3	L. Chalmers	*Mfg. Chem. Aerosol News* **38**, 37 (1967)	Chemistry and applications of organotin compounds	5 (45)
23.4	A. G. Davies	*Chem. Brit.* **4**, 403 (1968)	Organotin chemistry	4 (42)

(Continued)

Ref. No.	Authors	Reference	Title	No. of pages (No. of ref.)
23.5	J. D. Donaldson	*Progr. Inorg. Chem.* **8**, 287 (1967)	The chemistry of bivalent tin	70 (268)
23.6	F. W. Frey and H. Shapiro	*Fortschr. Chem. Forsch.* **16**, 243 (1971)	Commercial organolead compounds	55 (338)
23.7	M. Gielen	*Accounts Chem. Res.* **6**, 198 (1973)	From kinetics to the synthesis of chiral tetra-organotin compounds	5 (42)
23.8	M. Gielen	*Ind. Chim. Belge* **38**, 20, 138 (1973)	Synthesis and properties of tetraorganotin compounds (Fr.)	35+34 (525)
23.9	R. K. Ingham, S. D. Rosenberg, and H. Gilman	*Chem. Rev.* **60**, 459 (1960)	Organotin compounds	81 (924)
23.10	G. J. M. van der Kerk	*Chem. Ind.* (*London*) 644 (1970)	Present trends and perspectives in organotin chemistry	3 (11)
23.11	G. J. M. van der Kerk and J. G. A. Luijten	*Arzneim-Forsch.* **19**, 932 (1969)	Chemistry and applications of organotin compounds (Ger.)	3 (7)
23.12	G. J. M. van der Kerk, J. G. A. Luijten, J. C. van Egmond, and J. G. Noltes	*Chimia* **16**, 36 (1962)	Progress in organotin chemistry	7 (37)
23.13	G. J. M. van der Kerk, J. G. A. Luijten, and H. M. J. C. Creemers	*Chimia* **23**, 313 (1969)	Organotin chemistry. Application and research aspects (Ger.)	10 (52)
23.14	F. Klema	*Mitt. Chem. Forschunginst. Wirtsch. Oesterr.* **20**, 1 (1966)	Chemistry of organolead compounds I (Ger.)	6 (52)
23.15	F. Klema	*Mitt. Chem. Forschungist. Wirtsch. Oesterr.* **20**, 43 (1966)	Chemistry of organolead compounds II (Ger.)	4 (32)
23.16	F. Klema	*Chem. Ztg.* **90**, 106 (1966)	Organolead compounds (Ger.)	2 (18)

23.17	R. W. Leeper, L. Summers, and H. Gilman	*Chem. Rev.* **50**, 101 (1954)	Organolead compounds	66 (340)
23.18	J. G. A. Luijten	*Chem. Ind.* (*London*) 103 (1972)	Use of organotin compounds in organic synthesis	2 (13)
23.19	J. G. A. Luijten and G. J. M. van der Kerk	Tin Res. Inst., Greenford, Middlesex (1952)	A survey of the chemistry and applications of organotin compounds	30 (107)
23.20	J. G. A. Luijten	Tin Res. Inst., Greenford, Middlesex 436 (1972)	Toxicity of organotin compounds (a bibliography)	43 (195)
23.21	J. Nasielski	*Acad. Roy. Belg., Cl. Sci., Mem., Collect.* 8⁰ **39**, 59 (1971)	Organotin chemistry. Structure, mechanism, stereochemistry (Fr.)	59 (83)
23.22	W. P. Neumann and K. Kühlein	*Advan. Organometal. Chem.* **7**, 241 (1968)	Recent developments in the organic chemistry of lead; preparations and reactions of compounds with Pb–C, Pb–H, Pb–N and Pb–O bonds	70 (286)
23.23	W. P. Neumann	*Angew. Chem.* **75**, 225 (1963); *Angew. Chem., Int. Ed. Engl.*, **2**, 165 (1963)	Recent developments in organotin chemistry	11 (162)
23.24	J. G. Noltes and G. J. M. van der Kerk	Tin Res. Inst., Greenford, Middlesex (1958)	Functionally substituted organotin compounds	125 (146)
23.25	R. Okawara and M. Wada	*Advan. Organometal. Chem.* **5**, 137 (1967)	Structural aspects of organotin compounds	30 (98)
23.26	R. Piękoš	*Przem. Chem.* **48**, 255 (1969)	Organolead compounds: survey of their properties and application (Pol.)	6 (56)
23.27	S. Sakai and Y. Ishii	*Kagaku* (*Kyoto*) **26**, 142 (1971)	Organic synthesis utilizing organotin compounds (Japan)	7 (36)
23.28	G. Tagliavini, P. Zanella, and M. Fiorani	*Coordination Chem. Rev.* **1**, 249 (1966)	Pentacoordination of organotin compounds in non-aqueous solvents	6 (23)

(Continued)

Ref. No.	Authors	Reference	Title	No. of pages (No. of ref.)
23.29	M. Wada	*Kagaku (Kyoto)* **26**, 131 (1971)	Structure and properties of organotin compounds (Japan)	6 (35)
	See also: A108, A110, A111, A113–A115, A118–A122, A159, A161; B1–B4; E13.5, E16.3, E20; G4			
24.	*Tin and lead—M–H, M–N, M–O, and M–S compounds*			
24.1	A. G. Davies	*Synthesis* 56 (1969)	The potential of organotin oxides and alkoxides in organic synthesis	9 (36)
24.2	C. Giavarini	*Riv. Combust.* **24**, 172 (1970)	Tetraethyllead: industrial production process (Ital.)	9 (27)
24.3	P. G. Harrison	*Organometal. Chem. Rev. Sect. A* **4**, 379 (1969)	Metallostannoxanes and related compounds	99 (109)
24.4	K. Jones and M. F. Lappert	*Organometal. Chem. Rev.* **1**, 67 (1966)	Organic tin-nitrogen compounds	25 (81)
24.5	H. G. Kuivila	*Synthesis* 499 (1970)	Reduction of organic compounds by organotin hydrides	11 (63)
24.6	H. G. Kuivila	*Accounts Chem. Res.* **1**, 299 (1968)	Organotin hydrides and organic free radicals	7 (39)
24.7	H. G. Kuivila	*Advan. Organometal. Chem.* **1**, 47 (1964)	Reactions of organotin hydrides with organic compounds	40 (84)
24.8	E. Lindner	*Rev. Si, Ge, Sn, Pb Compounds* **1**, 35 (1972)	Sulfinato complexes of tin	23 (54)
24.9	H. Matsuda	*Kagaku (Kyoto)* **26**, 137 (1971)	Formation and cleavage of tin-carbon bonds (Japan)	4 (28)
24.10	Y. Nagai	*Kagaku No Ryoiki* **23**, 233 (1969)	Hydrostannation (Japan)	8 (76)
24.11	M. Wada and R. Okawara	*Kagaku No Ryoiki* **20**, 19 (1966)	Organic tin sulphides and related compounds (Japan)	6 (39)
	See also: A108, A110, A113, A115, A118–A122			

25. *Arsenic, antimony, bismuth, and tellurium*

No.	Author(s)	Reference	Title	Pages (Refs.)
25.1	G. Booth	*Advan. Inorg. Chem. Radiochem.* **6**, 1 (1964)	Complexes of the transition metals with phosphines, arsines and stibines	68 (460)
25.2	W. R. Cullen	*Advan. Organometal. Chem.* **4**, 145 (1966)	Organoarsenic chemistry	97 (509)
25.3	G. O. Doak and C. G. Long	*Trans. N.Y. Acad. Sci.* **28**, 402 (1966)	Structure of pentavalent antimony compounds	10
25.4	P. S. Elmes and B. O. West	*Coord. Chem. Rev.* **3**, 279 (1968)	Coordinating properties of pentamethylcyclopentaarsine	6 (12)
25.5	C. S. Hamilton and J. F. Morgan	*Org. React.* **2**, 415 (1944)	The preparation of aromatic arsenic and arsinic acids by the Bart, Bechamp and Rosenmund reactions	39 (231)
25.6	P. G. Harrison	*Organometal. Chem. Rev., Sect. A* **5**, 183 (1970)	Organobismuth chemistry	30 (214)
25.7	R. F. Hudson	*Chem. Ind.* (*Milan*) **54**, 335 (1972)	Chemistry of ylides (Ital.)	8 (59)
25.8	K. J. Irgolic and R. A. Zingaro	*Organometal. React.* **2**, 117 (1971)	Reactions of organotellurium compounds	218 (300)
25.9	G. Kamai and G. M. Usacheva	*Usp. Khim.* **35**, 1404 (1966; *Russ. Chem. Rev.* **35**, 601 (1966)	Stereochemistry of organic compounds of phosphorus and arsenic	8 (123)
25.10	F. G. Mann	*Progr. Org. Chem.* **4**, 217 (1958)	Heterocyclic derivatives of P, As and Sb	32 (65)
25.11	A. N. Nesmeyanov	*Probl. Org. Khim.* 65 (1970)	Phosphorus and arsenic ylides (Russ.)	8 (20)
25.12	N. Petragnani and M. de Moura Campos	*Organometal. Chem. Rev.* **2**, 61 (1967)	New topics in organotellurium chemistry	35 (118)
25.13	L. M. Venanzi	*Angew. Chem.* **76**, 621 (1964); *Angew. Chem., Int. Ed. Engl.* **3**, 453 (1964)	Complexes of tetradentate ligands containing phosphorus and arsenic	8 (27)

(*Continued*)

Ref. No.	Authors	Reference	Title	No. of pages (No. of ref.)
25.14	R. A. Zingaro	*Ann. N.Y. Acad. Sci.* **192**, 72 (1972)	Chemistry of selenium-bearing organometallic derivatives of group VA elements	18 (42)
	See also: A24, A123–A127; B4.2; G4			
26.	*Compounds containing metal–metal bonds*			
26.1	H. G. Ang and P. T. Lau	*Organometal. Chem. Rev., Sect. A* **8**, 235 (1972)	The chemistry of compounds containing silicon-to-transition metal bonds	66 (123)
26.2	M. C. Baird.	*Progr. Inorg. Chem.* **9**, 1 (1968)	Metal–metal bonds in transition metal compounds	159 (60)
26.3	R. J. Cross	*Organometal. Chem. Rev.* **2**, 97 (1967)	σ-Complexes of platinum (II) with hydrogen, carbon and other elements of group IV	43 (81)
26.4	C. S. Cundy, B. M. Kingston, and M. F. Lappert	*Advan. Organometal. Chem.* **11**, 253 (1973)	Organometal complexes with silicon-transition metal or silicon-carbon-transition metal bonds	78 (249)
26.5	E. I. Starovoitova and N. N. Motovilova	*Kauch. Rezina* **25**, 11 (1966)	Organosilicon compounds containing heteroatoms Sn, Ti, V, Al, Fe, Co, Ni, Cr, Al and P	6 (33)
26.6	N. S. Vyazankin, G. A. Razuvaev, and O. A. Kruglaya	*Z. Chem.* **11**, 53 (1971)	Organodimetallic compounds; new syntheses and reactions (Ger.)	8 (84)
26.7	N. S. Vyazankin, G. A. Razuvaev, and O. A. Kruglaya	*Organometal. Chem. Synth.* **1**, 205 (1971)	(Organosilyl and organogermyl)mercury derivatives for the synthesis of organometallic compounds of silicon and germanium	17 (66)
26.8	N. S. Vyazankin, G. A. Razuvaev, and O. A. Kruglaya	*Organometal. Chem. Rev., Sect. A* **3**, 323 (1968)	Organometallic compounds with metal–metal bonds between different metals	100 (396)
26.9	N. S. Vyazankin and O. A. Kruglaya	*Usp. Khim.* **35**, 1388 (1966); *Russ. Chem. Rev.* **35**, 593 (1966)	Covalent organodielementary compounds	16 (95)

26.10	N. S. Vyazankin, G. A. Razuvaev, and O. A. Kruglaya	*Organometal. React.* **5**.	Reactions of bimetallic organometallic compounds	
26.11	J. F. Young	*Advan. Inorg. Chem. Radiochem.* **11**, 92 (1968)	Transition metal complexes with group IVB elements	60 (254)
	See also: A126, A153, A184, A188; B3.1, B4.3			
27.	*Redistribution reactions*			
27.1	E. C. Ashby	*Ann. N.Y. Acad. Sci.* **159** (Part 1), 131 (1969)	Redistribution of organometallic compounds of groups IIA and B	12 (39)
27.2	A. G. Lee	*Organometal. Chem. Rev., Sect. A* **6**, 139 (1970)	The thermodynamics of redistribution reactions	12 (26)
27.3	J. C. Lockhart	*Chem. Rev.* **65**, 131 (1965)	Redistribution and exchange reactions in groups IIB–VIIB	21 (202)
27.4	K. Moedritzer	*Organometal. Chem. Rev.* **1**, 179 1966); for corrections see: A. R. Conrad, A. G. Leer, and K. Moedritzer, *Organometal. Chem. Rev., Sect. A* **3**, 135 (1968)	Redistribution reactions of organometallic compounds of Si, Ge, Sn and Pb	99 (187)
27.5	K. Moedritzer	*Advan. Organometal. Chem.* **6**, 171 (1967)	Redistribution equilibria of organometallic compounds	100 (336)
27.6	K. Moedritzer	*Organometal. React.* **2**, 1 (1971)	Redistribution reactions	115 (495)
27.7	W. P. Neumann	*Ann. N.Y. Acad. Sci.* **159** (Part 1), 56 (1969)	Substituent exchange equilibriums on Ge, Sn and Pb	17 (78)
27.8	J. R. van Wazer and K. Moedritzer	*Angew. Chem.* **78**, 401 (1966); *Angew. Chem., Int. Ed. Engl.* **5**, 341 (1966)	Scrambling equilibria and analysis of labile mixtures	11 (40)
27.9	D. R. Weyenberg, L. G. Mahone, and W. H. Atwell	*Ann. N.Y. Acad. Sci.* **159** (Part 1), 38 (1969)	Redistribution reactions in the chemistry of Si	18 (104)
	See also: A132–A136; C18.3; E13.4			

(Continued)

Ref. No.	Authors	Reference	Title	No. of pages (No. of ref.)
28. *Fluorocarbon compounds*				
28.1	R. E. Banks and R. N. Haszeldine	*Advan. Inorg. Chem. Radiochem.* **3**, 338 (1961)	Polyfluoroalkyl derivatives of metalloids and non-metals	95 (210)
28.2	M. I. Bruce and W. R. Cullen	*Fluorine Chem. Rev.* **4**, 79 (1969)	The chemistry of fluorinated acetylenes	39 (191)
28.3	R. D. Chambers and T. Chivers	*Organometal. Chem. Rev.* **1**, 279 (1968)	Perfluorophenyl-metal compounds	25 (65)
28.4	H. C. Clark	*Advan. Fluorine Chem.* **3**, 19 (1963)	Perfluoroalkyl derivatives of the elements	42 (180)
28.5	S. C. Cohen and A. G. Massey	*Advan. Fluorine Chem.* **6**, 185 (1970)	Polyfluoroaromatic derivatives of metals and metalloids	100 (332)
28.6	W. R. Cullen	*Fluorine Chem. Rev.* **3**, 73 (1969)	Fluoroalicyclic derivatives of metals and metalloids	56 (103)
28.7	W. R. Cullen	*Advan. Inorg. Chem. Radiochem.* **15**, 323 (1972)	Fluoroalicyclic derivatives of metals and metalloids	51 (220)
28.8	H. J. Emeléus	*Angew. Chem.* **74**, 189 (1962); *Angew. Chem., Int. Ed. Engl.* **1**, 129 (1962)	Recent studies on fluoroalkyls and related compounds	5 (36)
28.9	M. Fild and O. Glemser	*Fluorine Chem. Rev.* **3**, 129 (1969)	P, As, and Sb pentafluorophenyl compounds	15 (41)
28.10	N. Ishikawa and S. Hayashi	*Yuki Gosei Kagaku Kyokai Shi* **31**, 495 (1973)	Synthesis of metallic derivatives of poly-(fluoro)-aromatics (Japan)	13 (113)
28.11	I. L. Knunyants, R. N. Sterlin, and V. L. Isaev	*Zh. Vses. Khim Obshchest.* **15**, 25 (1970)	Perfluorovinyl derivatives of some elements (Russ.)	9 (96)
28.12	J. J. Lagowski	*Quart. Rev.* **13**, 232 (1959)	Perfluoroalkyl derivatives of metals and non-metals	32 (153)

28.13	C. Tamborski	*Trans. N.Y. Acad. Sci.* **28**, 601 (1966)	Perfluoroorganometallic compounds	10 (30)
	See also: A142, A146, A188; index 4.8			
29.	*Stereochemistry*			
29.1	B. J. Aylett	*Progr. Stereochem.* **4**, 213 (1969)	The sterochemistry of main group IV elements	59 (300)
29.2	D. C. Bradley	*Progr. Stereochem.* **3**, 1 (1962)	The stereochemistry of some elements of group III	52 (180)
29.3	F. G. Mann	*Progr. Stereochem.* **2**, 196 (1958)	The stereochemistry of the group V elements	231 (130)
294.	V. I. Sokolov and O. A. Reutov	*Usp. Khim.* **34**, 3 (1965); *Russ. Chem. Rev.* **34**, 1 (1965)	The asymmetric non-carbon atom	12 (170)
29.5	L. H. Sommer	*Angew. Chem.* **74**, 176 (1962); *Angew. Chem., Int. Ed. Engl.* **1**, 143 (1962)	The stereochemistry of organosilicon compounds	6 (18)
	See also: A70			
30.	*Kinetics and mechanism—heterolyses of C–M bonds*			
30.1	I. P. Beletskaya, K. P. Butin, A. N. Ryabtsev, and O. A. Reutov	*J. Organometal. Chem.* **59**, 1 (1973)	Stability of organo-mercury, -thallium, -tin and -lead complexes with anionic and neutral ligands	44 (187)
30.2	I. P. Beletskaya, K. P. Butin, and O. A. Reutov	*Organometal. Chem. Rev., Sect. A* **7**, 51 (1971)	S_E1 (N) mechanism in organometallic chemistry	28 (79)
30.3	R. E. Dessy and W. Kitching	*Advan. Organometal. Chem.* **4**, 268 (1966)	Organometallic reaction mechanisms	84 (250)
30.4	H. Fischer and D. Rewicki	*Progr. Org. Chem.* **7**, 116 (1968)	Acidic hydrocarbons	46 (155)
30.5	H. Hashimoto	*Yuki Gosei Kagaku Kyokai Shi* **24**, 1156 (1966)	Some reactions of σ-bonded organometallic compounds (Japan)	14 (87)

(Continued)

Ref. No.	Authors	Reference	Title	No. of pages (No. of ref.)
30.6	C. K. Ingold	*Helv. Chim. Acta* **47**, 1191 (1964)	Organo-metal substitution	13 (0)
30.7	J. R. Jones	*Quart. Rev.* **25**, 365 (1971)	Acidities of carbon acids	24 (47)
30.8	R. D. W. Kemmitt and J. Burgess	*Inorg. React. Mech.* **2**, 247 (1972)	Organometallic compounds. Substitution	19 (98)
30.9	W. Kitching	*Pure Appl. Chem.* **19**, 1 (1969)	Some carbonium-ion- and carbanion-producing heterolyses of carbon-metal bonds	16 (57)
30.10	G. Köbrich	*Angew. Chem.* **74**, 453 (1962); *Angew. Chem., Int. Ed. Engl.* **1**, 382 (1962)	Electrophilic substitutions at saturated carbon atoms	12 (135)
30.11	D. S. Matteson	*Organometal. Chem. Rev., Sect. A* **4**, 263 (1969)	Mechanism of electrophilic displacements of metals at saturated carbon	42 (142)
30.12	O. A. Reutov	*Pure Appl. Chem.* **17**, 79 (1968)	On the importance of coordination in electrophilic substitution reactions of organometallic compounds	16 (41)
30.13	O. A. Reutov	*Fortschr. Chem. Forsch.* **8**, 61 (1967)	Electrophilic substitution with organomercury compounds	30 (59)
30.14	O. A. Reutov	*Usp. Khim.* **36**, 414 (1967); *Russ. Chem. Rev.* **36**, 163 (1967)	Mechanism of reactions of electrophilic substitution at a saturated carbon atom	12 (78)
30.15	U. Schöllkopf	*Angew. Chem.* **82**, 795 (1970); *Angew. Chem., Int. Ed. Engl.* **9**, 763 (1970)	Recent results in carbanion chemistry	11 (80)
30.16	A. Streitwieser, Jr. and J. H. Hammons	*Progr. Phys. Org. Chem.* **3**, 41 (1965)	Acidity of hydrocarbons	40 (86)

See also: A38, A132–A135; B5; C18

31. *Kinetics and mechanism—homolytic processes, oxidation and autoxidation, organometallic peroxides*

31.1	Yu. A. Alexandrov	*J. Organometal. Chem.* **55**, 1 (1973)	Some advances in the liquid phase autoxidation of organic compounds of the non-transitional elements	40 (186)
31.2	K. C. Bass	*Organometal. Chem. Rev.* **1**, 391 (1966)	Homolytic decompositions of organometallic compounds in solution. Part 1: organic compounds of Hg	42 (114)
31.3	T. G. Brilkina and V. A. Shushunov	*Usp. Khim.* **35**, 1430 (1966); *Russ. Chem. Rev.* **35**, 613 (1966)	Progress in the study of the oxidation of organometallic compounds	9 (168)
31.4	I. M. T. Davidson	*Quart. Rev.* **25**, 111 (1971)	Silicon radical chemistry	23 (88)
31.5	A. G. Davies	*Chem. Ind. (London)* 832 (1972)	A radical approach to organometallic chemistry	3 (4)
31.6	A. G. Davies	*Org. Peroxides* **2**, 337 (1971)	Formation of organometallic peroxides by autoxidation	18 (78)
31.7	A. G. Davies and B. P. Roberts	*Accounts Chem. Res.* **5**, 387 (1972)	Bimolecular homolytic substitution at a metal centre	5 (53)
31.8	N. J. Friswell and B. G. Gowenlock	*Advan. Free Radical Chem.* **1**, 39 (1965)	Inorganic hydrogen and alkyl-containing free radicals Part I, groups II, III and IV	36 (200)
31.9	N. J. Friswell and B. G. Gowenlock	*Advan. Free Radical Chem.* **2**, 1 (1967)	Inorganic hydrogen and alkyl-containing free radicals Part II, groups V and VI	44 (300)
31.10	A. Hudson	*Electron Spin Reson.* **1**, 253 (1972)	Organometallic radicals	10 (53)
31.11	K. Ichikawa and S. Uemura	*Bull. Inst. Chem. Res., Kyoto Univ.* **50**, 225 (1972)	Oxidation involved in organometallic reactions	14 (32)
31.12	R. A. Jackson	*Advan. Free Radical Chem.* **3**, 231 (1969)	Group IVB radical reactions	57 (200)
31.13	R. A. Jackson	*Chem. Soc. Spec. Publ. No.* **24**, 295 (1970)	Si, Ge, Sn and Pb radicals	27 (117)

(Continued)

Ref. No.	Authors	Reference	Title	No. of pages (No. of ref.)
31.14	V. J. Karnojitzky	*Chem. Ind., Genie Chim.* **102**, 258 (1969)	Organometallic hydroperoxides (Fr.)	5 (20)
31.15	B. J. McClelland	*Chem. Rev.* **64**, 301 (1964)	Anionic free radicals	15 (149)
31.16	E. B. Milovskaya	*Usp. Khim.* **42**, 881 (1973); *Russ. Chem. Rev.* **42**, 384 (1973)	Homolytic substitution of the metal atom of organometallic compounds	8 (75)
31.17	I. Moritani and T. Hosokawa	*Kagaku No Ryoiki, Zokan No.* **93**, 183 (1970)	Photoreactions in organometallic compounds (Japan)	39 (81)
31.18	Y. Nagai	*Intra-Sci. Chem. Rep.* **4**, 115 (1970)	Chemistry of organosilicon free radicals	10 (102)
31.19	Y. Nagai	*Kagaku (Kyoto)* **25**, 544 (1970)	Reactions of halosilanes (Japan)	6 (50)
31.20	G. A. Razuvaev, V. A. Shushunov, V. Dodonov, and T. G. Brilkina	*Org. Peroxides* **3**, 141 (1972)	Acyl and alkyl peroxides and hydroperoxides with organometallic compounds	132 (271)
31.21	H. Sakurai	*Nippon Kagaku Zasshi* **91**, 885 (1970)	Reactions of free silyl radicals in the liquid phase (Japan)	13 (61)
31.22	L. M. Smorgonskii and A. B. Bruker	*Usp. Khim.* **15**, 81 (1946)	Free radicals in the Grignard reaction	20
31.23	G. Sosnovsky and J. M. Brown	*Chem. Rev.* **66**, 529 (1966)	The chemistry of organometallic and organometalloid peroxides	39 (212)
31.24	N. S. Vyazankin and O. A. Kruglaya	*Tr. Khim. Khim. Tekhnol.* 3 (1966)	Homolytic reactions of organometallic compounds in liquid phase	14 (126)

See also: A131, A136, A138–A139

32. *Kinetics and mechanism—carbenes, carbenoids*

32.1	E. Chinoporos	*Chem. Rev.* **63**, 235 (1963)	Carbenes: reactive intermediates containing divalent carbon	21 (153)
32.2	F. Ebel	*Fortschr. Chem. Forsch.* **12**, 387 (1969)	Structure and reactivity of carbanions and carbenes	57 (351)
32.3	W. Kirmse	*Progr. Org. Chem.* **6**, 164 (1964)	Carbenes	50 (275)
32.4	W. Kirmse	*Angew. Chem.* **77**, 1 (1965); *Angew. Chem., Int. Ed. Engl.* **4**, 1 (1965)	Intermediates of α-eliminations	10 (125)
32.5	G. Köbrich	*Angew. Chem.* **77**, 75 (1965); *Angew. Chem., Int. Ed. Engl.* **4**, 49 (1965)	Eliminations from olefins	20 (169)
32.6	G. Köbrich	*Angew. Chem.* **79**, 15 (1967); *Angew. Chem., Int. Ed. Engl.* **6**, 41 (1967)	Chemistry of stable α-halogenorganolithium compounds and the mechanism of carbenoid reactions	12 (52)
32.7	G. Köbrich	*Bull. Soc. Chim. Fr.* 2712 (1969)	Recent results in the chemistry of stable α-halolithium organyls (carbenoids)	9 (23)
32.8	G. Köbrich	*20 Jahre Fonds Chem. Ind., Beitr. Wiss. Veranst.* 101 (1970)	Thermolabile alkali metal organic compounds (Ger.)	11 (57)
32.9	G. Köbrich	*Angew. Chem.* **84**, 557 (1972); *Angew. Chem., Int. Ed. Engl.* **11**, 473 (1972)	The chemistry of carbenoids and other thermolabile organolithium compounds	13 (67)
32.10	O. M. Nefedov and M. N. Manakov	*Angew. Chem.* **78**, 1039 (1966); *Angew. Chem., Int. Ed. Engl.* **5**, 1021 (1966)	Inorganic, organometallic and organic analogues of carbenes	17 (185)
32.11	H. Nozaki and R. Noyori	*Yuki Gosei Kagaku Kyokai Shi* **24**, 519 (1966)	Chemistry of carbenoids I. Li and Zn carbenoids (Japan)	13 (86)
32.12	H. Nozaki and R. Noyori	*Yuki Gosei Kagaku Kyokai Shi* **24**, 632 (1966)	Chemistry of carbenoids II. (Decomposition of Hg carbenoids) (Japan)	14 (111)

(Continued)

Ref. No.	Authors	Reference	Title	No. of pages (No of ref.)
32.13	G. G. Rozantsev, A. A. Fainzil'berg, and S. S. Novikov	*Usp. Khim.* **34**, 177 (1965); *Russ. Chem. Rev.* **34**, 69 (1965)	Advances in the chemistry of carbenes	12 (231)
	See also: A137; E1.4, E2.2, E2.3			
33.	*Kinetics and mechanism—benzynes, hetarynes*			
33.1	H. Heaney	*Chem. Rev.* **62**, 81 (1962)	The benzyne and related intermediates	17 (135)
33.2	H. J. den Hertog and H. V. van der Plas	*Advan. Heterocyc. Chem.* **4**, 121 (1965)	Hetarynes	23 (49)
33.3	Th. Kaufmann	*Angew. Chem.* **77**, 557 (1965); *Angew. Chem., Int. Ed. Engl.* **4**, 543 (1965)	Hetarynes	15 (95)
33.4	Th. Kaufmann and R. Wirthwein	*Angew. Chem.* **83**, 21 (1971); *Angew. Chem., Int. Ed. Engl.* **10**, 20 (1971)	Progress in the hetaryne field	13 (72)
33.5	G. Wittig	*Angew. Chem.* **77**, 752 (1965); *Angew. Chem., Int. Ed. Engl.* **4**, 731 (1965)	1,2-Dehydrobenzene	7 (47)
33.6	G. Wittig	*Angew. Chem.* **74**, 479 (1962); *Angew. Chem., Int. Ed. Engl.* **1**, 415 (1962)	Small rings with carbon–carbon triple bonds	5 (27)
	See also: A184; E10.2			
34.	*Catalysis and related topics, polymerization, organometallic polymers*			
34.1	K. A. Andrianov, I. Haiduc, and L. M. Khananashvili	*Usp. Khim.* **34**, 27 (1965); *Russ. Chem. Rev.* **34**, 13 (1965)	Ability of elements to form polymers with inorganic chain molecules	18 (109)

34.2	C. H. Bamford	*J. Appl. Chem.* **13**, 525 (1963)	Recent advances in polymer science	13 (77)
34.3	C. E. H. Bawn and A. Ledwith	*Quart. Rev.* **16**, 361 (1962)	Stereoregular addition polymerisation	74 (218)
34.4	C. Beerman and H. Bestian	*Angew. Chem.* **71**, 618 (1959)	Organometallic titanium compounds as polymerisation catalysts	6 (16)
34.5	M. N. Berger, G. Boocock, and R. N. Howard	*Advan. Catal.* **19**, 211 (1969)	The polymerization of olefins by Ziegler catalysts	30 (38)
34.6	H. Bestian and K. Clauss	*Angew. Chem.* **75**, 1068 (1963); *Angew. Chem., Int. Ed. Engl.* **2**, 704 (1963)	Reactions of olefins with the titanium-carbon bond	11 (17)
34.7	H. Bestian, K. Clauss, H. Jensen, and E. Prinz	*Angew. Chem.* **74**, 955 (1962); *Angew. Chem., Int. Ed. Engl.* **2**, 32 (1963)	Low temperature polymerisation of ethylene	10 (26)
34.8	G. Bier	*Angew. Chem.* **73**, 186 (1961)	High molecular weight olefin mixed polymers prepared using mixed Ziegler catalysts	12 (35)
34.9	J. Boor	*Macromol. Rev.* **2**, 115 (1967)	Nature of the active site in the Ziegler-type catalyst	151 (266)
34.10	J. Boor	*Ind. Eng. Chem. Prod. Res. Develop.* **9**, 437 (1970)	Review of recent literature in Ziegler-type catalysts	18 (128)
34.11	N. Calderon	*J. Macromol. Sci., C* **7**, 105 (1972)	Ring opening polymerisation of cyclo-olefins	55 (86)
34.12	W. Cooper and G. Vaughan	*Progr. Polymer Sci.* **1**, 91 (1967)	Recent developments in the polymerisation of conjugated dienes	70 (272)
34.13	H. W. Coover, R. L. McConnell, and F. B. Joyner	*Macromol. Rev.* **1**, 91 (1966)	Relationship of catalyst composition to catalytic activity for the polymerisation of α-olefins	28 (59)

(*Continued*)

Ref. No.	Authors	Reference	Title	No. of pages (No. of ref.)
34.14	R. C. P. Cubbon and D. Margerison	*Progr. React. Kinet.* **3**, 403 (1965)	The kinetics of polymerisation of vinyl monomers by lithium alkyls	45 (101)
34.15	S. L. Davydova, N. A. Plate, and V. A. Kargin	*Usp. Khim.* **39**, 2256 (1970); *Russ. Chem. Rev.* **39**, 1082 (1970)	Synthesis and chemical reactions of metal-containing macromolecules	17 (111)
34.16	J. Furukawa	*Pure Appl. Chem.* **26**, 153 (1971)	Copolymerisation of vinyl chloride by organometallic compounds	19 (8)
34.17	A. Gumboldt	*Fortschr. Chem. Forsch.* **16**, 299 (1971)	Organometallic compounds as catalysts of olefin polymerisation	30 (70)
34.18	G. Henrici-Olivé and S. Olivé	*Angew. Chem.* **79**, 764 (1967); *Angew. Chem., Int. Ed. Engl.* **6**, 790 (1967)	The active species in homogeneous Ziegler–Natta catalysts for the polymerisation of ethylene	9 (28)
34.19	N. L. Holy and J. D. Marcum	*Angew. Chem.* **83**, 132 (1971); *Angew. Chem., Int. Ed. Engl.* **10**, 115 (1971)	Radical anion intermediates in organic chemistry	8 (80)
34.20	S. Inoue	*Nippon Kagaku Zasshi* **91**, 297 (1970)	Polymerisation by organometallic compounds (Japan)	16 (49)
34.21	B. R. James	"Homogeneous Hydrogenation," Chapter 15. Wiley, New York, 1973	Ziegler-type catalysts	21 (*ca.* 60)
34.22	J. P. Kennedy and T. Otsu	*J. Macromol. Sci., C* **6**, 237 (1972)	Hydrogen transfer polymerisation with anionic catalysts and the problem of anionic isomerisation polymerisation	47 (68)
34.23	M. H. Lehr	*Survey Progr. Chem.* **3**, 183 (1966)	Stereoregular polymers	12 (250)
34.24	A. R. Lyons	*Macromol. Rev.* **6**, 251 (1972)	Polymerisation of vinyl ketones	43 (116)

34.25	V. P. Mardykin, A. M. Antipova, and P. N. Gaponik	*Usp. Khim.* **40**, 24 (1971); *Russ. Chem. Rev.* **40**, 13 (1971)	Three-component complex organometallic catalysts for polymerisation of olefinic hydrocarbons	11 (236)
34.26	E. B. Milovskaya, L. V. Zamoiskaya, and E. L. Kopp	*Usp. Khim.* **38**, 420 (1969); *Russ. Chem. Rev.* **38**, 420 (1969)	The mechanism of the initiation of the radical polymerisation of polar monomers in systems comprising organometallic compounds	13 (135)
34.27	M. Morton and L. J. Fetters	*Macromol. Rev.* **2**, 71 (1967)	Homogeneous anionic polymerisation of unsaturated monomers	43 (347)
34.28	E. M. Natanson and M. T. Bryk	*Usp. Khim.* **41**, 1465 (1972); *Russ. Chem. Rev.* **41**, 671 (1972)	Metallopolymers	16 (175)
34.29	G. Natta, G. Dall'Asta, G. Mazzini, U. Giannini, and S. Cesca	*Angew. Chem.* **70**, 496 (1958)	Stereospecific polymerisation of vinyl ethers	6 (18)
34.30	G. Natta	*Makromol. Chem.* **35**, 94 (1960)	Advances in stereospecific polymerisation	38 (37)
34.31	G. Natta, G. Dall'Asta, and G. Mazzanti	*Angew. Chem.* **76**, 765 (1964); *Angew. Chem., Int. Ed. Engl.* **3**, 723 (1964)	Stereospecific homopolymerisation of cyclopentene	7 (19)
34.32	E. W. Neuse and H. Rosenberg	*J. Macromol. Sci., C* **4**, 1 (1970)	Metallocene polymers	145 (297)
34.33	F. Patat and H. Sinn	*Angew. Chem.* **70**, 496 (1958)	The result of low pressure polymerisation of α-olefins: polymerisation by complexes	5 (28)
34.34	H. Sinn and F. Patat	*Angew. Chem.* **76**, 805 (1964); *Angew. Chem., Int. Ed. Engl.* **3**, 93 (1964)	The action mechanism of organometallic catalysts	8 (18)
34.35	M. Szwarc	*Advan. Chem. Phys.* **2**, 147 (1959)	Recent advances in polymer chemistry—survey of anionic polymerisation	39 (57)
34.36	M. Szwarc and J. Smid	*Progr. React. Kinet.* **2**, 217 (1964)	The kinetics and propagation of anionic polymerisation and copolymerisation	68 (180)

(Continued)

Ref. No.	Authors	Reference	Title	No. of pages (No. of ref.)
34.37	M. Szwarc	*Progr. Phys. Org. Chem.* **6**, 323 (1968)	Chemistry of radical-ions	116 (328)
34.38	M. Szwarc	*Accounts Chem. Res.* **5**, 169 (1972)	Radical anions and carbanions as donors in electron-transfer processes	8 (42)
34.39	A. V. Topchiev, B. A. Krentsel', and L. L. Stotskaya	*Usp. Khim.* **30**, 462 (1961); *Russ. Chem. Rev.* **30**, 192 (1961)	Organometallic complexes: olefin polymerisation catalysts	16 (80)
34.40	H. Weber	*Fortschr. Chem. Forsch.* **16**, 329 (1971)	Organometallic compounds as catalysts for the preparation of stereopolymers	35 (138)
34.41	E. A. Youngman and J. Boor	*Macromol. Rev.* **2**, 33 (1967)	Syndiotactic polypropylene	37 (74)
	See also: A163–A180; C19, C20; E4, E13.9, E13.10, E19.4			
35. *Physical methods—infrared and Raman spectroscopy*				
35.1	H. Bürger	*Organometal. Chem. Rev., Sect. A* **3**, 425 (1968)	Infrared spectra of trimethylsilyl compounds (Ger.)	43 (147)
35.2	N. A. Chemaevskii	*Usp. Khim.* **32**, 1152 (1963); *Russ. Chem. Rev.* **32**, 509 (1963)	Vibrational spectra of compounds containing elements of the carbon sub-group	11 (188)
35.3	D. K. Huggins and H. D. Kaesz	*Progr. Solid State Chem.* **1**, 417 (1964)	Use of IR and Raman spectroscopy in the study of organometallic compounds	95 (314)
35.4	T. G. Spiro	*Progr. Inorg. Chem.* **11**, 1 (1970)	Vibrational spectra and metal–metal bonds	51 (145)
35.5	T. Tanaka	*Organometal. Chem. Rev., Sect. A* **5**, 1 (1970)	Vibrational spectra of organotin and organolead compounds	51 (280)
	See also: A152–154; B3.2; C16.1; E18.1; G9, G10			
36. *Physical methods—nuclear magnetic resonance*				
36.1	J. W. Akitt	*Annu. Rev. NMR Spectrosc.* **5A**, 466 (1972)	NMR spectroscopy in liquids containing compounds of aluminium and gallium	91 (298)

36.2	T. L. Brown	*Accounts Chem. Res.* **1**, 23 (1968)	NMR studies of organometallic exchange processes	10 (26)
36.3	J. W. Emsley and L. Phillips	*Progr. NMR Spectrosc.* **7**, 1 (1971)	Fluorine chemical shifts	524 (220)
36.4	L. A. Fedorov	*Usp. Khim.* **39**, 1389 (1970); *Russ. Chem. Rev.* **39**, 655 (1970)	NMR spectroscopy of organometallic allyl compounds	18 (171)
36.5	N. S. Ham and T. Mole	*Progr. NMR Spectrosc.* **4**, 91 (1969)	The application of NMR to organometallic exchange reactions	102 (185)
36.6	W. G. Henderson and E. F. Mooney	*Annu. Rev. NMR Spectrosc.* **2**, 219 (1969)	^{11}B NMR spectroscopy	73 (192)
36.7	H. Haraguchi	*Kagaku No Ryoiki* **24**, 802 (1970)	Study of aluminium complexes by ^{27}Al-magnetic resonance (Japan)	7 (43)
36.8	K. Jones and E. F. Mooney	*Annu. Rev. NMR Spectrosc.* **3**, 340 (1970)	^{19}F NMR spectroscopy	31 (60)
36.9	K. Jones and E. F. Mooney	*Annu. Rev. NMR Spectrosc.* **4**, 450 (1970)	^{19}F NMR spectroscopy	20 (50)
36.10	M. L. Maddox, S. L. Stafford, and H. D. Kaesz	*Advan. Organometal. Chem.* **3**, 1 (1965)	Applications of nuclear magnetic resonance to the study of organometallic compounds	179 (380)
36.11	E. F. Mooney and P. H. Winson	*Annu. Rev. NMR Spectrosc.* **1**, 280 (1966)	^{19}F NMR spectroscopy	14 (30)
36.12	E. F. Mooney and P. H. Winson	*Annu. Rev. NMR Spectrosc.* **2**, 174 (1967)	^{13}C NMR spectroscopy	3 (6)
	See also: A155; B3.2, B6; C16.1; E18.3			
37. *Physical methods—mass spectroscopy*				
37.1	M. I. Bruce	*Advan. Organometal. Chem.* **6**, 273 (1967)	Mass spectra of organometallic compounds	61 (303)
37.2	D. B. Chambers, F. Glockling, and J. R. C. Light	*Quart. Rev.* **22**, 317 (1968)	Mass spectra of organometallic compounds	21 (51)

(Continued)

Ref. No.	Authors	Reference	Title	No. of pages (No. of ref.)
37.3	J. Müller	*Angew. Chem.* **84**, 725 (1972); *Angew. Chem., Int. Ed. Engl.* **11**, 653 (1972)	Decomposition of organometallic complexes in the mass spectrometer	12 (112)
	See also: A156–A157; B3.3; C16.1; E17.2			
38. *Physical methods—Mössbauer spectroscopy*				
38.1	G. M. Bancroft and P. H. Platt	*Advan. Inorg. Chem. Radiochem.* **15**, 59 (1972)	Mössbauer spectra of inorganic compounds: bonding and structure	100 (577)
38.2	V. I. Goldanskii, V. V. Khrapov, and R. A. Stukin	*Organometal. Chem. Rev., Sect. A* **4**, 225 (1969)	Application of the Mössbauer effect in the study of organometallic compounds	36 (129)
38.3	R. H. Herber	*Progr. Inorg. Chem.* **8**, 1 (1967)	Chemical applications of Mössbauer spectroscopy	41 (108)
38.4	R. V. Parish	*Progr. Inorg. Chem.* **15**, 101 (1972)	The interpretation of ^{119}Sn Mössbauer spectra	100 (142)
38.5	P. S. Smith	*Organometal. Chem. Rev., Sect. A* **5**, 373 (1970)	Mössbauer parameters of organotin compounds	28 (86)
38.6	J. J. Zuckerman	*Advan. Organometal. Chem.* **9**, 22 (1970)	Applications of ^{119}Sn Mössbauer spectroscopy to the study of organotin compounds	112 (397)
	See also: B3.2; C6.1; E18.2			
39. *Physical methods—photoelectron spectroscopy*				
39.1	H. Bock and B. G. Ramsey	*Angew. Chem.* **85**, 773 (1973); *Angew. Chem., Int. Ed. Engl.* **12**, 734 (1973)	Photoelectron spectra of nonmetal compounds and their interpretation by MO methods	19 (106)
	See also: B.3.2			
40. *X-ray and electron-diffraction methods*				

40.1	M. Keeton	*Coord. Chem. Rev.* **8**, 260 (1972)	X-ray bibliography	29 (160)
	See also: B3.1; C16.1; E17.1, G7			
41. *Energetics*				
41.1	W. F. Lautsch, A. Tröber, H. Körner, K. Wagner, R. Kaden, and S. Blase	*Z. Chem.* **3**, 415 (1963)	Energetic data for metalorganic compounds. 1. Enthalpies of combustion and formation	7 (56)
41.2	W. Lautsch *et al.*	*Z. Chem.* **4**, 441 (1964)	Energetic data for metalorganic compounds. 2. Enthalpies of vapourisation and formation	14 (155)
41.3	W. Lautsch *et al.*	*Z. Chem.* **6**, 171 (1966)	Energetic data for metalorganic compounds. 3. Specific methods of measurement and consideration of errors	11 (20)
41.4	H. A. Skinner	*Advan. Organometal. Chem.* **2**, 49 (1964)	The strengths of metal-to-carbon bonds	65 (182)
	See also: A147–A151; 11.5			
42. *Miscellaneous topics*				
42.1	B. Bock, K. Flatau, H. Junge, M. Kuhr, and H. Musso	*Angew. Chem.* **83**, 239 (1971); *Angew. Chem., Int. Ed. Engl.* **10**, 225 (1971)	Bond character of β-diketone metal chelates	11 (114)
42.2	K. A. R. Mitchell	*Chem. Rev.* **69**, 157 (1969)	The use of outer d-orbitals in bonding	22 (251)
42.3	B. Saville	*Angew. Chem.* **79**, 966 (1967); *Angew. Chem., Int. Ed. Engl.* **6**, 928 (1967)	The concept of hard and soft acids and bases as applied to multi-centre chemical reactions	12 (84)
42.4	R. S. Tobias	*Organometal. Chem. Rev., Sect. A* **1**, 93 (1966)	σ-bonded organometallic cations in aqueous solutions and crystals	36 (136)

(*Continued*)

Ref. No.	Authors	Reference	Title	No. of pages (No. of ref.)
42.5	M. J. Taylor	*Essays Chem.* **4**, 19 (1972)	Variable valence as a property of the main group metals—an account of low oxidation states in the Zn, Al, Sn and Sb sub-groups	40 (6)
42.6	G. Wittig	*Quart. Rev.* **20**, 191 (1966)	The role of ate complexes as reaction-determining intermediates	20 (35)
42.7	G. Wittig and H. Reiff	*Angew. Chem.* **80**, 8 (1968); *Angew. Chem., Int. Ed. Engl.* **7**, 7 (1968)	Directed aldol condensations	7 (16)
42.8	R. Blackburn A. Kabi	*Rad. Res. Rev.* **2**, 103 (1968)	Ionising radiations and organometallic compounds	28 (163)

Author Index[1]

Reviews of Main Group Elements

[1] Author Index for appendix (pp. 486–542) to the article "The Organometallic Chemistry of the Main Group Elements—A Guide to the Literature" by J. D. Smith and D. R. M. Walton.

Subject Index

A

R

S

T

V

W

Y

Z

Cumulative List of Contributors

Cumulative List of Titles

A 5
B 6
C 7
D 8
E 9
F 0
G 1
H 2
I 3
J 4